Lecture Notes in Networks and Systems

Volume 586

The series "Lecture Notes in Networks and Systems" publishes the latest developments in Networks and Systems—quickly, informally and with high quality. Original research reported in proceedings and post-proceedings represents the core of LNNS.

Volumes published in LNNS embrace all aspects and subfields of, as well as new challenges in, Networks and Systems.

The series contains proceedings and edited volumes in systems and networks, spanning the areas of Cyber-Physical Systems, Autonomous Systems, Sensor Networks, Control Systems, Energy Systems, Automotive Systems, Biological Systems, Vehicular Networking and Connected Vehicles, Aerospace Systems, Automation, Manufacturing, Smart Grids, Nonlinear Systems, Power Systems, Robotics, Social Systems, Economic Systems and other. Of particular value to both the contributors and the readership are the short publication timeframe and the world-wide distribution and exposure which enable both a wide and rapid dissemination of research output.

The series covers the theory, applications, and perspectives on the state of the art and future developments relevant to systems and networks, decision making, control, complex processes and related areas, as embedded in the fields of interdisciplinary and applied sciences, engineering, computer science, physics, economics, social, and life sciences, as well as the paradigms and methodologies behind them.

Indexed by SCOPUS, INSPEC, WTI Frankfurt eG, zbMATH, SCImago.

All books published in the series are submitted for consideration in Web of Science.

For proposals from Asia please contact Aninda Bose (aninda.bose@springer.com).

Massimo Tistarelli · Shiv Ram Dubey ·
Satish Kumar Singh · Xiaoyi Jiang
Editors

Computer Vision and Machine Intelligence

Proceedings of CVMI 2022

Springer

Editors
Massimo Tistarelli
Computer Vision Laboratory
University of Sassari
Alghero, Sassari, Italy

Satish Kumar Singh
Computer Vision and Biometrics Lab
Department of Information Technology
Indian Institute of Information Technology
Allahabad, India

Shiv Ram Dubey
Computer Vision and Biometrics Lab
Department of Information Technology
Indian Institute of Information Technology
Allahabad
Prayagraj, India

Xiaoyi Jiang
University of Münster
Münster, Germany

ISSN 2367-3370 ISSN 2367-3389 (electronic)
Lecture Notes in Networks and Systems
ISBN 978-981-19-7866-1 ISBN 978-981-19-7867-8 (eBook)
https://doi.org/10.1007/978-981-19-7867-8

This Springer imprint is published by the registered company Springer Nature Singapore Pte Ltd.
The registered company address is: 152 Beach Road, #21-01/04 Gateway East, Singapore 189721, Singapore

Organization

Patron

Prof. Bidyut Baran Chaudhuri, Indian Statistical Institute (ISI), Kolkata, India
Prof. A. G. Ramakrishnan, Indian Institute of Sciences (IISc), Bangalore, India
Prof. P. Nagabhushan, Former Director, IIIT, Allahabad, India
Prof. R. S. Verma, Director, IIIT, Allahabad, India

General Chairs

Prof. Massimo Tistarelli, University of Sassari, Italy
Prof. Bhabatosh Chanda, Indian Statistical Institute (ISI), Kolkata, India
Prof. Balasubramanian Raman, IIT Roorkee, India
Prof. Shekhar Verma, IIIT, Allahabad, India

General Co-chairs

Prof. Peter Peer, University of Ljubljana, Slovenia
Prof. Abdenour Hadid, University Polytechnique Hauts-de-France, France
Prof. Wei-Ta Chu, National Cheng Kung University, Taiwan
Prof. Pavan Chakraborty, IIIT, Allahabad, India

Conference Chairs

Prof. K. C. Santosh, University of South Dakota, USA
Dr. Satish Kumar Singh, IIIT, Allahabad, India
Dr. Shiv Ram Dubey, IIIT, Allahabad, India
Dr. Mohammed Javed, IIIT, Allahabad, India

Technical Program Chairs

Prof. Vrijendra Singh, IIIT, Allahabad, India
Dr. Kiran Raja, NTNU, Norway

Conference Publication Chairs

Prof. Massimo Tistarelli, University of Sassari, Italy
Prof. Xiaoyi Jiang, University of Münster, Germany
Dr. Satish Kumar Singh, IIIT, Allahabad, India
Dr. Shiv Ram Dubey, IIIT, Allahabad, India

Conference Conveners

Dr. Navjot Singh, IIIT, Allahabad, India
Dr. Anjali Gautam, IIIT, Allahabad, India

Local Organizing Committee

Prof. Pritish Varadwaj (Chair), IIIT, Allahabad, India
Dr. Satish Kumar Singh, IIIT, Allahabad, India
Mr. Rajit Ram Yadav, IIIT, Allahabad, India
Mr. Ajay Tiwary, IIIT, Allahabad, India
Mr. Deep Narayan Das, IIIT, Allahabad, India
Mr. Sanjay Kumar, IIIT, Allahabad, India

Conference Hospitality Chairs

Dr. Akhilesh Tiwari, IIIT, Allahabad, India
Dr. Rajat Kumar Singh, IIIT, Allahabad, India

Conference Registration and Certification Chair

Dr. Anjali Gautam, IIIT, Allahabad, India

International Advisory Committee

Prof. Bidyut Baran Chaudhuri, Indian Statistical Institute (ISI), Kolkata
Prof. A. G. Ramakrishnan, Indian Institute of Sciences (IISc), Bangalore
Prof. P. Nagabhushan, IIIT, Allahabad, India
Prof. Xiaoyi Jiang, University of Münster, Germany
Prof. Gaurav Sharma, University of Rochester, USA
Prof. Massimo Tistarelli, University of Sassari, Italy
Prof. K. C. Santosh, University of South Dakota, USA
Prof. Mohan S. Kankanhalli, National University of Singapore, Singapore
Prof. Daniel P. Lopresti, Lehigh University, USA
Prof. Gian Luca Foresti, University of Udine, Italy
Prof. Raghavendra Ramachandra, NTNU, Norway
Prof. Paula Brito, University of Porto, Portugal
Prof. Peter Peer, University of Ljubljana, Slovenia
Prof. Abdenour Hadid, University Polytechnique Hauts-de-France, France
Prof. Wei-Ta Chu, National Cheng Kung University, Taiwan
Prof. Abdelmalik Taleb-Ahmed, Université Polytechnique Hauts-de-France, France
Prof. Lee Hwee-Kuan, A*STAR, Singapore
Prof. Michal Haindl, Czech Academy of Sciences, Czech Republic
Dr. Kiran Raja, NTNU, Norway
Dr. Ajita Rattani, Wichita State University, USA
Dr. Alireza Alaei, Southern Cross University, Australia
Prof. Bhabatosh Chanda, ISI, Kolkata, India
Prof. Umapada Pal, ISI, Kolkata, India
Prof. B. M. Mehtre, IDRBT, Hyderabad, India
Prof. Sri Niwas Singh, IIT Kanpur, India
Prof. G. C. Nandi, IIIT, Allahabad, India
Prof. D. S. Guru, University of Mysore, India
Prof. O. P. Vyas, IIIT, Allahabad, India
Prof. Anupam Agrawal, IIIT, Allahabad, India

Prof. Balasubramanian Raman, IIT Roorkee, India
Prof. Shekhar Verma, IIIT, Allahabad, India
Prof. Sanjay Kumar Singh, IIT BHU, India
Prof. Nishchal K. Verma, IIT Kanpur, India
Prof. Sushmita Gosh, Jadavpur University, India
Prof. Vrijendra Singh, IIIT, Allahabad, India
Prof. Pavan Chakraborty, IIIT, Allahabad, India
Prof. Ashish Khare, University of Allahabad, India
Dr. B. H. Shekhar, Mangalore University, India
Prof. Pritee Khanna, IIITDM Jabalpur, India
Dr. Swagatam Das, ISI, Kolkata, India
Dr. Omprakash Kaiwartya, Nottingham Trent University, UK
Dr. Partha Pratim Roy, IIT Roorkee, India
Dr. P. V. VenkitaKrishnan, ISRO, Bangalore, India
Dr. Surya Prakash, IIT Indore, India
Dr. M. Tanveer, IIT Indore, India
Dr. Puneet Goyal, IIT Ropar, India
Dr. Sharad Sinha, IIT Goa, India
Dr. Krishna Pratap Singh, IIIT, Allahabad, India
Dr. Sachin Kumar, South Ural State University, Russia
Dr. Balakrishna Pailla, Reliance Pvt. Ltd., India
Dr. Hemant Aggarwal, GE Healthcare Bangalore, India
Dr. Bunil Kumar Balabantaray, NIT Meghalaya, India
Dr. K. R. Udaya Kumar Reddy, Dayananda Sagar College of Engineering, Bengaluru, India
Dr. V. Chandra Sekhar, Samsung Pvt. Ltd., India

Preface

The present world is witnessing rapid advancements in the field of Information Technology, specifically, in the areas of Computer Vision and Machine Intelligence, cherished by the society and industry. The amount of Research and Development activities across the globe has been drastically scaled up from the past one decade in these research areas. Hence, a new state-of-the-art Computer Vision and Machine Intelligence (CVMI) Conference is conceived by Computer Vision and Biometrics Laboratory (CVBL), Department of Information Technology, Indian Institute of Information Technology, Allahabad, India, for the researchers to disseminate their research outcomes. The worldwide leading research laboratories, including Computer Vision Laboratory, Faculty of Computer and Information Science, University of Ljubljana, Slovenia, Multimedia and Computer Vision Laboratory, Department of Computer Science and Information Engineering, National Cheng Kung University, Taiwan, University Polytechnique Hauts-de-France, France, and Department of Computer Science, University of Münster, Germany, are the organizing partners. The CVMI 2022 conference was held in hybrid mode at IIIT, Allahabad, Prayagraj, India. The CVMI 2022 conference is "Endorsed by International Association for Pattern Recognition (IAPR)" and "Technically Sponsored by IEEE Signal Processing Society Uttar Pradesh Chapter".

CVMI 2022 received submissions on topics such as biometrics, forensics, content protection, face, iris, emotion, sign language and gesture recognition, image/video processing for autonomous vehicles, 3D image/video processing, image enhancement/super resolution/restoration, action and event detection/recognition, motion and tracking, medical image and video analysis, image/video retrieval, vision-based human GAIT analysis, document and synthetic visual processing, remote sensing, multispectral/hyperspectral image processing, datasets and evaluation, segmentation and shape representation, image/video scene understanding, image/video security, human–computer interaction, visual sensor hardware, document image analysis, compressed image/video analytics, other computer vision applications, machine learning, deep learning, computational intelligence, optimization techniques, explainable AI, fairness, accountability, privacy, transparency and ethics, brain–computer interaction, hand sensor-based intelligence, robotic intelligence, lightweight intelligent systems, limited data intelligence, hardware realization of intelligent systems, multimedia intelligent systems, supervised intelligence, unsupervised intelligence, self-supervised intelligence, transfer learning, multi-task learning, fooling intelligent systems, robustness of intelligent systems, and other topics of machine intelligence.

CVMI 2022 received 187 submissions from all over the world from countries including India, the United States of America, Japan, Germany, China, France, Sweden, and Bangladesh. All submissions were rigorously peer reviewed and selected 60 high-quality papers were presented at CVMI 2022. The program committee selected all the presented papers to be included in this volume of Computer Vision and Machine Intelligence (CVMI) proceedings published by Springer Nature. The conference advisory committee, technical program committee, and researchers of the Indian Institute of Information Technology, Allahabad, Prayagraj, India, made a significant effort to guarantee the success of the conference. We would like to thank all the members of the program committee and the referees for their commitment to help in the review process and for spreading our call for papers. We would like to thank Mr. Aninda Bose from Springer Nature for his helpful advice, guidance, and continuous support in publishing the proceedings. Moreover, we would like to thank all the authors for supporting CVMI 2022; without all their high-quality submissions, the conference would not have been possible.

Allahabad, India
August 2022

Shiv Ram Dubey

Contents

Editors and Contributors

About the Editors

Prof. Massimo Tistarelli received the Ph.D. in Computer Science and Robotics in 1991 from the University of Genoa. He is Full Professor in Computer Science and Director of the Computer Vision Laboratory at the University of Sassari, Italy. Since 1986, he has been involved as Project Coordinator and Task Manager in several projects on computer vision and biometrics funded by the European Community. Professor Tistarelli is Founding Member of the Biosecure Foundation, which includes all major European research centers working in biometrics. His main research interests cover biological and artificial vision, pattern recognition, biometrics, visual sensors, robotic navigation, and visuo-motor coordination. He is one of the world-recognized leading researchers in the area of biometrics, especially in the field of face recognition and multimodal fusion. He is Co-author of more than 150 scientific papers in peer-reviewed books, conferences, and international journals. He is Principal Editor for the Springer books *Handbook of Remote Biometrics* and *Handbook of Biometrics for Forensic Science*. Professor Massimo organized and chaired several world-recognized several scientific events and conferences in the area of Computer Vision and Biometrics, and he has been Associate Editor for several scientific journals including *IEEE Transactions on PAMI*, *IET Biometrics*, *Image and Vision Computing and Pattern Recognition Letters*. Since 2003, he is Founding Director for the International Summer School on Biometrics (now at the 17th edition). He is Fellow Member of the IAPR, Senior Member of IEEE, and Vice President of the IEEE Biometrics Council.

Dr. Shiv Ram Dubey has been with IIIT Allahabad since July 2021, where he is currently Assistant Professor of Information Technology. He was with the IIIT Sri City as Assistant Professor from December 2016 to July 2021 and Research Scientist from June 2016 to December 2016. He received the Ph.D. degree in IIIT Allahabad in 2016. Before that, from August 2012 to February 2013, he was Project Officer in the CSE at IIT Madras. Currently, Dr. Dubey is executing the research project funded by Global Innovation and Technology Alliance (GITA)—India–Taiwan project. He has also executed the projects funded by DRDO Young Scientist Lab in Artificial Intelligence (DYSL-AI) and Science and Engineering Research Board (SERB). He was Recipient of several awards including Best Ph.D. Award in Ph.D. Symposium, IEEE-CICT2017 at IIITM Gwalior, and NVIDIA GPU Grant Award Twice from NVIDIA. He received the Outstanding Certificate of Reviewing Award from Information Fusion, Elsevier, in 2018. He was also involved in the organization of Springer's CVIP conference in 2020 and 2021. He is serving as Associate Editor in *SN Computer Science Journal*. He is also involved in reviewing papers in top-notch journals, such as *IEEE TNNLS*, *IEEE TIP*, *IEEE SPL*, *IEEE TAI*, *IEEE TMM*, *IEEE TGRS*, *MTAP*, and *SiVP* and conferences such as WACV, ACMMM, ICME, ICVGIP, and CVIP. His research interest includes computer vision, deep learning, convolutional neural networks, generative adversarial networks, stochastic gradient descent optimizers, image retrieval, image-to-image transformation, etc.

Dr. Satish Kumar Singh is with the Indian Institute of Information Technology Allahabad India, as Associate Professor at the Department of Information Technology from 2013 and heading the Computer Vision and Biometrics Lab (CVBL). Before joining the IIIT Allahabad, he served the Department of Electronics and Communication Engineering, Jaypee University of Engineering and Technology Guna, India, from 2005 to 2012. His areas of interest include image processing, computer vision, biometrics, deep learning, and pattern recognition. He is Senior Member of IEEE. Presently, Dr. Singh is Section Chair IEEE Uttar Pradesh Section. Dr. Singh has also been involved as Editor in several journals and conferences, including *Springer's Neural Computing and Applications*, *Springer Nature Computer Science*, *IET-Image Processing*, *Springer's CVIP 2020*, CICT 2018, UPCON 2015, etc. Dr. Singh is also Technical Committee Affiliate of IEEE SPS IVMSP and MMSP and presently Chairperson, IEEE Signal Processing Society Chapter of Uttar Pradesh Section.

Prof. Xiaoyi Jiang received the Bachelor's degree from Peking University, Beijing, China, and the Ph.D. and Venia Docendi (Habilitation) degrees from the University of Bern, Bern Switzerland, all in Computer Science. He was Associate Professor with the Technical University of Berlin, Berlin Germany. Since 2002, he has been Full Professor with the University of Münster, Münster Germany, where he is currently Dean of the Faculty of Mathematics and Computer Science. His current research interests include biomedical imaging, 3D image analysis, and structural pattern recognition. Dr. Jiang is Editor-in-Chief of the *International Journal of Pattern Recognition and Artificial Intelligence*. He also serves on the Advisory Board and the Editorial Board of several journals, including *IEEE Transactions on Medical Imaging* and *International Journal of Neural Systems*. He is Senior Member of IEEE and Fellow of IAPR.

List of Contributors

Shajahan Aboobacker National Institute of Technology Karnataka, Surathkal, Karnataka, India

Wincy Abraham Department of Computer Science, Mangalore University, Konaje, India

Anuja Kumar Acharya School of Computer Engineering, KIIT Deemed to be University, Bhubaneswar, Odisha, India

Shyam Prasad Adhikari Applied Research Swiggy, Bangalore, India

N. G. Aditya Department of Computer Science, PES University, Bengaluru, Karnataka, India

Deepak Agarwal JPNATC, All India Institute of Medical Sciences, New Delhi, India

Rashi Agarwal Harcourt Butler Technical University, Kanpur, India

Supriya Agrahari Motilal Nehru National Institute of Technology Allahabad, Prayagraj, India

Manav Kamlesh Agrawal Department of IT, IIIT Allahabad, Prayagraj, U.P, India

R. K. Agrawal School of Computer & Systems Sciences, Jawaharlal Nehru University, Delhi, India

Shyamal Krishna Agrawal International Institute of Information Technology, Chhattisgarh, India

Sridhatta Jayaram Aithal IIIT Sricity, Sathyavedu, India

Ashly Ajith Department of Computer Science and Engineering, Amrita Vishwa Vidyapeetham, Amritapuri, Kerala, India

Asfak Ali Electronics and Telecommunication Engineering, Jadavpur, Kolkata, West Bengal, India

M. Anand Kumar Department of Information Technology, National Institute of Technology Karnataka, Surathkal, Mangalore, India

Wazib Ansar A. K. Choudhury School of IT, University of Calcutta, Kolkata, India

P. Arun Raj Kumar Department of Computer Science and Engineering, National Institute of Technology Calicut, Kozhikode, India

Saumya Asati Computer Science and Engineering, NIT, Hamirpur, Himachal Pradesh, India

Shubham Bagwari Thapar Institute of Engineering and Technology, Patiala, Punjab, India

Bunil Kumar Balabantaray Department of Computer Science, National Institute of Technology Meghalaya, Shillong, India

Kenneth Barner University of Delaware, Newark De, USA

Samet Bayram University of Delaware, Newark De, USA

Shrutilipi Bhattacharjee Department of Information Technology, National Institute of Technology Karnataka, Surathkal, Mangalore, India

Milan Bhuva Department of IT, IIIT Allahabad, Prayagraj, U.P, India

M. K. Bhuyan Department of Electronics and Electrical Engineering, Indian Institute of Technology Guwahati, Guwahati, India;
Department of Computer Science, Chubu University, Kasugai, Japan

Mrinal Bisoi Department of Computer Science, National Institute of Technology Meghalaya, Shillong, India

Michael Breuß Brandenburg University of Technology, Cottbus, Germany

Amlan Chakrabarti A. K. Choudhury School of IT, University of Calcutta, Kolkata, India

Basabi Chakraborty Iwate Prefectural University, Takizawa, Japan

Chandni Department of CSE, I.K. Gujral Punjab Technical University, Kapurthala, India

Sheli Sinha Chaudhuri Electronics and Telecommunication Engineering, Jadavpur, Kolkata, West Bengal, India

Prakash Chandra Chhipa Machine Learning Group, EISLAB, Luleå Tekniska Universitet, Luleå, Sweden

Kanika Choudhary Thapar Institute of Engineering and Technology, Patiala, Punjab, India

S. Sumam David National Institute of Technology Karnataka, Surathkal, Karnataka, India

Kanjar De Machine Learning Group, EISLAB, Luleå Tekniska Universitet, Luleå, Sweden

Ram Dewangan Thapar Institute of Technology, Patiyala, India

Gulshan Dhasmana Graphic Era Deemed to be University, Dehradun, India

P. B. Dhruval Department of Computer Science, PES University, Bengaluru, Karnataka, India

Enmao Diao Duke University, Durham, NC, USA

Jie Ding University of Minnesota-Twin Cities, Minneapolis, MN, USA

Shiv Ram Dubey Computer Vision and Biometrics Laboratory, Indian Institute of Information Technology, Allahabad, Prayagraj, India

Pavan Sai Santhosh Ejurothu Indian Institute of Information Technology, Sri City, India

Daniel L. Elliott 2AI: Applied AI Research Lab—Computer Science, University of South Dakota, Vermillion, SD, USA

Mahmoud Eltaher Brandenburg University of Technology, Cottbus, Germany; Al-Azhar University, Cairo, Egypt

K. V. Gangadharan Department of Mechanical Engineering, National Institute of Technology Karnataka, Surathkal, Mangalore, India

Neerja Mittal Garg CSIR-Central Scientific Instruments Organisation, Chandigarh, India;
Academy of Scientific and Innovative Research (AcSIR), Ghaziabad, India

Anmol Gautam National Institute of Technology Meghalaya, Shillong, India

Mrinmoy Ghorai Department of Electronics and Communication Engineering, Indian Institute of Information Technology, Chittoor, Andhra Pradesh, India

G. Gopakumar Department of Computer Science and Engineering, Amrita Vishwa Vidyapeetham, Amritapuri, India

Bibek Goswami Department of Electronics and Electrical Engineering, Indian Institute of Technology Guwahati, Guwahati, India

Saptarsi Goswami Department of Computer Science, Bangabasi Morning College, Kolkata, India

Mahendra Kumar Gourisaria School of Computer Engineering, KIIT Deemed to be University, Bhubaneswar, Odisha, India

Srirupa Guha Indian Institute of Science Bangalore, Bangalore, India

Rajeev Gupta Electronics and Communication Engineering Department, Motilal Nehru National Institute of Technology Allahabad, Prayagraj, India

Shresth Gupta International Institute of Information Technology, Chhattisgarh, India

Shubham Gupta PES University, Bangaluru, KA, India

Vibha Gupta Machine Learning Group, EISLAB, Luleå Tekniska Universitet, Luleå, Sweden

Vishal Gupta Centre for Development of Telematics, Telecom Technology Centre of Government of India, New Delhi, India

Habtu Hailu Mangalore University, Mangalagangothri, Karnataka, India

Pichhika Hari Chandana Indian Institute of Information Technology, Sri City, Chittoor, Sri City, Chittoor, AP, India

S. Hariharan University of Madras, Madras, India

Sohini Hazra GahanAI, Bengaluru, India

R. Hebbar Regional Remote Sensing Centre–South, ISRO, Bengaluru, KA, India

Yuji Iwahori Department of Electronics and Electrical Engineering, Indian Institute of Technology Guwahati, Guwahati, India;
Department of Computer Science, Chubu University, Kasugai, Japan

S. R. S. Iyengar Indian Institute of Technology Ropar, Ropar, Punjab, India

Krishna Chaitanya Jabu Department of Electronics and Communication Engineering, Indian Institute of Information Technology, Chittoor, Andhra Pradesh, India

Arti Jain Computer Science and Engineering, Jaypee Institute of Information Technology, Noida, Uttar Pradesh, India

Anamika Jain Centre for Advanced Studies, AKTU, Lucknow, India;
MIT-WPU, Pune, India

Upma Jain Graphic Era Deemed to be University, Dehradun, India

Mohammed Javed Department of IT, IIIT Allahabad, Prayagraj, U.P, India

Rahul Kala Center of Intelligent Robotics Indian Institute of Information Technology Allahabad Prayagraj, UP, India

Ranjith Kalingeri Center of Intelligent Robotics Indian Institute of Information Technology Allahabad Prayagraj, UP, India

Akash Kalluvilayil Venugopalan Department of Computer Science and Engineering, Amrita Vishwa Vidyapeetham, Amritapuri, India

Vipashi Kansal Graphic Era Deemed to be University, Dehradun, India

Ravi Ranjan Prasad Karn Department of Computer Science and Engineering, Indian Institute of Information Technology, Sri City, Chittoor, India

Srinivas Katharguppe Department of Computer Science, PES University, Bengaluru, Karnataka, India

Jyoti R. Kini Department of Pathology, Kasturba Medical College, Mangalore, India;
Manipal Academy of Higher Education, Manipal, Karnataka, India

Kisalaya Kishore Department of IT, IIIT Allahabad, Prayagraj, U.P, India

Ashwini Kodipalli Department of Artificial Intelligence and Data Science, Global Academy of Technology, Bangalore, India

Arnav Kotiyal Graphic Era Deemed to be University, Dehradun, India

György Kovács Machine Learning Group, EISLAB, Luleå Tekniska Universitet, Luleå, Sweden

Jadhav Sai Krishna Department of Computer Science and Engineering, Indian Institute of Information Technology Nagpur, Nagpur, Maharashtra, India

Manthan Milind Kulkarni Department of Computer Science and Engineering, Indian Institute of Information Technology Nagpur, Nagpur, Maharashtra, India

Akhilesh Kumar Defence Institute of Psychological Research (DIPR), DRDO, Delhi, India

Aman Kumar International Institute of Information Technology, Chhattisgarh, India

Basant Kumar Electronics and Communication Engineering Department, Motilal Nehru National Institute of Technology Allahabad, Prayagraj, India

Dhirendra Kumar Department of Applied Mathematics, Delhi Technological University, Delhi, India

Kamal Kumar National Institute of Technology, Srinagar, Uttarakhand, India

Mohit Kumar National Institute of Technology Hamirpur, Hamirpur, H.P., India

Mukesh Kumar Computer Science and Engineering Department, University Institute of Engineering and Technology, Panjab University, Chandigarh, India

Puneet Kumar School of Computer & Systems Sciences, Jawaharlal Nehru University, Delhi, India

Sanjay Kumar South Asian University, Chanakyapuri, New Delhi, Delhi, India

Saurav Kumar Department of Computer Science and Engineering, Indian Institute of Information Technology, Guwahati, India

Vikash Kumar Electronics and Telecommunication Engineering, Jadavpur, Kolkata, West Bengal, India

Kanya Kurauchi University of Fukui, Fukui, Japan

Alok Kumar Singh Kushwaha Department of CSE, Guru Ghasidas Vishwavidyalaya, Bilaspur, India

Vandana Kushwaha Center of Intelligent Robotics Indian Institute of Information Technology Allahabad Prayagraj, UP, India

Ketan Lambat Indian Institute of Information Technology, Sri City, Chittoor, India

Sidharth Lanka Department of Information Technology, National Institute of Technology Karnataka, Surathkal, Mangalore, India

Foteini Liwicki Machine Learning Group, EISLAB, Luleå Tekniska Universitet, Luleå, Sweden

Pallab Maji GahanAI, Bengaluru, India

Richa Makhijani Department of Computer Science and Engineering, Indian Institute of Information Technology Nagpur, Nagpur, Maharashtra, India

Priyank Makwana Department of Information Technology, IIIT Allahabad, Prayagraj, India

Muntasir Mamun 2AI: Applied AI Research Lab—Computer Science, University of South Dakota, Vermillion, SD, USA

Subhojit Mandal Indian Institute of Information Technology, Sri City, India

Philipp Mascha University of Augsburg, Augsburg, Germany

Tojo Mathew Department of Computer Science and Engineering, National Institute of Technology Karnataka, Surathkal, India;
Department of Computer Science and Engineering, The National Institute of Engineering, Mysuru, India

Sumit Mighlani Thapar Institute of Engineering and Technology, Patiala, Punjab, India

Hemant Misra Applied Research Swiggy, Bangalore, India

Kaushik Mitra Indian Institute of Technology, Madras, India

Hamam Mokayed Machine Learning Group, EISLAB, Luleå Tekniska Universitet, Luleå, Sweden

Soumen Moulik Department of Computer Science, National Institute of Technology Meghalaya, Shillong, India

Sonu Kumar Mourya Department of Computer Science and Engineering, Indian Institute of Information Technology Nagpur, Nagpur, Maharashtra, India

Susanta Mukhopadhyay Indian Institute of Technology (Indian School of Mines) Dhanbad, Jharkhand, India

Lakhindar Murmu International Institute of Information Technology, Chhattisgarh, India

Sandra V. S. Nair Department of Computer Science and Engineering, National Institute of Technology Calicut, Kozhikode, India

G. C. Nandi Center of Intelligent Robotics Indian Institute of Information Technology Allahabad Prayagraj, UP, India

S. Niyas Department of Computer Science and Engineering, National Institute of Technology Karnataka, Surathkal, India

Reena Oswal Department of Computer Science and Engineering, National Institute of Technology Karnataka, Surathkal, India

Akira Ouchi Department of Gastroenterological Surgery, Aichi Cancer Center Hospital, Nagoya, Japan

Bharathi Pilar Department of Computer Science, University College, Mangalore, India

Y. Raja Vara Prasad Department of Electronics and Communication Engineering, Indian Institute of Information Technology, Chittoor, Andhra Pradesh, India

Shraddha Priya Department of Computer Science and Engineering, National Institute of Technology Karnataka, Surathkal, India

R. Priyadarshini Department of Information Technology, National Institute of Technology Karnataka, Surathkal, Mangalore, India

U. Pruthviraj Department of Water and Ocean Engineering, National Institute of Technology Karnataka, Surathkal, Mangalore, India

Suresh Raikwar Thapar Institute of Engineering and Technology, Patiala, Punjab, India

Jeny Rajan Department of Computer Science and Engineering, National Institute of Technology Karnataka, Surathkal, India

Bulla Rajesh Department of IT, IIIT Allahabad, Prayagraj, U.P, India; Department of CSE, Vignan University, Guntur, A.P, India

Mayank Rajpurohit National Institute of Technology Karnataka, Surathkal, Karnataka, India

Sumit Rakesh Machine Learning Group, EISLAB, Luleå Tekniska Universitet, Luleå, Sweden

Prashant Singh Rana Thapar Institute of Engineering and Technology, Patiala, Punjab, India

Seema Rani Computer Science and Engineering Department, University Institute of Engineering and Technology, Panjab University, Chandigarh, India

Nicholas Rasmussen 2AI: Applied AI Research Lab—Computer Science, University of South Dakota, Vermillion, SD, USA

Reshma Rastogi South Asian University, Chanakyapuri, New Delhi, Delhi, India

Bytasandram Yaswanth Reddy Department of Computer Science and Engineering, Indian Institute of Information Technology, Sri City, Chittoor, India

Koppula Bhanu Prakash Reddy Department of Computer Science and Engineering, Indian Institute of Information Technology Nagpur, Nagpur, Maharashtra, India

Manali Roy Indian Institute of Technology (Indian School of Mines) Dhanbad, Jharkhand, India

Monika Sachdeva Department of CSE, I.K. Gujral Punjab Technical University, Kapurthala, India

Sumanth Sadu Computer Vision Group, Department of Computer Science and Engineering, Indian Institute of Information Technology, Sri City, Andhra Pradesh, India

Poorti Sagar Centre for Advanced Studies, AKTU, Lucknow, India

Shubham Sagar Department of Computer Science and Engineering, Indian Institute of Information Technology Nagpur, Nagpur, Maharashtra, India

Vishal Kumar Sahoo School of Computer Engineering, KIIT Deemed to be University, Bhubaneswar, Odisha, India

Rajkumar Saini Machine Learning Group, EISLAB, Luleå Tekniska Universitet, Luleå, Sweden;
Department of CSE, IIT Roorkee, Roorkee, India

Rishabh Samra Indian Institute of Technology, Madras, India

Rakesh Kumar Sanodiya Department of Computer Science and Engineering, Indian Institute of Information Technology, Sri City, Chittoor, India

Gaurav Santhalia Indian Institute of Information Technology, Allahabad, India

KC Santosh 2AI: Applied AI Research Lab—Computer Science, University of South Dakota, Vermillion, SD, USA

N. Shankaranarayan Department of Information Technology, National Institute of Technology Karnataka, Mangalore, India

Apurva Sharma CSIR-Central Scientific Instruments Organisation, Chandigarh, India;
Academy of Scientific and Innovative Research (AcSIR), Ghaziabad, India

Prasan Shedligeri Indian Institute of Technology, Madras, India

B. H. Shekar Department of Computer Science, Mangalore University, Konaje, India

Yasuhiro Shimizu Department of Gastroenterological Surgery, Aichi Cancer Center Hospital, Nagoya, Japan

S. S. Shylaja Department of Computer Science, PES University, Bengaluru, Karnataka, India

Tarandeep Singh CSIR-Central Scientific Instruments Organisation, Chandigarh, India;
Academy of Scientific and Innovative Research (AcSIR), Ghaziabad, India

Anil Kumar Singh Motilal Nehru National Institute of Technology Allahabad, Prayagraj, India

Arun Singh Bhadwal National Institute of Technology, Srinagar, Uttarakhand, India

Shrish Kumar Singhal Department of Electronics and Electrical Engineering, Indian Institute of Technology Guwahati, Guwahati, India

Anurag Singh International Institute of Information Technology, Chhattisgarh, India

Dinesh Singh Computer Science & Engineering, DCRUST, Sonepat, India

Maninder Singh Electronics and Communication Engineering Department, Motilal Nehru National Institute of Technology Allahabad, Prayagraj, India

Pragya Singh Indian Institute of Information Technology, Allahabad, India

Vinayak Singh School of Computer Engineering, KIIT Deemed to be University, Bhubaneswar, Odisha, India

Satish Kumar Singh Computer Vision and Biometrics Laboratory, Indian Institute of Information Technology, Allahabad, Prayagraj, India

S. Sowmya Kamath Department of Information Technology, National Institute of Technology Karnataka, Surathkal, Mangalore, India

S. R. Sreeja Department of Computer Science and Engineering, Indian Institute of Information Technology, Sri City, Andhra Pradesh, India

Saraswathy Sreeram Kasturba Medical College Mangalore, Manipal Academy of Higher Education, Manipal, Karnataka, India

Priyambada Subudhi Indian Institute of Information Technology, Sri City, Chittoor, Sri City, Chittoor, AP, India

B. Sudhakara Department of Information Technology, National Institute of Technology Karnataka, Surathkal, Mangalore, India

Pooja K. Suresh Kasturba Medical College Mangalore, Manipal Academy of Higher Education, Manipal, Karnataka, India

Kanji Tanaka University of Fukui, Fukui, Japan

Vahid Tarokh University of Minnesota-Twin Cities, Minneapolis, MN, USA

Kartikey Tewari National Institute of Technology Hamirpur, Hamirpur, H.P., India

Mainak Thakur Indian Institute of Information Technology, Sri City, India

Rohit Tripathi Department of Computer Science and Engineering, Indian Institute of Information Technology, Guwahati, India

D. Uma PES University, Bangaluru, KA, India

Richa Upadhyay Machine Learning Group, EISLAB, Luleå Tekniska Universitet, Luleå, Sweden

Raja Vara Prasad Yerra Indian Institute of Information Technology, Sri City, Chittoor, Sri City, Chittoor, AP, India

Rakshita Varadarajan Department of Information Technology, National Institute of Technology Karnataka, Surathkal, Mangalore, India

Gaurang Velingkar Department of Information Technology, National Institute of Technology Karnataka, Surathkal, Mangalore, India

Rishabh Verma GahanAI, Bengaluru, India

Deepu Vijayasenan National Institute of Technology Karnataka, Surathkal, Karnataka, India

Guoqiang Wang The Second Research Institute of Civil Aviation Administration of China, Chengdu, Sichuan, China

Honggang Wu The Second Research Institute of Civil Aviation Administration of China, Chengdu, Sichuan, China

Arun Kumar Yadav National Institute of Technology Hamirpur, Hamirpur, H.P., India

Divakar Yadav National Institute of Technology Hamirpur, Hamirpur, H.P., India

Ryogo Yamamoto University of Fukui, Fukui, Japan

Ruijing Yang University of Electronic Science and Technology of China, Chengdu, Sichuan, China;
Yangtze Delta Region Institute (Quzhou), University of Electronic Science and Technology of China, Quzhou, Zhejiang, China

Mitsuki Yoshida University of Fukui, Fukui, Japan

Xiang Zhang University of Electronic Science and Technology of China, Chengdu, Sichuan, China;
Yangtze Delta Region Institute (Quzhou), University of Electronic Science and Technology of China, Quzhou, Zhejiang, China

Efficient Voluntary Contact-Tracing System and Network for COVID-19 Patients Using Sound Waves and Predictive Analysis Using K-Means

Gaurav Santhalia and Pragya Singh

Abstract Patient tracking and contact mapping have been a challenge for government and hi-tech companies around the world in terms of precision after WHO declared COVID-19 as a pandemic on March 12, 2020. Our proposed method discloses a voluntary contact-tracing network using smartphone application by means of GPS using a microphone along with short-range communication like Bluetooth. The amalgamation of Bluetooth with a microphone (soundwaves) helps in identifying the distance of two people more accurately as Bluetooth alone does not calculate correct distances. The smartphone application transmits the exact location data captured by GPS and user's proximity distance data to an intelligent cloud platform using the Internet if the patient is diagnosed with COVID-19 after user content via smartphone. The smartphone captures the signals of nearby phones when in proximity and stores the connections between them in a local database. The data sync to intelligent cloud platform from each smartphone will have the unique mapping of each COVID-19 patient with exact contact tracing with location data once the patient is diagnosed with COVID-19. Hence, if a person tests positive for the COVID-19, the person can notify using an application that they have been infected, and the cloud system notifies other people whose phones came in close contact in the preceding days. The precision of contact mapping using sound wave technology along with Bluetooth will help the people as well as the local administration to deal with pandemics in an effective way. Moreover, the intelligent cloud platform will be enabled with an AI-supported algorithm that generates predictive insights from time to time on smartphones.

Keywords COVID-19 · Bluetooth · Sound waves · Microphone · GPS · K-means · Cluster · Cloud

---

G. Santhalia (✉) · P. Singh
Indian Institute of Information Technology, Allahabad, India
e-mail: gauravsanthalia@gmail.com

P. Singh
e-mail: pragyabhardwaj@iiita.ac.in

M. Tistarelli et al. (eds.), *Computer Vision and Machine Intelligence*,
Lecture Notes in Networks and Systems 586,
https://doi.org/10.1007/978-981-19-7867-8_1

1 Introduction

As the COVID-19 pandemic rages, technologists around the world have been rushing to build apps or services for contact tracing [1] and notifying all those who come in contact with a COVID-19-infected person. Small groups of coders are producing some services locally, while others are vast, global operations. The motivation is to allow society's greater freedom of movement by allowing quick tracing and breakdown of the infected chains. The infected person's past contact should be alerted quickly by the cloud system so that contacts can quarantine themselves more quickly. A robust system is required to break chains of infection by protecting the contacts of our contacts. Hence, a precision-oriented "COVID app" with predictive analysis using machine learning is required in the present situation.

Human activities change the signal propagation path and generate a change in said signal finally recognizing an act of human beings. In general, the signal received by a receiver is categorized into four categories such as vision [2], light [3], radio frequency (RF) [4], and acoustic/sound signal [5]. Table 1 which gives a brief overview of the signal with the recognition method. Hence, the selection of the model should be based on extrinsic conditions such as environment, climate, and weather.

For precision contact tracing to calculate the exact distance between the two contacts, the audio signal used along with Bluetooth (RF) seems to be a perfect model based on above Table 1. The microphone captures the nearby audio voices and will exchange the signal via Bluetooth. The audio extraction algorithm in the smartphone will calculate the distance discussed in detail in the methodology section. Since Bluetooth has its own limitation due to obstruction, amalgamation of these two signals creates the perfect model for contact tracing. Moreover, the requirement of storing the personal data of users can also be neglected using the above proposed system design. Hence, other signals like image and lights cannot suit for contact tracking as image and light require manual input, whereas RF and audio can be automatically captured and transmitted.

Table 1 Signal versus recognition method

Method	Signal	Recognition
Vision	Imge	Human authentication or activity recognition
Light	Light	Activity recognition
RF	RFID	Sign detection or activity recognition or human authentication
	Radar	
	OFDM	
Audio/Sound	Audible sound	Lip reading, contact tracking, activity recognition or human authentication
	Ultrasonic	

2 Literature Review

As of now, most of the developed countries, such as Europe and the USA, have developed applications, which trace contact using Bluetooth technology. In these countries, GPS cannot be used, as privacy of data is a legal issue [6]. Moreover, if GPS is enabled all the time, the battery is drained very fast. So, most of the smartphone applications that are developed in developed countries do not support GPS tracking. Google and Apple have recently been developing applications to be realized in the month of May, which does not use GPS for location tracing due to the above-listed reasons [7]. Moreover, these applications, which are currently available in the market as of date, do not support any predictive analysis based on AI algorithms. However, the United Kingdom National Health Service (NHS) in England has rolled out the contract to develop an AI-based system on collected data [8]. These country's systems are designed in such a way to contact profiles (frequent contacts patterns) or location profiles (location tracking) cannot be identified intentionally or unintentionally. No personal user data should be captured or stored such as mobile numbers and social media accounts as a matter of principle.

Google or Alphabet CEO, Sunder Pichai on April 10, 2020, tweeted "To help public health officials slow the spread of #COVID19, Google & @Apple are working on a contact tracing approach designed with strong controls and protections for user privacy. @tim_cook and I are committed to working together on these efforts". Hence, these countries will not allow the use of GPS for location tracking.

However, Asian counties like China and India where data privacy law is not so strong use GPS along with Bluetooth in the COVID app without any predictive analysis tool at the backend [9–11]. It is logically correct to use GPS along with Bluetooth in these counties due to the size of the population and degraded health infrastructure in remote areas. Counties due to the size of the population and degraded health infrastructure in remote areas. China used small AI support, later on, to support police administration to have insights of captured data [12].

In India, the lockdown was declared prematurely [13] and the government of India knew the gaps of infectivity of COVID virus from other countries; hence, they developed the Aarogya Setu application, which had 10 million users within 5 days of launch [14]. The application developed by the government of India is enabled with both GPS and Bluetooth for contact tracking. The application is been upgraded day-by-day based on identified gaps and government guidelines. The Aarogya Setu application does not have any predictive analysis based on the AI platform [15, 16] but has basic insights into data.

Based on the above background information available in the public domain, none of these countries (USA, Europe, India, and China) thought to fulfill the flaws of Bluetooth signal. Bluetooth signal may be obstructed due to any object or if kept in pocket or bag and other issues. Hence, Bluetooth fails to generate the exact distance between the two smartphones. None of these applications gives predictive analysis having a strong AI-based tool at the back end to support user and administration with new analysis on daily basis.

The proposed solution discloses a reliable solution to combine audio (soundwave) with Bluetooth so that the exact distance between two parties using a microphone can be identified, as the microphone has the capability of recognizing the sound signal, which even human cannot hear. Moreover, a smart AI K-means algorithm [17] is placed at the cloud platform to categorize the data and generate new predictive analysis, which is to be broadcasted to smartphone users at regular intervals.

Hence, the use of audio signals along with Bluetooth signals is the novelty aspect of the proposed solution. Further, K-means is the most efficient algorithm [18] to cluster similar data in such scenarios compared to other algorithms such as support vector machine (SVM) and genetic algorithms.

3 System Design or Methodology

When two proximity smartphones receive audio signals from microphones, the captured signal is processed to realize human activity. The proposed system in Fig. 1 extracts three types of information from the captured signal such as Doppler effect, phase, and time of flight (ToF) to determine the distance between two contacts. Once the distance is determined between the contacts, the said distance is stored in the local database of phones with anonymous identifiers exchanged by Bluetooth. The stored distance, key, and user identification are further synchronized to the cloud platform once the patient declares himself/herself as COVID.

Step 1: Audio Signal Acquisition Module: Sound signal generated by human activity is captured by nearby phone along with anonymous identifier exchanged by Bluetooth. The microphone receives the sound signal, and the processing unit

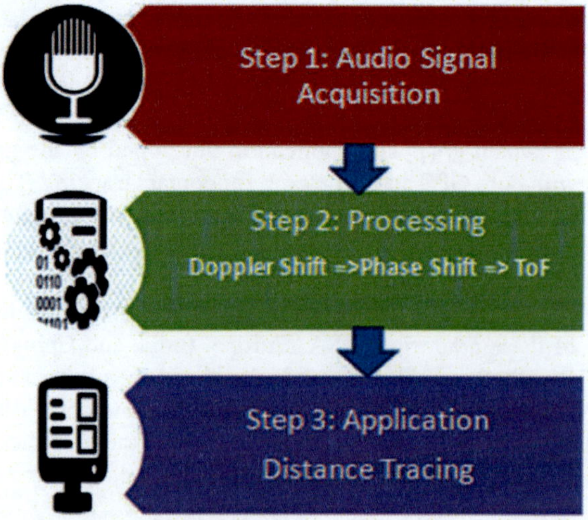

Fig. 1 High-level system design

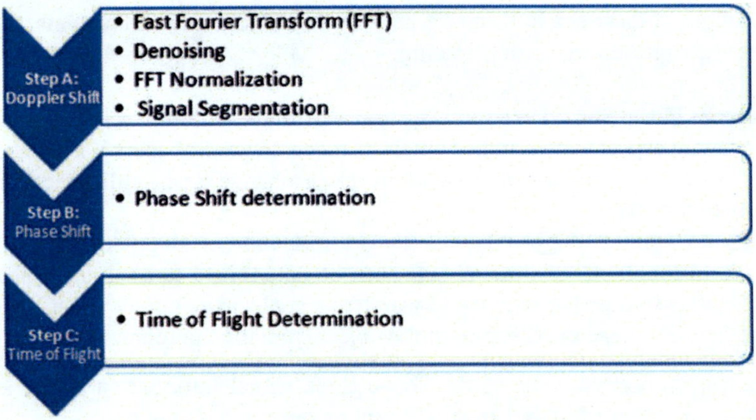

Fig. 2 Algorithm to determine Doppler shift

further processes the captured signal to extract the information to determine the contact distance.

Step 2: Processing: The raw sound signal captured by the microphone cannot be used to calculate the contact distance. Hence, processing unit consists of three major processing units to determine Doppler shift, phase shift, and time of flight (ToF) based on Fig. 2.

Step A: Doppler Shift: The Doppler effect (or the Doppler shift) is the change in frequency of a wave in relation to an observer who is moving relative to the wave source. According to this phenomenon, human activity can be determined using sound. The frequency shift is calculated using Doppler shift using the speed of audio sound in the air after the signal is preprocessed based on Fig. 2.

Before the Doppler shift is calculated, a preprocessing of the signal is performed so that signal is noise free. Some of these preprocessing steps are as follows:

- Fast Fourier Transform (FFT): The captured audio sound is a time-domain signal, and the signal is a transformed frequency-domain signal. Since sufficient information cannot be obtained from the time-domain signal, FFT transformation is essential.
- Denoising: Once the signal is transformed to frequency domain, the noise is removed from the signal using a bandpass filter. The outside noise occurs at very high frequency or low frequency which can be removed by using filters.
- FFT Normalization: To obtain noise-free signal still contains signal drift due to hardware. FFT normalization normalizes the amplitude of frequency to eliminate signal drift.
- Signal Segmentation: This method will divide the signal into a single motion element if the signal has the same motion for a continuous time period. For

example, if there is a human conversation sound for a continuous time, then this is to be captured for contact tracing.

Step B: Phase Shift: Determining phase shift information of a sound wave is an important aspect to determine contact distance. In general, there is a change in the signal phase for moving objects. Hence, phase change can be utilized to determine human activity.

Step C: Time of Flight (ToF): After determination of Doppler shift and Phase shift information, the time of flight can be determined to track the object. The difference in time between the transmitted signal and captured reflected signal is used to determine the distance of the object from the microphone.

Hence, the determining of the above three signal information is the key for determining contact distance in an efficient manner.

Step 3: Application: After processing the sound signal, contact information is extracted and stored the data in the local database of the phone. When a COVID patient declares himself/herself as COVID-19 infected on application, these stored data are synchronized to an intelligent cloud platform. The cloud platform used K-means algorithms and alerts the users with various analytics as discussed in the predictive analysis section.

4 Proposed Network Protocol

The proposed network protocol focuses to achieve two major objectives. The first objective is to accurately determine the distance between two smartphones by using audio along with Bluetooth. The second objective is to create an intelligent cloud platform, which gives predictive insights based on captured data.

4.1 Solution to Objective 1: Communication Protocol

Based on the above literature survey of different countries, we have proposed two types of contact-tracing solutions based on the above problems, which were identified in recent times. The proposed solutions are as follows:

1. **Contact tracing using GPS and Microphone along with Bluetooth**: This protocol suits countries like India and China, where personal data security law is not so strong and the population of these two countries is enormous. Hence, GPS can be used in this type of scenario along with a microphone and Bluetooth. As described in Fig. 3, the smartphone having a "COVID app" will enable GPS, microphone, and Bluetooth when the user registers for the first time. User identification can be the Email ID or cell phone number of the user while registering. Once the registration process is completed, the application starts storing

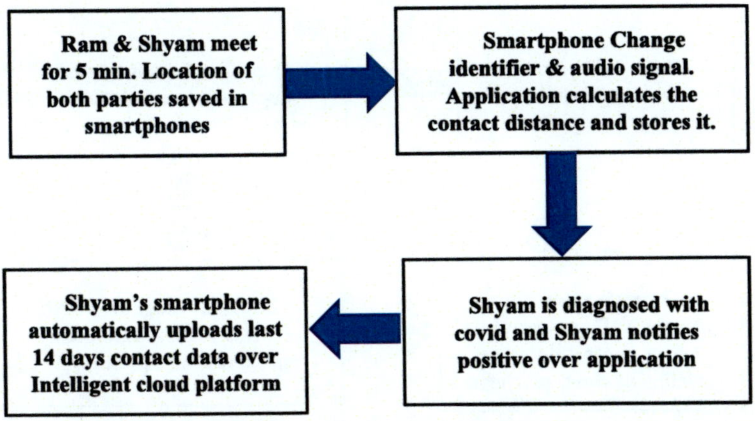

Fig. 3 Proposed protocol using GPS and microphone along with Bluetooth for contact tracing

location data with the help of GPS after every 10 min along with the timestamp. The application starts exchanging the anonymous identifier beacon using Bluetooth. Further, the microphone calculates the sound frequency of the phone and calculates the exact distance between two cell phones knowing that the speed of sound is 343 m per second. Hence, these captured data are initially stored in the local database of mobile phones and thereafter synchronized to the aggregator cloud platform if the patient is diagnosed with COVID-19 based on user consent. Today's smartphones have the capability to calculate the distance for certain ranges, and hence, such utility can be used for contact tracing. Hence, sound waves can detect proximity/close contacts even if the smartphones are in a bag, pocket, or any other obstacle without affecting distance measurement.

2. **Contact Tracing only using Microphone along with Bluetooth**: This protocol suits countries like Europe and the United States of America, where personal data security law is strong and the population of these two countries less in terms of India and China. Hence, location data using GPS is not preferred. In this scenario, the proposed protocol uses a microphone and Bluetooth for contact tracing. As described in Fig. 4, the smartphone having a "COVID app" will enable a microphone and Bluetooth when the user registers the first time. User identification can be the Email ID or cell phone number of the user while registering. Once the initial registration process is completed, the application starts exchanging the anonymous identifier beacon using Bluetooth. Further, the microphone calculates the sound frequency of the phone and calculates the exact distance between two cell phones knowing that the speed of sound is 343 m per second. Hence, these captured data are initially stored in the local database of mobile phones and thereafter synchronized to the aggregator cloud platform if the patient is diagnosed with COVID-19.

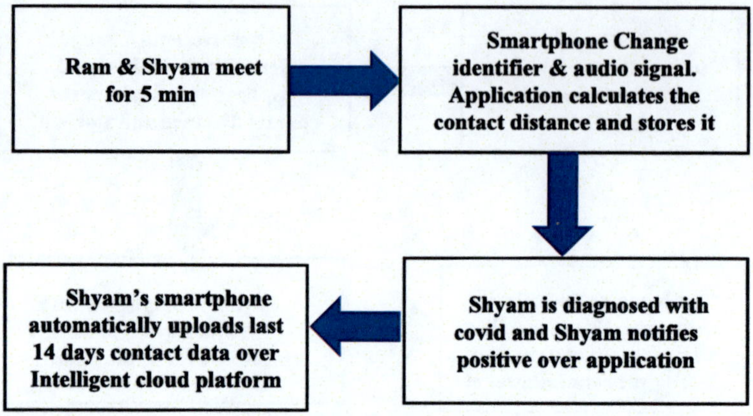

Fig. 4 Proposed protocol using microphone along with Bluetooth for contact tracing

4.2 *Solution to Objective 2: Predictive Analysis*

Once the patient is diagnosed with COVID-19, the application automatically synchronizes the last 14 days' location and contact data to an intelligent cloud server having an AI-based K-means algorithm. Typically, the location, contact, and distance data are voluntarily and periodically collected and stored privately for each individual and may be anonymously sent to the server without identifying the individual such that the data can be used to perform analysis. When a person marks himself as COVID-19 positive, the related temporal, spatial, contact, and personal data are sent to the intelligent cloud platform, where data analysis and machine learning can be performed at the back end to gain insights and notify the various parties. The personal data of the individual is crucial in recognizing the previous health conditions and symptoms associated with the infected person.

It is important to understand that such data analytics is essential to identify, track, and mitigate the spread of infection. This further helps in reducing pressure on the people who are engaged in combating this virus by providing technical assistance and awareness. Thus, it can aid in warning people about the risk of contracting the disease, but also lessen disruptions caused by panic and poor awareness. The infection prevention and treatment initiatives can be better aimed at the people who are more vulnerable to risk by providing real-time data on population behavior to the concerned authorities.

The obtained data is used for cluster analysis, for example using the K-means algorithm, to segregate people based on various features and identify alerting conditions. When data points form a collection due to certain similarities, it is termed a cluster. The K-means algorithm aims toward grouping related data points together and uncovering underlying patterns. To achieve this, K-means looks for a fixed number (k) of clusters in a data set. There are various flavors of the algorithm, which fulfill the gap and improve upon the basic K-means. This process can be applied by building

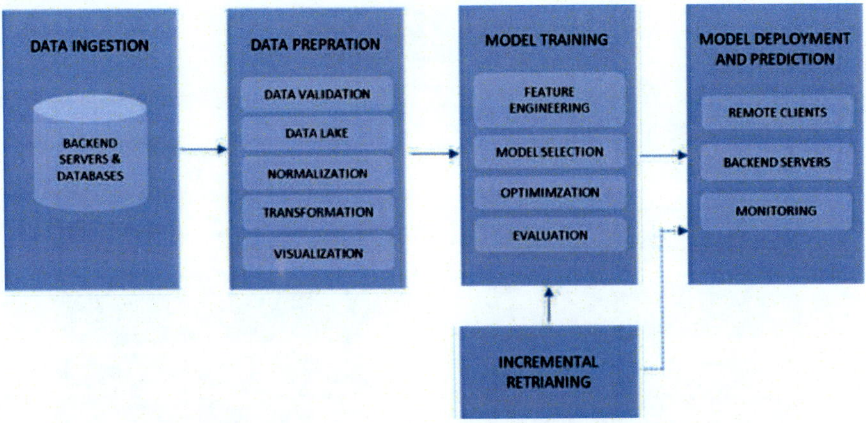

Fig. 5 Data modeling pipeline

a machine learning pipeline. First, we have to train the machine learning model by feeding it the input data, analyzing the data which includes cleaning and outlier detection, exploring and visualizing the data to recognize significant relationships, statistical measures, and certain insights, preprocessing the data, and then finally applying the clustering algorithm to classify the people based on their data. This training is consistently done to ensure the inclusion of each new incoming data point. Whenever a new data sample comes, it runs through the trained model and is assigned a cluster based on the characteristics of the model. Below Fig. 5 illustrates the modeling pipeline.

The data is ingested from the cloud servers and database storage. This data needs to be validated so that the model does not get trained on anomalous data. After getting the validated data, the data is normalized, transformed, outliers detected, and visualized to highlight important trends. To train the model, features are selected and data is clustered to determine the different clustering tendencies of the data and then optimize it continuously to reach a satisfactory level of threshold. As the new data keeps coming, it is important to continuously retrain the model to accommodate the new trends. The findings are then deployed to the various stakeholders in pre-defined formats.

The above model implementation can reveal various important insights. These insights are explained in detail below in Fig. 6.

- The data can be segregated based upon the location and distance data such that people coming within a few meters of each other can be identified. This helps officials ensure the practice of social distancing. It can track the spread of infection by identifying the moments of contact, i.e., instances when an infected person was within a few meters of another person. There can be cases of indirect spread of infection which can be tracked using the location and distance data of the people. Many people can come in indirect contact with a person who was in direct contact

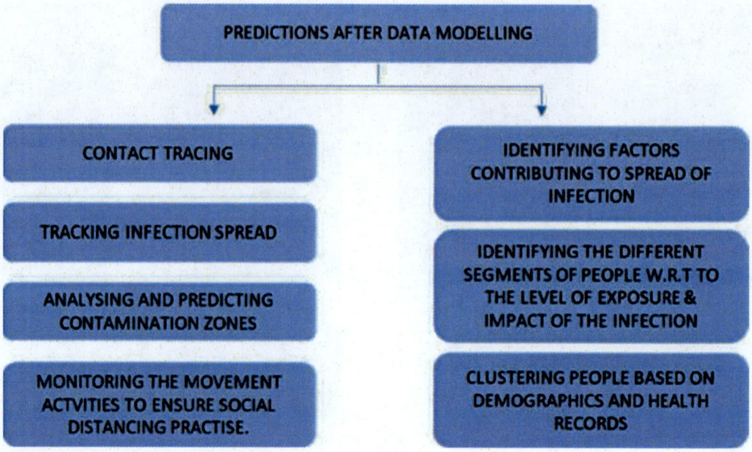

Fig. 6 Insights from data modeling

with an infected carrier. Thus, people can be notified immediately if a person who came in their contact within the past few days is now virus infected or was in contact with another virus-infected person. These interactions can be graphically represented to various stakeholders.

- A geographical area can be identified as being highly contaminated in real time. The locations can be segregated into safe, unsafe, or highly contaminated by analyzing the presence of infected people or the people who were in contact/vicinity of the infected people. Thus, people can be warned immediately about risk-prone areas and prevent them from getting infected.
- Users can also be clustered based on their previous health records, current symptoms, age, gender, location, and movement activities. The travel history of a person is a significant factor in deciding if they can be a potential virus carrier.
- The significant features identified while clustering the data can be further used to derive potential factors which contribute to spreading the virus. For example, age can be a factor, and old aged people are at higher risk of contracting the disease. Thus, the location and movement activity of elderly people can be strictly monitored such that they do not come in contact with many people.

5 Opportunity and Threat Analysis

Predictive modeling of the collected data provides various opportunities while fighting to control the spread of infection through COVID-19, and the management can be kept updated about the movement and locations of the citizens and if people are carefully practicing social distancing or not. People can be immediately notified if they are near an infected person or if they came in contact with an infected person.

The data is highly helpful in contact tracing. However, it is to be kept in mind that these things come with their own privacy concerns, but we have to weigh these against the benefits of being able to exit from the adversity. Further, these applications will be most helpful if everyone cooperates and install them in their smart devices, thus making predictions more accurate.

6 Limitations of Study

The most important limitation for proposed system design and protocol depends on the heterogeneity of the smartphones as the smartphones are of different brands and use various operating systems. So, while designing the algorithm, the heterogeneity is to be kept in mind so that every smartphone is able to process the data. The second limitation is the standardizing data sets for predictive modeling as different smartphones will have their own format of data set which may create inconsistency in a centralized database.

7 Conclusion and Future Scope

The proposed system design, protocol, and predictive modeling can be implemented based on country-specific requirements with high precision data. If distance/contact and location data are not accurately captured, the AI system may give wrong insights and the infection chain might now get collapsed. Further, the cloud platform is embedded with the powerful AI tool K-means, which is cluster similar type of data and presents insights into various forms. These insights can be very useful for both individuals as well for the government to further decide the next course of action.

The future work can lead to detection of automatic disease diagnostic, vital sign recognition, enhancing the security of smartphones for crowdsourcing applications and developing open-source tools for predictive modeling for epidemics like COVID-19.

References

1. HT Tech.: Republicans think Bill Gates will use Covid-19 vaccine to implant tracking chips, https://tech.hindustantimes.com/tech/news/republicans-have-a-conspiracy-theory-about-bill-gates-71590479295168.html
2. Rautaray, S.S., Agrawal, A.: Vision based hand gesture recognition for human computer interaction—a survey, pp. 1–54 (2015)
3. Li, T., Liu, Q., Zhou, X.: Practical human sensing in the light. In: 14th Proceeding International Conference Mobile System Application Services, pp. 71–84 (2016)

4. Ali, K., Liu, A.X., Wang, W., Shahzad, M.: Keystroke recognition using Wi-Fi signals. In: 21th Proceedings International Conference of Mobile Computation, pp. 90–102 (2015)
5. Cai, C., Zheng, R., Hu, M.: A survey on acoustic sensing (2019)
6. Hasan Chowdhury, Matthew Field, Margi Murphy. "NHS track and trace app: how will it work and when can you download it?" https://www.telegraph.co.uk/technology/2020/05/09/nhs-contact-tracing-app-what-how-coronavirus-download
7. Reuters.: Apple, Google bans location tracking in apps using their jointly-built contact tracing technology, https://www.thehindu.com/news/international/apple-google-ban-use-of-location-tracking-in-contact-tracing-apps/article31506222.ece
8. Andrea Downey.: NHS partners with tech giants to develop Covid-19 data platform, https://www.digitalhealth.net/2020/04/nhs-partners-with-tech-giants-to-develop-covid-19-data-platform
9. Sharma, S.: Governments tracking app tells if you were near COVID-19 patient
10. The Hindu.: Watch|How does the Aarogya setuApp works, https://www.thehindu.com/news/national/how-does-the-aarogya-setu-app-work/article31532073.ece
11. Ivan Mehta.: China's coronavirus detection app is reportedly sharing citizen data with police, https://thenextweb.com/china/2020/03/03/chinas-covid-19-app-reportedly-color-codes-people-and-shares-data-with-cops
12. Economic Times.: How to use Aarogya Setu app and find out if you have coronavirus symptoms, https://economictimes.indiatimes.com/tech/software/how-to-use-aarogya-setu-app-and-find-out-if-you-have-covid-19-symptoms/articleshow/75023152.cms
13. Natarajan, D.: Lockdown, Shutdown, Breakdown: India's COVID Policy Must Be Driven by Data, Not Fear, https://thewire.in/health/lockdown-shutdown-breakdown-indias-covid-policy-must-be-driven-by-data-not-fear
14. Choudhary, S.:How a mobile app helped China to contain the spread of covid-19, https://www.livemint.com/news/world/how-a-mobile-app-helped-china-to-contain-the-spread-of-covid-19-11585747212307.html
15. Bhatnagar, V., Poonia, R.C., Nagar, P., Kumar, S., Singh, V., Raja, L., Dass, P.: Descriptive analysis of covid-19 patients in the context of India. J. Interdisc. Math. (2020)
16. Singh, V., Poonia, R.C., Kumar, S., Dass, P., Agarwal, P., Bhatnagar, V., Raja, L.: Prediction of COVID-19 corona virus pandemic based on time series data using Support Vector Machine. J. Discrete Math. Sci. Crypt. (2020)
17. Li, Y., Wu, H.: A Clustering method based on K-means algorithm. Elsevier, 1104–1109 (2012)
18. Sujatha, S., Kumari, P.: Smart farming using K-means clustering and SVM classifier in image processing, IJSETR (2017)

Direct De Novo Molecule Generation Using Probabilistic Diverse Variational Autoencoder

Arun Singh Bhadwal and Kamal Kumar

Abstract In recent decades, there has been a significant increase in the application of deep learning in drug design. We present a basic method for generating diverse molecules using variational autoencoder. The variational autoencoder is based on recurrent neural network. A string representation SMILES of molecules is used for training the proposed model. A parameter that can be tweaked determines the level of diversity. Interpolation is also performed between two drug molecules in the latent space. The diverse variational autoencoder (dVAE) shows superior results as compare with other state-of-the-art methods. Thus, it can be used to generate controllable diverse molecules.

Keywords Variational autoencoder · Molecule generation · Diversity · SMILES · Interpolation

1 Introduction

Drug development is a time-consuming and expensive process. It take 10 to 15 years and cost up to $ 500 million. In recent years, deep learning (DL) has proven to be a highly effective strategy for drug design in recent years, allowing researchers to focus their efforts on a specific area of interest [1–3]. Inverse molecular design is one of the difficult problems that deep networks are attempting to solve [4]. Deep generative models have a revolutionary effect on a variety of content production, estimation, and prediction areas. Specially, the ability of neural networks to generate news headlines [5], synthesis music [6], or poetry [7] and produce photorealistic paintings [8], etc., has been demonstrated. It also have accelerated bio-science innovation by predicting bioactivity and synthesis [9], segmenting biological pictures [10], and designing new drug molecules [11]. The "simplified molecular-input line-entry system" (SMILES) [12] is the popularly recommended representation of

A. Singh Bhadwal (✉) · K. Kumar
National Institute of Technology, Srinagar, Uttarakhand, India
e-mail: arunsingh.phd19@nituk.ac.in

M. Tistarelli et al. (eds.), *Computer Vision and Machine Intelligence*,
Lecture Notes in Networks and Systems 586,
https://doi.org/10.1007/978-981-19-7867-8_2

chemicals [13]. Because SMILES strings are alphanumeric, they are compatible with NLP algorithms, such as RNNs, that do sequence modeling and generation. It has been shown that RNNs trained with SMILES strings can cover a very large chemical space [14].

1.1 Contributions

In this paper, a novel data-driven approach is proposed termed dVAE. Encoder and decoder are constructed using recurrent neural networks in the dVAE. A value of one (=1) is used as the diversity parameter (D=1) during the training process. In the generation mode, only the decoder portion of dVAE is active. The dVAE is capable of generating a proper balance between valid and diverse molecules. The level of diversity in the generated molecules is controlled by a tunable parameter.

1.2 Organization

The remaining sections of the paper are organised as follows: Related work is discussed in Sect. 2. Dataset and representation of molecules are discussed in Sect. 3. Architecture of dVAE model, training parameters, interpolation, and diversity is discussed in Sect. 4. The results of dVAE are compared with baseline model in Sect. 5. Finally, Sect. 6 concludes the article.

2 Related Work

High-dimensional discrete molecules can be reduced to low-dimensional hidden space by using NLP influential sequence-based DL models. However, models based on natural language processing (NLP) frequently produce invalid strings and sequences that do not correspond to valid molecules. To specialize the prior probabilistic neural network with smaller datasets, reinforcement and transfer learning can be used. Many assessments on more complex architecture [15] like autoencoder [16] which uses two concurrent neural networks for transforming input in or out of hidden representation are performed. The quality of hidden representation with Bayesian optimization [16] and "particle swam optimization" [17] directed randomized SMILES strings are also used for improving quality of latent representation [18]. Another approaches that eliminate optimization iterations learning techniques to precondition learning techniques use input vector of "variational autoencoder" (VAE) with SMILES and desired properties of molecules [19].

We address the issue of designing molecules that are similar to drugs like molecules. There is a lack of diversity in the molecules generated by trained vanilla

VAE, as demonstrated in Sect. 5. We concentrate on a novel approach and generate molecules with desirable amount of diversity. To generate a wide variety of molecules, a tunable parameter (D) is employed. As a result, we propose an approach based on VAE to control the generation of drug molecules in a controllable way.

3 Representation and Dataset

3.1 Molecule Representation

It is necessary to know how chemicals are represented in order to make the connection between chemistry and languages. Most of the time, they are shown in molecular graphs, which are called Lewis structures in material science. Atoms are vertices in molecular structures. The bonds between atoms are represented by the edges, which are labeled with the bond number (for example, single, double, or triple). So one could imagine a model that analyze and creates graphs. Several popular chemical systems encode molecules in such a way. The input and output of models for NLP are often sequences of individual letters, strings, or words. Therefore, we use the SMILES format, which represents molecular structures as human-readable sequences in a compact and comprehensible way. In SMILES, molecules are described using a vocabulary of characters, such as the letters O for oxygen, C and c for aliphatic and aromatic carbon atoms, respectively, and single, double, and triple bonds are denoted by −, = and #, respectively. It is necessary to add a number to recognize rings at the bonded atoms where they have been closed. For example, the chemical symbol for Caffeine in SMILES notation is CN1c2ncn(C)c2C(=O)N(C)C1=O. Round brackets are used to denote the presence of branches. To construct valid SMILES, the generator network must learn the SMILES syntax, which comprises rings and parentheses.

3.2 Dataset

For generation of molecules, the chemical data-driven model is trained on SMILES strings representation of molecules. ZINC database comprises commercially accessible molecules for virtual screening. A large amount of data is required for DL-based model, and the performance of model increases as the amount of data increases. Experimentally, we examine the result convergence with the dataset size. 500k SMILES are optimal number of SMILES for training proposed model. All SMILES strings were canonicalized by RDKit python package [20]. Thus, we use 500k molecules, 75% being used for training, and the remaining being used for testing. The mean of molecular weight (mw), LogP, and TPSA properties of training data is 333.20, 2.60, and 64.11, respectively. The standard deviation of mw, LogP, and TPSA of training data is 61.84, 1.43, and 22.93, respectively.

A total of 32 different symbols constitutes the vocabulary of dataset. The number of unique symbols in vocabulary set as hyperparameter in training; thus, only molecules with these 32 different symbols are generated during generation of molecules.

4 Experimental Setup

4.1 Model Architecture

The fundamental autoencoder (AE) transforms the compounds into a hidden space, and the decoder recreates the molecules from samples extracted from generated hidden space. However, using fundamental architecture, generalized continuous representation of molecules is not possible. With large number of parameters in AE and relatively less amount of training data, the decoder is unlikely to learn an effective mapping of training data and therefore cannot interpret random points in the hidden (continuous) space. Variational autoencoder (VAE) improves AE by introducing Bayesian inferences that allow generative model to generate continuous latent space.

The proposed dVAE is a diverse probabilistic generative model and is a extension of VAE. In the dVAE, the SMILES notation of molecules is first converted into embedding matrices S, then acts as input for the dVAE. The encoder part of the dVAE denoted as $p_\theta(h/S)$ convert the embedding matrices into a continuous hidden space h. Samples from the hidden space pass through the diverse layer. The output of diverse layer acts as input to the decoder denoted as $q_\phi(S/h)$ that convert them into samples similar to the real samples. The architecture of the dVAE is shown in the Fig. 1. Only decoder part of the architecture is active during the generation of the molecules. ϕ and θ are learnable parameters, and their optimal values can be inferred during training of dVAE. The decoder reconstruction error is minimized during training by optimizing the log-likelihood. The encoder is also regularized to minimize the Kullback-Leibler divergence in order to estimate the latent variable distribution. In case of mean 0 and unit variance of Gaussian distribution, the loss function for the dVAE can be expressed as:

$$\text{Loss}(\theta, \phi) = -D_{KL}(p_\theta(h/S) \parallel \mathcal{N}(0, 1)) + E[\log_{q_\phi}(S/h)] \quad (1)$$

4.2 Training of dVAE

Diverse drug-like molecules generation is the main objective of dVAE. Convergence is achieved after 100 iterations of training the model. The loss function's trend during training and validation is depicted in the Fig. 2. The loss function is less then 0.2 in

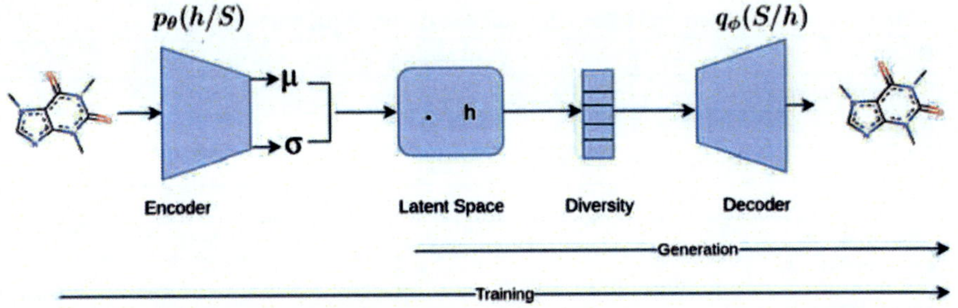

Fig. 1 Architecture of proposed dVAE. A diversity layer is introduced in generation mode. Decoder part is active in generation model of dVAE

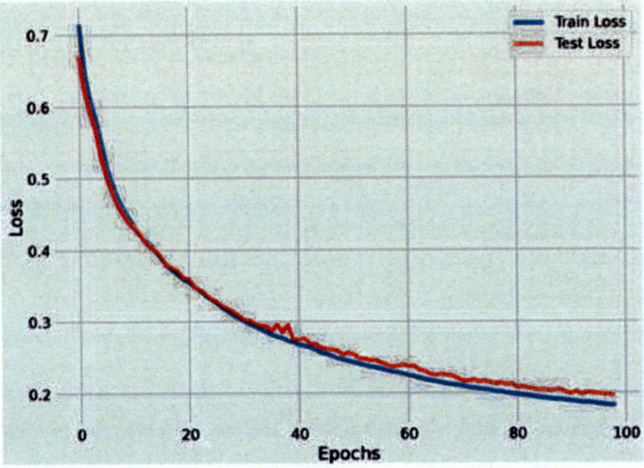

Fig. 2 Loss function of dVAE

both cases. To investigate the chemicals in the hidden space that are close to the parent molecule, we introduce diversity D to the hidden vector z. During training of dVAE, the diversity parameter is set to 1. All weights are initialized using a random gaussian distribution. All parameters that can be trained are optimized using the Adam optimizer algorithm. The SMILES strings are transformed into embedding matrices before being fed into dVAE. A teaching force mechanism is used to train the dVAE. The hyperparameters that were used in dVAE are listed in the Table 1.

4.3 Interpolation and Diversity

Intuitively, we want to explore the chemical space around the given molecule. Controllable noise is added to the latent vector using tunable parameters. This allows us to explore the latent space around a given molecule with varying degrees of diversity. Diversification corresponds to the variability of the molecules generated.

Table 1 Hyperparameters and there values used in the NC-VAE implementation

Parameters	Values
Molecule length	120
Batch size	128
Latent size	200
Mean	0.0
Std. deviation	1.0
Multi-RNN cell	512
No. of layers	3

Table 2 Evaluation of dVAE with different diversity and baseline for validity, uniqueness, and novelty of generated molecules

Model	Valid%	Unique%	FCD
Vanilla-VAE	88.01	60.11	3.01
Seq2Seq	77.12	65.31	3.40
dVAE, $D = 1$	96.60	95.87	3.44
dVAE, $D = 2$	81.35	96.17	3.98
dVAE, $D = 3$	63.84	98.81	4.38

Random sampling is used for injecting diversity in the generated molecules. We examine the impact of diversity layers on dVAE. The Table 2 compares dVAE with no diversity (i.e., D=1) to dVAE with a higher diversity value (i.e., D = 2 or 3). The results demonstrate the effect of the diversity parameter on the generated molecule library.

Interpolation is a methodology for assessing the qualities of latent space. The latent space representations of asprin and caffine are considered to be the starting and ending points (i.e., extreme points) for interpolation. 1280 molecules are generated from samples that fall somewhere between asprin and caffeine. Several generated molecules with interpolation are shown in Fig. 3. The properties of these generated molecules are nearly equivalent or lies in between the value of properties of asprin and caffeine molecules.

5 Results and Discussion

We were using the ZINC dataset discussed in Sect. 3.2 to train and validate dVAE. The proposed model results are compared with the models discussed in Sect. 5.1 and effectiveness of models are evaluated using the measures discussed in Sect. 5.2.

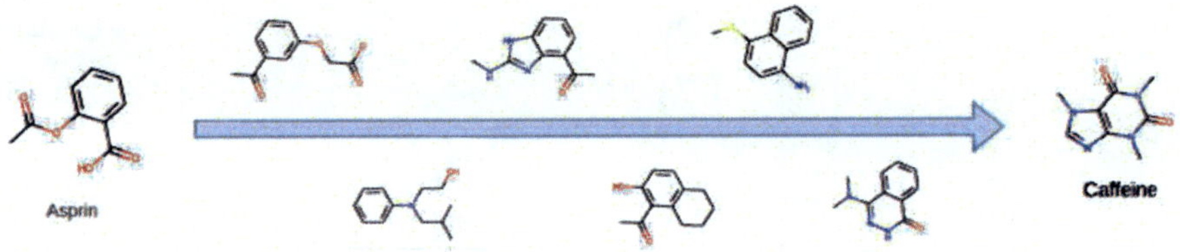

Fig. 3 Compounds reconstructed from the hidden vectors that emerged in the interpolation between two hidden vectors. The starting and finishing positions are the hidden vectors of Aspirin and caffeine, respectively

5.1 Compared Model

- Vanilla [21] VAE model makes new molecules by taking random samples from units of Gaussian randomness and making new molecules out of them.
- The SeqtoSeq [22] architecture can be used to predict sequence data. The RNN cells are used to build the encoder and decoder for the seq-to-seq model.

5.2 Evaluation Matrices

- **Validity:** It is defined as ratio between the number of valid molecules and the total number of generated molecules. Python package RDKit is used for checking the validity of molecules.
- **Uniqueness:** It is defined as ratio between the number of molecules that are not repeated and the total number of valid generated molecules.
- **Fréchet ChemNet Distance (FCD):** [23] To determine the discrepancies between the generated and training distributions of molecules, FCD is employed. The generated and actual molecules have a higher degree of similarity as the FCD metric decreases.

Table 2 shows the results of dVAE with different level of diversity and compare dVAE model with the state-of-the-art methods. It also shows the trade off between the diversity parameter and validity of generated molecules. As the diversity parameter increases, the value of FCD metric also increases indicating the generation of novel molecule.

The training molecules' behavior is compared to the generated molecules' behavior. A comparison of the molecular weights, LogP, and TPSA properties of molecules generated with varying levels of diversity is shown in the Figs. 4, 5, and 6, respectively. At $D = 1$, the generated molecules exhibit a high degree of similarity to training molecules. With increased diversity level through tunable parameter (i.e., D), the model generates molecules with properties that are different from those of the training molecules.

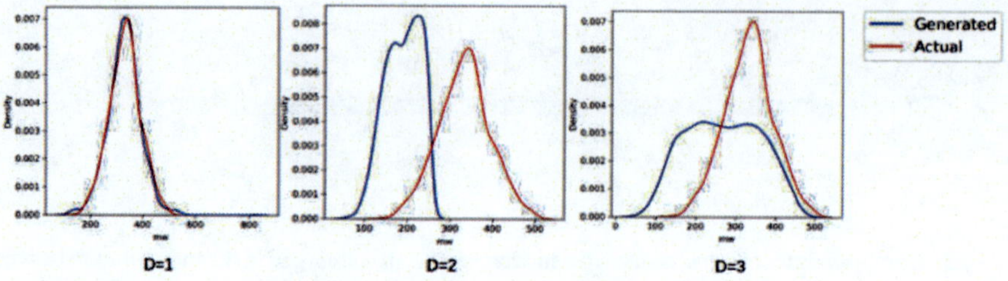

Fig. 4 Comparison of molecular property molecular weight between actual molecules and generated molecules at diversity level 1, 2, and 3

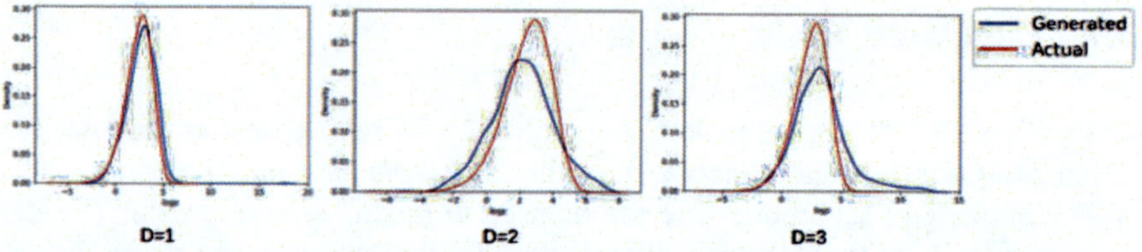

Fig. 5 Comparison of molecular property LogP between actual molecules and generated molecules at diversity level 1, 2 and 3

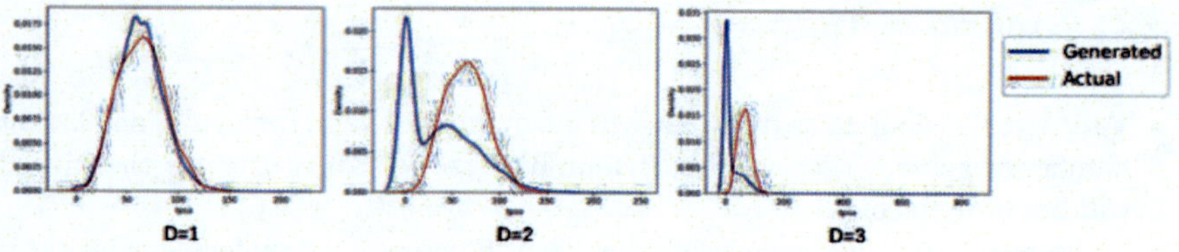

Fig. 6 Comparison of molecular property TPSA between actual molecules and generated molecules at diversity level 1, 2, and 3

The concept of diversity allows dVAE to generate molecules with diverse characteristics while maintaining high novelty and uniqueness at the same time.

6 Conclusion

Drug discovery is a way of finding optimal candidate for the target molecule. Common approaches use high throughput virtual screening (HTVS) for identifying optimal candidate. Further, in this approach, chemist supervision is also needed. State-of-the-art approaches of probabilistic generative algorithms of DL are focused in non-controllable molecules generation. These approaches have some limitation for generating valid and diverse molecules. In the proposed research work, we extend the use of VAE by introducing diversity in the generation. We extract samples around the known target molecules and generate molecules. Interpolation is also performed in the latent space for analyzing trend of molecules in the space. As the generated molecules around the target, molecule posses similar properties. A tunable diversity parameter is added in the generation mode to add diversity in the generated molecules.

References

1. Xu, Y., et al.: Deep learning for molecular generation. Future Med. Chem. **11**(6), 567–597 (2019)
2. Elton, D.C., et al.: Deep learning for molecular design-a review of the state of the art. Mol. Syst. Des. Eng. **4**(4), 828–849 (2019)
3. Vamathevan, J., et al.: Applications of machine learning in drug discovery and development. Nature Rev. Drug Discovery **18**(6), 463–477 (2019)
4. Sanchez-Lengeling, Benjamin, Aspuru-Guzik, Aláin.: Inverse molecular design using machine learning: generative models for matter engineering. Science **361**(6400), 360–365 (2018)
5. Lopyrev, K.: Generating news headlines with recurrent neural networks. arXiv preprint arXiv:1512.01712 (2015)
6. Briot, J.-P., Hadjeres, G., Pachet, F.-D.: Deep learning techniques for music generation. Springer (2020)
7. Wang, Z., He, W., Wu, H., Wu, H., Li, W., Wang, H., Chen, E.E.: Chinese poetry generation with planning based neural network. arXiv preprint arXiv:1610.09889 (2016)
8. Elgammal, A., et al.: Can: creative adversarial networks, generating art by learning about styles and deviating from style norms. arXiv preprint arXiv:1706.07068 (2017)
9. Segler, M.H., Preuss, M., Waller, M.P.: Planning chemical syntheses with deep neural networks and symbolic AI. Nature **555**(7698), 604–610 (2018)
10. Ronneberger, O., Fischer, P., Brox, T.: U-net: convolutional networks for biomedical image segmentation. In: International Conference on Medical Image Computing and Computer-Assisted Intervention, pp. 234–241. Springer
11. Chen, H., et al.: The rise of deep learning in drug discovery. Drug Discovery Today **23**(6), 1241–1250 (2018)
12. Weininger, D: SMILES, a chemical language and information system. 1. Introduction to methodology and encoding rules. J. Chem. Inf. Comput. Sci. **28**(1), 31–36 (1988)
13. Schwalbe-Koda, D., Gómez-Bombarelli, R.: Generative models for automatic chemical design. In: Machine Learning Meets Quantum Physics, pp. 445–467. Springer, Cham (2020)
14. Arús-Pous, J., et al.: Exploring the GDB-13 chemical space using deep generative models. J. Cheminformatics **11**(1), 1–14 (2019)
15. Polykovskiy, D., et al.: Molecular sets (MOSES): a benchmarking platform for molecular generation models. Front. Pharmacol **11**, 1931 (2020)
16. Gómez-Bombarelli, R., et al.: Automatic chemical design using a data-driven continuous representation of molecules. ACS Central Sci. **4**(2), 268-276 (2018)
17. Winter, R., et al.: Efficient multi-objective molecular optimization in a continuous latent space. Chem. Sci. **10**(34), 8016–8024 (2019)
18. Jannik Bjerrum, E., Sattarov, B.: Improving chemical autoencoder latent space and molecular De novo generation diversity with heteroencoders. arXiv e-prints: arXiv-1806 (2018)
19. Lim, J., et al.: Molecular generative model based on conditional variational autoencoder for de novo molecular design. J. Cheminformatics **10**(1), 1–9 (2018)
20. Landrum, G.: RDKit: Open-source cheminformatics. (Online). http://wwwrdkit.org. Accessed 3 Jan 2022, 2012 (2006)

21. Kingma, D.P., Welling, M.: Auto-encoding variational bayes. arXiv preprint arXiv:1312.6114 (2013)
22. Sutskever, I., Vinyals, O., Le, Q.V.: Sequence to sequence learning with neural networks. Adv. Neural Inf. Process. Syst. **27** (2014)
23. Preuer, K., et al.: Fréchet ChemNet distance: a metric for generative models for molecules in drug discovery. J. Chem. Inf. Model. **58**(9), 1736–1741 (2018)

Automated Molecular Subtyping of Breast Cancer Through Immunohistochemistry Image Analysis

S. Niyas, Shraddha Priya, Reena Oswal, Tojo Mathew, Jyoti R. Kini, and Jeny Rajan

Abstract Molecular subtyping has a significant role in cancer prognosis and targeted therapy. However, the prevalent manual procedure for this has disadvantages, such as deficit of medical experts, inter-observer variability, and high time consumption. This paper suggests a novel approach to automate molecular subtyping of breast cancer using an end-to-end deep learning model. Immunohistochemistry (IHC) images of the tumor tissues are analyzed using a three-stage system to determine the subtype. A modified Res-UNet CNN architecture is used in the first stage to segregate the biomarker responses. This is followed by using a CNN classifier to determine the status of the four biomarkers. Finally, the biomarker statuses are combined to determine the specific subtype of breast cancer. For each IHC biomarker, the performance of segmentation models is analyzed qualitatively and quantitatively. In addition, the patient-level biomarker prediction results are also assessed. The findings of the suggested technique demonstrate the potential of computer-aided techniques to diagnose the subtypes of breast cancer. The proposed automated molecular subtyping approach can accelerate pathology procedures, considerably reduce pathologists' workload, and minimize the overall cost and time required for diagnosis and treatment planning.

Keywords Biomarkers · Breast cancer · Deep learning · Immunohistochemistry · Molecular subtyping

---

S. Niyas (✉) · S. Priya · R. Oswal · T. Mathew · J. Rajan
Department of Computer Science and Engineering, National Institute of Technology Karnataka, Surathkal, India
e-mail: niyas.187co504@nitk.edu.in

T. Mathew
Department of Computer Science and Engineering, The National Institute of Engineering, Mysuru, India

J. R. Kini
Department of Pathology, Kasturba Medical College, Mangalore, India

Manipal Academy of Higher Education, Manipal, Karnataka, India

M. Tistarelli et al. (eds.), *Computer Vision and Machine Intelligence*,
Lecture Notes in Networks and Systems 586,
https://doi.org/10.1007/978-981-19-7867-8_3

1 Introduction

Global Cancer Statistics 2020 [23] reports that female breast cancer has surpassed lung cancer to become the most prevalent cancer type. In the year 2020, 2.3 million new cases and 0.68 million deaths were reported for breast cancer. Early diagnosis and targeted treatment are two important aspects to bring down the mortality rate of cancer. Targeted treatment considers the root cause of the malignancy to choose an appropriate treatment modality, and this can lead to better outcome. Molecular subtyping is a classification of cancer to classify the disease based on the genetic alterations that resulted in cancer. Gene expression profiling and immunohistochemistry (IHC) analysis are two ways of identifying the molecular subtypes of cancer. Since genomic analysis is costly and not routinely available, IHC surrogates are commonly used for molecular subtyping. Breast cancer has four common subtypes, namely Luminal A, Luminal B, HER2 enriched, and triple negative. These subtypes are determined based on the assessment of four biomarkers in the tumor tissues through IHC analysis. Since the presence and extent of these biomarkers are driven by the basic gene mutations, molecular subtyping via IHC analysis is an effective alternative for gene expression profiling [8].

In the routine pathology procedure for IHC analysis, the tissue samples extracted from the tumor regions are applied with appropriate antibody reagents and analyzed through a microscope by experienced pathologists. Color sensitivity of the biomarkers to corresponding antibodies indicates the extent of immunopositive cells. In the case of breast cancer molecular subtyping, the status of the four biomarkers (estrogen receptor (ER), progesterone receptor (PR), human epidermal growth factor receptor 2 (HER2), and antigen Ki67) are assessed, and the findings are aggregated to identify the cancer subtype. Figure 3a shows the samples of digitized slides of biomarkers used in the breast cancer molecular subtyping, and Table 1 shows the four primary breast cancer molecular subtypes and corresponding biomarker responses.

Molecular subtyping requires responses of all four biomarkers collected from a patient, and the manual procedure of analyzing these images is tedious and time-consuming. This research explores the feasibility of automated molecular subtyping through IHC image analysis using an IHC image dataset consisting of all four biomarker images. The main contributions in the proposed IHC analysis are listed below.

1. A fully automated molecular subtyping using four biomarkers: ER, PR, Ki67, and HER2 using deep learning techniques.
2. A modified CNN segmentation architecture is used to detect the cell components in each biomarker, followed by a CNN-based classifier for molecular subtyping.

2 Related Works

Automated analysis of the individual biomarkers used in molecular subtyping of breast cancer has been independently attempted before. To our knowledge, there are relatively few approaches in the literature that use automated IHC image processing to define the molecular subtypes of breast cancer. This section discusses the existing methodologies for automated molecular subtyping using biomarkers ER, PR, Ki67, and HER2, as these biomarkers form the constituent factors for molecular subtyping.

IHC biomarker images of ER and PR share some common characteristics in appearance. As a result, some of the works in literature have considered both these biomarkers for analysis whereas others deal with either of these biomarkers. A feasibility study on ER assessment [12] using digital slides shows that the automated assessment results are highly correlated with manual assessment. A Web application namely Immunoratio for quantitatively analyzing ER and PR is proposed in [24]. Manual evaluation of ER and PR is compared with auto-evaluation by Immunoratio in the work by Vijayashree et al. [27]. Segmentation of nuclei from ER slide images followed by classification into +ve or −ve class using fuzzy C-means is applied in the approach by Oscanoa et al. [15] for assessing the ER response. Rexhepaj et al. [18] proposed one of the initial methods for automated ER and PR analysis that used digitized tissue microarrays (TMA) as the dataset to find optimal thresholds for nuclei count to designate ER +ve or −ve status. Mouelhi et al. [13] proposed a method for segmentation of nuclei in IHC images that is based on color deconvolution and morphology operations. A four-class classifier CNN system for cell classification from whole slide images was presented by Jamaluddin et al. [5]. Saha et al. [19] presented a convolutional neural network (CNN) model made of a segmentation module followed by a scoring component for ER and PR scoring.

The membrane-bound biomarker HER2 accelerates tumor growth by facilitating faster cell growth and division. HER2 status in tumor tissues is also an essential factor of molecular subtyping [16]. Automated analysis of IHC images for HER2 assessment has also been attempted before. Normally, one of the three levels namely 'HER2 positive', 'HER2 equivocal', and 'HER2 negative' is assigned as a result of HER2 image analysis. The borderline cases of equivocal status are further analyzed using fluorescent in situ hybridization (FISH) method that is more costly and time-consuming. Reliability of IHC image analysis for ER and HER2 is studied by Lloyd et al. [10]. The results of automated and manual analysis are found to match substantially.

Due to wide adoption of IHC and consequent availability of datasets, most of the automation attempts are reported for IHC image-based HER2 scoring systems. Tuominen et al. [25] developed a Web application for HER2 scoring. American society of clinical oncology (ASCO) guidelines [28] for HER2 score computation is used in this application. Isolation of cell membrane and analyzing the membrane quantitatively is the strategy adopted by Hall et al. [4] for HER2 scoring.

Vandenberghe et al. [26] presented a CNN framework that uses nuclei-based image patches of 44 × 44 to train a CNN. In the method [17], the CNN model is trained using patches of size 128 × 128. Counting on the classification structure of the retrieved patches, the HER2 score is then given. Her2Net proposed by Saha et al. [20] uses CNN with trapezoidal long short-term memory (TLSTM) units for membrane segmentation and scoring.

The presence of molecular biomarker Ki67 indicates the various stages of cell division [2]. The number of cells/nuclei that give Ki67 positive expression shows how aggressively the tumor is growing. Ki67 assessment is also an essential component of molecular subtyping, and there are a few methods in literature to automate Ki67 scoring for breast cancer. An automated Ki67 scoring system for breast cancer proposed by Abubakar et al. [1] involves detection of nuclei in TMA images and training a TMA specific classifier and a universal classifier. Shi et al. [22] proposed a Ki67 scoring method for nasopharyngeal carcinoma using IHC images. The process pipeline involves preprocessing, feature extraction, segmentation, and post-processing stages. A dictionary learning model for counting Ki67 nuclei in images of neuroendocrine carcinoma was presented by Xing e al. [29]. Konsti et al. [7] assessed Ki67 positive nuclei in breast cancer TMA images and studied its prognostic value using a sample set of 1931 patients.

Khan et al. [6] use perceptual clustering to locate Ki67 hotspots from digitized slides of neuroendocrine tumors. Saha et al. [21] applied deep learning based on nuclei patches extracted from IHC images for Ki67 scoring of breast cancer. Segmentation of Ki67 positive nuclei using a U-Net-based architecture is proposed by Lakshmi et al. [9] for bladder cancer. Ki67 score is a vital prognostic parameter for various types of cancer, namely breast, nasopharyngeal, neuroendocrine, bladder, etc., as indicated by the methods that are proposed for these types of cancers whereas ER, PR, and HER2 assessment methods mostly deal with breast cancer. Mathew et al. [11] presented a CNN-based molecular subtyping classification using patch-wise analysis of cell elements from ER, PR, HER2, and Ki67 biomarkers.

The survey of the literature carried out for various biomarkers shows that most of the existing works focus on one or two biomarkers related to breast cancer subtyping. Since molecular subtyping is a patient-level procedure that involves the assessment of multiple biomarkers together, the existing works on individual biomarker analysis cannot lead to an effective solution. Moreover, any effort to that end is constrained by the lack of any public dataset of all the biomarkers collected patient-wise. We overcome this constraint by using a patient-wise dataset of the four biomarkers prepared by our collaborating pathology department to implement a comprehensive automated system for molecular subtyping of breast cancer. Our proposed method is elaborated in the following section.

3 Proposed Method

The proposed molecular subtyping uses a CNN-based analysis over the ER, PR, Ki67, and HER2 biomarker images. The complete workflow of the proposed method is presented in Fig. 1. It consists of a segmentation stage to extract nuclei and cell

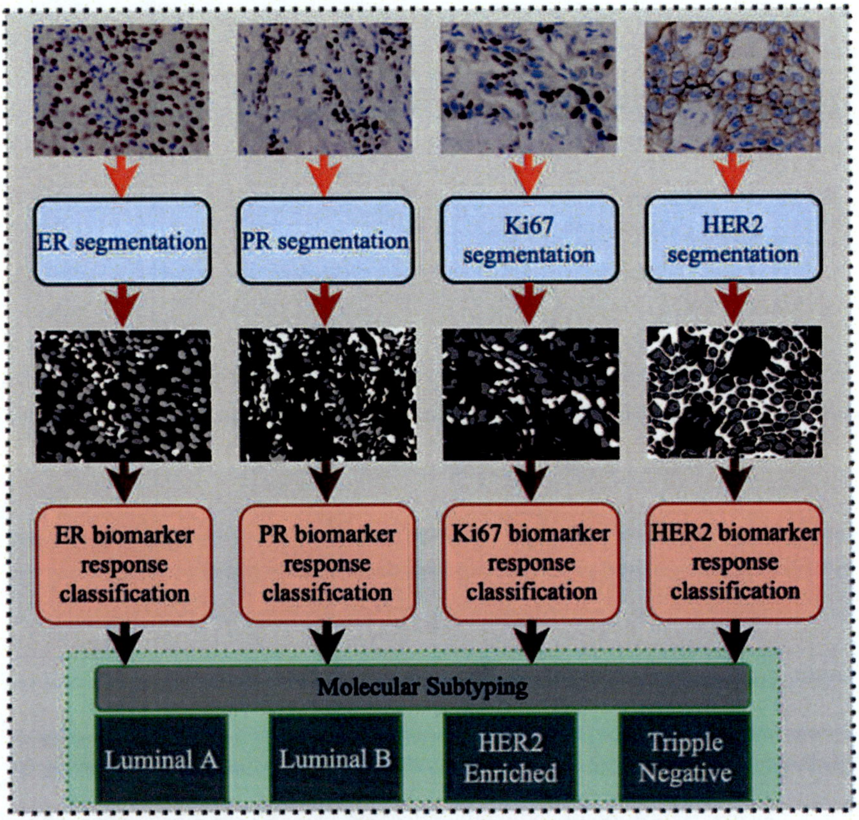

Fig. 1 Graphical overview of the proposed method

membrane, followed by a classification stage to predict the biomarker response from the segmented images. Image samples from each biomarker are analyzed individually, and molecular subtyping is determined from the overall biomarker responses according to the clinical guidelines. The biomarker response assessment from the four different IHC images depends upon distinct characteristics of the image samples. For instance, immunopositive cell nuclei are significant in ER, PR, and Ki67, while HER2 analysis mostly checks for the cell membrane of the immunopositive cells. Hence, separately trained segmentation and classification models are used for the individual biomarker analysis. The biomarker status from the classification stage is consolidated to determine the molecular subtype as specified in Table 1. The dataset used and the details of various stages of the proposed method are described in the following sections.

Table 1 Molecular subtypes of breast cancer and the corresponding biomarker status

Molecular Subtype	Biomarker response
Luminal A	ER+, PR+/−, HER2−, low Ki67
Luminal B	ER+, PR+/−, HER2+/−, high Ki67
HER2 Enriched	ER−, PR−, HER2+, high Ki67
Triple −ve	ER−, PR−, HER2−, high Ki67

3.1 Dataset

The proposed method uses a private dataset of digitized biopsy slides of the ER, PR, HER2, and Ki67 biomarkers captured at 40X magnification. The dataset is collected from Kasturba Medical College, Mangalore, India, and consists of 600 IHC images from 15 breast cancer patients. There are 150 images per biomarker, and each image has a spatial resolution of 1920×1440. Since the images are processed using a supervised segmentation scheme, ground truth masks corresponding to the image samples are also created under the supervision of an expert pathologist.

3.2 Segmentation

In this stage, relevant cell elements from the biomarker images are segmented. The size of the individual images (1920×1440) is too large for optimal processing with CNN models. Hence, non-overlapping slices of size 480×480 have been created from each image and then resized to 240×240 to get a better trade-off between the segmentation accuracy and the computation overhead for training. This way 12 sliced image patches are created from every IHC image, to make 1800 patches in total from 150 samples per biomarker. This way the training data requirement of the proposed CNN models is met.

Modified Res-UNet architecture: The proposed Res-UNet architecture is inspired by the Res-UNet model proposed by Zhang et al. [30]. The model possesses the advantages of both residual connections and the UNet architecture. The cascaded architecture of the segmentation and classification architecture is shown in Fig. 2. The architecture consists of an encoder and a decoder path. The encoder consists of four residual blocks, and each block consists of two convolution layers. There are 16 filters each in the convolution layers of the first encoder block, increasing by a factor of two in the subsequent layers. Strided convolution is used instead of max-pooling to downsample the feature space. The decoder is also made of four residual blocks, where each block consists of only a single convolution layer. In the decoder, the spatial dimensions of the feature map are doubled after each level, whereas the number of filters is reduced by a factor of two. The features from each encoder level are also concatenated with the corresponding decoder levels via skip connections.

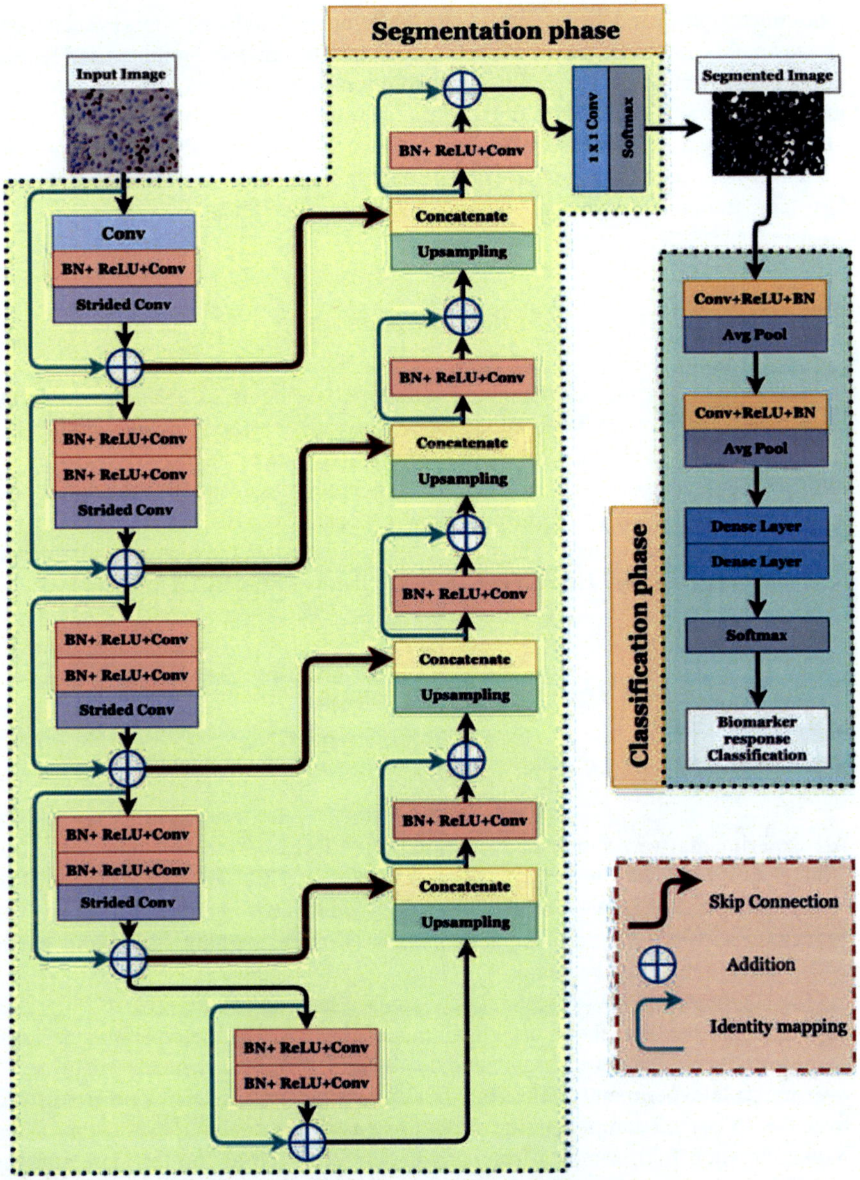

Fig. 2 CNN architecture of the segmentation and classification stages

When there are a series of convolution layers with N filters in each layer, the number of computations and memory cost is $N + (l - 1)N^2$, where l is the number of convolution layers [14]. This quadratic effect in the computation cost and memory requirement can be reduced by increasing the layers with reduced filters in each layer. However, this makes the model deeper and can lead to vanishing gradient and feature degradation problems. The residual convolution layers can avoid these vanishing gradient problems and feature degradation as it uses identity mapping. Hence, the network can act as a superset of multiple networks, and the training is possible by adaptively skipping irrelevant layers. The deeper architecture in the Res-UNet helps to use the trainable parameters by extracting multi-scale features.

3.3 Classification and Molecular Subtyping

Following the segmentation of immunopositive cell elements, a LeNet-based classification model predicts the segmented images' biomarker response. The classification model is shown in Fig. 2. The input to the classification stage is constructed by multiplying the original images with the segmentation map and then resizing it to a resolution of 240×240. Once the biomarker responses from ER, PR, HER2, and Ki67 are predicted, the molecular subtype is estimated based on the recommendations of *St. Gallen International Expert Consensus* [3] that is summarized in Table 1.

4 Experimental Results and Discussion

4.1 Experimental Setup

All experiments were conducted on NVIDIA® DGX-1® server with Canonical Ubuntu OS, Dual 20-Core Intel® Xeon E5-2698 v4 CPU @2.2 GHz, 512 GB of RAM, and 8X NVIDIA® Tesla® V100 GPU with 32GB graphics memory. The implementations are done using a Python-based deep learning framework Keras with TensorFlow as the backend.

The segmentation stage uses cropped non-overlapping patches of size 240×240. The convolutional kernel size used in all models is 3×3. The hyperparameters, such as the number of filters in each layer and depth, are selected based on the performance over multiple experiments. Leave-one-out cross-validation (LOOCV) is performed over the 15 patient samples in the dataset to ensure an unbiased evaluation of the model. In each fold, samples from one patient are used as the test data and the remaining 14 patients' samples for the training. The average of 15 folds is taken as the summarized performance and is done for all four biomarkers.

For the segmentation models, Tversky loss function is used as the loss function to alleviate the class imbalance issue within the pixel classes. Best performance is observed using a batch size of 4 and the dropout of 0.1 in the decoding layers. $L2$ regularization is used to avoid overfitting, and Adam optimizer is used with a learning rate of 0.001. The He normal initializer is used for initializing kernel weights

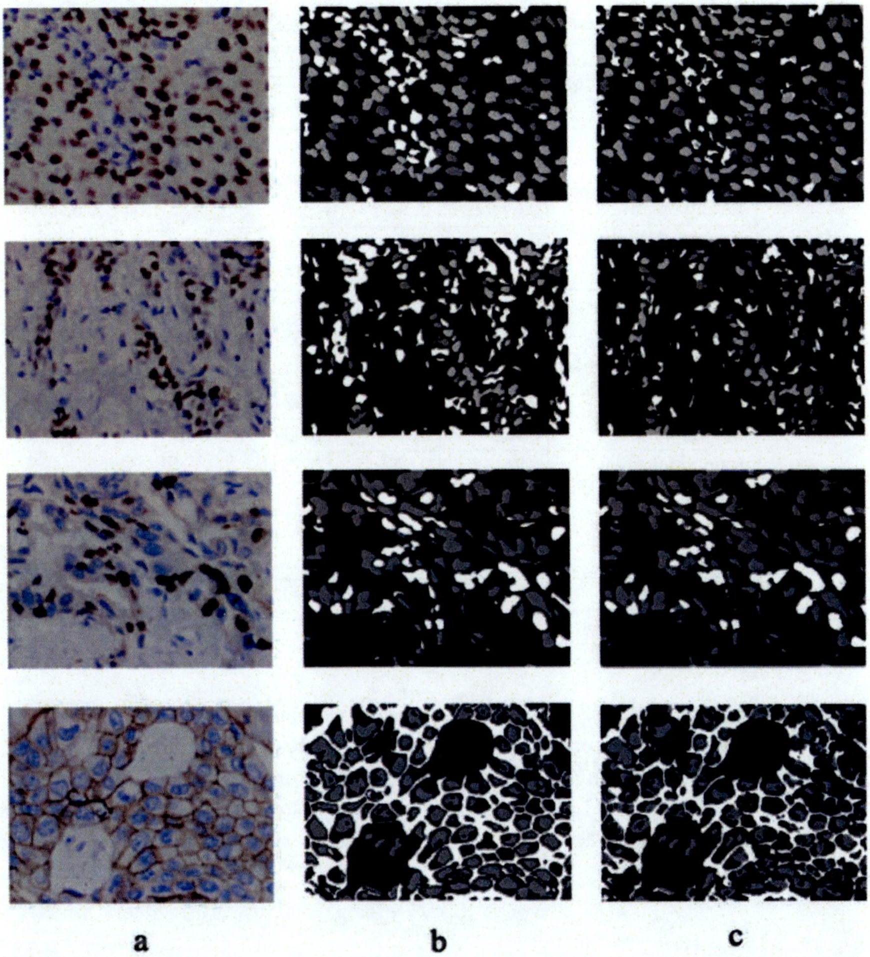

Fig. 3 Qualitative analysis of the immunopositive cell segmentation: **a** Biomarker image samples of ER, PR, Ki67 & HER2 (from top), **b** Ground truth images, and **c** Predicted images

in all segmentation models, and each model is trained from scratch for 100 epochs. Segmentation uses 4-class classification in ER and PR while 3-class segmentation is used in HER2 and Ki67 analysis. In the classification models, the first two layers of the convolution model use 16 filters, each with a kernel size of 5×5, while the following fully connected layers use 120, 84, and 2 neurons, respectively. ReLU activation function is used in all layers in segmentation and classification stages except in the final classification layer, while Softmax activation is used in the end layer.

The performance of the proposed method has been analyzed in all three stages of the methodology. Figure 3 shows the qualitative analysis of the immunopositive

Table 2 Results of the segmentation phase and biomarker response prediction for the four biomarkers

Biomarker type	Segmentation performance				Biomarker Response Accuracy (%)
	Pixel class	Precision	Recall	Dice	
ER	Background	0.97	0.98	0.98	100
	Strong positive	0.67	0.70	0.68	
	Intermediate positive	0.47	0.40	0.43	
	Weak positive	0.30	0.09	0.14	
PR	Background	0.98	0.99	0.99	96.7
	Strong positive	0.62	0.69	0.65	
	Intermediate positive	0.60	0.54	0.57	
	Weak positive	0.25	0.08	0.12	
Ki67	Background	0.92	0.95	0.93	100
	Immunopositive	0.76	0.64	0.69	
	Immunonegative	0.83	0.83	0.83	
HER2	Background	0.90	0.94	0.92	88
	Nuclei	0.75	0.64	0.69	
	Cell membrane	0.94	0.87	0.90	

cell segmentation. The false predictions across the subclasses are relatively less in all biomarker types. Table 2 represents the quantitative analysis of both segmentation and biomarker classification stages. The performance while predicting the final molecular subtyping using individual biomarker responses is also analyzed and obtained an overall patient-level accuracy of 87%.

5 Conclusion

This study proposed an automated molecular subtyping approach that mimics the standard manual analysis. The tri-stage analysis: immunopositive cell segmentation, biomarker response classification, and the final decision-making stage together contribute a robust end-to-end analysis by assessing four protein biomarkers. The CNN-based analysis using an improved Res-UNet segmentation model and a customized LeNet classification model helps to improve the overall prediction performance to make use in real-world cases. The proposed model automates molecular subtyping using IHC image analysis by considering all four biomarkers: ER, PR, HER2, and Ki67. The approach uses a cascaded CNN-based segmentation stage followed by classification. We see tremendous potential for future improvements in this domain. The main shortcoming of the proposed segmentation approach is the subpar segmentation results in pixel classes with high data imbalance. Using more training data and enhanced deep learning architectures to address the data imbalance, it is conceivable to further improve the molecular subtyping analysis and build automated assistive technologies for clinical trials.

6 Compliance with Ethical Standards

Institute scientific committee approval is obtained, and institutional ethical committee approval is exempted.

References

1. Abubakar, M., Howat, W.J., Daley, F., Zabaglo, L., McDuffus, L.A., Blows, F., Coulson, P., Raza Ali, H., Benitez, J., Milne, R., et al.: High-throughput automated scoring of ki67 in breast cancer tissue microarrays from the breast cancer association consortium. J. Pathol. Clin. Res. **2**(3), 138–153 (2016)
2. Gerdes, J., Li, L., Schlueter, C., Duchrow, M., Wohlenberg, C., Gerlach, C., Stahmer, I., Kloth, S., Brandt, E., Flad, H.D.: Immunobiochemical and molecular biologic characterization of the cell proliferation-associated nuclear antigen that is defined by monoclonal antibody ki-67. Am. J. Pathol. **138**(4), 867 (1991)
3. Goldhirsch, A., Winer, E.P., Coates, A., Gelber, R., Piccart-Gebhart, M., Thürlimann, B., Senn, H.J., Albain, K.S., André, F., Bergh, J., et al.: Personalizing the treatment of women with early breast cancer: highlights of the st gallen international expert consensus on the primary therapy of early breast cancer 2013. Ann. Oncol. **24**(9), 2206–2223 (2013)
4. Hall, B.H., Ianosi-Irimie, M., Javidian, P., Chen, W., Ganesan, S., Foran, D.J.: Computer-assisted assessment of the human epidermal growth factor receptor 2 immunohistochemical assay in imaged histologic sections using a membrane isolation algorithm and quantitative analysis of positive controls. BMC Med. Imaging **8**(1), 1–13 (2008)
5. Jamaluddin, M.F., Fauzi, M.F., Abas, F.S., Lee, J.T., Khor, S.Y., Teoh, K.H., Looi, L.M.: Cell classification in er-stained whole slide breast cancer images using convolutional neural network. In: 2018 40th Annual International Conference of the IEEE Engineering in Medicine and Biology Society (EMBC). pp. 632–635. IEEE (2018)
6. KHAN NIAZI, M.K., Yearsley, M.M., Zhou, X., Frankel, W.L., Gurcan, M.N.: Perceptual clustering for automatic hotspot detection from ki-67-stained neuroendocrine tumour images. J. Microsc. **256**(3), 213–225 (2014)
7. Konsti, J., Lundin, M., Joensuu, H., Lehtimäki, T., Sihto, H., Holli, K., Turpeenniemi-Hujanen, T., Kataja, V., Sailas, L., Isola, J., et al.: Development and evaluation of a virtual microscopy application for automated assessment of ki-67 expression in breast cancer. BMC Clin. Pathol. **11**(1), 1–11 (2011)
8. Kornegoor, R., Verschuur-Maes, A.H., Buerger, H., Hogenes, M.C., De Bruin, P.C., Oudejans, J.J., Van Der Groep, P., Hinrichs, B., Van Diest, P.J.: Molecular subtyping of male breast cancer by immunohistochemistry. Mod. Pathol. **25**(3), 398–404 (2012)
9. Lakshmi, S., Vijayasenan, D., Sumam, D.S., Sreeram, S., Suresh, P.K.: An integrated deep learning approach towards automatic evaluation of ki-67 labeling index. In: TENCON 2019-2019 IEEE Region 10 Conference (TENCON). pp. 2310–2314. IEEE (2019)
10. Lloyd, M.C., Allam-Nandyala, P., Purohit, C.N., Burke, N., Coppola, D., Bui, M.M.: Using image analysis as a tool for assessment of prognostic and predictive biomarkers for breast cancer: How reliable is it? J. Pathol. Inform. **1** (2010)

11. Mathew, T., Niyas, S., Johnpaul, C., Kini, J.R., Rajan, J.: A novel deep classifier framework for automated molecular subtyping of breast carcinoma using immunohistochemistry image analysis. Biomed. Signal Process. Control **76**, 103657 (2022)
12. Mofidi, R., Walsh, R., Ridgway, P., Crotty, T., McDermott, E., Keaveny, T., Duffy, M., Hill, A., O'Higgins, N.: Objective measurement of breast cancer oestrogen receptor status through digital image analysis. Eur. J. Surg. Oncol. (EJSO) **29**(1), 20–24 (2003)
13. Mouelhi, A., Sayadi, M., Fnaiech, F.: A novel morphological segmentation method for evaluating estrogen receptors' status in breast tissue images. In: 2014 1st International Conference on Advanced Technologies for Signal and Image Processing (ATSIP). pp. 177–182. IEEE (2014)
14. Niyas, S., Vaisali, S.C., Show, I., Chandrika, T., Vinayagamani, S., Kesavadas, C., Rajan, J.: Segmentation of focal cortical dysplasia lesions from magnetic resonance images using 3d convolutional neural networks. Biomed. Signal Process. Control **70**, 102951 (2021)
15. Oscanoa, J., Doimi, F., Dyer, R., Araujo, J., Pinto, J., Castaneda, B.: Automated segmentation and classification of cell nuclei in immunohistochemical breast cancer images with estrogen receptor marker. In: 2016 38th Annual International Conference of the IEEE Engineering in Medicine and Biology Society (EMBC). pp. 2399–2402. IEEE (2016)
16. Perez, E.A., Cortés, J., Gonzalez-Angulo, A.M., Bartlett, J.M.: Her2 testing: current status and future directions. Cancer Treat. Rev. **40**(2), 276–284 (2014)
17. Pitkäaho, T., Lehtimäki, T.M., McDonald, J., Naughton, T.J., et al.: Classifying her2 breast cancer cell samples using deep learning. In: Proc. Irish Mach. Vis. Image Process. Conf., 1–104 (2016)
18. Rexhepaj, E., Brennan, D.J., Holloway, P., Kay, E.W., McCann, A.H., Landberg, G., Duffy, M.J., Jirstrom, K., Gallagher, W.M.: Novel image analysis approach for quantifying expression of nuclear proteins assessed by immunohistochemistry: application to measurement of oestrogen and progesterone receptor levels in breast cancer. Breast Cancer Res. **10**(5), 1–10 (2008)
19. Saha, M., Arun, I., Ahmed, R., Chatterjee, S., Chakraborty, C.: Hscorenet: A deep network for estrogen and progesterone scoring using breast ihc images. Pattern Recogn. **102**, 107200 (2020)
20. Saha, M., Chakraborty, C.: Her2net: A deep framework for semantic segmentation and classification of cell membranes and nuclei in breast cancer evaluation. IEEE Trans. Image Process. **27**(5), 2189–2200 (2018)
21. Saha, M., Chakraborty, C., Arun, I., Ahmed, R., Chatterjee, S.: An advanced deep learning approach for ki-67 stained hotspot detection and proliferation rate scoring for prognostic evaluation of breast cancer. Sci. Rep. **7**(1), 1–14 (2017)
22. Shi, P., Zhong, J., Hong, J., Huang, R., Wang, K., Chen, Y.: Automated ki-67 quantification of immunohistochemical staining image of human nasopharyngeal carcinoma xenografts. Sci. Rep. **6**(1), 1–9 (2016)
23. Sung, H., Ferlay, J., Siegel, R.L., Laversanne, M., Soerjomataram, I., Jemal, A., Bray, F.: Global cancer statistics 2020: Globocan estimates of incidence and mortality worldwide for 36 cancers in 185 countries. CA: A Cancer J. Clin. **71**(3), 209–249 (2021)
24. Tuominen, V.J., Ruotoistenmäki, S., Viitanen, A., Jumppanen, M., Isola, J.: Immunoratio: a publicly available web application for quantitative image analysis of estrogen receptor (er), progesterone receptor (pr), and ki-67. Breast Cancer Res. **12**(4), 1–12 (2010)
25. Tuominen, V.J., Tolonen, T.T., Isola, J.: Immunomembrane: a publicly available web application for digital image analysis of her2 immunohistochemistry. Histopathology **60**(5), 758–767 (2012)
26. Vandenberghe, M.E., Scott, M.L., Scorer, P.W., Söderberg, M., Balcerzak, D., Barker, C.: Relevance of deep learning to facilitate the diagnosis of her2 status in breast cancer. Sci. Rep. **7**(1), 1–11 (2017)
27. Vijayashree, R., Aruthra, P., Rao, K.R.: A comparison of manual and automated methods of quantitation of oestrogen/progesterone receptor expression in breast carcinoma. J. Clin. Diagn. Res.: JCDR **9**(3), EC01 (2015)
28. Wolff, A.C., Hammond, M.E.H., Schwartz, J.N., Hagerty, K.L., Allred, D.C., Cote, R.J., Dowsett, M., Fitzgibbons, P.L., Hanna, W.M., Langer, A., et al.: American society of clinical

oncology/college of american pathologists guideline recommendations for human epidermal growth factor receptor 2 testing in breast cancer. Arch. Pathol. Lab. Med. **131**(1), 18–43 (2007)
29. Xing, F., Su, H., Neltner, J., Yang, L.: Automatic ki-67 counting using robust cell detection and online dictionary learning. IEEE Trans. Biomed. Eng. **61**(3), 859–870 (2013)
30. Zhang, Z., Liu, Q., Wang, Y.: Road extraction by deep residual u-net. IEEE Geosci. Remote Sens. Lett. **15**(5), 749–753 (2018)

Emotions Classification Using EEG in Health Care

Sumit Rakesh, Foteini Liwicki, Hamam Mokayed, Richa Upadhyay, Prakash Chandra Chhipa, Vibha Gupta, Kanjar De, György Kovács, Dinesh Singh, and Rajkumar Saini

Abstract Online monitoring of mental well-being and factors contributing to it is vital, especially in pandemics (like COVID-19), when physical contact is discouraged. As emotions are closely connected to mental health, the monitoring and classification of emotions are also vital. In this paper, we present an emotion recognition framework recognizing emotions (*angry, happiness, neutral, sadness, and scare*) from neurological signals. Subjects can monitor their own emotions; the recognition outcome can also be transferred to healthcare professionals online for further investi-

S. Rakesh · F. Liwicki · H. Mokayed · R. Upadhyay · P. C. Chhipa · V. Gupta (✉) · K. De · G. Kovács · R. Saini
Machine Learning Group, EISLAB, Luleå Tekniska Universitet, Luleå, Sweden
e-mail: vibha.gupta@associated.ltu.se

S. Rakesh
e-mail: Sumit.Rakesh@ltu.se

F. Liwicki
e-mail: Foteini.Liwicki@ltu.se

H. Mokayed
e-mail: Hamam.Mokayed@ltu.se

R. Upadhyay
e-mail: Richa.Upadhyay@ltu.se

P. C. Chhipa
e-mail: Prakash.Chandra.Chhipa@ltu.se

K. De
e-mail: Kanjar.De@ltu.se

G. Kovács
e-mail: Gyorgy.Kovacs@ltu.se

R. Saini
e-mail: rsaini@cs.iitr.ac.in

D. Singh
Computer Science & Engineering, DCRUST, Murthal, Sonepat, India

R. Saini
Department of CSE, IIT Roorkee, Roorkee, India

M. Tistarelli et al. (eds.), *Computer Vision and Machine Intelligence*, Lecture Notes in Networks and Systems 586,
https://doi.org/10.1007/978-981-19-7867-8_4

gation. For this, we use the brain's electrical responses, containing the affective state of mind. These electrical responses or brain signals are recorded with electroencephalography (EEG) sensors. The EEG sensors are well established in their utility in non-medical and medical applications, such as biometrics or epilepsy detection. Here, we extract features from the EEG sensors and feed them as input to support vector machine (SVM) and random forest (RF) classifiers. Recognition rates of 76.32% and 79.18% have been recorded with SVM and RF classifiers, respectively. These results, along with the results attained using prior methods, demonstrate the efficacy of the proposed method.

Keywords Electroencephalography (eeg) · Emotion recognition · Frequency bands · Random forest (rf) · Support vector machine (svm)

1 Introduction

Emotions impact health and quality of life in general. Different sensor devices have been used to understand the behavioral and physiological nature of emotions. In the past few decades, emotion studies based on body gesture [11], facial expressions [23], and eye movement [34] were in the focus of e-healthcare systems. For example, Tivatansakul et al. [29] proposed an emotional classification system based on facial expressions. Using SVM, the authors were able to classify six different emotions based on facial expressions: {*sadness, surprise, happiness, fear, disgust, anger*}. However, the similarity of the facial expressions elicited by various emotions can make classifying certain emotions difficult. Due to this difficulty and the poor emotion classification performance of the above-mentioned techniques with a single modality, the research community has started focusing on electroencephalography (EEG) based neuro-sensing techniques to classify emotions.

As emotional changes impact mental health, these sensors capture slight changes in behavioral patterns and help healthcare professionals investigate the emotional states of the patients. Different techniques like electroencephalography (EEG) [6], magnetic resonance imaging (MRI), functional magnetic resonance imaging (fMRI) have been used to study the neural activities of the brain. However, due to their easy portability, low-cost EEG sensors are the most commonly used. One tool applicable here is the neuroheadset, an award-winning device equipped with various sensors and electrodes. The electrical impulses known as EEG signals are detected by a headset placed over a person's scalp. These electrical signals are subsequently sent through bluetooth technology to the connected computer or smartphone device. In order to understand the behavioral functioning which occurs due to frequent mood changes, researchers have started using EEG sensor-based emotion classification systems to support their evidence. The studies that utilized EEG in prediction [13, 19, 20], among others, include healthcare applications, such as real-time seizure detection [32], stroke detection [31], the assessment of mental stress [1] the identification of mental burden [10], detecting abnormalities in the brain [22], as well as home automation studies [25] and person identification [9, 17, 26]. For emotion classification using EEG data, Lan et al. [21] proposed a subject-dependent algorithm. They looked at four different emotions: pleasant, happy, frightened, and angry. The EEG signal data for these experiments was gathered over the course of eight days in two sessions. Their results demonstrated that an emotion identification system might be utilized to assess people's emotional states based on EEG data. These experiments, however, were carried out for only five subjects. In the work of Jenke et al. [15], the International Affective Picture System (IAPS) pictures database was used to show emotions. The data was collected from 16 subjects. From the collected EEG data, Hilbert–Huang spectrum (HHS), higher-order spectra (HOS), and higher-order crossings (HOC) features were extracted. The five different emotions (*happy, curious, angry, sad, quiet*), were then classified using a quadratic discriminant analysis (QDA), and accuracy of 36.8% was achieved.

Syahril et al. [28] in their proposed methodology investigated sad, fear, happiness, and disgust emotions. The data from 15 subjects was collected, and spectral features were extracted from it. Despite the fact that the authors worked on the alpha and beta wavebands, the results have been reported using only $\{Fp_i \mid i = 1, 2, 3, 4\}$ channels. Bahari et al. [3] used three different class emotions, valence, arousal, and liking, for classification. The nonlinear features were extracted using recurrence quantification analysis (RQA) which were further classified using KNN (K-Nearest Neighbor) classifier. The mean accuracies of 58.05, 64.56, and 67.42% have been reported for all the classes. The authors in [24] used only two primary emotions (happiness and sadness) to classify the data collected from six subjects after exploring new methods of selecting subject-specific frequency bands rather than fixed bands. An average accuracy of 74.17% has been achieved after classification using a common spatial pattern and support vector machine.

Bird et al. [4] in their research extracted discriminative EEG-based features to categorize three emotional states relaxing, neutral, and concentrating. Although the accuracy of 87% has been achieved using three different classifiers on the data from five subjects. The data for two different emotions, happiness and sadness, has been collected using the IAPS images in [2]. For the classification of emotions, artificial neural network (ANN) has been used in the paper where an accuracy of 81.8% is reported with only two electrodes, i.e., FP1 and FP2. EEG data is generally divided into multiple frequency rhythms to evaluate the brain's reaction to various stimuli. These frequency rhythm names (frequency range) are gamma (32Hz—above), beta (16–32Hz), alpha (8–16Hz), theta (4–8Hz), and delta (0.5–4Hz). Srinivas et al. [27] used DEAP dataset for emotion recognition. Authors employed wavelet transform (WT) and discrete Fourier transform (DFT) to distinguish emotions. Different frequency bands are extracted using WT and DFT. The beta and gamma rhythms improve performance when categorized with the SVM's radial basis function (RBF), as per the data obtained from 11 people utilizing video stimuli. They only evaluated three channels when reporting the accuracy.

Different from above work where authors mainly focused on using handcrafted features, Lei et al. [16] investigate the feasibility of of EEG signals for automatic emotion recognition during reminiscence therapy for older people. They have col-

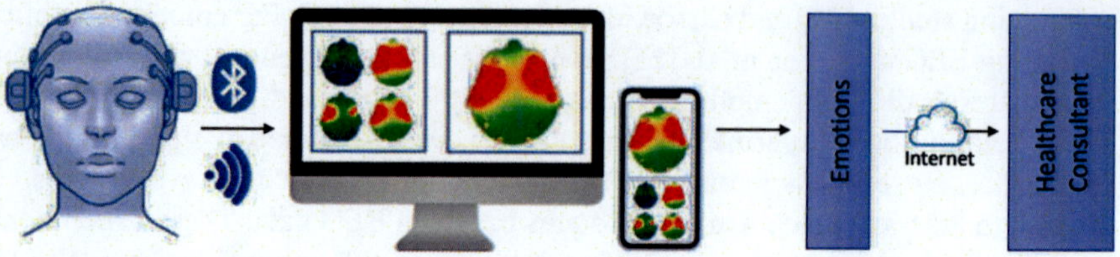

Fig. 1 The flow of the information from emotion recognition to the health care using Internet

lected data from eleven older people (mean 71.25, SD 4.66) and seven young people (mean 22.4, SD 1.51), and used depth models like LSTM and Bi-LSTM to extract complex emotional features for automatic recognition of emotions. In the work [35], authors proposed 2D CNN uses two convolutional kernels of different sizes to extract emotion-related features along both the time direction and the spatial direction. The public emotion dataset DEAP is used in experiments. Sofien et al. [12] proposed ZTWBES algorithm to identify the epochs, pre-selecting the electrodes that successfully identified the epochs and for every emotional state, determining relevant electrodes in every frequency band. Different classification schemes were defined using QDC and RNN and evaluated using the DEAP database.

In this study, we propose an emotion recognition system (depicted in Fig. 1) where the brain signals are first analyzed to detect the emotions of the subject. Then, the recognized emotions are sent to the healthcare centers via the Internet for further analysis.

The remainder of the paper is structured as follows: In Sect. 2, we present the proposed framework. Section 2.2 discusses wavelets and feature extraction. The description of classifiers used in the proposed work can be found in Sect. 2.3. The results of our experiments are reported in Sect. 3. Lastly, Sect. 4 reports the conclusions and plans for future work.

2 Proposed System

In this section, we discuss our entire proposed framework (see, Fig. 2), including data collection, the process of feature extraction, and the classification methods used.

2.1 The EEG Headset

The data for our experiments is captured after the subject watched 2–3 minute-long video clips conveying different emotions, namely sadness, happiness, scare, and anger. The data for different emotions is acquired by an EEG neuroheadset called Emotiv Epoc+. This headset consists of 14 electrodes, i.e., $\{AFi|i = 3,\ 4\}$, $\{Fi|i =$

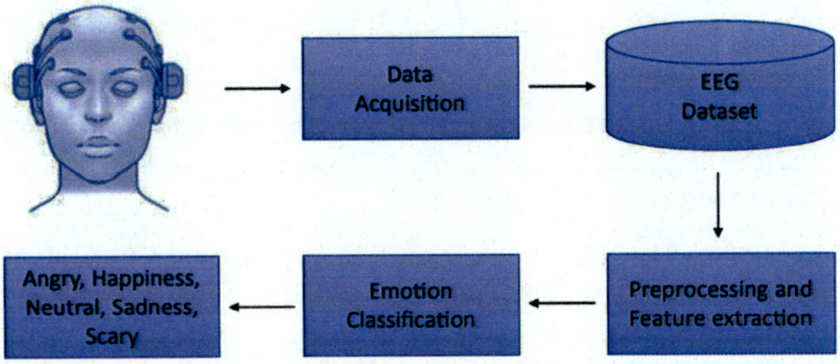

Fig. 2 Framework representing classification of happiness, anger, neutral, scary, and sadness emotions

3, 4, 7, 8}, $\{FCi|i = 5, 6\}$, $\{Pi|i = 7, 8\}$, $\{Ti|i = 7, 8\}$, $\{Oi|i = 1, 2\}$, and two references CMS and DRL with a sampling rate 128 Hz[1]. The headset records the data in EDF file format (European Data Format)[2] from which raw signals are extracted. The sensor comes with a saline water to activate electrodes for better conductivity.

The data recorded is sent to the connected computer via Bluetooth, using a USB dongle. The resulting data can be analyzed using the Emotiv Xavier Test Bench 3.1 tool [10]. This tool allows the presentation of EEG data in a time-dependent manner. For the classification of emotions, data from all fourteen channels is processed. Emotion analysis is also performed on different brain lobes, i.e., front-lobe, parietal-lobe, temporal-lobe, occipital-lobe, front, and rear-lobe. Furthermore, the gamma frequency waveband is used to extract statistical features from preprocessed data followed by classification. The gamma frequency band is extracted using discrete wavelet transform (DWT).

2.2 *Wavelet Decomposition and Feature Extraction*

To analyze EEG signals in the proposed framework, we use the discrete wavelet transform (DWT). Apart from time-scale analysis, it uses signal decomposition and signal compression techniques to decompose the EEG signal into different bands [24]. Researchers have used the DWT for various application, such as gaming [14], seizure detection [30], and security [26]. Daubechies wavelet (db4) has been used to perform the decomposition of input EEG signals to get different frequency bands, namely alpha, beta, gamma, delta, and theta. Their frequency ranges are (gamma: 32–100 HZ), (beta: 16–32 Hz), (alpha: 8–16 Hz), (theta: 4–8Hz), and (delta: 0–4 Hz). The general equation of the mother wavelet transformation is defined in Eq. 1 which is obtained by continuous shifting and dilating of the signal [18].

[1] https://www.emotiv.com/epoc/.

[2] https://en.wikipedia.org/wiki/European_Data_Format.

$$\psi_{r,s}(k) = \frac{1}{\sqrt{|r|}}\psi\left(\frac{k-s}{r}\right) r,\ s \in W,\ i \neq 0, \tag{1}$$

where mother wavelet is obtained by $\psi_{r,s}(k)$. Here, r denotes the scaling parameter measuring the degree of compression, whereas s represents time location. The wavelet space is indicated by the variable W.

Feature Extraction: The statistical features, namely, mean (M_n), standard deviation (STD), and root mean square (RMS) are extracted for the experiments. They are calculated as defined in Eq. 2, 3, and 4, respectively.

Arithmetic Mean (M): We divide the sum total of the data by the number of data points in the signal to get the arithmetic mean (or average) as shown in Eq. 2, where x_k signifies the kth sample of the signal X and s specifies the signal's length.

$$M = \frac{1}{s}\sum_{k=1}^{s} x_k \tag{2}$$

Standard Deviation (STD): STD is used to calculate the variability in EEG signals as defined in Eq. 3, where x_k ($k = 1, 2, \ldots, s$) denotes the kth value, s is the length of the signal and mean of the signals is defined by M.

$$STD = \sqrt{\frac{1}{s-1}\sum_{k=1}^{s}(x_k - M)^2} \tag{3}$$

Root Mean Square (RMS): The square root of the arithmetic mean of the squared data can be used to calculate RMS as defined in Eq. 4, where x_k signifies the kth sample of the signal X and s specifies the signal's length. It is also known as the quadratic mean of the signal.

$$\text{RMS} = \sqrt{\frac{1}{s}\sum_{k=1}^{s}(x_k)^2} \tag{4}$$

The M, STD, and RMS are concatenated to create final feature vector f_v from the signal as shown in Eq. 5.

$$f_v = \{M, \text{STD}, \text{RMS}\} \tag{5}$$

2.3 *Classification of Emotions Using Different Classifiers*

Random Forest (RF): Proposed initially by Breiman et al. [5], the random forest is an ensemble of decision trees. Each tree contributes to the final decision by designating the test input a class label. Here, trees are grown randomly, and their posterior distribution is used to estimate the leaf nodes over several classes. The root node is the starting point for constructing a random tree; in this, splitting of the training data is done based on features, and it continues for each feature value. On the other hand, test root node splitting and selection is based on the information gain.

Support Vector Machines (SVM): This classifier can accurately predict results with a large number of vectors and thus make it an appropriate classifier for various biomedical use cases. It divides several classes by mapping them onto a high-dimensional space with the widest possible margin and provides different kernel functions using which calculations can be performed. SVM is covered in more depth in [7]. The experiments have been done using radial basis function kernel [7], but we also examined the performance of linear and polynomial kernels. In our study, we used the LIBSVM [7] implementation of the classifier.

3 Experiments and Results

In this section, we describe the dataset that has been used for the classification of five emotions. The accuracy from the data is evaluated and achieved using RF and SVM classifiers. The experiments are performed on a 64- bit Windows 7 operating system with Intel®Core™ i5 CPU @2.30 GHz and 8 GB of RAM. We have used Matlab 2016a to conduct RF and SVM classifications. The details are as follows:

3.1 *Dataset Description and Experimental Protocol*

For the emotion classification, the study is conducted on a real-time EEG signal dataset collected by [17] from 10 subjects. The 10 s data was captured after (after each video) the subject watched 2–3 minute-long video clips conveying different emotions, namely sadness, happiness, scare, and anger; eyes were closed while recording the data. The data used in this work is the part of the dataset proposed in [17] subjected to emotion analysis. Emotiv Epoc+ neuroheadset with 14 channels and a sample rate 128 Hz was used to record EEG signals. The raw signals are were extracted from EDF files for further processing and feature extraction. The emotion data was captured for four different emotions: sadness, happiness, scare, and anger after a video of 2–3 min was shown to the subjects. A valence-arousal model has been followed to incite the proper emotions. The data for the emotional state was captured in different sessions. The neutral emotion data is captured while subjects sit in a closed-eyes

Table 1 True positive and false positives obtained using RF classifier

	Angry	Scare	Happiness	Sadness	Neutral
Angry	0.76	0.12	0.02	0.04	0.06
Scare	0.10	0.79	0.06	0.0	0.04
Happiness	0.08	0.14	0.71	0.02	0.04
Sadness	0.02	0.02	0.02	0.88	0.06
Neutral	0.12	0.02	0.0	0.4	0.82

resting position. The performance of the proposed framework has been evaluated using fivefold cross-validation, where 80% of the dataset has been used in training EEG signals and the rest 20% used for testing. The EEG data has also been analyzed from various lobes of the brain, i.e., parietal-lobe (P7, P8), occipital-lobe (O1, O2), left-lobe (AF3, F7, F3, FC5), right-lobe (AF4, F4, F8, FC6), temporal-lobe (T7, T8), and rear-lobe (P7, P8, O1, O2). The data recorded is then segmented into multiple files of different seconds. After the experiments, it has been observed that maximum accuracy is achieved on 4-sec EEG emotion data for all subjects.

3.2 Emotion Classification Results Using RF

The emotion classification experiments were first conducted using an RF classifier following extensive experimentation for the optimization of hyper-parameters. The tree size used in this study ranged from one to fifty. The maximal accuracy of 79.18% was obtained at 39 trees considering all 14 electrodes when tested with features, i.e., M, STD, and RMS. Table 1 illustrates the accuracies obtained form RF (with 39 trees). The diagonal elements of table shows the performance for respective classes; Angry (76%), Scare (79%), Happiness (71%), Sadness (88%), and Neutral (72%). The off diagonal elements depicts the miss-classification. From the table, it is observed that the model achieves the highest accuracy for the emotion of sadness and the lowest for the emotion of happiness.

3.3 Comparative Analysis of SVMs Using Different Kernels

This section presents the results achieved using different SVM kernels. The results attained with different kernels are presented in Table 2. As can be seen in Table 2, the RBF kernel achieves the highest emotion recognition accuracy score (76.32%), and the linear kernel achieves the lowest (37.55%). .

Table 2 Emotion classifications using different kernels of SVM with different features

Features	SVM (linear)	SVM (polynomial)	SVM (RBF)	RF
M, STD	35.1	58.77	76.32	73.87
M, RMS	36.32	58.76	75.91	75.91
STD, RMS	37.55	60	75.91	78.77
M, STD, RMS	37.55	64.08	76.32	79.18

Table 3 Classifications performance considering different lobes with SVM

Lobes	SVM (linear)	SVM (polynomial)	SVM (RBF)	RF
Frontal	33.06	42.44	54.29	55.51
Left-frontal	28.98	30.20	41.63	45.30
Right-frontal	22.85	28.97	46.12	48.57
Occipital	26.53	25.71	37.96	39.18
Temporal	24.48	26.53	47.32	50.20
Parietal	24.90	26.53	35.10	33.47
Rear	28.98	32.25	52.25	52.25

3.4 Comparative Analysis at Different Brain Lobes

This section presents the comparative analysis performed on data from different lobes. These include the left-lobe (AF3, F7, F3, FC5), the right-lobe (AF4, F4, F8, FC6), the parietal-lobe (P7, P8), the occipital-lobe (O1, O2), the temporal-lobe (T7, T8), and the rear-lobe (P7, P8, O1, O2). A comparison of the resulting accuracies is presented in Table 3. Different performances were noticed for different lobes with different classifiers. From the table, it is noticed that frontal lobe is dominate the emotion activity. This dominance is also noticed across different kernel of SVM.

3.5 Comparative Analysis by Varying Time

The proposed system was evaluated by varying the duration (seconds) of the EEG signals to analyze the effect of time. The accuracy scores attained are depicted in Fig. 3; it can be seen that the highest accuracy has been achieved for 4-sec data.

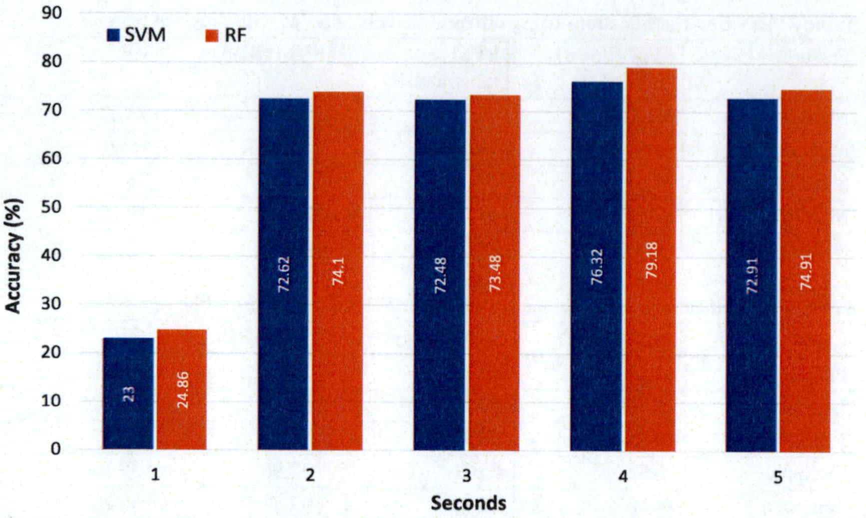

Fig. 3 Performance of RF and SVM at different time intervals

Table 4 Summary of the quantitative results

	#Subjects	#Channels	#Emotions	Classifier (accuracy)%
Daly et al. [8], 2016	8	32	3	SVM (53.96)
Zhuang et al. [36], 2017	32	8	2	SVM (69.10)
Wang et al. [33], 2018	32	6	4	SVM (68.59)
Proposed work	10	14	5	RF (79.18) SVM (76.32)

3.6 Comparative Performance Analysis

A summary of the prior research has been made in this section, as shown in Table 4. Daly et al. [8] proposed a real-time emotion monitoring system based on three emotions (happiness, calmness, and stress). An accuracy of 53.96% was achieved when data was recorded from 8 subjects while listening to music. Two emotional classes, valence (positive and negative) and arousal (high and low) have been used to classify emotions. Authors used music video clips as stimuli [36]. An accuracy of 69.10% has been reported with the SVM classifier. In [33], authors used the same protocol to record data for four emotions fun, happy, sad, and terrible. An accuracy of 68.59% has been achieved while with six channels. When compared to existing methods, it is found that the proposed methodology outperforms raw signal findings with ten subjects, fourteen channels, and five emotions.

4 Conclusion and Future Scope

Mental health is as important as physical health, and EEG signals can be used to identify different emotions to analyze mental health. The consistent negative emotions (sadness, anger, scare) could seriously harm the subjects' health. Therefore, emotion analysis can assist medical care professionals in caring for their patients. Five different emotions have been considered in this paper: happiness, sadness, scare, anger, and neutral. The preprocessed signals were used to extract features that describe the different emotions. The wavelet decomposition is done, and gamma band is extracted. The recorded signals were classified using RF and SVM methods, achieving an accuracy of 79.18% and 76.32%, respectively. Such systems can be used in mental health care. With the help of the Internet and IoT, healthcare professionals directly monitor the mental health and emotions of the subjects. They can recommend prescriptions, making them suitable for pandemic situations like Covid19. In the future, we aim to increase the number of emotions and evaluate their impact on the brain using different classifiers. Furthermore, efficient inter-subject emotion recognition and similar applications can be realized with the availability of massive datasets. The handcrafted features can also be tested with deep learning-based approaches on more extensive datasets. The proposed method is beneficial for health care as it helps in finding negative emotions (sadness, anger, scare) which could seriously harm the subjects' health. Moreover, handcrafted features designed based on domain knowledge showed promising results, and hence possible to consider in real-time application because of their low complexity.

References

1. Al-Shargie, F., Tang, T.B., Badruddin, N., Kiguchi, M.: Towards multilevel mental stress assessment using SVM with ECOC: an EEG approach. Med. Biol. Eng. Comput. **56**(1), 125–136 (2018)
2. Ang, A.Q.X., Yeong, Y.Q., Wee, W.: Emotion classification from EEG signals using time-frequency-dwt features and ANN. J. Comput. Commun. **5**(3), 75–79 (2017)
3. Bahari, F., Janghorbani, A.: Eeg-based emotion recognition using recurrence plot analysis and k nearest neighbor classifier. In: 2013 20th Iranian Conference on Biomedical Engineering (ICBME). pp. 228–233. IEEE (2013)
4. Bird, J.J., Manso, L.J., Ribeiro, E.P., Ekart, A., Faria, D.R.: A study on mental state classification using eeg-based brain-machine interface. In: 2018 International Conference on Intelligent Systems (IS). pp. 795–800. IEEE (2018)
5. Breiman, L.: Random forests. Mach. Learn. **45**(1), 5–32 (2001)
6. Chakladar, D.D., Kumar, P., Roy, P.P., Dogra, D.P., Scheme, E., Chang, V.: A multimodal-Siamese neural network (mSNN) for person verification using signatures and EEG. Inf. Fusion **71**, 17–27 (2021)
7. Chang, C.C., Lin, C.J.: LIBSVM: a library for support vector machines. ACM Trans. Intell. Syst. Technol. (TIST) **2**(3), 1–27 (2011)
8. Daly, I., Williams, D., Kirke, A., Weaver, J., Malik, A., Hwang, F., Miranda, E., Nasuto, S.J.: Affective brain-computer music interfacing. J. Neural Eng. **13**(4), 046022 (2016)

9. Das, B.B., Kumar, P., Kar, D., Ram, S.K., Babu, K.S., Mohapatra, R.K.: A spatio-temporal model for EEG-based person identification. Multimedia Tools Appl. **78**(19), 28157–28177 (2019)
10. Di Stasi, L.L., Diaz-Piedra, C., Suárez, J., McCamy, M.B., Martinez-Conde, S., Roca-Dorda, J., Catena, A.: Task complexity modulates pilot electroencephalographic activity during real flights. Psychophysiology **52**(7), 951–956 (2015)
11. Faust, O., Hagiwara, Y., Hong, T.J., Lih, O.S., Acharya, U.R.: Deep learning for healthcare applications based on physiological signals: a review. Comput. Methods Programs Biomed. **161**, 1–13 (2018)
12. Gannouni, S., Aledaily, A., Belwafi, K., Aboalsamh, H.: Emotion detection using electroencephalography signals and a zero-time windowing-based epoch estimation and relevant electrode identification. Sci. Rep. **11**(1), 1–17 (2021)
13. Gauba, H., Kumar, P., Roy, P.P., Singh, P., Dogra, D.P., Raman, B.: Prediction of advertisement preference by fusing EEG response and sentiment analysis. Neural Netw. **92**, 77–88 (2017)
14. Hazarika, J., Kant, P., Dasgupta, R., Laskar, S.H.: Neural modulation in action video game players during inhibitory control function: An EEG study using discrete wavelet transform. Biomed. Signal Process. Control **45**, 144–150 (2018)
15. Jenke, R., Peer, A., Buss, M.: Feature extraction and selection for emotion recognition from EEG. IEEE Trans. Affect. Comput. **5**(3), 327–339 (2014)
16. Jiang, L., Siriaraya, P., Choi, D., Kuwahara, N.: Emotion recognition using electroencephalography signals of older people for reminiscence therapy. Front. Physiol., 2468 (2022)
17. Kaur, B., Singh, D., Roy, P.P.: A novel framework of EEG-based user identification by analyzing music-listening behavior. Multimedia Tools Appl. **76**(24), 25581–25602 (2017)
18. Kaur, B., Singh, D., Roy, P.P.: Age and gender classification using brain-computer interface. Neural Comput. Appl. **31**(10), 5887–5900 (2019)
19. Khurana, V., Gahalawat, M., Kumar, P., Roy, P.P., Dogra, D.P., Scheme, E., Soleymani, M.: A survey on neuromarketing using EEG signals. IEEE Trans. Cognitive Develop. Syst. (2021)
20. Kumar, P., Scheme, E.: A deep spatio-temporal model for eeg-based imagined speech recognition. In: ICASSP 2021-2021 IEEE International Conference on Acoustics, Speech and Signal Processing (ICASSP). pp. 995–999. IEEE (2021)
21. Lan, Z., Sourina, O., Wang, L., Liu, Y.: Real-time EEG-based emotion monitoring using stable features. Vis. Comput. **32**(3), 347–358 (2016)
22. Liu, T., Chen, Y., Lin, P., Wang, J.: Small-world brain functional networks in children with attention-deficit/hyperactivity disorder revealed by EEG synchrony. Clin. EEG Neurosci. **46**(3), 183–191 (2015)
23. Muhammad, G., Alsulaiman, M., Amin, S.U., Ghoneim, A., Alhamid, M.F.: A facial-expression monitoring system for improved healthcare in smart cities. IEEE Access **5**, 10871–10881 (2017)
24. Pan, J., Li, Y., Wang, J.: An eeg-based brain-computer interface for emotion recognition. In: 2016 International Joint Conference on Neural Networks (IJCNN). pp. 2063–2067. IEEE (2016)
25. Roy, P.P., Kumar, P., Chang, V.: A hybrid classifier combination for home automation using EEG signals. Neural Comput. Appl. **32**(20), 16135–16147 (2020)
26. Saini, R., Kaur, B., Singh, P., Kumar, P., Roy, P.P., Raman, B., Singh, D.: Don't just sign use brain too: a novel multimodal approach for user identification and verification. Inf. Sci. **430**, 163–178 (2018)
27. Srinivas, M.V., Rama, M.V., Rao, C.: Wavelet based emotion recognition using RBF algorithm. Int. J. Innov. Res. Electr. Electron., Instrum. Control Eng. **4** (2016)
28. Syahril, S., Subari, K.S., Ahmad, N.N.: EEG and emotions: α-peak frequency as a quantifier for happiness. In: 2016 6th IEEE International Conference on Control System, Computing and Engineering (ICCSCE). pp. 217–222. IEEE (2016)
29. Tivatansakul, S., Ohkura, M., Puangpontip, S., Achalakul, T.: Emotional healthcare system: Emotion detection by facial expressions using japanese database. In: 2014 6th Computer Science and Electronic Engineering Conference (CEEC). pp. 41–46. IEEE (2014)

30. Tzimourta, K., Tzallas, A., Giannakeas, N., Astrakas, L., Tsalikakis, D., Tsipouras, M.: Epileptic seizures classification based on long-term eeg signal wavelet analysis. In: International Conference on Biomedical and Health Informatics. pp. 165–169. Springer (2017)
31. Van Kaam, R.C., Van Putten, M.J., Vermeer, S.E., Hofmeijer, J.: Contralesional brain activity in acute ischemic stroke. Cerebrovascular Diseases **45**(1–2), 85–92 (2018)
32. Vidyaratne, L.S., Iftekharuddin, K.M.: Real-time epileptic seizure detection using EEG. IEEE Trans. Neural Syst. Rehabil. Eng. **25**(11), 2146–2156 (2017)
33. Wang, C.l., Wei, W., LI, T.y.: Emotion recognition based on EEG using IMF energy moment. DEStech Trans. Comput. Sci. Eng. **pcmm** (2018)
34. Wang, Y., Lv, Z., Zheng, Y.: Automatic emotion perception using eye movement information for e-healthcare systems. Sensors **18**(9), 2826 (2018)
35. Wang, Y., Zhang, L., Xia, P., Wang, P., Chen, X., Du, L., Fang, Z., Du, M.: EEG-based emotion recognition using a 2d CNN with different kernels. Bioengineering **9**(6), 231 (2022)
36. Zhuang, N., Zeng, Y., Tong, L., Zhang, C., Zhang, H., Yan, B.: Emotion recognition from EEG signals using multidimensional information in EMD domain. BioMed Res. Int. **2017** (2017)

Moment Centralization-Based Gradient Descent Optimizers for Convolutional Neural Networks

Sumanth Sadu, Shiv Ram Dubey, and S. R. Sreeja

Abstract Convolutional neural networks (CNNs) have shown very appealing performance for many computer vision applications. The training of CNNs is generally performed using stochastic gradient descent (SGD)-based optimization techniques. The adaptive momentum-based SGD optimizers are the recent trends. However, the existing optimizers are not able to maintain a zero mean in the first order moment and struggle with optimization. In this paper, we propose a moment centralization-based SGD optimizer for CNNs. Specifically, we impose the zero mean constraints on the first order moment explicitly. The proposed moment centralization is generic in nature and can be integrated with any of the existing adaptive momentum-based optimizers. The proposed idea is tested with three state-of-the-art optimization techniques, including Adam, Radam, and Adabelief on benchmark CIFAR10, CIFAR100, and TinyImageNet datasets for image classification. The performance of the existing optimizers is generally improved when integrated with the proposed moment centralization. Further, the results of the proposed moment centralization are also better than the existing gradient centralization. The analytical analysis using the toy example shows that the proposed method leads to a shorter and smoother optimization trajectory. The source code is made publicly available at https://github.com/sumanthsadhu/MC-optimizer.

S. Sadu (✉)
Computer Vision Group, Department of Computer Science and Engineering, Indian Institute of Information Technology, Sri City, Andhra Pradesh, India
e-mail: venkatasaisumanth.s18@iiits.in

S. R. Dubey
Computer Vision and Biometrics Laboratory, Indian Institute of Information Technology, Allahabad, Uttar Pradesh, India
e-mail: srdubey@iiita.ac.in

S. R. Sreeja
Department of Computer Science and Engineering, Indian Institute of Information Technology, Sri City, Andhra Pradesh, India
e-mail: sreeja.sr@iiits.in

M. Tistarelli et al. (eds.), *Computer Vision and Machine Intelligence*,
Lecture Notes in Networks and Systems 586,
https://doi.org/10.1007/978-981-19-7867-8_5

1 Introduction

Deep learning has become very prominent to solve the problems in computer vision, natural language processing, and speech processing [17]. Convolutional neural networks (CNNs) have been exploited to utilize deep learning in the computer vision area with great success to deal with the image and video data [1, 9, 15, 19, 22, 24]. The CNN models are generally trained using the stochastic gradient descent (SGD) [3]-based optimization techniques where the parameters of the model are updated in the opposite direction of the gradient. The performance of any CNN model is also sensitive to the chosen SGD optimizer used for the training. Hence, several SGD optimizers have been investigated in the literature.

The vanilla SGD algorithm suffers from the problem of zero gradients at local minimum and saddle regions. The SGD with momentum (SGDM) uses the accumulated gradient, i.e., momentum, to update the parameters instead of the current gradient [25]. Hence, it is able to update the parameters in the local minimum and saddle regions. However, the same step size is used by SGDM for all the parameters. In order to adapt the step size based on the gradient consistency, AdaGrad [8] divides the learning rate by the square root of the accumulated squared gradients from the initial iteration. However, it leads to a vanishing learning rate after some iteration as the squared gradient is always positive. In order to tackle the learning rate diminishing problem of AdaGrad, RMSProp [10] uses a decay factor on the accumulated squared gradient. The idea of both SGDM and RMSProp is combined in a very popular Adam gradient descent optimizer [13]. Basically, Adam uses a first order moment as the accumulated gradient for parameter update and a second order moment as the accumulated squared gradient to control the learning rate. A rectified Adam (Radam) [18] performs the rectification in Adam to switch to SGDM to improve the precision of the convergence of training. Recently, Adabelief optimizer [27] utilizes the residual of gradient and first order moment to compute the second order moment which improves the training at saddle regions and local minimum. Other notable recent optimizers include diffGrad [6], AngularGrad [20], signSGD [2], Nostalgic Adam [11], and AdaInject [7].

The above-mentioned optimization techniques do not utilize the normalization of the moment for smoother training. However, it is evident that normalization plays a very important role in the training of deep learning models, such as data normalization [23], batch normalization [12], instance normalization [5], weight normalization [21], and gradient normalization [4]. The gradient normalization is also utilized with SGD optimizer in [26]. However, as the first order moment is used for the parameter updates in adaptive momentum-based optimizers, we proposed to perform the normalization on the accumulated first order moment. The contributions of this paper can be summarized as follows:

- We propose the concept of moment centralization to normalize the first order moment for the smoother training of CNNs.
- The proposed moment centralization is integrated with Adam, Radam, and Adabelief optimizers.

- The performance of the existing optimizers is tested with and without the moment centralization.

The rest of the paper is structured as follows: Sect. 2 presents the proposed moment centralization-based SGD optimizers; Sect. 3 summarizes the experimental setup, datasets used, and CNN models used; Sect. 4 illustrates the experimental results, comparison, and analysis; and finally, Sect. 5 concludes the paper.

2 Proposed Moment Centralization-Based SGD Optimizers

In this paper, a moment centralization strategy is proposed for the adaptive momentum-based SGD optimizers for CNNs. The proposed approach imposes the explicit normalization over the first order moment in each iteration. The imposed zero mean distribution helps the CNN models for a smoother training. The proposed moment centralization is generic and can be integrated with any adaptive momentum based SGD optimizer. In this paper, we use it with state-of-the-art optimizers, including Adam [13], Radam [18], and Adabelief [27]. The Adam optimizer with moment centralization is termed AdamMC. Similarly, the Radam and Adabelief optimizers with moment centralization are referred to as RadamMC and AdabeliefMC, respectively. In this section, we explain the proposed concept with Adam, i.e., AdamMC.

Let a function f represents a CNN model having parameters θ for image classification. Initially, the parameters θ_0 are initialized randomly and updated using backpropagation of gradients in the subsequent iterations of training. During the forward pass of training, the CNN model takes a batch of images (I_B) containing B images as the input and returns the cross-entropy loss as follows:

$$L_{CE} = -\sum_{i=1}^{B} \log(p_{c_i}) \tag{1}$$

where p_{c_i} is the softmax probability of given input image I_i for the correct class level c_i. Note that the class level is represented by the one-hot encoding in the implementation. During the backward pass of the training, the parameters are updated based on the gradient information. Different SGD optimizers utilize the gradient information differently for the parameter update. The adaptive momentum-based optimizers such as Adam are very common in practice which utilizes the first order and second order momentum which is the exponential moving average of gradient and squared gradient, respectively. Hence, we explain the proposed concept below with the help of Adam optimizer.

Consider g_t to be the gradient of an objective function w.r.t. the parameter θ, i.e., $g_t = \nabla_\theta f_t(\theta_{t-1})$ in the tth iteration of training, where θ_{t-1} represents the parameter values obtained after $(t-1)$th iteration. The first order moment (m_t) in the tth iteration is computed as follows:

$$m_t = \beta_1 m_{t-1} + (1 - \beta_1) g_t \tag{2}$$

where m_{t-1} is the first order moment in $(t-1)$th iteration and β_1 is a decay hyperparameter. Note that the distribution of first order moments is generally not zero centric which leads to the update in a similar direction for the majority of the parameters. In order to tackle this problem, we propose to perform the moment centralization by normalizing the first order moment with zero mean as follows:

$$m_t = m_t - \text{mean}(m_t). \tag{3}$$

The second order moment (v_t) in the tth iteration is computed as follows:

$$v_t \leftarrow \beta_2 v_{t-1} + (1 - \beta_2) g_t^2 \tag{4}$$

where v_{t-1} is the second order moment in $(t-1)$th iteration and β_2 is a decay hyperparameter. Note that we do not perform the centralization of second order moments as it may lead to a very inconsistent effective learning rate.

Algorithm 1: Adam Optimizer [13]

Initialize: $\theta_0, m_0 \leftarrow 0, v_0 \leftarrow 0, t \leftarrow 0$ **Hyperparameters:** $\alpha, \beta_1, \beta_2, T$
For $\theta_t = [1,...,T]$ **do**
$g_t \leftarrow \nabla_\theta f_t(\theta_{t-1})$ ▷ *Gradient Computation*
$m_t \leftarrow \beta_1 m_{t-1} + (1 - \beta_1) g_t$ ▷ *First Order Moment*
$v_t \leftarrow \beta_2 v_{t-1} + (1 - \beta_2) g_t^2$ ▷ *Second Order Moment*
$\widehat{m_t} \leftarrow m_t/(1 - \beta_1^t), \quad \widehat{v_t} \leftarrow v_t/(1 - \beta_2^t)$ ▷ *Bias Correction*
$\theta_t \leftarrow \theta_{t-1} - \alpha\widehat{m_t}/(\sqrt{\widehat{v_t}} + \epsilon)$ ▷ *Parameter Update*
return θ_T

Algorithm 2: Adam + Moment Centralization (AdamMC) Optimizer

Initialize: $\theta_0, m_0 \leftarrow 0, v_0 \leftarrow 0, t \leftarrow 0$ **Hyperparameters:** $\alpha, \beta_1, \beta_2, T$
For $\theta_t = [1,...,T]$ **do**
$g_t \leftarrow \nabla_\theta f_t(\theta_{t-1})$ ▷ *Gradient Computation*
$m_t \leftarrow \beta_1 m_{t-1} + (1 - \beta_1) g_t$ ▷ *First Order Moment*
$m_t \leftarrow m_t - mean(m_t)$ ▷ *Moment Centralization*
$v_t \leftarrow \beta_2 v_{t-1} + (1 - \beta_2) g_t^2$ ▷ *Second Order Moment*
$\widehat{m_t} \leftarrow m_t/(1 - \beta_1^t), \quad \widehat{v_t} \leftarrow v_t/(1 - \beta_2^t)$ ▷ *Bias Correction*
$\theta_t \leftarrow \theta_{t-1} - \alpha\widehat{m_t}/(\sqrt{\widehat{v_t}} + \epsilon)$ ▷ *Parameter Update*
return θ_T

The first and second order moments suffer in the initial iteration due to very small values leading to very high effective step size. In order to cope up with this bias, Adam [13] has introduced a bias correction as follows:

$$\widehat{m_t} \leftarrow m_t/(1 - \beta_1^t), \quad \widehat{v_t} \leftarrow v_t/(1 - \beta_2^t) \tag{5}$$

where $\widehat{m_t}$ and $\widehat{v_t}$ are the bias corrected first and second order moment, respectively. Finally, the parameter update in t^{th} iteration is performed as follows:

$$\theta_t \leftarrow \theta_{t-1} - \alpha\widehat{m_t}/(\sqrt{\widehat{v_t}} + \epsilon) \tag{6}$$

where θ_t is the updated parameter, θ_{t-1} is the parameter after $(t-1)$th iteration, α is the learning rate, and ϵ is a very small constant number used for numerical stability to avoid the division by zero. The steps of Adam optimizer without and with the proposed moment centralization concept are summarized in Algorithm 1 (Adam) and Algorithm 2 (AdamMC), respectively. The changes in the AdamMC are highlighted in blue color. Similarly, we also incorporate the moment centralization concept with Radam [18] and Adabelief [27] optimizers. The steps for Radam, RadamMC, Adabelief, and AdabeliefMC are illustrated in Algorithm 3, 4, 5, and 6, respectively.

3 Experimental Setup

All experiments are conducted on Google Colab GPUs using the Pytorch 1.9 framework. We want to emphasize that our moment centralization (MC) approach does not include any additional hyperparameters. To incorporate MC into the existing optimizers, only one line of code is required to be included in the code of existing adaptive momentum-based SGD optimizers, with all other settings remaining untouched.

Algorithm 3: Radam Optimizer [18]

Initialize: $\theta_0, m_0 \leftarrow 0, v_0 \leftarrow 0$ **Hyperparameters:** $\alpha, \beta_1, \beta_2, T$

For $\theta_t = [1,...,T]$ **do**

- $g_t \leftarrow \nabla_\theta f_t(\theta_{t-1})$ ▷ *Gradient Computation*
- $m_t \leftarrow \beta_1 m_{t-1} + (1-\beta_1)g_t$ ▷ *First Order Moment*
- $v_t \leftarrow \beta_2 v_{t-1} + (1-\beta_2)g_t^2$ ▷ *Second Order Moment*
- $\rho_\infty \leftarrow 2/(1-\beta_2) - 1$
- $\rho_t = \rho_\infty - 2t\beta_2^t/(1-\beta_2^t)$
- **If** $\rho_t \geq 5$ ▷ *Check if the variance is tractable*
  - $\rho_u = (\rho_t - 4) \times (\rho_t - 2) \times \rho_\infty$
  - $\rho_d = (\rho_\infty - 4) \times (\rho_\infty - 2) \times \rho_t$
  - $\rho = \sqrt{(1-\beta_2) \times \rho_u/\rho_d}$ ▷ *Variance rectification term*
  - $\alpha_1 = \rho \times \alpha/(1-\beta_1^t)$ ▷ *Rectified learning rate*
  - $\theta_t \leftarrow \theta_{t-1} - \alpha_1 m_t/(\sqrt{v_t} + \epsilon)$ ▷ *Update parameters with rectification*
- **Else**
  - $\alpha_2 = \alpha/(1-\beta_1^t)$ ▷ *Bias correction*
  - $\theta_t \leftarrow \theta_{t-1} - \alpha_2 m_t$ ▷ *Update parameters without rectification*

return θ_T

Algorithm 4: Radam + Moment Centralization (RadamMC) Optimizer

Initialize: $\theta_0, m_0 \leftarrow 0, v_0 \leftarrow 0$ **Hyperparameters:** $\alpha, \beta_1, \beta_2, T$

For $\theta_t = [1,...,T]$ **do**

$g_t \leftarrow \nabla_\theta f_t(\theta_{t-1})$ ▷ *Gradient Computation*

$m_t \leftarrow \beta_1 m_{t-1} + (1-\beta_1) g_t$ ▷ *First Order Moment*

$m_t \leftarrow m_t - mean(m_t)$ ▷ *Moment Centralization*

$v_t \leftarrow \beta_2 v_{t-1} + (1-\beta_2) g_t^2$ ▷ *Second Order Moment*

$\rho_\infty \leftarrow 2/(1-\beta_2) - 1$

$\rho_t = \rho_\infty - 2t\beta_2^t/(1-\beta_2^t)$

If $\rho_t \geq 5$ ▷ *Check if the variance is tractable*

$\rho_u = (\rho_t - 4) \times (\rho_t - 2) \times \rho_\infty$

$\rho_d = (\rho_\infty - 4) \times (\rho_\infty - 2) \times \rho_t$

$\rho = \sqrt{(1-\beta_2) \times \rho_u/\rho_d}$ ▷ *Variance rectification term*

$\alpha_1 = \rho \times \alpha/(1-\beta_1^t)$ ▷ *Rectified learning rate*

$\theta_t \leftarrow \theta_{t-1} - \alpha_1 m_t/(\sqrt{v_t} + \epsilon)$ ▷ *Update parameters with rectification*

Else

$\alpha_2 = \alpha/(1-\beta_1^t)$ ▷ *Bias correction*

$\theta_t \leftarrow \theta_{t-1} - \alpha_2 m_t$ ▷ *Update parameters without rectification*

return θ_T

Algorithm 5: Adabelief Optimizer [27]

Initialize: $\theta_0, m_0 \leftarrow 0, v_0 \leftarrow 0, t \leftarrow 0$ **Hyperparameters:** $\alpha, \beta_1, \beta_2, T$

For $\theta_t = [1,...,T]$ **do**

$g_t \leftarrow \nabla_\theta f_t(\theta_{t-1})$ ▷ *Gradient Computation*

$m_t \leftarrow \beta_1 m_{t-1} + (1-\beta_1) g_t$ ▷ *First Order Moment*

$v_t \leftarrow \beta_2 v_{t-1} + (1-\beta_2)(g_t - m_t)^2$ ▷ *Second Order Moment*

$\widehat{m_t} \leftarrow m_t/(1-\beta_1^t)$, $\widehat{v_t} \leftarrow v_t/(1-\beta_2^t)$ ▷ *Bias Correction*

$\theta_t \leftarrow \theta_{t-1} - \alpha\widehat{m_t}/(\sqrt{\widehat{v_t}} + \epsilon)$ ▷ *Parameter Update*

return θ_T

3.1 CNN Models Used

We use VGG16 [22] and ResNet18 [9] CNN models in the experiments to validate the performance of the proposed moment centralization-based optimizers, including AdamMC, RadamMC, and AdabeliefMC. The VGG16 is a plain CNN model with sixteen learnable layers. It uses three fully connected layers toward the end of the network. It is one of the popular CNN models utilized for different computer vision tasks. The ResNet18 is a directed acyclic graph-based CNN model which utilizes the identity or residual connections in the network. The identity connection improves the gradient flow in the network during backpropagation and helps in the training of a deep CNN model.

Algorithm 6: Adabelief + Moment Centralization (AdabeliefMC) Optimizer

Initialize: $\theta_0, m_0 \leftarrow 0, v_0 \leftarrow 0, t \leftarrow 0$ **Hyperparameters:** $\alpha, \beta_1, \beta_2, T$

For $\theta_t = [1,...,T]$ **do**

$g_t \leftarrow \nabla_\theta f_t(\theta_{t-1})$ ▷ *Gradient Computation*

$m_t \leftarrow \beta_1 m_{t-1} + (1 - \beta_1) g_t$ ▷ *First Order Moment*

$m_t \leftarrow m_t - mean(m_t)$ ▷ *Moment Centralization*

$v_t \leftarrow \beta_2 v_{t-1} + (1 - \beta_2)(g_t - m_t)^2$ ▷ *Second Order Moment*

$\widehat{m_t} \leftarrow m_t/(1 - \beta_1^t), \quad \widehat{v_t} \leftarrow v_t/(1 - \beta_2^t)$ ▷ *Bias Correction*

$\theta_t \leftarrow \theta_{t-1} - \alpha\widehat{m_t}/(\sqrt{\widehat{v_t}} + \epsilon)$ ▷ *Parameter Update*

return θ_T

3.2 *Datasets Used*

We test the performance of the proposed moment centralization optimizers on three benchmark datasets, including CIFAR10, CIFAR100 [14], and TinyImageNet[1] [16]. The CIFAR10 dataset consists of 60,000 images from 10 object categories with 6000 images per category. The 5000 images per category are used for the training, and the remaining 1000 images per category are used for testing. Hence, the total number of images used for training and testing in CIFAR10 data is 50,000 and 10,000, respectively. The CIFAR100 contains the same 60,000 images of CIFAR10 but divides the images into 100 object categories which is beneficial to test the performance of the optimizers for fine-grained classification. The TinyImageNet dataset is part of the full ImageNet challange. It contsists of 200 object categories. Each class has 500 images for training. However, the test set consists of 10,000 images. The dimension of all the images is 64 × 64 in color in TinyImageNet dataset.

3.3 *Hyperparameter Settings*

All the optimizers in the experiment share the following settings. The decay rates of first and second order moments $\beta 1$ and $\beta 2$ are 0.9 and 0.999, respectively. The first and second order moments (m, v) are initialized to 0. The training is performed for 100 epochs with 0.001 learning rate for the first 80 epochs and 0.0001 for the last 20 epochs. The weight initialization is performed using random numbers from a standard normal distribution.

[1] http://cs231n.stanford.edu/tiny-imagenet-200.zip.

Table 1 Results comparison of Adam [13], Radam [18], and Adabelief [27] optimizers with the gradient centralization [26] and the proposed moment centralization on CIFAR10 dataset using VGG16 [22] and ResNet18 [9] CNN models

CNN model	VGG16 model			ResNet18 model		
Optimizer	Adam	AdamGC	AdamMC	Adam	AdamGC	AdamMC
Run1	92.49	92.43	92.49	93.52	91.82	93.42
Run2	92.52	90.64	92.26	93.49	93.43	93.37
Run3	92.70	92.15	92.56	93.95	91.18	93.3
Mean±Std	**92.57**±0.11	91.74±0.96	92.44±0.16	**93.65**±0.26	92.14±1.16	93.36±0.06
Optimizer	Radam	RadamGC	RadamMC	Radam	RadamGC	RadamMC
Run1	92.79	92.16	93.33	94.02	93.67	94.00
Run2	93.30	92.92	93.33	92.85	93.76	93.69
Run3	92.80	92.58	93.36	94.06	93.7	94.08
Mean±Std	92.96±0.29	92.55±0.38	**93.34**±0.02	93.64±0.69	93.71±0.05	**93.92**±0.21
Optimizer	Adabelief	AdabeliefGC	AdabeliefMC	Adabelief	AdabeliefGC	AdabeliefMC
Run1	92.91	92.71	93.07	93.86	93.74	93.78
Run2	92.67	93.04	92.94	92.31	93.88	93.72
Run3	92.85	92.83	92.91	93.98	93.73	93.56
Mean±Std	92.81±0.12	92.86±0.17	**92.97**±0.09	93.38±0.93	**93.78**±0.08	93.69±0.11

Note that the experiments are repeated three times with independent weight initialization

Table 2 Results comparison of Adam [13], Radam [18], and Adabelief [27] optimizers with the gradient centralization [26] and the proposed moment centralization on CIFAR100 dataset using VGG16 [22] and ResNet18 [9] CNN models

CNN model	VGG16 model			ResNet18 model		
Optimizer	Adam	AdamGC	AdamMC	Adam	AdamGC	AdamMC
Run1	67.74	67.68	68.47	71.4	73.74	74.48
Run2	67.51	67.73	69.43	71.47	73.34	73.92
Run3	67.96	68.26	68.01	71.64	73.58	74.64
Mean±Std	67.74±0.23	67.89±0.32	**68.64**±0.72	71.5±0.12	73.55±0.2	**74.35**±0.38
Optimizer	Radam	RadamGC	RadamMC	Radam	RadamGC	RadamMC
Run1	69.56	70.31	70.89	73.54	73.75	74.31
Run2	70.05	69.81	70.57	73.15	73.07	74.61
Run3	69.82	70.26	70.35	73.41	73.88	74.48
Mean±Std	69.81±0.25	70.13±0.28	**70.60**±0.27	73.37±0.2	73.57±0.44	**74.47**±0.15
Optimizer	Adabelief	AdabeliefGC	AdabeliefMC	Adabelief	AdabeliefGC	AdabeliefMC
Run1	70.84	70.85	71.44	74.05	73.69	74.38
Run2	70.71	71.17	70.83	74.11	73.95	74.86
Run3	70.37	70.56	70.84	74.3	73.74	74.74
Mean±Std	70.64±0.24	70.86±0.31	**71.04**±0.35	74.15±0.13	73.79±0.14	**74.66**±0.25

Note that the experiments are repeated three times with independent weight initialization

Table 3 Results comparison of Adam [13], Radam [18], and Adabelief [27] optimizers with the gradient centralization [26] and the proposed moment centralization on TinyImageNet dataset using VGG16 [22] and ResNet18 [9] CNN models

CNN model	VGG16 model			ResNet18 model		
Optimizer	Adam	AdamGC	AdamMC	Adam	AdamGC	AdamMC
Run1	42.38	45.02	43.50	49.08	53.90	53.02
Run2	41.92	43.00	41.06	49.00	53.60	53.62
Run3	42.72	44.42	39.22	49.00	52.54	54.10
Mean±Std	42.34±0.4	**44.15**±1.04	41.26±2.15	49.03±0.05	53.35±0.71	**53.58**±0.54
Optimizer	Radam	RadamGC	RadamMC	Radam	RadamGC	RadamMC
Run1	43.86	45.94	47.22	49.84	51.62	53.40
Run2	44.10	45.86	47.48	50.50	51.54	52.72
Run3	45.18	45.80	47.26	50.78	51.92	52.88
Mean±Std	44.38±0.70	45.87±0.07	**47.32**±0.14	50.37±0.48	51.69±0.20	**53.00**±0.36
Optimizer	Adabelief	AdabeliefGC	AdabeliefMC	Adabelief	AdabeliefGC	AdabeliefMC
Run1	47.32	47.84	47.22	51.82	52.02	53.04
Run2	49.12	47.60	52.86	46.78	51.80	53.08
Run3	47.16	47.94	48.34	51.04	51.56	53.78
Mean±Std	47.87±1.09	47.79±0.17	**49.47**±2.99	49.88±2.71	51.79±0.23	**53.30**±0.42

Note that the experiments are repeated three times with independent weight initialization

4 Experimental Results and Analysis

In order to demonstrate the improved performance of the optimizers with the proposed moment centralization, we conduct the image classification experiments using VGG16 [22] and ResNet18 [9] CNN models on CIFAR10, CIFAR100, and TinyImageNet datasets [14] and report the classification accuracy. The results are compared with the corresponding optimization method without using moment centralization. Moreover, the results are also compared with the corresponding optimization method with the gradient centralization [26]. We repeat all the experiments three times with independent initializations and consider the mean and standard deviation for the comparison purpose. All results are evaluated with the same settings as described above.

The classification accuracies on the CIFAR10, CIFAR100, and TinyImagenet datasets are reported in Tables 1, 2 and 3, respectively. It is noticed from these results that the optimizers with moment centralization (i.e., AdamMC, RadamMC, and AdabeliefMC) outperform the corresponding optimizers without moment centralization (i.e., Adam, Radam, and Adabelief, respectively) and with gradient centralization (i.e., AdamGC, RadamGC, and AdabeliefGC, respectively) in most of the scenario. The performance of the proposed AdamMC optimizer is also comparable with Adam on the CIFAR10 dataset, where AdamGC (i.e., Adam with gradient centralization [26]) fails drastically. It is also worth mentioning that the classification accuracy using the proposed optimizers is very consistent in different trials with a

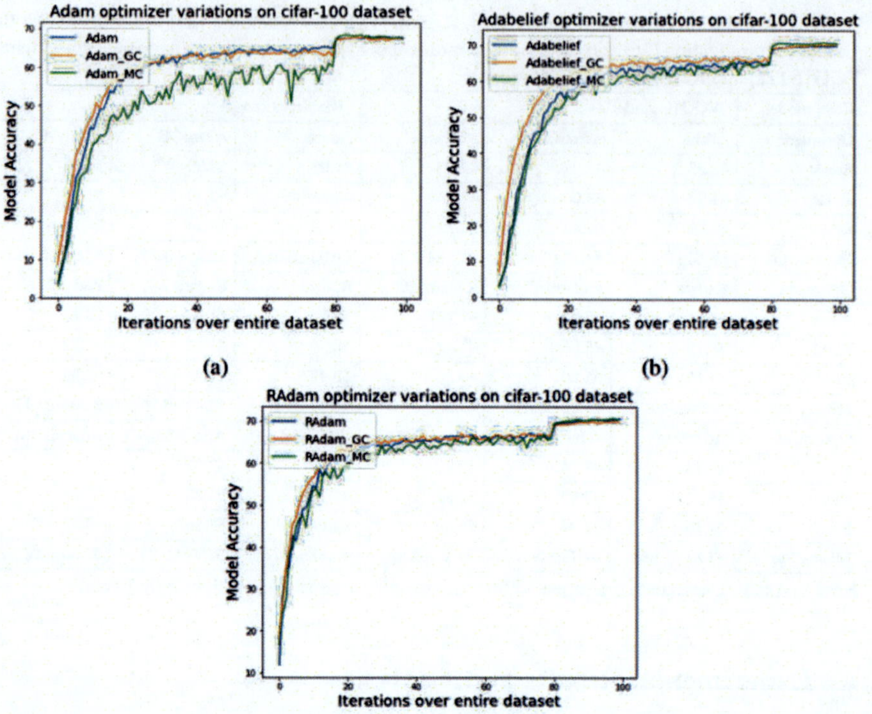

(c)

Fig. 1 Test accuracy vs epoch plots using **a** Adam, Adam_GC & Adam_MC, **b** Adabelief, Adabelief_GC & Adabelief_MC, and **c** Radam, Radam_GC & Radam_MC optimizers on CIFAR100 dataset

reasonable standard deviation in the results, except on TinyImageNet dataset using VGG16 model.

The accuracy plot w.r.t. epochs is depicted in Fig. 1 on CIFAR100 dataset for different optimizers. Note that the learning rate is set to 0.001 for first 80 epochs and 0.0001 for last 20 epochs. It is noticed in all the plots that the performance of the proposed optimizers boosts significantly and outperforms other optimizers when the learning rate is dropped. It shows that the proposed moment centralization leads to better regularization and reaches closer to minimum.

In order to justify the improved performance of the proposed Adam_MC method, we show the convergence plot in terms of the optimization trajectory in Fig. 2 with random intializations for following toy example:

$$f(x, y) = -2e^{\frac{-((x-1)^2+y^2)}{2}} - 3e^{\frac{-((x+1)^2+y^2)}{2}} + x^2 + y^2. \tag{7}$$

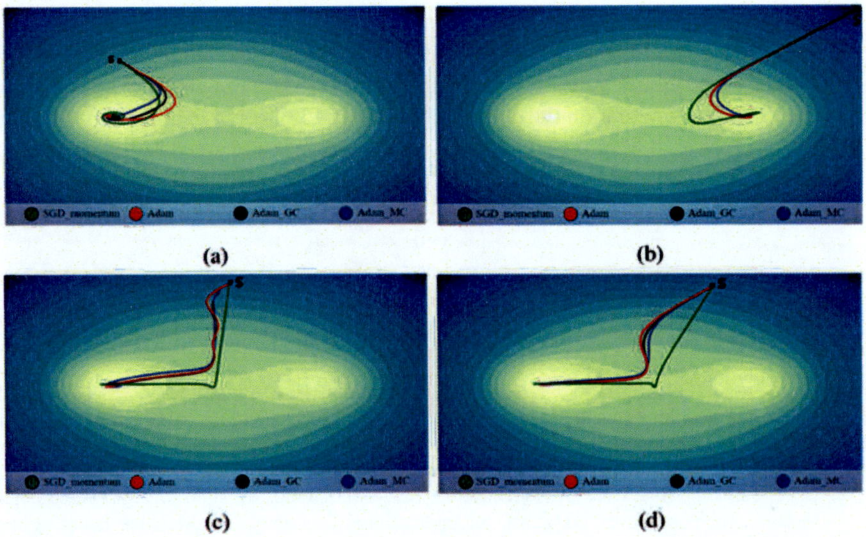

Fig. 2 Convergence of different optimization methods on a toy example with random initializations. S is the starting point

It is a quadratic 'bowl' with two gaussian looking minima at $(1, 0)$ and $(-1, 0)$, respectively. It can be observed in Fig. 2a that the Adam_MC optimizer leads to a shorter path and faster convergence to reach minimum as compared to other optimizers. Figure 1 illustrates how the moment centralization for Adam, RAdam, and Adabelief achieved higher accuracy after the 80th epoch than without. This demonstrates that optimizers with MC variation have faster convergence speed than without. The SGD shows much oscillations near the local minimum. In Fig. 2b, Adam and Adam_GC depict fewer turns as compared to SGD with momentum, while Adam_MC exhibits smoother updates. In Fig. 2c, d, Adam_MC leads to less oscillations throughout its course as compared to other optimizers. Hence, it is found that the proposed optimizer leads to shorter and smoother path as compared to other optimizers.

5 Conclusion

In this paper, a moment centralization is proposed for the adaptive momentum-based SGD optimizers. The proposed approach explicitly imposes the zero mean constraints on the first order moment. The moment centralization leads to better training of the CNN models. The efficacy of the proposed idea is tested with state-of-the-art optimizers, including Adam, Radam, and Adabelief on CIFAR10, CIFAR100, and TinyImageNet datasets using VGG16 and ResNet18 CNN models. It is found that the performance of the existing optimizers is improved when integrated with the proposed moment centralization in most of the cases. Moreover, the moment centralization outperforms the gradient centralization in most of the cases. Based on the findings from the results, it is concluded that the moment centralization can be used very effectively to train the deep CNN models with improved performance. It is also observed that the proposed method leads to shorter and smoother optimization trajectory. The future work includes the exploration of the proposed idea on different types of computer vision applications.

References

1. Basha, S.S., Ghosh, S., Babu, K.K., Dubey, S.R., Pulabaigari, V., Mukherjee, S.: Rccnet: An efficient convolutional neural network for histological routine colon cancer nuclei classification. In: 2018 15th International Conference on Control, Automation, Robotics and Vision (ICARCV). pp. 1222–1227. IEEE (2018)
2. Bernstein, J., Wang, Y.X., Azizzadenesheli, K., Anandkumar, A.: signsgd: Compressed optimisation for non-convex problems. In: International Conference on Machine Learning. pp. 560–569 (2018)
3. Bottou, L.: Large-scale machine learning with stochastic gradient descent. In: Proceedings of the COMPSTAT, pp. 177–186 (2010)
4. Chen, Z., Badrinarayanan, V., Lee, C.Y., Rabinovich, A.: Gradnorm: Gradient normalization for adaptive loss balancing in deep multitask networks. In: International Conference on Machine Learning. pp. 794–803. PMLR (2018)
5. Choi, S., Kim, T., Jeong, M., Park, H., Kim, C.: Meta batch-instance normalization for generalizable person re-identification. In: Proceedings of the IEEE/CVF Conference on Computer Vision and Pattern Recognition. pp. 3425–3435 (2021)
6. Dubey, S.R., Chakraborty, S., Roy, S.K., Mukherjee, S., Singh, S.K., Chaudhuri, B.B.: diffgrad: an optimization method for convolutional neural networks. IEEE transactions on neural networks and learning systems **31**(11), 4500–4511 (2019)
7. Dubey, S., Basha, S., Singh, S., Chaudhuri, B.: Curvature injected adaptive momentum optimizer for convolutional neural networks. arXiv preprint arXiv:2109.12504 (2021)
8. Duchi, J., Hazan, E., Singer, Y.: Adaptive subgradient methods for online learning and stochastic optimization. J. Mach. Learn. Res. **12**(Jul), 2121–2159 (2011)
9. He, K., Zhang, X., Ren, S., Sun, J.: Deep residual learning for image recognition. In: Proceedings of the IEEE conference on computer vision and pattern recognition. pp. 770–778 (2016)
10. Hinton, G., Srivastava, N., Swersky, K.: Neural networks for machine learning. Lecture 6a overview of mini-batch gradient descent course (2012)
11. Huang, H., Wang, C., Dong, B.: Nostalgic adam: Weighting more of the past gradients when designing the adaptive learning rate. In: Proceedings of the Twenty-Eighth International Joint Conference on Artificial Intelligence. pp. 2556–2562 (2019)
12. Ioffe, S., Szegedy, C.: Batch normalization: Accelerating deep network training by reducing internal covariate shift. In: International conference on machine learning. pp. 448–456. PMLR (2015)
13. Kingma, D.P., Ba, J.: Adam: A method for stochastic optimization. In: International Conference on Learning Representations (2015)
14. Krizhevsky, A., Hinton, G., et al.: Learning multiple layers of features from tiny images. Tech Report (2009)
15. Krizhevsky, A., Sutskever, I., Hinton, G.E.: Imagenet classification with deep convolutional neural networks. Adv. Neural Inf. Process. Syst. **25**, 1097–1105 (2012)
16. Le, Y., Yang, X.: Tiny imagenet visual recognition challenge. CS 231N **7**(7), 3 (2015)

17. LeCun, Y., Bengio, Y., Hinton, G.: Deep learning. Nature **521**(7553), 436–444 (2015)
18. Liu, L., Jiang, H., He, P., Chen, W., Liu, X., Gao, J., Han, J.: On the variance of the adaptive learning rate and beyond. In: International Conference on Learning Representations (2019)
19. Repala, V.K., Dubey, S.R.: Dual cnn models for unsupervised monocular depth estimation. In: International Conference on Pattern Recognition and Machine Intelligence. pp. 209–217. Springer (2019)
20. Roy, S., Paoletti, M., Haut, J., Dubey, S., Kar, P., Plaza, A., Chaudhuri, B.: Angulargrad: A new optimization technique for angular convergence of convolutional neural networks. arXiv preprint arXiv:2105.10190 (2021)
21. Salimans, T., Kingma, D.P.: Weight normalization: a simple reparameterization to accelerate training of deep neural networks. Adv. Neural Inf. Process. Syst. **29**, 901–909 (2016)
22. Simonyan, K., Zisserman, A.: Very deep convolutional networks for large-scale image recognition. In: International Conference on Learning Representations (2015)
23. Singh, D., Singh, B.: Investigating the impact of data normalization on classification performance. Appl. Soft Comput. **97**, 105524 (2020)
24. Srivastava, Y., Murali, V., Dubey, S.R.: Hard-mining loss based convolutional neural network for face recognition. In: International Conference on Computer Vision and Image Processing. pp. 70–80. Springer (2020)
25. Sutskever, I., Martens, J., Dahl, G., Hinton, G.: On the importance of initialization and momentum in deep learning. In: Proceedings of the International Conference on Machine Learning. pp. 1139–1147 (2013)
26. Yong, H., Huang, J., Hua, X., Zhang, L.: Gradient centralization: a new optimization technique for deep neural networks. In: European Conference on Computer Vision. pp. 635–652. Springer (2020)
27. Zhuang, J., Tang, T., Ding, Y., Tatikonda, S.C., Dvornek, N., Papademetris, X., Duncan, J.: Adabelief optimizer: Adapting stepsizes by the belief in observed gradients. Adv. Neural Inf. Process. Syst. **33** (2020)

Context Unaware Knowledge Distillation for Image Retrieval

Bytasandram Yaswanth Reddy, Shiv Ram Dubey, Rakesh Kumar Sanodiya, and Ravi Ranjan Prasad Karn

Abstract Existing data-dependent hashing methods use large backbone networks with millions of parameters and are computationally complex. Existing knowledge distillation methods use logits and other features of the deep (teacher) model and as knowledge for the compact (student) model, which requires the teacher's network to be fine-tuned on the context in parallel with the student model on the context. Training teacher on the target context requires more time and computational resources. In this paper, we propose context unaware knowledge distillation that uses the knowledge of the teacher model without fine-tuning it on the target context. We also propose a new efficient student model architecture for knowledge distillation. The proposed approach follows a two-step process. The first step involves pre-training the student model with the help of context unaware knowledge distillation from the teacher model. The second step involves fine-tuning the student model on the context of image retrieval. In order to show the efficacy of the proposed approach, we compare the retrieval results, no. of parameters, and no. of operations of the student models with the teacher models under different retrieval frameworks, including deep cauchy hashing (DCH) and central similarity quantization (CSQ). The experimental results confirm that the proposed approach provides a promising trade-off between the retrieval results and efficiency. The code used in this paper is released publicly at https://github.com/satoru2001/CUKDFIR.

B. Y. Reddy (✉) · R. K. Sanodiya · R. R. P. Karn
Department of Computer Science and Engineering, Indian Institute of Information Technology, Sri City, Chittoor, India
e-mail: yaswanthreddy.b18@iiits.in

R. K. Sanodiya
e-mail: rakesh.s@iiits.in

R. R. P. Karn
e-mail: raviranjanprasad.k@iiits.in

S. R. Dubey
Computer Vision and Biometrics Laboratory, Indian Institute of Information Technology, Allahabad, India
e-mail: srdubey@iiita.ac.in

M. Tistarelli et al. (eds.), *Computer Vision and Machine Intelligence*,
Lecture Notes in Networks and Systems 586,
https://doi.org/10.1007/978-981-19-7867-8_6

Keywords Knowledge distillation · Image retrieval · CNN model · Model compression

1 Introduction

In this age of big data, where voluminous data are generated from various sources with very fast speed, for image retrieval-based applications, parallel to the indexing methods [1], hashing methods have shown promising results. In the hashing methods, high-dimension media data like images/video are compressed into a low-dimension binary code (hash) such that media with similar data items has identical hash. Deep learning has become successful in various fields in the past few years. Many state-of-the-art models [2, 3] emerged with a large number of trainable parameters, making them good at learning complex patterns from data and computationally costly to run on low-end devices. Many model compression and acceleration techniques were introduced to decrease the computational complexity, like parameter tuning/quantization, transferred/compact convolution filters, low-rank factorization, and knowledge distillation [4]. Knowledge distillation is a model compression and acceleration method that helps the compact (student) model to perform nearly equal to or better than the deep (teacher) models by learning knowledge gained by deep (teacher) models. In vanilla Knowledge distillation [5, 6], logits of the pre-trained teacher model on the context are used as knowledge for the student network.

Knowledge distillation can be categorized into three types [7]. First, response-based knowledge distillation in which the logits of teachers are used as knowledge for students. Hilton et al. [5] used soft targets, which are soft-max probabilities of classes in which input is predicted to be ascertained by the teacher model as knowledge to distill. Second, feature-based knowledge distillation in which both the features of the last and intermediate layers are used as knowledge [8]. Third, relation-based knowledge, instead of learning from the features of intermediate layers/logits like in the previous models, relation-based knowledge distillation tries to distill the relationship between different layers or data samples. Yim et al. [9] used a Gram matrix between two layers that are calculated using the inner products between features from both the layers. It summarizes the relationship between the two layers and tries to distill this knowledge into the student. There are two main schemas of training during knowledge distillation, such as online and offline schemas. In the offline schema, we have fine-tuned teacher networks, and the student network will learn the teacher's knowledge along with the downstream task. In an online schema, the teacher and the student are trained simultaneously on the downstream task and knowledge distillation. A great deal of comprehensive survey is conducted on this in [7].

It has been seen that, recently, deep learning to hash methods enables end-to-end representation learning. They can learn complex non-linear hash functions and achieve state-of-the-art retrieval performance. In particular, they proved that the networks could learn the representations that are similarity-preserved and can quantize representations to be binary codes [10–16]. Deep learning-based hashing can be categorized into various buckets based on the training approach, namely supervised, unsupervised, semi-supervised, weakly supervised, pseudo supervised, etc. The backbone architectures used for these different training modes contain CNN, auto-encoders, Siamese, triplet networks, GAN, etc., and use different descriptors for representing hash code, namely binary, where hashes are a combination of 0's and 1's, real-valued, and aggregation of both binary and real-valued. A comprehensive survey on image retrieval is conducted in [17] which can be referred for holistic understanding.

Although most existing deep learning methods for hashing are tailored to learn efficient hash codes, the used backbone models become computationally costly with millions of parameters. On the other hand, most existing knowledge distillation modes are carried out in two steps; the first step consists of fine-tuning the teacher model on the context, followed by the training of the student model with knowledge distillation and contextual losses. However, constantly, fine-tuning teacher models for each context might be computationally expensive since they tend to be deeper networks.

In this work, we present the findings on the approach for context unaware knowledge distillation in which the knowledge is transferred from a teacher network that is not fine-tuned on the context/downstream task. This approach is carried out in two steps; first, we distill the knowledge from an un-trained teacher (on context) to a student on a specific dataset making the student mimic the output of the teacher for that dataset. Then, we fine-tune our student on any context of the same dataset, thus decreasing the computation overhead incurred from fine-tuning the teacher network for each context on the dataset. We experimented our approch in the context of image retrieval, then compared the results of teacher and student by using each of them as a backbone network in CSQ [11] and DCH [10]. The experiments are conducted on the image retrieval task on two different datasets with multiple hash lengths.

2 Related Work

The supervised deep learning to hash methods such as DCH [10], CSQ [11], DTQ [14], DQN [13], and DHN [12] is successful in learning non-linear hash function for generating hash of different bit-sizes (16, 32, 48, 64...) and achieved state-of-the-art results in image retrieval. DQN [13] uses a Siamese network that uses pairwise cosine loss for better linking the cosine distance of similar images and a quantization loss for restricting the bits to binary. DTQ [14] introduces the group hard triplet selection module for suitable mining triplets (anchor, positive, and negative) in real-time. The concept behind it is to divide the training data into many groups at random, then pick one hard negative sample for each anchor-positive pair from each group at random. A specified triplet loss is used for pulling together anchor and positive pairs and moving away anchor and negative pairs, as well as a quantization loss for monitoring the efficiency and restricting hash bits to be binary.

DCH [10] exploits the lack of the capability of concentrating relevant images to be within a small hamming distance of existing hashing methods. Instead, it uses Cauchy distribution for distance calculation between similar and dissimilar image pairs instead of traditional sigmoid. Similar to the previous methods, DCH [10] includes quantization loss for controlling hash quality. CSQ [11] replaces the low efficiency in creating image datasets while using pairwise or triplet loss by introducing a new concept of "hash centers". Hash centers are unique K dimension vectors (K refers to desired hash size), one for each class. They are defined as K-dimensional binary points (0/1) in hamming space with an average pairwise distance greater than or equal to $K/2$ between any two hash centers. Now, they use these hash centers as a target, similar to a multi-class classification problem.

DTQ [15] uses a compact student network for fast image retrieval. The training is carried out in three phases. In the first phase, a modified teacher is trained on the classification task. The teacher and student models have a fully connected layer of N neurons ($FC@N$) (where N indicates desired hash length) before the classification layer. The second phase consists of knowledge distillation between teacher and student, where the output of the teacher's $FC@N$ is used as knowledge. The loss function for the student network includes knowledge distillation loss (a regression loss between the teacher and student's output of $FC@N$) and the classification loss. In the last phase, the full precision student model is quantized to get a ternary model (where weights of each layer are represented in only three states) and fine-tuned with knowledge distillation to find the best ternary model.

3 Context Unaware Knowledge Distillation

In this section, we first present an overview of ResNet followed by the architecture of two student models V1 and V2, one for each teacher, namely ResNet50 [18] and AlexNet [19], respectively. We then present the process of knowledge distillation between teacher and student.

3.1 ResNet Overview

The building block of ResNet is the residual block which is one of the significant advancements in deep learning. The problem with general plain deep learning architectures, which constitute a sequence of convolution layers and other layers like batch normalization, etc., is diminishing gradient during backpropagation. Weights of specific layers cannot be updated since no gradient affects the model's learning and degrades its performance. Residual blocks tackle this problem by adding an identity connection which acts as a way to avert the vanishing gradient problem. Let us denote $H(x)$ as the desired mapping function where x is the input of the residual block. We make the residual building block fit the mapping function $F(x)$ such that $F(x) = H(x) - x$. It is demonstrated in the left subfigure of Fig. 1.

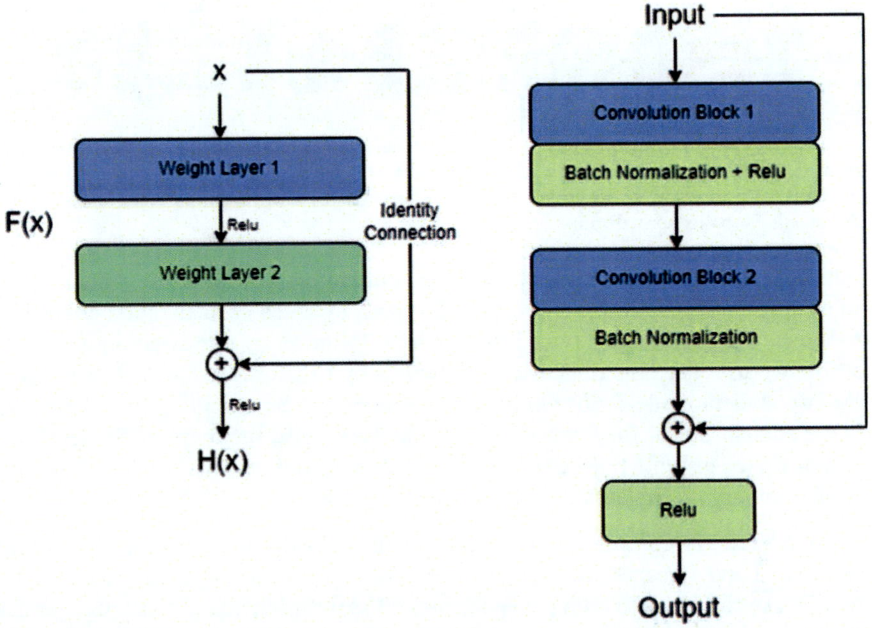

Fig. 1 (left) Residual block of ResNet architecture [18] and (right) building block of student architecture

3.2 *Student Model*

Each student model consists of two building blocks, namely basic block, which is inspired by residual blocks [18]. These blocks are stacked to form a layer. These layers are then stacked together to form the student network.

Basic Block and Layer The architecture of the basic block is demonstrated in the right subfigure of Fig. 1. Each convolution layer contains a kernel of size 3×3 and a stride of one. The input dimension is retained throughout the basic block to support the identity connection. Each student model contains five layers, and each of the five layers contains 2, 3, 5, 3, and 2 basic blocks stacked together, respectively. The dimension of the input feature is retained in a layer. Each layer (except the 5th) is followed by a convolution block with a 3×3 kernel and stride of two to reduce the dimension of the features by half, then followed by a batch normalization layer. The architecture of the student is demonstrated in Fig. 2. Each layer and the following convolution block contain an equal number of filters.

Student Network Since the dimensions of flattened output features before classification layer of ResNet50 [18] are 2048 and that of AlexNet [19] is 4096, we made two different student models **StudentV1** and **StudentV2**. Before passing input to the first layer, we tried to reduce the dimension of the input similar to ResNet [18]. The

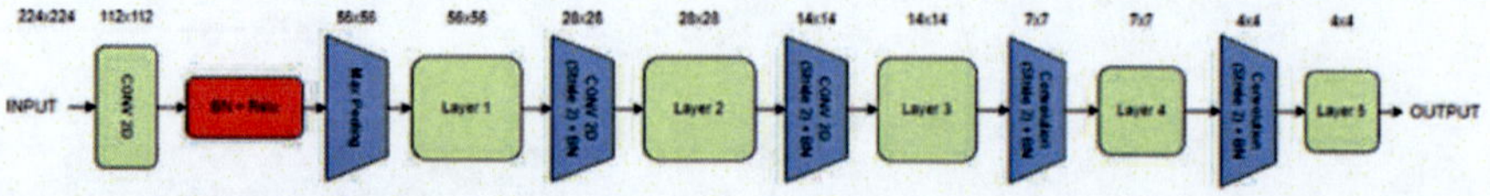

Fig. 2 represents student model, $X \times X$ on top of each blocks represents the dimension of output from that layer (excluding channel dimension). Each layer is followed by a convolution layer (CONV2D) with stride of two to decrease the dimensions by half instead of using max-pooling. BN stands for batch normalization

initial module of student consists of a convolution block with a 7×7 filter, stride of two, and padding of three. It is then followed by batch normalization and ReLU to reduce the dimensionality of input by half, followed by a max-pooling layer of stride two to reduce the dimension of features by half further. The architectures of the two student models differ only in the number of filters in 5th layer. **StudentV1** is the student model for ResNet50 [18]. The layer wise summary of StudentV1 is demonstrated in Table 1. **StudentV2** is the student model for AlexNet [19]. The layer wise summary of StudentV2 is demonstrated in Table 2.

Comparison The comparison of teacher and their respective student models is done in Tables 3 and 4, respectively. With less trainable parameters and fewer FLOPs, the model takes less time to train per epoch as well as for inference. We can observe that 85.46% reduction in the number of trainable parameters in the ResNet50–StudentV1 pair and a 91.16% reduction in the number of trainable parameters in the AlexNet–StudentV2 pair. In the AlexNet–StudentV2 pair, as StudentV2 contains identity connection, whereas AlexNet doesn't, it results in higher FLOPs for student despite having lower trainable parameters. In the ResNet50–StudentV1 pair, since both networks contain identity connections, fewer trainable parameters in student resulted in lower FLOPs.

Table 1 Layerwise summary of StudentV1 with input size of $224 \times 224 \times 3$

Layer	Number of filters	Output shape
InputLayer	0	$224 \times 224 \times 3$
Initial module (Conv 2D + BN + ReLU)	64	$112 \times 112 \times 64$
Max Pool	64	$56 \times 56 \times 64$
Layer_1	64	$56 \times 56 \times 64$
Conv_2d_1 (Conv 2D + BN)	64	$28 \times 28 \times 64$
Layer_2	64	$28 \times 28 \times 64$
Conv_2d_2 (Conv 2D + BN)	128	$14 \times 14 \times 128$
Layer_3	128	$14 \times 14 \times 128$
Conv_2d_3 (Conv 2D + BN)	128	$7 \times 7 \times 128$
Layer_4	128	$7 \times 7 \times 128$
Conv_2d_4 (Conv 2D + BN)	128	$4 \times 4 \times 128$
Layer_5	128	$4 \times 4 \times 128$
Flatten	NA	2048

BN batch normalization

3.3 Knowledge Distillation

We can view the process of knowledge distillation from teacher to student network as a regression problem. Hence, we can use the loss function of regression problems like L1 (mean absolute error, mae), L2 (mean squared error, mse), or smooth L1 loss. We consider mean square error (MSE) as our loss function as most of the activation values of the last layer are less than 1, which makes smooth L1 loss perform similar to L2 loss. Here, we use the output features of the last layer of the teacher as knowledge to train the student. The equation is given as follows:

Table 2 Layerwise summary of StudentV2 with input size of $224 \times 224 \times 3$

Layer	Number of filters	Output shape
InputLayer	0	$224 \times 224 \times 3$
Initial module (Conv 2D + BN + ReLU)	64	$112 \times 112 \times 64$
Max Pool	64	$56 \times 56 \times 64$
Layer_1	64	$56 \times 56 \times 64$
Conv_2d_1 (Conv 2D + BN)	64	$28 \times 28 \times 64$
Layer_2	64	$28 \times 28 \times 64$
Conv_2d_2 (Conv 2D + BN)	128	$14 \times 14 \times 128$
Layer_3	128	$14 \times 14 \times 128$
Conv_2d_3 (Conv 2D + BN)	128	$7 \times 7 \times 128$
Layer_4	128	$7 \times 7 \times 128$
Conv_2d_4 (Conv 2D + BN)	128	$4 \times 4 \times 128$
Layer_5	256	$4 \times 4 \times 256$
Flatten	NA	4096

BN batch normalization

Table 3 Comparison between ResNet50 and StudentV1. FLOPs are calculated on image of dimension $224 \times 224 \times 3$

Model	Trainable parameters	FLOPs
ResNet50	23, 639, 168	4.12 Giga
StudentV1	3, 437, 568	1.10 Giga

Table 4 Comparison between AlexNet and StudentV2. FLOPs are calculated on image of dimension $224 \times 224 \times 3$

Model	Trainable parameters	FLOPs
AlexNet	57, 266, 048	0.72 Giga
StudentV2	5, 060, 352	1.119 Giga

$$L_2(T, S) = \frac{1}{N}\sum_{i=1}^{N}\sum_{j=1}^{K}(S_{ij} - T_{ij})^2 \quad (1)$$

where N represents the number of images in the dataset, K represents the dimensionality of features which is 2048 and 4096 for ResNet50 [18] and AlexNet [19], respectively, T_{ij} represents the feature vectors of the last layer of the teacher network, and S_{ij} represents the feature vector of the last layer of the student network. We fix the teacher weights and only update the student weights during backpropagation as portrayed in Fig. 3.

4 Experiments

As discussed above, the first step includes training the teacher–student pair with knowledge distillation loss. We then fine-tune the student model using the pre-trained student model as the backbone network instead of their respective teacher on image retrieval task under the retrieval frameworks of CSQ [11] and DCH [10].

4.1 Datasets and Evaluation Metrics

We use the CIFAR10 [21] and NUS-WIDE [20] datasets for experiments. CIFAR10 [21] contains images from 10 different classes (categories), and each class includes 6,000 images. For knowledge distillation, we use the entire dataset for training teacher–student pairs. For fine-tuning on image retrieval task, we randomly select

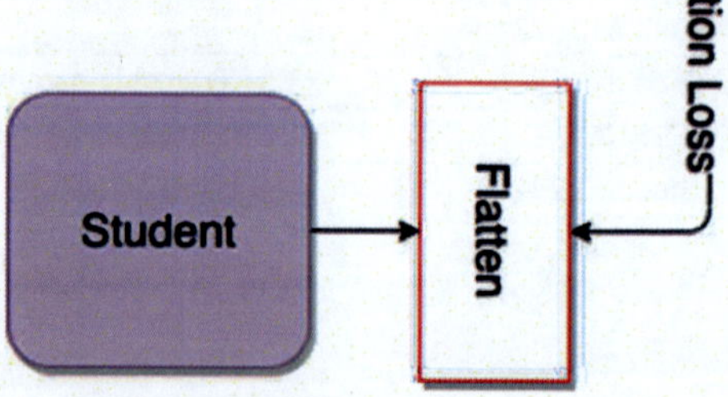

Fig. 3 Training process for knowledge distillation in which we freeze the weights of teacher model and train student model with MSE-based knowledge distillation loss

Table 5 Comparison of mAP of different bits under CSQ retrieval framework [11]

Model	NUS-WIDE (mAP@5000)			CIFAR10 (mAP@5000)		
	16 bit	32 bit	64 bit	16 bit	32 bit	64 bit
ResNet50	0.812	0.833	0.839	0.834	0.851	0.849
StudentV1	0.779	0.807	0.819	0.824	0.822	0.840
AlexNet	0.762	0.794	0.808	0.784	0.778	0.787
StudentV2	0.765	0.798	0.812	0.763	0.747	0.767

Table 6 Comparison of mAP of different bits under DCH retrieval framework [10]

Model	NUS-WIDE (mAP@5000)			CIFAR10 (mAP@5000)		
	16 bit	32 bit	48 bit	16 bit	32 bit	48 bit
ResNet50	0.778	0.784	0.780	0.844	0.868	0.851
StudentV1	0.766	0.781	0.782	0.819	0.828	0.845
AlexNet	0.748	0.76	0.758	0.757	0.786	0.768
StudentV2	0.743	0.755	0.762	0.754	0.769	0.754

1000 images (100 images per class) as the query set and 5000 images as the training set (500 images per class), with the remaining images as database images as done in the works of DCH [10] and CSQ [11]. NUS-WIDE [20] is a public Web image dataset that contains 2,69,648 images. We use the subset of NUS-WIDE in which there are only 21 frequent categories. We use the entire NUS-WIDE dataset to train teacher–student pairs for knowledge distillation. For fine-tuning on the image retrieval task, we randomly choose 2100 images (100 images per class) as a test set and 10,500 images (500 images per class) as a training set, leaving the rest 1,49,736 as a database.

We use mean average precision (mAP) as the evaluation metric for image retrieval. To calculate $mAP@N$ for a given set of queries, we first calculate average precision $(AP)@N$ for each set of the query as specified as follows:

$$AP@N = \frac{\sum_{i=1}^{N} P(i)\alpha(i)}{\sum_{i=1}^{N} \alpha(i)} \tag{2}$$

where $P(i)$ is precision of ith retrieved image and $\alpha(i) = 1$ if the retrieved image is a neighbor (belongs to same class) and $\alpha(i) = 0$ otherwise. N denotes number of images in the database. mAP is calculated as mean of each query average precision and is represented as follows:

$$mAP@N = \frac{\sum_{i=1}^{Q} AP@N(i)}{Q} \tag{3}$$

where Q represents a number of query images. We use $mAP@5000$ as our evaluation metric for the image retrieval for both datasets.

Table 7 Comparison between the training time per epoch (in seconds) of students and their respective teachers on NUS-WIDE dataset

Loss	ResNet50	StudentV1	AlexNet	StudentV2
CSQ 64	118	37	40	38
DCH 48	120	39	39	38

All the experiments are conducted on Tesla T4 GPU and Intel Xeon CPU

4.2 Training and Results

Both the teacher networks are initialized with the pre-trained weights of ImageNet [22] classification task. During knowledge distillation training, the Adam optimizer is used with learning rate (LR) of 1e-4 for the ResNet50–StudentV1 pair and LR of 3e-6 for the training of the AlexNet–StudentV2 pair. We train teacher–student pairs for 160 and 120 epochs for CIFAR10 [21] and NUS-WIDE [20] datasets, respectively. We add a fully connected layer with n neurons (where n is the number of desired hash bits such as 16,32,48,64) to the student and teacher models for image retrieval. We then train it under the retrieval frameworks of CSQ [11] and DCH [10]. For training on image retrieval, we use the RMSProp optimizer with a learning rate of 1e-5. Tables 5 and 6 represent the results with backbone networks as teachers (i.e., ResNet50 and AlexNet), students (i.e., StudentV1 and StudentV2) for different hash bits under CSQ and DCH retrieval frameworks, respectively. It can be seen that the performance of StudentV1 and StudentV2 models is either better or very close to ResNet50 and AlexNet teacher models, respectively, in spite of having significantly reduced number of parameters. Moreover, the no. of FLOPS of StudentV1 model is also significantly as compared to the ResNet50 model. It is also noted that the StudentV2 model outperforms the AlexNet model on NUS-WIDE dataset under CSQ framework. The performance of the proposed student models is better for 48 bit hash codes on NUS-WIDE dataset under DCH framework. Table 7 represents the training time per epoch of teacher with their respective students (in seconds) on NUS-WIDE dataset for CSQ 64-bit configuration and DCH 48 bit configuration, and we can observe a drop of nearly $3.1X$ and $1.05X$ times training time per epoch in the case of StudentV1 and StudentV2, respectively.

For a given query image from the test set, Figs. 4 and 5 represent the top five retrieved images from the database of 1, 49, 736 images based on Hamming distance. The query image belongs to three categories: Building, Cloud, and Sky. All the retrieved images contain at least two categories, i.e., Building and Sky making them relevant retrieved images.

Fig. 4 Top row shows the retrieved images using ResNet50 [18] as the backbone in CSQ [11] with a 64-bit hash, and the bottom row shows the retrieved images using StudentV1 as a backbone in CSQ with a 64-bit hash. The left subfigure in each row represents the query image from the test set of NUS-WIDE [20], and the following subfigures represent the top 5 similar images from the database

Fig. 5 Top row shows the retrieved images using AlexNet [19] as the backbone in CSQ [11] with a 64-bit hash, and the bottom row shows the retrieved images using StudentV2 as a backbone in CSQ with a 64-bit hash. The left subfigure in each row represents the query image from the test set of NUS-WIDE [20], and the following subfigures represent the top 5 similar images from the database

5 Conclusions

Deep learning to hash is an active research area for image retrieval tasks where one uses deep learning algorithms to act as hash functions with appropriate loss functions such that images with similar content have a similar hash. Most present-day, deep learning algorithms have deep convolution neural networks as backbone models, which are computationally expensive. In general, compact (student) models with less trainable parameters are less computationally complex but do not perform well as deep (teacher) models on the tasks. Most existing knowledge distillation methods require the teacher model to be fine-tuned on the task, requiring more training time and computational resources. In this work, we propose a two-fold solution to increase the performance of the student model using knowledge from a teacher which is not trained on the context.

We observe that the student model performed in equal terms to teacher models with a maximum of only a 4% drop in mAP and a maximum of 0.4% gain in mAP compared to their respective teacher models. At the same time, number of trainable parameters is reduced by 85.4% in the case of StudentV1 and 91.16% in the case of StudentV2. A decrease in the number of trainable parameters leads to significant reduction in training time per epoch. We got a nearly $3.1X$ times drop in training time per epoch in the case of StudentV1 and a $1.05X$ times drop in training time per epoch in the case of StudentV2. Once the student–teacher knowledge distillation is done on a dataset, we can reuse our student model for any fine-tuning task on the same dataset without repeating the knowledge distillation step and without training the teacher model on the fine-tuning task. This solution can be used in diverse applications where model compression is required.

References

1. Lew, M.S. , Sebe, N., Djeraba, C., Jain, R.: Content-based multimedia information retrieval: state of the art and challenges. ACM Trans. Multimedia Comput. Commun. Appl. (TOMM) **2**(1), pp. 1–19 (2006)
2. Dubey, S.R., Singh, S.K., Chu, W.T.: Vision transformer hashing for image retrieval. IEEE International Conference on Multimedia and Expo (2022)
3. Singh, S.R., Yedla, R.R., Dubey, S.R., Sanodiya, R., Chu, W.T.: Frequency disentangled residual network. arXiv preprint arXiv:2109.12556 (2021)
4. Cheng, Y., Wang, D., Zhou, P., Zhang, T.: A survey of model compression and acceleration for deep neural networks. arXiv preprint arXiv:1710.09282 (2017)
5. Hinton, G., Vinyals, O., Dean J.: Distilling the knowledge in a neural network. arXiv preprint arXiv:1503.02531 (2015)
6. Ba, L.J., Caruana, R.: Do deep nets really need to be deep?" CoRR, Vol. abs/1312.6184 (2013). [Online]. Available: http://arxiv.org/abs/1312.6184
7. Gou, J., Yu, B., Maybank, S.J., Tao, D.: Knowledge distillation: A survey, CoRR, Vol. abs/2006.05525 (2020). [Online]. Available: https://arxiv.org/abs/2006.05525
8. Romero, A., Ballas, N., Kahou, S.E., Chassang, A., Gatta, C., Bengio, Y.: Fitnets: Hints for thin deep nets, arXiv preprint arXiv:1412.6550 (2014)
9. Yim, J., Joo, D., Bae, J., Kim, J.: A gift from knowledge distillation: fast optimization, network minimization and transfer learning. In: Proceedings of the IEEE Conference on Computer Vision and Pattern Recognition, pp. 4133–4141 (2017)
10. Cao, Y., Long, M., Liu, B., Wang, J.: Deep cauchy hashing for hamming space retrieval. In: Proceedings of the IEEE Conference on Computer Vision and Pattern Recognition, pp. 1229–1237 (2018)
11. Yuan, L., Wang, T., Zhang, X., Tay, F.E., Jie, Z., Liu, W., Feng, J.: Central similarity quantization for efficient image and video retrieval. In: Proceedings of the IEEE/CVF Conference on Computer Vision and Pattern Recognition, pp. 3083–3092 (2020)
12. Zhu, H., Long, M., Wang, J., Cao, Y.: Deep hashing network for efficient similarity retrieval. In: Proceedings of the AAAI Conference on Artificial Intelligence, vol. 30 (2016)

13. Cao, Y., Long, M., Wang, J., Zhu, H., Wen, Q.: Deep quantization network for efficient image retrieval. In: AAAI (2016)
14. Liu, B., Cao, Y., Long, M., Wang, J., Wang, J.: Deep triplet quantization. In: Proceedings of the 26th ACM international conference on Multimedia, pp. 755–763 (2018)
15. Zhai, H., Lai, S., Jin, H., Qian, X., Mei, T.: Deep transfer hashing for image retrieval. IEEE Trans. Circuits Syst. Video Technol. **31**(2), 742–753 (2020)
16. Cao, Z., Long, M., Wang, J., Yu, P.S.: Hashnet: deep learning to hash by continuation. In: Proceedings of the IEEE International Conference on Computer Vision, pp. 5608–5617 (2017)
17. Dubey, S. R.: A decade survey of content based image retrieval using deep learning. IEEE Trans. Circuits Syst. Video Technol. (2021)
18. He, K., Zhang, X., Ren, S., Sun, J.: Deep residual learning for image recognition. In: Proceedings of the IEEE Conference on Computer Vision and Pattern Recognition, pp. 770–778 (2016)
19. Krizhevsky, A., Sutskever, I., Hinton, G.E.: Imagenet classification with deep convolutional neural networks. Adv. Neural Inf. Process. Syst. **25**, 1097–1105 (2012)
20. Chua, T.-S., Tang, J., Hong, R., Li, H., Luo, Z., Zheng, Y.: Nus-wide: a real-world web image database from national university of Singapore. In Proceedings of the ACM International Conference on Image and Video Retrieval, pp. 1–9 (2009)
21. Krizhevsky, A.: Learning multiple layers of features from tiny images, pp. 32-33 (2009). [Online]. Available: https://www.cs.toronto.edu/kriz/learning-features-2009-TR.pdf
22. Deng, J., Dong, W., Socher, R., Li, L.-J.: Kai Li and Li Fei-Fei, "ImageNet: A large-scale hierarchical image database." IEEE Conf. Comput. Vis. Pattern Recognit. **2009**, 248–255 (2009). https://doi.org/10.1109/CVPR.2009.5206848

Detection of Motion Vector-Based Stegomalware in Video Files

Sandra V. S. Nair and P. Arun Raj Kumar

Abstract Cybercriminals are increasingly using steganography to launch attacks on devices. The cyberattack is more threatening as steganography hides the embedded malware, if any, making it harder to detect by various anti-virus tools. Such malware is called stegomalware. Since video files are larger and have a complex structure, they also have a high capacity for hiding malware. Motion vector (MV)-based steganography techniques do not cause much distortion in video files. Therefore, it remains to be one of the leading video steganography techniques. This paper deals with a lightweight solution for MV-based stegomalware detection in video files. Our model is compatible with state-of-the-art video coding standards having variable macroblock sizes and different motion vector resolutions. The proposed method obtained an accuracy of 95.8% on testing H.264 videos with various embedding rates. The 81-D spatial and temporal features result in the high performance of the proposed model.

Keywords Stegomalware · Steganography · Stego video · Motion vector

1 Introduction

Steganography is the process of embedding secret information (stego object) in an ordinary file (cover object) to avoid detection. Steganography techniques for image, video, audio, and text files are gaining popularity. These methods exploit the limitations of the human visual or auditory system. A little distortion in audio or image data is unrecognizable by humans. For example, a steganography algorithm creates

S. V. S. Nair (✉) · P. Arun Raj Kumar
Department of Computer Science and Engineering, National Institute of Technology Calicut, Kozhikode, India
e-mail: sandravsnair@gmail.com

P. Arun Raj Kumar
e-mail: park@nitc.ac.in

M. Tistarelli et al. (eds.), *Computer Vision and Machine Intelligence*,
Lecture Notes in Networks and Systems 586,
https://doi.org/10.1007/978-981-19-7867-8_8

an embedded image visually similar to the original image. Efficient steganography algorithms embed data with the least possible distortion of the original content.

The tool used for analyzing and detecting stegomalware is called a steganalyzer. There are two types of steganalyzers: quantitative and qualitative. A qualitative steganalyzer typically searches for the presence of a secret message, whereas a quantitative steganalyzer estimates information such as its embedding rate and content. Similarly, there are two types of steganalysis methods: specific (targeted) and blind (universal). Specific steganalysis detects a particular steganography method. Blind steganalysis detects the presence of a group of steganography algorithms.

Stegomalware is a malware type that uses steganography to escape its detection. Video, audio, and text files shared through the Internet hide stegomalware. The users downloading the particular embedded file will be affected. Finally, the attacker gets access to the user's data without authorization.

The study on the stegomalware history points out that images are used the most for embedding malware. Image steganography techniques include embedding secret information in pixel data, discrete cosine transform (DCT) coefficients, discrete wavelet transform (DWT) coefficients, etc. According to [2], most stegomalware methods hide malware settings or configuration files in ordinary files. The URLs inside the configuration files are utilized for downloading further resources. Finally, the malicious file enters the victim's system.

The malware TeslaCrypt is a ransomware that utilizes steganography. It redirects the users to a malicious Web page. The most appropriate attack method for the user is selected. Then, a malicious executable is downloaded via an HTTP page with hidden C&C commands. The malware Lokibot hides its malicious source code in a PNG image file. In the case of Cerber ransomware, the victim downloads a malicious word document attached to a phishing email. The word document contains a macro code. It downloads a steganographic image that hides a malicious executable code. The malicious code is executed, and the user data is encrypted. Stegoloader is a malware that hid encrypted URLs in BMP and PNG images. These URLs point to additional components required by the malware. The above-listed malware utilized steganography to hide malicious codes, URLs, or commands. However, Duqu malware implemented steganography to embed the stolen user data into an image before sending it to the controller server to avoid detection.

Currently, video steganography techniques are attaining more popularity. Because of the impact of social media, the usage and download rate of videos are high. Thus, video files are a perfect target for embedding malware. Video steganography methods include embedding data into motion vectors (MVs), DCT coefficients, variable-length codes, etc. Video coding does not compress motion vector information. Also, manipulation of MVs does not cause visible visual distortion. Therefore, motion vector steganography is the best choice to hide secret bits. This paper implements an MV-based blind qualitative steganalyzer for video files.

Summing up, the contributions of this paper are follows:

- A review of steganalysis techniques in video files.
- A lightweight solution for MV-based stegomalware detection in video files that considers both spatial and temporal features of the video. This paper introduces an 81-dimensional feature set.
- Evaluation of the model with real-world malware and 95.8% detection accuracy.
- To the best of our knowledge, this paper is the first to study the steganalysis of stego videos with malware embedded in them.

2 Literature Survey

2.1 Steganography in Video Files

There are two types of frames in a video: intra-frame and inter-frame. An intra-frame does not depend on other video frames for compression (no reference frame required). Meanwhile, an inter-frame depends on additional video frames for the compression process. The performance of intra-frame steganography techniques is not better than image steganography techniques. Hence, inter-frame steganography has more popularity among video steganography techniques. The literature describes various ways of inter-frame video steganography methods.

Steganography uses motion vector attributes such as its magnitude and phase angle. Xu et al. [21] proposed a method for video steganography by selecting motion vectors with higher magnitude and embedding bits to them. Similarly, Jue et al. [8] chose motion vectors with higher magnitude for steganography. Since motion vectors with high magnitude may not have high prediction error, [1] chose motion vectors for embedding based on the prediction error.

Discrete cosine transform (DCT), quantized DCT (QDCT), and discrete wavelet transform (DWT) coefficients are used for video steganography [7, 9]. Similarly, embedding techniques on variable-length codes [11] are in practice.

2.2 Steganalysis in Video Files

The following literature survey discusses the existing techniques used for stego video detection. It is worth noting that none of the papers have embedded malware in the videos.

Deep Learning Deep learning is used increasingly for stegomalware detection. Huang et al. [6] modify SRNet—steganalysis residual network for images and formed video steganalysis residual network (VSRNet) for feature extraction. The quantitative steganalyzer solution has two input channels that use the motion vector and the prediction residual features, respectively. It outputs a 512-dimensional feature vector. The solution includes M feature extraction modules for different embedding rates $P_1, P_2, \ldots, P_M$. The final feature vector will be a $M \times 512$-dimensional vector. It is the input for an embedding rate estimation module.

Liu et al. [12] consider embedding as adding a low amplitude noise to the image content. The proposed method is a convolutional neural network (CNN). It is a universal steganalyzer for MV and intra-prediction mode steganography. In video files, spatial features reveal the relationship among different macroblocks or pixels in the same video frame. Temporal features signify the temporal correlation among different video frames. This particular solution considers only the spatial features of videos.

Motion Vector-Based Motion vector steganography is the most popular method of video steganography due to the minimal visual distortion. Motion vectors are critical elements used in the process of video compression. During video compression, to encode a macroblock B of size $n \times n$, the encoder uses a frame R as the reference frame and searches for the best matching block R_B within R. The spatial displacement (motion vector) between B and R_B is determined. The prediction error representing the difference of pixel values in B and R_B is also coded and transmitted.

According to Wu et al. [18], joint probability matrices with values specifying the probability of difference between the motion vector of a central macroblock and its neighboring macroblocks act as the feature input of a LIBSVM classifier module. If the difference between an MV and its neighboring MV is more, the correlation between two MVs is probably less and vice versa.

Ghamsarian et al. [5] suggest an improvement over the previous experiment. The spatial and temporal features of motion vectors form a 54-dimensional feature set in this paper. It detects MV-based stego videos blindly. The spatial features used in this method depend on the probability of differences between the current motion vector and neighboring motion vectors. The temporal features exploit the local optimality feature of motion vectors.

Pixel Intensity-Based According to Chakraborty et al. [4], pixels in a frame have the property of Markov random field (MRF). The image intensity of a pixel depends only on its neighborhood in the MRF model. Images are locally smooth except for a few discontinuities like edges and region boundaries. The MRF model contains cliques. Cliques are a set of points or pixels which are neighbors to each other. The proposed method computes the clique potential for each clique of each frame. Then, the sum of the clique potentials (Gibbs energy) is calculated. Finally, Gibbs energy distribution is determined. The method calculates standard deviation S for all the frames. The average value of S is much higher for tampered videos. S is calculated for every frame in this method. The difference of S for every consecutive frame is determined. The video is malicious if the difference is more than a predefined threshold.

Optimal Matching Property The decoded motion vector will be different if the original motion vector is manipulated to hide information. Only the slightest changes are made to the motion vector to reduce the detection rate of steganography. The embedding of secret information replaces the correct motion vector with one of its neighbors. Hence, the original MV is one of the neighbors of the decoded MV. It will be the optimal MV compared to the decoded one.

Ren et al. [15] propose a solution considering this particular property. The method calculates the sum of absolute difference (SAD) value between a particular macroblock and all the macroblocks in the neighborhood (Say $N_{5\times5}$) of its reference macroblock. If the SAD value is minimum between the macroblock and the reference macroblock, then the motion vector is said to be optimally matching. Otherwise, it is not optimally matching. The proposed method considers this feature of each macroblock in the video. The subtractive probability of optimal matching (SPOM) denotes the difference between the probability of optimal matching of calibrated (recompressed) video and the original video. It acts as the feature that distinguishes benign and malicious video.

Wang et al. [17] propose a solution for the motion vector-based steganography detection. MV steganography method modifies the associated prediction error values in addition to the motion vectors. Therefore, distortion in video frames will be minimal. But in the case of stego videos, some MVs will not be locally optimal. The proposed method defines an operation called adding-or-subtracting-one (AoSO). The method computes an MV matrix and a SAD matrix using the MV and SAD values of the current macroblock with its reference macroblock and eight neighbors. These matrices are utilized to extract the feature set required for training the classifier.

Zhang et al. [22] introduced a 36-D feature set for the motion vector steganography detection. It uses Lagrangian cost function for feature extraction. Recent video coding standards implement Lagrangian cost function to find the best matching macroblock. The prediction error is weighed against the number of bits required to represent the MV in this function. The macroblock having the minimal Lagrangian cost function is chosen as the reference macroblock. The proposed method computes SAD-based and sum of absolute transposed differences (SATD)-based Lagrangian costs for decompressed MV and its eight neighbors. The feature set consists of four types of features extracted from these values.

Statistical Feature-Based Su et al. [16] introduced a steganalysis method that considers both spatial and temporal features of videos. The video frames are segmented into blocks of size 8×8. The feature set is based on the histogram distribution for local motion intensity and texture complexity. This method uses support vector machine (SVM) classifier. The dataset contains MPEG-2 and H.264 videos.

Li et al. [10] proposed a steganalysis method for HEVC videos. A macroblock can be partitioned into 25 different partition modes in HEVC video coding standard. There are steganography algorithms that hide information in the partition modes. This technique mainly deals with detecting such steganography algorithms. The 25-dimensional feature set consists of the probability of each partition mode. The dimensionality of the feature set is further reduced to three. SVM classifier detects the stego videos.

3 Video Compression and Decompression

3.1 Motion Estimation and Motion Compensation

Consecutive video frames may contain the same objects in similar positions. This redundancy helps to reduce the storage space required for video files. Each video frame consists of several macroblocks. The encoder uses an already compressed frame R as the reference frame to encode a macroblock B of size $n \times n$ during video compression. The best matching block R_B is searched within R. The motion vector represents the spatial displacement between B and R_B in horizontal and vertical directions. Although the macroblocks are similar, the pixel intensities are different. The prediction error matrix represents the difference in pixel values between B and R_B. Discrete cosine transformation is applied on to the prediction error matrix. Finally, the prediction error matrix is quantized and entropy coded. This process is called motion estimation.

Motion vector information, prediction error matrix, and the reconstructed reference frame are required for decompression to reconstruct a particular video frame. This process is called motion compensation.

3.2 Block Matching Algorithms

An efficient algorithm is required for finding the best possible matching macroblock from the reference frame. It is impractical to search the whole reference frame for a matching macroblock. One of the first algorithms used for macroblock matching is the full search or the exhaustive search. The algorithm compares the current macroblock with each macroblock within a search window of the reference frame. Thus, the best matching macroblock is selected. Although this algorithm ensures the best image quality, it is computationally intensive. Suppose the search window is of size 'P' pixels on all four sides. Then, the algorithm searches a total of $(2P+1)\times(2P+1)$ positions. Therefore, other search algorithms like three-step search (TSS), binary search (BS), etc., are used for macroblock matching.

3.3 Variable-Sized Macroblocks

In older video coding standards, the macroblock size used to be 16×16. In the newer video coding standards, a macroblock can be further subdivided into 16×8, 8×16, 8×8, 8×4, 4×8, 4×4, etc. It helps in achieving a better compression ratio. Hence, current video compression standards can have variable-sized macroblocks. Figure 1 illustrates variable-sized macroblocks in a frame.

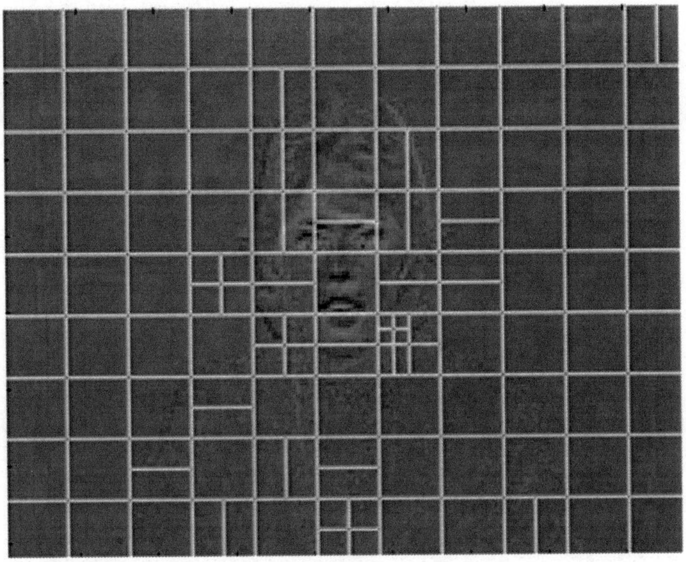

Fig. 1 Variable-sized macroblocks in a video frame

3.4 Sub-pixel Motion Estimation

Since the consecutive video frames are similar, the spatial displacement between the macroblocks in them can be half-pixel, 1/4 pixel, or even 1/8 pixel. The pixel values at integer locations are interpolated to form half-pixel and quarter-pixel values. It is called sub-pixel motion estimation. Block matching algorithms give better accuracy with sub-pixel motion estimation. Current compression standards such as H.264 implement such interpolation techniques. H.264 uses quarter-pixel motion vector resolution. Figure 2 shows integer-pixel, half-pixel, and quarter-pixel locations in a video frame.

4 Proposed Design

A video is a sequence of images. It has both spatial and temporal features. Image steganalysis methods consider only the spatial features. The proposed method considers both the spatial and temporal features of video frames.

Current video compression standards can have variable-sized macroblocks. The new standards also support sub-pixel motion accuracy. Both the features (spatial and temporal) used in this method can accommodate it. Most of the existing techniques have higher computational complexity for feature extraction. Therefore, the proposed design is a lightweight solution that computes the required features from the motion vector matrix.

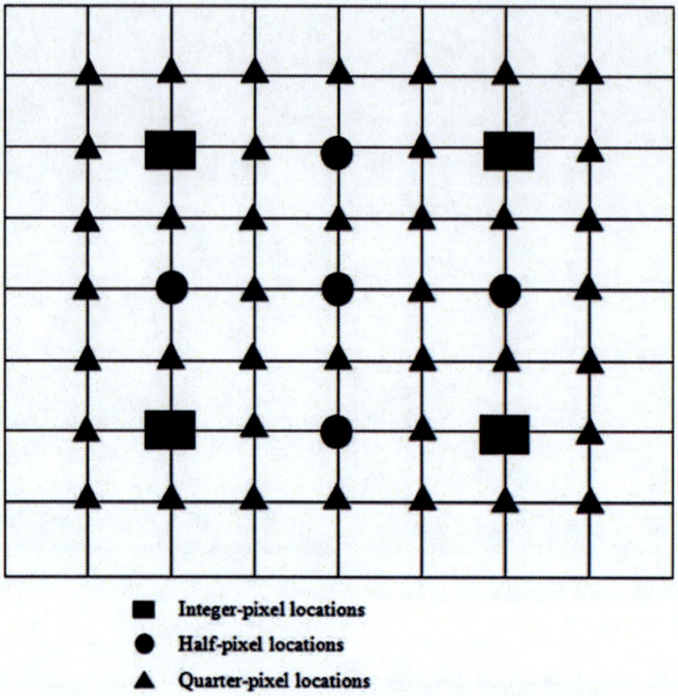

Fig. 2 Integer-pixel, half-pixel, and quarter-pixel locations in a video frame

4.1 Dataset and Preprocessing

The dataset [19] is a set of YUV video sequences of resolution 176×144. The *Y* in YUV stands for luminance (brightness). *U* and *V* provide color difference signals. *U* stands for blue minus luminance, and *V* stands for red minus luminance. YUV sequences are raw and uncompressed. The luma channel (Y) of YUV video sequences is converted into frames to prepare the dataset for feature extraction.

Since the dataset of videos embedded with malware is unavailable, the malware bits are embedded during the compression stage to create a set of stego videos for training the classifier. YUV sequence from the available dataset is converted into frames to create a dataset of stego videos. Further, the video frames are compressed. The steganography algorithm select motion vectors that satisfy particular criteria. These motion vectors are called candidate motion vectors (CMVs). The malware bits are embedded in the CMVs. Then, the prediction error matrix corresponding to the modified motion vector is calculated and compressed.

4.2 Feature Extraction

The feature set consists of spatial features and temporal features of a video. The proposed design is motivated by the spatial feature set of [5] and temporal feature set of [18].

The first feature utilized in this design is the spatial feature of the video. Consider a macroblock B in a video frame. There exist eight neighboring macroblocks for a fixed-size macroblock in a video frame. Eight neighboring pixels are selected as illustrated in [5, Fig. 4] to accommodate coding standards with variable macroblock sizes. The current macroblock starts at position (i_0,j_0) and has width BS_x and height BS_y. The pixels $P_1 - P_8$ represent the eight neighboring pixels. Each of these pixels is part of a macroblock. Each of these macroblocks has a motion vector associated with it.

It is possible to find the differences between the neighboring motion vectors and the motion vector of B. A constant called truncation threshold is selected to make it more compact. If the truncation threshold is 4, the differences will be limited to the set $\{-4, -3, -2, -1, 0, 1, 2, 3, 4\}$. The difference values greater than 4 or lesser than −4 are rounded to 4 and −4, respectively.

If the difference between the MVs is more, the two MVs are probably unrelated and vice versa. The proposed method computes the spatial feature set as shown in (1) for classification purposes.

$$f_s^p(K,L) = \begin{cases} P(d^p/\text{mvr} \le -T) & \text{if } L = -T \\ P(d^p/\text{mvr} = L) & \text{if } -(T-1) \le L \le (T-1) \\ P(d^p/\text{mvr} \ge T) & \text{if } L = T \end{cases} \tag{1}$$

MV_0 represents the motion vector of the current macroblock. MV_{1-8} represents the motion vectors of the eight neighboring macroblocks. P denotes the probability function. Each motion vector has a horizontal component and a vertical component. p can be v (difference in vertical component) or h (difference in horizontal component). $K \in \{1, 2, 3, 4, 5, 6, 7, 8\}$ denotes the neighboring macroblocks, whereas $L \in \{-4, -3, -2, -1, 0, 1, 2, 3, 4\}$ represents the truncated difference values. T represents the truncation threshold and is equal to 4. The mvr is equal to the motion vector resolution. In this experiment, mvr is 0.25. d denotes the difference between MVs. It is calculated as per (2).

$$d^p = MV_0^p - MV_K^p \tag{2}$$

Finally, combining all the features, the resultant vector contains 8×2×9 features (8 neighboring motion vectors, 2 components, and 9 difference values).

Similarly, a macroblock depends on the neighbors in its reference frame. The temporal feature of the design is based on this property.

Consider a macroblock B at position (i_0,j_0) in a frame F with width BS_x and height BS_y. Suppose the reference frame of B is R. There exists a macroblock R_B

at position (i_0, j_0) with width BS_x and height BS_y in R. The motion vectors of its eight neighboring macroblocks (MV_{1-8}) are considered (similar to [5, Fig. 4]). The motion vector of R_B is MV_9. The proposed method computes the temporal feature set as shown in (3) for classification purposes.

$$f_t^p(K, L) = \begin{cases} P(d^p/\text{mvr} \leq -T) & \text{if } L = -T \\ P(d^p/\text{mvr} = L) & \text{if } -(T-1) \leq L \leq (T-1) \\ P(d^p/\text{mvr} \geq T) & \text{if } L = T \end{cases} \quad (3)$$

Here, $K \in \{1, 2, 3, 4, 5, 6, 7, 8, 9\}$ denotes the neighboring macroblocks. The resultant vector contains 9×2×9 features (9 motion vectors from R, 2 components, and 9 difference values).

4.3 Dimensionality Reduction

The large dimension of the spatial feature may lead to overfitting of the classifier and the requirement of a huge training set. Hence, the spatial feature dimension is reduced from 144-D to 36-D, and the temporal feature is reduced from 162-D to 45-D. The methods used for dimensionality reduction are motivated by [5]. The following methods (eqs. (4)-(7)) are used for dimensionality reduction.

1. Averaging the features of two neighboring pixels each.

$$\begin{aligned} f_{sH}^{p}(L) &= (f_s^p(4, L) + f_s^p(5, L))/2 \\ f_{sV}^{p}(L) &= (f_s^p(2, L) + f_s^p(7, L))/2 \\ f_{sRD}^{p}(L) &= (f_s^p(3, L) + f_s^p(6, L))/2 \\ f_{sLD}^{p}(L) &= (f_s^p(1, L) + f_s^p(8, L))/2 \end{aligned} \quad (4)$$

$$\begin{aligned} f_{tH}^{p}(L) &= (f_t^p(4, L) + f_t^p(5, L))/2 \\ f_{tV}^{p}(L) &= (f_t^p(2, L) + f_t^p(7, L))/2 \\ f_{tRD}^{p}(L) &= (f_t^p(3, L) + f_t^p(6, L))/2 \\ f_{tLD}^{p}(L) &= (f_t^p(1, L) + f_t^p(8, L))/2 \\ f_{tB}^{p}(L) &= f_t^p(9, L) \end{aligned} \quad (5)$$

2. Grouping the values of horizontal and vertical components together.

$$\begin{aligned} f_{sH}^{vh}(L) &= exp(f_{sH}^{v}(L) + f_{sH}^{h}(L)) \\ f_{sV}^{vh}(L) &= exp(f_{sV}^{v}(L) + f_{sV}^{h}(L)) \\ f_{sLD}^{vh}(L) &= exp(f_{sLD}^{v}(L) + f_{sRD}^{h}(L)) \\ f_{sRD}^{vh}(L) &= exp(f_{sRD}^{v}(L) + f_{sRD}^{h}(L)) \end{aligned} \quad (6)$$

Fig. 3 Video frame before and after embedding 100 malware bits into the candidate motion vectors using Xu method

$$
\begin{aligned}
f^{vh}_{tH}(L) &= exp(f^{v}_{tH}(L) + f^{h}_{tH}(L)) \\
f^{vh}_{tV}(L) &= exp(f^{v}_{tV}(L) + f^{h}_{tV}(L)) \\
f^{vh}_{tLD}(L) &= exp(f^{v}_{tLD}(L) + f^{h}_{tRD}(L)) \\
f^{vh}_{tRD}(L) &= exp(f^{v}_{tRD}(L) + f^{h}_{tRD}(L)) \\
f^{vh}_{tB}(L) &= exp(f^{v}_{tB}(L) + f^{h}_{tB}(L))
\end{aligned}
\qquad (7)
$$

Finally, the results of (6) and (7) are used as the feature vector input for the classifier.

5 Implementation Results

This experiment follows video compression and decompression methods of the H.264/AVC standard. Group of pictures (GOP) size is 12. The dataset consists of non-overlapping video sequences of 60 frames each. A macroblock of size 16×16 can be further subdivided into smaller macroblocks. The implementation uses quarter-pixel motion vector resolution. The compression of YUV sequences is carried out using two macroblock matching algorithms: exhaustive search and three-step search.

A malware executable that performs reverse TCP attacks is created using msfvenom and Kali Linux. The executable is converted into binary bits and is embedded into the motion vectors. Two popular motion vector steganography methods are selected to embed malware into the videos: Xu method [21] and Aly method [1]. Embedding rates are calculated using bits per frame (bpf). Stego videos with the embedding rates of 50 bpf, 80 bpf, and 100 bpf are generated using each steganography method. The dataset includes 720 samples in total. The ratio of training data to testing data is equal to 80:20. SVM classifier with polynomial kernel acts as the classifier for the experiment. Figure 3 shows a video frame before and after embedding malware bits into the CMVs using Xu method [21].

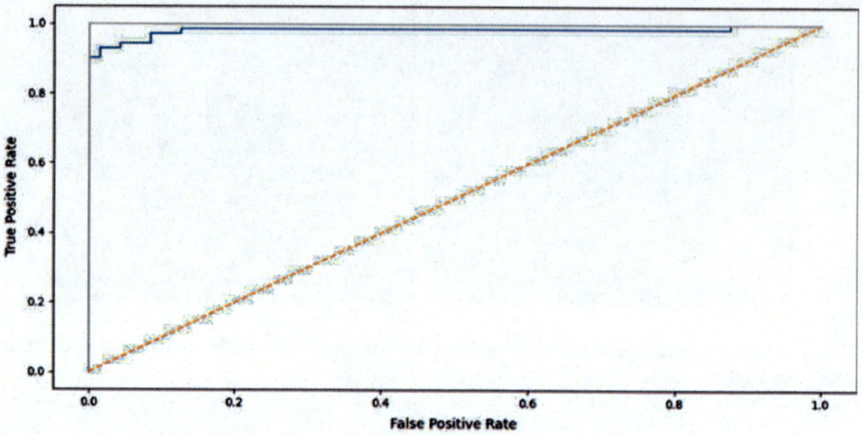

Fig. 4 Receiver operating characteristic (ROC) curve of the proposed method

The confusion matrix values obtained after testing the model are as follows: True positives (TP) = 67
True negatives (TN) = 71
False positives (FP) = 1
False negatives (FN) = 5

The blue curve in Fig. 4 shows the receiver operating characteristic (ROC) curve of the proposed solution. The ROC curve depicts how much a model is capable of differentiating between the classes. The ROC curve plots true positive rate (TPR) against the false positive rate (FPR) at different threshold values. The TPR and FPR are calculated as per (8) and (9).

$$\mathrm{TPR} = \frac{\mathrm{TP}}{\mathrm{TP} + \mathrm{FN}} \tag{8}$$

$$\mathrm{FPR} = \frac{\mathrm{FP}}{\mathrm{TN} + \mathrm{FP}} \tag{9}$$

The accuracy of the proposed solution is 95.8%. To verify the effectiveness of the model, we have extended the testing process using four different real-world malware that are not used to train the classifier. The malware used in this experiment is as follows:

1. Cerber ransomware [3]
2. XP Antivirus rogue [20]
3. MeltingScreen worm [13]
4. NotPetya [14]

Table 1 Detection accuracy for real-world malware

Malware	Accuracy (%)	Number of detected videos/Total number of videos
Cerber ransomware	100	12/12
XP Antivirus rogue	100	12/12
MeltingScreen worm	91.6	11/12
NotPetya	91.6	11/12
Total accuracy	95.8	46/48

Table 2 True positive rates of the JPMF model and our proposed model for different embedding rates

Bits per frame (bpf)	JPMF	Proposed model
	TPR (%)	TPR (%)
100	96.4	100
60	85.5	94.4
40	74.5	86.1

Each of the above-listed malware is embedded into a set of 12 videos. Table 1 shows the detection accuracy of the classifier in this experiment. The classifier detected 46/48 stego videos. It proves that our model works effectively in a practical scenario.

The feature set of our proposed model is motivated by the joint probability mass function (JPMF) model [18]. Both the methods compute the probability of motion vector difference for the feature set. To prove the effectiveness of our proposed method, we compare its results with that of JPMF. Table 2 shows the true positive rates obtained by the model with stego videos of three different embedding rates—100 bpf, 60 bpf, and 40 bpf. The dimensionality of our feature set is lesser than that of JPMF. It leads to the higher performance of our model. Similarly, while the JPMF model considers only three neighboring macroblocks simultaneously, our method considers eight neighboring macroblocks. The experimental results demonstrate the better performance of our features relative to that of JPMF.

6 Conclusion

This paper presents a lightweight solution for motion vector-based stegomalware detection in video files. The proposed solution detects the stegomalware with an accuracy of 95.8%, a precision of 98.5%, and a recall of 93%. The area under curve (AUC) is 0.98. Experiment results of testing the model with malware such as Cerber ransomware, XP Antivirus rogue, MeltingScreen worm, and NotPetya have shown that the proposed model is effective in a real-world scenario. In contrast to the previous steganalysis methods, our feature set is adaptable to various video coding standards having different configuration settings such as sub-pixel motion estimation, and variable-sized macroblocks. It considers both the spatial and temporal features of video files. The 81-D steganalysis feature set is readily extractable from the motion vector information. Therefore, the computational complexity of feature extraction is low. For future work, the proposed model can be further extended to include stego videos that use recent MV-based steganography techniques and different macroblock matching algorithms.

References

1. Aly, H.A.: Data hiding in motion vectors of compressed video based on their associated prediction error. IEEE Trans. Inf. Forensics Secur. **6**(1), 14–18 (2010)
2. Caviglione, L., Choraś, M., Corona, I., Janicki A., Mazurczyk, W., Pawlicki, M., Wasielewska, K.: Tight arms race: overview of current malware threats and trends in their detection. IEEE Access **9**, 5371–5396 (2021)
3. Cerber ransomware, https://www.f-secure.com/v-descs/trojan_w32_cerber.shtml
4. Chakraborty, S., Das, A.K., Sinha, C., Mitra, S.: Steganalysis of videos using energy associated with each pixel. In: Fourth International Conference on Image Information Processing (ICIIP) (2017)
5. Ghamsarian, N., Schoeffmann, K., Khademi, M.: Blind MV-based video steganalysis based on joint inter-frame and intra-frame statistics. Multimedia Tools Appl. **80**(6), 9137–9159 (2020)
6. Huang, X., Hu, Y., Wang, Y., Liu, B., Liu, S.: Deep learning-based quantitative steganalysis to detect motion vector embedding of HEVC videos. In: IEEE Fifth International Conference on Data Science in Cyberspace (DSC), pp. 150–155 (2020)
7. Idbeaa, T.F., Samad, S.A., Husain, H.: An adaptive compressed video steganography based on pixel-value differencing schemes. In: International Conference on Advanced Technologies for Communications (ATC), pp. 50–55 (2015)
8. Jue, W., Min-Qing, Z., Juan-Li, S.: Video steganography using motion vector components. In: Communication Software and Networks (ICCSN), 2011 IEEE 3rd International Conference, pp. 500–503 (2011)
9. Li, Y., Chen, H.X., Zhao, Y.: A new method of data hiding based on h.264 encoded video sequences. In: International Conference on Signal Processing Proceedings ICSP, pp. 1833–1836 (2010)
10. Li, Z., Meng, L., Xu, S., Li, Z., Shi, Y., Liang, Y.: A HEVC video steganalysis algorithm based on PU partition modes. Secur. Commun. Netw. **59**, 563–574 (2019)
11. Liao, K., Lian, S., Guo, Z., Wang, J.: Efficient information hiding in h.264/avc video coding. Telecommun. Syst., 261–269 (2012)
12. Liu, P., Li, S.: Steganalysis of intra prediction mode and motion vector-based steganography by noise residual convolutional neural network. IOP Conf. Ser.: Mater. Sci. Eng. **719** (2020)
13. MeltingScreen worm, https://www.f-secure.com/v-descs/melting.shtml
14. NotPetya, https://en.wikipedia.org/wiki/Petya_and_NotPetya
15. Ren, Y., Zhai, L., Wang, L., Zhu, T.: Video steganalysis based on subtractive probability of optimal matching feature. Multimedia Tools Appl., 83–99 (2015)
16. Su, Y., Yu, F., Zhang, C.: Digital video steganalysis based on a spatial temporal detector. KSII Trans. Internet Inf. Syst. **11**, 360–373 (2017)
17. Wang, K., Zhao, H., Wang, H.: Video steganalysis against motion vector-based steganography by adding or subtracting one motion vector value. In: IEEE Trans. Inf. Forensic Secur. **9**(5) (2014)

18. Wu, H.-T., Liu, Y., Huang, J., Yang, X.-Y.: Improved steganalysis algorithm against motion vector based video steganography. In: IEEE International Conference on Image Processing (ICIP), pp. 5512–5516 (2014)
19. Xiph.org (1999). https://media.xiph.org/video/derf
20. XP Antivirus rogue, https://www.f-secure.com/v-descs/rogue_w32_xpantivirus.shtml
21. Xu, C., Ping, X., Zhang, T.: Steganography in compressed video stream. In: First International Conference on Innovative Computing, Information and Control-Volume I (ICICIC'06), vol. 1. IEEE, pp. 269–272 (2006)
22. Zhang, H., Cao, Y., Zhao, X.: A steganalytic approach to detect motion vector modification using near-perfect estimation for local optimality. IEEE Trans. Inf. Forensics Secur. **12**(2), pp. 465–478 (2017)

Unsupervised Description of 3D Shapes by Superquadrics Using Deep Learning

Mahmoud Eltaher and Michael Breuß

Abstract The decomposition of 3D shapes into simple yet representative components is a very intriguing topic in computer vision as it is very useful for many possible applications. Superquadrics may be used with benefit to obtain an implicit representation of the 3D shapes, as they allow to represent a wide range of possible forms by few parameters. However, in the computation of the shape representation, there is often an intricate trade-off between the variation of the represented geometric forms and the accuracy in such implicit approaches. In this paper, we propose an improved loss function, and we introduce beneficial computational techniques. By comparing results obtained by our new technique to the baseline method, we demonstrate that our results are more reliable and accurate, as well as much faster to obtain.

Keywords Implicit shape representation · Superquadrics · Deep learning · 3D shape description

1 Introduction

The decomposition of 3D data into compact low-dimensional representations is a classic problem in shape analysis. Path planning and grasping, as well as identification, detection, and shape manipulation, might benefit from such representations. To this end, shape primitives such as 3D polyhedral forms [22], generalized cylinders [3], geons [2], as well as superquadrics [19] were studied in the early days of computer vision. However, because of a lack of computer power and data at that time, extracting such representations proved highly challenging.

M. Eltaher (✉) · M. Breuß
Brandenburg University of Technology, Cottbus, Germany
e-mail: Mahmoud.Eltaher@b-tu.de

M. Breuß
e-mail: breuss@b-tu.de

M. Eltaher
Al-Azhar University, Cairo, Egypt

M. Tistarelli et al. (eds.), *Computer Vision and Machine Intelligence*,
Lecture Notes in Networks and Systems 586,
https://doi.org/10.1007/978-981-19-7867-8_9

Let us consider representations of 3D data by modern deep learning (DL) approaches. The surely most simple 3D representation may be obtained by exploring the underlying Cartesian grid structure of 3D space. The Cartesian grid makes it possible to apply DL paradigms in a straightforward way to 3D voxel data while maintaining the standard 2D convolution approach. As examples of such techniques, let us mention [4, 14, 24] where rigid shapes with minimal deformations have been compared, or the extension of the 2D techniques in [30] where the 3D voxel data of basic objects has been utilized directly.

On the other hand, the representation of many complex 3D structures may be very simple if cast into a format not ideally modeled over a Cartesian grid, exploring, e.g., local highly adaptive descriptions as in mentioned classic works. However, such representations lack the natural grid array structure. As a result, typical DL methods are difficult to apply, yet such representations may be beneficial for assessing non-rigid objects in many applications, such as human body model segmentation [13], or point-to-point correspondence [8, 15, 28]. To widen the scope of DL architectures and make DL models more applicable to the analysis of 3D data, it is thus desirable to explore corresponding non-Cartesian approaches to 3D data representations.

Our Contributions In this paper, we build upon and extend the superquadrics baseline DL model proposed in [18]. Our new model varies from the latter and other current models, as it implements the following contributions *(i)*–*(iv)*. The arguably most apparent contribution is concerned with the fundamental modeling, namely *(i)* we re-engineer the loss function which results in more accurate shape representations. In doing this, our aim is also to improve the balancing between the proper representation of a given shape versus the accuracy, which appears to be a delicate issue in approaches working with implicit shape descriptions as in our work. Let us stress that a careful design of the loss function is very important for obtaining reasonable results since the approach is unsupervised. The other advances we propose are concerned with computational efficiency in several respects. We propose *(ii)* to employ PointNet as a feature detector layer, which makes working with point cloud datasets much easier than in [18], where a volumetric CNN was employed which is computationally much more intensive. Furthermore, *(iii)* we employ the cyclic learning rate [25], which allows to get better results in fewer epochs while at the same time it helps to avoid the loss plateau difficulty. Finally, *(iv)* we make use of the Kaiming weight initialization process [9], which produces a model with improved stability.

2 Related Work

In this section, we review in some detail the most important related DL approaches that have the same subject as this paper into two categories. Let us note that all the mentioned approaches as well as our method are unsupervised, which distinguishes the complete approach from other possible decomposition methods, such as, e.g., [10, 12, 31].

CNN-Based Methods By learning to build shapes using 3D volumetric primitives, Tulsiani et al. [27] examined the learning framework for abstracting complicated forms. This enables automated discovery and utilization of consistent structures, in addition to creating geometrically interpretable descriptions of 3D objects.

Following that, Paschalidou et al. [18] enhance the technique from [27]. They show how to decompose a 3D shape by utilizing superquadrics as atomic elements. As an important aspect of [18] which serves as a motivation for us to proceed in that line, the use of the superquadrics may produce more expressive 3D shape parses than standard 3D cuboid representations; moreover, they may offer advantages for the learning process.

Deng et al. [7] propose an auto-encoder network design to represent a low-dimensional family of convex polytopes, as any solid object may be broken down into a group of convex polytopes. A convex object may then be interpreted formally as a mesh formed by computing the vertices of a convex hull, or implicitly as a collection of half-space constraints, respectively, support functions.

RNN-Based Methods 3D-PRNN is a generative recurrent neural network for form synthesis that uses repeating primitive-based abstractions, proposed by Zou et al. [31]. The method employs a low-parametric model to represent more complicated forms, allowing to model shapes using few training samples.

The neural star domain (NSD), proposed by Kawana et al. [11], is a unique representation for learning primitive shapes in the star domain, represented as continuous functions over the sphere. This enables implicit as well as explicit form representation.

Paschalidou et al. [16] describe a learning-based technique for recovering the geometry of a 3D object as a set of primitives, as well as their underlying hierarchical structure. This approach describes objects as a binary tree of primitives, with few primitives representing simple object parts and more components representing the more intricate sections. However, a ground truth is employed to tackle situations where primitive points not belong to the input shape.

To tackle the trade-off between reconstruction quality and amount of parts, Paschalidou et al. [17] adopt neural parts, a 3D representation built again over the sphere that generates primitives using an invertible neural network. The latter allows to compute both a primitive's implicit surface function and its mesh. This model may be understood as a distinct extension of the superquadrics approach from [18], however, the simplicity and descriptive potential of the superquadrics has not been explored in it.

3 Superquadrics and Our New Model

In general, the network's purpose is to represent a 3D input shape by a collection of primitives. As indicated, the loss function is very important for the network's performance. In our unsupervised setting, there is no ground truth, and the network is confined to the shape primitives it can predict.

Let us briefly recall the basic definition of the superquadrics that are used as primitives before we proceed with further descriptions. In 1981, the superquadrics were introduced by Barr [1] in the field of computer graphics. Technically, they represent an extension of quadric surfaces derived from conic sections. A superquadric can be described in an implicit way via the equation

$$f(x,y,z)=1 \quad \text{with} \quad f(x,y,z)=\left(\left[\frac{x}{\alpha_1}\right]^{\frac{2}{\epsilon_2}}+\left[\frac{y}{\alpha_2}\right]^{\frac{2}{\epsilon_2}}\right)^{\frac{\epsilon_2}{\epsilon_1}}+\left[\frac{z}{\alpha_3}\right]^{\frac{2}{\epsilon_1}} \tag{1}$$

The size of the superquadric is determined by $\alpha = [\alpha_1, \alpha_2, \alpha_3]$, while the general shape of the superquadric is determined by $\epsilon = [\epsilon_1, \epsilon_2]$. We will make use of the property that a 3D point with coordinates (x_0, y_0, z_0) is included in a superquadric if $f(x_0, y_0, z_0) \leq 1$ and outside of it for $f(x_0, y_0, z_0) > 1$.

Superquadrics are often considered as a good choice for expressing a wide range of forms by a relatively simple parameterization, compare Fig. 1 for a few examples. Despite the fact that non-linear least squares may be used to fit superquadrics in a variety of ways [6, 26], Paschalidou et al. [18] appear to be the first to train a deep network to predict superquadrics from 3D inputs.

3.1 *Setup of Our Network*

Our model will now be described. We start by using the PointNet [21] as an encoder layer, which allows us to work directly with point cloud data. The encoder layer's output is then transmitted to a fully connected layer, which predicts the five parameters that form a superquadric. Figure 2 depicts a high-level representation of our model. Other PointNet-based methods in 3D are [23, 29].

The encoder uses a number of layer combinations to turn the point cloud input into a single channel 1×1024 feature vector. The feature vector is then passed through four fully connected layers before being fed into one of the four regressors, which determines the final value of the predicted parameters. Each of the four regressors

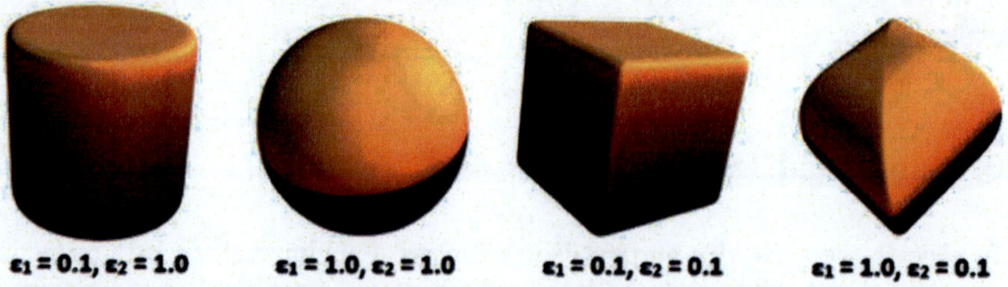

Fig. 1 Superquadrics with different values of ϵ_1, ϵ_2 that allow to control the shape curvature. Size parameters $\alpha_1, \alpha_2, and\alpha_3$ are kept constant

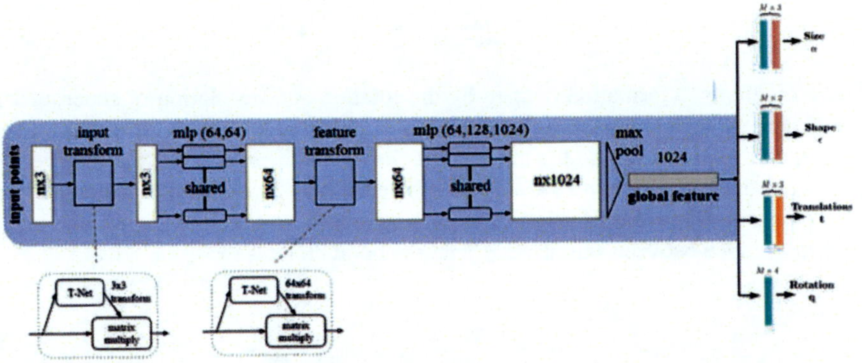

Fig. 2 Overview of our strategy. We utilize a CNN to predict basic parameters from the input point cloud X. The anticipated parameters distinguished by four subnetworks specify converted primitives whose combination produce a predicted shape

begins with a completely connected layer. The last layer of this size, the form regressor, uses sigmoid gates because the output of a sigmoid gate shall be between 0 and 1.

To avoid singularities, a small constant is added to the sigmoid gate of the alpha regressor, and the output of the fully connected layer of the epsilon regressor is added and scaled to fit within the needed values. The shape range in this study is 0.4 to 1.1. The tanh-gate at the conclusion of the translation regressor produces a number between -1 and 1. The completely connected layer is the only regressor that provides the value of the quaternion that yields the rotation of the superquadric.

3.2 *New Model for the Loss Function*

The loss function is the function that the network will try to optimize by utilizing gradient descent to find a suitable local minimum. It is a key part of developing an unsupervised neural network since there is no right or wrong response by examples. As a result, if the loss function is not carefully created, even a minor loss that may occur may yield results that are apparently far from optimal. In our setting, the goal of the loss function is to resolve the trade-off between represented form structure and accuracy.

Max-Primitive to Point Cloud Loss This loss is used to make sure the primitive shape is as similar to the input shape as possible. For computation, we sample the continuous surface of primitive number m, so that it is represented by a set of points $Y_m = \{y_k^m\}_{k=1}^{K}$, see Sect. 4 for some more details on sampling. The input point cloud is given by $X = \{x_i\}_{i=1}^{N}$. Then, the shortest distance between every point k on superquadric m and transformed input point cloud is computed as follows:

$$\Delta_k^m = \min_{i=1,\ldots,N} \|T_m(x_i) - \{y_k^m\}\|_2 \tag{2}$$

$T_m(\cdot)$ is the transformation defined by the rotation and translation of superquadric m. Then, we take the maximum value of the all shortest distances for each sampled point on a superquadric to the input point cloud, and after this, we take the average between the results over all primitives, to give the max-primitive to point cloud loss.

This part of the loss function thus gives the ability to consider the average maximum distance between primitives P and the point cloud X:

$$L_{P\to X}^m(P, X) = \frac{1}{M} \sum_{m=1}^{M} \max_{i=1,\ldots,k} \Delta_m^k \tag{3}$$

Outside-to-Primitive Loss While the loss in (3) means that we care about the distance of all input points to the primitive points, we found that it is in addition mandatory to consider the minimum distance of any point of point cloud P to all the primitives, where M is again the number of primitives:

$$\delta_i^m = \min_{m=1,\ldots,M} \min_{k=1,\ldots,k} \|T_m(x_i) - \{y_k^m\}\|_2 \tag{4}$$

Also, we found that we should refine this distance, taking care for the input points that belong to the input shape but are not contained in the primitives. This can be checked by the condition $f(x, y, z) > 1$. Thus in practice, we set up a mask described by the primitives, and set the distance inside this mask to zero, which leads us to define:

$$\tilde{\Delta}_i^m = \begin{cases} \delta_i^m & \text{if} \quad f(x, y, z) > 1 \\ 0 & else \end{cases} \tag{5}$$

Putting these considerations together, the loss will be, where N is the number of outside points given by set O only:

$$L_{O\to P}^m(O, P) = \frac{1}{N} \sum_{x_i \in O} \tilde{\Delta}_i^m \tag{6}$$

Dynamic Weights Rather than weighting the contributions from (3) and (6) by a user defined constant, we suggest to train these two weights. To this end, we define *(i)* the outside-to-primitive weight w_1, where $\|O\|$ is the number of outside points only, and N is total number of input points, and *(ii)* the primitive to point cloud weight w_2, with $\|\tilde{P}_{\text{inside}}\|$ as defined below and K denoting the overall number of primitive points

$$w_1 = 1 - \frac{\|O\|}{N} \quad \text{and} \quad w_2 = 1 - \frac{\|P_{\text{inside}}\|}{K} \tag{7}$$

Thereby, we say that a sampled point of a primitive for which we compute $\tilde{\Delta}_k^m$ as in (5) is in P_{inside} if $\tilde{\Delta}_k^m < \text{mean}(\tilde{\Delta}_k^m)$, where the mean is given by the arithmetic average of all values of $\tilde{\Delta}_k^m$.

3.3 Comparison to the Previous Model for the Loss Function

Let us recall that the network in [18] incorporates two loss functions: *(i)* A primitive to point cloud type loss for accuracy that strives to maintain the predicted primitives' surfaces near to the surface of the input form, and *(ii)* for structural integrity, a point cloud to primitive type loss that ensures that every region of the input's surface is characterized by at least one primitive.

At first glance, this appears to be similar as in our work. However, as pointed out already, in an unsupervised approach, the concrete choice of the loss function is quite delicate, and the advances we propose relate to the work [18] as follows.

The first mentioned loss from [18] used for accuracy is calculated by the shortest distance between each point on the primitives to the input point cloud. So the starting point for this measure is given by the primitive points. The computation gives a set of shortest distances corresponding to the set of primitives. Then, the arithmetic average of all these distances computed for the set of primitives is employed in [18]. We modify this averaging in (3) to taking the maximum of the shortest distances in that set in our model instead.

The second mentioned loss from [18] formulated for structural integrity ensures that each input point of a given point cloud is represented by at least one primitive. As in [18] and many other works in the field, it is conjectured that a minimal distance between each point of a given point cloud to the set of computed primitives gives a measure for this. So the starting point in this measure is a considered point of the point cloud, and the minimal distance between each such point and each primitive is calculated in this loss.

However, we discovered that the network is sometimes misled by distances belonging to points inside the primitives, in the sense that this sometimes does not satisfy the indirect optimization goal of structural representation. Therefore, we utilize the mask as in (5) to take into account just the points outside the primitives for optimization. Let us note in this context that input points are understood as represented by a primitive if they are inside the primitive, in our work as well as in [18].

When thinking of the complete modeling approach made up by the loss functions, it is in total in our approach more apparent, how the geometric accuracy is enforced by penalizing maximal deviations, while at the same time, it is more explicit how structural accuracy is understood and optimized. Note also that the dynamic weights help to enforce the balancing between the losses in our model.

4 Experimental Evaluation

To initialize weight, we use the Kaiming initialize weight technique [9] since it has a significant influence on training reliability. The cyclic learning rate [25] is also used since it helps a network to reach convergence in less epochs.

For evaluating the loss, we sample points on the superquadric surface. To achieve a uniform point distribution, we sample η and ω as proposed in [20]. In all experiments, we take uniformly 200 points from the surface of every superquadric, while the target point cloud is represented by 1000 points.

Evaluation metrics Let us discuss in some detail the evaluation metrics we consider. In the work [18], the primitive to point cloud distance (denoted here as prim-to-pcl, smaller is better) is employed, which evaluates the obtained accuracy. For assessing in addition to accuracy also the structural representation properties of the two methods, we propose here two additional evaluation metrics that have not been used in [18]. These two measures are designed to give accounts of the relative number of represented points of the point cloud and the relative number of primitive points that do not serve the representation.

The percentage of input points that is represented by the output primitives is the structural accuracy (larger is better). Let us note here again that this means we evaluate the percentage of points located inside the primitives. Therefore, this measure is calculated by dividing the number of input points that are located within the computed primitives by the total number of input points of the given point cloud.

The last measure we consider is the primitive accuracy (larger is better) in the following sense. The input point cloud is converted to a watertight mesh of the given shape, and then, we check if a primitive point is within the generated closed volumetric shape representation or not. The primitive accuracy measure is then given by the percentage of these points relative to the total number of primitive points.

Let us note that in the work [18] also another error measure is considered, based on volume of the considered shapes. However, while this appears to be natural in the volumetric CNN framework explored there, in contrast, our method is entirely based on use of point clouds, so that we refrain to consider that error measure here.

Experiments The results we now present are based on the 3D ShapeNet dataset [5]. We consider the code provided for the baseline method [18] to compare with our results.

We do our experiment on the chair, airplane, and table categories with 20 samples for each category, for simplicity and to decrease method execution time. Let us note that this number of experiments is often considered adequate since run times for unsupervised learning are considerable.

By Table 1, we can see that the two evaluated models appear at first glance close in evaluation metrics. However, while the primitive to point cloud accuracy, which represents overall approximation accuracy, is virtually identical, there is a small but still notable advantage of our model compared to the baseline method in terms of both structural accuracy measures. This may seen as experimental proof that distance is not the only important element influencing the outcome. Moreover, since

Table 1 Quantitative comparison between baseline superquadrics method [18] and our new model. For fair comparison, the baseline method has been employed here with Kaiming initialization (which makes it more stable) as is done in our model

Category	Method	Prim-to-pcl	Structural accuracy	Primitive accuracy
Chair	Baseline	0.0003	0.99	0.71
	Our result	0.0003	0.99	0.76
Airplane	Baseline	0.0001	0.96	0.68
	Our result	0.0001	0.99	0.71
Table	Baseline	0.0004	0.95	0.72
	Our result	0.0004	0.99	0.76

Results are averaged over 20 input shapes for each of the categories

our model performs consistently better in the structural evaluations without losing approximation quality, we conjecture that our model is balanced in a favorable way.

It makes sense to complement the quantitative results from Table 1 as well as our advances with a visual evaluation. The Fig. 3 shows evidence that the results obtained by our method are much more stable than for the original baseline method without Kaiming initialization. Our results displayed in second row are consistent with the overall chair structure, and it shows similar results at any of the evaluated epochs, while the original baseline method gives inconsistent results here and even does not converge in this example.

We can observe the beneficial impact of the Kaiming intiliazation for both evaluated techniques in Fig. 4. Moreover, we employ here for our method also the cyclic learning rate so that our method requires just about 1000 epochs to converge, while the baseline method takes about 5000 epochs to achieve a reasonable result. Moreover, one may observe that the baseline method [18] still generates visually inaccurate predictions after 50,000 epochs, such as the chair-back being bigger here than the original one. Our method corrects such structural errors, giving us more precise shape representations.

The Fig. 5 confirms that our results give a more reasonable account of structural characteristics compared to the baseline method. One may observe this, e.g., at the arm of the chair in column two or the structure of the table in columns three and four.

5 Conclusion

We have considered the problem of abstracting 3D shapes into a collection of primitives, as an extension of the baseline method from [18]. We have proposed a new model with two loss functions and dynamic weighting. The purpose of the new model is to minimize the gap between expected shape descriptions and input forms without relying on ground truth, which is typically costly and difficult to obtain. Experimental comparisons confirm that by the new developments, our model enables more accurate

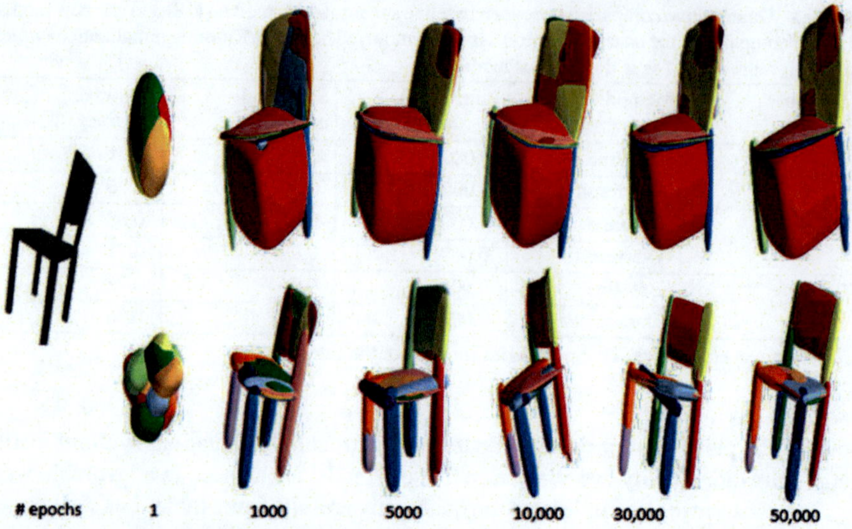

Fig. 3 Comparison between baseline method without Kaiming initialization (first row) and our solution (second row). As can be seen, our new model gives qualitatively reliable results and appears to provide the same result every evaluated epoch, yet the baseline method is inconsistent and in this example it does not converge

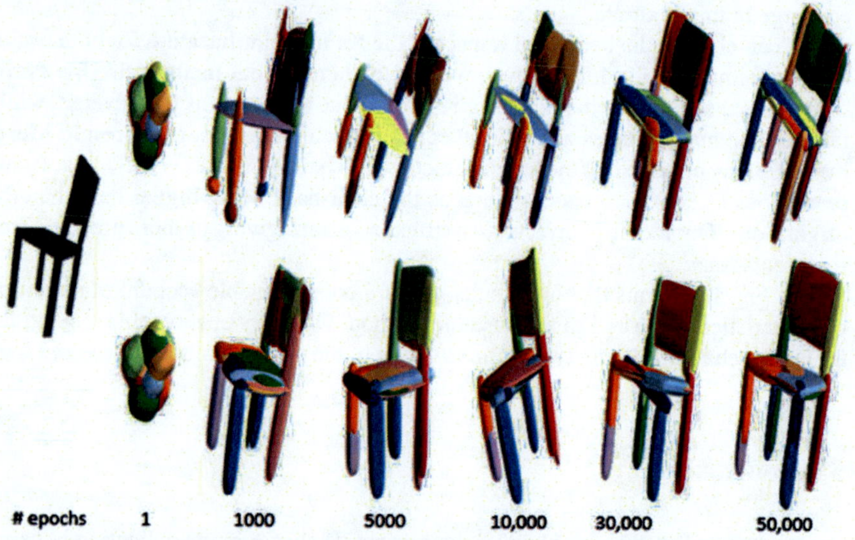

Fig. 4 Comparison of the superquadrics baseline method with Kaiming initialization (first row) with our discovery (second row). The results clearly confirm the benefit of the Kaiming initialization. Still, in comparison, our model converges much faster to more accurate representations

Fig. 5 Some more results on ShapeNet examples; (first row) input point cloud, (second row) baseline result with Kaiming initialization, (third row) our new method. This confirms the overall higher reliability of our method

structure representation than the baseline method, while keeping the accuracy. Our contributions are both in terms of modeling and improved computational efficiency.

In ongoing and future work, the aim is to obtain even better structural accuracy, since in some representations an object part tends to be represented by several superquadrics. This seems to happen especially in some cases when the object part is bent to some degree. Therefore, we aim at enhancing the representation by including deformations such as tapering and bending.

Acknowledgements The current work was supported by the European Regional Development Fund, EFRE 85037495. Furthermore, the authors acknowledge the support by BTU Graduate Research School (STIBET short-term scholarship for international PhD Students sponsored by the German Academic Exchange Service (DAAD) with funds of the German Federal Foreign Office).

References

1. Barr, A.H.: Superquadrics and angle-preserving transformations. IEEE Comput. Graph. Appl. **1**(1), 11–23 (1981)
2. Biederman, I.: Human image understanding: Recent research and a theory. Comput. Vis. Graph. Image Process. **32**(1), 29–73 (1985)
3. Binford, I.: Visual perception by computer. In: IEEE Conference of Systems and Control (1971)
4. Brock, A., Lim, T., Ritchie, J.M., Weston, N.: Generative and discriminative voxel modeling with convolutional neural networks. arXiv preprint arXiv:1608.04236 (2016)

5. Chang, A.X., Funkhouser, T., Guibas, L., Hanrahan, P., Huang, Q., Li, Z., Savarese, S., Savva, M., Song, S., Su, H., et al.: Shapenet: An information-rich 3d model repository. arXiv preprint arXiv:1512.03012 (2015)
6. Chevalier, L., Jaillet, F., Baskurt, A.: Segmentation and superquadric modeling of 3d objects (2003)
7. Deng, B., Genova, K., Yazdani, S., Bouaziz, S., Hinton, G., Tagliasacchi, A.: Cvxnet: Learnable convex decomposition. In: Proceedings of the IEEE/CVF Conference on Computer Vision and Pattern Recognition. pp. 31–44 (2020)
8. Fey, M., Lenssen, J.E., Weichert, F., Müller, H.: Splinecnn: Fast geometric deep learning with continuous b-spline kernels. In: Proceedings of the IEEE Conference on Computer Vision and Pattern Recognition. pp. 869–877 (2018)
9. He, K., Zhang, X., Ren, S., Sun, J.: Delving deep into rectifiers: Surpassing human-level performance on imagenet classification. In: Proceedings of the IEEE international conference on computer vision. pp. 1026–1034 (2015)
10. Huang, J., Gao, J., Ganapathi-Subramanian, V., Su, H., Liu, Y., Tang, C., Guibas, L.J.: Deep-primitive: Image decomposition by layered primitive detection. Comput. Vis. Media **4**(4), 385–397 (2018)
11. Kawana, Y., Mukuta, Y., Harada, T.: Neural star domain as primitive representation. arXiv preprint arXiv:2010.11248 (2020)
12. Li, C., Zeeshan Zia, M., Tran, Q.H., Yu, X., Hager, G.D., Chandraker, M.: Deep supervision with shape concepts for occlusion-aware 3d object parsing. In: Proceedings of the IEEE Conference on Computer Vision and Pattern Recognition. pp. 5465–5474 (2017)
13. Maron, H., Galun, M., Aigerman, N., Trope, M., Dym, N., Yumer, E., Kim, V.G., Lipman, Y.: Convolutional neural networks on surfaces via seamless toric covers. ACM Trans. Graph. **36**(4), 71–1 (2017)
14. Maturana, D., Scherer, S.: Voxnet: A 3d convolutional neural network for real-time object recognition. In: 2015 IEEE/RSJ International Conference on Intelligent Robots and Systems (IROS). pp. 922–928. IEEE (2015)
15. Monti, F., Boscaini, D., Masci, J., Rodola, E., Svoboda, J., Bronstein, M.M.: Geometric deep learning on graphs and manifolds using mixture model cnns. In: Proceedings of the IEEE conference on computer vision and pattern recognition. pp. 5115–5124 (2017)
16. Paschalidou, D., Gool, L.V., Geiger, A.: Learning unsupervised hierarchical part decomposition of 3d objects from a single RGB image. In: Proceedings of the IEEE/CVF Conference on Computer Vision and Pattern Recognition. pp. 1060–1070 (2020)
17. Paschalidou, D., Katharopoulos, A., Geiger, A., Fidler, S.: Neural parts: Learning expressive 3d shape abstractions with invertible neural networks. In: Proceedings of the IEEE/CVF Conference on Computer Vision and Pattern Recognition. pp. 3204–3215 (2021)
18. Paschalidou, D., Ulusoy, A.O., Geiger, A.: Superquadrics revisited: Learning 3d shape parsing beyond cuboids. In: Proceedings of the IEEE/CVF Conference on Computer Vision and Pattern Recognition. pp. 10344–10353 (2019)
19. Pentland, A.: Parts: Structured descriptions of shape. In: AAAI. pp. 695–701 (1986)
20. Pilu, M., Fisher, R.B.: Equal-distance sampling of superellipse models (1995)
21. Qi, C.R., Su, H., Mo, K., Guibas, L.J.: Pointnet: deep learning on point sets for 3d classification and segmentation. In: Proceedings of the IEEE conference on computer vision and pattern recognition. pp. 652–660 (2017)
22. Roberts, L.G.: Machine perception of three-dimensional solids. Ph.D. thesis, Massachusetts Institute of Technology (1963)
23. Shi, W., Rajkumar, R.: Point-gnn: Graph neural network for 3d object detection in a point cloud. In: Proceedings of the IEEE/CVF conference on computer vision and pattern recognition. pp. 1711–1719 (2020)
24. Sinha, A., Bai, J., Ramani, K.: Deep learning 3d shape surfaces using geometry images. In: European Conference on Computer Vision. pp. 223–240. Springer (2016)
25. Smith, L.N.: Cyclical learning rates for training neural networks. In: 2017 IEEE winter conference on applications of computer vision (WACV). pp. 464–472. IEEE (2017)

26. Solina, F., Bajcsy, R.: Recovery of parametric models from range images: The case for superquadrics with global deformations. IEEE Trans. Pattern Anal. Mach. Intell. **12**(2), 131–147 (1990)
27. Tulsiani, S., Su, H., Guibas, L.J., Efros, A.A., Malik, J.: Learning shape abstractions by assembling volumetric primitives. In: Proceedings of the IEEE Conference on Computer Vision and Pattern Recognition. pp. 2635–2643 (2017)
28. Verma, N., Boyer, E., Verbeek, J.: Feastnet: Feature-steered graph convolutions for 3d shape analysis. In: Proceedings of the IEEE conference on computer vision and pattern recognition. pp. 2598–2606 (2018)
29. Wu, B., Liu, Y., Lang, B., Huang, L.: Dgcnn: Disordered graph convolutional neural network based on the gaussian mixture model. Neurocomputing **321**, 346–356 (2018)
30. Wu, Z., Song, S., Khosla, A., Yu, F., Zhang, L., Tang, X., Xiao, J.: 3d shapenets: A deep representation for volumetric shapes. In: Proceedings of the IEEE conference on computer vision and pattern recognition. pp. 1912–1920 (2015)
31. Zou, C., Yumer, E., Yang, J., Ceylan, D., Hoiem, D.: 3d-prnn: generating shape primitives with recurrent neural networks. In: Proceedings of the IEEE International Conference on Computer Vision. pp. 900–909 (2017)

Multimodal Controller for Generative Models

Enmao Diao, Jie Ding, and Vahid Tarokh

Abstract Class-conditional generative models are crucial tools for data generation from user-specified class labels. Existing approaches for class-conditional generative models require nontrivial modifications of backbone generative architectures to model conditional information fed into the model. This paper introduces a plug-and-play module, named 'multimodal controller' to generate multimodal data without introducing additional learning parameters. In the absence of the controllers, our model reduces to non-conditional generative models. We test the efficacy of multimodal controllers on CIFAR10, COIL100, and Omniglot benchmark datasets. We demonstrate that multimodal controlled generative models (including VAE, PixelCNN, Glow, and GAN) can generate class-conditional images of significantly better quality when compared with conditional generative models. Moreover, we show that multimodal controlled models can also create novel modalities of images.

Keywords Image generation · Computer vision

1 Introduction

In recent years, many generative models based on neural networks have been proposed and achieved remarkable performance. The main backbones of generative models include autoencoder, autoregression, normalization flow, and adversarial generative models. Perhaps the most well-known representatives of them are ariational autoencoder (VAE) [14], PixelCNN [26], Glow [15], and generative adversarial network (GAN) [8], respectively. VAE learns a parametric distribution over an encoded

E. Diao (✉)
Duke University, Durham, NC 27708, USA
e-mail: enmao.diao@duke.edu

J. Ding · V. Tarokh
University of Minnesota-Twin Cities, Minneapolis, MN 55455, USA
e-mail: dingj@umn.edu

V. Tarokh
e-mail: vahid.tarokh@duke.edu

M. Tistarelli et al. (eds.), *Computer Vision and Machine Intelligence*,
Lecture Notes in Networks and Systems 586,
https://doi.org/10.1007/978-981-19-7867-8_10

latent space, samples from this distribution, and then constructs generations from decoded samples. PixelCNN uses autoregressive connections to factorize the joint image distribution as a product of conditionals over sub-pixels. Glow optimizes the exact log-likelihood of the data with a deterministic and invertible transformation. GAN was introduced as a generative framework where intractable probabilistic distributions are approximated through adversarial training.

In many application scenarios, we are interested in constructing generations based on a conditional distribution. For instance, we may be interested in generating human face images conditional on some given characteristics of faces such as hair color, eye size, gender, etc. A systematic way to incorporate conditional information may enable us to control the data generating process with more flexibility. In this direction, conditional generative models including conditional variational autoencoder (CVAE) [31], conditional generative adversarial network (CGAN) [21], and conditional PixelCNN (CPixelCNN) [25] have been proposed which model conditional information by learning the associated embeddings. The learned features are usually concatenated or added with non-conditional features at various network layers. Conditional Glow (CGlow) learns a class-conditional prior distribution and an optional auxiliary classifier.

In this paper, we propose a plug-and-play module named multimodal controller (MC) to allocate uniformly sampled subnetwork for each mode of data. Instead of introducing additional learning parameters to model conditional information, multimodal controlled generative models generate each mode of data from its corresponding unique subnetwork. Our main contributions of this work are three-fold.

- We introduce a novel method to transform non-conditional generative models into class-conditional generative models, by simply attaching a multimodal controller at each layer. Unlike existing methods, our method does not introduce additional learning parameters, and it can be easily incorporated into existing implementations.
- We empirically demonstrate that our method outperforms various well-known conditional generative models for datasets with small intra-class variation and a large number of data modalities. We achieve this advantage by allocating specialized subnetworks for corresponding data modalities.
- We show that multimodal controlled generative models can create novel data modalities by allocating un-trained subnetworks from genetic crossover or resampling of a codebook.

We experiment with CIFAR10, COIL100, and Omniglot datasets [16, 17, 23]. We compare our method with the conditional generative models with different backbone generative models such as VAE, PixleCNN, Glow, and GAN.

The rest of the paper is organized as follows. In Sect. 2, we review the related work. In Sect. 3, we introduce our proposed multimodal controller. In Sect. 4, we provide experimental results demonstrating the performance of our approach. Finally, we make our concluding remarks in Sect. 5.

2 Related Work

We use four distinct backbone generative models, including variational autoencoder (VAE) [14], PixelCNN [26], Glow [15], and generative adversarial network (GAN) [8] to demonstrate general compatibility of our method. VAE is a directed generative model with probabilistic latent variables. PixelCNN is an autoregressive generative model that fully factorizes the joint probability density function of images into a product of conditional distributions over all sub-pixels. Glow is a flow-based generative model that enables exact and tractable log-likelihood and latent space inference by ensuring a bijective neural network mapping. GAN consists of a generator (G) network and a discriminator (D) network that trains to find a Nash equilibrium for generating realistic-looking images.

Conditional generative models treat class-conditional information h as input to the model. Typically, modeling such h requires additional learning parameters, and the resulting objective is the conditional distribution $p_\theta(x \mid h)$. Conditional VAE (CVAE) [31] and conditional GAN (CGAN) [21] concatenate trainable embeddings to both encoder (discriminator) and decoder (generator). Apart from trainable embeddings, this approach also requires additional learning parameters on the backbone generative models to incorporate the embeddings. Instead of concatenating, conditional PixelCNN (CPixelCNN) [26] and conditional Glow (CGlow) [15] add trainable embeddings to the features. This method requires the size of embedding to match the channel size of features. There also exist many other ways of modeling h. ACGAN [24] introduces an auxiliary classifier with a class-conditional objective function to model h. The conditional normalization [3, 5] learns class-conditional affine parameters, which can be considered another way to incorporate embeddings. [22] proposed a projection discriminator to mode h by measuring the cosine similarity between features and a learned embedding. A hybrid approach that combines the previous two methods was used in BigGAN [1]. Recently, StyleGAN [12] enables scale-specific synthesis by transforming style information with fully connected layers into affine parameters for conditional normalization. MSGAN [20] models h to generate a mode-seeking regularization term. STGAN [18] enables image attribute editing by modeling the difference between the source and target h. StyleGAN2 addresses the artifact problem caused by the adaptive instance normalization by demodulation operation [13]. Latest variants of StyleGAN show that auxiliary classifier improves the performance of class-conditional generation [11, 30]. Transformer-based generative models leverage style vector similar to StyleGAN [27]. Score-based generative models modulate the generation process by conditioning on information not available during training [32].

We aim to introduce a novel and generic alternative method to generate data class-conditionally. Intuitively, training separate models for each mode of data can guarantee class-conditional data generation. However, we do not want the model complexity to grow with the number of data modalities.

Recently, a few works pay attention to subnetworks. The lottery ticket hypothesis [7, 35] states that there may exist a subnetwork that can match the performance of the original network when trained in isolation. PathNet [6] demonstrates successful transfer learning through migration from one subnetwork to the other. Piggyback and some other works in continual learning literature also adopt similar methods to train subnetworks for various tasks of data [19, 28, 34]. HeteroFL [4] utilizes subnetworks in federated learning to reduce the computation and communication costs.

Multimodal controlled generative models uniformly sample subnetworks from a backbone generative model and allocate a unique computational path for each data modality. In our paper, we refer data modality as the class-conditional modality of data rather than the type of data, i.e., language and image. Our method is different from the existing weight masking approach [19, 28, 34] because our mask is applied to the network representations of networks rather than the model parameters. Specifically, our approach is able to train data from multiple data modalities in the same batch of data simultaneously. However, existing weight masking methods can only optimize one data modality at a time. This paper empirically justifies the following claim. Uniformly sampled subnetworks through masking network representations can well-represent a substantial number of data modalities in class-conditional generative models by allocating a unique computational path for each mode of data.

3 Multimodal Controller

Suppose that there is a dataset X with C data modalities. Each mode of data $X_c = \{x_c^i\}_{i=1}^{N_c}$ consists of N_c i.i.d. samples of a (continuous or discrete) random variable. Given a set of learning model parameters $\theta \in \mathbb{R}^D$ with size D, each mode of data is modeled with a random subset $\theta_c \subset \theta$. For notational convenience, we will interchangeably use the notions of subset and subvector. In this way, the allocated parameters for each mode will represent both the inter-mode association and intra-mode variation, thanks to parameter sharing and specialization.

Next, we discuss technical details in the specific context of neural networks. Suppose a uniformly sampled subnetwork takes input $X_c \in \mathbb{R}^{N_c \times K_c}$, where N_c and K_c are the batch size and the input channel size for data mode c. Suppose that the subnetwork is parameterized by a weight matrix $W_c \in \mathbb{R}^{D_c \times K_c}$ and bias vector $b_c \in \mathbb{R}^{D_c}$, where D_c is the output channel size for data mode c, then we have output $y_c \in \mathbb{R}^{N_c \times D_c}$ where

$$y_c = \phi(\mathrm{BN}_c(X_c \times W_c^T + b_c))$$

$\mathrm{BN}_c(\cdot)$ denotes the batch normalization (BN) [10] with affine parameters for corresponding data mode c, and $\phi(\cdot)$ is the activation function.

Existing methods [19] allocate subnetworks by masking out model parameters, i.e., $W_c = W \odot e_c$ where e_c is a binary mask, and $\odot$ indicates Hadamard product. However, the computation cost of the above formulation increases with the number of data modalities, as we need to sequentially backward the subnetwork of each data modality.

Therefore, we propose a nonparametric module, named **multimodal controller (MC)** using a masking method similar to dropout [33], which allocates subnetworks by masking out network representations. We choose to uniformly sample the codebook because we want to create a large number of unique subnetworks. The number of unique subnetworks we can generate with MC is 2^D. We do not train the codebook, because it is possible for some data modalities to have the same binary masks.

We uniformly draw C unique modality codewords $e_c \in \mathbb{F}_2^D$ to construct a modality codebook $e \in \mathbb{F}_2^{C \times D}$, where $\mathbb{F}_2$ denotes the binary field. Note that each row of the codebook is a binary mask allocated for each mode of data. Let $\times$ denote the usual matrix multiplication. Let $X \in \mathbb{R}^{N \times K}$ denotes the output from the previous layer. Suppose that the original network is parameterized by a weight matrix $W \in \mathbb{R}^{D \times K}$ and bias vector $b \in \mathbb{R}^D$. Then for a specific mode of data c, we have multimodal controlled output $y \in \mathbb{R}^{N \times D}$ where

$$\begin{aligned}
\hat{W}_c &= W \odot e_c, \quad \hat{b}_c^D = b \odot e_c, \\
\hat{\text{BN}}_c &= \text{BN} \odot e_c, \quad \hat{\phi}_c(\cdot) = \phi(\cdot) \odot e_c \\
y &= \hat{\phi}_c(\hat{\text{BN}}_c(X \times \hat{W}_c^T + \hat{b}_c)) \\
&= \phi(\text{BN}(X \times W^T + b)) \odot e_c
\end{aligned}$$

Note that $\hat{W}_c, \hat{b}_c^D, \hat{\text{BN}}_c$, and $\hat{\phi}_c$ are uniformly masked from original network. Because we are masking out the network representations, we can factorize e_c to avoid interfering with the calculation of running statistics and activation function because $(e_c)^n = e_c$. Suppose we have class-conditional information $h \in \mathbb{F}_2^{N \times C}$ where each row is a one-hot vector. The multimodal controller can optimize all data modalities of one batch of data in parallel as follows

$$y = \phi(\text{BN}(X \times W^T + b)) \odot (h \times e).$$

Suppose that the above formulation is for an intermediate layer. Then, X is the output masked from the multimodal controller in the previous layer. Denote $\tilde{e}_c \in \mathbb{F}_2^K$ as the codeword for data mode c in the previous multimodal controller, then the effective subnetwork weight matrix $\tilde{W}$ is $W \odot (e_c \times \tilde{e}_c)$. Since each codeword is uniformly sampled, the size of $\tilde{W}$ is approximately $\frac{1}{4}$ the original W. Indeed, the multimodal controller is a nonparametric computation module modulating model architecture and trading memory space for speedup in runtime. Although the above formulation only describes the interaction of multimodal controller and linear network. It can be readily extended to other parametric modules, such as convolution layers. We demonstrate the suggested way of attaching this plug-and-play module in practice in Fig. 1.

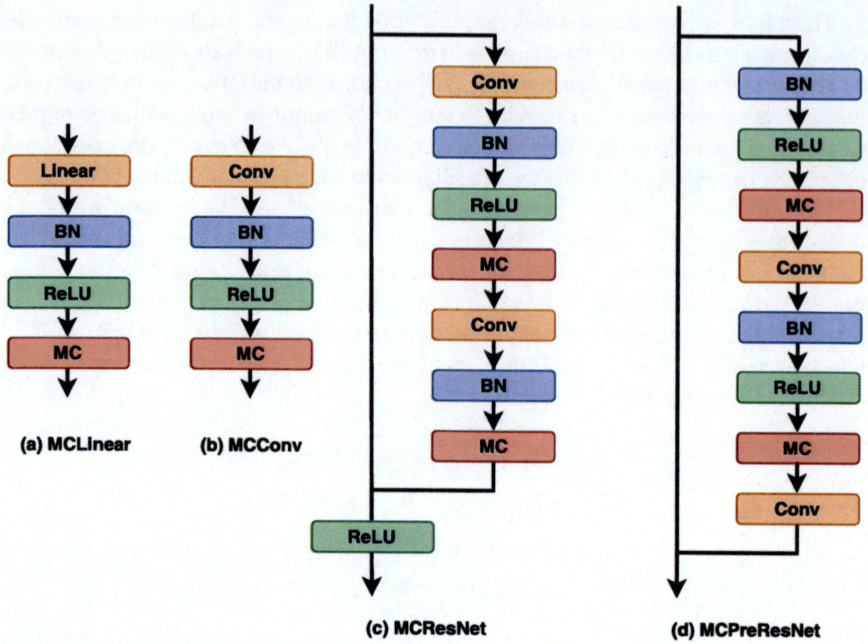

Fig. 1 Multimodal Controlled Neural Networks

4 Experiments

This section demonstrates applications of multimodal controlled generative models to data generation and creation. We compare the result of our proposed method with that of conditional generative models. We illustrate our results for four different types of multimodal controlled generative models, including VAE, PixelCNN, Glow, and GAN on CIFAR10, COIL100, and Omniglot datasets [16, 17, 23]. Due to our proposed method's random nature, we conduct 12 random experiments for each generative model on each dataset. The standard errors in parentheses show that uniformly sampled subnetworks are robust enough to produce stable results for either a small or large number of data modalities.

4.1 Image Generation

In this section, we present quantitative and qualitative results of generations from conditional and multimodal controlled generative models. More results are included in the supplementary document. Based on our results, multimodal controlled generative models can generate samples of comparable or better fidelity and diversity com-

Table 1 Inception score (IS) and Fréchet Inception distance (FID) for conditional and multimodal controlled generative models

	CIFAR10		COIL100		Omniglot	
	IS	FID	IS	FID	IS	FID
CVAE	3.4 (0.07)	133.7 (3.13)	89.4 (1.94)	37.6 (2.30)	539.8 (21.61)	367.5 (14.16)
MCVAE	**3.4 (0.05)**	**128.6 (1.32)**	**95.2 (0.50)**	**29.5 (1.91)**	**889.3 (16.94)**	**328.5 (9.82)**
CPixelCNN	**5.1 (0.06)**	**70.8 (1.78)**	94.2 (0.68)	8.1 (0.51)	**1048.5 (160.33)**	**23.4 (5.38)**
MCPixelCNN	4.8 (0.04)	75.2 (1.60)	**98.1 (0.75)**	**4.7 (0.69)**	762.4 (86.07)	43.3 (6.20)
CGlow	4.4 (0.05)	**63.9 (1.34)**	78.3 (2.25)	**35.1 (2.67)**	616.9 (6.71)	**40.8 (1.28)**
MCGlow	**4.8 (0.05)**	65.2 (1.21)	**89.6 (1.63)**	42.0 (5.19)	**998.5 (29.10)**	47.2 (2.69)
CGAN	**8.0 (0.10)**	**18.1 (0.75)**	97.9 (0.78)	24.4 (9.86)	677.3 (40.07)	51.0 (12.65)
MCGAN	7.9 (0.13)	21.4 (0.83)	**98.8 (0.09)**	**7.8 (0.44)**	**1288.8 (5.38)**	**23.9 (0.73)**

pared with conditional counterparts. We report our quantitative results in Table 1 with inception score (IS) [29] and Fréchet inception distance (FID) [9] which are perhaps the two most common metrics for comparing generative models. Both conditional and multimodal controlled generative models share the same backbone generative models for a fair comparison.

Quantitative results shown in Table 1 demonstrate that multimodal controlled generative models perform considerably better than the conditional generative models for COIL100 and Omniglot datasets and perform comparably with conditional generative models for the CIFAR10 dataset. The major difference among them is that CIFAR10 dataset has sufficient shots for a small number of data modalities while COIL100 and Omniglot datasets only have a few shots for a large number data modalities. Note that uniformly sampled subnetworks are using approximately a quarter of the number of learning parameters as the original network. Therefore, it is difficult for a small subnetwork to learn one mode of data with a high intra-class variation. When intra-class variation is small, and the number of data modalities is large as in COIL100 and Omniglot datasets, our proposed method has a significant advantage because we can specialize subnetworks for their corresponding data modalities by allocating unique computational paths. It is worth mentioning that CPixelCNN for Omniglot outperforms MCPixelCNN because it has almost twice the learning parameters as the later. The additional parameters are used for learning conditional embeddings, which increase with the number of data modalities.

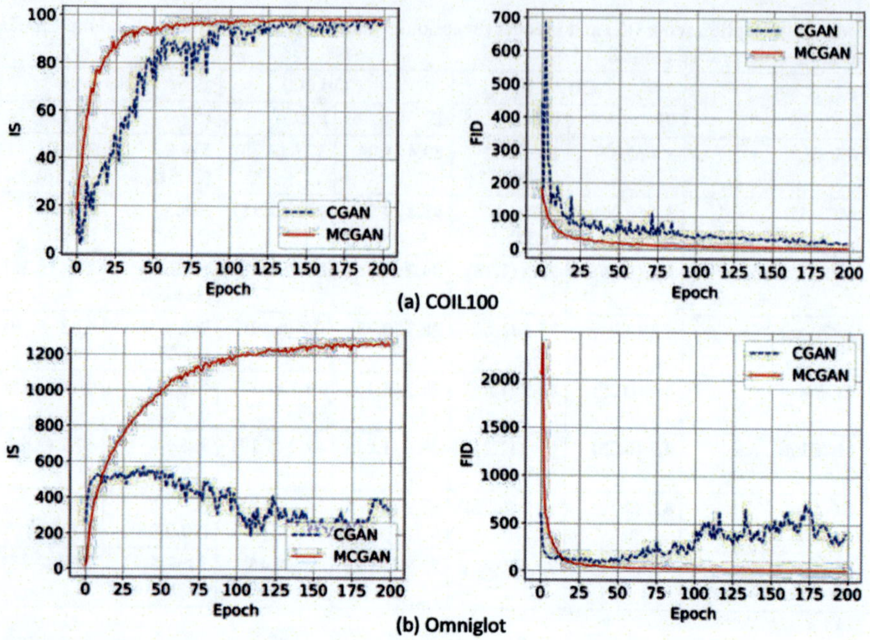

Fig. 2 Learning curves of Inception score (IS) of CGAN and MCGAN for COIL100 and Omniglot datasets

We illustrate the learning curves of of MCGAN and CGAN for COIL100 and Omniglot datasets in Fig. 2. The learning curves also show that MCGAN consistently outperforms CGAN. The results show that conditional generative models fail to capture and inter-class variation when the number of data modalities is large. Both our quantitative and qualitative results demonstrate the efficacy and advantage of our proposed multimodal controller for datasets with small intra-class variation and a large number of data modalities.

Qualitative results regarding multimodal controlled GAN (MCGAN) and conditional GAN (CGAN) for CIFAR10, COIL100, and Omniglot datasets are shown in Figs. 3, 4, and 5, respectively. For CIFAR10 datasets, MCGAN and CGAN generate similar qualitative results. For the COIL100 dataset, the rotation of different objects can be readily recognized by MCGAN. However, CGAN fails to generate objects of diverse rotations, as shown in red lines. For the Omniglot dataset, the intra-class variation and inter-class distinction can also be identified by MCGAN. However, CGAN fails to generate some data modalities, as shown in images squared in red lines.

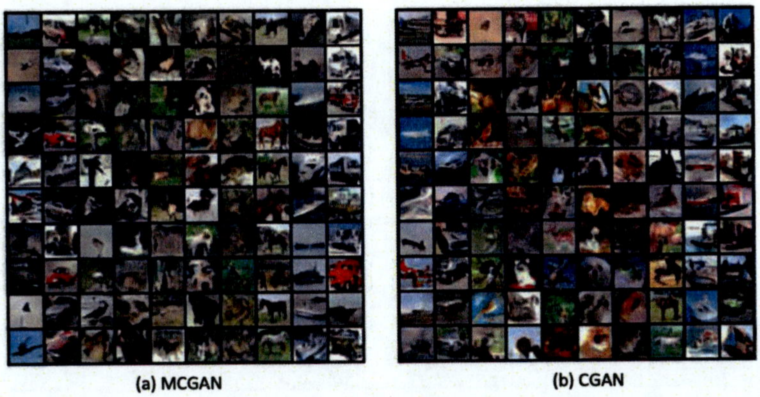

Fig. 3 **a** MCGAN, **b** CGAN trained with CIFAR10 dataset. Generations in each column are from one data modality

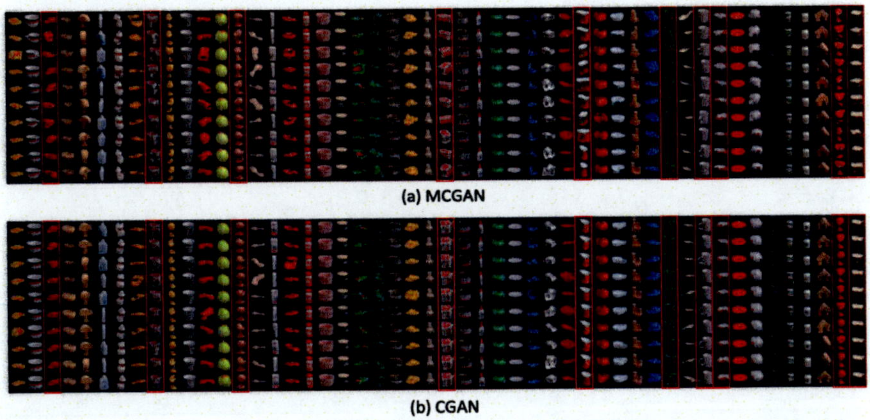

Fig. 4 **a** MCGAN, **b** CGAN trained with COIL100 dataset. Generations in each column are from one data modality

4.2 Image Creation from Novel Data Modalities

In this section, we provide quantitative and qualitative results of data creation from conditional and multimodal controlled generative models. We show that multimodal controlled generative models can class-conditionally synthesize from a novel data modality not prescribed in the training dataset. We propose an unbiased way of creating novel data modalities for our proposed method. Because the pre-trained codewords are uniformly sampled binary masks, we can naturally create unbiased novel data modalities by uniformly resampling the codebooks of multimodal controllers plugged in each layer. To compare with conditional generative models, we

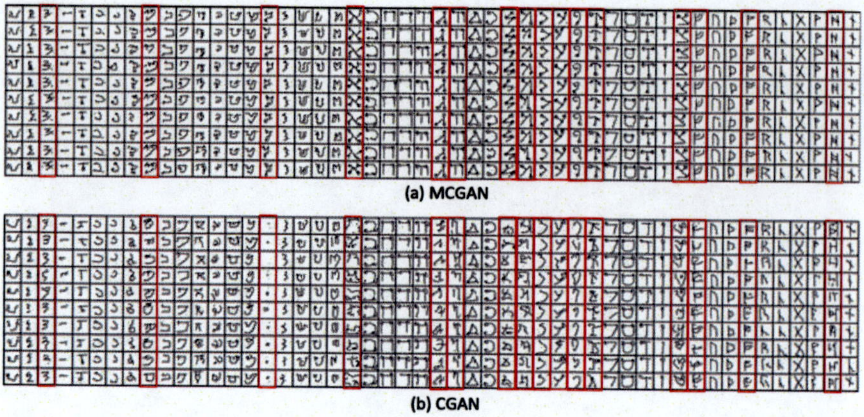

Fig. 5 **a** MCGAN **b** CGAN trained with Omniglot dataset. Generations in each column are from one data modality

Table 2 Davies-Bouldin Index (DBI) for conditional and multimodal controlled generative models on the uniformly created datasets. The created dataset has the same number of modalities as the raw dataset. Small DBI values indicate that data creations are closely clustered on novel data modality

	CIFAR10	COIL100	Omniglot
Raw dataset	12.0	2.7	5.4
CVAE	37.9 (5.10)	15.5 (0.60)	8.5 (0.04)
MCVAE	**2.1 (0.17)**	**1.8 (0.06)**	**3.0 (0.03)**
CPixelCNN	27.6 (1.84)	17.9 (0.59)	9.0 (0.19)
MCPixelCNN	**3.8 (0.34)**	**4.8 (0.18)**	**4.6 (0.19)**
CGlow	40.3 (5.09)	14.3 (0.87)	8.0 (0.02)
MCGlow	**5.4 (0.46)**	**2.8 (0.12)**	**5.1 (0.14)**
CGAN	33.2 (3.46)	10.4 (2.88)	7.8 (0.08)
MCGAN	**2.0 (0.30)**	**1.5 (0.07)**	**3.6 (0.02)**

uniformly create new data modalities for conditional generative models by sampling the weights of a convex combination of pre-trained embeddings from a Dirichlet distribution.

Quantitative results are shown in Table 2. We evaluate the quality of uniform data creation with Davies-Bouldin Index (DBI) [2]. Small DBI values indicate that data creations are closely clustered on novel data modalities. Because novel data modalities are parameterized by resampled subnetworks, data created by our proposed method can be closely clustered together. Data created by conditional generative models are not closely clustered together and have much higher DBI. It shows that a random convex combination of embeddings is not enough to create unbiased novel data modalities from pre-trained class-conditional generative models.

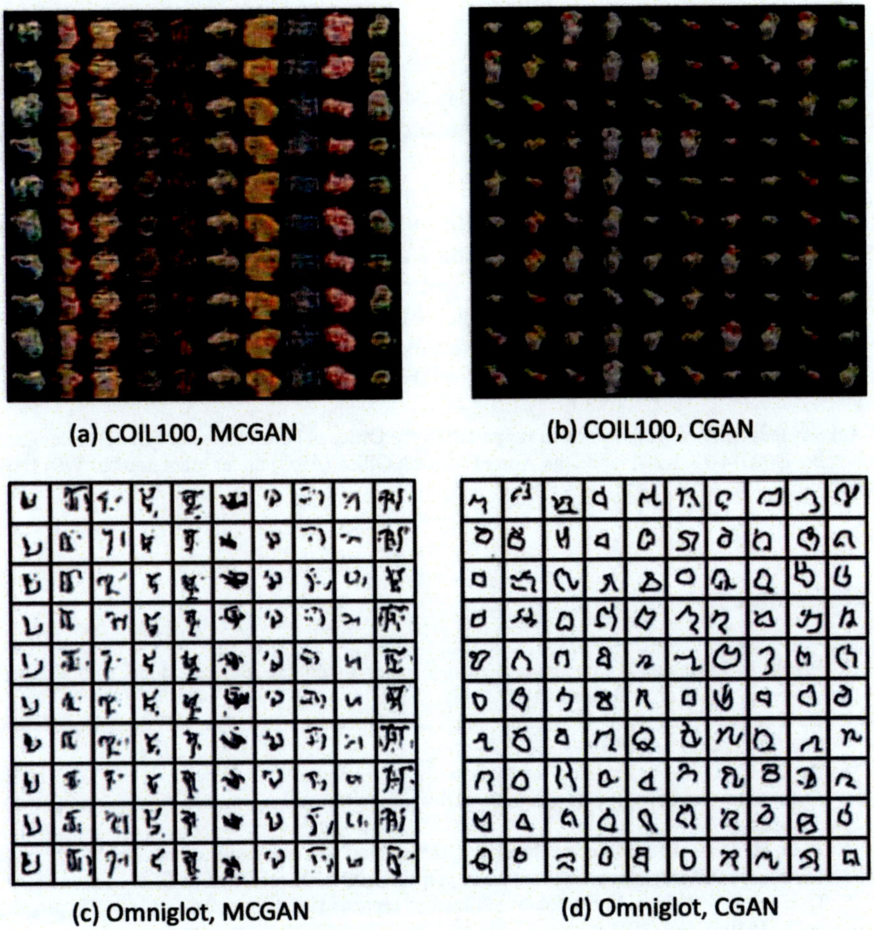

Fig. 6 a,b COIL100 and (c,d) Omniglot datasets trained with MCGAN and CGAN. (a,c). We uniformly create new data modalities (each column) from pre-trained data modalities

Qualitative results are shown in Fig. 6. The results demonstrate that subnetworks can create unbiased novel data modalities that have never been trained before. For COIL100 and Omniglot datasets, our proposed method can create novel data modalities with high fidelity and diversity because both datasets have a large number of data modalities. In particular, the learning parameters of those resampled subnetworks have been sufficiently exploited by a large number of pre-trained subnetworks.

5 Conclusion

In this work, we proposed a plug-and-play nonparametric module, named multimodal controller (MC), to equip generative models with class-conditional data generation. Unlike classical conditional generative models that introduce additional learning parameters to model class-conditional information, our method allocates a unique computation path for each data modality with a uniformly sampled subnetwork. The multimodal controller is a general method applicable to various well-known backbone generative models, and it works particularly well for a substantial number of modalities (e.g., the Omniglot challenge). Multimodal controlled generative models are also capable of creating novel data modalities. We believe that this work will shed light on the use of subnetworks for large-scale and multimodal deep learning.

Acknowledgements This work was supported by the Office of Naval Research (ONR) under grant number N00014-18-1-2244, and the Army Research Office (ARO) under grant number W911NF-20-1-0222.

References

1. Brock, A., Donahue, J., Simonyan, K.: Large scale gan training for high fidelity natural image synthesis. arXiv preprint arXiv:1809.11096 (2018)
2. Davies, D.L., Bouldin, D.W.: A cluster separation measure. IEEE Trans. Pattern Anal. Mach. Intell. **2**, 224–227 (1979)
3. De Vries, H., Strub, F., Mary, J., Larochelle, H., Pietquin, O., Courville, A.C.: Modulating early visual processing by language. In: Advances in Neural Information Processing Systems. pp. 6594–6604 (2017)
4. Diao, E., Ding, J., Tarokh, V.: Heterofl: Computation and communication efficient federated learning for heterogeneous clients. arXiv preprint arXiv:2010.01264 (2020)
5. Dumoulin, V., Shlens, J., Kudlur, M.: A learned representation for artistic style. arXiv preprint arXiv:1610.07629 (2016)
6. Fernando, C., Banarse, D., Blundell, C., Zwols, Y., Ha, D., Rusu, A.A., Pritzel, A., Wierstra, D.: Pathnet: Evolution channels gradient descent in super neural networks. arXiv preprint arXiv:1701.08734 (2017)
7. Frankle, J., Carbin, M.: The lottery ticket hypothesis: Finding sparse, trainable neural networks. arXiv preprint arXiv:1803.03635 (2018)
8. Goodfellow, I., Pouget-Abadie, J., Mirza, M., Xu, B., Warde-Farley, D., Ozair, S., Courville, A., Bengio, Y.: Generative adversarial nets. In: Advances in neural information processing systems. pp. 2672–2680 (2014)
9. Heusel, M., Ramsauer, H., Unterthiner, T., Nessler, B., Hochreiter, S.: Gans trained by a two time-scale update rule converge to a local nash equilibrium. In: Advances in neural information processing systems. pp. 6626–6637 (2017)
10. Ioffe, S., Szegedy, C.: Batch normalization: Accelerating deep network training by reducing internal covariate shift. arXiv preprint arXiv:1502.03167 (2015)
11. Kang, M., Shim, W., Cho, M., Park, J.: Rebooting acgan: Auxiliary classifier gans with stable training. Adv. Neural Inf. Process. Syst. **34**, 23505–23518 (2021)
12. Karras, T., Laine, S., Aila, T.: A style-based generator architecture for generative adversarial networks. In: Proceedings of the IEEE Conference on Computer Vision and Pattern Recognition. pp. 4401–4410 (2019)

13. Karras, T., Laine, S., Aittala, M., Hellsten, J., Lehtinen, J., Aila, T.: Analyzing and improving the image quality of stylegan. In: Proceedings of the IEEE/CVF conference on computer vision and pattern recognition. pp. 8110–8119 (2020)
14. Kingma, D.P., Welling, M.: Auto-encoding variational bayes. arXiv preprint arXiv:1312.6114 (2013)
15. Kingma, D.P., Dhariwal, P.: Glow: Generative flow with invertible 1x1 convolutions. In: Advances in Neural Information Processing Systems. pp. 10215–10224 (2018)
16. Krizhevsky, A., et al.: Learning multiple layers of features from tiny images (2009)
17. Lake, B.M., Salakhutdinov, R., Tenenbaum, J.B.: Human-level concept learning through probabilistic program induction. Science **350**(6266), 1332–1338 (2015)
18. Liu, M., Ding, Y., Xia, M., Liu, X., Ding, E., Zuo, W., Wen, S.: Stgan: A unified selective transfer network for arbitrary image attribute editing. In: Proceedings of the IEEE Conference on Computer Vision and Pattern Recognition. pp. 3673–3682 (2019)
19. Mallya, A., Davis, D., Lazebnik, S.: Piggyback: Adapting a single network to multiple tasks by learning to mask weights. In: Proceedings of the European Conference on Computer Vision (ECCV). pp. 67–82 (2018)
20. Mao, Q., Lee, H.Y., Tseng, H.Y., Ma, S., Yang, M.H.: Mode seeking generative adversarial networks for diverse image synthesis. In: Proceedings of the IEEE Conference on Computer Vision and Pattern Recognition. pp. 1429–1437 (2019)
21. Mirza, M., Osindero, S.: Conditional generative adversarial nets. arXiv preprint arXiv:1411.1784 (2014)
22. Miyato, T., Koyama, M.: cgans with projection discriminator. arXiv preprint arXiv:1802.05637 (2018)
23. Nene, S.A., Nayar, S.K., Murase, H., et al.: Columbia object image library (coil-100)
24. Odena, A., Olah, C., Shlens, J.: Conditional image synthesis with auxiliary classifier gans. In: Proceedings of the 34th International Conference on Machine Learning-Volume 70. pp. 2642–2651. JMLR. org (2017)
25. Van den Oord, A., Kalchbrenner, N., Espeholt, L., Vinyals, O., Graves, A., et al.: Conditional image generation with pixelcnn decoders. In: Advances in neural information processing systems. pp. 4790–4798 (2016)
26. Oord, A.v.d., Kalchbrenner, N., Kavukcuoglu, K.: Pixel recurrent neural networks. arXiv preprint arXiv:1601.06759 (2016)
27. Park, J., Kim, Y.: Styleformer: Transformer based generative adversarial networks with style vector. In: Proceedings of the IEEE/CVF Conference on Computer Vision and Pattern Recognition. pp. 8983–8992 (2022)
28. Rajasegaran, J., Hayat, M., Khan, S., Khan, F.S., Shao, L.: Random path selection for incremental learning (2019)
29. Salimans, T., Goodfellow, I., Zaremba, W., Cheung, V., Radford, A., Chen, X.: Improved techniques for training gans. In: Advances in Neural Information Processing Systems. pp. 2234–2242 (2016)
30. Sauer, A., Schwarz, K., Geiger, A.: Stylegan-xl: Scaling stylegan to large diverse datasets. arXiv preprint arXiv:2202.00273 (2022)
31. Sohn, K., Lee, H., Yan, X.: Learning structured output representation using deep conditional generative models. In: Advances in neural information processing systems. pp. 3483–3491 (2015)
32. Song, Y., Sohl-Dickstein, J., Kingma, D.P., Kumar, A., Ermon, S., Poole, B.: Score-based generative modeling through stochastic differential equations. arXiv preprint arXiv:2011.13456 (2020)
33. Srivastava, N., Hinton, G., Krizhevsky, A., Sutskever, I., Salakhutdinov, R.: Dropout: a simple way to prevent neural networks from overfitting. J. Mac. Learn. Res. **15**(1), 1929–1958 (2014)
34. Wortsman, M., Ramanujan, V., Liu, R., Kembhavi, A., Rastegari, M., Yosinski, J., Farhadi, A.: Supermasks in superposition. arXiv preprint arXiv:2006.14769 (2020)
35. Zhou, H., Lan, J., Liu, R., Yosinski, J.: Deconstructing lottery tickets: Zeros, signs, and the supermask. arXiv preprint arXiv:1905.01067 (2019)

TexIm: A Novel Text-to-Image Encoding Technique Using BERT

Wazib Ansar, Saptarsi Goswami, Amlan Chakrabarti, and Basabi Chakraborty

Abstract Often when we read some text, it leaves an impression in our mind. This perception imbibes the knowledge conveyed, the context, and the lexical information. Although there has been abundant research on the representation of text, research on devising techniques for visualization of embedded text is absent. Thus, we propose a novel "text-to-image" (TexIm) encoding enabling visualization of textual features. The proposed TexIm extracts the contextualized semantic and syntactic information present in the text through BERT and generates informed pictorial representations through a series of transformations. This unique representation is potent enough to assimilate the information conveyed, and the linguistic intricacies present in the text. Additionally, TexIm generates concise input representation that reduces the memory footprint by 37%. The proposed methodology has been evaluated on a hand-crafted dataset of Cricketer Biographies for the task of pair-wise comparison of texts. The conformity between the similarity of texts and the corresponding generated representations ascertain its fruitfulness.

Keywords BERT · Image processing · NLP · PCA · Word embedding

W. Ansar (✉) · A. Chakrabarti
A. K. Choudhury School of IT, University of Calcutta, Kolkata, India
e-mail: waakcs_rs@caluniv.ac.in

A. Chakrabarti
e-mail: acakcs@caluniv.ac.in

S. Goswami
Department of Computer Science, Bangabasi Morning College, Kolkata, India
e-mail: sgakc@caluniv.ac.in

B. Chakraborty
Iwate Prefectural University, Takizawa, Japan
e-mail: basabi@iwate-pu.ac.jp

M. Tistarelli et al. (eds.), *Computer Vision and Machine Intelligence*,
Lecture Notes in Networks and Systems 586,
https://doi.org/10.1007/978-981-19-7867-8_11

1 Introduction

With the advent of natural language processing (NLP), the ability of computing devices to process unstructured text has improved manifold [1]. Besides, the amount of text available online is massive and expanding at a breakneck pace. As of December 2021, the indexed Web contains above 3.15 billion Webpages.[1] Every minute, more than 474,000 tweets are posted on Twitter,[2] 510,000 comments are posted on Facebook[3], and on Google,[4] over 2,400,000 search queries are executed.[5] This ushers the need to represent text effectively and efficiently.

Conventionally, a text is represented in memory as a stream of bytes corresponding to the characters [2]. Such representation is devoid of the syntactic, lexical as well as contextual characteristics present in the text. To mitigate this, fixed-length sparse encodings like bag-of-words [3] to vector embeddings like Word2Vec [4, 5] and global vectors for word representation (GloVe) [6] were devised. Further advancements include contextualized word embedding techniques such as Context2Vec [7], embeddings from language models (ELMo) [8] and bidirectional encoder representations from transformers (BERT) [9]. Besides effective representation, efficient representation of text is becoming a key factor to reduce the memory footprint. Various approaches have been proposed to compress the text without significant information loss like Huffman encoding [10–12], Lempel-Ziv-Welch algorithm [13, 14], and Hahn's encoding technique [15]. However, these approaches faltered in distinguishing between the meanings of polysemous words in varying contexts [2]. This highlights the dearth of efficient textual representation techniques in terms of memory footprint retaining the information conveyed and the linguistic details.

Despite the abundance of textual information, visual content accounts for 90% of the information processed by the brain.[6] To bridge the gap between text and image, a few research works have been conducted to substitute text with similar illustrations [16] or to generate transformed representations of text [17–19]. However, the progress made in this domain is still in its nascent stage and necessitates devising techniques for visualization of text retaining all its characteristics with the added benefit of reduced memory footprint.

As an effort to address the above-mentioned research gaps, we hereby propose an innovative methodology which automates visualization of textual features through a novel "text-to-image" (TexIm) encoding technique. The proposed methodology captures the features present in the text using BERT embeddings and transforms them into a pictorial representation through an innovative process. Besides, it compresses the text into a concise form so that subsequent processing operations become more efficient. The proposed methodology provides dual advantage of tapping the contex-

[1] https://www.worldwidewebsize.com/.

[2] https://twitter.com.

[3] https://www.facebook.com.

[4] https://www.google.com.

[5] https://blog.microfocus.com/how-much-data-is-created-on-the-internet-each-day/.

[6] https://www.t-sciences.com/news/humans-process-visual-data-better.

tualized syntactic and lexical information in the text along with subsequently generating pictorial representations for enhanced visualization, optimal processing, and less memory consumption. The principal contributions in this paper are as follows:

1. A novel "text-to-image" (TexIm) encoding technique has been proposed which automates visualization of textual features.
2. The proposed TexIm captures the contextualized syntactic and lexical features in the text through an innovative feature representation utilizing BERT.
3. An efficacy of the proposed methodology has been evaluated for the task of pair-wise comparison of texts on a hand-crafted dataset comprising of biographies of renowned international cricketers.
4. The generated pictorial representations bring down the memory consumption by 37.370% compared to normal text which optimizes processing.

The compendium to this paper has been ushered herein. The prior contributions in this domain are surveyed in Sect. 2. The proposed methodology is enunciated in Sect. 3. Section 4 outlines the experimental setup to implement and analyze the efficacy of the proposed methodology. Section 5 presents the results obtained based on a variety of parameters and metrics followed by a discussion in Sect. 6. Finally, in the conclusion section, the inferences are drawn with a commentary on the future avenues for this work.

2 Related Works

A review of how prior works and their inherent flaws pave the way for the conceptualization of the proposed methodology has been offered in this section.

2.1 Representation of Text

Effective representation of text is the key to churning out optimal performance from a model in NLP [20]. Fixed-length sparse encodings like bag-of-words despite being pretty naive as well as effective were deficient in semantic and contextual information [3]. To alleviate this drawback, vector-based techniques such as Word2Vec were developed. It is comprised of two single-hidden layer neural networks, i.e., continuous bag-of-words (CBoW) and skip-gram (SG) [4, 5]. Although Word2Vec deftly captured the local context, it faltered in comprehending the global context [6]. Global vectors for word representation (GloVe) solved this limitation through matrix factorization of the word-context co-occurrence matrix using stochastic gradient descent (SGD) [6]. Besides syntactic and lexical information, contextual information is crucial to distinguish between the information conveyed by a word in varying contexts. For instance, the word "band" can refer to a group of musicians or an adornment worn around the wrist. This can be achieved through contextualized word embedding techniques such as Context2Vec [7], ELMo [8], and BERT [9].

2.2 Compression of Text

Apart from the efficacy of the input representation, the efficiency in terms of memory footprint is also significant. Taking cognizance of the magnanimity of the data being fed to the contemporary deep learning models, reducing the memory consumption and computational overhead has become a prime concern [21]. For text compression, Huffman [10] proposed variable-length encoding for characters based on their frequency. Ziv and Lempel [13, 14] proposed an algorithm to convert input strings of variable lengths into codes of fixed length by substituting common patterns of frequently occurring n-grams with an unused byte for the next string of incremental length. Hahn [15] encoded non-blank characters in fixed-length groups as distinct numbers with fixed precision. Recent advances include Huffman trees variations like Zopfli by Google [11] and techniques deploying quaternary trees [12]. A common drawback of these approaches was their inability to deal with polysemy and failure to distinguish between possible syntactic as well as semantic variations of identical words in different contexts [2]. Therefore, the need arises to devise compression techniques to preserve the semantic and lexical information in the text.

2.3 Visualization and Analysis of Text

In NLP, a few studies have been conducted for the visualization of textual content. Zakraoui et al. [16] surveyed works to substitute a piece of text with illustrations or animations from a repository having the highest probability. Here, the works did not enable retrieval of text from the generated images. Nataraj et al. [17] targeted malware detection through transforming text into grayscale images by converting characters into intensity values. They asserted their approach made the model resilient to code obfuscation and API injections. He et al. [18] took this study forward by designing a CNN model with spatial pyramid pooling (SPP) [22] to accommodate variable-size inputs. They inferred that conversion to RGB images would enhance the performance compared to grayscale images. Petrie and Julius [19] created a string-map representation for pair-wise comparison of patent-inventor records using AlexNet [23]. Each string-map was assigned a single color, and during comparison,it was superimposed upon the other. It was restricted to comparison purposes only as no significant information could be obtained from individual representations. From the above-mentioned contributions, it can be inferred that limited research has been performed to generate pictorial representations from texts. Furthermore, works focusing on compression of the text capturing the semantic and syntactic dependencies are absent.

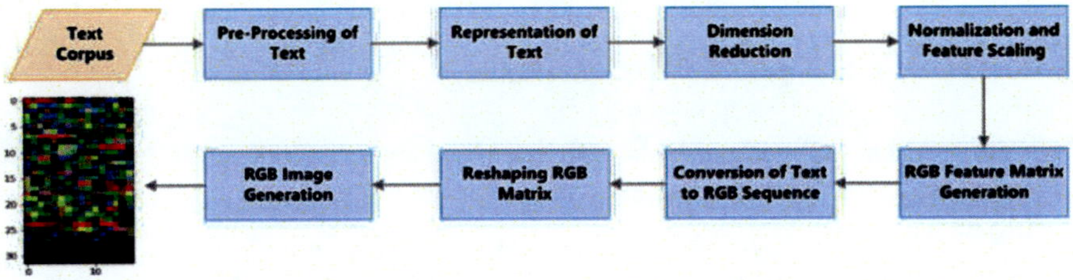

Fig. 1 Illustration of the proposed methodology

3 Proposed Methodology

Utilizing the proposed TexIm, the text sequence is transformed into its pictorial representation through a series of steps described in this section. The flow of the proposed methodology has been depicted in Fig. 1.

3.1 Pre-processing of Text

Firstly, the text is subjected to cleaning operations to filter out insignificant portions. It includes removal of line breaks, tab sequences, URL patterns, digits, non-ASCII characters, punctuation marks, and white-space standardization. After that lemmatization is performed. This aids in normalizing a word through the elimination of morphological derivatives from it [24]. Finally, certain frequent but contextually insignificant words also known as stop words are removed.

3.2 Representation of Text

The text representation is prepared as per the specifications of the BERT model consisting of token embeddings, segment embeddings, and position embeddings [9]. The text is tokenized using WordPiece tokenizer which decomposes the words into one or more subwords or morphemes [25]. After that, for sequences longer than the maximum sequence length sl, a special truncation operation is carried out such that the most lengthy sentence present in the sequence is truncated instead of chopping off the beginning or the end of the sequence. Alternatively, sequences shorter than sl are zero-padded to ensure uniform sequence length. To distinguish between the tokens and the padding, mask ID $M[j]$ for jth position is incorporated as enunciated in Eq. (1).

$$M[j] = \begin{cases} 1 & \text{for token} \\ 0 & \text{for padding} \end{cases}, \forall j \in [0, sl) \tag{1}$$

The segment embedding deals with assigning segment indices to the sentences. For the jth token in the kth sentence, the segment embedding $S_v[j, k]$ is generated using Eq. (2).

$$S_v[j, k] = k \tag{2}$$

The position embedding is applied on the sequence to establish ordering of tokens. For the sequence, $S = \{t_1, t_2, t_3, ..., t_{sl}\}$, the position embedding $P[t_i]$ of the ith token, $t_i \in S$ is calculated through Eq. (3) in which the positional indices are incremented by δ with β being the offset.

$$P[t_i] = \beta + i\delta, \forall i \in [0, sl) \tag{3}$$

Finally, the contextualized embedding vectors are generated through fine-tuning a pre-trained BERT model. Here, a combination of the hidden layers is selected which captures best the significant information present in the text. In the proposed methodology, for the ith token t_i, the word vector $WV[t_i]$ is obtained through concatenation of last four hidden layers $H_t^i[-4:-1]$ as depicted in Eq. (4).

$$WV[t_i] = \text{Concat}(H_t^i[-4:-1]) \tag{4}$$

3.3 Dimension Reduction

After computing the word embeddings, the next step is to reduce the dimensions of the word embeddings. This is essential as ultimately for RGB representation, only three components would be required, i.e., red, green, and blue channels. In the proposed methodology, PCA has been selected due to its ability to capture the maximum variability in the data through the selection of orthogonal components in decreasing order of variability [26].

3.4 Normalization and Feature Scaling

The three-dimensional embeddings obtained from the previous step need further processing to serve as RGB channels. To accomplish this, the components are normalized within the bounds of [0, 1] as presented in Eq. (5). It is followed by feature scaling the normalized embeddings between legitimate ranges of the RGB channels, i.e., $[0 - 255]$ as shown in Eq. (6).

$$E'_t[d] = \frac{E_t[d] - \min(E_t[d])}{\max(E_t[d]) - \min(E_t[d])}, \forall d \in [0, 3) \text{ and } t \in V \quad (5)$$

$$E^s_t[d] = E'_t[d] * 255, \forall d \in [0, 3) \text{ and } t \in V \quad (6)$$

where $E_t[d]$ denotes the dth embedding component for token t in the vocabulary V. $E'_t[d]$ and $E^s_t[d]$ denote the corresponding normalized and the scaled embedding components, respectively. Here, normalization has been chosen over standardization [27] to avoid negative values and obtain a well-distributed representation within the range [0–255].

3.5 RGB Feature Matrix Generation

The outcome of this step is to map the RGB embeddings generated in the previous step with their corresponding tokens. Here, an RGB feature matrix is generated such that the ith index contains the RGB embedding corresponding to the token having index i. The matrix once generated serves as a lookup table for generating the TexIm embeddings and also enables lossless reconstruction of text from the pixels. This approach has been described in Algorithm 1.

Algorithm 1 RGB feature matrix generation

Require: Vocabulary V , Scaled Embedding Vector E^s
Ensure: RGB Feature matrix Fm_{RGB}
  $l_V \leftarrow size(V)$
  for $i \leftarrow 0$ to $(l_V - 1)$ **do**
    $Fm_{RGB}[i] \leftarrow \{V_i : E^s[V_i]\}$ ▷ V_i is the ith token in V
  end for

3.6 Text to RGB Sequence Conversion

After obtaining the RGB feature matrix, the text sequences in the corpus need to be converted into equivalent RGB sequences. For each word, the corresponding RGB embedding is obtained through looking up the RGB feature matrix Fm_{RGB} as portrayed in Algorithm 2. A point to ponder here is that a single word may be tokenized into one or more subwords. In that case, each of those subwords will be mapped to a unique RGB value in Fm_{RGB}. It enables the proposed methodology to preserve the morphological constructs present in a word and diminish the out-of-vocabulary (OOV) problem.

Algorithm 2 Text to RGB mapping

Require: Text Sequence S , RGB Feature matrix Fm_{RGB}
Ensure: RGB Embedding Sequence Es_{RGB}

```
S_WP ← WordPiece_Tokenizer(S)                ▷ S_WP is the tokenized sequence
l_WP ← size(S_WP)
for i ← 0 to l_WP − 1) do
    Es_RGB[i] ← {Fm_RGB[t_i]}                ▷ t_i is the i^th token in S_WP
end for
```

3.7 *Reshaping RGB Matrix and Image Generation*

The RGB sequence obtained from the previous step is one-dimensional, but a typical image has two dimensions. To convert this one-dimensional sequence to two-dimensional form, reshaping operation is applied as depicted in Algorithm 3. Similar to the input sequence length, the dimensions of the RGB matrix are predefined to ensure uniformity in the dimensions of pictorial representations for all inputs. Finally, the image is generated from the RGB matrix. Here, each vector in the RGB embedding matrix is cast into (3 * 8 = 24) bit unsigned int ('uint8') notation. This helps to reduce the memory footprint compared to the usual 32-bit floating point ('float32') notation.

Algorithm 3 Reshaping RGB Matrix

Require: RGB Embedding Sequence Es_{RGB} of length l
Ensure: $r * c$ RGB Embedding Matrix Em_{RGB}

```
k ← 0
for i ← 0 to (r − 1) do
    for j ← 0 to (c − 1) do
        Em_RGB[i, j] ← Es_RGB[k]
        k ← k + 1
    end for
end for
```

4 Experimental Setup

In this section, the experimental setup consisting of the dataset description, implementation details, hyperparameters as well as metrics to analyze the efficacy of the proposed methodology has been enunciated.

Table 1 BERT hyperparameters

Hyperparameter	Value
Number of transformer blocks	12
Number of attention heads	12
Number of parameters	110 M
Hidden layer size	768
Batch size	128

4.1 Data

The proposed methodology has been implemented on a hand-crafted dataset compiled by us. It consists of 200 biographies of international cricketers representing nations around the globe extracted from Wikipedia.[7] The fields in the data include name, country, and biography. The length of each biography ranges between 100 and 600 words occupying about 2.16 KB of memory on average. Furthermore, the dataset has been augmented with 20 biographies of renowned personalities apart from cricketers to test the ability of the proposed methodology to handle data from different domains.

4.2 Implementation Details and Hyperparameters

The proposed methodology presented in Sect. 3 has been implemented on Python 3 Google Compute Engine provided by Google Colab[8] with 8 TPU cores, 12.69 GB RAM, and 107.72 GB Disk. For encoding of text, an uncased BERT_Base model has been fine-tuned with hyperparameters listed in Table 1. The overall time taken to fine-tune the BERT model upon our dataset is 5 min and 33 s. The maximum sequence length of the input has been fixed as 512. The dimensions of the embeddings obtained using BERT are (4 * 768) = 3072, while RGB embeddings generated using the proposed methodology have three dimensions corresponding to the red, green, and blue channels. Finally, the RGB sequence has been represented as a two-dimensional image of size (32 * 16) pixels.

---

[7] https://www.wikipedia.org.

[8] https://colab.research.google.com.

5 Results and Analysis

In this section, the results obtained using the proposed methodology have been presented based on a variety of parameters and metrics.

5.1 Demonstration of TexIm

As part of a demonstration of the proposed methodology, the text and its corresponding pictorial representation generated through TexIm have been presented in Table 2 and Fig. 2 respectively. In Fig. 2, the colored pixels represent the TexIm embeddings corresponding to the tokens in the sequence, while the black region indicates the padding applied to ensure uniform dimensions of the generated images. At first sight, the generated RGB representation may not appear informative to the naked eye. But, it imbibes the linguistic intricacies present in the text such that similar hues indicate contextual as well as semantic similarity of words as elucidated in Sect. 5.2. An interesting observation is that the text contains 458 words containing 2780 characters occupying 2780 bytes of memory, while the generated RGB image is of (512 * 3 = 1536) bytes achieving 44.75% memory compression.

5.2 Representation of the TexIm Embeddings

Figure 3a represents the entire vocabulary in three-dimensional space, with axes denoting the red, green, and blue channels of the embeddings and having its range between [0, 255]. Here, each point has been color-coded indicating its embedded RGB value. Furthermore, Fig. 3b portrays the RGB feature representation of words

Fig. 2 RGB image

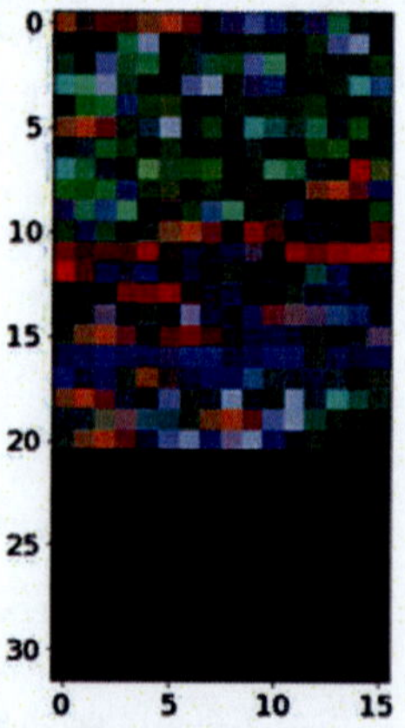

Table 2 Demonstration of the proposed methodology

	Text: *Sachin Ramesh Tendulkar is a former Indian international cricketer who captained the Indian squad. He is the only player to score 100 international centuries, the first batsman to score a double century in a One-Day International (ODI), the all-time leading run scorer in both Test and ODI cricket, and the only player to reach 30,000 runs in international cricket. In 2013, he was the only Indian cricketer nominated to an all-time Test World XI to commemorate Wisden Cricketers' Almanack's 150th anniversary. "Little Master" or "Master Blaster" are some of his nicknames. Tendulkar began playing cricket at the age of eleven, making his Test debut against Pakistan in Karachi on November 15, 1989, at the age of sixteen, and going on to represent Mumbai domestically and internationally for nearly twenty-four years. He was voted the second-greatest Test batsman of all time, behind Don Bradman, and the second-greatest ODI batsman of all time, behind Viv Richards, by Wisden Cricketers' Almanack in 2002. Tendulkar was a member of the Indian squad that won the 2011 World Cup, which was his first win for India in six World Cup appearances. In the 2003 edition of the event, held in South Africa, he was named "Player of the Tournament". Tendulkar got the Arjuna Medal for excellent sports accomplishment in 1994, the Rajiv Gandhi Khel Ratna award in 1997, India's highest sporting honor, and the Padma Shri and Padma Vibhushan awards in 1999 and 2008, India's fourth and second highest civilian awards, respectively. On November 16, 2013, a few hours after his final match, the Prime Minister's Office announced that he would be awarded the Bharat Ratna, India's highest civilian honor. He is the award's youngest recipient to date, as well as the first sportsperson to receive it. At the 2010 ICC awards, he also received the Sir Garfield Sobers Trophy for cricketer of the year. Tendulkar was elected to the Rajya Sabha, India's highest chamber of parliament, in 2012. He was also the first sportsperson and the first person without an aviation background to be honored by the Indian Air Force with the honorary rank of group captain. In 2012, he was inducted into the Order of Australia as an Honorary Member. Sachin was named one of the "Most Influential People in the World" in Time magazine's annual Time 100 list in 2010. Tendulkar declared his retirement from One-Day Internationals in December 2012. In October 2013, he retired from Twenty20 cricket, then on November 16, 2013, he retired from all forms of cricket after playing his 200th Test match against the West Indies in Mumbai's Wankhede Stadium. Tendulkar appeared in 664 international cricket matches, scoring 34,357 runs in the process. Tendulkar was inducted into the ICC's Cricket Hall of Fame in 2019.*

similar to the word "cricket". An interesting characteristic observed here is that in the given three-dimensional space, the proximity of the words is based on their mutual similarity. Apart from this, the similarity is also reflected in the hues corresponding to the words. For instance, the words "cricket" and "domestic" have similar hues. On the other hand, the word "premier" seems slightly out of context indicated by its contrasting hue.

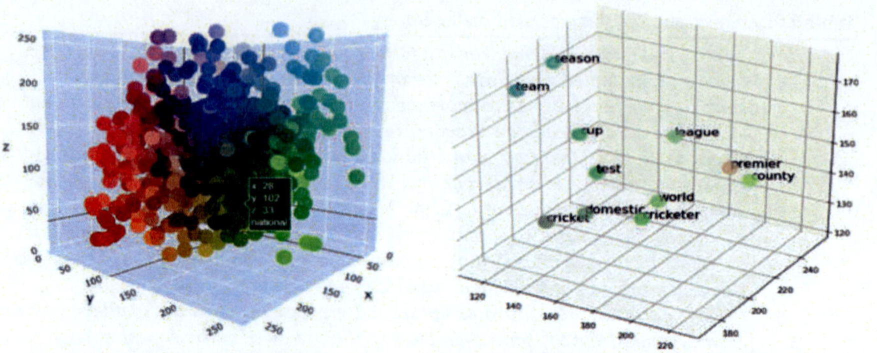

(a) Entire vocabulary (b) Words similar to "cricket"

Fig. 3 RGB feature representation of the TexIm embeddings

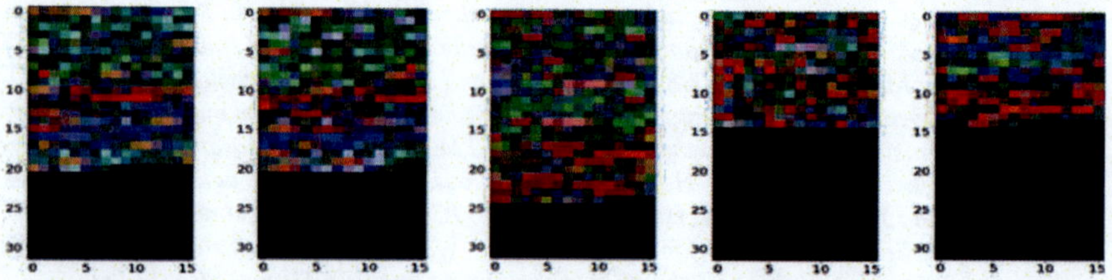

(a) T1 image (b) T2 image (c) T3 image (d) T4 image (e) T5 image

Fig. 4 Pictorial representations generated through the proposed methodology

5.3 Evaluation of Efficacy

The efficacy of the proposed TexIm in retaining the information present in the input text has been analyzed through the task of pair-wise comparison of texts. For this, four distinct comparisons have been studied as follows:

1. **Text Comparison 1 (TC1)**: A cricketer's biography (T1) is compared with its paraphrased text (T2).
2. **Text Comparison 2 (TC2)**: A cricketer's biography (T1) is compared with another cricketer's biography (T3) both representing the same country.
3. **Text Comparison 3 (TC3)**: A cricketer's biography (T1) is compared with another cricketer's biography (T4) representing different countries.
4. **Text Comparison 4 (TC4)**: A cricketer's biography (T1) is compared with a politician's biography (T5).

Figure 4 depicts the pictorial representations of these five texts generated through the proposed methodology. Figure 5 shows the comparison among the histograms of the pictorial representations. Figure 6 presents the comparisons among the texts, the generated RGB images as well as the histograms of the images based on evalu-

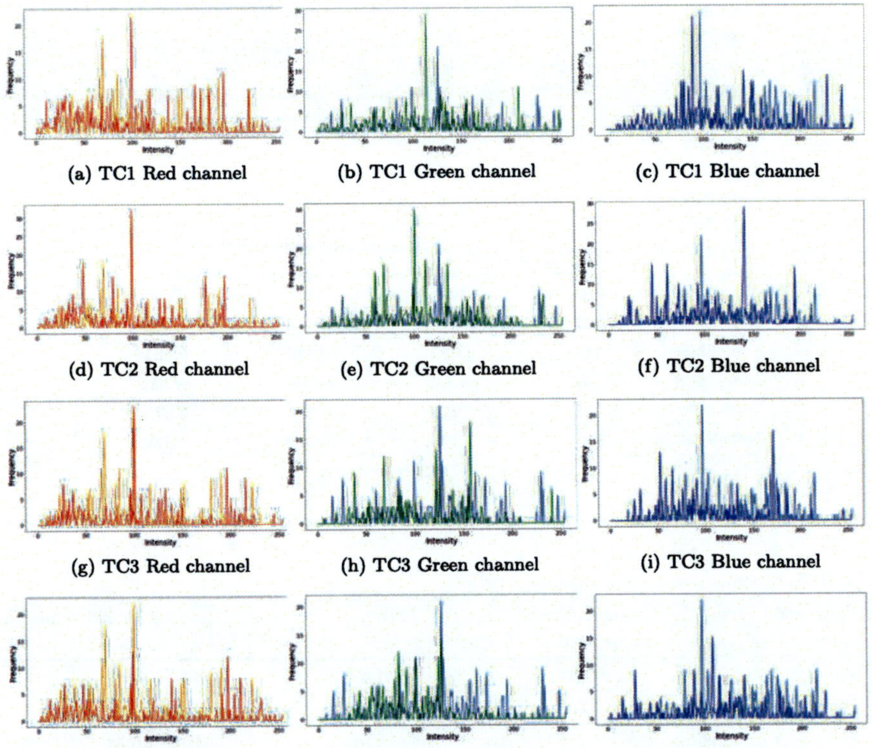

(a) TC1 Red channel (b) TC1 Green channel (c) TC1 Blue channel

(d) TC2 Red channel (e) TC2 Green channel (f) TC2 Blue channel

(g) TC3 Red channel (h) TC3 Green channel (i) TC3 Blue channel

(j) TC4 Red channel (k) TC4 Green channel (l) TC4 Blue channel

Fig. 5 Pair-wise comparison of histograms of the pictorial representations

ation metrics like mean square error ($MS\epsilon$), root mean square error ($RMS\epsilon$), peak signal-to-noise-ratio (PSNR) [28], structural-similarity index measure (SSIM) [29], and word mover distance ($WM\delta$) [30]. From the trend between the $WM\delta$ and other similarity metrics, it can be inferred that the similarity among the texts is in accordance with the similarity among the images and their histograms. This ascertains that the generated images retain the contextualized semantic as well as syntactic information of the text.

6 Discussion

Based on the results obtained, the efficacy, efficiency as well as limitations of the proposed TexIm have been discussed in this section.

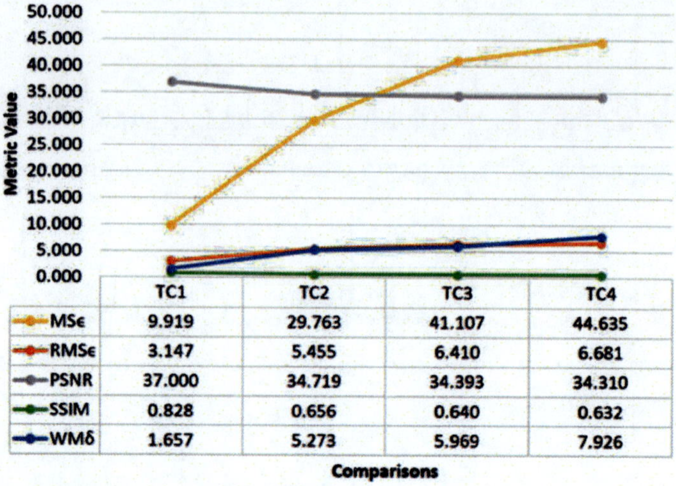

Fig. 6 Illustration of pair-wise similarity trend

Table 3 Comparison of compression capability

Parameter	Conventional approach	TexIm	Compression achieved
Average word length	4.79 bytes	3 bytes	37.370%
Encoding data type	float32	uint8	75%
Encoding dimensions	3072	3	1024 times

6.1 Compression of Text

In TexIm, 24 bits are used to encode each word irrespective of the number of characters present in it. Thus, the proposed methodology achieves text compression for all words greater than three characters assuming each character is comprised of 8 bits. Given the fact that the average word length in the English language is 4.79 characters with just 0.706% distinct words having less than or equal to two characters,[9] one can even deduce that the proposed methodology achieves approximately 37.370% compression of text. Besides, the selection of "uint8" notation for embedding representation consumes 75% less memory compared to widely used "float32" notation. The compression capability of the proposed TexIm has been demonstrated in Table 3.

[9] http://www.norvig.com/mayzner.html.

6.2 Maximum Supported Vocabulary Size

Theoretically, the proposed representation is capable of providing unique embeddings for 256^3 =16,777,216 distinct words. Practically, a vocabulary of this size cannot be fully mapped as having all possible combinations of embedding dimensions may not be possible. It is to be noted that as per the Oxford English Dictionary, 171,476 words are currently in use.[10] Whereas, according to Wiktionary, 1,317,179 total entries are present in the English language.[11] This accounts for only 7.85% of the vocabulary limit of the proposed TexIm.

6.3 Dealing with Unseen Text

The proposed TexIm utilizes the WordPiece tokenizer which decomposes the words into one or more subwords or morphemes. This serves three purposes—Firstly, to deal with unseen textual content through partially understanding out-of-vocabulary (OOV) words by decomposing it into identifiable subwords. Secondly, to establish similarity among words with identical constituent subwords. Finally, by carrying out "meaning preserving" subword splits, it accomplishes a reduction in the size of the vocabulary as multiple words may be constituted from the same subword. For example, words like "snow", "snowman", "fire", and "fireman" can be tokenized into ["snow", "##man", "fire"].

7 Conclusion

In this paper, a novel methodology automating visualization of textual content, i.e., TexIm has been proposed. The proposed TexIm generates pictorial representations from text retaining the linguistic intricacies with the additional benefit of information compression. Furthermore, an extensive analysis accompanied with an experiment to study the performance of the proposed methodology for the task of pair-wise comparison of texts has been conducted on a hand-crafted Cricketer Biography dataset. The results obtained ascertain its utility as well as efficacy. Despite the advantages discussed above, the proposed methodology has a few limitations too. Firstly, it uses PCA for dimension reduction that might lead to loss of information. Secondly, the RGB feature matrix needs to be recomputed whenever there is an alteration in the vocabulary. In future, efforts can be directed toward addressing these limitations. For instance, other dimensionality reduction measures like t-distributed stochastic-neighbor-embedding (t-SNE), linear-discriminant analysis (LDA), and so on may

[10] https://www.lexico.com/explore/how-many-words-are-there-in-the-english-language.

[11] https://en.wikipedia.org/wiki/List_of_dictionaries_by_number_of_words#cite_note-11.

be explored to obtain better results. Furthermore, comparison of the efficacy of the proposed methodology for diverse NLP tasks can be undertaken in future.

References

1. Chowdhary, K.R.: Natural language processing. In: Fundamentals of Artificial Intelligence, pp. 603–649. Springer, New Delhi (2020)
2. Ainon, R.N.: Storing text using integer codes. In: Coling 1986 Volume 1: The 11th International Conference on Computational Linguistics (1986)
3. Harris, Z.S.: Distributional structure. Word **10**(2–3), 146–162 (1954)
4. Mikolov, T., Chen, K., Corrado, G., Dean, J.: Efficient estimation of word representations in vector space. arXiv preprint arXiv:1301.3781 (2013)
5. Mikolov, T., Sutskever, I., Chen, K., Corrado, G.S., Dean, J.: Distributed representations of words and phrases and their compositionality. In: Advances in Neural Information Processing Systems, pp. 3111–3119 (2013)
6. Pennington, J., Socher, R., Manning, C.: Glove: global vectors for word representation. In: Proceedings of the 2014 Conference on Empirical Methods in Natural Language Processing (EMNLP), pp. 1532–1543 (2014)
7. Melamud, O., Goldberger, J., Dagan, I.: Context2vec: learning generic context embedding with bidirectional LSTM. In: Proceedings of the 20th SIGNLL Conference on Computational Natural Language Learning, pp. 51–61 (2016)
8. Peters, M.E., Neumann, M., Iyyer, M., Gardner, M., Clark, C., Lee, K., Zettlemoyer, L.: Deep contextualized word representations. arXiv preprint arXiv:1802.05365 (2018)
9. Devlin, J., Chang, M.-W., Lee, K., Toutanova, K.: Bert: pre-training of deep bidirectional transformers for language understanding. arXiv preprint arXiv:1810.04805 (2018)
10. Huffman, D.A.: A method for the construction of minimum-redundancy codes. Proc. IRE **40**(9), 1098–1101 (1952)
11. Alakuijala, J., Vandevenne, L.: Data Compression Using Zopfli. Tech. Rep, Google (2013)
12. Habib, A., Jahirul Islam, M., Rahman, M.S.: A dictionary-based text compression technique using quaternary code. Iran J. Comput. Sci. **3**(3), 127–136 (2020)
13. Ziv, J., Lempel, A.: A universal algorithm for sequential data compression. IEEE Trans. Inf. Theory **23**(3), 337–343 (1977)
14. Ziv, J., Lempel, A.: Compression of individual sequences via variable-rate coding. IEEE Trans. Inf. Theory **24**(5), 530–536 (1978)
15. Hahn, B.: A new technique for compression and storage of data. Commun. ACM **17**(8), 434–436 (1974)
16. Zakraoui, J., Saleh, M., Ja'am, A.: Text-to-picture tools, systems, and approaches: a survey. Multimedia Tools Appl. **78**(16), 22833–22859 (2019)
17. Nataraj, L., Karthikeyan, S., Jacob, G., Manjunath, B.S.: Malware images: visualization and automatic classification. In: Proceedings of the 8th International Symposium on Visualization for Cyber Security, pp. 1–7 (2011)
18. He, K., Kim, D.-S.: Malware detection with malware images using deep learning techniques. In: 2019 18th IEEE International Conference on Trust, Security and Privacy in Computing and Communications/13th IEEE International Conference on Big Data Science and Engineering (TrustCom/BigDataSE) (2019). https://doi.org/10.1109/TrustCom/BigDataSE.2019.00022
19. Petrie, S.M., Julius, T.D.: Representing text as abstract images enables image classifiers to also simultaneously classify text. arXiv preprint arXiv:1908.07846 (2019)
20. Zhu, L., Li, W., Shi, Y., Guo, K.: SentiVec: learning sentiment-context vector via kernel optimization function for sentiment analysis. IEEE Trans. Neural Networks Learn. Syst. **32**(6), 2561–2572 (2020)

21. Strubell, E., Ganesh, A., McCallum, A.: Energy and policy considerations for deep learning in NLP. arXiv preprint arXiv:1906.02243 (2019)
22. Hu, W., Tan, Y.: Black-box attacks against RNN based malware detection algorithms. In: Workshops at the Thirty-Second AAAI Conference on Artificial Intelligence (2018)
23. Krizhevsky, A., Sutskever, I., Hinton, G.E.: Imagenet classification with deep convolutional neural networks. In: Advances in Neural Information Processing Systems, vol. 25 (2012)
24. Plisson, J., Lavrac, N., Mladenic, D.: A rule based approach to word lemmatization. Proc. IS **3**, 83–86 (2004)
25. Wu, Y., Schuster, M., Chen, Z., Le, Q.V., Norouzi, M., Macherey, W., Krikun, M., et al.: Google's neural machine translation system: bridging the gap between human and machine translation. arXiv preprint arXiv:1609.08144 (2016)
26. Clark, N.R., Ma'ayan, A.: Introduction to statistical methods to analyze large data sets: principal components analysis. Sci. Signal. **4**(190) (2011)
27. Patro, S., Sahu, K.K.: Normalization: a preprocessing stage. arXiv preprint arXiv:1503.06462 (2015)
28. Hore, A., Ziou, D.: Image quality metrics: PSNR versus SSIM. In: 2010 20th International Conference on Pattern Recognition, pp. 2366–2369. IEEE (2010)
29. Wang, Z., Bovik, A.C., Sheikh, H.R., Simoncelli, E.P.: Image quality assessment: from error visibility to structural similarity. IEEE Trans. Image Process. **13**(4), 600–612 (2004)
30. Kusner, M., Sun, Y., Kolkin, N., Weinberger, K.: From word embeddings to document distances. In: International Conference on Machine Learning, pp. 957–966. PMLR (2015)

ED-NET: Educational Teaching Video Classification Network

Anmol Gautam, Sohini Hazra, Rishabh Verma, Pallab Maji, and Bunil Kumar Balabantaray

Abstract Convolutional neural networks (CNNs) are widely used in computer vision-based problems. Video and image data are dominating the Internet. This has led to extensive use of deep learning (DL)-based models in solving tasks like image recognition, image segmentation, video classification, etc. Encouraged by the enhanced performance of CNNs, we have developed ED-NET in order to classify videos as teaching videos or non-teaching videos. Along with the model, we have developed a novel dataset, Teach-VID, containing teaching videos. The data is collected through our e-learning platform Gyaan, an online end-to-end teaching platform developed by our organization, GahanAI. The purpose is to make sure we can restrict non-teaching videos from being played on our portal. The models proposed along with the dataset provide benchmarking results. There are two models presented one that makes use of 3D-CNN and the other uses 2D-CNN and LSTM. The results suggest that the models can be used in real-time settings. The model based on 3D-CNN has reached an accuracy of 98.87%, and the model based on 2D-CNN has reached an accuracy of 96.34%. The loss graph of both models suggests that there is no issue

Supported by Organization GahanAI, Bengaluru, India

A. Gautam (✉) · B. K. Balabantaray
National Institute of Technology Meghalaya, Shillong, India
e-mail: t20cs004@nitm.ac.in

B. K. Balabantaray
e-mail: bunil@nitm.ac.in

S. Hazra · R. Verma · P. Maji
GahanAI, Bengaluru, India
e-mail: sohini.hazra@gahanai.com
URL: https://gahanai.com/

R. Verma
e-mail: rishabh.verma@gahanai.com
URL: https://gahanai.com/

P. Maji
e-mail: pallab.maji@gahanai.com
URL: https://gahanai.com/

M. Tistarelli et al. (eds.), *Computer Vision and Machine Intelligence*,
Lecture Notes in Networks and Systems 586,
https://doi.org/10.1007/978-981-19-7867-8_12

of overfitting and underfitting. The proposed model and dataset can provide useful results in the field of video classification regarding teaching versus non-teaching videos.

Keywords Video classification · Deep learning · CNN · LSTM

1 Introduction

The amount of audio–video data has increased exponentially in recent times. This has led researchers to develop methods and datasets working directly in video modality. Video classification has seen strong progress in the recent past. Advancements in deep learning (DL)-based models and techniques have led to the introduction of powerful models that have solved video-based vision problems. But, model architecture development is only part of the problem. To develop a solution, a dataset for solving the specific problem is necessary. Large-sized datasets like ImageNet were key in the development of DL-based solutions for image-related tasks.

Currently, convolutional neural networks (CNNs)-based models have been extensively used for video classification. The information in video data has a spatiotemporal aspect along with sequential information. This also makes handling video data directly a challenging task. The video is composed of sequential images with overlapping contents.

Due to a shift in teaching mode, in the wake of the global pandemic, we have witnessed a new form of data modality that started dominating the Internet. The online shift of teaching institutions ranging from schools to graduate colleges. Though, YouTube and other related platforms had online content related to education and training categories. But, this recent shift has led to entire classes being held on platforms like Google Meet, Microsoft Teams, and some custom-made applications like the e-learning portal developed by our organization. A new problem that has emerged in online teaching mode is whether the video being played on the platform belongs to the teaching category or not. In this regard, we have developed a system that can classify the video being played on our e-learning platform as teaching or not. If the video is categorized as a non-teaching category, then the stream will stop immediately.

In this paper, we introduce Teach-VID, a dataset containing online teaching videos which will be useful for video classification tasks. In our study, we have treated video classification as producing relevant labels given video clips. Some frames of a video clip from the Teach-VID dataset can be seen in Fig. 1.

Along with the dataset, we have also proposed a model that can classify the given video as teaching or non-teaching. The baseline model can be used for benchmarking purposes. Our contributions can be summarized as follows:

- Teach-VID dataset, a novel dataset containing online video clips. Each clip is of sixty-second duration and is extracted from the videos uploaded on our e-learning platform.

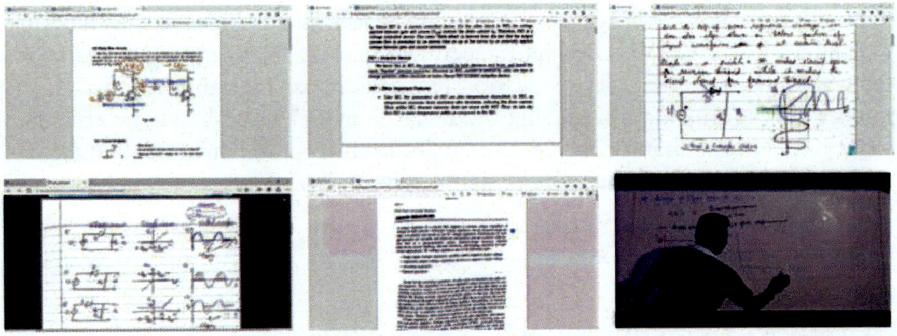

Fig. 1 Sample frames from Teach-VID dataset

- We propose two models for the video classification task. One is using 3D convolutional layers for feature extraction and dense layers for classification. Another model is developed using 2D convolutional layers for frame-wise feature extraction and using the LSTM layer to incorporate sequential information. In our study, both models have shown competitive results and can be used as benchmarking models for further work on the Teach-VID dataset.

2 Related Work

The task of video analysis is computationally expensive due to complicated data modalities. The sequence information has to be processed efficiently to get the most out of it. In this regard, Bhardwaj et al. [2] have proposed a teacher–student architecture that is computationally efficient. They utilized the idea of knowledge distillation. The network makes use of a fraction of frames at inferencing time. The student network performs on a similar level to the teacher network on the YouTube-8M [1] dataset. Xu et al. [15] have utilized networks with two streams that learn static and motion information. They have utilized the late fusion approach in order to reduce overfitting risk. Peng et al. [10] have used spatiotemporal attention along with spatial attention to obtain discriminative features to achieve better results. Appearance and relation network was introduced by Wang et al. [14] to learn video representation in an end-to-end manner. The SMART block introduced in the network separates out spatiotemporal information in spatial modeling and temporal modeling. Tran et al. [13] have developed channel separated convolutional network which factorizes 3D convolutions. This helps to reduce the computation along with boosting the results. Long et al. [8] have used the idea of using pure attention networks for video classification. They have proposed pyramid-pyramid attention clusters that incorporate channel attention and temporal attention. Pouyanfar et al. [11] have extracted spatiotemporal features from video sequences using residual attention-based bidirectional long short-term memory. Further, to handle data imbalance, they have used weighted SVM. NeXtVLAD is introduced by Lin et al. [7], to aggregate frame-wise features into a compact feature vector. They decompose high-dimensional features into a group of smaller dimensional vectors. The model was found to be efficient in aggregating temporal information. Li et al. [6] have proposed SmallBig Net that uses two branches to incorporate core semantics through one and contextual semantics through another.

3 Dataset Development and Description

The Teach-VID dataset contains 60-s clips of teaching videos uploaded on our e-learning platform developed by our organization. Our Web-based product is an e-learning platform that aims to provide an online medium to conduct educational and e-learning activities. This platform is available to institutes, coaching, schools, and individuals who want to create and manage classes in an online manner. The video uploaded on our platform is used to create the dataset. Each video was preprocessed to remove identification markers like visual clues and audio clues to ensure privacy. To train the models for the classification task, we have used the entire dataset of 746 clips of 60 s along with teaching videos taken from platforms like YouTube. We are also releasing a smaller version of the dataset to be used on our Web site after making sure the faces of either students or teachers are removed due to privacy concerns. The total number of clips that is released are 58 and the audio stream is removed from them. The dataset can be accessed at the link https://gahanai.com/dataset.php. The frames of the clips have been resized to the dimension of (256, 256, 3). In order to make the model more robust, further collection of data is ongoing. To make sure we had enough diversity, we trained the model by taking videos of teaching categories from open platforms like YouTube www.youtube.com. The remaining videos will be released in due course of time as they contained the faces of teachers or students. In order to ensure the anonymity and privacy of the users of our portal, the said video has been kept private. After taking the necessary steps to ensure privacy, the remaining videos will be released to be used openly.

4 Methodology

This section focuses on the proposed ED-Net model and its components used for the video classification task. The methodology we followed consists of several steps which are discussed herewith. First, we extracted twenty frames from each clip of the Teach-VID dataset and from UCF-50 dataset [12] as well. This allowed us to represent a given clip in a four-dimensional NumPy array. This is done for all the clips. As it is a binary classification problem, we formed two classes 'teaching' and 'non-teaching'. Now, our two proposed models perform classification using these labeled training data. In the first method which makes use of 3D-CNN, the architecture itself takes into consideration the spatial and temporal information, as discussed in the following section. In the second model, we have used 2D-CNN to first extract features from each frame; therefore, for 20 frames per video clip, we extract 20 feature maps and two LSTM layers that combine the temporal information present. Therefore, in the first method, spatial and temporal information is processed simultaneously, and in the second, first, we extract spatial features and then temporal features. Finally, a classifier is used to classify the extracted features.

4.1 Problem Formulation

Let $X \in \mathbb{R}^{h \times i \times j \times k}$ is input video and Y is its corresponding label. Y belongs to binary class, i.e., teaching videos will be encoded as label '0' and non-teaching videos will be labeled as '1'. The proposed model after training classifies each given video clip as belonging to Class 0 or Class 1.

4.2 Model Architectures

We have proposed two different architectures to explore the usability and efficacy of the task of teaching video classification. The proposed model, ED-Net Model-A, uses 3D convolution to take volumetric data and use it to extract temporal and spatial features in an end-to-end manner. Another model, ED-Net Model-B, uses 2D convolutional layers first to extract features from each frame and finally uses an LSTM layer to incorporate sequential information. Both models have performed well on the task at hand. A detailed description of each model is given in the following sections. The goal behind developing simple models is to give benchmarking results on the dataset at hand for the task of video classification.

Attention Module We have used triplet attention (TA) [9] and a modified version of it which can be used along with 3D-CNN; the module provides to refine the feature map representation by adding little cost to the network. Triplet attention improves the overall performance of the model as the ablation study done by authors [9] has established this. The TA module uses three paths to calculate the interaction between the three dimensions, height, width, and channel. The three paths are represented as $(H \times W \times C)$, $(H \times C \times W)$, and $(W \times H \times C)$ where the input tensor is represented as $(H \times W \times C)$.

Each path is represented as O_1, O_2 and O_3.

X is the input tensor, and X' is the output tensor.

$f_{\text{perm}}(.)$ represents permutation of the tensor along an axis. $f_{\text{BN}}(.)$ represents batch normalization (BN), $f_{k \times k}(.)$ is $k \times k$ convolution and $f_{\text{concat}}(.)$ is concatenation operation on the input tensors along respective axis. σ represents sigmoid activation function. We have placed the BN layer to avoid internal covariate shift due to three paths in the triplets.

$$O_1 = \sigma\left(f_{\mathrm{BN}}\left(f^{7\times7}\left(f_{\mathrm{concat}}\left(\sqsubset F^{S}_{\mathrm{avg}}, F^{S}_{\mathrm{max}} \sqsupset\right)\right)\right)\right) \tag{1}$$

$$O_1{}' = X \odot O_1 \tag{2}$$

$$O_2 = \sigma\left(f_{\mathrm{BN}}\left(f^{7\times7}\left(f_{\mathrm{concat}}\left(\sqsubset F'_{\mathrm{avg}}, F'_{\mathrm{max}} \sqsupset\right)\right)\right)\right) \tag{3}$$

$$O_2{}' = X' \odot O_2 \tag{4}$$

$$O_3 = \sigma\left(f_{\mathrm{BN}}\left(f^{7\times7}\left(f_{\mathrm{concat}}\left(\sqsubset F''_{\mathrm{avg}}, F''_{\mathrm{max}} \sqsupset\right)\right)\right)\right) \tag{5}$$

$$O_3{}' = X' \odot O_3 \tag{6}$$

$$X' = X \odot \left(f_{\mathrm{BN}}\left(f_{1\times1}\left(\sqsubset O'_1; O'_2; O'_3 \sqsupset\right)\right)\right) \tag{7}$$

ED-Net Model-A (3D-CNN Based Model) The proposed model takes M frames of $N \times N \times 3$ dimensions as input and outputs a corresponding label value. The 3D-CNN-based architecture using 3D convolutions is gaining popularity to extract features from video. They use 3D kernels that slide along three dimensions to extract spatial and temporal features. The architecture of the proposed model is composed of the following layers:

- 3D-Convolution Layer: This layer makes use of 3D convolutions or filters which slide across the 3-axis to extract low-level features. The output of the filter is a 3D-tensor, through which spatial and temporal information is extracted.
- 3D-Max Pooling Layer: This layer is used to down-sample the input tensor along the specified axis.
- Batch-Normalization Layer: This is used to make the network faster and converge quickly and avoids internal covariate shift.
- Attention Layer: This layer helps to focus on the most informative aspect of input data during training. This allows the model to focus on what and where to look for the task at hand.
- Flatten Layer: This layer is used to take the input tensor and reshape it into a 1D tensor which can be further given to the classifier for the classification task.
- Dense Layer: It is used to build a multi-layer perceptron which will be acting as the classifier for the proposed model.

The architecture is built using 5 3D convolutional layers and 4 3D maxpooling layers of filter sizes as given in Fig. 2. The model consists of 5 blocks for feature extraction. The classifier is built using 2 fully connected layers. Finally, the output layer is a sigmoid layer with a single neuron.

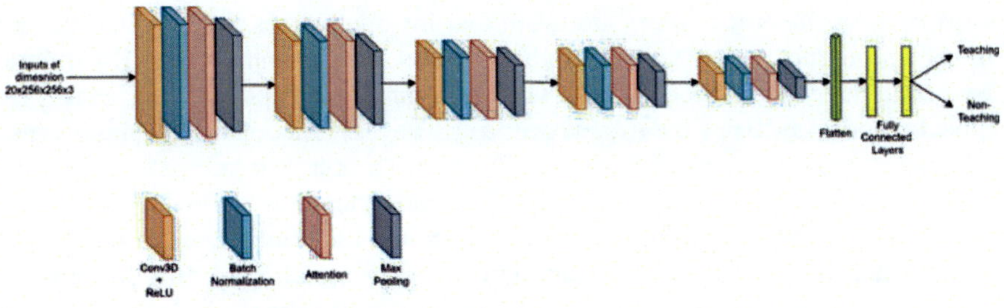

Fig. 2 Proposed model architecture using 3D-CNN

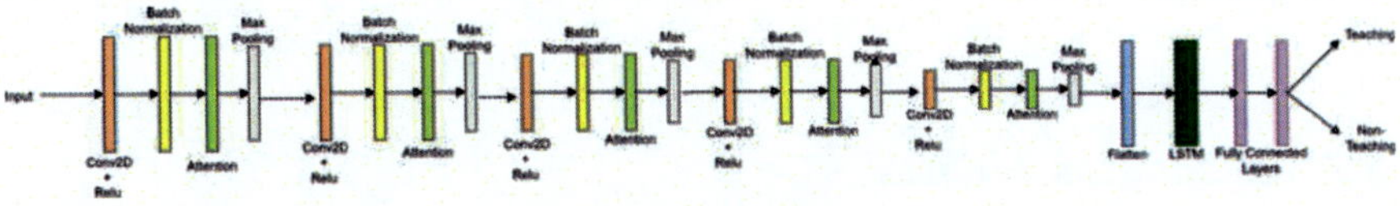

Fig. 3 Proposed model architecture using 2D-CNN and LSTM

Inspired by the work of [9] we have used triplet attention in our model to extract richer and better feature representations. The attention module is computationally efficient in extracting channel and spatial attention. The module takes in the input tensor and returns a transformed tensor of a similar shape. The output of the module is element-wise multiplied by the input tensor to produce the final attention map.

ED-Net Model-B (CNN-LSTM Based Model) Along with the 3D-CNN model, we also propose another CNN-LSTM-based model in order to compare the inference time between two different types of architecture. The proposed CNN-LSTM architecture is built using 5 2D convolutional layers with ReLU activation. For feature down-sampling, maxpooling is used with a filter size of 2. The dropout layer is used during the feature extraction process to introduce regularization. After the feature extraction, the tensor is flattened, and to incorporate the sequential information, LSTM layer is used before giving it to the classifier. The details of the model are given in Fig. 3. To incorporate the temporal aspect, the TimeDistributed layer is used as provided in the Keras framework to apply the said layer to every temporal slice of the input. In this way, we apply the feature extraction using Conv2D on the temporal slices of N frames consecutively.

5 Experimental Setup

5.1 Implementation Details

The models are implemented using the Keras framework and TensorFlow back-end. The data for the 'non-teaching' class is taken from different datasets including UCF-50 [12], Sports-1M [5], and YouTube-8M [1] to add to the diversity of videos. The clips from open-source places are taken that include classes like gaming, boxing, traveling, product review videos, football, cricket, etc., to name a few. The training data contains no data imbalance between 'teaching' and 'non-teaching' class. The model is optimized using the Adam optimizer. The loss function used is binary cross-entropy as the task at hand is binary classification. The model is trained for 20 epochs, and the results are presented in Sect. 6. The early stopping and reducing learning rate on the plateau are used. The model is trained on Nvidia Quadro RTX 6000, with a batch size of 16 only. The dataset is divided into an 8:1:1 ratio for training, validation, and testing. Apart from testing on the dataset, the model is also tested on random videos taken from open-source platforms to see the generalized capability of the model.

5.2 *Evaluation Metrics*

For evaluation and monitoring of our training and generalizability of the model, we have used precision, recall, and accuracy as our collective metrics. The mathematical equation of each metric is given in the following equations. True positive (TP) measures how many pixels the model is labeling to its correct class. False positive (FP) measures how many pixels labeled to positive class belonged to the negative class. True negative (TN) measures how many pixels classified as a negative class were actually from the negative class (Figs. 4, 5, 6 and 7).

$$\text{Precision} = \frac{\text{TP}}{\text{TP} + \text{FP}} \tag{8}$$

$$\text{Recall} = \frac{\text{TP}}{\text{TP} + \text{FN}} \tag{9}$$

$$\text{Accuracy} = \frac{\text{TP} + \text{TN}}{\text{TP} + \text{TN} + \text{FP} + \text{FN}} \tag{10}$$

6 Result and Discussion

The quantitative results are presented in Table 1. Model-A reaches the highest accuracy of 98.87% while Model-B reaches an accuracy of 96.34%. Figures 4 and 6 represent the loss curve for Model-A and Model-B. The graph shows that the model doesn't overfit or underfit. We have also tested our dataset using pre-trained models like ResNet-152 [4], Xception [3], and multi-head attention mechanism using implementation given at https://keras.io/examples/vision/video_transformers/. Figures 5 and 7 represent the accuracy graph for Model-A and Model-B. For qualitative anal-

Fig. 4 Training and validation loss for ED-Net Model-A

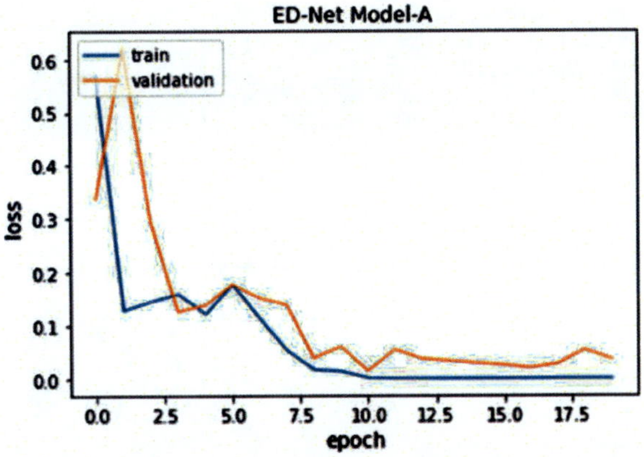

Fig. 5 Training and validation accuracy for ED-Net Model-A

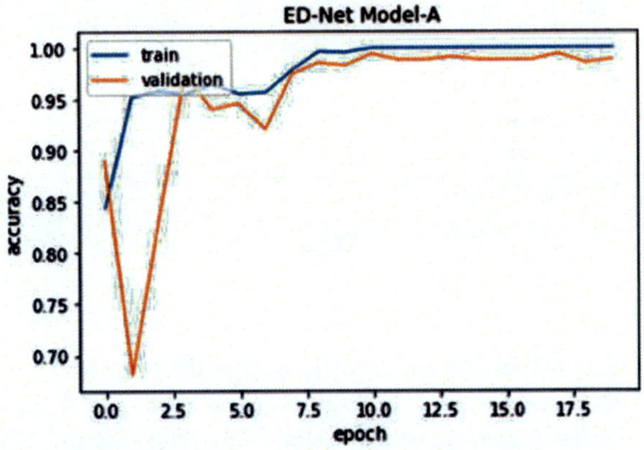

Fig. 6 Training and validation loss for ED-Net Model-B

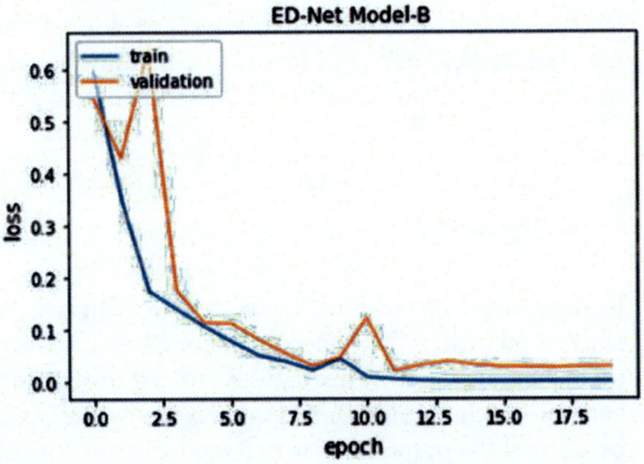

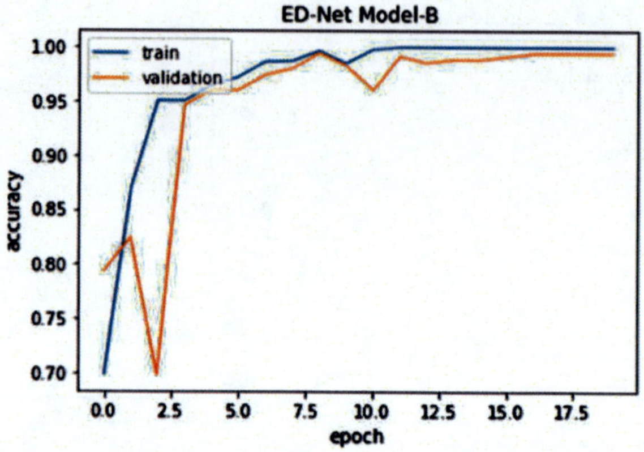

Fig. 7 Training and validation accuracy for ED-Net Model-B

Table 1 Results on Teach-VID dataset

Model	Precision (%)	Recall (%)	Accuracy (%)
ResNet-152 + LSTM	96.09	96.41	97.77
Xception + LSTM	99.15	98.33	99.53
ViT	98.37	98.37	98.83
ED-Net Model-A	99.22	72.11	98.87
ED-Net Model-B	97.35	92.45	96.34

ysis, the results are checked using videos from the test set. In this, the prediction is performed on each individual frame of the input video. The results suggest possible failure cases of the models as well. The teaching videos generally have two characteristics, written texts, and a teacher or some annotation tool. The videos having texts or sentences in the videos which do not belong to the teaching category were misclassified as teaching videos. In our future work, we will try to fix this issue by including the audio information in feature extraction and classification tasks.

7 Conclusion

In this paper, we have presented a novel dataset, Teach-VID, that contains video clips of 60 s each. The video clips are taken from an e-learning platform developed by our organization. This dataset is used to develop a system to classify teaching and non-teaching videos. Along with the dataset, we have proposed two models for benchmarking purposes. The first model is developed using 3D-CNN, and the second model is developed using 2D-CNN and LSTM.

References

1. Abu-El-Haija, S., Kothari, N., Lee, J., Natsev, P., Toderici, G., Varadarajan, B., Vijayanarasimhan, S.: Youtube-8m: a large-scale video classification benchmark. arXiv preprint arXiv:1609.08675 (2016)
2. Bhardwaj, S., Srinivasan, M., Khapra, M.M.: Efficient video classification using fewer frames. In: Proceedings of the IEEE/CVF Conference on Computer Vision and Pattern Recognition, pp. 354–363 (2019)
3. Chollet, F.: Xception: deep learning with depthwise separable convolutions. In: Proceedings of the IEEE Conference on Computer Vision and Pattern Recognition, pp. 1251–1258 (2017)
4. He, K., Zhang, X., Ren, S., Sun, J.: Deep residual learning for image recognition. In: Proceedings of the IEEE Conference on Computer Vision and Pattern Recognition, pp. 770–778 (2016)
5. Karpathy, A., Toderici, G., Shetty, S., Leung, T., Sukthankar, R., Fei-Fei, L.: Large-scale video classification with convolutional neural networks. In: Proceedings of the IEEE Conference on Computer Vision and Pattern Recognition, pp. 1725–1732 (2014)
6. Li, X., Wang, Y., Zhou, Z., Qiao, Y.: Smallbignet: integrating core and contextual views for video classification. In: Proceedings of the IEEE/CVF Conference on Computer Vision and Pattern Recognition, pp. 1092–1101 (2020)
7. Lin, R., Xiao, J., Fan, J.: Nextvlad: an efficient neural network to aggregate frame-level features for large-scale video classification. In: Proceedings of the European Conference on Computer Vision (ECCV) Workshops (2018)
8. Long, X., De Melo, G., He, D., Li, F., Chi, Z., Wen, S., Gan, C.: Purely attention based local feature integration for video classification. IEEE Trans. Pattern Anal. Mach. Intell. (2020)
9. Misra, D., Nalamada, T., Arasanipalai, A.U., Hou, Q.: Rotate to attend: convolutional triplet attention module. In: Proceedings of the IEEE/CVF Winter Conference on Applications of Computer Vision, pp. 3139–3148 (2021)
10. Peng, Y., Zhao, Y., Zhang, J.: Two-stream collaborative learning with spatial-temporal attention for video classification. IEEE Trans. Circuits Syst. Video Technol. **29**(3), 773–786 (2018)
11. Pouyanfar, S., Wang, T., Chen, S.C.: Residual attention-based fusion for video classification. In: Proceedings of the IEEE/CVF Conference on Computer Vision and Pattern Recognition Workshops (2019)
12. Reddy, K.K., Shah, M.: Recognizing 50 human action categories of web videos. Mach. Vision Appl. **24**(5), 971–981 (2013)
13. Tran, D., Wang, H., Torresani, L., Feiszli, M.: Video classification with channel-separated convolutional networks. In: Proceedings of the IEEE/CVF International Conference on Computer Vision, pp. 5552–5561 (2019)
14. Wang, L., Li, W., Li, W., Van Gool, L.: Appearance-and-relation networks for video classification. In: Proceedings of the IEEE Conference on Computer Vision and Pattern Recognition, pp. 1430–1439 (2018)
15. Xu, X., Wu, X., Wang, G., Wang, H.: Violent video classification based on spatial-temporal cues using deep learning. In: 2018 11th International Symposium on Computational Intelligence and Design (ISCID), vol. 1, pp. 319–322. IEEE (2018)

Detection of COVID-19 Using Machine Learning

Saurav Kumar and Rohit Tripathi

Abstract In this paper, we aimed at experimenting and developing suitable machine learning algorithms along with some deep learning architectures, to achieve the task of COVID-19 classification. We experimented with various supervised learning algorithms for the classification of COVID-19 images and normal images, from X-ray images and radiography images. We evaluated the performance of our models on performance metrics like accuracy, precision, recall, $F1$ score, and Cohen Kappa score. The proposed model achieved the accuracy of 98.5%, $F1$ score of 97%, precision of 95%, and Cohen Kappa score of 0.88. In this experiment, we jotted down the values of performance of various performance metrics, that were obtained after tuning various parameters of the model and on different image resolutions of input image. We performed linear regression on the results to study the behavior of various performance metrics when we tuned our model with different parameters with different input image resolutions.

Keywords COVID-19 · Machine learning · Supervised learning · Classification

1 Introduction

COVID-19 virus outbreak started from Wuhan city of China in December 2019. The virus belongs to the family of SARSCov2 and causes many respiratory problems in humans. It spread across the globe very rapidly, and World Health Organization (WHO) declared it a pandemic. The number of infected people due to the COVID-19 virus was increasing day by day. In 2019, there was no proper medication available to fight this deadly virus. The symptoms of the novel corona virus-infected people

S. Kumar (✉) · R. Tripathi
Department of Computer Science and Engineering, Indian Institute of Information Technology, Guwahati, India
e-mail: saurav.kumar@iiitg.ac.in

R. Tripathi
e-mail: rohit@iiitg.ac.in

M. Tistarelli et al. (eds.), *Computer Vision and Machine Intelligence*,
Lecture Notes in Networks and Systems 586,
https://doi.org/10.1007/978-981-19-7867-8_13

were similar to people who caught cough and cold initially [20]. Lockdowns were imposed by the governments of many countries; air travel with other countries was stopped, and other means of transportation were stopped to slow the rate of spread of the virus due to human mobility. Detecting COVID-19 is also and challenge because its symptoms are similar to cold and flu. Correct diagnosis of COVID-19 will result in the correct medication and treatment of the infected patients. Different diagnostic approaches are used for the detection of COVID-19 [3, 7]. These approaches are [3] molecular diagnostic tools: RT-PCR, computed tomography (CT) scans, laboratory serological testing. Machine learning and artificial intelligence methods are previously used for the detection and classification of diseases. Machine learning methods can be used for detection and diagnosis of COVID-19. It is used globally for classification task due to its higher prediction accuracy. Machine learning approach for COVID-19 detection from radiography images will provide supportive tool to the existing tools for detecting COVID-19 patients.

The rest of the paper is organized as follows: Sect. 2 describes related work. Section 3 method describes about the dataset followed by the algorithms that we implemented as a part of our work. Section 4 proposed model describes about the proposed model that we used to achieve the goals. Section 5 results and discussion tells about the various performance metrics we used to evaluate our model and performance evaluation of our model, followed by results of linear regression, and the paper is finally concluded in Sect. 6.

2 Related Work

Corona virus is from the family of respiratory virus [15]. These viruses exhibit disorders like common cold and other respiratory disorders [23]. A new corona virus strain came into existence in 2019, termed as COVID-19 (Corona Virus Diseases 19) [6, 16]. Diagnosis of COVID-19 at an early stage is important to minimize the person-to-person transmission and to provide proper medical care to the patients. Machine learning (ML) and artificial intelligence (AI) were used previously on detecting other diseases. The researchers are exploring different ways to detect and diagnose COVID-19 infected patients using ML and AI techniques. Researchers observed that apart from the results of RT-PCR test and antigen test, COVID-19 infected patients also have computed tomography scans, and chest X-rays with them. The Chinese researchers successfully diagnosed the para-clinical and clinical activities of COVID-19 using radiography images [1, 13, 18] (Table 1).

Research works and papers were published related to COVID-19 show that different machine learning paradigms such as supervised learning (SL), unsupervised learning (UL), and reinforcement learning (RL) were used for the task of detecting COVID-19 infected people [20]. Supervised learning can be used for classification and regression analysis; unsupervised learning can be used for cluster analysis and dimensionality reduction, and reinforcement learning can do classification and control. For the task of detection and classification, most of the work and researches were

Table 1 Algorithms for classification

Author	ML algorithms
Khanday et al. [19]	Logistic regression and Naive Bayes
Burdick et al. [8]	Logistic regression
DAN et al. [12]	ANN
Constantin et al. [3]	CNN

done using supervised learning techniques. The dataset available in most of the published works is labeled data, and supervised learning techniques were considered an optimal way to achieve the task. Many researchers had used deep learning architecture frameworks for feature extraction, in order to get higher accuracy rate features. The task of classification is divided into two phases: feature extraction and classification. Many traditional machine learning algorithms were also used to achieve the classification task like Naive Bayes, logistic regression, KNN algorithms as shown in Table 2.

Brunese et al. [7] published their work, that uses KNN algorithm to detect COVID-19 in X-ray images. The experiment was performed on 90% training dataset and 10% of test dataset. The proposed model achieved FP rate of 0.068, precision of 0.965, recall of 0.965, and F-measure of 0.964. Many of the published works used deep learning framework to achieve the task of detecting COVID-19. Apostolopoulus et al. [4] performed COVID-19 detection on X-ray images using transfer learning techniques. They considered pre-trained networks like VGG-19, MobileNet V2, Inception ResNet V2, and other pre-trained networks on X-ray images and considered performance evaluation metrics like area under curve, mean sensitivity, mean specificity, and mean accuracy. AI tools produced stable results for in applications that are either images or other type of data [11, 22, 24]. For achieving the goals related to this experiment, we referred to the publications listed in Tables 1 and 2. The table contains list of algorithms which were used for the classification task. In our work, we used multiple algorithms, deep learning frameworks, and evaluated the performance of the model on balanced dataset and imbalanced dataset. We performed linear regression on the results obtained after training the model on different dataset on different hyper-parameters.

3 Method

In past years, machine learning and artificial intelligence technologies and methods were used for prediction and classification of various diseases. We experimented with four different machine models, in combination with deep learning frameworks like CNN and VGG-16. We implemented machine learning classification algorithms like KNN [7, 14], decision trees, support vector machines. Each of the experiment is performed on two datasets, Kaggle and COVID-19 radiography database. We also evaluated the performance metrics on different input image size for different datasets on different models. The reason behind doing this is to find the behavior of machine learning models on different input image pixels [21].

3.1 Dataset

We trained and tested our model on two datasets, which were used previously for this work. For doing the implementation, we had chosen Kaggle Dataset and COVID-19 radiography database [10]. The COVID-19 radiography dataset is balanced dataset which contains 1200 X-ray images of COVID-19 infected patients and 1341 patients who are not infected by COVID-19, where original size of each input image is 400×300 pixels. Similarly, the Kaggle Dataset contains 60 chest X-rays of COVID-19 infected patients and 880 chest X-rays of patients who were not infected by COVID-19.

3.2 K-Nearest Neighbors Algorithm

KNN algorithm is a supervised machine learning algorithm. It is used for doing classification. The core concept behind this algorithm is that similar or identical data points or entities exist in close proximity. As a part of our experimentation, we prepared the experimental setup proposed by Brunese et al. [7], later we implemented the classification model using our parameters and tested it on two datasets. The working of the model is divided into two parts. In the first part, the task is to extract features from the input images. This task was achieved by using a colored layout descriptor. The extracted feature is used as an input for the KNN algorithm. In the second part of the experimentation, we had taken the input parameters the same as proposed. We had taken $K = 4$ and Euclidean distance for doing proximity calculations. The metrics are considered for the evaluation of the performance of the classifier: accuracy, precision, $f1$ score, and recall. The result obtained with this experimental setup is depicted in Table 3. We implemented the existing models and then implemented my own classification model. We had taken different value for K. In our research, we observed that choosing correct value of K is very crucial because if we take K value very less then the classification model will exhibit overfitting. If K value is very high, then it may lead to under-fitting will be very high. We analyzed results obtained after different values of K and reached to a conclusion that for binary classification the value of K should be taken always odd. I got satisfactory results when I took $K = 5$. We used Manhattan distance [5] instead of Euclidean distance in the algorithm.

$$\text{Manhattan distance} = \sum_{i=1}^{n} |X_i - Y_i| \tag{1}$$

3.3 Support Vector Machine

Support vector machine is a supervised machine learning method that can perform achieve classification and regression. Most commonly, SVM is used for classification. In this algorithm, we plot the data points into n-dimensional space, where n is the number of features. Each feature indicates a coordinate in the n-dimensional space. To do classification, we try to find the correct hyper-plane which divides the two classes. We used CNN for feature extraction and SVM for achieving the task of COVID-19 classification. We implemented the model on two datasets. In CNN, we used ReLU activation function in convolution layer. We used hinge loss as a loss function and "Adam" optimizer for doing regularization. The feature extracted from CNN is given as input to SVM, where linear activation function used for classification because the task was to do binary classification. After the classification task, we implemented same task with resolution of images from the same datasets. The idea behind is to study and evaluate the performance of models on different resolution of input images. Also, as a part of this experiment, we performed linear regression analysis on the results that are obtained after implementation of model on different resolution of images and different number of images in the dataset. We evaluated performance of the proposed model on metrics: accuracy, precision, recall, and $F1$ score.

3.4 Decision Trees

A decision tree is a graph or a model of a decision. It is a supervised learning method used for classification and regression. In most cases, it is used for doing classification. The decision tree represents the step-by-step decision. Following these decisions, we reach the conclusion which is stored in leaf nodes. A decision tree will have a lot of internal nodes. In a decision tree, each internal node represents the outcome of the test, and each branch represents the outcome of the test, and each leaf node represents the output class labels. Each internal node splits into another internal node or leaf node. The leaf node should have the answer that the model can predict. The best decision tree should require fewer steps to reach a decision at its minimum. To do so, at each level of the decision tree, correct feature should be chosen for splitting. Like previous two proposed model, I evaluated the performance of model using accuracy, recall, $f1$ score, and precision. The detailed description of results obtained on Kaggle dataset is jotted down on table, and the results of COVID-19 radiography dataset are jotted down on table.

3.5 XGBoost

XGBoost is widely used for image classification problems. It was developed by Chen and Guetrin [9], and is a sound algorithm for doing regression and classification. It is applied to win many Kaggle competitions. It uses gradient boosting framework. It adds new decision tree to fit the value with multiple iteration. This is how XGBoost improves efficiency.

4 Proposed Model

To achieve better results, we used a new machine learning model in combination with deep learning framework. We used VGG-16 architecture for extracting the features from the images and XGBoost algorithm for classification task. We evaluated the model based on performance metrics like accuracy, precision, recall, $f1$ score. And, we also evaluated the performance of model based on Cohen Kappa Score. We used Cohen Kappa Score for evaluating the model because, it can measure the performance on the basis of agreement of two rater which classify items in the dataset into two categories. We put the images into the VGG-16 framework, and extracted the features. We give the extracted to the XGBoost classifier. We found improved accuracy, $f1$ score, precision, recall and Cohen Kappa score for both balanced dataset and imbalanced datasets. We also applied linear regression on the results obtained from the classification on various

5 Results and Discussions

The classification was performed on different input image resolution of both the dataset and also with different depth value of decision tree of the classification algorithm. We performed this because we want to know the correct input image size for our model. For medical images, if we give input image of lower resolution, many important information will be lost and hence performance of model won't be efficient, due to information loss [21]. Similarly, the model will show overfitting for certain input image resolution. In this section, the prediction results of machine learning models: K-nearest neighbors, SVM, decision trees and XGBoost, are summarized.

5.1 Performance Metrics

The performance of machine learning models is calculated using some performance measures. In our work, we evaluated our model based on performance metrics like accuracy, $F1$ score, precision, recall, and Cohen Kappa score.
Accuracy: A prediction model generates four types of results which are categorized as true positive (TP), false positive (FP), true negative (TN), and false negative (FN). Accuracy is the ratio of total number of correct predictions made by the model to the

total number of data points [2, 5].

$$\text{Accuracy} = \frac{\text{TP} + \text{TN}}{\text{TP} + \text{TN} + \text{FN} + \text{FP}} \tag{2}$$

Recall: It is the ratio of patients who were correctly detected or diagnosed with COVID-19 to all the patients who are diagnosed with COVID-19 [2, 5].

$$\text{Recall} = \frac{\text{TP}}{\text{TP} + \text{FN}} \tag{3}$$

Precision: It is the ratio of number of correctly predicted COVID-19 cases from the actual COVID-19 cases in the dataset [2, 5].

$$\text{Precision} = \frac{\text{TP}}{\text{TP} + \text{FP}} \tag{4}$$

***F*1 Score**: It is the harmonic mean of precision and recall [5].

$$F1\ \text{Score} = 2 * \frac{\text{Precision} * \text{Recall}}{\text{Precision} + \text{Recall}} \tag{5}$$

Cohen Kappa Score: It is a statistical measure which is used to evaluate the performance of classification model [5, 17]. It checks the consistency of the model. It compares the predicted results with actual results. The value of Kappa (K) lies between 0 and 1.

$$K = \frac{P_0 - P_e}{1 - P_e} \tag{6}$$

5.2 *Performance Evaluation*

We mentioned in the previous sections that we tested the proposed model performance on different image resolution. For achieving the goal, the minimum image resolution is 32 × 32 pixels because VGG-16 can't take input resolution below 32 × 32. The maximum input image resolution is 300 × 300 pixels for images in COVID-19 radiography database and 256 × 256 pixels for images in Kaggle dataset. We used XGBoost algorithm for classification, which is a decision tree. We aimed at getting better results with less depth of tree shown in Figs. 1 and 2.

From Fig. 1, we can observe that the performance metrics values reach to overfitting condition. Whereas, in Fig. 2, after 224 pixel image size, we are not getting excellent performance. The model performs well on both dataset, on input image

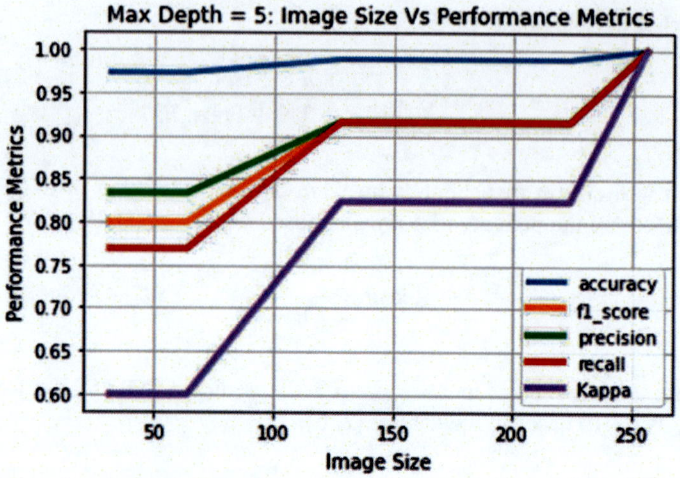

Fig. 1 Performance on Kaggle dataset

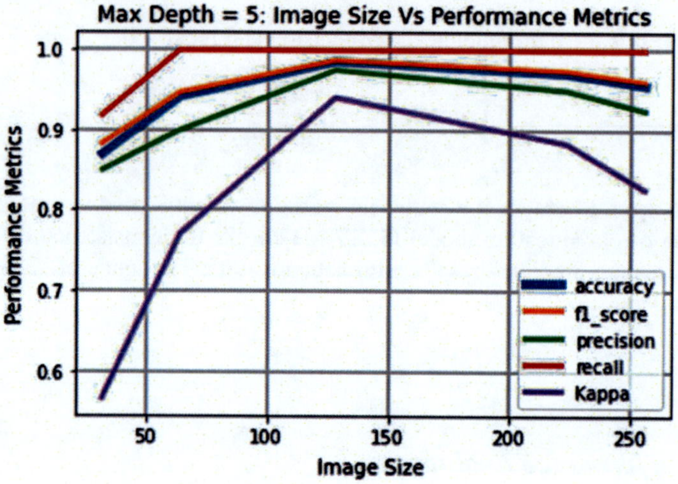

Fig. 2 Performance on radiography dataset

size of 224×224 pixel, with maximum depth 5 on both decision tree, shown in Table 4. The results depicted in Table 3, if we compare the Cohen Kappa score of the previously implemented models like K-nearest neighbors, CNN with SVM and CNN with decision tree, with the proposed model, then the Cohen Kappa score of the current model is improved for the datasets as shown in Figs. 3 and 4.

Table 2 Classification results using XGBoost

Dataset	Precision	Recall	$F1$ score	Accuracy	Kappa
Kaggle	0.91666	0.91666	0.91666	0.98936	0.82397
Radiography	0.95	1	0.9743589	0.9710144	0.88324

Table 3 Regression results using XGBoost on Kaggle dataset

y	m_1	m_2	m_3	m_4	c	R^2
Accuracy	1.016106	1	1	6.661186	0.971016	0.87
$F1$ score	0.00076304	0	0	0.0002435	0.779395	0.88
Precision	0.00063142	0	0	0.0010088	0.805115	0.90
Recall	0.00088454	0	0	−0.00046292	0.7556538	0.89
Kappa	0.00152119	0	0	0.00082212	0.5527714	0.89

Table 4 Regression results using XGBoost on radiography dataset

y	m_1	m_2	m_3	m_4	c	R^2
Accuracy	3.24736	1	8.3278	1.9575	0.8874	0.87
$F1$ score	2.9424	1	7.9959	1.87950	0.89798	0.88
Precision	3.303295	1	1.6192	3.806137	0.85287	0.86
Recall	2.50343565	1	−1.4843632	−3.48908	0.94818	0.89
Kappa	1.011974	1	3.19374	7.507095	0.612551	0.85

5.3 Linear Regression

We performed linear regression on the results obtained after implementing the models. The reason behind performing linear regression on the features and output variable is to determine the dependency of the output features on the particular input features. The higher the value of the coefficient of the feature, the higher will be dependency of output variable on that particular feature. In our experiment, we tried to figure out the important features that will affect the results by varying the values of m. For our proposed model, we choose size of input image, number of images in the dataset, ratio of COVID-19 infected cases and normal patients, maximum depth of the tree as independent features for the linear regression. We considered accuracy, precision, recall, $F1$ score, and Cohen Kappa score as dependent features for performing the linear regression, as shown in equation below.

$$y = m_1 * \text{size} + m_2 * \text{num of image} + m_3 * \text{ratio} + m_4 * \text{max depth} + c \quad (7)$$

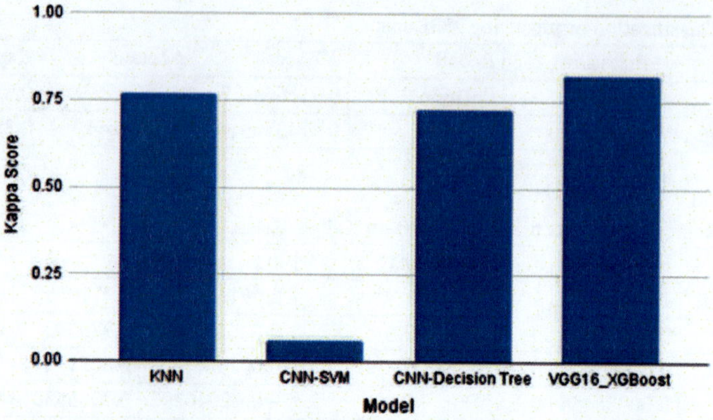

Fig. 3 Performance of various models Kaggle dataset

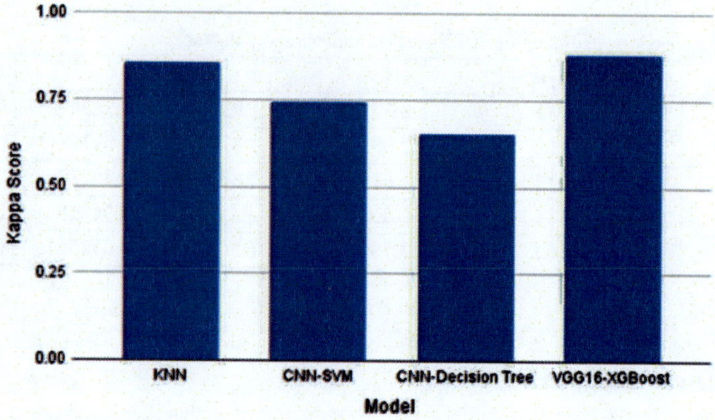

Fig. 4 Performance of various models radiography dataset

To understand the behavior of the proposed model, we used all the independent features and dependent features one at a time. The coefficient obtained after performing linear regression is jotted on Table 3 for Kaggle dataset and on Table 4 for radiography dataset. We also evaluated the performance of our linear regression model on the basis of R^2 and obtained a satisfactory R^2 values for regression model.

6 Conclusion

We draw conclusion from Table 3 that, we should not evaluate the performance of model only on accuracy, precision, recall, and $f1$ score only. The dataset which we used for training will be imbalanced in some cases, and in such cases, accuracy and other performance metrics will perform better, but overall, the model will not do good. Therefore, we used Kappa score as well which tell the overall performance of classification model. The Kappa score is satisfactory both types of dataset, that is, balanced and unbalanced dataset. The results obtained after performing the linear regression in Table 4 indicates that for imbalanced dataset, the accuracy of the proposed model highly dependent on maximum depth of the XGBoost. Similarly, the results of linear regression on the balanced dataset, i.e., radiography dataset, indicate that accuracy and $F1$ score are highly dependent on the ratio of positive and negative image in the dataset, which is highly true, because comparing it with results of regression on Kaggle dataset, we found that coefficient m_3 is 0 in imbalanced dataset. When precision is considered as output variable for the linear regression, we can observe that for balanced dataset, it depends on size of the dataset. The more will the images in the dataset, the more will be the precision of the model. Cohen Kappa value in Table 5 is higher for the feature maximum depth of the XGBoost tree. On comparing this with value of feature m_4 in Table 4, for Cohen Kappa score as dependent feature, we observed that for imbalanced dataset, the values is very less, due to imbalance in the dataset.

References

1. Ai, T., Yang, Z., Hou, H., Zhan, C., Chen, C., Lv, W., Tao, Q., Sun, Z., Xia, L.: Correlation of chest CT and RT-PCR testing for coronavirus disease 2019 (COVID-19) in china: a report of 1014 cases. Radiology **296**(2), E32–E40 (2020)
2. Alpaydin, E.: Introduction to Machine Learning. MIT Press (2020)
3. Anastasopoulos, C., Weikert, T., Yang, S., Abdulkadir, A., Schmülling, L., Bühler, C., Paciolla, F., Sexauer, R., Cyriac, J., Nesic, I., et al.: Development and clinical implementation of tailored image analysis tools for COVID-19 in the midst of the pandemic: the synergetic effect of an open, clinically embedded software development platform and machine learning. Eur. J. Radiol. **131**, 109233 (2020)
4. Apostolopoulos, I.D., Mpesiana, T.A.: COVID-19: automatic detection from X-ray images utilizing transfer learning with convolutional neural networks. Phys. Eng. Sci. Med. **43**(2), 635–640 (2020)
5. Bonaccorso, G.: Machine Learning Algorithms. Packt Publishing Ltd (2017)
6. Brooks, W.A.: Bacterial pneumonia. In: Hunter's Tropical Medicine and Emerging Infectious Diseases, pp. 446–453. Elsevier (2020)
7. Brunese, L., Martinelli, F., Mercaldo, F., Santone, A.: Machine learning for coronavirus COVID-19 detection from chest X-rays. Proc. Comput. Sci. **176**, 2212–2221 (2020)
8. Burdick, H., Lam, C., Mataraso, S., Siefkas, A., Braden, G., Dellinger, R.P., McCoy, A., Vincent, J.L., Green-Saxena, A., Barnes, G., et al.: Prediction of respiratory decompensation in COVID-19 patients using machine learning: the ready trial. Comput. Biol. Med. **124**, 103949 (2020)
9. Chang, Y.C., Chang, K.H., Wu, G.J.: Application of extreme gradient boosting trees in the construction of credit risk assessment models for financial institutions. Appl. Soft Comput. **73**, 914–920 (2018)
10. Chowdhury, M.E., Rahman, T., Khandakar, A., Mazhar, R., Kadir, M.A., Mahbub, Z.B., Islam, K.R., Khan, M.S., Iqbal, A., Al Emadi, N., et al.: Can AI help in screening viral and COVID-19 pneumonia? IEEE Access **8**, 132665–132676 (2020)
11. Dai, S., Li, L., Li, Z.: Modeling vehicle interactions via modified LSTM models for trajectory prediction. IEEE Access **7**, 38287–38296 (2019)

12. Deng, X., Shao, H., Shi, L., Wang, X., Xie, T.: A classification-detection approach of COVID-19 based on chest X-ray and CT by using Keras pre-trained deep learning models. Comput. Model. Eng. Sci. **125**(2), 579–596 (2020)
13. Fang, Y., Zhang, H., Xie, J., Lin, M., Ying, L., Pang, P., Ji, W.: Sensitivity of chest CT for COVID-19: comparison to RT-PCR. Radiology **296**(2), E115–E117 (2020)
14. Hamed, A., Sobhy, A., Nassar, H.: Accurate classification of COVID-19 based on incomplete heterogeneous data using a KNN variant algorithm. Arab. J. Sci. Eng. 1–12 (2021)
15. Holmes, K.V.: Sars-associated coronavirus. New Engl. J. Med. **348**(20), 1948–1951 (2003)
16. Huang, C., Wang, Y., Li, X., Ren, L., Zhao, J., Hu, Y., Zhang, L., Fan, G., Xu, J., Gu, X., et al.: Clinical features of patients infected with 2019 novel coronavirus in Wuhan, China. Lancet **395**(10223), 497–506 (2020)
17. Jeni, L.A., Cohn, J.F., De La Torre, F.: Facing imbalanced data-recommendations for the use of performance metrics. In: 2013 Humaine Association Conference on Affective Computing and Intelligent Interaction, pp. 245–251. IEEE (2013)
18. Jin, X., Lian, J.S., Hu, J.H., Gao, J., Zheng, L., Zhang, Y.M., Hao, S.R., Jia, H.Y., Cai, H., Zhang, X.L., et al.: Epidemiological, clinical and virological characteristics of 74 cases of coronavirus-infected disease 2019 (COVID-19) with gastrointestinal symptoms. Gut **69**(6), 1002–1009 (2020)
19. Khanday, A.M.U.D., Rabani, S.T., Khan, Q.R., Rouf, N., Din, M.M.U.: Machine learning based approaches for detecting COVID-19 using clinical text data. Int. J. Inf. Technol. **12**(3), 731–739 (2020)
20. Kwekha-Rashid, A.S., Abduljabbar, H.N., Alhayani, B.: Coronavirus disease (COVID-19) cases analysis using machine-learning applications. Appl. Nanosci. 1–13 (2021)
21. Luke, J.J., Joseph, R., Balaji, M.: Impact of image size on accuracy and generalization of convolutional neural networks (2019)
22. Ozsahin, I., Sekeroglu, B., Mok, G.S.: The use of back propagation neural networks and 18F-Florbetapir PET for early detection of Alzheimer's disease using Alzheimer's disease neuroimaging initiative database. PLoS One **14**(12), e0226577 (2019)
23. Pyrc, K., Jebbink, M., Vermeulen-Oost, W., Berkhout, R., Wolthers, K., Wertheim-van, P.D., Kaandorp, J., Spaargaren, J., Berkhout, B., et al.: Identification of a new human coronavirus. Nat. Med. **10**(4), 368–373 (2004)
24. Yılmaz, N., Sekeroglu, B.: Student performance classification using artificial intelligence techniques. In: International Conference on Theory and Application of Soft Computing, Computing with Words and Perceptions, pp. 596–603. Springer (2019)

A Comparison of Model Confidence Metrics on Visual Manufacturing Quality Data

Philipp Mascha

Abstract After ground-breaking achievements through the application of modern deep learning, there is a considerable push towards using machine learning systems for visual inspection tasks part of most industrial manufacturing processes. But whilst there exist a lot of successful proof-of-concept implementations, productive use proves problematic. Whilst missing interpretability is one concern, the constant presence of data drift is another. Changes in pre-materials or process and degradation of sensors or product redesigns impose constant change towards statically trained machine learning models. To handle these kind of changes, a measurement of system confidence is needed. Since pure model output probabilities often lack in this concern better solutions are required. In this work, we compare and contrast several pre-existing methods used to describe model confidence. In contrast to previous works, they are evaluated on a large set of real-world manufacturing data. It is shown that utilizing an approach based on auto-encoder reconstruction error proves to be most promising in all scenarios tested.

1 Introduction

Since recent advances in deep learning made it possible to use machine learning for safety critical applications, for example, in the medical field [29] or even for autonomous driving [11]. It allows a lot of tasks, especially from the cognitive area, to be automated where it was previously impossible to do so. Thus, it is apparent that there is a strong push towards bringing deep learning into industrial manufacturing, where automation is key.

Most need arises usually in quality assurance, where manufactured parts may have to undergo manual single part inspections, which is costly, slow, and error prone.

This work was supported by OSRAM Automotive.

P. Mascha (✉)
University of Augsburg, Universitätsstraße 2, 86159 Augsburg, Germany
e-mail: philipp.mascha@ams-osram.com; p.mascha@gmx.net

M. Tistarelli et al. (eds.), *Computer Vision and Machine Intelligence*,
Lecture Notes in Networks and Systems 586,
https://doi.org/10.1007/978-981-19-7867-8_14

Modern machine learning approaches for visual manufacturing quality assurance exist in various forms [20, 27, 32, 33]. Even commercial tools and applications exist from manufacturers like Halcon or Cognex.

But whilst initial proof-of-concepts may work fine, industrial data usually differentiates in a few key aspects from usual machine learning problems:

- Since a plant may manufacture thousands of parts per day, huge amounts of data exist. But labelling might be costly as it requires experts that have knowledge of the underlying process and quality guidelines.
- Classes are usually very imbalanced, since faulty products should be much less frequent than acceptable ones. Some defects might even only arise in a handful of products.
- There exists only low variance between samples and even classes. This is desired behaviour since in the best case, each product manufactured should look exactly the same.
- Observed data may change from the samples trained on due to changing processes, pre-materials, or product design as well as sensor drift.

The last point is especially important. Whilst pictures found in the typical evaluation datasets like MNIST or CIFAR do not change over time, machine learning models deployed in automatic optical inspection (AOI) systems may degrade in performance over time.

This may lead to cases where images during live production are too different from the samples used during training. The tendency of deep models to rate wrong-classified samples with high confidence is well documented [10, 26] and has advanced to one of the main problems in AI security [2].

From own experience, the mixture of low data variance, high class imbalance, and apparent data drift often lead to unknown samples labelled as fulfilling quality requirements. This is the worst case scenario, where defect parts might be supplied to the customer.

In machine learning, derivation from the training data can take various forms. Samples may be drawn from a completely different distribution, resulting in out-of-distribution (OOD) samples [34]. Closely related are anomalies [6], which may be either defects of the device under test or failures in measurement. Parameters may also change continuously over time which is called data drift.

Whilst one may chose from a wide array of confidence metrics, their evaluation is usually done on standard classification sets like MNIST, CIFAR, or SVHN. These sets do not provide any data that mirrors the criteria introduced for industrial images, as they all provide high class-balance, high inter-class variance, and no inherent data drift.

But confidence in measuring systems is essential for a serious application of machine learning inside production quality assurance. Contemporary manufacturing tools like statistical process control (SPC) rely on a feedback from the measuring system to determine if a process still works as expected [8]. Since plain model output cannot provide an adequate metric and more sophisticated approaches are not tried on production data, empiric evidence is needed to justify their application in this domain.

Thus, the goal of this work is to close this gap by testing existing metrics on novel data which was gathered from a real-world industrial use-case. An ideal method should both detect drift and separate the cases where the models output is still based on interpolation from training data and where it is extrapolating or simply 'guessing'.

2 Related Work

There exists a wide array of methods to achieve a reliable confidence metric. Most can be classified by either post-hoc evaluation of either data or model or utilizing regularization during training [7]. The later shall not be considered inside the scope of this work.

2.1 Plain Network Output

The most intuitive and straight-forward way to specify uncertainty is to use the output of the model. Calculating the softmax function on the models output $f(x)$ gives a good baseline for confidence, as is stated in [15]. The amount of uncertainty $d(x)$ is simply measured by $d_{\text{softmax}}(x) := \max(f(x))$, assuming softmax activation for the model's output.

2.2 Intermediate Layer Distributions

Whilst the softmax may give an appropriate baseline, there exist many cases where the model might be assigning high output scores to wrongly classified samples. Thus, a more sophisticated approach is desired.

Research in adversarial learning has shown that models may be easily manipulated into providing wrong results with high output confidence [5]. To circumvent this, there are many techniques that compare the distribution of features inside intermediate layers $f_l(x)$ between the training set and during inference.

To do so [21] used the Mahanalobis distance on each layer. Another possibility is to use a k-nearest neighbour approach as was done in [28]. The methods key component can be calculated as follows:

$$d_{\text{kNN}}(x) := \frac{1}{N}\sum_{l}^{N} \frac{|\{c \in \Gamma_{l,k} : c \neq \arg\max(f(x))\}|}{k} \tag{1}$$

where $\Gamma_{l,k}$ is the labels of the k-nearest samples to the embedding layer $f_l(x)$ drawn from the calibration set. This approach is very similar to [25] which uses additional regularization on the embedding layer.

This metric was further modified in [16] by calculating only the ratio of distance d between the k-th out-of-class ($\hat{\gamma}_{l,k}$) and the k-th in-class ($\gamma_{l,k}$) sample:

$$d_{\text{dist}}(x,l) := \frac{d(f_l(x), \hat{\gamma}_{l,k})}{d(f_l(x), \gamma_{l,k})} \tag{2}$$

It is only calculated on one of the last layers of the network.

2.3 *Distribution of Input Data*

Another, completely different approach is to compare the data used for training with the one during inference. This is especially well researched in the context of data drift [4]. State-of-the-art approaches check the distribution of input values utilizing tools like the Kolmogorov-Smirnov test [30], p-values [18], or MD3 [31]. Also, specifically tested on IoT data streams, an offline ensemble method was applied in [23].

These methods are usually not well suited for visual data, since the data dimensionality is so high, and the relativity between dimensions is far more important than individual values. To handle such high amounts of complexity, auto-encoders are a possible solution to the problem. These are models consisting of an encoder $enc(x)$ and a decoder $dec(x)$ with the goal to reconstruct the input x so that $x \approx dec(enc(x))$. To prevent the encoder from simply learning the identity function, the dimensionality of the latent space $enc(x)$ is usually much smaller than the input dimension.

The fact that auto-encoders are very good at reconstructing images similar to their training set whilst failing to do so on OOD samples can be easily exploited. Their reconstruction error $\mathbb{E}$ is then used as a measurement of the grade of dissimilarity. By using the mean squared logarithmic error (MSLE), we derive as follows:

$$d_{\text{AE}}(x) := \frac{1}{N}\sum_{i}^{N}(\log(x_i+1) - \log(dec(enc(x))_i+1))^2 \tag{3}$$

Many possibilities exist to design an auto-encoder. A widely used and well-researched approach is principal component analysis (PCA, [17]) which has been applied in both 1D [24] and 2D quality control [22].

More sophisticated approaches are deep convolutional auto-encoders [12]. These consist of an encoder formed by stacking convolutional and max-pooling layers with a final dense layer at the bottom. The decoder part utilizes both convolutional and up-sampling layers, which repeat each pixel in x and y dimension. To improve feature distribution of the latent space, additional regularization may be applied to create a variational auto-encoders [3].

It is also possible to compare the distribution of the latent space to detect drift in input data, for example, in [1, 35]. Also, reconstruction can be done by searching for existing samples that have a low distance in latent space [9].

It should be noted that the training of auto-encoders is unsupervised. This accommodates the high amount of unlabelled data present in manufacturing quality data.

3 Experimental Setup

Data drift detectors were tested on real-world manufacturing data. It was gathered from an automatic optical inspection (AOI) of a welding process. An existing and fully automated manufacturing line was retro-fitted with two cameras observing the welding point. A detection model has to determine whether the lead-in wire is properly attached to the soldering lug based on the acquired images (see Fig. 1).

In the past, evaluation of these images was done by classical image processing using a combination of edge detection, segmentation, and pattern matching. Due to an unstable measurement process (inconsistent lighting and focus, blur caused by mechanical slack), various lighting artefacts accuracy on this system was low, and manual re-evaluation was needed on a significant amount of products.

A new evaluation based on a 101-layer deep ResNet [13] yielded far superior results with 99.7% evaluation accuracy, thus making manual re-evaluation obsolete. During the implementation period, it became apparent that the images varied greatly over time (see Fig. 2), degrading model performance. Additionally, single images could be completely un-rateable due to wrong timing of the camera, a loose camera carriage or an under- or overexposing light source (see Fig. 3).

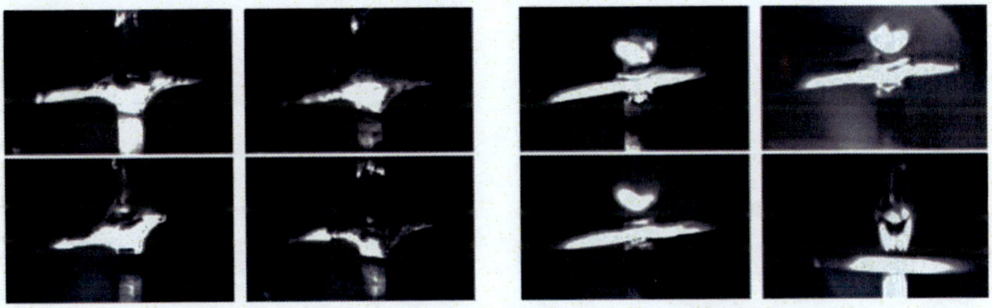

(a) Ok (b) Defect

Fig. 1 Examples of good and bad welds

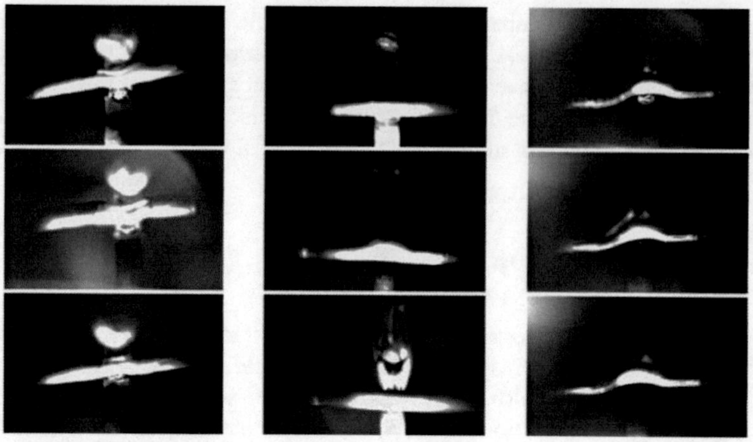

(a) Day 1 (b) Day 2 (c) Day 3

Fig. 2 Images varying over time

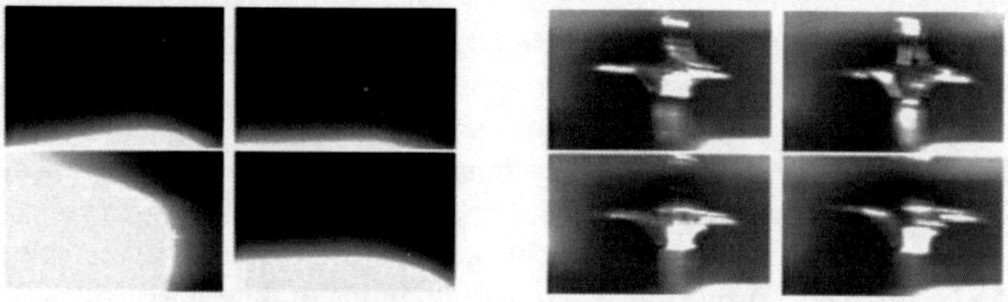

(a) Lighting over-exposition (b) Blurry images

Fig. 3 Images that are un-rateable

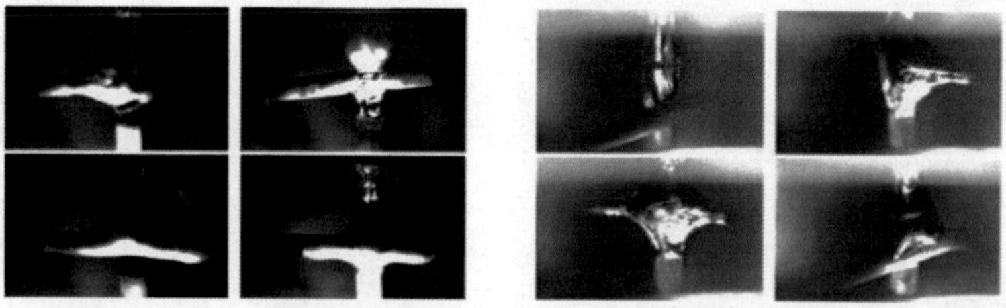

(a) Left camera (b) Right camera

Fig. 4 Difference between left and right camera pictures

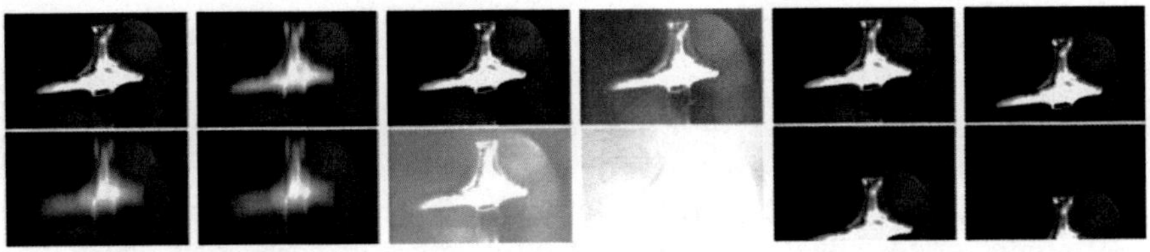

(a) Artificial blur (b) Artificial lighting (c) Artificial movement

Fig. 5 Examples of manipulated images

3.1 Instantaneous Change Scenario

To test which method is suitable for real-world application, they were tested on OOD samples for a realistic scenario. The existing AOI consists of two cameras where each observes a different welding joint. They differ in quality of lighting as well as the angle of welding. An exemplary subset for each camera can be seen in Fig. 4.

Each confidence indicator was tested whether it is able to distinguish the source of each image when calibrated only on images from the first camera. Since the data had not to be labelled manually, it was possible to use a wast dataset gathered over the course of a month. It consists of about 47,000 images per camera. These images were neither used for training the base model nor calibrating the drift detectors.

3.2 Continuous Drift Scenario

To also evaluate sensitivity towards continuous drift, an additional evaluation scenario was created. An ideal drift detector should show a change as soon as detection accuracy of the base model decreases under increased data drift. Thus, we define an ideal drift detector as one that correlates with the detection error rate the most.

Three kinds of continuous drift were tested: blurry images, lighting overexposure, and camera movement. All three may happen in the real world (see Fig. 3) and are easy to simulate trough image post-processing.

Blur was generated using a Gaussian blur kernel with increasing standard deviation σ. To simulate an unsuitable light source, the images pixel values $p \in [0, 255]$ were increased using the function $\min(p + \delta_L, 255)$. For artificial movement, the image was translated in the y direction by δ_M times the images height. Examples of the resulting images can be seen in Fig. 5.

Since it was vital to determine the detection rate on these images, only labelled data could be used for experiments. Of the available 14,500 images, about 8500 were used for training the base model, another 3000 to calibrate the drift detectors, and 3000 for evaluation.

It should be noted that, as it is usual in manufacturing data, the dataset is highly unbalanced with defect samples only making up around 4% of total images.

4 Selected Methods

Various methods presented in Sect. 2 were implemented and tested. As baseline, the models output was used (*softmax*, [15]).

For immediate layer distances, the k-nearest neighbour approach was used (*kNN*, [28]). In the original work, all layers of the network were considered, which itself were rather shallow. For a state-of-the-art ResNet, it was not possible to track all layers due to memory restrictions. Thus, only the final output of each residual block before pooling was used.

Additionally, the distance ratio towards the k-th in-class and out-off-class was tested (*k distance*, [16]) on both the global-pooling layer input and the output of the last residual block before the final down-scale layer, which was more performant. Results are only displayed for the later. The method originally suggests an additional parametrized filter which was omitted since it did not prove useful in experiments. k was set to the suggested value of 10 in both cases.

For auto-encoders PCA with 90%, explained variance (20 components) was used. Additionally, a custom designed deep convolutional auto-encoder was created with 6 stacked residual blocks of both 3 convolutional and batch normalization with a max-pooling/ up-sampling layer in-between. Both conventional training (*CAE*, [12]) as well as variational regularization (*VAE*, [3]) was used.

All methods that required training or calibration used a calibration set X_{Cal} which was drawn from the same distribution as and is complementary to the training set. To consider the high data imbalance in industrial data, the samples in X_{Cal} were class balanced. Each result was averaged using fivefold cross validation.

5 Results and Discussion

For the instant change scenario presented in 3.1, an ideal detector would assign low values to both the calibration set X_{Cal} and the data gathered on the left camera X_{c1}. The OOD set of right camera pictures X_{c2} should be distinguishable by a high score. It should be noted that there exists drift between X_{Cal} and X_{c1}, since they were gathered on different points in time, and no labels for training are available for X_{c1}.

Figure 6 shows the distribution of output values for each set. But whilst approaches working on the detection model itself only produce vaguely distinguishable distributions, all auto-encoder-based methods visibly excel at this task. An additional benefit is that their logarithmic scores form a close to normal distribution over each set. This makes them especially well suited for statistical process control (SPC) which is a key component of modern lean manufacturing environments [8].

The quantitative evaluation can be found in Table 1. The unregularized encoder performs the best, probably due to best reconstruction performance, but all methods based on reconstruction error are clearly suitable.

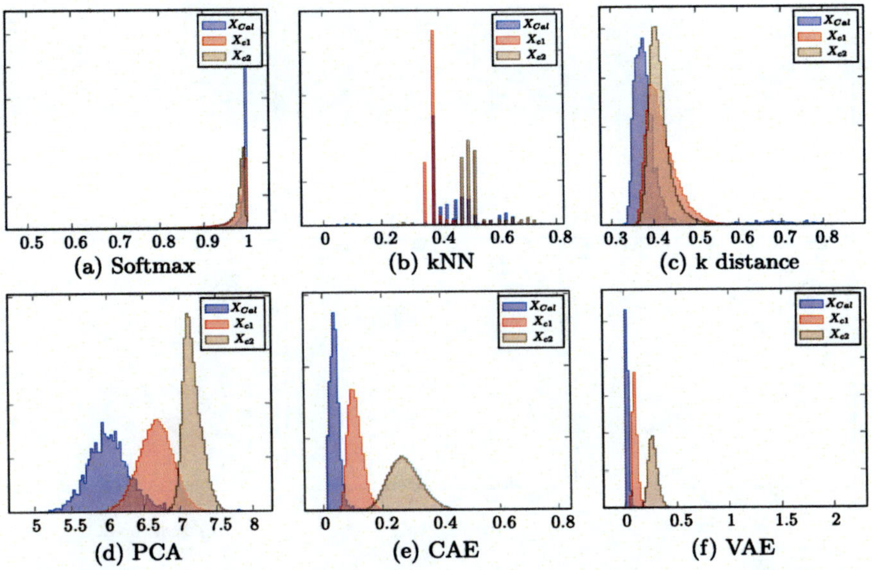

Fig. 6 Histogram of the output values for both the calibration set X_{cal} and productive set X_{Cam1} (both taken from the first camera) as well as the set from the second camera X_{Cam2}. Counts for each set are normalized to sum to 1

Table 1 Area under curve scores for OOD samples as well as correlation values for each manipulation scenario. Auto-encoders provide best or close-to-best results in any category

Method	ROC AUC	Correlation			
		Blur	Light	Move	Average
Softmax	0.31 ±0.13	−0.34 ±0.29	−0.47 ±0.19	0.25 ±0.84	−0.19
KNN	0.58 ±0.25	−0.93 ±0.09	−0.67 ±0.31	−0.88 ±0.11	−0.82
k distance	0.25 ±0.12	−0.95 ±0.07	−0.85 ±0.14	−0.96 ±0.06	−0.92
PCA	0.95 ±0.02	**0.78** ±0.07	−0.94 ±0.05	0.85 ±0.11	0.23
CAE	**0.99** ±0.01	0.75 ±0.13	0.31 ±0.50	**0.90** ±0.04	0.65
VAE	0.97 ±0.03	0.77 ±0.11	**0.87** ±0.14	0.85 ±0.18	**0.83**

Similar results can be seen in the continuous case: Fig. 7 shows the error rate and assigned score in correspondence with the degree of manipulation.

All of the approaches that supervise the detection model have very random or even strong negative correlation. They often correlate strongly with the percentage of samples assigned 'Ok' by the model, which are almost exclusively false negatives. This may stem from the large class imbalance in the training set and is rather problematic for industrial data, since false negatives in defect-detection might mean faulty products being shipped to the customer.

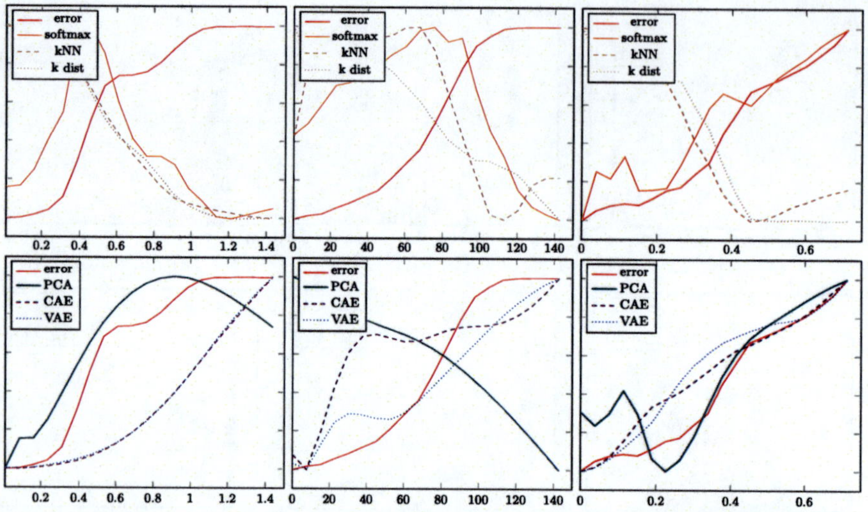

(a) Blur (b) Lighting (c) Movement

Fig. 7 Distance metrics compared to the error rate on various continuous manipulations. Values have been normalized for better visibility

Again, only the auto-encoders provide sufficient results. Correlation values seen in Table 1 show that with a few exceptions, these methods capture the apparent data drift (the result for CAE on lighting probably being an outlier). The fact of PCA to have better reconstruction with increased lighting may stem from its decoder being a linear combination.

This confirms the assumption done in [14] that ReLU-based networks always struggle with interpretable prediction values. If one does not want to regularize training or even change the model's architecture, auto-encoders seem to offer the best results, since they can also be trained independent from the model and do not require labelled data.

6 Conclusion and Further Research

Finding a robust metric for prediction confidence is of great importance to enable large-scale application of machine learning in industrial quality control. To verify whether different model confidence metrics are feasible, it is imperative to use real-world manufacturing data. In this work, a novel dataset with both real out-of-distribution samples as well as simulated data drift was used.

The experiments show that methods based on supervision of intermediate or final feature representations fail altogether, a notion that was already hinted at by [19]. At the same time, approaches construct an additional and unsupervised auto-encoder model excel at the tasks given. Thus, this research highly suggests that reconstruction error-based methods are applicable to calculate confidence metrics on visual manufacturing data.

This evaluation could be further improved upon by applying data augmentation during both auto-encoder construction and training of the final model. Thus, it could be examined if the expanded dataset is properly reflected by a more stable confidence metric.

In terms of industrial process engineering, it could be useful to test how well these drift detectors can be integrated into a continuous improvement scenario for the underlying manufacturing process. They could detect unstable processes, enabling process engineer to improve and stabilize them based on the confidence metric. Another application would be to use them for camera calibration to reconstruct the settings used during the collection of the training data.

References

1. Abati, D., Porrello, A., Calderara, S., Cucchiara, R.: Latent space autoregression for novelty detection. In: Proceedings of the IEEE/CVF Conference on Computer Vision and Pattern Recognition (CVPR) (2019)
2. Amodei, D., Olah, C., Steinhardt, J., Christiano, P., Schulman, J., Mané, D.: Concrete problems in AI safety (2016). https://doi.org/10.48550/ARXIV.1606.06565, https://arxiv.org/abs/1606.06565
3. An, J., Cho, S.: Variational autoencoder based anomaly detection using reconstruction probability. Spec. Lect. IE **2**(1), 1–18 (2015)
4. Barros, R.S.M., Santos, S.G.T.C.: A large-scale comparison of concept drift detectors. Inf. Sci. **451–452**, 348–370 (2018). https://doi.org/10.1016/j.ins.2018.04.014, http://www.sciencedirect.com/science/article/pii/S0020025518302743
5. Carlini, N., Wagner, D.: Towards evaluating the robustness of neural networks. In: 2017 IEEE Symposium on Security and Privacy, pp. 39–57. IEEE (2017)
6. Chalapathy, R., Chawla, S.: Deep learning for anomaly detection: a survey. arXiv preprint arXiv:1901.03407 (2019)
7. Du, X., Wang, Z., Cai, M., Li, Y.: VOS: learning what you don't know by virtual outlier synthesis. CoRR **abs/2202.01197**, https://arxiv.org/abs/2202.01197 (2022)
8. Forza, C.: Work organization in lean production and traditional plants: what are the differences? Int. J. Oper. Prod. Manage. (1996)
9. Gong, D., Liu, L., Le, V., Saha, B., Mansour, M.R., Venkatesh, S., Hengel, A.V.D.: Memorizing normality to detect anomaly: memory-augmented deep autoencoder for unsupervised anomaly detection. In: Proceedings of the IEEE/CVF International Conference on Computer Vision (ICCV) (2019)
10. Goodfellow, I.J., Shlens, J., Szegedy, C.: Explaining and harnessing adversarial examples. arXiv preprint arXiv:1412.6572 (2014)
11. Grigorescu, S., Trasnea, B., Cocias, T., Macesanu, G.: A survey of deep learning techniques for autonomous driving. J. Field Rob. **37**(3), 362–386 (2020)
12. Guo, X., Liu, X., Zhu, E., Yin, J.: Deep clustering with convolutional autoencoders. In: International Conference on Neural Information Processing, pp. 373–382. Springer (2017)
13. He, K., Zhang, X., Ren, S., Sun, J.: Identity mappings in deep residual networks (2016). https://doi.org/10.48550/ARXIV.1603.05027, https://arxiv.org/abs/1603.05027

14. Hein, M., Andriushchenko, M., Bitterwolf, J.: Why Relu networks yield high-confidence predictions far away from the training data and how to mitigate the problem. In: Proceedings of the IEEE/CVF Conference on Computer Vision and Pattern Recognition, pp. 41–50 (2019)
15. Hendrycks, D., Gimpel, K.: A baseline for detecting misclassified and out-of-distribution examples in neural networks (2018)
16. Jiang, H., Kim, B., Guan, M.Y., Gupta, M.R.: To trust or not to trust a classifier. In: NeurIPS, pp. 5546–5557 (2018)
17. Jolliffe, I.T., Cadima, J.: Principal component analysis: a review and recent developments. Phil. Trans. R. Soc. A Math. Phys. Eng. Sci. **374**(2065), 20150202 (2016)
18. Jordaney, R., Sharad, K., Dash, S.K., Wang, Z., Papini, D., Nouretdinov, I., Cavallaro, L.: Transcend: detecting concept drift in malware classification models. In: 26th USENIX Security Symposium (USENIX Security 17), pp. 625–642. USENIX Association, Vancouver, BC (2017). https://www.usenix.org/conference/usenixsecurity17/technical-sessions/presentation/jordaney
19. Laugel, T., Lesot, M.J., Marsala, C., Renard, X., Detyniecki, M.: The dangers of post-hoc interpretability: unjustified counterfactual explanations. arXiv preprint arXiv:1907.09294 (2019)
20. Lee, K.B., Cheon, S., Kim, C.O.: A convolutional neural network for fault classification and diagnosis in semiconductor manufacturing processes. IEEE Trans. Semicond. Manuf. **30**(2), 135–142 (2017)
21. Lee, K., Lee, K., Lee, H., Shin, J.: A simple unified framework for detecting out-of-distribution samples and adversarial attacks. In: Advances in Neural Information Processing Systems, vol. 31 (2018)
22. Li, D., Liang, L.Q., Zhang, W.J.: Defect inspection and extraction of the mobile phone cover glass based on the principal components analysis. Int. J. Adv. Manuf. Technol. **73**(9), 1605–1614 (2014)
23. Lin, C.C., Deng, D.J., Kuo, C.H., Chen, L.: Concept drift detection and adaption in big imbalance industrial IoT data using an ensemble learning method of offline classifiers. IEEE Access **7**, 56198–56207 (2019)
24. Malhi, A., Gao, R.X.: PCA-based feature selection scheme for machine defect classification. IEEE Trans. Instrum. Measur. **53**(6), 1517–1525 (2004)
25. Mandelbaum, A., Weinshall, D.: Distance-based confidence score for neural network classifiers. arXiv preprint arXiv:1709.09844 (2017)
26. Nguyen, A., Yosinski, J., Clune, J.: Deep neural networks are easily fooled: high confidence predictions for unrecognizable images. In: Proceedings of the IEEE Conference on Computer Vision and Pattern Recognition (CVPR) (2015)
27. Ozdemir, R., Koc, M.: A quality control application on a smart factory prototype using deep learning methods. In: 2019 IEEE 14th International Conference on Computer Sciences and Information Technologies (CSIT), vol. 1, pp. 46–49. IEEE (2019)
28. Papernot, N., McDaniel, P.: Deep k-nearest neighbors: towards confident, interpretable and robust deep learning. arXiv preprint arXiv:1803.04765 (2018)
29. Piccialli, F., Di Somma, V., Giampaolo, F., Cuomo, S., Fortino, G.: A survey on deep learning in medicine: why, how and when? Inf. Fusion **66**, 111–137 (2021)
30. Reis, D.M.d., Flach, P., Matwin, S., Batista, G.: Fast unsupervised online drift detection using incremental Kolmogorov-Smirnov test. In: Proceedings of the 22nd ACM SIGKDD International Conference on Knowledge Discovery and Data Mining, pp. 1545–1554 (2016)
31. Sethi, T.S., Kantardzic, M.: On the reliable detection of concept drift from streaming unlabeled data. Expert Syst. Appl. **82**, 77–99 (2017)
32. Villalba-Diez, J., Schmidt, D., Gevers, R., Ordieres-Meré, J., Buchwitz, M., Wellbrock, W.: Deep learning for industrial computer vision quality control in the printing industry 4.0. Sensors **19**(18), 3987 (2019)
33. Wankerl, H., Stern, M.L., Altieri-Weimar, P., Al-Baddai, S., Lang, K.J., Roider, F., Lang, E.W.: Fully convolutional networks for void segmentation in X-ray images of solder joints. J. Manuf. Process. **57**, 762–767 (2020)

34. Yang, J., Zhou, K., Li, Y., Liu, Z.: Generalized out-of-distribution detection: a survey. CoRR **abs/2110.11334**, https://arxiv.org/abs/2110.11334 (2021)
35. Zong, B., Song, Q., Min, M.R., Cheng, W., Lumezanu, C., Cho, D., Chen, H.: Deep autoencoding gaussian mixture model for unsupervised anomaly detection. In: International Conference on Learning Representations (2018). https://openreview.net/forum?id=BJJLHbb0-

High-Speed HDR Video Reconstruction from Hybrid Intensity Frames and Events

Rishabh Samra, Kaushik Mitra, and Prasan Shedligeri

Abstract An effective way to generate high dynamic range (HDR) videos is to capture a sequence of low dynamic range (LDR) frames with alternate exposures and interpolate the intermediate frames. Video frame interpolation techniques can help reconstruct missing information from neighboring images of different exposures. Most of the conventional video frame interpolation techniques compute optical flow between successively captured frames and linearly interpolate them to obtain the intermediate frames. However, these techniques will fail when there is a nonlinear motion or sudden brightness changes in the scene. There is a new class of sensors called event sensors which asynchronously measures per-pixel brightness changes and offer advantages like high temporal resolution, high dynamic range, and low latency. For HDR video reconstruction, we recommend using a hybrid imaging system consisting of a conventional camera, which captures alternate exposure LDR frames, and an event camera which captures high-speed events. We interpolate the missing frames for each exposure by using an event-based interpolation technique which takes in the nearest image frames corresponding to that exposure and the high-speed events data between these frames. At each timestamp, once we have interpolated all the LDR frames for different exposures, we use a deep learning-based algorithm to obtain the HDR frame. We compare our results with those of non-event-based interpolation methods and found that event-based techniques perform better when a large number of frames need to be interpolated.

Keywords HDR imaging · Event-based interpolation

R. Samra (✉) · K. Mitra · P. Shedligeri
Indian Institute of Technology, Madras, India
e-mail: samrarishabh@gmail.com

K. Mitra
e-mail: kmitra@ee.iitm.ac.in

P. Shedligeri
e-mail: ee16d409@ee.iitm.ac.in

M. Tistarelli et al. (eds.), *Computer Vision and Machine Intelligence*,
Lecture Notes in Networks and Systems 586,
https://doi.org/10.1007/978-981-19-7867-8_15

1 Introduction

Natural scenes have a much higher dynamic range than most conventional cameras with 8-bit sensors. The most common way is to capture the LDR images at multiple exposures and then fuse them [1]. Here, the final image combines information from multi-exposure images, and these methods are known as bracketed exposure techniques. But, for the general public, HDR video is not reachable since the approach requires specialized cameras, which are bulky and expensive for them as it needs complex sensors and optical systems [2, 3]. Recently [4] claimed HDR video can be reconstructed even by inexpensive off-the-shelf cameras. Kalantari and Ramamoorthi [4] proposed to capture the input low dynamic range sequences at alternating exposures and then recover the missing HDR details from neighboring frames of different exposures by training two sequential CNNs that do alignment and merging. But, when there is very fast motion in the scene, or there is large pixel displacement between successive frames, this method can lead to ghosting artifacts. So, it is required to resort to a good interpolation technique.

The problem of HDR video reconstruction is more challenging than HDR image reconstruction because in addition to producing artifact-free images it is required that the frames remain temporally coherent. Most of the non-event-based interpolation techniques compute optical flow between two frames and then linearly interpolate to obtain the intermediate frames. However, the scene motion need not be linear. This is likely to be true for fast camera or object motion. Also, optical flow-based methods will fail in case of sudden changes in scene brightness. For capturing very fast motion, there are novel sensors called event sensors which measure asynchronously per-pixel brightness changes and offer advantages like high dynamic range, high temporal resolution, low latency, etc. So, to get temporally coherent frames while doing the interpolation, we propose to use a hybrid imaging system consisting of a conventional camera, which captures LDR frames, and an event camera which captures high-speed events. This additional information from the event camera should lead to a more coherent HDR video output. Note that a hybrid imaging system, consisting of an LDR conventional camera and an event sensor, is already available commercially, known as the DAVIS camera [5]. However, we cannot change the exposure time at every frame in the DAVIS camera. Hence, we show all our results on the simulated data.

Since we capture only a single exposure image at each timestamp, the other exposure images are missing. So, we interpolate the other exposure frames at each timestamp using the nearest frames of corresponding exposure and event data. For this purpose, we used the event-based interpolation technique as described in [6]. Then, after getting multi-exposure LDR images at each time instant, we use a deep learning-based network [7] to obtain HDR frames. We compare our proposed method with non-event-based interpolation techniques for generating HDR frames. Our experiments show that our event-based interpolation technique has a clear edge over non-event-based HDR techniques. We show that as the number of frames to interpolate between two consecutive frames increases, event-based methods perform much better than non-event-based interpolation methods. Following are the main contributions of our paper:

1. To capture high-speed HDR video, we propose a hybrid imaging system consisting of an LDR conventional camera and an event camera.
2. We use an event-based interpolation technique to obtain intermediate frames from the nearest input LDR frames at the same exposure time and corresponding events. After that, we fuse them using a deep learning-based HDR technique to obtain HDR video frames.
3. We show that our proposed event-based HDR method performs better than those which use non-event-based interpolation techniques such as RRIN [8] and Super-Slomo [9]. The performance gap between our method and the non-event-based methods increases as the number of frames to interpolate between input LDR frames increases.

2 Related Works

2.1 *HDR Image-Based Reconstruction*

Previously, Debevec and Malik proposed an HDR imaging method [1] that merges photographs at different exposures. But, the resultant HDR reconstructed image can have ghosting artifacts that occur due to misalignment of different LDR images because of large camera motion or brightness changes. This led to research in methods to de-ghost the images caused due to misalignment [10–12].

Using optical flow [13] first aligned LDR images at multiple exposures and then used neural networks which reconstruct an HDR image after merging multiple LDR Images. Inverse tone mapping is one of the methods proposed in [14] to reconstruct a high dynamic range image, which is visually appealing, from a single low dynamic range image. There are also other methods that use CNNs for single image reconstruction [15–18]. But, they can't handle the noise in the low-exposure images which have dark regions and also tend to focus on hallucinating the saturated regions. CNNs are also applied to merge multi-exposure images [7, 19–21]. However, these methods cannot be applied directly for HDR video reconstruction from sequences of alternate exposures as it relies on a fixed exposure as a reference.

2.2 *HDR Video Reconstruction*

HDR imaging problems can be well-posed by using the computational approaches or by using the cameras that encode the high dynamic range of the scene. There are several methods using specialized hardware for capturing images at a high dynamic range. An HDR video system was implemented [3] that captures three images at different exposures simultaneously using a beam splitter, and then, they merged them to get a high dynamic range image. In another method, a photographic filter was placed over the lens, and a colored filter array was placed on the conventional camera sensor for doing single-shot HDR imaging [22]. Recently, Chen et al. [23] introduces a deep learning-based coarse to Fine network for HDR video reconstruction from alternate exposure sequences. First, it performs coarse video reconstruction of high dynamic range in image space and then removes ghosting artifacts by refining the previous predictions in feature space.

2.3 *Event-Based Interpolation*

Inspired by the human retina mechanism, event sensors detect the changes in scene radiance asynchronously. As compared to conventional image cameras, event sensors offer several advantages like low latency, high dynamic range, low power, etc. [24] Thus, for recovering a very high dynamic range of the scene, reconstructing an image from raw events can be more effective. Recently, Gehrig et al. [6] attempted to combine the method of synthesis-based approach with the flow-based approach to reconstruct high-speed images using both image data as well as corresponding event sensor data [6].

3 Hybrid Event-HDR

This section explains the overall procedure that was followed to obtain high-speed HDR video given the sequence of alternate exposure LDR images. The entire approach is described in Fig. 1.

In our algorithm, we take as input a sequence of LDR images that have been captured with alternating exposure durations.

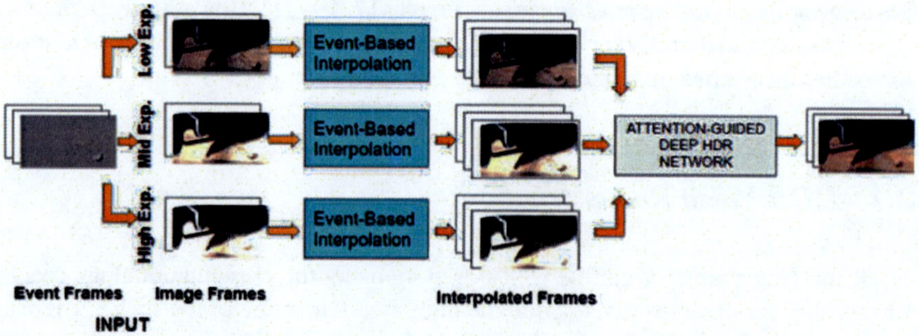

Fig. 1 Our proposed method where the input is a sequence of alternate exposure LDR frames. We perform hybrid event-based interpolation for all the exposures and after that fuse them using a deep-HDR network to obtain the HDR video frames

Specifically, we consider 3 different exposure times. In a video sequence of these LDR frames where the first frame is indexed 0, the frames $0, 3, 6, \ldots$ are captured with low-exposure time; the frames $1, 4, 7, \ldots$ are captured with mid-exposure time, and the frames $2, 5, 8, \ldots$ are captured with high-exposure time. To generate an HDR frame at each time instant, we require LDR frames with all three different exposure times. This can be achieved through interpolating the missing exposure LDR frame from the neighboring frames of the same exposure time. Here, we utilize motion information from the event sensor data that assist us in interpolating the LDR frames. Event sensor information additionally helps in interpolating at multiple intermediate temporal locations between successive input video frames. Thus, we are able to generate high-speed HDR videos from a sequence of alternating exposure LDR frames and corresponding event sensor data.

We specifically use Timelens [6], a recent state-of-the-art algorithm for frame interpolation from a hybrid event and conventional image sensor. Timelens takes as input, a low-frame rate video with LDR frames, and the corresponding high temporal resolution motion information acquired using event sensor data. We utilize Timelens to interpolate multiple frames between successive LDR frames of the same exposure time. We first consider frames acquired with low-exposure time (frames $0, 3, 6, \ldots$) and interpolate multiple frames in between. A similar procedure is followed to interpolate LDR frames corresponding to the mid- and high-exposure times. The number of frames to interpolate depends on the desired output frame rate of the HDR video. If we want to get an HDR video with the same frame rate as the input LDR frames, then we need to interpolate 2 frames between any successive input LDR frames of the same exposure. For getting a two times faster HDR video, we need to do 5-frame interpolation, and for three times faster HDR video, we need to do 8-frame interpolation.

After interpolation at each time instant, we obtain LDR frames with all three exposure times. To reconstruct an HDR frame, these three frames must be fused at each time instant. For this, we utilize a deep neural network proposed in [7]. The neural network proposed in [7] consists of 2 stages: (a) attention module and (b) merging module. The attention module attends to each of the input LDR images and fuses only the relevant information necessary for HDR reconstruction. The merging module which consists of dilated dense residual blocks then reconstructs a ghost-free HDR image. The output is further tonemapped to generate the final HDR image.

Algorithm 1 Hybrid Event-HDR Algorithm

Input: Sequence of alternate exposure LDR frames and corresponding simulated event sensor data.

Output: Reconstructed HDR Video frames.

for t in T **do**

- Interpolate missing exposure frames using the method explained in [6].
- Use the deep-HDR network [7], convert multi-exposure LDR frames to corresponding HDR frames.

end for

4 Experiments, Datasets, and Simulation

We obtain a dataset of HDR video sequences from [25]. Specifically, we use the following sequences from the HDR video dataset: Poker_Fullshot, Bistro, Cars_Longshot, Fishing_Longshot, Showgirl_2, Fireplace2. Following [4], we convert each of these HDR videos into an LDR video sequence where the respective LDR frames are obtained using alternating exposures values of $1.0, 0.25, 0.0625$. The spatial resolution of these RGB videos is 1050×1890. However, for our experiments, we reduce the image size by half along both dimensions. We also require the event-sensor data corresponding to these videos. Using the E-Sim simulator [26], we generated event sensor data from the input HDR videos. Note that we use HDR video sequences for generating events as event sensors can capture high dynamic range data [5]. For the E-Sim simulator, we set the output frame rate as 25 fps, exposure as 10 ms, and contrast threshold as 0.15. E-Sim outputs event sensor data as a sequence of tuple (x, y, t, p), where $[x, y] \in \mathbb{Z}^{h \times w}$ represents spatial location, $p \in [-1, 1]$ represent event-polarity, and t represents the timestamp at which the event was fired. We convert this sequence of tuple into event frames by accumulating events between two timestamps corresponding to the frames in the input HDR video.

If we want HDR video to have the same rate as the input LDR video, then we sample low-exposure LDR at $t = 0, 3, 6 \ldots$, mid-exposure LDR at $t = 1, 4, 7 \ldots$, and high-exposure LDR at $t = 2, 5, 8 \ldots$. Now, our task is to interpolate the two missing LDR frames at each time instant. For this, we only need to interpolate 2 LDR frames between successive LDR frames captured with the same (low, mid, or high) exposure. Similarly, if we wish to have the output HDR video at twice the frame rate of input LDR video, then we sample low-exposure LDR at $t = 0, 6, 12 \ldots$, mid-exposure LDR at $t = 2, 8, 14 \ldots$, and high-exposure LDR at $t = 4, 10, 16 \ldots$. This sequence leaves a gap of 1 frame between any successive low-mid or mid-high exposure frames. We again utilize the event-sensor data to interpolate 5 LDR frames between successive LDR frames captured with identical exposure time. This ensures that at each time instant $t = 0, 1, 2, \ldots$ we obtain LDR frames with all 3 different exposure values. Finally, to show output HDR video at 3 times the frame rate as the input LDR video, we sample low-exposure LDR at $t = 0, 9, 18 \ldots$, mid-exposure LDR at $t = 3, 12, 21 \ldots$, and high-exposure LDR at $t = 6, 15, 24 \ldots$. Here, we require interpolating 8 frames between successive LDR frames captured with the same exposure time. In summary, to obtain HDR videos at the same rate, twice the frame rate, and thrice the frame rate of the input LDR video, we need to interpolate 2, 5, and 8 frames, respectively, between successive LDR frames with the same exposure time.

We use Timelens [6], which utilizes information from both an event sensor and a conventional image sensor, to interpolate video frames. As Timelens is trained only on intensity images that have good exposure values, we finetune the Timelens network to interpolate LDR frames with low and high exposures. After separately interpolating the low, mid, and high-exposure videos, we obtain the information of all three exposure images at each time instant. These 3 frames are then fused using an attention-guided deep neural network proposed in [7], resulting in the final

HDR video sequence. Then, we compare our proposed algorithm with Superslomo [9] and RRIN [8]-based interpolation calling them Superslomo-HDR and RRIN-HDR, respectively. In Superslomo-HDR and RRIN-HDR, we replace our event-based LDR frame interpolation step with Superslomo and RRIN video interpolation, respectively. Once the LDR frames are interpolated using Superslomo (or RRIN), we fuse the three LDR frames at each time instant using [7]. Hence, compared to our proposed algorithm, Superslomo-HDR and RRIN-HDR differ only in the video interpolation technique used. While we use an event-based learned interpolation technique, Superslomo and RRIN use optical-flow-based learned video interpolation.

5 Results

We compare our hybrid event-HDR method with other HDR methods like RRIN-HDR and Slomo-HDR for various interpolation rates. For obtaining HDR video with the same frame rate as the input LDR video, we perform 2-frame interpolation, and

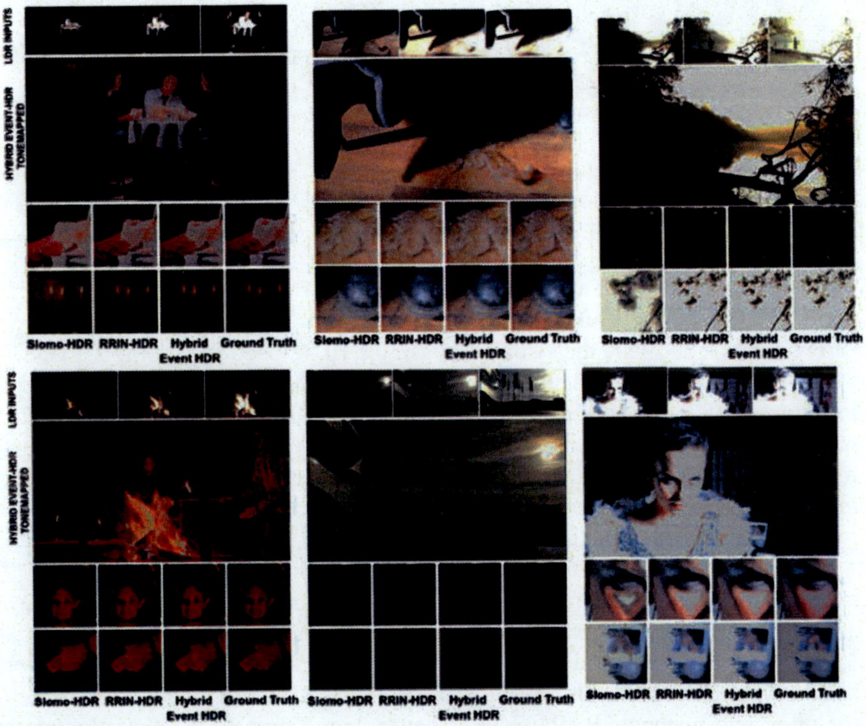

Fig. 2 Comparision of hybrid event-HDR with Slomo-HDR and RRIN-HDR for **2-frame interpolation**

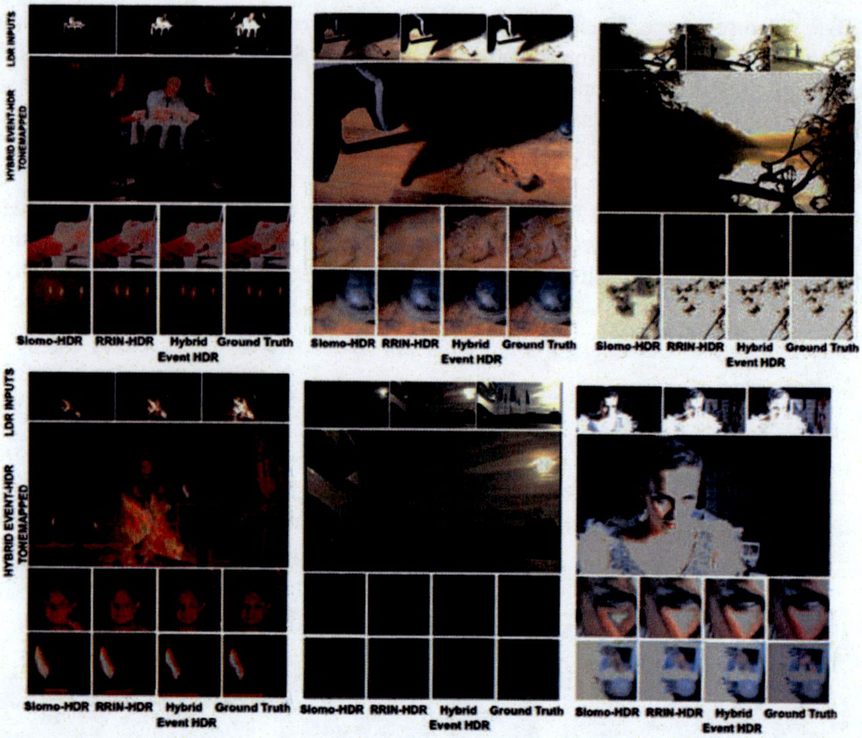

Fig. 3 Comparision of hybrid event-HDR with Slomo-HDR and RRIN-HDR for **5-frame interpolation**

for obtaining HDR video with twice (or thrice) the frame rate as that of input, we perform 5-frame (or 8-frame) interpolation.

Figures 2, 3, and 4 are the visual results of the HDR video after doing 2-frame, 5-frame, and 8-frame interpolation, respectively. For 2-frame interpolation, the output of Slomo-HDR is not good but hybrid event-HDR and RRIN-HDR performs much better and resembles the ground truth. From Fig. 3 which shows results for 5-frame interpolation, we can see that hybrid event-HDR performs even better than other methods. For 8-frame interpolation, we can note that the hybrid event-HDR performs much better than other methods and performance gap increases.

Table 1 shows the quantitative comparison of the above experiments. HDR-VDP2 is considered an ideal metric for evaluating the performance of HDR images[27]. The numbers in Tables 1 and 2 correspond to reconstructed frames shown in Figs. 2, 3, and 4. It can be inferred from Table 1 that with respect to hdr-vdp2 scores, our hybrid event-HDR performs better than both RRIN-HDR and Slomo-HDR. Among the other two methods, RRIN-HDR is better than Slomo-HDR. Note that as the number of frames to interpolate increases, hybrid event-HDR performs better than other techniques. We can infer from Table 2 that for 8-frame interpolation, hybrid event-HDR performs better than other techniques for all the datasets in terms of

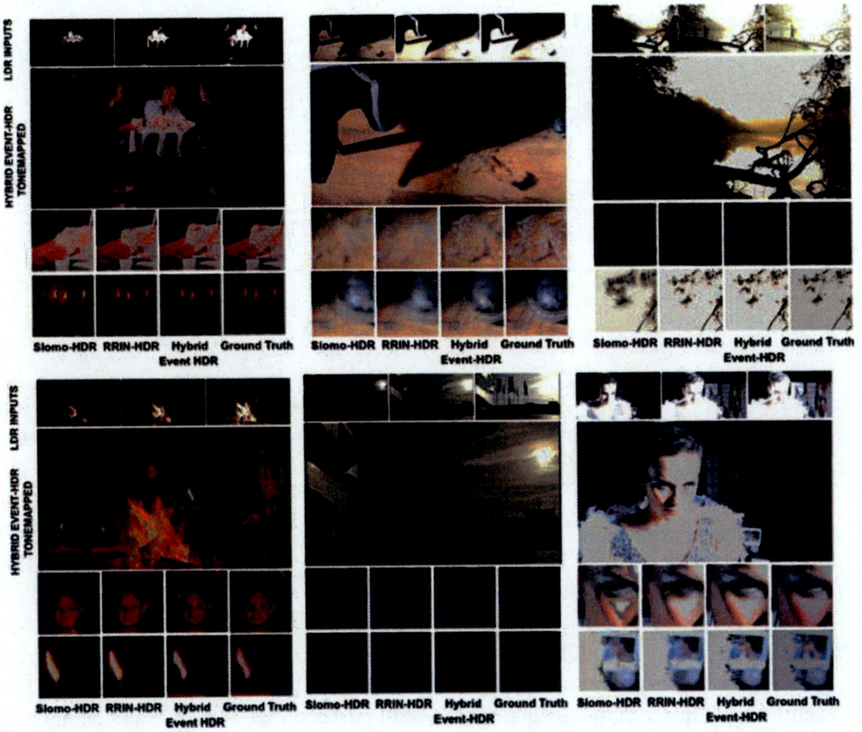

Fig. 4 Comparision of hybrid event-HDR with Slomo-HDR and RRIN-HDR for **8-frame interpolation**

Table 1 HDR-VDP2 scores for 2, 5 & 8-frame interpolation

Datasets	**2 frame interpolation**			**5-frame interpolation**			**8-frame interpolation**		
	Slomo HDR	**RRIN HDR**	**Hybrid event HDR**	**Slomo HDR**	**RRIN HDR**	**Hybrid event HDR**	**Slomo HDR**	**RRIN HDR**	**Hybrid event HDR**
Poker	67.72	74.51	**74.61**	67.63	73.05	**74.54**	67.44	71.59	**73.961**
Fireplace	65.21	70.41	**70.42**	56.96	57.25	**64.40**	57.013	57.200	**60.665**
Showgirl	66.33	67.79	**68.97**	64.83	64.62	**70.21**	64.29	62.898	**69.211**
Cars	68.86	**79.74**	77.44	66.56	**71.38**	68.333	66.785	62.614	**69.18**
Fishing	65.16	70.95	**71.26**	63.65	66.92	**70.56**	63.90	65.362	**69.663**
Bistro	66.37	68.70	**71.22**	64.08	63.36	**68.84**	63.866	63.534	**68.407**

Table 2 Quantitative metrics for 8-frame interpolation

Datasets	Slomo-HDR			RRIN-HDR			Hybrid Event HDR		
	PSNR	SSIM	HDR-VDP2	PSNR	SSIM	HDR-VDP2	PSNR	SSIM	HDR-VDP2
Poker_Fullshot	26.466	0.586	67.44	34.957	0.906	71.59	**35.982**	**0.910**	**73.961**
Fireplace_02	20.374	0.324	57.013	22.558	0.506	57.200	**26.322**	**0.560**	**60.665**
Showgirl_02	19.45	0.751	64.29	19.24	0.727	62.898	**22.806**	**0.915**	**69.211**
Cars_Longshot	24.164	0.910	66.785	24.722	0.942	62.614	**24.81**	**0.947**	**69.18**
Fishing_Longshot	22.296	0.821	63.90	21.286	0.813	65.362	**25.512**	**0.947**	**69.663**
Bistro_03	26.545	0.833	63.866	28.062	0.892	63.534	**32.830**	**0.930**	**68.407**

hdr-vdp2 score as well as PSNR/SSIM, and also, the performance gap for hybrid event-HDR, as compared to other methods, increases when a large number of frames are interpolated or when a higher output speed of HDR video is needed.

6 Conclusion

We proposed an algorithm for the reconstruction of HDR video sequences from an input of alternating exposure LDR frames and corresponding event-sensor data. This is achieved by first interpolating the necessary LDR frames at each time instant with the help of dense motion information obtained from the event-sensor data. Then, the LDR frames with different exposures are fused using an attention-based deep learning algorithm to generate an HDR frame at each time instant. We demonstrated HDR video reconstruction upto 3 times the frame rate of the input LDR video sequence. Due to the use of event-sensor data for video frame interpolation, we obtain more accurate reconstructions compared to previous state-of-the-art techniques, and the performance gap between our method and the non-event-based method increases as we move toward higher output video frame rate. As a future work, we would like to capture the real-world event data and will perform similar experiments with it.

References

1. Debevec, P.E., Malik, J.: Recovering high dynamic range radiance maps from photographs. In: Proceedings of the 24th Annual Conference on Computer Graphics and Interactive Techniques, USA, 1997, SIGGRAPH '97, pp. 369–378. ACM Press/Addison-Wesley Publishing Co (1997)
2. Zhao, H., Shi, B., Fernandez-Cull, C., Yeung, S.K., Raskar, R.: Unbounded high dynamic range photography using a modulo camera. In: 2015 IEEE International Conference on Computational Photography (ICCP), pp. 1–10. IEEE (2015)
3. Tocci, M.D., Kiser, C., Tocci, N., Sen, P.: A versatile HDR video production system. ACM Trans. Graph. (TOG) **30**(4), 1–10 (2011)

4. Kalantari, N.K., Ramamoorthi, R.: Deep HDR video from sequences with alternating exposures. Comput. Graph. Forum Wiley Online Libr. **38**, 193–205 (2019)
5. Brandli, C., Berner, R., Yang, M., Liu, S.-C., Delbruck, T.: A 240× 180 130 dB 3 μs latency global shutter spatiotemporal vision sensor. IEEE J. Solid-State Circ. **49**(10), 2333–2341 (2014)
6. Tulyakov, S., Gehrig, D., Georgoulis, S., Erbach, J., Gehrig, M., Li, Y., Scaramuzza, D.: Time lens: event-based video frame interpolation. In: Proceedings of the IEEE/CVF Conference on Computer Vision and Pattern Recognition, pp. 16155–16164 (2021)
7. Yan, Q., Gong, D., Shi, Q., van den Hengel, A., Shen, C., Reid, I., Zhang, Y.: Attention-guided network for ghost-free high dynamic range imaging. In: Proceedings of the IEEE/CVF Conference on Computer Vision and Pattern Recognition, pp. 1751–1760 (2019)
8. Li, H., Yuan, Y., Wang, Q.: Video frame interpolation via residue refinement. In: ICASSP 2020—2020 IEEE International Conference on Acoustics, Speech and Signal Processing (ICASSP), pp. 2613–2617. IEEE (2020)
9. Jiang, H., Sun, D., Jampani, V., Yang, M.-H., Learned-Miller, E., Kautz, J.: Super slomo: high quality estimation of multiple intermediate frames for video interpolation. In: Proceedings of the IEEE Conference on Computer Vision and Pattern Recognition, pp. 9000–9008 (2018)
10. Arif Khan, E., Oguz Akyuz, A., Reinhard, E.: Ghost removal in high dynamic range images. In: 2006 International Conference on Image Processing, pp. 2005–2008. IEEE (2006)
11. Oh, T.H., Lee, J.-Y., Tai, Y.-W., Kweon, I.S.: Robust high dynamic range imaging by rank minimization. IEEE Trans. Pattern Anal. Mach. Intell. **37**(6), 1219–1232 (2014)
12. Sen, P., Kalantari, N.K., Yaesoubi, M., Darabi, S., Goldman, D.B., Shechtman, E.: Robust patch-based HDR reconstruction of dynamic scenes. ACM Trans. Graph. **31**(6), 203 (2012)
13. Kalantari, N.K., Ramamoorthi, R., et al.: Deep high dynamic range imaging of dynamic scenes. ACM Trans. Graph. 36(4), 1–12, Art. No. 144 (2017)
14. Banterle, F., Ledda, P., Debattista, K., Chalmers, A.: Inverse tone mapping. In: Proceedings of the 4th International Conference on Computer Graphics and Interactive Techniques in Australasia and Southeast Asia, pp. 349–356 (2016)
15. Moriwaki, K., Yoshihashi, R., Kawakami, R., You, S., Naemura, T.: Hybrid loss for learning single-image-based HDR reconstruction. arXiv preprint arXiv:1812.07134 (2018)
16. Zhang, J., Lalonde, J.-F.: Learning high dynamic range from outdoor panoramas. In: Proceedings of the IEEE International Conference on Computer Vision, pp. 4519–4528 (2017)
17. Santana Santos, M., Ren, T.I., Khademi Kalantari, N.: Single image HDR reconstruction using a CNN with masked features and perceptual loss. arXiv preprint arXiv:2005.07335 (2020)
18. Eilertsen, G., Kronander, J., Denes, G., Mantiuk, R.K., Unger, J.: HDR image reconstruction from a single exposure using deep CNNs. ACM Trans. Graph. (TOG) **36**(6), 1–15 (2017)
19. Niu, Y., Wu, J., Liu, W., Guo, W., Lau, R.W.H.: HDR-GAN: HDR image reconstruction from multi-exposed LDR images with large motions. IEEE Trans. Image Process. **30**, 3885–3896 (2021)
20. Yan, Q., Zhang, L., Liu, Y., Sun, J., Shi, Q., Zhang, Y.: Deep HDR imaging via a non-local network. IEEE Trans. Image Process. **29**, 4308–4322 (2020)
21. Wu, S., Xu, J., Tai, Y.-W., Tang, C.-K.: Deep high dynamic range imaging with large foreground motions. In: Proceedings of the European Conference on Computer Vision (ECCV), pp. 117–132 (2018)
22. Hirakawa, K., Simon, P.M.: Single-shot high dynamic range imaging with conventional camera hardware. In: 2011 International Conference on Computer Vision, pp. 1339–1346. IEEE (2011)
23. Chen, G., Chen, C., Guo, S., Liang, Z., Wong, K.Y.K., Zhang, L.: HDR video reconstruction: a coarse-to-fine network and a real-world benchmark dataset. arXiv preprint arXiv:2103.14943 (2021)
24. Gallego, G., Delbruck, T., Orchard, G., Bartolozzi, C., Taba, B., Censi, A., Leutenegger, S., Davison, A., Conradt, J., Daniilidis, K., et al.: Event-based vision: a survey. arXiv preprint arXiv:1904.08405 (2019)
25. Froehlich, J., Grandinetti, S., Eberhardt, B., Walter, S., Schilling, A., Brendel, H.: Creating cinematic wide gamut HDR-video for the evaluation of tone mapping operators and HDR-displays. Digit. Photogr. X Int. Soc. Opt. Photonics **9023**, 90230X (2014)

26. Rebecq, H., Gehrig, D., Scaramuzza, D.: ESIM: an open event camera simulator. In: Conference on Robot Learning, pp. 969–982. PMLR (2018)
27. Mantiuk, R., Kim, K.J., Rempel, A.G., Heidrich, W.: HDR-VDP-2: a calibrated visual metric for visibility and quality predictions in all luminance conditions. ACM Trans. Graph. (TOG) **30**(4), 1–14 (2011)

Diagnosis of COVID-19 Using Deep Learning Augmented with Contour Detection on X-rays

Rashi Agarwal and S. Hariharan

Abstract The WHO has declared the infectious respiratory disease COVID-19 caused due to novel coronavirus an international pandemic on March 11, 2020. The raging pandemic created a colossal loss of human life and created economic and social disruptions for millions worldwide. As the illness is new, the medical system and infrastructure are presently inadequate to counter the condition. The situation demands innovating creatively and instituting countervailing measures to circumvent the crisis. Artificial intelligence and machine learning need to be the engine for leading the technological transformation across the healthcare industry amidst the pandemic. As human resources have stretched, automation is the key to tiding over the situation. Early computer-aided automated detection of the disease can provide the necessary edge in combating this deadly virus. Though the availability of a dataset with sufficient ground truth remains a challenge, a convolution neural network (CNN)-based approach can play a dominant role as a classifier solution for the chest X-rays. This paper explores the possibility of using the convolutional neural networks to classify the chest X-rays on the preprocessed images using thresholding followed by morphological processing for edge detection and contour detection. The accuracy of a stand-alone CNN network increases remarkably when preprocessed images are used as input.

Keywords COVID-19 · CNN · Otsu thresholding · Morphological processing · X-ray · Contours

---

R. Agarwal (✉)
Harcourt Butler Technical University, Kanpur 208002, India
e-mail: dr.rashiagrawal@gmail.com

S. Hariharan
University of Madras, Madras, India

M. Tistarelli et al. (eds.), *Computer Vision and Machine Intelligence*,
Lecture Notes in Networks and Systems 586,
https://doi.org/10.1007/978-981-19-7867-8_16

1 Introduction

COVID-19 is a rapidly spreading viral disease that infects humans and animals alike, which has created panic worldwide. The rapid escalation of deaths poses a significant challenge to humanity in containing this virus [3]. There have been virus diseases like severe acute respiratory syndrome (SARS), middle east respiratory syndrome (MERS), and flu. Still, they lasted for a few days or months [13]. The reverse transcription polymerase chain reaction (RTPCR) remains the reference standard for confirming the infection despite a high false-negative rate. The errors occur due to multiple preanalytical and analytical features like the absence of standardization, delay in delivery of samples, lack of quality storage conditions and substandard assays, non-adherence to the procedure, and presence of mutation [14].

Despite best efforts, several countries face an acute shortage of RTPCR testing kits and reagents for screening the illness among suspected patients. Thus, efforts to detect the COVID-19 infection using alternate methods like computed tomography (CT) chest images have indicated up to 91% accuracy during the diagnosis [9]. To expedite the detection of infection, the possibility of deep learning on the CT images was explored [20]. Despite this fact, the inherent disadvantage of CT is that it is expensive, and the image acquisition process is relatively more time-consuming than conventional X-rays. Moreover, due to the cost factor, the facility of CT imaging is not readily available for the needy in many underdeveloped and developing nations. On the contrary, X-ray imaging technology is faster, cheaper, and thus rapidly screening COVID-19 patients if adequately exploited. However, this aspect is not devoid of challenges due to the intervening ribs and non-availability of sufficient datasets in the structured format. The layered images in the CT and magnetic resonance imaging (MRI) give a higher signal-to-noise ratio making them easier to process than the X-ray and thus revealing more detail [7]. CNN is a good option for processing X-ray images as it is an end-to-end network. In the first step, we use CNN to classify infected and normal X-ray images. The images are preprocessed before being fed into the CNN to enhance accuracy in the next step. We preprocess images with Gaussian filtering, Otsu thresholding, and morphological segmentation to successfully improve classification accuracy on the existing CNN network. We use adaptive thresholding to provide better accuracy than the conventional thresholding technique [19].

This paper intends to demonstrate the accuracy levels achieved by CNN networks on chest X-ray images that are preprocessed using Gaussian filtering and Otsu thresholding combined with morphological operations for early rapid detection of COVID-19 patient screening.

1.1 Preprocessing Algorithms

1.1.1 Gaussian Filtering

Gaussian filters have been used in this setup for noise removal. They are two-dimensional (2D) convolution operators which work on the input, smooth them, and remove noise in which the Gaussian filter is used as a point spread function [5]. The Gaussian filters are low pass filters, with the center coefficients having the highest value and the peripheral values the lowest in a reducing form. The separable property of the filter is exploited to get higher computational speed.

1.1.2 Canny Edge Filtering

Canny edge filtering is used to detect the edges in image processing applications. The working principle is that these filters identify borders by reading the pixel intensity value in their neighborhood. If there is a drastic change in the pixel values, then it is identified as an edge. The Canny detector works by differentiating the image in two orthogonal directions and computing the gradient magnitude as the root sum square of derivatives. The gradient derivative is calculated as the arctangent of the ratio of the derivatives [11].

1.1.3 Otsu Thresholding

The Otsu thresholding is a method for automatic clustering, assuming that there are two groups of pixels in the foreground and background, by calculating the optimum threshold that results in the minimal intra-class variance and the maximum inter-class variance [18]. Then based on this threshold value, each pixel is classified if it belongs to the foreground or background. This threshold value is represented within the class of variances, as given below

$$\sigma^2 = w_0(t)\sigma_0(t) + w_1(t)\sigma_1^2(t) \tag{1}$$

w_0 and w_1 are the probabilities of both the classes and 't' is the threshold value. The σ_0^2 and σ_1^2 are the variance of the two classes. The probability is as given as

$$w_o(t) = \sum_{i=0}^{t-1} p(i) \tag{2}$$

$$w_1(t) = \sum_{l=t}^{l-1} p(i) \tag{3}$$

In the OTSU method, the inter-class variance maximization is same as intra-class variance minimization [6]. The total variance is given as

$$\sigma^2 = \underbrace{\sigma_w^2(t)} + \underbrace{w_0(t)[1 - w_0(t)][\mu_0(t) - \mu_1(t)]^2} \tag{4}$$

First part shows intra-class variance and second part shows inter-class variance.

Connected Components and Contours—The connected component analysis technique detects connected regions in a binary image and labels them. Connected component labeling is different from segmentation, albeit they appear similar. The labeling is based on heuristics and is performed after scanning the image from top to bottom and left to right. Contour is the bounding line that encloses or connects a shape or object based on extracted edge. The contour is defined as the segmentation of an image u as a finite set of rectifiable Jordan curves for the contours k. The region k is a connected component, and we assume that each contour meets the other. It is assumed that each contour separates two separate regions and that each tip is common to at least three contours, and the segmentation is expressed as—$p = \{R_i\}i \quad of \quad \Omega$ and p= partition $\{R_i\}$ is contour separating regions.

1.2 Convolution Neural Network

Neural networks are an alternative solution to the computational space wherein the solution is derived not from explicit instructions, but the computer learns these from a series of examples; the inspiration for this concept originally came from the method of functioning of the brain in processing information [4]. A feedforward artificial neural network is a mapping function that translates the input function into an output function based on self-derived rules. These mapping functions are nonlinear operations. This set of rules is based on the values of the weights of the parameters that have been derived automatically. The given equation can best describe a simple artificial network as per [15]—$a = \sum_{i=1}^{d} w_i x_i + w_0$.

Here, w_i is the weight of the corresponding input function and w_0 is the bias of the function.

Convolution Layer This layer is the first layer used to extract the feature from the input by running a filter of a predetermined size. The corresponding data points in the filter perform convolution operation with the input. The resultant of the convolution is fed into the subsequent layers. The output size is given by the formula given below

$$O = \left(\frac{W - F + 2 * P}{S}\right) + 1 \tag{5}$$

where O is the output width, W is the width of input size, F is the filter size, P is the padding, and S is the stride size. The overall convolution formula is given by [17] as

$$X_j^i = f < \sum_{i \epsilon M_j}^{l-1} * W_{ij}^l + b_j^l > \tag{6}$$

where the * is the convolution operator and $M_j = \{i | i \ \epsilon \ (l-1)^{th}\text{layer}\}$ map linked to the jth map in the lth layer.

Nonlinearity Layer This layer is associated with the activation function. This layer calculates the relationship between the parameters and the input and decides whether a feature should proceed to the next layer or be discontinued. Sigmoid was the most prevalent activation function, but the recent trend among the users is in favor of rectified linear output (ReLU). The definition for ReLU is given below [2].

$$\text{ReLU}(x) = \max(0, x) \tag{7}$$

$$\frac{\text{dRelu}(x)}{\text{d}x} = \begin{cases} 1, & \text{if} x > 0 \\ 0, & \text{otherwise} \end{cases} \tag{8}$$

Pooling Layer The main function of this component is to reduce the size of the feature map, or it may also be interpreted as a reduction in the image's resolution. This layer reduces the computational load of the next layers.

Fully Connected Layer Every node is connected to the node in the succeeding and the preceding layer in this layer. Nodes in the last frame of the pooling layer are connected as vectors to the first layers of the fully connected (FC) layer [8], and these are the widely used parameters; therefore, the computation is time-consuming. Since the number of parameters is more thus, efforts are made to eliminate a few nodes and connections to reduce computational complexity. Let the two consecutive layers be given as below

$$l^{k-1} \in \mathbb{R}^{m^{k-1}}, l^k \in \mathbb{R}^{m^k}$$

and if the weight of the matrix is—$w^k \in \mathbb{R}^{m^{k-1}m^k}$, where m^k is the number of neurons in layer, then the output of the FC layer is as given below

$$o^k = \psi_k * (x^{k-1}) = \sigma\left((x^{k-1})^T w^{(k)} + b^{(k)}\right) \tag{9}$$

2 Literature Survey

In [12], the authors have used CNN, deep feature extraction, end-to-end training, and fine-tuning of pre-trained neural networks to screen the COVID-19 chest X-ray. ResNet18, ResNet50, ResNet101, VGG16, and VGG19 have been utilized for this purpose. The support vector machine was used with linear, quadratic, Gaussian, and cubic functions as a classifier of the features. Deep transfer learning (fine-tuning combined with deep feature extraction) is used to overcome training limitations. The results indicated that the highest accuracy was achieved with a ResNet50 SVM classifier with a linear kernel filter of 94.6%.

In [18], the authors employ the technique of optimized CNN for identifying COVID-19 patients from their chest X-rays. One uses the ResNet50 with an error correcting code output of the three models. The other two use CNN optimized using the gray wolf algorithm (GWO) and whale optimization (WAO) using the BAT+ algorithm. The test images are scaled down to 448 × 448 × 1 size before being fed into the optimized CNN. The study results indicate that the optimized CNN gave better results than the un-optimized CNN. Meanwhile, the GWO setup achieved accuracy up to 98%, and the optimizer using WAO achieved 96% accuracy.

In [10], authors propose a sequential CNN model to detect COVID-19 patients from chest X-rays with a simplistic architecture. The tuned dataset is fed into a convolutional layer with a 3 × 3 kernel with ReLU activation followed by three 2D convolutional layers and max pooling. The sigmoid function, in the end, is used to classify the input as a COVID-19 patient or not. Three models of CNN are used for this setup, each with three, four, and five convolutional layers. The results demonstrate that the model's accuracy with four convolutional layers is highest (97%) than the other models with 03 and 05 convolutional layers (96%).

In [1], the authors propose a machine learning architecture with features extracted by the histogram of gradients (HOG) and classification by CNN to detect COVID-19 disease from chest X-rays utilizing modified anisotropic diffusion filtering to remove speckle noise. The independently extracted feature vectors are fused to generate a larger dataset. Out of the 7876 features fused, 1186 features were selected. The watershed mechanism of segmentation is utilized. Among the various CNN models tested, the VGG19 gave the highest levels of accuracy, nearing 99%.

In [16], the authors have proposed a model that is an end-to-end selection and classification employing 05 different CNN models and suggesting a high accuracy decision support system for detecting COVID-19 patients. Data augmentation was implemented with horizontal flipping enabled in the dataset. The CNN models used were ResNet50, ResNet101, ResNet152, Inception V3, and Inception-ResNetV52. ResNet50 had the highest level of accuracy compared to the other models. ResNet50 demonstrated an accuracy of 96% when performed for the binary class (COVID-19/normal). The model's performance increased up to 99.4% when performed on the binary class (COVID-19/viral pneumonia).

In [17], the authors propose a CNN model to classify the chest X-rays for early detection of COVID-19. The model specifically used in this setup is CVDNet. The proposed model uses two parallel convolutional layers of size 5 × 5 and 13 × 13 with a stride of 01. Finally, the output of the parallel convolution layers is concatenated to generate the final feature maps. The results show that the model achieved an average accuracy of 96.69% across fivefold classification.

3 Methodology

The dataset of the chest X-rays was taken from the Kaggle COVID-19 radiography database, which consisted of 10,192 normal chest X-rays and 3616 COVID-19 X-rays. The dataset was partitioned into two subsets with 20% data used to train and test the proposed CNN. The training dataset contained 2531 COVID-19 X-ray images and 7135 normal X-ray images, making 9666 X-ray images for training purposes. The

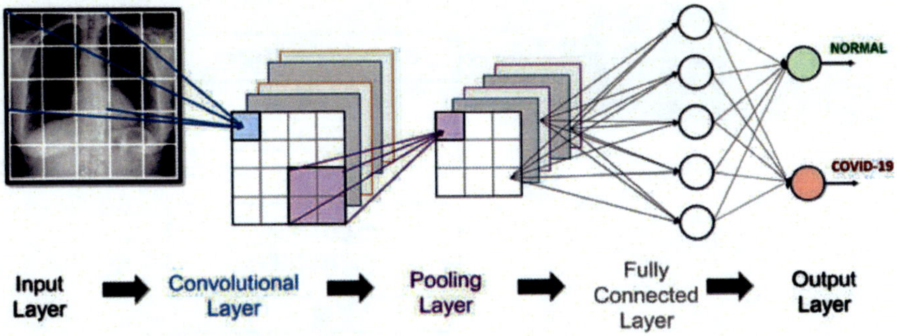

Fig. 1 CNN architecture used for classification

testing dataset similarly contained 2761 X-ray images, in which 723 X-ray images were from each class COVID-19 positive and 2038 normal. A 10% validation size is taken. A vanilla CNN, as shown in Fig. 1, was applied to all these images.

Model Used A vanilla model composed in a sequential organization. It starts with a batch normalization layer, two convolutional-pooling layers, and one dropout layer. Then, flattening is performed, and the classification is achieved at the end of two dense layers. The CNN model thus achieved an accuracy of 88% with the test data subset used from the processed dataset of this study with a precision of 88.58%, with the model shown in Fig. 1, preprocessed as per the flowchart shown in Fig. 2. Other important metrics have been adopted in this study to evaluate the overall performance and accuracy, including $F1$ score, precision, accuracy, and recall. The scores of these parameters are reported in Table 3.

The results were recorded, and further exploration was done on the images to perform preprocessing to derive contours from the X-rays. The raw images are filtered using Gaussian filtering. A low pass filter and additional weights to the pixels on the edge led to a significant reduction in edge blurring. The smoothened image is segmented using the Otsu method. Otsu algorithm takes in a noise-reduced image and outputs a binary image using a threshold value based on the image's histogram. The threshold value can segregate the pixels into foreground and background. This binary image is further processed through a Canny edge detector for identifying the edges. The Canny edge detector works on a multi-stage algorithm in which the spatial derivatives are derived. Non-maximal suppression is carried out on the ridges, making all the pixels equal zero, not on top of the ridge. This traversing output is controlled between two threshold values to prevent any noisy edges from being fragmented. The detected edges are then subjected to connected component analysis, which yields the contour outlay of an image. The processing was conducted in a Keras Tensor-flow environment on an i7 processor @ 3.4 GHz 64GB RAM. A sample image with the segmentation and contours is depicted in Fig. 5.

The contoured image for each image in the dataset is generated based on the pixel intensity values and the connectivity property, as shown in Fig. 4. These are

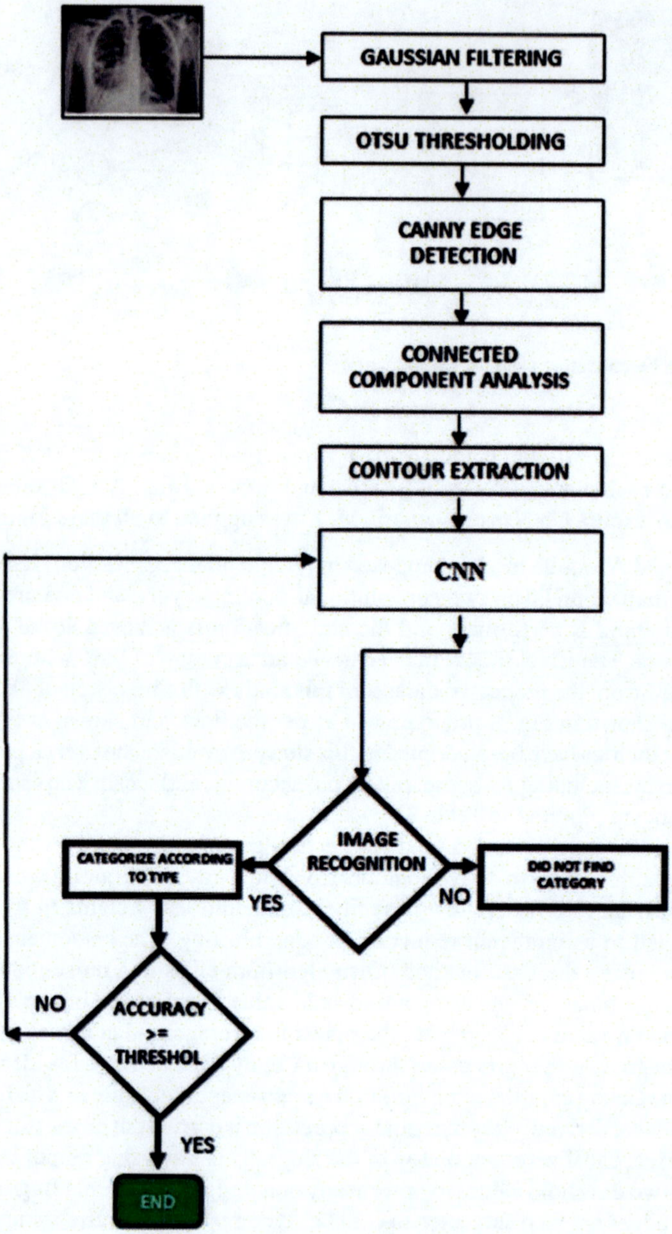

Fig. 2 Flow chart depicting the pipeline for image processing

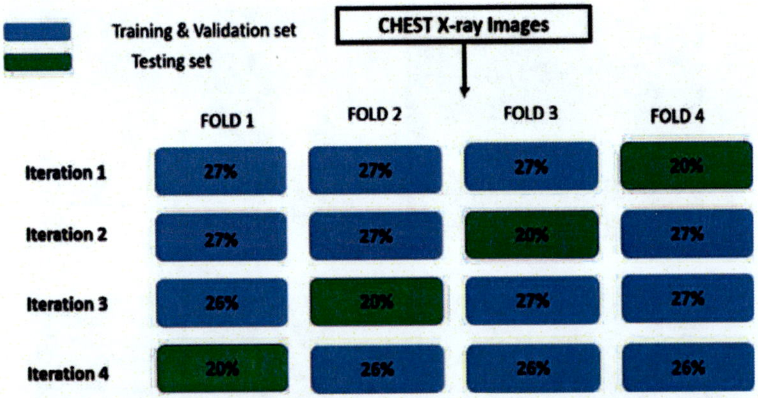

Fig. 3 Schematic diagram of training, testing, and validation in a fourfold cross-validation scheme

Fig. 4 Confusion matrix for the normal and COVID-19 without pre processing

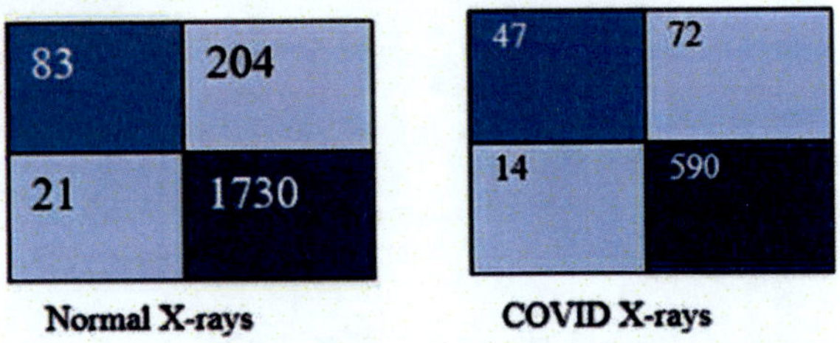

labeled according to their original class. We observed that the number of contours in COVID-infected chest X-rays was much larger than in uninfected chest X-rays (Figs. 3, 4 and 5).

The same vanilla CNN model used before is now used to study the effectiveness of the preprocessing operations performed on the X-rays. The CNN model now achieved an accuracy of 94.58% and a recall of 98.45%. To evaluate the overall performance and accuracy of the contour-based image CNN model, other important metrics have been adopted in this study, including $F1$ score, precision, accuracy, and recall. The scores of these parameters are reported in Table 5.

4 Results

The COVID-19 X-rays are characterized by ground-glass opacity, air space consolidation, broncho vascular thickening, and traction bronchiectasis [17]. These defects are more pertinent in the contour images as the number of contours increases drastically for such images (Table 1).

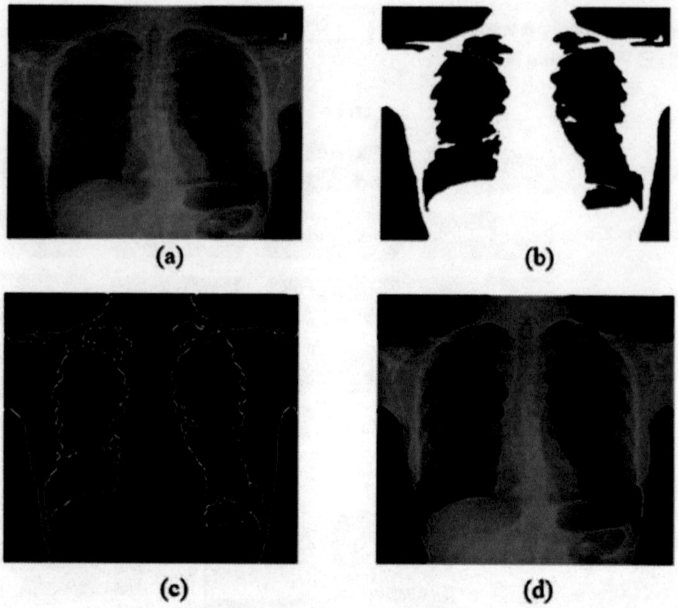

Fig. 5 **a** Normal X-ray. **b** Image after undergoing thresholding operation. **c** Image after segmentation. **d** Contours are used to form the connected components

Table 1 Details of training, validation, and test set

Class	Number of images	Training	Validation	Test
Normal	10192	7135	1019	2038
COVID-19	3616	2531	362	723

To study the efficiency of our new system, a fourfold cross-validation method was utilized to validate the test results and measure the system's efficiency. All the sample data were split into four groups.

One group was utilized to test the data, and the remaining three sets were utilized for training the data. 70% of the chest X-rays were used for training, the remaining 10% for validation, and the other 20% for testing data. Out of the 10,192 normal chest X-rays, 7135 were used for training, 1019 were used for validation, and 2038 images were used for testing. Out of the 3616 COVID-19 X-rays, 2531 X-rays were used for testing, 362 were used for validation, and 723 images were used for testing. All the selections were random. The images are subjected to fourfold validation with 04 iterations, as shown in Fig. 3.

The dataset images are contoured and then trained using the same vanilla CNN using fourfold cross-validation. The confusion matrix got is shown in Fig. 4. Other important metrics have been adopted in this study to evaluate the overall performance

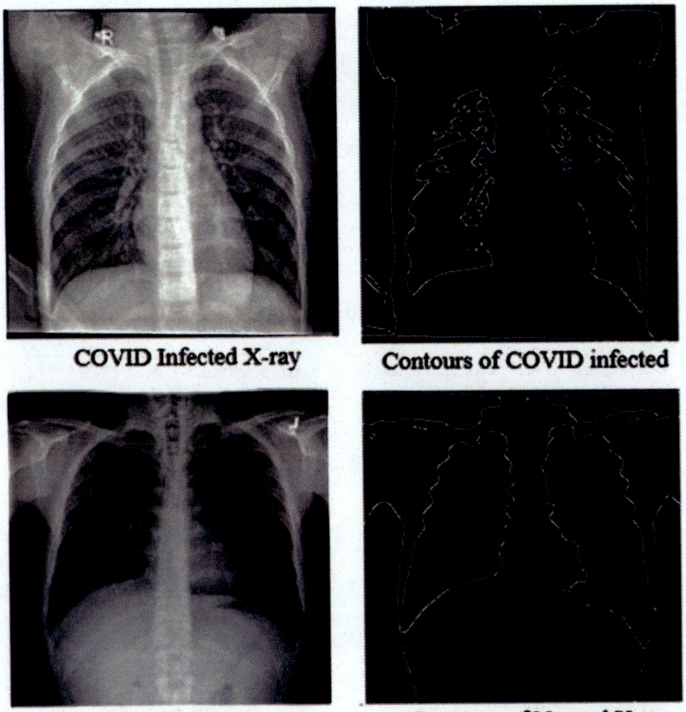

Fig. 6 Normal and COVID-19 X-rays are shown on the left and the processed images classified are given to the right of image

Table 2 Precision, accuracy, recall, and $F1$ score of CNN in fold-1, fold-2, fold-3, and fold-4 without preprocessing

Fold	Class	Precision (%)	Accuracy (%)	Recall (%)	$F1$ score (%)
Fold-1	Normal	87.47	87.24	98.69	92.741
	COVID-19	87.67	86.44	97.32	92.289
Fold-2	Normal	88.63	88.42	99.0	93.52
	COVID-19	88.78	87.94	97.82	93.08
Fold-3	Normal	88.72	88.37	98.9	93.53
	COVID-19	88.80	88.52	98.65	93.46
Fold-4	Normal	89.45	88.96	98.8	93.89
	COVID-19	89.12	88.10	97.6	93.16

and accuracy, including $F1$ score, precision, accuracy, and recall. The scores of these parameters are reported in Tables 2, 3, 4 and 5. A sample normal and COVID-infected X-ray with corresponding contour images are shown in Fig. 6.

Table 3 Performance of the CNN in each fold without preprocessing

Fold	Precision (%)	Accuracy (%)	Recall (%)	$F1$ score (%)
Fold-1	87.57	86.84	98.00	92.51
Fold-2	88.70	88.18	98.41	93.30
Fold-3	88.76	88.45	98.77	93.49
Fold-4	89.29	88.53	98.20	93.50
Average	88.58	88.00	98.35	93.20

Table 4 Precision, accuracy, recall, and $F1$ score of CNN in fold-1, fold-2, fold-3, and fold-4 with preprocessing

Fold	Class	Precision (%)	Accuracy (%)	Recall (%)	$F1$ score (%)
Fold-1	Normal	94.04	93.03	98.4	96.17
	COVID-19	94.7	94.19	98.5	96.56
Fold-2	Normal	95.01	93.81	98.24	96.59
	COVID-19	95.25	94.74	98.68	96.93
Fold-3	Normal	95.69	94.35	98.14	96.02
	COVID-19	95.77	95.15	98.70	97.21
Fold-4	Normal	95.93	95.29	98.96	97.42
	COVID-19	96.72	96.12	98.85	97.77

Table 5 Performance of the CNN in each fold with preprocessing

Fold	Precision (%)	Accuracy (%)	Recall (%)	$F1$ score (%)
Fold-1	94.37	93.61	98.45	96.37
Fold-2	95.13	94.28	98.46	96.76
Fold-3	95.73	94.75	98.42	96.62
Fold-4	96.33	95.71	98.91	97.60
Average	95.39	94.58	98.56	96.84

The results indicate that the metrics and efficiency substantially increase when CNN is combined with the preprocessing. As the system is functioning bereft of the preprocessing, the average accuracy is 87.69%, precision is 88.11%, recall is 87.10%, and $F1$ score is 86.45%. When the CNN is clubbed with the preprocessing technique, the precision is 98.10%, accuracy is 97.36%, recall is 97.42%, and $F1$ score is 96.91%. The model substantially improved in classifying the COVID-19 X-rays and the normal X-rays. Wrong classification in the CNN model with preprocessing occurred mainly because of poor image quality.

5 Conclusion

The present study proposed the method for detecting COVID-19 patients from chest X-rays. The method could distinguish clearly the COVID-19 patients from the normal with the aid of the X-ray. Our model is trained on a sufficient dataset sourced from public and open sources. The efficiency was measured using the fourfold cross-validation scheme. It was observed that the CNN model with preprocessing technique achieved an average precision of 98.10% and an accuracy of 97.36% compared to an average precision of 88.11% and an accuracy of 87.69% for the model using images without the preprocessing. These encouraging results are an indication that this is a promising tool that can be utilized by health workers combating the pandemic.

References

1. Ahsan, Md.M., Based, J.H., Kowalski, M., et al.: COVID-19 detection from chest X-ray images using feature fusion and deep learning. Sensors **21**(4), 1480 (2021)
2. Albawi, S., Mohammed, T.A., Al-Zawi, S.: Understanding of a convolutional neural network. In: 2017 International Conference on Engineering and Technology (ICET), pp. 1–6. IEEE (2017)
3. Bai, Y., Yao, L., Wei, T., Tian, F., Jin, D.-Y., Chen, L., Wang, M.: Presumed asymptomatic carrier transmission of COVID-19. Jama **323**(14), 1406–1407 (2020)
4. Bishop, C.M.: Neural networks and their applications. Rev. Sci. Instrum. **65**(6), 1803–1832 (1994)
5. Cabello, F., León, J., Iano, Y., Arthur, R.: Implementation of a fixed-point 2D Gaussian filter for image processing based on FPGA. In: 2015 Signal Processing: Algorithms, Architectures, Arrangements, and Applications (SPA), pp. 28–33. IEEE (2015)
6. Chithra, A.S., Renjen Roy, R.U.: Otsu's adaptive thresholding based segmentation for detection of lung nodules in CT image. In: 2018 2nd International Conference on Trends in Electronics and Informatics (ICOEI), pp. 1303–1307. IEEE (2018)
7. Dong, Y., Pan, Y., Zhang, J., Xu, W.: Learning to read chest X-ray images from 16000+ examples using CNN. In: 2017 IEEE/ACM International Conference on Connected Health: Applications, Systems and Engineering Technologies (CHASE), pp. 51–57. IEEE (2017)
8. Dumoulin, V., Visin, F.: A guide to convolution arithmetic for deep learning. arXiv preprint arXiv:1603.07285 (2016)
9. Fang, Y., Zhang, H., Xie, J., Lin, M., Ying, L., Pang, P., Ji, W.: Sensitivity of chest CT for COVID-19: comparison to RT-PCR. Radiology **296**(2), E115–E117 (2020)
10. Haque, K.F., Haque, F.F., Gandy, L., Abdelgawad, A.: Automatic detection of COVID-19 from chest X-ray images with convolutional neural networks. In: 2020 International Conference on Computing, Electronics & Communications Engineering (iCCECE), pp. 125–130. IEEE (2020)
11. Heath, M.D., Sarkar, S., Sanocki, T., Bowyer, K.W.: A robust visual method for assessing the relative performance of edge-detection algorithms. IEEE Trans. Pattern Anal. Mach. Intell. **19**(12), 1338–1359 (1997)
12. Ismael, A.M., Şengür, A.: Deep learning approaches for COVID-19 detection based on chest X-ray images. Expert Syst. Appl. **164**, 114054 (2021)
13. Jain, R., Gupta, M., Taneja, S., Jude Hemanth, D.: Deep learning based detection and analysis of COVID-19 on chest X-ray images. Appl. Intell. **51**(3), 1690–1700 (2021)

14. Lippi, G., Simundic, A.-M., Plebani, M.: Potential preanalytical and analytical vulnerabilities in the laboratory diagnosis of coronavirus disease 2019 (COVID-19). Clin. Chem. Lab. Med. (CCLM) **58**(7), 1070–1076 (2020)
15. McCulloch, W.S., Pitts, W.: A logical calculus of the ideas immanent in nervous activity. Bull. Math. Biophys. **5**(4), 115–133 (1943)
16. Narin, A., Kaya, C., Pamuk, Z.: Automatic detection of coronavirus disease (COVID-19) using X-ray images and deep convolutional neural networks. Pattern Anal. Appl. 1–14 (2021)
17. Ouchicha, C., Ammor, O., Meknassi, M.: CVDNet: a novel deep learning architecture for detection of coronavirus (COVID-19) from chest X-ray images. Chaos, Solitons Fractals **140**, 110245 (2020)
18. Pathan, S., Siddalingaswamy, P.C., Ali, T.: Automated detection of COVID-19 from chest X-ray scans using an optimized CNN architecture. Appl. Soft Comput. **104**, 107238 (2021)
19. Roy, P., Dutta, S., Dey, N., Dey, G., Chakraborty, S., Ray, R.: Adaptive thresholding: a comparative study. In: 2014 International Conference on Control, Instrumentation, Communication and Computational Technologies (ICCICCT), pp. 1182–1186. IEEE (2014)
20. Shi, F., Xia, L., Shan, F., Song, B., Dijia, W., Wei, Y., Yuan, H., Jiang, H., He, Y., Gao, Y., et al.: Large-scale screening to distinguish between COVID-19 and community-acquired pneumonia using infection size-aware classification. Phys. Med. Biol. **66**(6), 065031 (2021)

A Review: The Study and Analysis of Neural Style Transfer in Image

Shubham Bagwari, Kanika Choudhary, Suresh Raikwar, Prashant Singh Rana, and Sumit Mighlani

Abstract The transfer of artistic styles into the image has become prevalent in industry and academia. The neural style transfer (NST) is a method to transfer the style of an image to another image. The study and analysis of the NST methods are essential to obtaining realistic, stylized images efficiently. This study explored different methods to perform the style transfer and revealed the key challenges in the style transfer. Further, the specific research gaps have been identified in the field of NST. Moreover, an exhaustive analysis of the existing methods of the NST has been presented in this study. The qualitative and quantitative comparisons of the renowned methods of NST have been conducted and presented in this study. The COCO dataset has been utilized to compute the PSNR and SSIM values to compare the results. Further, the computation time of different methods of the NST has been discussed. Finally, the study has been concluded with future scope in the field of NST.

Keywords Convolution neural network · Deep learning · Neural networks · Neural style transfer

S. Bagwari (✉) · K. Choudhary · S. Raikwar · P. S. Rana · S. Mighlani
Thapar Institute of Engineering and Technology, Patiala, Punjab, India
e-mail: sbagwari_me20@thapar.edu

K. Choudhary
e-mail: kchoudhary_me20@thapar.edu

S. Raikwar
e-mail: suresh.raikwar@thapar.edu

P. S. Rana
e-mail: prashant.rana@thapar.edu

S. Mighlani
e-mail: smiglani@thapar.edu

M. Tistarelli et al. (eds.), *Computer Vision and Machine Intelligence*,
Lecture Notes in Networks and Systems 586,
https://doi.org/10.1007/978-981-19-7867-8_17

1 Introduction

The painting is a form of the art. The painter imparts his style in paint a scene. This style is unique for each painter. However, redrawing an image in the style of different painters is a challenge. The neural style transfer (NST) is a machine learning-based approach to transfer the style of a painter into the content of a scene [1].

The NST combines two images (a style image and a content image) to produce stylized image (the content image drawn in the form of style image) [1]. The NST takes a content image and a style image (artwork by any well-known artist) as inputs and produces stylized image by blending both of the input images, as shown in Fig. 1. It can be observed from Fig. 1 that a NST model takes style image and content image as input to produce stylized image as output. The stylized image is redrawn content image according to the pattern of input style image.

The NST can be utilized in variety of applications such as scene understanding recognition and speech recognition using speech style transfer [1, 2]. The NST can be used in commercial and industrial applications like Prisma, maestro (an application for music style transfer), and clinical practices (e.g., design of a framework to remove appearance shift for ultrasound image segmentation) [3].

1.1 Motivation

The number of style transfer methods has increased exponentially in the last few years. There are many commercial apps available in the market that most people like to use in performing style transfer. Thus, it is essential to understand the evolution of the style transfer methods for further advancements. Thus, the works of style transfer (from a traditional to neural-based style transfer) have been studied to present a detailed study and analysis. This study presents a general strategy of the style transfer and analyzed the NST specifically. Finally, we eventually limited our study to the application of the NST in images. The MS-COCO dataset [4] is used to present the comparison of the different methods of the NST.

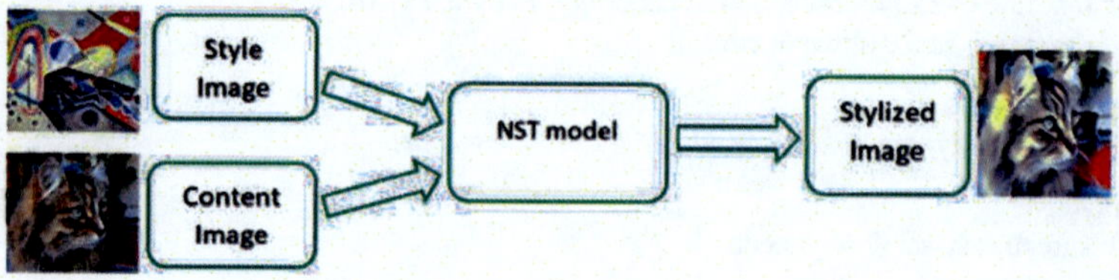

Fig. 1 The process to perform neural style transfer

1.2 Key Challenges

The use of neural networks in the field of NST has reduced the complexity in extracting feature for style transfer. But there are some challenges faced by the NST methods. The major key challenges faced by the methods of style transfer are listed below:

1. The production of natural effects in the style image is challenging due to different types of local impacts in the content image (like sky color in a natural image). While doing ST, geometric effect features should be avoided, like window grids should stay as grids and not be distorted.
2. The structure of the content image should be preserved after stylization. Real-world scenes have diversity in objects, such as in a cityscape, the appearance of buildings should mimic the appearance of other buildings, and the sky should match the sky.

The organization of this paper is as follows: Section 2 explained simple style transfer and the NST. Sect. 2.1 introduced the traditional methods of style transfer such as stroke-based rendering and region-based. In Sect. 2.2, a study of the methods in NST has been presented. Sect. 3 gives overview of the performance metrics to be used in the NST. The probable research gaps have been identified in Sect. 4. In Sect. 5, the key challenges and the issues have been explored. Finally, in Sect. 6, the paper has been concluded with a discussion on the future scope.

2 The Literature Survey

The different types of methods to perform the NST have been studied and to present a detailed literature review. The style transfer methods can be classified in two broad categories as follows:

- Simple Style Transfer
- Neural Style Transfer

2.1 Simple Style Transfer

The method in [5] is based on artistic rendering (AR), which is independent of neural networks and concentrates on the artistic stylization of images. The work in [5] has utilized artistic stylization and has a broad background. This work is significant because of its broad scope of application. Prior to the introduction of NST, non-photorealistic rendering is a field that grew out of research studies.

Stroke-Based Rendering (SBR) is a technique for reproducing an image using virtual strokes (e.g., stipples, tiles, and brush strokes) on a digital canvas [6]. The

SBR works by starting with a source image, gradually blending strokes to mimic the shot, and generating non-photorealistic graphics. These graphics resemble the original content image but have an artistic style. These types of techniques have an objective function to be optimized. The SBR method has been designed to reflect predefined aesthetics as accurately as possible. As a result, they are often good at imitating specific styles (e.g., sketches, watercolors, oil paintings.). However, each SBR method is specifically built for a single style and is not capable of replicating any other style, making it inflexible.

Techniques Based on Region is a method based on region segmentation, which allows rendering adaptation using region in the content image. The geometry of region is used to guide the placement of strokes in initial region-based Image Based-AR (IB-AR) method [7, 8]. This method can create distinct stroke patterns in various semantic regions of an image. Song et al. [9] introduced a region-based IB-AR method for manipulating shapes in the context of creative styles. Their system provides reduced shape rendering effects by substituting regions with various canonical shapes. The region-based methods are capable of focusing on minute details present in the small local region. However, region-based methods have the same inflexibility/drawbacks as SBR, including the incapability in imitating the different type of styles.

Example-Based Rendering (EBR) is based on finding out how a pair of exemplars map together. Hertzmann et al. are the pioneers of such a type of IB-AR approach. They presented a framework based on image analogies [10]. Image analogies aim to generate a supervised connection between two input images and stylized images. The taring set of image analogies consists of pairs of unstylized input images and their stylized counterparts in a specific style. When testing on input image, the image analogy method learns the analogous transformation from the example training pair and produces analogous stylized outputs. Image analogy can be used in numerous ways, such as to learn how to arrange strokes in a portrait painting. Image analogies are often efficient for a wide range of artistic genres. However, in practice, the pairs of training data are rarely accessible.

Image Filtering and Processing is a process of producing an artistic image, which aims at visual simplification and abstraction. Some methods in image filtering and processing use filters to generate output images with cartoon-like effects. Method in [11, 12] uses bilateral filtering and difference of Gaussian filters [13]. Image filtering-based rendering methods are often simple to design and effectual in practice compared to IB-AR approaches. Image filtering-based methods have a limited range of styles.

In summary, the IB-AR methods are competent to illustrate specific mandated styles authentically. These methods often exhibit style diversity, flexibility, and effective image structure extractions.

2.2 *Neural Style Transfer*

The CNN-based methods proved to be crucial in the NST. Gatys et al. [2] attempted to recreate famous artwork styles using natural content images. In the CNN methods, the content image adapts the artwork of style image. The main idea is to iteratively improve stylized image. The NST-based methods can be categorized into five classes as follows:

NST with Patch The methods of Li and Wand [14] and Li and Wand's [15] are based on Markovian network, which increased the computational efficiency due to adversarial network.

A patch-based non-parametric technique is similar to [15]. Because of their patch-based design, [14] method surpasses Johnson et al. [16] and Ulyanov et al. [17] methods in preserving intelligible textures in images. However, both of these networks do not consider semantics. These methods underperform with non-textured styles (such as face pictures) and lack in capturing details/differences in brush strokes, which are essential visual elements.

Method of Li and Wand [15] is less potent due to unstable training process. However, it has been observed that GAN-based style transfer (such as multicontent GAN and anime sketches via auxiliary classifier gan [18, 19]) is better in comparison with other existing methods.

[20] proposed a method to transfer style of universal face photo-sketch without being limited by drawing styles. As a result, this method can produce texture, lighting information, and shading in the synthesized results. They are useful for face recognition in law enforcement situations. For patch representation, these methods plan to use more advanced deep neural network architectures. For universal-style transmission, it is also conceivable to combine complementary representation skills of various deep networks. These techniques may additionally be utilized in order to face sketch-to-photo conversion by swapping roles of the sketches and photos.

Moreover, the Dynamic ResBlock (DRB) GAN for artistic style transfer is presented by Xu et al. [21]. "Style codes" are modeled as standard parameters in this model for Dynamic ResBlocks, connecting both style transfer and encoding network to close the gap between collection and arbitrary style transfer in a single model.

This method has presented attention discriminative style information and mechanism fully utilized it in target-style images. This enhanced the model's capability to transfer artistic style. DRG-GAN model performed very well and created high-quality synthetic style images.

Mallika et al. [22] presented an embedding technique based on the NST method and destyling GAN for recreating the embedded image. This technique suffices for two primary purposes: executing destyling and steganography for styled images rendered by the NST method. The novel goal was destyling and stego image generation of a styled image from supervised image-to-image translation. However, the above-mentioned method could improve spatial content, scale, and color using fast ST.

NST with Text Zhang et al. [23] looked into a more generic manner of representing style/content information, assuming that all styles/contents have the same feature.

This technique is shown to be more generalized. However, this strategy is only applicable to Chinese and English text, having limited applicability.

The architecture presented by Zhu et al. [24] enables the possibility of style transfer between several styles and languages. Given the appropriate reference sets, it can generate text in any style-content combination. As a result, it is predicted to generalize effectively to new language styles and contents. However, testing shows that this model does not perform well for font and texture generation on a distinct IR without fine-tuning. Future research will focus on finer-grained deep similarity fusion for improved text ST results across languages.

Chen et al. [25] style transfer network is the first to link back to the traditional text mapping methods, providing a fresh outlook on NST. However, the auto-encoder could include semantic segmentation as an additional layer of supervision in the region decomposition, resulting in a more spectacular region-specific transfer. Incorporating the proposed approach into video and stereoscopic applications is also intriguing.

NST on Real Time Adaptive Convolution (AdaConv) is an extension proposed by Chandran et al. [26] that allows for the real-time transfer of both structural and statistical styles. In addition to style transfer, their method may easily be broadened to style-based image production and other task.

Xu et al. [27] proposed VTNet, which is a real-time temporally coherent stylized video generator that is edge-to-edge trained from effectively unlimited untagged video data. The stylizing and temporal prediction branches of VTNet transmit the style of a reference image toward the source video frames.

NST with VGG Network Based on whitening and coloring transforms, Yoo et al. [28] adjusted the wavelet transformation. During stylization, it permits features to keep their spatial and statistical aspects of the VGG feature space. However, eliminating the need for semantic labels should be correct for a perfect performance.

For target styles, Li et al. [29] develops an N-dimensional one-hot vector as a selection unit for style selection. They include a learnable light matrix that may be used to optimize feature mapping and help in reducing function with an objective.

NST with Image Feature Johnson et al. [16] and Ulyanov et al. [17] proposed methods. Both methods are based on the alike idea: producing a stylized output including a single forward pass and pre-training feed-forward style-specific network during testing. They are primarily a difference in network architecture, with Johnson et al. approach broadly following Radford et al. [36] network but with fractionally strided convolutions along with residual blocks, and Ulyanov et al. generator network being a multi-scale architecture. Shortly after [16, 17] Ulyanov et al. [37] found that implement normalization into each individual image instead of a collection of images (specifically Batch Normalization) results in a large advancement in stylization attribute. At the time that batch size is set to 1, instance normalization (IN), which is the same as batch normalization, is used to normalize a single image. IN's style transfer network converges faster than Batch Normalization (BN) and produces better visual results. Instance Normalization is a type of style normalization in that style of each content image is directly normalized toward the appropriate style [30], according to one interpretation. As a result, the objective is simpler to understand because the rest of the network cares about content loss.

Wang et al. [38] proposed a sequence-level feature sharing technique for long-term temporal consistency, as well as a dynamic inter-channel filter to enhance the stylization impact ourselves. Temporal consistency further can be used in conjunction with GAN to improve performance.

Although the Per-Style-Per-Model (PSPM) approaches may generate stylized images two orders of magnitude quicker than earlier image optimization-based NST techniques, each style image necessitates the training of distinct generative networks, which is inflexible and long-drawn-out. Numerous artworks (e.g., impressionist paintings) have identical paint strokes however varying primarily in their color palettes, probably redundant to train a different network for each of them. Multiple-Style-Per-Model Neural Methods (MSPM) is offered to increase PSPM's versatility by combining various styles into a single model. Dealing with such problems has commonly two approaches: (i) associating each style to a little range of parameters in a network ([39, 40] and (ii) employing a single network, such as PSPM, but with inputs for both style and content [29, 31]. In contrast, for multiple styles, Li et al. [29] and Zhang and Dana's [31] algorithms use the same trainable network lights on a single network. The model size problem has been addressed. However, there appears to be some interplay between distinct styles, which has a minor impact on the stylization quality.

In both NPR and NST, an aesthetic appraisal is a key concern. Many academics in the subject of NPR stress the importance of aesthetic judgment [5, 41–44], e.g., in [5], researchers proposes two phases to investigate such problem. These issues are progressively analytical and precisely the area of NST along with NPR matured, explained in [5] that researchers require some decisive criteria to evaluate advantages of their suggested strategy over the prior art, as well as a method to analyze the appropriateness of one methodology to a certain case. Most NPR and NST articles, on the other hand, evaluate their suggested approach using metrics generated or instinctive visual comparisons from numerous user studies [32, 45].

This technique successfully avoids distortion and achieves appropriate Photorealistic Style Transfers (PST) in a wide range of settings, including mimicking aesthetic edits, the time of day, season, and others, according to Luan and Paris [46]. This study proposes the first solution to the problem. However, some other breakthroughs and enhancements can be made.

The breakthrough image stylization method developed by Cheng et al. [47] includes an additional structure representation. Firstly, the depth map represents the global structure. Secondly, the image edges represent the local structure details. It perfectly describes the spatial distribution of all elements in an image and the formation of notable items. The method provides terrific visual effects, especially when processing images hypersensitive to structural deformation, such as images having many items possibly at various depths or notable items with distinct structures, as demonstrated by testing results. Even so, there is scope for improvement.

With the Adversarial distillation learning technique, Qia et al. [33] are able to produce clear images in a short time. Although the network has been well trained, it still has unrealistic artifacts and a high computational cost. The difficulty of simultaneously improving efficiency in PST and visual quality remains unsolved.

For style transfer between unpaired datasets, Li et al. [48] efficiently handle and preserve the features on essential salient locations. Two new losses are proposed to improve the overall image perception quality by optimizing the generator and saliency networks. Furthermore, tasks where the saliency objects are modified, such as dog2cat translation, are not suited for the proposed SDP-GAN.

Ling et al. [49] use the background to expressly formulate the visual style, which they then apply to the foreground. Despite its progress, the proposed method still has two notable drawbacks. To begin with, it is unclear why employing region-aware adaptive instance normalization exclusively in the encoder yields such a low gain. Second, the model will reduce the aesthetic contrast and attenuate the sharp foreground object in examples with sharp foreground objects and dark backgrounds.

The mismatch between human sense of stylization quality and the classic AST style loss was studied by Cheng et al. [50]. During training, the core cause of the problem was identified as the style-agnostic cluster of sample-wise losses. They used a novel style-balanced loss with style-aware normalization to obtain theoretical limitations for the style loss. In the future, more substantial limitations for the style loss could be derived toward enhancing style-aware normalization.

Because of superior image-object retention, Lin et al. [51] outperform rival systems in terms of achieving higher nighttime vehicle detection accuracy. However, its uni-modality is a disadvantage. They want to try explicitly encoding a random noise vector to the structure-aware latent vector in the future to achieve model diversity when executing a unitary image-to-image translation.

Virtusio et al. [52] proposed a single-style input style transfer strategy that focuses on an H-AI-encouraged architecture which involves human control across the stylization procedure, allowing for a variety of outputs. This method is inspired by the various perceptual qualities that can be discovered in a sole style image. For example, it could include a variety of different colors and textures. However, neural style transfer is focused on accelerating the process. On the other hand, these works may only discover a limited amount of styles.

Xiao et al. [53] approach can efficiently handle the issues of blurring, poor diversity, and distortion of the output images relative to other data augmentation methods. Despite the lack of realistic images, extensive experimental findings show that this strategy can still be effective. They expanded the field of deep learning's application possibilities for simulated images. However, the image quality created is too low, style transfer takes a long time, and the quantity of augmented data produced unavoidably brings out the CNN over-fitting issue.

3 Experimental Results

The experiment is performed using the tools like TensorFlow-GPU 1.15, CUDA 10.2, and Python 3.7.6 on NVIDIA GeForce GTX 166Ti GPU. The methods in [2, 15–17, 29–33] have been implemented to conduct the analysis of the existing methods of NST by using COCO dataset [4]. This dataset consists of more than 80,000 real and artistic images of size 256 * 256 or large.

Further, the PSNR (peak signal-to-noise ratio) and SSIM (Structural Similarity Index) are used as the evaluation metrics. Both of these metrics compare two images for their perceptual and structural measurements. The PSNR is calculated using Eq. (1)

$$\text{PSNR} = 10 * \log_{10}\left(\frac{R^2}{\text{MSE}}\right) \tag{1}$$

where R is the maximum possible intensity value in the image and MSE is the mean squared error.

Next, the SSIM measures the structural quality of image. The SSIM is computed using Eq. (2).

$$\text{SSIM}(x, y) = \frac{(2\mu_x\mu_y + c_1)(2\sigma + c_2)}{(\mu_x^2 + \mu_y^2 + c_2)(\sigma_x^2 + \sigma_y^2 + c_2)} \tag{2}$$

where μ_x is the average of x;μ_y is the average of y; σ_x^2 is the the variance of x; σ_y^2 is the variance of y; σ_{xy} is the covariance of x and y; and $c_1 = (k_1L)^2, c_2 = (k_2L)^2$ are two variables to stabilize the division with ak denominator. The computed values of the PSNR and SSIM by using the methods in [2, 15–17, 29–33] have been shown in Table 2.

The experiment has been conducted using different style images as presented in Table 1. However, the PSNR and SSIM values have been computed between the con-

Table 1 Style images: done by famous artists, available in public domain

S. No.	Name and year	Author
1	Divan Japonais, 1893	Henri de Toulouse-Lautrec
2	Edith with Striped Dress (1915)	Egon Schiele
3	Head of a Clown (1907)	Georges Rouault
4	Landscape at Saint-Remy (1889)	Vincent van Gogh
5	Portrait of Pablo Picasso (1912)	Juan Gris
6	Ritmo plastico del 14 luglio (1913)	Severini Gino
7	The Tor of Babel (1563)	Pieter Bruegel the Elder
8	Three Fishing Boats (1886)	Claude Monet
9	Trees in a Lane (1847)	John Ruskin
10	White Zig Zags (1922)	Wassily Kandinsky

Fig. 2 The style image used to obtain stylized images for each content image shown in Fig. 3a

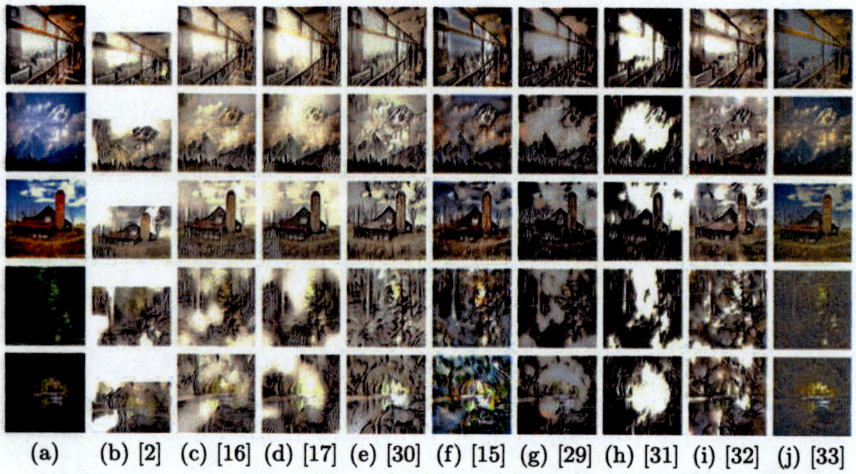

Fig. 3 Comparison between different models. One style image and different content images are taken for better visualization and understanding. Every model's produced stylized images

tent image (shown in Fig. 2a) and the stylized image (obtained by different methods in [2, 15–17, 29–33] by using style image, shown in Fig. 2) shown in Fig. 3b–j. The reason behind using the same style and content images is easily comparable (Table 2).

Figure 3a shows the content images. Each column in Fig. 3b–j presents stylized image obtained by methods in [2, 15–17, 29–33], respectively. In method [32], the stylized image has poor clarity and shapes, due to blending of minute details by VGG network. However, method in [15] generated visually pleasing stylized images. Methods [2, 16, 17] have generated much better stylized images due to texture analysis. Methods in [29, 31] have obtained dark stylized images. Further, method in [15] generated over-saturated stylized images, as shown in Fig. 3(i). The method in [33] is able to generate balanced and visually better stylized images, compared to other methods due to perceptual-aware distillation and pixel-aware distillation.

Table 2 Comparison of PSNR and SSIM between different models

S. No.	Method name	PSNR	SSIM
1	Gatys Style [2]	11.56602	0.28774
2	Johnson style [16]	11.90553	0.407876
3	Ulyanov style [17]	12.70022	0.542995
4	Huang style [30]	12.4078	0.309562
5	Li and Wand style [15]	12.25893	0.308762
6	Li Diverse style [29]	12.49095	0.383528
7	Zhang style [31]	12.47281	0.417546
8	Li Universal style [32]	11.46045	0.264825
9	Yingxu Style [33]	14.12685	0.684205

Table 3 Types of loss functions, utilized in different algorithms

Paper	Loss	Description
Li and Wand [15]	Adversarial loss	PatchGAN was used to calculate this result. Contextual correspondence is used to link patches together. In complex images, it is more effective in maintaining texture coherence
Johnson et al. [16]	Perceptual loss	The method of content loss based on perceptual similarity is widely used
Zhang and Dana [31]	Multi-scale gram loss	Gram loss is calculated using multi-scale characteristics. Getting rid of a few objects
Yingxu et al. [33]	Pixel-aware loss and perception-aware loss	The convenience of perceptual-aware distillation and pixel-aware distillation in image space in feature space are blended. The teacher network's sophisticated feature transforms may be trained using a simple fast-to-execute NN, that gives comparable results in a matter of seconds

Furthermore, the quantitative evaluation of the methods in [2, 15–17, 29–33] has been presented in Table 2. The method in [32] achieved lowest value of PSNR and SSIM as presented in Table 2. This happens due to blending of minute details by VGG network in [32]. However, method in [2] is better compared to method in [32] due to modified VGG network. The PSNR and SSIM have been improved by the method in [16] due to the use of perceptual loss in the training. The methods in [15, 17, 29–31] are based on texture analysis. Thus, they have achieved similar performance in terms of the PSNR and SSIM, as shown in Table 2. The method in [33] performed better compared to other methods due to perceptual-aware distillation and pixel-aware distillation in image space. This comparison infers that the methods based on perceptual-aware distillation and pixel-aware distillation have achieved state-of-the-art results.

Table 4 Timings of different models at different dimensions [33, 34]

Methods	Time (s)		
	256 × 256	512 × 512	1024 × 1024
Gatys et al. [2]	14.32	51.19	200.3
Li and Wand [15]	0.015	0.055	0.229
Johnson et al. [16]	0.014	0.045	0.166
Ulyanov et al. [17]	0.022	0.047	0.145
Zhang and Dana [31]	0.019	0.059	0.230
Li et al. [29]	0.017	0.064	0.254
Chen and Schmidt [35]	0.123	1.495	–
Li et al. [32]	0.620	1.139	2.947
Yingxu et al. [33]	0.060 (avg.)	0.1 (avg.)	0.256 (avg.)

Moreover, Table 3 presents different existing methods based on perceptual loss, adversarial loss, multi-scale gram loss, and pixel-aware loss. These methods have been proved to obtain pleasing results. In these, the perception-aware loss and pixel-aware loss-based methods are computationally fast compared to other methods, as presented in Table 4.

The computation time (in seconds) of methods in [2, 15–17, 29–33] has been presented in Table 4 using images with sizes $256 * 256$, $512 * 512$, and $1024 * 1024$. The computation time of the method in [2] is very high due to the use of VGG network. Methods in [15, 31] support statistics of storing encoded style, which speeds up the stylization process. The computation time of [31] is very high, and the GPU runs out of memory. The computation time of [15, 16, 29] is similar due to the utilization of same architecture by both of these methods. This comparison indicates that perceptual loss-based methods are computationally fast and produce visually pleasing results.

4 Research Gaps

The exhaustive evaluation and study of the presented literature identified the following research gaps.

- The style transfer in the textual images is challenges due to high representational variation in the languages [23].
- The stylized images suffer from unrealistic artifacts and heavy computational cost, due to erroneous estimation of the light parameters [24, 53].
- The training of the network with optimized loss is a challenging task due to the limited size of the available dataset [27, 50].
- The NST networks neglects the blurred features, which produce inaccurate stylized image [28, 49].

5 Challenges and Issues

Although existing methods are capable of performing effectively. But there are some issues and problems to be resolved in the NST. The NST faces challenges of NPR problems (as summarized in [5, 41–44]). Further, the following are specific issues in the NST.

1. Removal of semantic labels in stylized image is challenging for the flawless results [28].
2. The structure perseverance in the stylized image is a tedious task.
3. The evaluation of the NST methods is challenging due to the distinct structures of the content and stylized image.

6 Conclusion and Future Scope

The style transfer has been used in many applications such as scene understanding recognition and speech recognition using speech style transfer. The NST-based methods proved to be effective and style transfer. This research explores various methods of the NST, their pros/cons, and qualitative and quantitative evaluation. Comparing different NST methods shows that the perceptual loss-based NST methods are better compared to other methods. Moreover, the use of different loss functions has been explored in this study. The future work can be based on design of a more advanced deep neural architecture to obtain the stylized image efficiently.

References

1. Jing, Y., Yang, Y., Feng, Z., Ye, J., Yizhou, Y., Song, M.: Neural style transfer: a review. IEEE Trans. Vis. Comput. Graph. **26**(11), 3365–3385 (2020)
2. Gatys, L.A., Ecker, A.S., Bethge, M.: A neural algorithm of artistic style (2015)
3. Liu, Z., Yang, X., Gao, R., Liu, S., Dou, H., He, S., Huang, Y., Huang, Y., Luo, H., Zhang, Y., Xiong, Y., Ni, D.: Remove appearance shift for ultrasound image segmentation via fast and universal style transfer. In: 2020 IEEE 17th International Symposium on Biomedical Imaging (ISBI), pp. 1824–1828 (2020)
4. Lin, T.-Y., Maire, M., Belongie, S., Bourdev, L., Girshick, R., Hays, J., Perona, P., Ramanan, D., Lawrence Zitnick, C., Dollár, P.: Microsoft coco: common objects in context (2014)
5. Kyprianidis, J.E., Collomosse, J., Wang, T., Isenberg, T.: State of the "art": a taxonomy of artistic stylization techniques for images and video. IEEE Trans. Vis. Comput. Graph. **19**(5), 866–885 (2013)
6. Hertzmann, A.: Painterly rendering with curved brush strokes of multiple sizes. In: Proceedings of the 25th Annual Conference on Computer Graphics and Interactive Techniques, SIGGRAPH '98, pp. 453–460. Association for Computing Machinery, New York, NY, USA (1998)
7. Kolliopoulos, A.: Image segmentation for stylized non-photorealistic rendering and animation (2005)

8. Gooch, B., Coombe, G., Shirley, P.: Artistic vision: painterly rendering using computer vision techniques. In: NPAR Symposium on Non-photorealistic Animation and Rendering (2003)
9. Song, Y.-Z., Rosin, P.L., Hall, P.M., Collomosse, J.: Arty shapes. In: Proceedings of the Fourth Eurographics Conference on Computational Aesthetics in Graphics, Visualization and Imaging, Computational Aesthetics'08, pp. 65–72. Eurographics Association, Goslar, DEU (2008)
10. Hertzmann, A., Jacobs, C.E., Oliver, N., Curless, B., Salesin, D.H.: Image analogies. In: Proceedings of the 28th Annual Conference on Computer Graphics and Interactive Techniques, SIGGRAPH '01, pp. 327–340. Association for Computing Machinery, New York, NY, USA (2001)
11. Winnemoeller, H., Olsen, S., Gooch, B.: Real-time video abstraction. ACM Trans. Graph. **25**, 1221–1226 (2006)
12. Tomasi, C., Manduchi, R.: Bilateral filtering for gray and color images. In: Sixth International Conference on Computer Vision (IEEE Cat. No.98CH36271), pp. 839–846 (1998)
13. Gooch, B., Reinhard, E., Gooch, A.: Human facial illustrations: creation and psychophysical evaluation. ACM Trans. Graph. **23**(1), 27–44 (2004). Jan
14. Li, C., Wand, M.: Combining Markov random fields and convolutional neural networks for image synthesis. In: Proceedings of the IEEE Conference on Computer Vision and Pattern Recognition (CVPR) (2016)
15. Li, C., Wand, M.: Precomputed real-time texture synthesis with Markovian generative adversarial networks. In: European Conference on Computer Vision, pp. 702–716. Springer (2016)
16. Johnson, J., Alahi, A., Fei-Fei, L.: Perceptual losses for real-time style transfer and super-resolution. In: Leibe, B., Matas, J., Sebe, N., Welling, M. (eds.) Computer Vision—ECCV 2016, pp. 694–711. Springer International Publishing, Cham (2016)
17. Ulyanov, D., Lebedev, V., Vedaldi, A., Lempitsky, V.S.: Texture networks: feed-forward synthesis of textures and stylized images. In: ICML vol. 1, p. 4 (2016)
18. Azadi, S., Fisher, M., Kim, V., Wang, Z., Shechtman, E., Darrell, T.: Multi-content GAN for few-shot font style transfer (2017)
19. Zhang, L., Ji, Y., Lin, X.: Style transfer for anime sketches with enhanced residual u-net and auxiliary classifier GAN (2017)
20. Peng, Chunlei, Wang, Nannan, Li, Jie, Gao, Xinbo: Universal face photo-sketch style transfer via multiview domain translation. IEEE Trans. Image Process. **29**, 8519–8534 (2020)
21. Xu, W., Long, C., Wang, R., Wang, G.: DRB-GAN: a dynamic ResBlock generative adversarial network for artistic style transfer. In: Proceedings of the IEEE/CVF International Conference on Computer Vision (ICCV), pp. 6383–6392 (2021)
22. Mallika, Ubhi, J.S., Aggarwal, A.K.: Neural style transfer for image within images and conditional GANs for destylization. J. Vis. Commun. Image Representation **85**, 103483 (2022)
23. Zhang, Y., Zhang, Y., Cai, W.: A unified framework for generalizable style transfer: style and content separation. IEEE Trans. Image Process. **29**, 4085–4098 (2020)
24. Zhu, A., Lu, X., Bai, X., Uchida, S., Iwana, B.K., Xiong, S.: Few-shot text style transfer via deep feature similarity. IEEE Trans. Image Process. **29**, 6932–6946 (2020)
25. Dongdong Chen, L., Yuan, J.L., Nenghai, Y., Hua, G.: Explicit filterbank learning for neural image style transfer and image processing. IEEE Trans. Pattern Anal. Mach. Intell. **43**(7), 2373–2387 (2021)
26. Chandran, P., Zoss, G., Gotardo, P., Gross, M., Bradley, D.: Adaptive convolutions for structure-aware style transfer. In: Proceedings of the IEEE/CVF Conference on Computer Vision and Pattern Recognition (CVPR), pp. 7972–7981 (2021)
27. Kai, X., Wen, L., Li, G., Qi, H., Bo, L., Huang, Q.: Learning self-supervised space-time CNN for fast video style transfer. IEEE Trans. Image Process. **30**, 2501–2512 (2021)
28. Yoo, J., Uh, Y., Chun, S., Kang, B., Ha, J.-W.: Photorealistic style transfer via wavelet transforms. In: Proceedings of the IEEE/CVF International Conference on Computer Vision (ICCV) (2019)
29. Li, Y., Fang, C., Yang, J., Wang, Z., Lu, X., Yang, M.-H.: Diversified texture synthesis with feed-forward networks. In: Proceedings of the IEEE Conference on Computer Vision and Pattern Recognition, pp. 3920–3928 (2017)

30. Huang, X., Belongie, S.: Arbitrary style transfer in real-time with adaptive instance normalization. In: Proceedings of the IEEE International Conference on Computer Vision, pp. 1501–1510 (2017)
31. Zhang, H., Dana, K.: Multi-style generative network for real-time transfer. In: Proceedings of the European Conference on Computer Vision (ECCV) Workshops (2018)
32. Li, Y., Fang, C., Yang, J., Lu, X., Yang, M.-H.: Universal style transfer via feature transforms (2017)
33. Qiao, Y., Cui, J., Huang, F., Liu, H., Bao, C., Li, X.: Efficient style-corpus constrained learning for photorealistic style transfer. IEEE Trans. Image Process. **30**, 3154–3166 (2021)
34. Jing, Y., Yang, Y., Feng, Z., Ye, J., Yu, Y., Song, M.: Neural style transfer: a review. IEEE Trans. Vis. Comput. Graph. **26**(11), 3365–3385 (2020). Nov
35. Chen, T.Q., Schmidt, M.: Fast patch-based style transfer of arbitrary style. arXiv preprint arXiv:1612.04337 (2016)
36. Radford, A., Metz, L., Chintala, S.: Unsupervised representation learning with deep convolutional generative adversarial networks. arXiv preprint arXiv:1511.06434 (2015)
37. Ulyanov, D., Vedaldi, A., Lempitsky, V.: Improved texture networks: maximizing quality and diversity in feed-forward stylization and texture synthesis. In Proceedings of the IEEE Conference on Computer Vision and Pattern Recognition, pp. 6924–6932 (2017)
38. Wang, W., Yang, S., Jizheng, X., Liu, J.: Consistent video style transfer via relaxation and regularization. IEEE Trans. Image Process. **29**, 9125–9139 (2020)
39. Dumoulin, V., Shlens, J., Kudlur, M.: A learned representation for artistic style. CoRR, abs/1610.07629 (2016)
40. Chen, D., Yuan, L., Liao, J., Yu, N., Hua, G.: An explicit representation for neural image style transfer. Stylebank (2017)
41. Collomosse, J., Rosin, P.: Image and Video-Based Artistic Stylisation, vol. 42 (2013)
42. Gooch, A.A., Long, J., Ji, L., Estey, A., Gooch, B.S.: Viewing progress in non-photorealistic rendering through Heinlein's lens. In: Proceedings of the 8th International Symposium on Non-Photorealistic Animation and Rendering, NPAR '10, pp. 165–171. Association for Computing Machinery, New York, NY, USA (2010)
43. DeCarlo, D., Stone, M.: Visual explanations. In: Proceedings of the 8th International Symposium on Non-Photorealistic Animation and Rendering, NPAR '10, pp. 173–178. Association for Computing Machinery, New York, NY, USA (2010)
44. Hertzmann, A.: Non-photorealistic rendering and the science of art. In: Proceedings of the 8th International Symposium on Non-Photorealistic Animation and Rendering, NPAR '10, pp. 147–157. Association for Computing Machinery, New York, NY, USA (2010)
45. Mould, D.: Authorial subjective evaluation of non-photorealistic images. In: Proceedings of the Workshop on Non-Photorealistic Animation and Rendering, NPAR '14, pp. 49–56. Association for Computing Machinery, New York, NY, USA (2014)
46. Luan, F., Paris, S., Shechtman, E., Bala, K.: Deep photo style transfer. In: Proceedings of the IEEE Conference on Computer Vision and Pattern Recognition (CVPR) (2017)
47. Cheng, M.-M., Liu, X.-C., Wang, J., Shao-Ping, L., Lai, Y.-K., Rosin, P.L.: Structure-preserving neural style transfer. IEEE Trans. Image Process. **29**, 909–920 (2020)
48. Li, R., Chi-Hao, W., Liu, S., Wang, J., Wang, G., Liu, G., Zeng, B.: SDP-GAN: saliency detail preservation generative adversarial networks for high perceptual quality style transfer. IEEE Trans. Image Process. **30**, 374–385 (2021)
49. Ling, J., Xue, H., Song, L., Xie, R., Gu, X.: Region-aware adaptive instance normalization for image harmonization. In: Proceedings of the IEEE/CVF Conference on Computer Vision and Pattern Recognition (CVPR), pp. 9361–9370 (2021)
50. Cheng, J., Jaiswal, A., Wu, Y., Natarajan, P., Natarajan, P.: Style-aware normalized loss for improving arbitrary style transfer. In: Proceedings of the IEEE/CVF Conference on Computer Vision and Pattern Recognition (CVPR), pp. 134–143 (2021)
51. Lin, C.-T., Huang, S.-W., Yen-Yi, W., Lai, S.-H.: Gan-based day-to-night image style transfer for nighttime vehicle detection. IEEE Trans. Intell. Transp. Syst. **22**(2), 951–963 (2021)

52. Virtusio, J.J., Ople, J.J.M., Tan, D.S., Tanveer, M., Kumar, N., Hua, K.: Neural style palette: a multimodal and interactive style transfer from a single style image. IEEE Trans. Multimedia **23**, 2245–2258 (2021)
53. Xiao, Q., Liu, B., Li, Z., Ni, W., Yang, Z., Li, L.: Progressive data augmentation method for remote sensing ship image classification based on imaging simulation system and neural style transfer. IEEE J. Sel. Top. Appl. Earth Observations Remote Sens. **14**, 9176–9186 (2021)

A Black-Box Attack on Optical Character Recognition Systems

Samet Bayram and Kenneth Barner

Abstract Adversarial machine learning is an emerging area showing the vulnerability of deep learning models. Exploring attack methods to challenge state-of-the-art artificial intelligence (AI) models is an area of critical concern. The reliability and robustness of such AI models are one of the major concerns with an increasing number of effective adversarial attack methods. Classification tasks are a major vulnerable area for adversarial attacks. The majority of attack strategies are developed for colored or gray-scaled images. Consequently, adversarial attacks on binary image recognition systems have not been sufficiently studied. Binary images are simple—two possible pixel-valued signals with a single channel. The simplicity of binary images has a significant advantage compared to colored and gray-scaled images, namely computation efficiency. Moreover, most optical character recognition systems (OCRs), such as handwritten character recognition, plate number identification, and bank check recognition systems, use binary images or binarization in their processing steps. In this paper, we propose a simple yet efficient attack method, efficient combinatorial black-box adversarial attack (ECoBA), on binary image classifiers. We validate the efficiency of the attack technique on two different data sets and three classification networks, demonstrating its performance. Furthermore, we compare our proposed method with state-of-the-art methods regarding advantages and disadvantages as well as applicability.

Keywords Adversarial examples · Black-box attack · Binarization

S. Bayram (✉) · K. Barner
University of Delaware, Newark De 19716, USA
e-mail: sbayram@udel.edu

K. Barner
e-mail: barner@udel.edu

M. Tistarelli et al. (eds.), *Computer Vision and Machine Intelligence*,
Lecture Notes in Networks and Systems 586,
https://doi.org/10.1007/978-981-19-7867-8_18

1 Introduction

The existence of adversarial examples has drawn significant attention to the machine learning community. Showing the vulnerabilities of machine learning algorithms has opened critical research areas on the attack and robustness areas. Studies have shown that adversarial attacks are highly effective on many existing AI systems, especially on image classification tasks [1–3]. In recent years, a significant number of attack and defense algorithms were proposed for colored and gray-scaled images [4–10]. In contrast, AI binary image adversarial attacks and defenses are not well studied. Existing attack algorithms are inefficient or not well suited to binary image classifiers because of the binary nature of such images. We explain the inefficiency of existing attack methods under the Related Works section.

Binary image classification and recognition models are widely used in daily image processing tasks, such as license plate number recognizing, bank check processing, and fingerprint recognition systems. Critically, binarization is a pre-processing step for OCR systems, such as Tesseract [11]. The fundamental difference between binary and color/grayscale images, regarding generating adversarial examples, is their pixel value domains. Traditional color and grayscale attacks do not lend themselves to binary images because of their limited black/white pixel range. Specifically, color and grayscale images have a large range of pixel values, which allows crafting small perturbations to affect the desired (negative) classification result. Consequently, it is possible to generate imperceptive perturbations for color and grayscale images. However, in terms of perception, such results are much more challenging for binary images because there are only two options for the pixel values. Thus, a different approach is necessary to create attack methods for binary image classifiers. Moreover, the number of added, removed, or shifted pixels should be constrained to minimize the visual perception of attack perturbations.

In this study, we introduce a simple yet efficient attack method in black-box settings for binary image classification models. Black-box attack only requires access to the classifier's input and output information. The presented results show the efficiency and performance of the attack method on different data sets as well as on multiple binary image classification models.

1.1 Related Works

Szegedy et al. [3] show that even small perturbations in input testing images can significantly change the classification accuracy. Goodfellow et al. [4] attempt to explain the existence of adversarial examples and propose one of the first efficient attack algorithms in white-box settings. Madry et al. [12] proposed projected gradient descent (PGD) as a universal first-order adversarial attack. They stated that the network architecture and capacity play a big role in adversarial robustness. One extreme case of an adversarial attack was proposed by [13]. In their study, they only changed the value of a single-pixel of an input image to mislead the classifier. TramÃ¨r et al. [14] show the transferability of black-box attack among different ML models. Balkanski et al. [15] propose an attack method, referred to as scar, on binary image recognition systems. Scar resembles one of our perturbation models, namely additive perturbations. In this attack, it adds perturbation in the background of characters. Scar tries to hide the perturbations by placing them close to the character. However, this requires more perturbations to mislead the classifier.

Inefficiency of Previous Attack Methods Attacking the binary classifiers should not be a complex problem at first sight. The attack method can only generate white or black pixels. However, having only two possible pixel values narrow downs the attack ideas. State-of-the-art methods such as PGD or FGSM create small perturbations to make adversarial examples look like the original input image. Those attack methods are inefficient on binary images because the binarization process wipes the attack perturbations in the adversarial example before itâL™s fed to the binary image classifier. This method, binarizing the input image, is considered a simple defense method against state-of-the-art adversarial attacks. Wang et al. [16] proposed a defense method against adversarial attacks by binarizing the input image as a pre-processing step before the classification. They achieved 91.2% accuracy against white-box attacks on MNIST digits. To illustrate this phenomenon, we apply

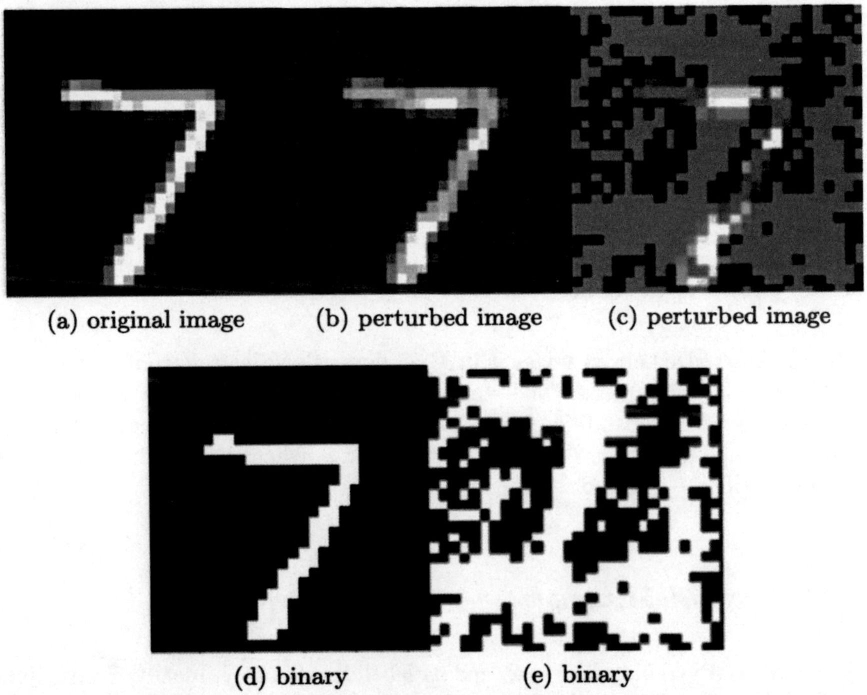

Fig. 1 Effect of binarization on adversarial examples created by the PGD method. Perturbations in **b** are smaller than the binarization threshold, and perturbations in **c** are more significant than the threshold. **d** and **e** are the binary versions of **b** and **c**, respectively

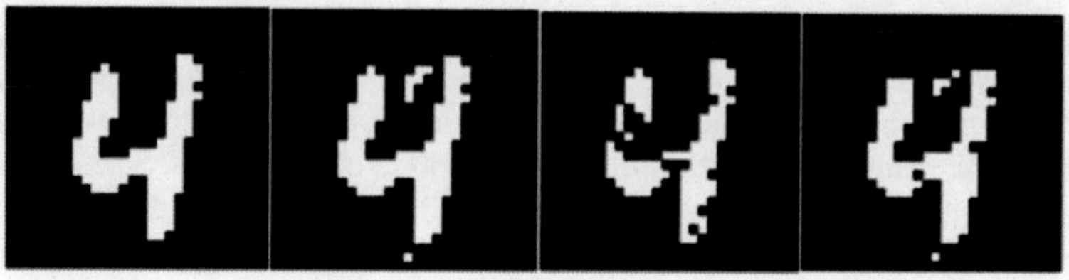

Fig. 2 From left to right: original binary image, adversarial examples after only additive perturbations, only erosive perturbations, and final adversarial example with the proposed method

PGD on a gray-scaled digit image whose ground truth label is seven. The PGD attack fools the gray-scaled digit classifier resulting output of three. However, after the binarization process of the same adversarial example, the perturbations generated by PGD are removed, and the image is classified as seven, as illustrated in Fig. 1. For this reason, the state-of-the-art methods that generate perturbations less than the binarization threshold is inefficient when the image is converted to binary form.

Adversarial perturbations created by the PGD attack method is disappeared when the perturbations are smaller than the binarization threshold. After the binarization process, the image is classified correctly for the case where perturbations are smaller than the threshold. On the other hand, adversarial perturbations can go through binarization when perturbations are bigger than the threshold. However, the final adversarial example contains an excessive amount of perturbations that ruin the main character (digit or letter), which is an unwanted situation for creating adversarial examples. Our proposed method produces a few white and black perturbations that can pass through the binarization process and misleads the classifiers. We show those perturbations in Fig. 2.

2 Problem Definition

Let $\boldsymbol{x}$ be a rasterized binary image with $d \times 1$ dimension. Each element of $\boldsymbol{x}$ is either 0 (black) or 1 (white). A trained multi-class binary image classifier F takes $\boldsymbol{x}$ as input and gives $\boldsymbol{n}$ probabilities for each class. The label with highest probability, y, is the predicted label of $\boldsymbol{x}$. Thus, $y = \text{argmax}_i F(\boldsymbol{x})_i$, where $F(\cdot)_i$ defines the binary image classifier, $\boldsymbol{i} \in \boldsymbol{n}$, and $\boldsymbol{n}$ is the number of classes.

2.1 *Adversarial Example*

An adversarial variation of $\boldsymbol{x}$ is $\tilde{\boldsymbol{x}}$, and its label denoted as $\tilde{y}$. Ideally, $\tilde{\boldsymbol{x}}$ resembles $\boldsymbol{x}$ as much as possible (metrically and/or perceptually), while $y \neq \tilde{y}$. The classical mini–max optimization is adopted in this setting. That is, we want to maximize the similarity between the original input and adversarial example while minimizing the confidence of the true label y. While minimizing the confidence is generally straightforward, hiding the perturbations in adversarial samples is challenging in the binary image case, especially in low (spatial) resolution images.

3 Proposed Method

Here, we propose a black-box adversarial attack on binary image classifiers. The ECoBA consists of two important components: additive perturbations and erosive perturbations. We separate perturbations into two categories to have full control over whether to apply the perturbations to the character. The additive perturbations occur in the character's background, while the erosive perturbations appear on the character. Since preventing the visibility of attack perturbations is impossible for the binary image case, it is important to damage the character as less as possible while fooling the classifier successfully. Since the images are binary, we assume, without loss of generality, that white pixels represent the characters in the image, and black pixels represent the background. Proposed attack algorithms change the pixel value based on the decline in classification accuracy. We define this change as adversarial error, ϵ_i, for the flipped ith pixel of input $\boldsymbol{x}$. For instance, $\boldsymbol{x} + \boldsymbol{w}_i$ means image $\boldsymbol{x}$ with ith pixel is flipped, from black to white. Thus, the adversarial example is $\tilde{\boldsymbol{x}} = \boldsymbol{x} + \boldsymbol{w}_i$, and the adversarial error is simply $\epsilon_i = \boldsymbol{x}_i - \tilde{\boldsymbol{x}}_i$, for the flipped ith pixel.

3.1 Additive Perturbations

To create additive perturbations, an image is scanned, flipping each background (black) pixel, one by one, in an exhaustive fashion. The performance of the classifier is recorded for each potential pixel flip, and the results are ordered and saved in a dictionary, D_{AP}, with the corresponding pixel index that causes the error. Pixels switched from black to white are denoted as $\boldsymbol{w}_i$, where i represents the pixel index. The procedure is repeated, with k indicating the number of flipped pixels, starting with the highest error in the dictionary, and continuing until the desired performance level or the number of flipped pixels is achieved. That is, notionally

$$\arg\min_i F(\boldsymbol{x} + \boldsymbol{w}_i) \text{ where } \|\tilde{\boldsymbol{x}} - \boldsymbol{x}\|_0 \leq k. \tag{1}$$

The confidence of the adversarial example applied to the classifier is recorded after each iteration. If $\epsilon_i > 0$, then the ith pixel index is saved. Otherwise, the procedure is repeated, skipping to the next pixel. The procedure is completed by considering each pixel in the image.

3.2 Erosive Perturbations

In contrast to additive perturbations, creating erosive perturbations is the mirror procedure. That is, pixels on the character (white pixels) are identified that cause the most significant adversarial error and thus flipped. Although previous works [15] utilize perturbing around or on the border of the character in an input image, erosive perturbations occur directly on (or within) the character. This approach can provide some advantages regarding the visibility of perturbations and maximizes the similarity between the original image and its adversarial example. Similarly, the sorted errors are saved in a dictionary, D_{EP}, with the corresponding pixel index that causes the error. Pixels flipped from white to black are denoted as $\boldsymbol{b}_i$. The optimization procedure identifies that the pixels that cause the most considerable decrease in confidence are flipped. That is, notionally

$$\arg\min_i F(\boldsymbol{x} + \boldsymbol{b}_i) \text{ where } \|\tilde{\boldsymbol{x}} - \boldsymbol{x}\|_0 \leq k. \tag{2}$$

3.3 ECoBA: Efficient Combinatorial Black-Box Adversarial Attack

The ECoBA can be considered as a combination, in concert, of both additive and erosive perturbations. The errors and corresponding pixel numbers are stored in D_{AP} and D_{EP}, merging them in a composite dictionary, D_{AEP}. For example, the top row of the D_{AEP} contains the highest ϵ values for $\boldsymbol{w}_i$ and $\boldsymbol{b}_i$. For $k = 1$, two pixels are flipped, corresponding to $\boldsymbol{w}_1$ and $\boldsymbol{b}_1$, resulting in no composite change in the number of black (or white) pixels. That is, there is no change in the L_0 norm. Accordingly, we utilize k as the iteration index, corresponding to the number of flipped pixel pairs and the number of perturbations.

The detailed steps of the proposed attack method are shown in Algorithm 1.

Algorithm 1 ECoBA

procedure ADV(x) ▷ Create adversarial example of input image x
  $\tilde{x} \leftarrow x$
  while $\arg\min_i F(x + w_i)$ where $\|\tilde{x} - x\|_0 \leq k$ **do**
    $w_i \leftarrow \arg\max_i F(\tilde{x})$
    $\epsilon_i \leftarrow F(x_i) - F(\tilde{x}_i)$
    $D_{AP'} \leftarrow w_i, \epsilon_i$ ▷ Dictionary with pixel index and its corresponding error
  $D_{AP} \leftarrow sort(D_{AP'})$ ▷ Sort the index of pixels starting from max error.
  while $\arg\min_i F(x + b_i)$ where $\|\tilde{x} - x\|_0 \leq k$ **do**
    $b_i \leftarrow \arg\max_i F(\tilde{x})$
    $\epsilon_i \leftarrow F(x_i) - F(\tilde{x}_i)$
    $D_{EP'} \leftarrow b_i, \epsilon_i$
  $D_{EP} \leftarrow sort(D_{EP'})$ ▷ Sort the index of pixels starting from max error.
  $D_{AEP} \leftarrow stack(D_{AP}, D_{EP})$ ▷ Merge dictionaries into one.
  $\tilde{x} \leftarrow x + D_{AEP_i}$ ▷ add perturbation couples from the merged dictionary
  return $\tilde{x}$

The amount of perturbations is controlled by k, which will be the step size in the simulations. Figure 2 shows an example of the input image and the effect of perturbations.

4 Simulations

We present simulations over two data sets and three different neural network-based classifiers in order to obtain comprehensive performance evaluations of the attack algorithms. Since the majority of optical characters involve with numbers and letters, we chose one data set for handwritten digits and another data set for handwritten letters.

4.1 Data Sets

Models were trained and tested on the handwritten digits MNIST [17] and letters EMNIST [18] data sets. Images in the data sets are normalized between 0 and 1 as grayscale images are binarized using a global thresholding method with the threshold of 0.5. Both data sets consist of 28×28 pixel images. MNIST and EMNIST have 70,000 and 145,000 examples, respectively. We use the split of 85%–15% of each data set for training and testing.

Table 1 Training performance of models

Top-1 training accuracy			
Data set	MLP-2	LENET	CNN
MNIST	0.97	0.99	0.99
EMNIST	0.91	0.941	0.96

4.2 Models

Three classifiers are employed for the training and testing. The simplest classifier, MLP-2, consists of only two fully connected layers with 128 and 64 nodes, respectively. The second classifier, a neural network architecture, is LeNet [19]. Finally, the third classifier is a two-layer convolutional neural network (CNN), with 16 and 32 convolution filters of kernel size 5×5. Training accuracies of each model on both data sets are shown in Table 1.

The highest training accuracy was obtained with the CNN classifier, then LeNet and MLP-2, respectively. Training accuracies for both data sets with all classifiers are high enough to evaluate with testing samples.

4.3 Results

We evaluate the results of the proposed attacking method on three different neural network architectures over two different data sets. Figure 3 shows the attack performance over images from MNIST and EMNIST data sets. Ten input images are selected among correctly classified samples for the attack. The Y-axis of plots represents the averaged classification accuracy of input images, while the X-axis represents the number of iterations (number of added, removed, or shifted pixels).

An observation of Fig. 3 shows that all approaches yield successful attacks, with the proposed method generating the most successful attacks in all cases. Moreover, the classifier results on the adversarial examples yield very high confidence levels. The attack perturbations are applied even after the classifier gives the wrong label as a classification result to observe the attack strength. For instance, obtained average step size of ECoBA for misleading the MLP-2 classifier on the digit data set is six. This means that changing the six pixels of the input image was enough to mislead the classifier. The averaged confidence level of the ground truth labels drops to zero when the attack perturbations are intensified on MLP-2. On the other hand, the proposed method generated more perturbations to mislead CNN classifier. We show the average step sizes for a successful attack for different attack types with respect to data sets in Table 2.

Another important outcome of the simulations is an observation of interpolations between classes, as reported earlier in [20]. As we increase the number of iterations,

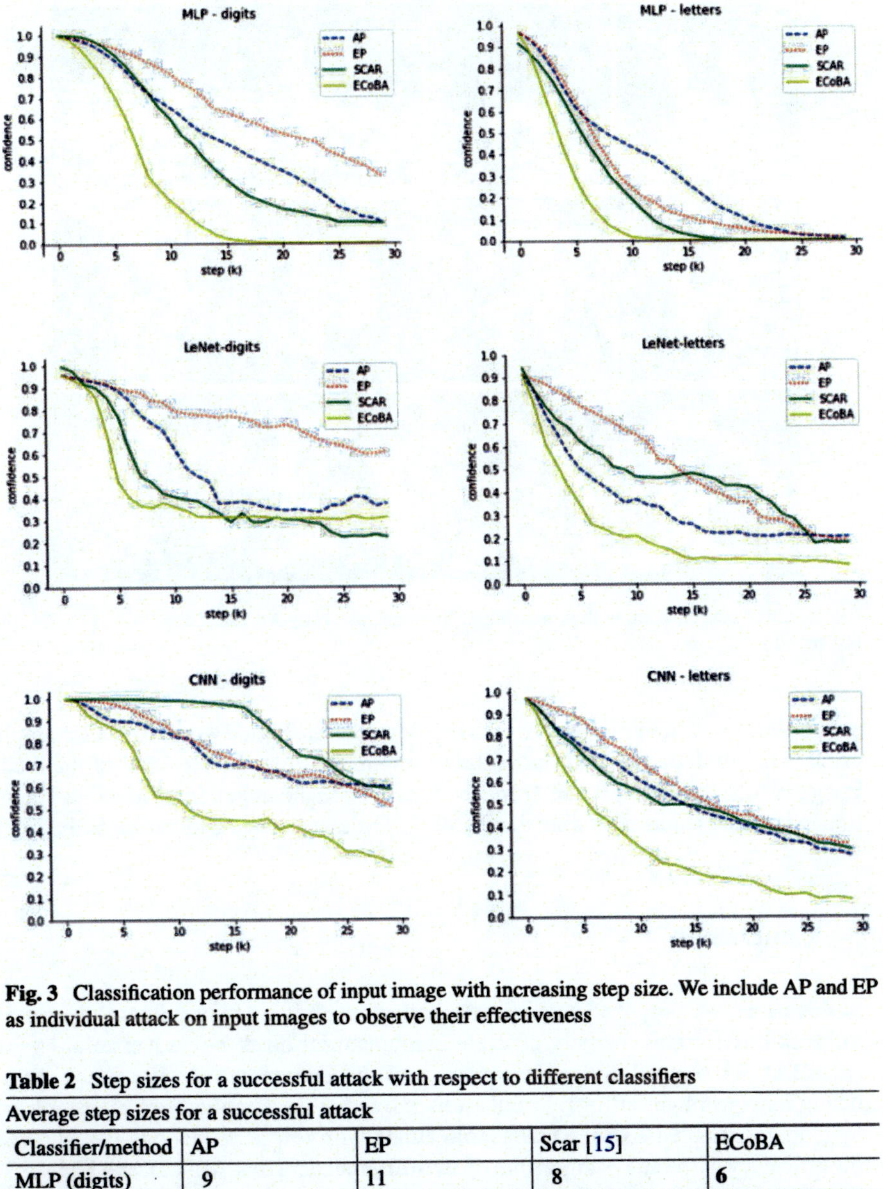

Fig. 3 Classification performance of input image with increasing step size. We include AP and EP as individual attack on input images to observe their effectiveness

Table 2 Step sizes for a successful attack with respect to different classifiers

Average step sizes for a successful attack

Classifier/method	AP	EP	Scar [15]	ECoBA
MLP (digits)	9	11	8	**6**
MLP (letters)	9	9	7	**5**
LeNet (digits)	10	18	8	**6**
LeNet (letters)	8	12	10	**5**
CNN (digits)	12	13	17	**8**
CNN (letters)	9	11	9	**7**

Bold numbers represents the smallest step size of attack method

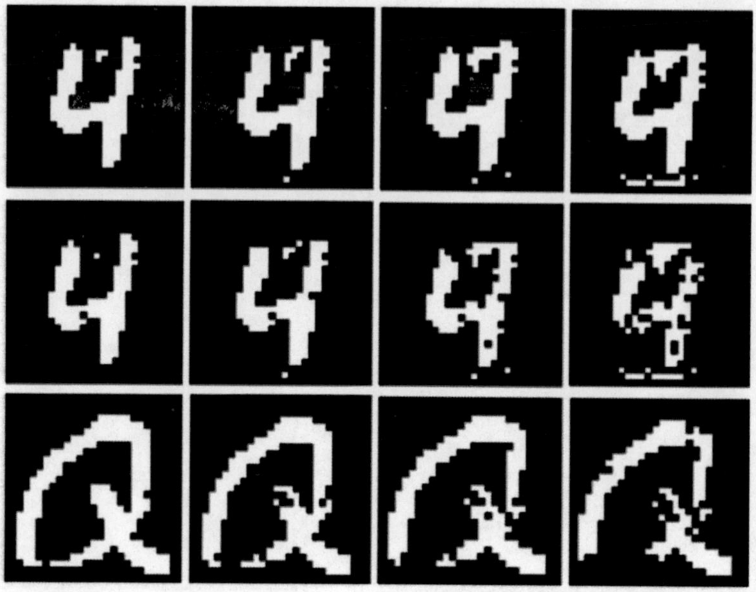

Fig. 4 Class interpolation with increasing k. The first row: only AP, the second row: ECoBA, the last row: EP

the original input image interpolates to the closest class. Figure 4 provides an example of class interpolation. In this particular example, the ground truth label of the input image is four for the digit and Q for the letter. Once the attack intensifies, the input image is classified as digit nine and letter P, while the image evolves visually.

5 Conclusion

In this paper, we proposed an adversarial attack method on binary image classifiers in black-box settings, namely efficient combinatorial black-box adversarial attack (ECoBA). We showed the inefficiency of most benchmark adversarial attack methods in binary image settings. Simulations show that the simplicity of the proposed method has enabled a strong adversarial attack with few perturbations. We showed the efficiency of the attack algorithm on two different data sets, MNIST and EMNIST. Simulations utilizing the MLP-2, LENET, and CNN networks show that even a small number of perturbations are enough to mislead classifiers with very high confidence.

References

1. Dalvi, N., Domingos, P., Mausam, Sanghai, S., Verma, D.: Adversarial classification. In: Proceedings of the Tenth ACM SIGKDD International Conference on Knowledge Discovery and Data Mining, KDD '04, pp. 99–108. Association for Computing Machinery, New York, NY, USA (2004). ISBN 1581138881. https://doi.org/10.1145/1014052.1014066
2. Biggio, B., Roli, F.: Wild patterns: ten years after the rise of adversarial machine learning. Pattern Recogn. **84**, 317–331 (2018.) ISSN 0031-3203. https://doi.org/10.1016/j.patcog.2018.07.023
3. Szegedy, C., Zaremba, W., Sutskever, I., Bruna, J., Goodfellow, I., Fergus, R.: Intriguing properties of neural networks. Dumitru Erhan (2014)
4. Goodfellow, I., Shlens, J., Szegedy, C.: Explaining and harnessing adversarial examples. In: International Conference on Learning Representations (2015). https://doi.org/10.48550/arXiv.1412.6572
5. Carlini, N., Wagner, D.: Towards evaluating the robustness of neural networks. In: 2017 IEEE Symposium on Security and Privacy (SP), pp. 39–57 (2017). https://doi.org/10.1109/SP.2017.49
6. Nguyen, A., Yosinski, J., Clune, J.: High confidence predictions for unrecognizable images: deep neural networks are easily fooled (2015)
7. Moosavi-Dezfooli, S.-M., Fawzi, A., Frossard, P.: DeepFool: a simple and accurate method to fool deep neural networks (2016)
8. Sharif, M., Bhagavatula, S., Bauer, L., Reiter, M.K.: Accessorize to a crime: real and stealthy attacks on state-of-the-art face recognition. In: Proceedings of the 2016 ACM SIGSAC Conference on Computer and Communications Security, CCS '16, pp. 1528–1540. Association for Computing Machinery, New York, NY, USA (2016.) ISBN 9781450341394. https://doi.org/10.1145/2976749.2978392
9. Kurakin, A., Goodfellow, I.J., Bengio, S.: Adversarial examples in the physical world. CoRR (2016). https://doi.org/10.48550/arXiv.1607.02533
10. Eykholt, K., Evtimov, I., Fernandes, E., Li, B., Rahmati, A., Xiao, C., Prakash, A., Kohno, T., Song, D.: Robust physical-world attacks on deep learning models (2018)
11. Smith, R.: An overview of the tesseract OCR engine. In: Proceedings of the Ninth International Conference on Document Analysis and Recognition (ICDAR), pp. 629–633 (2007)
12. Madry, A., Makelov, A., Schmidt, L., Tsipras, D., Vladu, A.: Towards deep learning models resistant to adversarial attacks (2019)
13. Su, J., Vargas, D.V., Sakurai, K.: One pixel attack for fooling deep neural networks. IEEE Trans. Evol. Comput. **23**(5), 828–841 (2019). ISSN 1941-0026. https://doi.org/10.1109/TEVC.2019.2890858
14. Tramér, F., Papernot, N., Goodfellow, I., Boneh, D., McDaniel, P.: The space of transferable adversarial examples (2017)
15. Balkanski, E., Chase, H., Oshiba, K., Rilee, A., Singer, Y., Wang, R.: Adversarial attacks on binary image recognition systems (2020)
16. Wang, Y., Zhang, W., Shen, T., Hui, Y., Wang, F.-Y.: Binary thresholding defense against adversarial attacks. Neurocomputing **445**, 61–71 (2021). https://doi.org/10.1016/j.neucom.2021.03.036
17. LeCun, Y., Cortes, C., Burges, C.J.: MNIST handwritten digit database. ATT Labs [Online]. Available: http://yann.lecun.com/exdb/mnist, 2 (2010)
18. Cohen, G., Afshar, S., Tapson, J., van Schaik, A.: EMNIST: an extension of MNIST to handwritten letters. arXiv preprint arXiv:1702.05373 (2017)
19. Lecun, Y., Bottou, L., Bengio, Y., Haffner, P.: Gradient-based learning applied to document recognition. Proc. IEEE **86**(11), 2278–2324 (1998). https://doi.org/10.1109/5.726791
20. Tsipras, D., Santurkar, S., Engstrom, L., Turner, A., Madry, A.: Robustness may be at odds with accuracy (2019)

Segmentation of Bone Tissue from CT Images

Shrish Kumar Singhal, Bibek Goswami, Yuji Iwahori, M. K. Bhuyan, Akira Ouchi, and Yasuhiro Shimizu

Abstract Segmentation of the bone structures in computed tomography (CT) is crucial for research as it plays a substantial role in surgical planning, disease diagnosis, identification of organs and tissues, and analysis of fractures and bone densities. Manual segmentation of bones could be tedious and not suggested as there could be human bias present. In this paper, we evaluate some existing approaches for bone segmentation and present a method for segmenting bone tissues from CT images. In this approach, the CT image is first enhanced to remove the artifacts surrounding the bone. Subsequently, the image is binarized and outliers are removed to get the bone regions. The proposed method has a Dice index of 0.9321, Jaccard index (IoU) of 0.8729, a precision of 0.9004, and a recall of 0.9662.

Keywords Bone segmentation · computed tomography (CT) · Image processing

1 Introduction

The human skeletal structure, which forms the basic framework, consists of various individual bones and cartilages distributed throughout the body, having varying characteristics such as size, shape, and composition depending on the individual. Simple wear and tears in bones, apparent anomalies in bone positioning, and inflammations of bones can be and have been detected over the years using X-rays. Conventional X-ray images do not help with diminutive changes in the bone structure. Therefore,

S. K. Singhal (✉) · B. Goswami · Y. Iwahori · M. K. Bhuyan
Department of Electronics and Electrical Engineering, Indian Institute of Technology Guwahati, Guwahati 781039, India
e-mail: shrishkumar@iitg.ac.in

Y. Iwahori · M. K. Bhuyan
Department of Computer Science, Chubu University, Kasugai 487-8501, Japan

A. Ouchi · Y. Shimizu
Department of Gastroenterological Surgery, Aichi Cancer Center Hospital, Nagoya 464-8681, Japan

M. Tistarelli et al. (eds.), *Computer Vision and Machine Intelligence*,
Lecture Notes in Networks and Systems 586,
https://doi.org/10.1007/978-981-19-7867-8_19

computed tomography is being widely used to detect those minuscule changes leading to the diagnosing of various diseases and medical conditions which sometimes go undiagnosed with conventional clinical diagnostic methods and tools.

CT scan incorporates a sequence of X-rays taken at different angles around the body and operates on computer processing to produce cross-sectional images (slices) of the bones and blood vessels, enhancing the delicate tissue visibility. With CT images, we can pinpoint tumor locations, blood clots, and infections, detect internal bleeding, and measure the effectiveness of specific treatments.

Manual or semi-automatic bone segmentation has the disadvantage of being time-consuming, leading to delays, and therefore is infrequently used in clinical practice. Hence, in this paper, we present a reliable and entirely automatic bone segmentation method in whole-body CT images of patients.

2 Literature Review

In image analysis, the process of isolation of object of interest from the background is termed segmentation. The paper suggested the demarcation of the bone tissue from its surrounding from a compendium of CT images. Thresholding-based techniques are commonly used in image segmentation [1]. Bone tissue can usually be separated from the soft tissue to a certain extent using thresholding-based methods since it is of high-intensity levels in CT images, especially the outer cortical tissue. A semi-automatic and manual method to segment bone interactively was proposed by Tomazevic et al. [2]. Tanssani et al. [3] segmented fractured bone tissue using a global fixed threshold value. However, intensity variation over the slices poses a problem in identifying such a threshold. A threshold-based segmentation technique has also been proposed to implement binarization of CT images having long humerus bones [4]. Recurrently, authors use two-dimensional [5, 6] and three-dimensional [7, 8] region growing techniques in addition to different variations of thresholding-based segmentation to segment and label bones. A multi-region growing approach, where the entire image was scanned and several pixels with intensity above a specified threshold were identified as seed points was proposed by Lee et al. [9]. Bone regions were extracted using a constrained 3D region growing algorithm by Huang et al. [10]. However, a pertinent threshold value is necessary to be recognized via the time-consuming and expensive iteration process for every slice to avoid over-segmentation. Pérez-Carrasco et al. used successive max-flow optimization as well as histogram-based energy diminishment [11]. Klein et al. used a deep convolutional neural network influenced by U-Net [12]. Zhang et al. used a DFM-Net which applied Dense Block as the basic module to extract features and amplify the feature transfer [13]. Xiong et al. proposed a deep convolutional neural network (CNN) approach which combined localization for the automated segmentation of pelvis in computed tomography (CT) scans [14]. Moreover, deformable models [15, 16], probabilistic watershed transform [17], registration-based models [18] have also been applied to fragment the bone pieces from CT images.

To achieve the desired output, most of the methods described above require a lot of manual intervention, such as providing the seed points or threshold value. The design and development of the proposed algorithm are described in particular in the consecutive sections.

3 Dataset

We evaluated our method on the dataset used by Peréz-Carrasco [11]. The data used in this work and the ground truth associated with it have been made publicly available [19]. The dataset utilized in the experiments contained 27 CT volumes (as MATLAB files) resembling different body regions from 20 patients. All the CT volumes were manually annotated in harmony by two experts and were taken as ground truth. Only the bone tissue segments are considered in this work. The patients' age was from 16 to 93, with an average age of 50. The number of male patients was 8, and that of females was 12. Informed consent was obtained from all the individuals participating in the study. Helical CT by Philips Medical Systems was used for the acquisition device, with a slice size of 512 $\times$512 pixels at 0.781 mm/pixel and a slice thickness of 5mm. In all the experiments, a conventional PC (Intel ® Core^{TM} i7-10510U, CPU @ 2.30 GHz, 16GB RAM) was used.

4 Proposed Methodology

The proposed algorithm consists of first enhancing the input image using an image processing technique contrast stretching; secondly, segmentation is carried out to extract the bone tissues using another image processing technique known as region growing; and finally, removing the outliers present in the segmented image. The flow diagram of the proposed algorithm is shown in Fig. 1 and would be discussed in detail in this section.

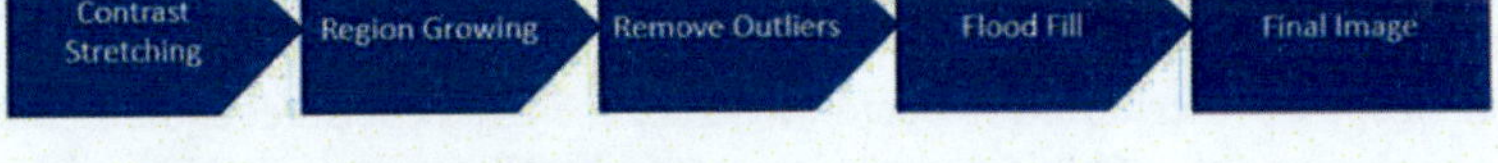

Fig. 1 Flow diagram of proposed methodology

4.1 Contrast Stretching

A large portion of the CT image is covered by the soft tissue surrounding the bone. It is vital to develop a method for removing undesired artifacts from an image without compromising the bone tissue to develop an accurate bone detection and segmentation system. For this purpose, we have implemented contrast stretching which also enhances the bone tissue, making segmentation in subsequent steps more efficient.

When a histogram is plotted after converting the image into grayscale, it is observed that the CT images have a low contrast as most of the pixels fall under the small range of 70–150 pixels [20]. Contrast stretching is thus performed to enhance the image, the gray level scope of the input image is stretched to fit the complete dynamic range of 0–255 pixels using the formula in Eq. 1.

$$
\begin{aligned}
I_{\text{out}} &= (I_{\text{in}} - i_{\text{il}}) * \left(\frac{\text{Expected Range}}{\text{Actual Range}}\right) \\
\text{Expected Range} &= i_{\text{ih}} - i_{\text{il}} \\
\text{ActualRange} &= i_{\text{oh}} - i_{\text{ol}}
\end{aligned}
\tag{1}
$$

where I_{out} is the output image, I_{in} is the input image, i_{il} denotes the input low intensity, i_{ih} denotes the input high intensity, i_{ol} denotes desired output low intensity (0 in this case), and i_{oh} denotes desired output high intensity (255 in this case).

The actual low and high-intensity values of 15 and 150 are stretched to 0–255 as desired low and high-intensity values, respectively, to cover the entire dynamic range. A wide margin is taken over the narrow range for bones 70–150 pixels to get close to 100% sensitivity at the cost of a few false positives as shown in Fig. 2(b). Pixels with intensity below 15 are set to 0 and above 150 are set to 255. When the input low intensity is increased to 30, it is observed that artifacts surrounding the bone are rejected at the cost of removal of bone tissue as well.

4.2 Region Growing

Region growing is a simple pixel-based image segmentation method frequently used in image segmentation. The adjacent pixels of initial seed points are reviewed against a specified criterion to decide whether the adjacent pixel should be counted in the region or not, and this process is repeated until convergence. The algorithm is described as follows:

- An arbitrary seed pixel is chosen, and its neighbors are compared.
- The neighboring pixel is added to the region in case it is found to be similar.
- Another seed pixel that has not yet been categorized into any region is chosen when the growth of one region stops and the process starts again.
- This whole process continues until all pixels belong to some region.

The initial seed points are selected on some user-defined criterion. Depending on the criteria defined, the regions are then grown from these seed points to nearby points. For instance, the criterion may be color, grayscale texture, or pixel intensity based on the application and problem being addressed.

In this work, we started with 10 random seed points in the image obtained after contrast stretching and we limit the variance to 4 for region growing. The result is shown in Fig. 2(c).

4.3 *Outlier Removal*

A lot of noise is observed in images obtained after the region growing algorithm is applied. To get rid of it without compromising on the details inside the bone tissue, we developed an ingenious algorithm as follows:

- Generate limits: For each row in the image, we determine the leftmost and the rightmost bright pixel to determine the limits within which the bone region lies. Similarly, we determine the topmost and the bottom-most bright pixel for each column.
- For each column, upward-facing square brackets are formed of varying sizes centered at the topmost white pixel. If all the pixels on the edges of all the brackets are found to be "dark", all pixels inside the largest bracket are set to "dark".

This is carried out from all directions and iterated until convergence. The result is shown in Fig. 2(d).

5 Performance Metrics

The following metrics were computed to compare the different methods: Dice index, Jaccard index, and sensitivity. These metrics measure agreement between the ground truth and the prediction of false-positive, false-negative, true-negative, and true-positive counts. In our approach, true positives (TP) correspond to pixels correctly labeled as belonging to the category under analysis (bone). False positives (FP) are pixels that have been incorrectly identified as belonging to the category under investigation. True negatives (TN) are pixels that are correctly classified as not belonging to the relevant category, while false negatives (FN) are pixels that are incorrectly classified.

(a) Original Image

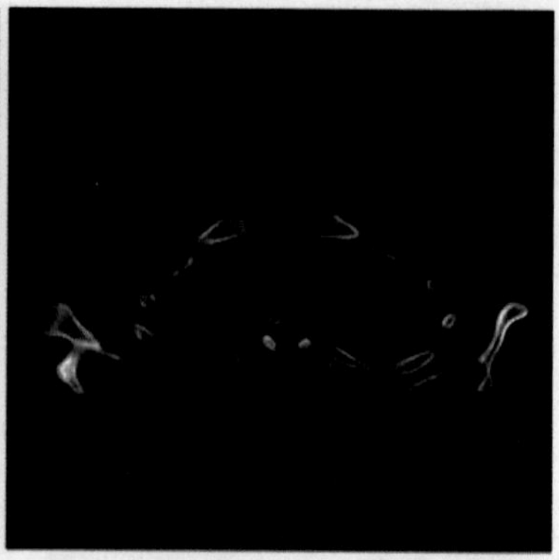

(b) Contrast Stretched Image

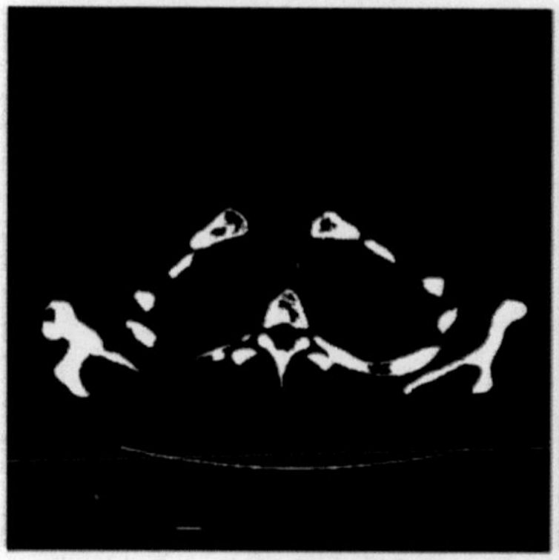

(c) Region Growing Applied Image

(d) Final Output

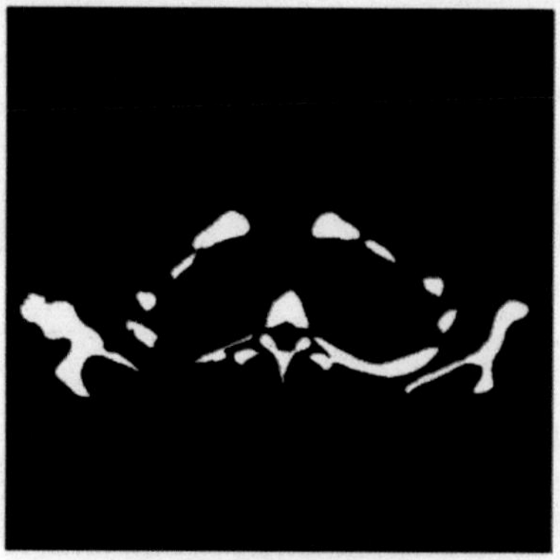

(e) Ground Truth

Fig. 2 Output images at different stage of the algorithm

The different parameters were computed as follows:

$$\text{Dice Index} = \frac{(2 * \text{TP})}{2 * \text{TP} + \text{FP} + \text{FN}}$$
$$\text{Jaccard Index} = \frac{\text{TP}}{\text{TP} + \text{FP} + \text{FN}}$$
$$\text{Sensitivity} = \frac{\text{TP}}{\text{TP} + \text{FN}}$$
$$\text{PPV} = \frac{\text{TP}}{\text{TP} + \text{FP}}. \quad (2)$$

The Dice and Jaccard coefficients are used to compare the similarity of the two segments. Positive Predictive Value (PPV) or Precision computes the likelihood that a voxel categorized as positive in a specific category truly belongs to that category, whereas sensitivity or recall computes the fraction of true positives accurately recognized by the algorithm.

6 Result and Inference

Results obtained had clear demarcation of the bone tissue in CT images when the described approach was applied to the dataset. Figure 2(a) shows the original CT image. Figure 2(b) shows the image obtained when contrast stretching was applied. Figure 2(c) shows the outcome obtained by applying region growing to this image. On comparing our results with the image labeled by the experts as shown in Fig. 2(e), it is evident that our approach provides optimal results. We get detailed bone segmentation as shown in Fig. 2(d). However, some artifacts surrounding the bone are also marked as bone. The proposed algorithm returned a Dice index of 0.9321, a Jaccard index or Intersection over Union (IoU) of 0.8729, a precision of 0.9004 and a recall of 0.9662.

From Table 1, we infer that the proposed algorithm achieves better scores in performance metrics: Jaccard index and recall when compared to the state-of-the-art

Table 1 Comparison of performance metrics with state-of-the-art algorithms

Algorithm	Dice index	Jaccard index	Precision	Recall
Max-flow optimization [11]	0.91		**0.97**	
U-Net [12]	0.92	0.85		
DFM-Net [13]	0.8897	0.8040	0.9398	0.8485
CNN [14]	**0.9396**			
Proposed algorithm	0.9321	**0.8729**	0.9004	**0.9662**

Bold depicts the highest attained value of that metric across all the frameworks algorithms. When considering the Dice index, the proposed algorithm outperforms all the algorithms but, lacks from the state-of-the-art algorithm [14] by 0.0075 which is very less compared to the computational expense and a large amount of data required for training the algorithm proposed in [14]. It is known that deep learning algorithms serving as the state-of-the-art algorithms need vigorous training and are resource-dependent. Whereas, this methodology is almost commensurable with the existing state-of-the-art methods and can be implemented in real-time applications as the processing speed stands at an average of 66 secs per image.

7 Conclusion

This study presents a method for segmentation of bone tissues from CT images, which comprises artifacts removal and segmentation. The image is cleaned by simple yet efficient contrast stretching technique, which removes unwanted artifacts surrounding the bone tissue to a large extent and enhances the bone region. An iterative region growing algorithm is then applied, which further refines our segmentation process and extracts the bone region giving optimal results. The noise is removed through ingenious techniques, and the bone is accurately segmented. The overall performance is acceptable with a Dice index of 0.9321, Jaccard index (IoU) of 0.8729, a precision of 0.9004, and a recall of 0.9662. The proposed method outperformed the state-of-the-art algorithms in this domain in performance with limited resources and data. It is been implemented in real time having significantly less processing time of an average of 66 s for each image.

8 Limitations and Future Work

The proposed algorithm requires user input at some point of time to decide on the initial seed in the region growing algorithm. We will try to create a complete automatic bone segmentation algorithm with better performance metrics in future work. We would also explore the application areas of this algorithm and try to provide an ideal solution to solve the existing problems of the domain.

Acknowledgements This research is supported by JSPS Grant-in-Aid for Scientific Research (C) (20K11873) and Chubu University Grant.

References

1. Pham, D.L., Xu, C., Prince, J.L.: Current methods in medical image segmentation. Annu. Rev. Biomed. Eng. **2**, 315–37 (2000). https://doi.org/10.1146/annurev.bioeng.2.1.315
2. Tomazevic, M., Kreuh, D., Kristan, A., Puketa, V., Cimerman, M.: Preoperative planning program tool in treatment of articular fractures process of segmentation procedure. In: Bamidis, P.D., Pallikarakis, N. (eds.) XII Mediterranean Conference on Medical and Biological Engineering and Computing 2010. IFMBE Proceedings, vol. 29. Springer, Berlin, Heidelberg (2010). https://doi.org/10.1007/978-3-642-13039-7_108
3. Tassani, S., Matsopoulos, G.K., Baruffaldi, F.: 3D identification of trabecular bone fracture zone using an automatic image registration scheme: a validation study. J. Biomech. **45**(11), 2035–2040 (2012)
4. He, Y., Shi, C., Liu, J., Shi, D.: A segmentation algorithm of the cortex bone and trabecular bone in Proximal Humerus based on CT images. In: 2017 23rd International Conference on Automation and Computing (ICAC), pp. 1–4 (2017). https://doi.org/10.23919/IConAC.2017.8082093
5. Paulano, F., Jiménez, J.J., Pulido, R.: 3D segmentation and labeling of fractured bone from CT images. Vis. Comput. **30**, 939–948 (2014). https://doi.org/10.1007/s00371-014-0963-0
6. Fornaro, J., Székely, G., Harders, M.: Semi-automatic segmentation of fractured Pelvic bones for surgical planning. In: Bello, F., Cotin, S. (eds.) Biomedical Simulation. ISBMS 2010. Lecture Notes in Computer Science, vol. 5958. Springer, Berlin, Heidelberg (2010). https://doi.org/10.1007/978-3-642-11615-5_9
7. Kyle Justice, R., Stokely, E.M., Strobel, J.S., Ideker M.D.R.E., Smith, W.M.: Medical image segmentation using 3D seeded region growing. In: Proceedings of SPIE 3034, Medical Imaging 1997: Image Processing (1997). https://doi.org/10.1117/12.274179
8. Kaminsky, J., Klinge, P., Rodt, T., Bokemeyer, M., Luedemann, W., Samii, M.: Specially adapted interactive tools for an improved 3D-segmentation of the spine. Comput. Med. Imag. Graph. **28**(3), 119–27 (2004). https://doi.org/10.1016/j.compmedimag.2003.12.001
9. Lee, P.-Y., Lai, J.-Y., Hu, Y.-S., Huang, C.-Y., Tsai, Y.-C., Ueng, W.-D.: Virtual 3D planning of pelvic fracture reduction and implant placement. Biomed. Eng. Appl. Basis Commun. **24**, 245–262 (2012). https://doi.org/10.1142/S101623721250007X
10. Huang, C.-Y., Luo, L.-J., Lee, P.-Y., Lai, J.-Y., Wang, W.-T.: Efficient segmentation algorithm for 3D bone models construction on medical images. J. Med. Biol. Eng. (2011). https://doi.org/10.5405/jmbe.734
11. Pérez-Carrasco, J.A., Acha-Piñero, B., Serrano, C.: Segmentation of bone structures in 3D CT images based on continuous max-flow optimization. In: Progress in Biomedical Optics and Imaging—Proceedings of SPIE (2015). https://doi.org/10.1117/12.2082139
12. Klein, A., Warszawski, J., Hillengaß, J.: Automatic bone segmentation in whole-body CT images. Int. J. CARS **14**, 21–29 (2019). https://doi.org/10.1007/s11548-018-1883-7
13. Zhang, J., Qian, W., Xu, D., Pu, Y.: DFM-Net: a contextual inference network for T2-weighted image segmentation of the pelvis. In: Thirteenth International Conference on Graphics and Image Processing (ICGIP 2021). SPIE 12083 (2021). https://doi.org/10.1117/12.2623470
14. Xiong, X., Smith, B.J., Graves, S.A., Sunderland, J.J., Graham, M.M., Gross, B.A., Buatti, J.M., Beichel, R.R.: Quantification of uptake in pelvis F-18 FLT PET-CT images using a 3D localization and segmentation CNN. Med. Phys. **49**(3), 1585–1598 (2022). https://doi.org/10.1002/mp.15440
15. Gangwar, T., Calder, J., Takahashi, T., Bechtold, J., Schillinger, D.: Robust variational segmentation of 3D bone CT data with thin cartilage interfaces. Med. Image Anal. **47** (2018). https://doi.org/10.1016/j.media.2018.04.003
16. Sebastian, T.B., Tek, H., Crisco, J.J., Kimia, B.B.: Segmentation of carpal bones from CT images using skeletally coupled deformable models. Med. Image Anal. **7**(1), 21–45 (2003). https://doi.org/10.1016/s1361-8415(02)00065-8

17. Shadid, W., Willis, A.: Bone fragment segmentation from 3D CT imagery using the probabilistic watershed transform. Proc. IEEE Southeastcon **2013**, 1–8 (2013). https://doi.org/10.1109/SECON.2013.6567509
18. Moghari, M.H., Abolmaesumi, P.: Global registration of multiple bone fragments using statistical atlas models: feasibility experiments. Ann. Int. Conf. IEEE Eng. Med. Biol. Soc. **2008**, 5374–7 (2008). https://doi.org/10.1109/IEMBS.2008.4650429
19. USevillabonemuscle Dataset. http://grupo.us.es/grupobip/research/research-topics/segmentation-of-abdominal-organs-and-tumors/
20. Ruikar, D.D., Santosh, K.C., Hegadi, R.S.: Automated fractured bone segmentation and labeling from CT images. J. Med. Syst. **43**(3), 60 (2019). https://doi.org/10.1007/s10916-019-1176-x

Fusion of Features Extracted from Transfer Learning and Handcrafted Methods to Enhance Skin Cancer Classification Performance

B. H. Shekar and Habtu Hailu

Abstract Nowadays, medical imaging has become crucial for detecting several diseases using machine learning and deep learning techniques. Skin cancer has been the most common of several diseases. If it has not treated early, a severe illness may cause in the patients, and this difficulty may lead to death. Several automated detection methods have been explored in the area, but the performance still does not reach the level needed by the medical sector. The common factors that affect the performance of the detection process are limited dataset, the method used for feature extraction, classification, hyperparameter tuning, and the like. This work proposed a fused feature extraction technique containing a transfer learning of the DenseNet-169 model and six handcrafted methods to capture richer and more detailed features. We applied a well-known machine learning algorithm called gradient boosting machine (GBM) for classification. We have used a publicly available dataset to train and evaluate our method: ISIC Archive datasets. The result shows that the proposed method improves the performance of skin cancer classification. GBM with the fused feature extraction technique is the highest performer with 87.91% of accuracy. Moreover, we use a few recent and best previously worked methods for comparison, using the same dataset. Our proposed method outperforms all of the previously worked algorithms by most of the evaluation metrics.

Keywords Feature Extraction · Skin cancer · Transfer Learning · Hand crafted feature

1 Introduction

Over the years, many computer-aided systems have been developed to diagnose and detect human disease by using medical images. The most recent imaging modalities rely on high-resolution imaging to provide radiologists with multi-oriented representations, which helps them with a clinical diagnosis and perform accurate predictions

B. H. Shekar · H. Hailu (✉)
Mangalore University, Mangalagangothri, Karnataka, India
e-mail: habtuhailu@gmail.com

M. Tistarelli et al. (eds.), *Computer Vision and Machine Intelligence*,
Lecture Notes in Networks and Systems 586,
https://doi.org/10.1007/978-981-19-7867-8_20

and treatment for a patient. Ultrasound, magnetic resonance imaging (MRI), X-ray computed tomography, and endoscopy are some of the most common modalities for medical imaging [11]. Due to the fast development of medical imaging technologies, the field of digital pathology, which concentrates on the analysis and management of image information produced by these technologies, is expanding quickly [24]. Digital pathology with machine learning and deep learning technologies holds considerable potential for medical practices (diagnostic medicine and disease prediction) and research related to biomedical [12, 14].

The health industry actively seeks new and accurate technology to diagnose, detect, and control several diseases with better performance. As a result, AI has become the most promising area to solve this problem because of its automatic feature extraction, representation, and classification ability. AI includes machine learning and deep learning techniques as a sub-area, which consists of many different algorithms that will apply to solve several complex tasks and then provide intelligent models [4]. These algorithms require an extensive amount of training data for better generalization ability. Nevertheless, publicly, getting a sufficient amount of data to implement these techniques is very difficult. Transfer learning which focuses on the sharing of knowledge that learns from a very complex domain to the other new task becomes an appropriate technique to overcome this issue [2, 28, 32]. It has become the most popular and efficient technique to automate clinical applications and replace radiologists' expensive and tiresome labor work due to its superior performance and flexibility.

One of the global public issues is skin cancer, caused by the abnormal growth of melanocyte cells in the human body. It is the most dangerous and deadly disease if it is not treated in the early stage. However, skin cancer is curable if identified and treated in its beginning stage. The traditional way of diagnosing skin cancer is by examining physical and biopsy. Even though the methods are the simplest ways to detect skin cancer, the methodology is burdensome and unreliable. To improve these methods, recently, dermatologists have been assisted by macroscopic and dermoscopy images to diagnose skin cancer [1]. Dermoscopy images are the best choice for dermatologists to improve the capability of skin cancer detection due to their high-resolution images and capturing the deeper structure of the skin [10]. However, skin cancer detection accurately is a challenging issue for dermatologists, even with the high-resolution dermoscopy image.

To assist dermatologists and overcome their challenges of detecting skin cancer disease, several researchers have attempted to develop a computer-aided diagnosis (CAD) system [15, 26, 27]. Nevertheless, predicting a skin cancer with high performance using a dermoscopy image is a challenging task. The scarcity of dermoscopy images, the lower resolution of the images, the efficiency of the methods used for feature extraction and classification, etc., are some of the challenges in the area. This work reveals a feature extraction mechanism that uses a fusion of handcrafted and deep transfer learning feature extraction mechanisms to enhance skin cancer image classification. For the handcrafted techniques, we apply six different techniques: speeded up robust features (SURF), geometric moment, local binary pattern (LBP), Haralick, Zernike moment, and color histogram. These techniques extract the image feature from different aspects. On the other hand, we used a DenseNet-169 pre-trained model for extracting the deep feature. Finally, a machine learning algorithm, gradient boosting machine (GBM), is applied to classify the extracted features. After detailed experimentation, the classification performance is compared between each method of feature extraction (handcrafted feature, deep learning feature, and the fusion of the two). Also, the best-proposed method is compared to the recent state-of-the-art models, which are done on the same dataset (ISIC Archive dataset).

2 Literature

Any organ of the human body is subjected to cancer disease if there is an unmanageable abnormal growth of cells. Cancer primarily affects the human body's skin, lungs, breasts, stomach, prostate, and liver. According to the WHO, the rate of global spread is increasing at an alarming rate. To help physicians for the detection and diagnosis of cancer, the following works are attempted.

Ain et al. [3] employ a genetic programming technique for feature selection and construction operations extracted from skin cancer images by using local binary pattern methods. An Internet of health and things (IoHT) framework, which is based on the deep transfer learning method, is presented by [21] for the classification of skin lesions. For feature extraction and classification, they apply the pre-trained algorithms InceptionV3, SqueezeNet, VGG19, and ResNet50. Saba et al. [30] introduce a deep neural network strategy for skin lesion classification that uses fast local Laplacian filtering to enhance contrast, CNN to extract the wound's boundaries, and the InceptionV3 algorithm to extract the lesion feature in-depth. The entropy-controlled technique is used to select the most significant features. They used the PH2 and ISIC-2017 datasets for training and testing the model. To categorize skin cancer images into malignant and benign, Shahin Ali et al. [6] developed a deep convolutional neural network method with a noise removal technique. To address the challenges posed by the scarcity of image datasets, Zhao et al. [35] propose a style-based GAN dataset augmentation approach with a combination of DenseNet201. Monika et al. [23] apply the SVM algorithm to classify dermoscopic images with several images preprocessing techniques such as dull razor for hair removal, Gaussian filter for image smoothing, median filter for filtering the noise, and k-mean for segmentation purposes.

Most recently, to safeguard against cancer disease for the classification of melanoma tumors, Indraswari et al. [20] employ a widely used type of deep learning, MobileNetV2 architecture. They test the chosen architecture on four different melanoma image datasets. Sharma et al. [31] established a binary classification based on histopathology images using a pre-trained Xception algorithm to detect breast cancer. In order to improve the performance of the classifier, the researchers utilized a handcrafted feature extraction method. To detect skin cancer from the dermoscopic image, Kamrul Hasan et al. [17] propose a dermoscopic expert framework by using a hybrid convolutional neural network. The framework consists of three distinct modules for extracting features. Lesion segmentation, augmentation, and class rebalancing operations are applied in the preprocessing stage on the selected datasets, namely ISIC-2016, ISIC-2017, and ISIC-2018. Ali et al. [5] suggest different preprocessing approaches such as hair removal, augmentation, and resizing techniques to enhance the performance of the classifier. They use a transfer learning method for classification on the HAM10000 dataset. Zhang et al. [34] proposed a technique that is currently an emerging technique for detecting breast cancer, namely Romana spectroscopy with an SVM algorithm.

Even if researchers explore several works in the literature, the performance of the proposed method in all the metrics is not satisfactory. High-end performance accuracy is the main requirement of designing a computer-aided diagnosis (CAD) for identifying skin cancer. The authors of this work determined that getting a better classification performance is highly dependent on feature extraction and attempted to improve this technique. Most of the existing works apply a single or combination of pre-trained deep learning (DL) algorithms for feature extraction. Although DL is powerful for extracting features and classification of the image, it has some limitations because the algorithms require a large training dataset which is difficult in medical image tasks. Moreover, it takes the weights that learn from the ImageNet dataset, which is not directly related to the dataset used for our work. We fuse features extracted from DL and handcrafted methods to overcome these limitations. A pre-trained DenseNet-169 algorithm, the best performer in the ImegeNet dataset and previous works, is used as a DL feature extractor. For handcrafted features, a variety of six different algorithms: SURF, geometric moment, LBP, Haralick, Zernike moment, and color histogram are utilized. In addition, we employ an GBM algorithm for the classification, which had exemplary achievements in the previous works using the cross-validation technique on the train set to classify the image datasets. Figure 1 illustrates the architecture of the proposed method.

3 Proposed Methodology

Several research studies have been accomplished to detect skin cancer from dermoscopy images. The primary goal was to obtain the desired performance by training different models on a given dataset. We have used a fused feature extraction method and an GBM classification technique to support this issue to detect skin cancer disease.

Figure 1 presents a clear general overview and the steps included in the proposed methodology. As described in the figure, the proposed method starts with image acquisition; then, features extraction is done by two different techniques, such as the handcrafted and pre-trained deep learning method. Next, feature vector generation is done by concatenating the features extracted from these two techniques and fetched to the classifier. Finally, the GBM classifier model performs the prediction, and evaluation is conducted. The following sections explain the details of the dataset used, feature extraction, and classification techniques in the proposed method.

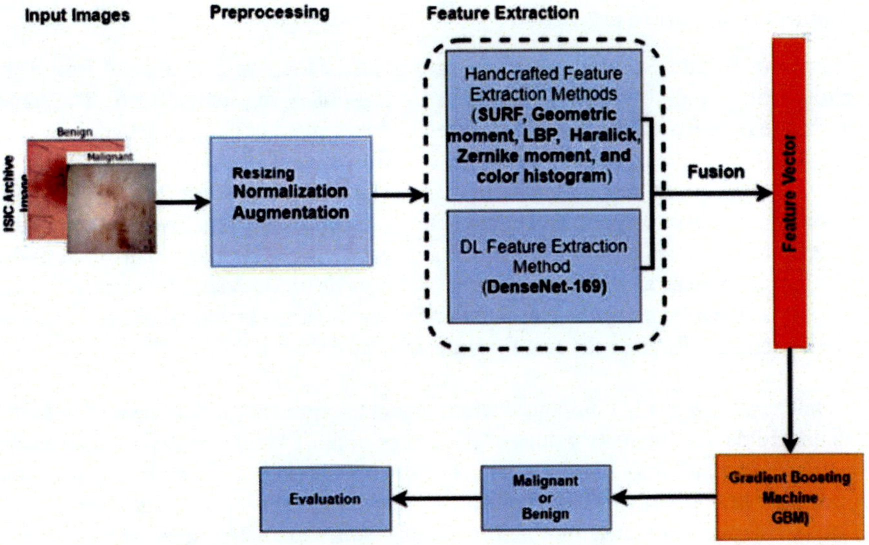

Fig. 1 Proposed architecture

3.1 Preprocessing

In the proposed method, we apply the basic preprocessing operations using the Keras ImageDataGenerator built-in functions. First, we resized the image dataset to 224 × 224 pixel size to fit the selected pre-trained deep learning model, DenseNet-169. The normalization operation is applied by dividing the pixel value by 255, to make the range of pixel values between 0 and 1. Finally, we applied an augmentation operation to increase the dataset size. These preprocessing techniques help the proposed method to enhance its generalization capability.

3.2 Feature Extraction Techniques

Feature extraction is the most crucial part of image classification tasks, affecting the model's prediction performance. For this work, we used DL and handcrafted techniques to extract features. To extract DL features, we applied a DenseNet-169 algorithm which is well-known and highly performed algorithm in the previous work. SURF, color histogram, LBP, Haralick, Zernike moment, and geometric moment are employed to extract local features like global shape, texture, color, etc. The fusion of features is done by a python concatenation operation hstack(). First, the features are extracted for each handcrafted and deep learning method, then all the handcrafted features are fused. Finally, the combined handcrafted features are merged with the deep learning features. The techniques are briefly explained in the following sections.

Handcrafted feature extraction

Handcrafted features refer to the characteristics of the image emanated from it by using different algorithms that are directly present in the image itself. The handcrafted feature extraction algorithms used in this paper are described as follows:

Speeded Up Robust Features (SURF) Bay et al. [8] introduce a feature extraction algorithm based on keypoints detector and descriptor, which is invariant to the displacement deformation of photometric and geometric. The Laplacian-of-Gaussian (LoG) approximation with the box filter technique is used to detect the image's interest points. Then by using the Hessian matrix, the detected keypoints are represented. The dimensional size of the feature generated by SURF is 64 or 128.

Geometric Moments In computer vision, object recognition, pattern recognition, and related fields, the intensity values of the image pixels are represented by the specific weighted average (moment), which is usually chosen to have some appealing interpretation. Image moments are helpful in describing objects in image data [9]. Hu [18] develops a geometric moment invariant feature extraction method that focuses on shape detection tasks from image data. The method extracts features via the rotation scale translation (RST) invariant. This means that the features obtained through this method are not altered for the variation of translation, rotation, and scaling. To represent the image by using geometric moments, Hu derives seven invariants. They are invariant for similarity, translation, rotation, and reflection.

Zernike Moments To enhance geometric invariants for rotation, Teague [33] presents Zernike moment invariant. This algorithm works by allocating image data onto a set of Zernike polynomial. Zernike moment describes the characteristics of an image without information overlapping between the moments because Zernic polynomials are orthogonal to each other. In addition, the orthogonal effects of the algorithm make it easier to utilize in the reconstruction process. To compute Zernike moment, Z_{nm}, from a digital image with current pixel $P(x, y)$ that has $M \times N$ size, Eq. 1 is used.

$$Z_{nm} = \frac{n+1}{\pi} \sum_{x} \sum_{y} P(x, y) V_{nm}^{*}(x, y) \tag{1}$$

where $x^2 + y^2 \leq 1$, n is a called order which is non-negative integer, m is repetition which is an integer, $0 \leq |m| \leq n$, and V_{nm} is an orthogonal Zernike polynomials.

Haralick Texture Feature Haralick textural feature proposed by [16] is a feature extraction technique in image processing tasks by capturing helpful information regarding the patterns that appear in the image data. It uses gray-level co-occurrence matrix (GLCM) method to find the features. GLCM is a statistical approach for investigating an image's textural feature, which considers a spatial interrelationship of the pixel's gray level. A GLCM matrix is generated by computing how frequently a pixel with intensity value i appears in a particular spatial association to a j value pixel. The distribution of the properties of GLCM in the matrix will depend on the distance and directions, such as diagonal, vertical, and horizontal relationships between the pixels [7]. Using the GLCM, Haralick extracted 28 different types of textural features; contrast, correlation, entropy, energy, and homogeneity are some examples. For grayscale pixels, i and j, the GLCM, which is termed as the number of the co-occurrence matrix in different directions, is computed as Eq. 2.

$$P(i, j|d, \theta) = \frac{p(i, j|d, \theta)}{\sum_i \sum_j p(i, j|d, \theta)} \tag{2}$$

where d and θ are the given distance and direction between i and j, respectively.

Linear Binary Pattern (LBP) Ojala et al. [25] introduced a feature extraction algorithm called LBP, a nonparametric operator with invulnerable to light variations to represent the local structure around each image pixel. As illustrated in Fig. 2, the LBP algorithm compares each image pixel with its eight neighbors by finding the difference from the central pixel's value. If the difference is a positive value, it is encoded to 1, and if the result is negative, the encoding value is 0. Then, a binary number is formulated by concatenating these binary codes starting from the top-left value in a clockwise direction. The decimal equivalence for this binary number is used for labeling. The LBP codes are these emanated binary numbers. Finally, the decimal expression form of LBP for a given pixel at (x_c, y_c) is given by Eq. 3 and function $s(x)$ is defined as Eq. 4.

$$\mathrm{LBP}_{P,R}(x_c, y_c) = \sum_{p=0}^{p-1} s(i_P - i_c)2^P \tag{3}$$

where

- i_c = the central pixel gray-level value
- i_P = the P surrounding pixels gray-level value
- R = radius

$$s(x) = \begin{cases} 1 \text{ if } x \geq 0 \\ 0 \text{ if } x < 0 \end{cases} \tag{4}$$

Color Histogram Color histogram is a straightforward, most widely used, and low-level technique of color descriptor from the image data [9]. It uses color bins to represent the frequency of color distribution by counting the pixels with similar values. It is not used spatial relations within regions of the image; this shows the invariance of translation and rotation. The color histogram can be represented for any color space, although the most common way used for this work is three-dimensional space. We select the hue-saturation-value (HSV) space, which is in one-to-one relation with the red-green-blue (RGB) space. The full-color space for an RGB color image S

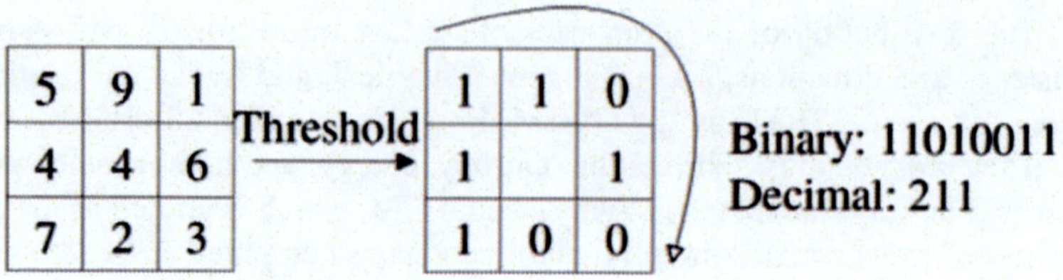

Fig. 2 Example of how LBP operator works

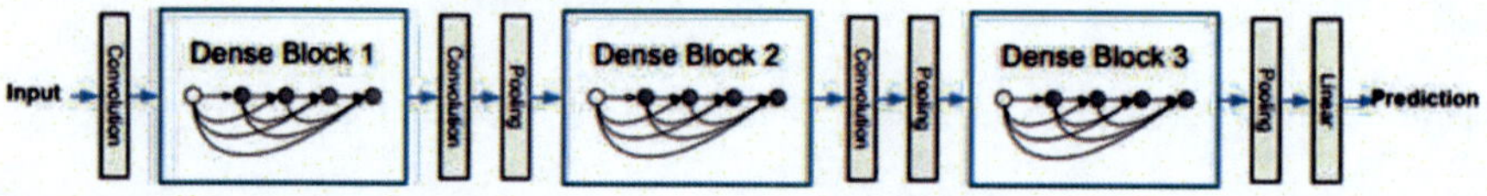

Fig. 3 DenseNet-169 architecture

can be equally divided into $N \times N \times N$ histogram bins. Then, the number of pixels included in each bin is used to compute the color histogram, and the histogram vector can be generated as $(h_{111}, h_{112}, ...h_{NNN})$.

Deep Learning Based Feature Extraction

Recently, the predominant approach for image processing operations like feature extraction, segmentation, and classification is convolutional neural networks (CNN). CNN has several forms starting from the original LeNet [22] which has only five layers. In this paper, we use one of the best and well-performed variants of CNN, DenseNet-169 [19], which contains 169 layers (165 convolution + 3 transition + 1 classification) grouped in four dense blocks. DenseNet-169 is chosen because it is easier and faster to train with no loss of accuracy due to the improved gradient flow. This algorithm is formed by connecting all layers directly to each other to secure the flow of the highest information between layers. Each layer receives input from all prior layers and gives its output feature-map to all successive layers. Figure 3 illustrates the layout of the DenseNet algorithm. We extract features from the skin lesion image dataset by using transfer learning a pre-trained DenseNet-169. We remove the top layer and freeze the other layers. The prediction output from the last dropout layer is taken as the feature vector.

3.3 Classification Technique

The study used a gradient boosting machine (GBM) learning classification algorithm as a classifier which is the best performer in the previous works. GBM is a robust model created from many decision trees by ensembling them. GBM can be applied for classification and regression tasks. Meanwhile, it compensates for the errors of the previous tree by forming a sequential tree, and it is described in Eq. 5

$$G(x) = g_1(x) + g_2(x) + g_3(x) + \cdots \quad (5)$$

Here, $g_n(x)$ is generated by reducing the error of $g_{n-1}(x)$ with each successively generated tree.

The grid search hyperparameter tuning is conducted for the optimal performance of the models. This hyperparameter tuning helps the classifier obtain the best parameters for training. A Sklearn library function GridSearchCV() is implemented for the grid search.

4 Experimental Setup and Result Analysis

The current study applied the K-fold cross-validation ($K = 5$) configuration to train the model to obtain the best achievement. This approach reduced bias and over-fitting by splitting the input dataset into five equally disconnected stratified subsets. The approach is more suitable for small datasets. The training is performed by using one of the subgroups as a validation dataset and the other four subgroups as a training dataset. This process continues until all the subgroups are used as a validation dataset and the model is trained and validated $k = 5$ times.

In addition, the grid search technique was utilized for tuning the hyperparameters for the classifier algorithm applied in the study. Hyperparameter tuning helps choose the optimal combination of hyperparameters for the learning algorithm. We have specified the possible hyperparameter values for the learning algorithm used in the study. The following subsections describe the details of evaluation metrics, the dataset used and the experimental result.

4.1 Evaluation Metrics

The study used several performance evaluation metrics to evaluate the model's effectiveness, including accuracy (Ac), precision (Pr), recall (Re), $F1$-score, and Cohen's Kappa. Moreover, for the visual representation of the performance of the models, we applied the confusion matrix and area under the receiving operating characteristic (ROC) curve. A confusion matrix summarizes the models' prediction outcomes in the classification issue, and it is used to numerically summarize the correctly and incorrectly predicted test images by their count value. On the other hand, the ROC curve is a graph of two parameters: True Positive Rate and False Positive Rate, to visualize the performance of the classification architecture at all successive prediction thresholds. The area under the curve (AUC) is the summarizing form of the ROC curve that estimates the usefulness of the classifier model to distinguish between classes.

4.2 Dataset Description

We use a publicly available image dataset for training, validating, and testing the proposed work: ISIC Archive [13]. The ISIC Archive dataset contains 3297 images of skin cancer, 1800 of which are benign and 1497 of which are malignant. This dataset is stored the training and testing images separately as 2637 for training and 660 for testing. As we can see, in Table 1, the class distribution of the dataset is somewhat balanced for each class. Sample images are illustrated in Fig. 4 for all classes. The train and test data are in the ratio of 80:20.

4.3 Experimental Results and Analysis

Table 2 shows the comparison of the performance of the GBM classifier on the different feature extraction methods using accuracy, precision, recall, F1-score, Cohen's kappa, and AUC. The fusion of handcrafted and deep learning features obtains the highest achievement. It performed 87.91%, 84.46%, 90.98%, 87.11%, 75.72%, and 95.19% for accuracy, precision, recall, $f1$-score, Cohen's kappa, and AUC, respectively. The second performer is the deep learning feature, which scores 85.66% of accuracy, and the handcrafted feature performs the most diminutive, with an accuracy of 51.00%.

Table 1 Class distribution of the dataset used for this work

Class	Total image	Training image	Testing image
Benign	1800	1440	360
Malignant	1497	1197	300
Total	3297	2637	660

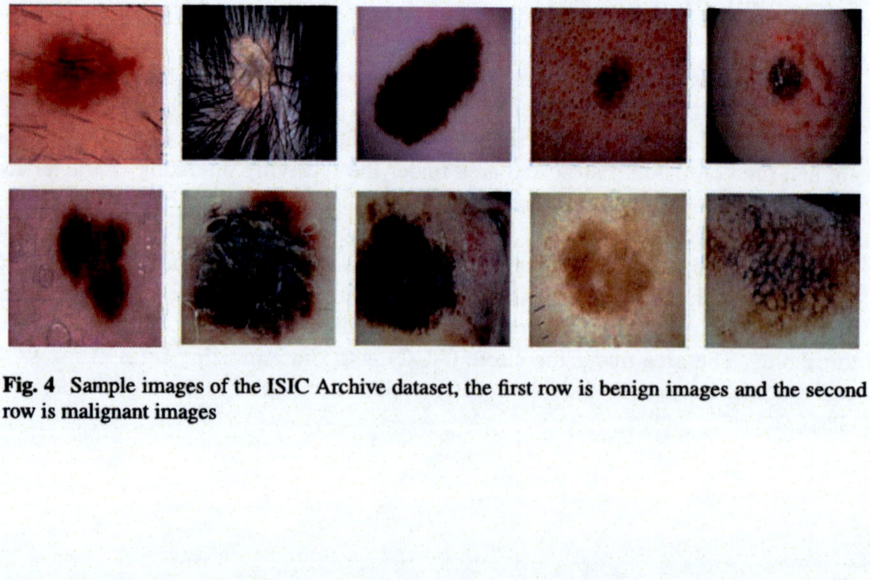

Fig. 4 Sample images of the ISIC Archive dataset, the first row is benign images and the second row is malignant images

Table 2 Performance of the proposed method using different feature extraction techniques

Feature type	Accuracy	Precision	Recall	F1-score	Kappa	AUC
Hand crafted	51.00	45.71	40.32	42.85	3.47	53.15
Deep learning	85.66	82.31	87.08	84.64	71.17	93.97
Fusion of HC and DL	87.91	84.46	91.67	87.11	75.72	95.19

The other measurement tool we have used to evaluate the proposed method is the confusion matrix used to visualize the number of samples classified correctly or wrongly. The confusion matrices of the proposed method for all utilized feature

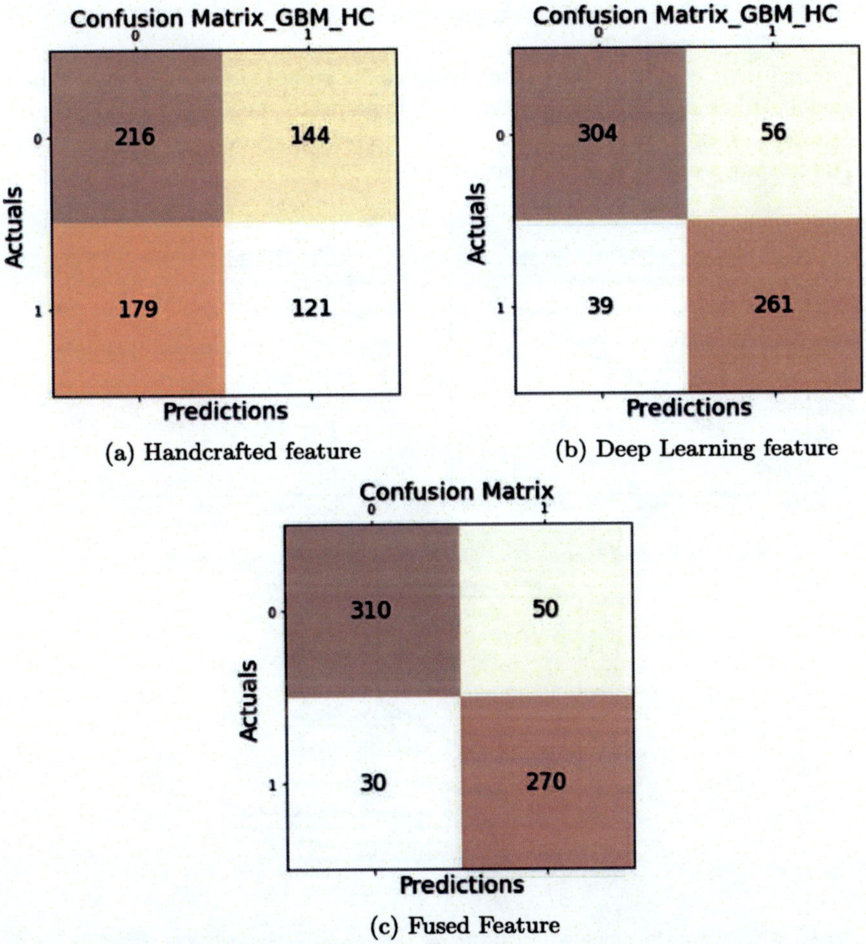

Fig. 5 Confusion matrices obtained for proposed method for each feature extraction technique on the ISIC Archive dataset

extraction mechanisms on the ISIC Archive dataset are depicted in Fig. 5. The number of misclassified samples for the fused feature extraction mechanism is much less compared to the other techniques. Furthermore, the number of misclassified samples is slightly higher than the other techniques for a handcrafted technique. For the best performer model, a fused feature extraction method, the number of misclassified test samples is 80. For deep learning and handcrafted features, 95 and 323 images are incorrectly classified.

In addition, we have used the ROC curve to overview the performance of the proposed method. The AUC also indicates for each technique to compare them quantitatively. Figure 6 illustrates the ROC curve with the corresponding AUC values for the proposed method on the ISIC Archive dataset. In Fig. 6, we observe that the fused feature extraction technique can attain 95.19% for the evaluation metric AUC. The handcrafted and the deep learning methods scores for the AUC were 53.15% and 93.97%, respectively.

Finally, we conduct a comparison between the proposed method and the existing models. To compare the performance of the proposed model on the ISIC Archive dataset, we cannot find enough works that were previously done on the same dataset. The existing works that we can see are the following.

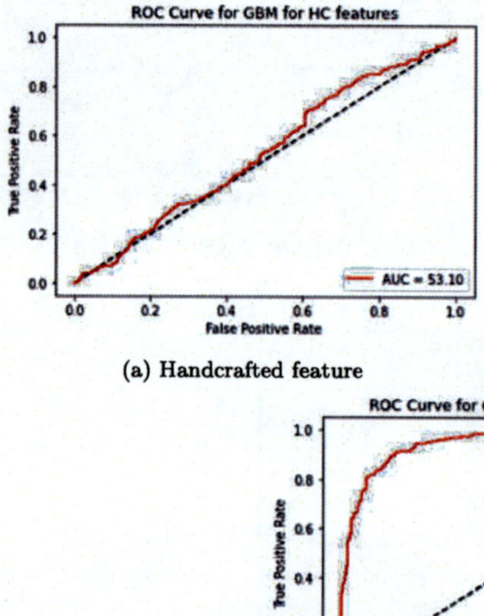

(a) Handcrafted feature

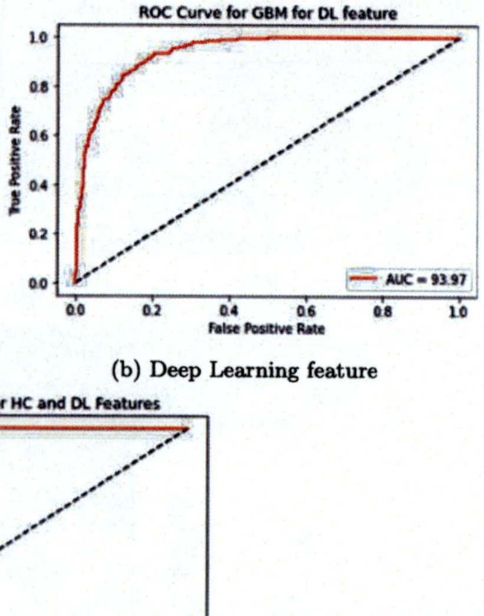

(b) Deep Learning feature

(c) Fused Feature

Fig. 6 ROC obtained for proposed method for each feature extraction technique on the ISIC archive dataset

Table 3 Evaluation results of the proposed method and the models proposed in previous works for comparison on ISIC Archive dataset

Work	Accuracy	Precision	Recall	$F1$-score	Kappa
Indraswari et al. [20]	85.00	83.00	85.00	–	–
Rokhana et al. [29]	84.76	–	**91.97**	–	–
Proposed model	**87.91**	**84.46**	90.98	**87.11**	**75.72**

The bold indicates the highest score in the column

- **Indraswari et al.** [20] employ a widely used type of deep learning, MobileNetV2 architecture, as a base model. They customize their work by adding several layers to the base model called the head model. The head model consists of a global pooling layer, two fully-connected or dense layers, and finally, the output layer that consists of two nodes.
- **Rokhana et al.** [29] suggest a deep CNN architecture for classifying skin images into benign and malignant. Their architecture includes a collection of convolution layers, max-pooling layers, dropout layers, and fully-connected layers.

The proposed method outperforms the state-of-the-art works on the ISIC Archive dataset in most of the metrics used. As shown in Table 3, Rokhana et al. [29] score a better result in recall 91.97% among the existing works. But the method achieves much less accuracy, and it is not specified the results in other metrics.

5 Conclusion

In this work, we propose a feature extraction model to optimize skin cancer detection using image datasets. The feature extraction method fuses handcrafted features and deep learning features. The handcrafted features contain a concatenation of six feature extractors that extract the image's shape, texture, and color, and we use a pre-trained DenseNet-169 for deep learning features. Then, the experimentation is conducted on ISIC Archive dataset for each type of feature: handcrafted, deep learning, and fused features. As a result, the fused feature extraction technique outperforms the other types and the handcrafted method scores much less. The deep learning feature extraction technique scores a comparative score to the fused feature extraction method. Finally, we compare the proposed method with the models developed in the related works that use the same dataset, as shown in Table 3. The result in the table verifies the superiority of the proposed method. In the future, we plan to evaluate the proposed method's robustness by using a more variety of medical image datasets with multi-class and improve the achievement.

References

1. Abbas, Q., Emre Celebi, M., Garcia, I.F., Ahmad, W.: Melanoma recognition framework based on expert definition of ABCD for dermoscopic images. Skin Res. Technol. **19**(1), e93–e102 (2013)
2. Ahmad, N., Asghar, S., Gillani, S.A.: Transfer learning-assisted multi-resolution breast cancer histopathological images classification. Vis. Comput. 1– 20 (2021)
3. Ain, Q.U., Xue, B., Al-Sahaf, H., Zhang, M.: Genetic programming for feature selection and feature construction in skin cancer image classification. In: Pacific Rim International Conference on Artificial Intelligence, pp. 732–745. Springer (2018)
4. Alafif, T., Tehame, A.M., Bajaba, S., Barnawi, A., Zia, S.: Machine and deep learning towards COVID-19 diagnosis and treatment: survey, challenges, and future directions. Int. J. Environ. Res. Public Health **18**(3), 1117 (2021)
5. Ali, K., Shaikh, Z.A., Khan, A.A., Laghari, A.A.: Multiclass skin cancer classification using efficientnets—a first step towards preventing skin cancer. Neurosci. Inform. 100034 (2021)
6. Ali, Md.S., Miah, Md.S., Haque, J., Rahman, Md.M., Islam, Md.K.: An enhanced technique of skin cancer classification using deep convolutional neural network with transfer learning models. Mach. Learn. Appl. **5**, 100036 (2021)
7. Bagri, N., Johari, P.K.: A comparative study on feature extraction using texture and shape for content based image retrieval. Int. J. Adv. Sci. Technol. **80**(4), 41–52 (2015)
8. Bay, H., Ess, A., Tuytelaars, T., Van Gool, L.: Speeded-up robust features (SURF). Comput. Vis. Image Underst. **110**(3), 346–59 (2008)
9. Chadha, A., Mallik, S., Johar, R.: Comparative study and optimization of feature-extraction techniques for content based image retrieval. arXiv preprint arXiv:1208.6335 (2012)
10. Chaturvedi, S.S., Tembhurne, J.V., Diwan, T.: A multi-class skin cancer classification using deep convolutional neural networks. Multimedia Tools Appl. **79**(39), 28477–28498 (2020)
11. Chowdhary, C.L., Acharjya, D.P.: Segmentation and feature extraction in medical imaging: a systematic review. Proc. Comput. Sci. **167**, 26–36 (2020)
12. Fan, J., Lee, J.H., Lee, Y.K.: A transfer learning architecture based on a support vector machine for histopathology image classification. Appl. Sci. **11**(14), 6380 (2021)
13. Fanconi, C.: Skin cancer: malignant versus benign-processed skin cancer pictures of the ISIC archive (2019). [Online]. Available: https://www.kaggle.com/fanconic/skin-cancer-malignant-vs-benign
14. Foucart, A., Debeir, O., Decaestecker, C.: SNOW: semi-supervised, noisy and/or weak data for deep learning in digital pathology. In: IEEE 2019 IEEE 16th International Symposium on Biomedical Imaging (ISBI 2019), pp. 1869–1872 (2019)
15. Haggenmüller, S., Maron, R.C., Hekler, A., Utikal, J.S., Barata, C., Barnhill, R.L., Beltraminelli, H., Berking, C., Betz-Stablein, B., Blum, A., Braun, S.A., Carr, R., Combalia, M., Fernandez-Figueras, M.T., Ferrara, G., Fraitag, S., French, L.E., Gellrich, F.F.: Skin cancer classification via convolutional neural networks: systematic review of studies involving human experts. Eur. J. Cancer **156**, 202–216 (2021)
16. Haralick, R.M., Shanmugam, K., Dinstein, I.H.: Textural features for image classification. IEEE Trans. Syst. Man Cybern. **6**, 610–621 (1973)
17. Hasan, Md.K., Elahi, Md.T.E., Alam, Md.A., Jawad, Md.T., Martí, R.: Dermoexpert: skin lesion classification using a hybrid convolutional neural network through segmentation, transfer learning, and augmentation. Inform. Med. Unlocked 100819 (2022)
18. Hu, M.-K.: Visual pattern recognition by moment invariants. IRE Trans. Inf. Theory **8**(2), 179–187 (1962)
19. Huang, G., Liu, Z., Van Der Maaten, L., Weinberger, K.Q.: Densely connected convolutional networks. In: Proceedings of the IEEE Conference on Computer Vision and Pattern Recognition, pp. 4700–4708 (2017)
20. Indraswari, R., Rokhana, R., Herulambang, W.: Melanoma image classification based on mobilenetv2 network. Proc. Comput. Sci. **197**, 198–207 (2022)

21. Khamparia, A., Singh, P.K., Rani, P., Samanta, D., Khanna, A., Bhushan, B.: An internet of health things-driven deep learning framework for detection and classification of skin cancer using transfer learning. Trans. Emerg. Telecommun. Technol. 32(7), e3963 (2021)
22. LeCun, Y., Bottou, L., Bengio, Y., Haffner, P.: Gradient-based learning applied to document recognition. Proc. IEEE **86**(11), 2278–2324 (1998)
23. Monika, M.K., Vignesh, N.A., Kumari, Ch.U., Kumar, M.N.V.S.S., Lydia, E.L.: Skin cancer detection and classification using machine learning. Mater. Today Proc. **33**, 4266–4270 (2020)
24. Mormont, R., Geurts, P., Marée, R.: Comparison of deep transfer learning strategies for digital pathology. In: Proceedings of the IEEE Conference on Computer Vision and Pattern Recognition Workshops, pp. 2262–2271 (2018)
25. Ojala, T., Pietikainen, M., Maenpaa, T.: Multiresolution gray-scale and rotation invariant texture classification with local binary patterns. IEEE Trans. Pattern Anal. Mach. Intell. **24**(7), 971–987 (2002)
26. Oliveira, R.B., Papa, J.P., Pereira, A.S., Tavares, J.M.R.S.: Computational methods for pigmented skin lesion classification in images: review and future trends. Neural Comput. Appl. **29**(3), 613–636 (2018)
27. Pathan, S., Prabhu, K.G., Siddalingaswamy, P.C.: Techniques and algorithms for computer aided diagnosis of pigmented skin lesions—a review. Biomed. Signal Process. Control **39**, 237–262 (2018)
28. Penatti, O.A.B., Nogueira, K., Dos Santos, J.A.: Do deep features generalize from everyday objects to remote sensing and aerial scenes domains? In: Proceedings of the IEEE Conference on Computer Vision and Pattern Recognition Workshops, pp. 44–51 (2015)
29. Rokhana, R., Herulambang, W., Indraswari, R.: Deep convolutional neural network for melanoma image classification. In: IEEE 2020 International Electronics Symposium (IES), pp. 481–486 (2020)
30. Saba, T., Khan, M.A., Rehman, A., Marie-Sainte, S.L.: Region extraction and classification of skin cancer: a heterogeneous framework of deep CNN features fusion and reduction. J. Med. Syst. **43**(9), 1–19 (2019)
31. Sharma, S., Kumar, S.: The Xception model: a potential feature extractor in breast cancer histology images classification. ICT Express (2021)
32. Shin, H.-C., Roth, H.R., Gao, M., Lu, L., Xu, Z., Nogues, I., Yao, J., Mollura, D., Summers, R.M.: Deep convolutional neural networks for computer-aided detection: CNN architectures, dataset characteristics and transfer learning. IEEE Trans. Med. Imag. **35**(5), 1285–1298 (2016)
33. Teague, M.R.: Image analysis via the general theory of moments (a). J. Opt. Soc. Am. **(1917–1983)**(69), 1468 (1979)
34. Zhang, L., Li, C., Peng, D., Yi, X., He, S., Liu, F., Zheng, X., Huang, W.E., Zhao, L., Huang, X.: Raman spectroscopy and machine learning for the classification of breast cancers. Spectrochimica Acta Part A Mol. Biomol. Spectro. **264**, 120300 (2022)
35. Zhao, C., Shuai, R., Ma, L., Liu, W., Hu, D., Wu, M.: Dermoscopy image classification based on stylegan and densenet201. IEEE Access **9**, 8659–8679 (2021)

Investigation of Feature Importance for Blood Pressure Estimation Using Photoplethysmogram

Shyamal Krishna Agrawal, Shresth Gupta, Aman Kumar, Lakhindar Murmu, and Anurag Singh

Abstract Continuous blood pressure monitoring is essential for persons at risk of hypertension and cardiovascular disease. This work presents the analysis of different temporal features of photoplethysmogram (PPG) useful for estimating cuffless blood pressure by utilizing several statistical test approaches to identify the features' contribution to this estimation and their correlation with a target mean arterial pressure values. The regression is performed using a random forest regressor to estimate mean arterial pressure (MAP) with temporal features, and statistical analysis with a ranking of features is done after estimation using p-value, correlation, and z-test. The significant ranking temporal features are selected and used to estimate MAP, DBP, and SBP.

Keywords Analysis · Blood pressure · Photoplethysmogram signal · Features · Random forest

1 Introduction

Health issues are also increasing in this highly modern world and the era of technological advancement. Nowadays, many people suffering from blood pressure can be seen and at each moment; measuring blood pressure by the traditional method

S. K. Agrawal · S. Gupta (✉) · A. Kumar · L. Murmu · A. Singh
International Institute of Information Technology, Naya Raipur, Chhattisgarh, India
e-mail: shresth@iiitnr.edu.in

S. K. Agrawal
e-mail: shyamal19101@iiitnr.edu.in

A. Kumar
e-mail: aman19102@iiitnr.edu.in

L. Murmu
e-mail: anurag@iiitnr.edu.in

A. Singh
e-mail: lakhindar@iiitnr.edu.in

M. Tistarelli et al. (eds.), *Computer Vision and Machine Intelligence*,
Lecture Notes in Networks and Systems 586,
https://doi.org/10.1007/978-981-19-7867-8_21

is impossible. Additionally, the procedure—in which the cuff is placed around the shoulder blade, exposed and stretched out, and then inflated until blood cannot flow via the brachial artery—requires both time and effort to complete. After that, the air is gradually let out of the cuff. At the same time, we may even assert that it is challenging to use a stethoscope to auscultate the brachial artery and detect the emergence, bafflement, or disappearance of the Korotkoff sounds, which stand for SBP and DBP [1].

The idea behind this paper is to use a photoplethysmogram (PPG) to predict blood pressure with fewer temporal features, so the prediction can be as fast as possible; hence, we have ranked the features in Table 3. As such, a device can be easily installed in a small device like a wristband or watch [2–5]. Using photoplethysmogram (PPG) waveforms, we can represent arterial oxygenation with respect to time. It is also a time-dependent depiction of blood volume. The signal we get is called PPG if we can figure out the optical way for sensing the plethysmogram signal. PPG was recorded using a technique that involved shining a red or infrared light on an organ and recording the reflected light. After two pulses of a typical PPG signal, the arterial oxygen level is determined using the amount of light that is reflected. After receiving the signal, we can forecast the values of the systolic blood pressure (SBP), diastolic blood pressure (DBP), and mean arterial pressure using the machine learning model (MAP). MAP is the average arterial pressure experienced by a patient throughout a single cardiac cycle. In contrast to DBP, SBP is caused by an infection of the ascitic fluid when there are no intra-abdominal, surgically curable sources of infection. We can determine the blood pressure using all three of them.

In Hasanzadeh et al. [6], they used the PPG signal morphological features to estimate BP. Tenfold cross-validation and AdaBoost are used in the following paper, while the outputs where SBP, DBP, MAP, and the dataset have a total of 7 columns extracted, and five columns are used for preprocessing.

The blood pressure and HRV parameters were calculated by Mafi et al. [7]. They implemented and realized a wearable device using synchronous recordings of ECG and PPG signals. In order to estimate blood pressure as a function of PAT value, they try to find the relationship between PAT and BP obtained with a typical device by computing the pulse arrival time (PAT) in two modes (PAT1 and PAT2).

In Podaru et al. [8], both electrocardiogram (ECG) and PPG are used to drive pulse arrival time (PAT), and then, it is used to predict SBP and DBP.

Kumar et al. [9] have used ECG, PPG, and pulse transit time (PTT) to predict SBP and DBP.

Li et al. [10], in this machine learning algorithm, are used multiple linear regression (MLR), a support vector machine (SVM), decision tree for predicting SBP and DBP. The minimum MAE achieved is about 7.44 for SBP, while for DBP, it is 5.09. This work is intended to investigate the importance of pre-existing features using different statistical tests. This can help to efficiently develop BP-estimation-based algorithms with the most significant and correlated temporal features with target BP values.

2 Database

We have utilized a subset of MIMIC-II is also available at University of California Irvine (UCI) Machine Learning Repository [11] and has also been used in BP estimation works.

3 Methodology

The proposed analysis framework broadly shown in Fig. 1 involves preprocessing of raw PPG signal followed by feature extraction that is based on morphological key points available in PPG contour. The obtained temporal feature set is used further in different regression models with ABP signal as ground truth available in the database. The estimation obtained from regression models provides the prediction of BP values, and in order to see the contribution of each feature, we utilized feature analysis and feature ranking approach. A detailed description of each block is provided below in subsections.

3.1 Preprocessing

Baseline wandering and high-frequency noise, which are discovered to be primarily confined in time and even have greater amplitudes, corrupt the normalized raw PPG [12]. In order to eliminate the coefficients responsible for undesired noise, wavelet denoising is used to decompose normalized PPG at the 10-level using a Db8 mother wavelet. After that, gentle thresholding is used on the remaining coefficients,

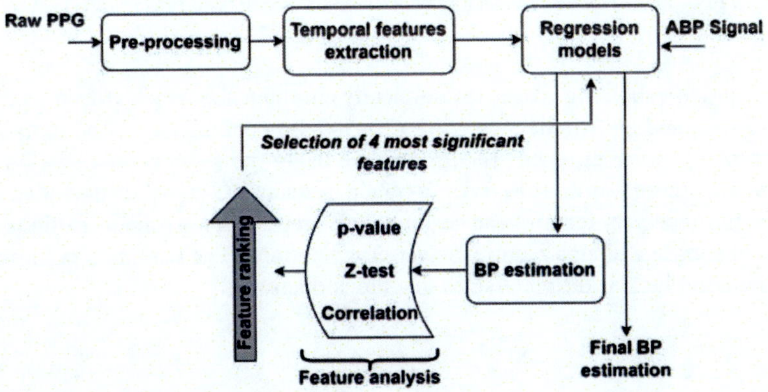

Fig. 1 Proposed framework for feature analysis and ranking

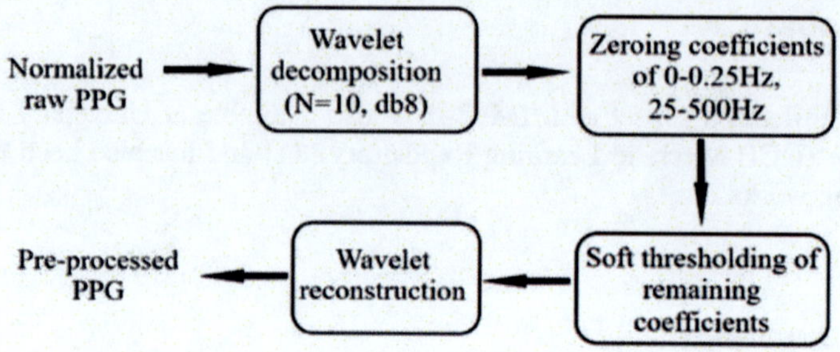

Fig. 2 Preprocessing framework

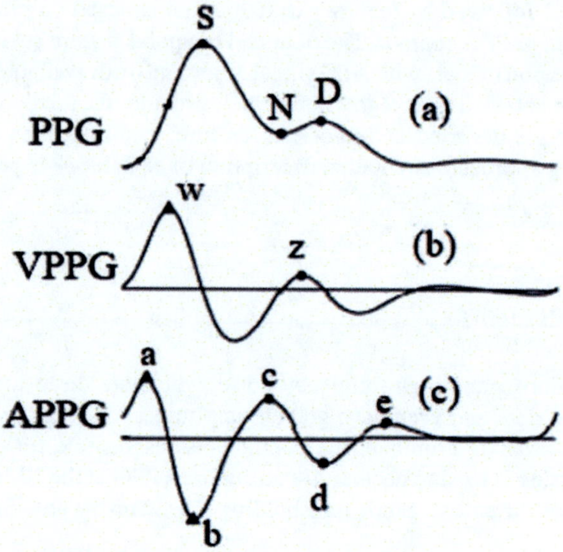

Fig. 3 **a** PPG, **b** Differentiated/velocity PPG (VPPG), **c** Accelerated PPG (APPG)

while hard thresholding is used to completely eliminate the very low-frequency noise (0–0.25Hz) and high-frequency noise (25–500Hz) coefficients. After applying soft thresholding, wavelet reconstruction is done using the same coefficients on which soft thresholding was done in order to obtain a clean PPG signal using which extraction of the necessary features can be done. This preprocessing scheme exhibits a good phase response and also requires fewer computations. The complete preprocessing is shown in Fig. 2 as different steps in a block diagram.

3.2 Feature Extraction

The PPG episodes after preprocessing are segmented to obtain single PPG cycles, as shown in Fig. 3a, to extract features. Hence, detection of onset and end points are performed to capture one single cycle to extract temporal features. There are different state-of-the-art features of PPG that correlate with mean arterial pressure for the estimation of cuffless blood pressure. The description of the temporal features is given below.

1. The amplitude of systolic peak: The maximum amplitude of the systolic peak is taken as an important temporal feature. The amplitude is measured from the base of the cycle to the highest detected peak in the PPG cycle.
2. Notch Distance: Distance of notch point from the origin.
3. Differentiated peak: Peak obtained from a differentiated version of PPG as shown in Fig. 3b, also called velocity PPG.
4. Augmentation Index (AI): The augmentation pressure (AP) determines how much the reflected wave boosts the systolic arterial pressure. The reflected wave from the boundary to the center is used to calculate it. When elastic artery compliance decreases, the reflected wave returns earlier and occurs in systole rather than diastole. This early wave return results in an unbalanced rise in systolic pressure, an increase in pulse pressure, a commensurate rise in left ventricular afterload, a decrease in DBP, and poor coronary perfusion. Takazawa et al. [13] define the AI as the ratio of P to Q as follows:

$$AI = P/Q - 1 \tag{1}$$

   When P is the late systolic peak height and Q is the first systolic peak in the pulse. Padilla et al. [14] used the RI as a reflection index as follows:

$$RI = P/Q \tag{2}$$

   Rubins et al. [15] used the reflection index and found an alternative augmentation index as follows:

$$AI = P - Y/Q \tag{3}$$

5. Dicrotic notch: This crucial physiological point in the PPG contour depicts the upstroke that follows in the descending half of a pulse trace and corresponds to the momentary rise in aortic pressure following the closure of the aortic valve.
6. Large artery stiffness index (LASI): Stiffness of the major arteries is frequently caused by a continuous increase in blood pressure, especially when additional risk factors are present. As a result of the increased stiffness, hypertension is worse by an increase in SBP, which can lead to cardiac hypertrophy and arterial lesions.
7. Crest Time: The period of time between the PPG waveform's base and its peak is referred to as the crest time. A crucial factor in categorizing cardiovascular

Table 1 Comparison with other works for regression

Paper	SBP		DBP		MAP	
	MAE	STD	MAE	STD	MAE	STD
Kachuee et al. [1]	11.17	–	5.35	–	5.92	–
Hasanzade et al. [6]	8.22	10.38	4.17	4.22	4.58	5.53
Li et al. [10]	5.09	5.66	7.44	7.37	–	–
Mase et al. [17]	4.24	4.72	4.53	5.72	5.75	6.67
Random forest	**4.99**	**5.13**	**2.56**	**3.06**	**3.58**	**4.02**
Decision tree	5.43	5.51	2.71	3.32	3.68	4.13

The bold values signifies the result obtained from the proposed approach

illnesses is the crest time. They developed a mechanism to categorize people into high and low-pulse wave velocity using data taken from the PPG (corresponding to high and low cardiovascular disease risk). Peak-to-peak time (T), crest time (CT), and stiffness index (SI =h/T) were the most effective criteria for the first derivative of the PPG's accurate classification of cardiovascular diseases. They used a combination of these traits to classify things.

8. S1: This is the region of the PPG signal curve that is closest to its highest slope.
9. S2: This is the region of the PPG signal's curve that extends up to the systolic peak.
10. S3: This is the region of the PPG signal's curve that extends up to the diastolic peak.
11. S4: The entire area is under the PPG signal waveform.
12. Peak distance 1: This is the distance between the origin and the systolic peak.
13. Peak distance 2: This is the separation between the origin and the diastolic peak.
14. RATIO b_a: According to Takazawa et al. [13] research, the b/a ratio rises with age as a result of a rise in vascular stiffness. Imanaga et al. [16] discovery that the peripheral artery's distensibility is correlated with the magnitude of the APPG's (shown in Fig. 3c) b/a suggests that the magnitude of b/a is a useful non-invasive indicator of atherosclerosis and altered arterial distensibility.
15. RATIO e_a: According to Takazawa et al. [13], the e/a ratio decreases with aging, and a higher e/a ratio suggests reduced arterial stiffness.

3.3 Blood Pressure Estimation Using Regression Models

For the regression purpose, we have found the values of MAP, SBP, and DBP by using the two algorithms, i.e., decision tree and random forest regression, in this paper. After that, we have classified whether it is high, low, or normal blood pressure. Decision trees are a nonparametric supervised learning method. They can be used for classification as well as regression. A decision tree algorithm uses a series of decisions drawn from the data and its behavior to make decisions. Boosting and bagging algorithms were created as ensemble models based on the basic premise of decision trees modified slightly. The judgments in the decision tree are based on any of the features' conditions. Random forest is a supervised learning method. Both classification and regression are done using them [14]. Its basis is the concept of ensemble learning. The classification performance is increased for this database by the random forest approach, which takes the averages of a series of decision trees for various subsets of the dataset for making a prediction. Instead of depending just on one decision tree, random forests use numerous predictors to combine the predictions from each tree to produce the final outcome or prediction. As a result, the random forest method consists of two phases. The first method creates a random forest by combining n trees, while the second algorithm calculates forecasts and means for each tree produced in the first stage. The predicted value of each tree is used to predict the final output.

Random forest and decision trees were giving low mean absolute error and decent accuracy with all 15 temporal features; hence, we used them to get predictions with 4 most significant features, i.e., amplitude, dicrotic, S4, ratio_e_a, the mean absolute error value for predicted values are shown in Table 2.

Table 2 BP estimation results with 15 temporal features

Algorithm	SBP		DBP		MAP	
	MAE	STD	MAE	STD	MAE	STD
Random forest	4.01	4.12	3.5	4.18	3.89	4.65
Decision tree	4.85	4.69	3.88	4.79	5.12	5.06
Support vector machine (SVM)	5.2	5.13	4.32	5.08	5.65	5.53

3.4 Feature Analysis

For the selection of features for the prediction of MAP, the following statistics are used on data:- We have started with 15 temporal features and predicted the MAP. Then, we used the P-value, correlation coefficient, and Z-test to refine the number of features. We kept thresholds as follows: The threshold for the correlation coefficient was set to be 0.001, and the features above the threshold value were used to predict MAP. Then, the p-value of the threshold was set to be 0.5, and the features above the threshold value were again used for the prediction purpose. Similarly, for the z-value, the threshold was set to be 0.75 and then used for prediction. Finally, we concluded with four features found by using the z-test, and the features were amplitude, dicrotic, S4, and RATIO_E_A, which can be regarded as the most significant temporal features, and using them, we get the best result.

1. Correlation: It is one of the concepts used in statistics to indicate how closely two variables move in unison with each other. The two variables have a negative correlation coefficient when they move in the opposite direction. They have a positive correlation coefficient if they travel in the same direction. For N observable variables, the difference might have happened purely by chance. The correlation can be seen for each of the columns of the dataset in Fig. 4.

$$r = \frac{\sum(x_i - \bar{x})(y_i - \bar{y})}{\sqrt[2]{\sum(x_i - \bar{x})^2 \sum(y_i - \bar{y})^2}} \tag{4}$$

   where r=correlation coefficient, x_i = values of the feature in a sample, $\bar{x}$ = mean of the values of the feature, y_i =values of the MAP in a sample , $\bar{y}$=mean of the values of the MAP

2. P-value: A p-value is an observed difference that could have occurred by random chance. It is a measure of probability. The p-value can be seen for each of the columns of the dataset in Fig. 5.
3. Z-statistic: It is a statistical test used to determine how well the sample data generalize with the entire population by checking whether the means of two populations are different when the sample size is large and the variances are known. The Z-test can be seen for each of the columns of the dataset in Fig. 6.

4 Result

We have used machine learning models like SVR, decision tree, and random forest for blood pressure estimation. By seeing the other graphs like the ones in the feature analysis section, we found four features of top rank can be used, but to check the

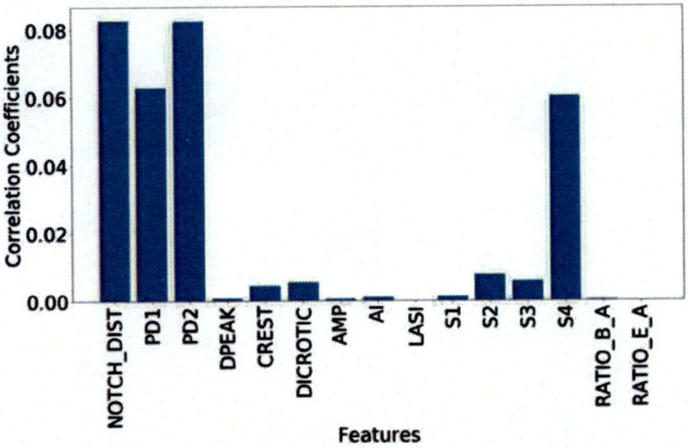

Fig. 4 Individual correlation coefficient with MAP for 15 temporal features

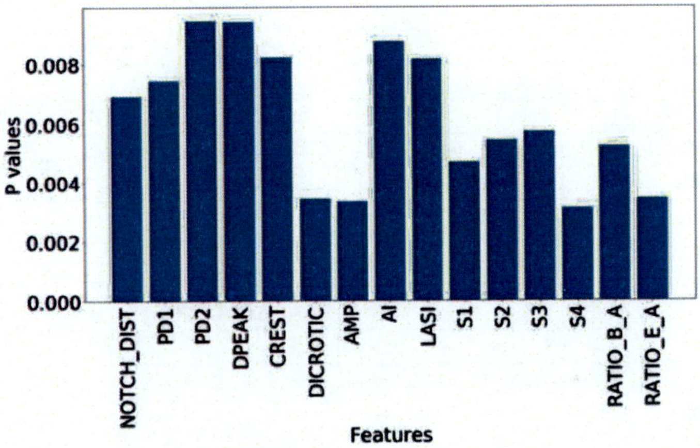

Fig. 5 P value for each feature corresponding to MAP

same, we even used feature reduction, i.e., half the number of features or the floor of the value. After ranking the features, we reduced the number of features in the same order as ranking from 15 to 7 (half the number of features and the floor of the value) and then applied the three models again. We found that change in mean absolute error value was too low when the number of features was changed from 15 to 7, so the number of features was decreased in the same manner, i.e., now, for three features, we used the three algorithms. Still, a significant gap was found in mean absolute error by changing the number of features from feature 7 to 3, so we tried with four features and found that the loss for mean absolute error is adjustable for decreasing the features. So, as a result, we found four features with a low-value change of mean absolute error and can decrease the complexity of the model. We used to calculate

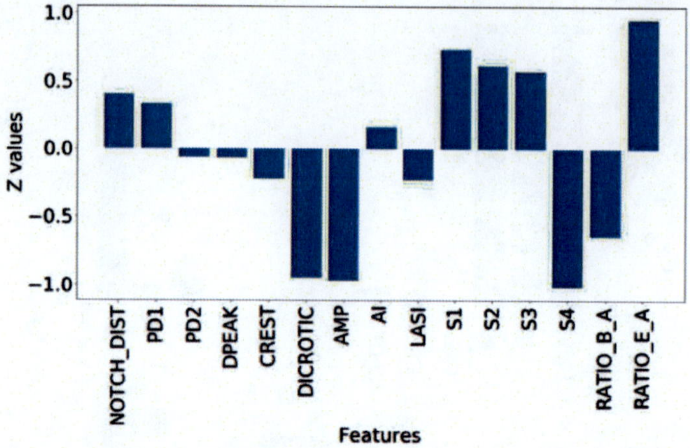

Fig. 6 Z value for each feature corresponding to MAP

Table 3 Temporal features ranking

Feature	Rank	Feature	Rank
S4	1	NOTCH DISTANCE	9
RATIO_E_A	2	LASI	10
AMPLITUDE	3	CREST	11
DICROTIC	4	AUG INDEX	12
S1	5	DIFFRENTIATED PEAK	13
RATIO_B_A	6	PEAK DISTANCE2	14
S2	7	PEAK DISTANCE1	15
S3	8		

the MAP value per the method mentioned above, but it was not able to provide a good result for predicting SBP and DBP; hence, we used predicted MAP as a feature for them and achieved a good result. Now, referring to Table 2, we can say that the random forest is a giving a low mean absolute error and better accuracy than the decision tree with only four features, while we can see the same with 15 features in Table 1. By considering Table 3, we can say the features amplitude, dicrotic, S4, and ratio_e_a are an essential feature for estimation of blood pressure using PPG.

5 Conclusion

In this paper, we have applied decision tree and random forest machine learning techniques, and after statistical feature analysis again, we estimated the blood pressure using the most significant features found after our test. Here, the random forest was the one to give better accuracy and low mean absolute error. Our obtained mean absolute error was 4.99 for SBP, 2.56 for DBP, and 3.58 for MAP. The number of features contributed to this low mean absolute error is four, and the features are amplitude, dicrotic, S4, and ratio_e_a. Further, the contribution of the work lies in both estimation of BP and analysis of significant features using various statistical tests. However, our investigation was limited to temporal features only. In the future, we are planning to analyze the other domain PPG features useful for the estimation of blood pressure and hypertension.

References

1. Kachuee, M., Kiani, M.M., Mohammadzade, H., Shabany, M.: Cuffless blood pressure estimation algorithms for continuous healthcare monitoring. IEEE Trans. Biomed. Eng. **64**(4), 859–869 (2017)
2. Zhu, S., Tan, K., Zhang, X., Liu, Z., Liu, B.: MICROST: A mixed approach for heart rate monitoring during intensive physical exercise using wrist-type PPG signals. In: 2015 37th Annual
3. Essalat, M., Mashhadi, M.B., Marvasti, F.: Supervised heart rate tracking using wrist-type photoplethysmographic (PPG) signals during physical exercise without simultaneous acceleration signals. IEEE Glob Conf. Signal Inform. Process. (GlobalSIP) **2016**, 1166–1170 (2016). https://doi.org/10.1109/GlobalSIP.2016.7906025
4. Zhang, Z., Pi, Z., Liu, B.: TROIKA: A general framework for heart rate monitoring using wrist-type photoplethysmographic signals during intensive physical exercise. IEEE Trans. Biomed. Eng. **62**(2), 522–531 (2015). https://doi.org/10.1109/TBME.2014.2359372
5. Gupta, Shresth, Singh, Anurag, Sharma, Abhishek: Dynamic large artery stiffness index for cuffless blood pressure estimation. IEEE Sens. Lett. **6**(3), 1–4 (2022)
6. Hasanzadeh, N., Ahmadi, M.M., Mohammadzade, H.: Blood pressure estimation using photoplethysmogram signal and its morphological features. IEEE Sens. J. **20**(8), 4300–4310, 15 Apr 15, 2020, https://doi.org/10.1109/JSEN.2019.2961411.
7. Mafi, M., Rajan, S., Bolic, M., Groza, V.Z., Dajani, H.R.: Blood pressure estimation using maximum slope of oscillometric pulses. Annu. Int. Conf. IEEE Eng. Med. Biol. Soc. **2012**, 3239–3242 (2012). https://doi.org/10.1109/EMBC.2012.6346655
8. Podaru, A.C., David, V.: Blood pressure estimation based on synchronous ECG and PPG recording. Int. Conf. Exposition Electr. Power Eng. (EPE) **2020**, 640–645 (2020). https://doi.org/10.1109/EPE50722.2020.9305544
9. Kumar, S., Ayub, S.: Estimation of blood pressure by using electrocardiogram (ECG) and photo-plethysmogram (PPG). Fifth Int. Conf. Commun. Syst. Netw. Technol. **2015**, 521–524 (2015). https://doi.org/10.1109/CSNT.2015.99
10. Li, P., Laleg-Kirati, T.-M.: Central blood pressure estimation from distal PPG measurement using semi classical signal analysis features. IEEE Access **9**, 44963–44973 (2021). https://doi.org/10.1109/ACCESS.2021.3065576
11. Kachuee, M., Kiani, M.M., Mohammadzade, H., Shabany, M.: Cuff-less high-accuracy calibration-free blood pressure estimation using pulse transit time. In: IEEE International Symposium on Circuits and Systems (ISCAS'15) (2015)
12. Gupta, S., Singh, A., Sharma, A.: Photoplethysmogram based mean arterial pressure estimation using LSTM. In: 2021 8th International Conference on Signal Processing and Integrated Networks (SPIN), pp. 806–811. IEEE (2021)

13. Takazawa, K.T.N., Fujita, M., Matsuoka, O., Saiki, T., Aikawa, M., Tamura, S., Ibukiyama, C.: Assessment of vasoactive agents and vascular aging by the second derivative of photoplethysmogram waveform. Hypertension **32**, 365–370 (1998)
14. Padilla, J.M., Berjano, E.J., Saiz, J., Facila, L., Diaz, P., Merce, S.: Assessment of relationships between blood pressure, pulse wave velocity and digital volume pulse. Comput. Cardiol. 893–896 (2006)
15. Rubins, U., Grabovskis, A., Grube, J., Kukulis, I.: Photoplethysmography Analysis of Artery Properties in Patients with Cardiovascular Diseases. Springer, Berlin (2008)
16. Imanaga, I., Hara, H., Koyanagi, S., Tanaka, K.: Correlation between wave components of the second derivative of plethysmogram and arterial distensibility. Jpn Heart J. **39**, 775–784 (1998)
17. Masé, M., Mattei, W., Cucino, R., Faes, L., Nollo, G.: Feasibility of cuff-free measurement of systolic and diastolic arterial blood pressure. J. Electrocardiol. **44**(2), 201–207 (2011)

Low-Cost Hardware-Accelerated Vision-Based Depth Perception for Real-Time Applications

N. G. Aditya, P. B. Dhruval, S. S. Shylaja, and Srinivas Katharguppe

Abstract Depth estimation and 3D object detection are critical for autonomous systems to gain context of their surroundings. In recent times, compute capacity has improved tremendously, enabling computer vision and AI on the edge. In this paper, we harness the power of CUDA and OpenMP to accelerate ELAS (a stereoscopic vision-based disparity calculation algorithm) and 3D projection of the estimated depth while performing object detection and tracking. We also examine the utility of Bayesian inference in achieving real-time object tracking. Finally, we build a drive-by-wire car equipped with a stereo camera setup to test our system in the real world. The entire system has been made public and easily accessible through a Python module.

Keywords Robot vision · Stereoscopic vision · Rgb-d perception · 3d object tracking · Disparity · Depth estimation · CUDA · OpenMP · Bayesian inference · Parallel computing · Deep Q-learning

1 Introduction

Due to the ever-increasing dependence on robots for automation, they need to be able to perceive their environment with reliable accuracy. Depth information is crucial in such systems. A photograph is a projection of a three-dimensional scene onto a two-dimensional plane (losing out on depth). With just one more image of the

N. G. Aditya (✉) · P. B. Dhruval · S. S. Shylaja · S. Katharguppe
Department of Computer Science, PES University, Bengaluru, Karnataka 560085, India
e-mail: adityang5@gmail.com

P. B. Dhruval
e-mail: dhruvalpb@pesu.pes.edu

S. S. Shylaja
e-mail: shylaja.sharath@pes.edu

S. Katharguppe
e-mail: srinivasks@pes.edu

M. Tistarelli et al. (eds.), *Computer Vision and Machine Intelligence*,
Lecture Notes in Networks and Systems 586,
https://doi.org/10.1007/978-981-19-7867-8_22

Fig. 1 Our system in action on the KITTI dataset. The process of going from a stereo pair to 3D environment map involves disparity estimation, object bounding box prediction, and back projection into 3D. We have accelerated this entire pipeline using CUDA and OpenMP

same scene taken from a slightly different perspective, we can extract depth, allowing us to visualize the scene in three dimensions: width, height, and depth. This is called stereoscopic vision where depth is measured as the parallax error between two images from a stereoscopic source, much like the human visual system. It is critical to emphasize that a depth map alone is insufficient for gaining context of the environment; we also need to detect the objects within it. Various object recognition neural networks enable this. Once the objects have been detected in the depth map, it is advantageous for a moving system to be able to follow and anticipate their positions in the near future. This data can be used by path planning algorithms (Fig. 1).

In this paper, we accelerate ELAS (Efficient LArge-scale Stereo Matching) [1] by parallelizing it on both the CPU (via OpenMP [2]) and the GPU (via CUDA [3]) in order to achieve real-time disparity calculation. We have also projected the estimated depth map, obtained from the disparity, onto 3D space using the parallelism offered by GPUs and visualized it using OpenGL [4]. To sum up, our system receives video feeds from a pair of calibrated stereo cameras and generates a 3D depth map with bounding boxes drawn around objects detected as well as their expected future positions.

We then compare our vision-based depth perception system with LiDAR in a real-time simulated environment by training a Deep Q Learning agent to complete an autonomous driving task. Motivated by the promising results obtained, we built a drive-by-wire car equipped with a stereo camera setup to put our system to the test in the real world. We also investigate the efficacy of Bayesian inference in achieving real-time object tracking. The entire system is now open to the public and easily accessible via a Python module (Github Link).

2 Related Work

Path planning for a robot demands an accurate depth map of its environment. Using an array of Time of Flight (ToF) sensors, such as LiDAR, which involves illuminating a target scene with a laser and measuring differences in laser return times and wavelengths to reconstruct the scene in three dimensions, has been the traditional approach. Vision-based depth perception, on the other hand, necessitates accurate and fast disparity calculation methods. Due to increased compute power, using purely vision-based approaches has recently become viable and is seen as the less expensive option of the two. The paper "Pseudo-LiDAR from Visual Depth Estimation" [5] made strides in 3D object detection from 2D stereo images by performing object detection on 3D depth data rather than disparity, thus mimicking the LiDAR signal. Such research and advancements have increased trust in vision-based depth mapping approaches. There are two primary approaches to vision-based depth perception: feature engineering and neural networks.

2.1 *Feature Engineering Methods*

Feature engineering methods use image features such as edges, gradient changes, and shapes to match blocks of pixels in one image to another, hence the name block matching. Efficient LArge-scale Stereo [1] abbreviated as ELAS is a Bayesian approach to stereo matching. It builds a prior on the disparities by forming a triangulation on a set of support points that can be robustly matched, reducing the matching ambiguities of the remaining points. This allows for efficient exploitation of the disparity search space, yielding accurate dense reconstruction without the need for global optimization. We parallelized ELAS using OpenMP [2] and CUDA [3] to achieve faster processing times.

2.2 *Learned Approach (Neural Networks)*

Various neural network architectures like convolutional neural networks (CNNs) [6], variational auto-encoders (VAEs) [7], recurrent neural networks (RNNs) [8], and generative adversarial networks (GANs) [9] have manifested their effectiveness to address the problem of depth estimation. Monocular (depth predicted from a single input image) and binocular (depth predicted from a pair of images fed as input to the neural network) are the two methods in this approach. A more detailed and well-written overview of monocular depth estimation approaches can be found in [10]. Furthermore, binocular learned approaches like [11–14] can produce more accurate depth maps as they take advantage of epipolar constraints on the possible disparity values. While these methods have improved accuracy, they come at a cost in compute time and/or hardware requirements. It becomes impossible to run such methods in real time on an embedded device that may not be powerful enough without specialized hardware like FPGAs.

3 System Design

The design of our implementation is shown in Fig. 2. The system receives images from a calibrated camera pair as input. ELAS disparity generation is run on the left image with respect to the right image. At the same time, YOLOv4-Tiny [15] object detection is run on the left image to generate 2D bounding boxes. The disparity map and 2D bounding boxes of the detected objects are then projected onto 3D space using the stereo camera pair's calibration parameters to obtain a point cloud. We then use an OpenGL graphing tool that we built to render this point cloud in 3D.

3.1 Point Cloud Generation

For a given scene viewed through a set of cameras, once we compute its disparity map, a corresponding 3D point cloud can be constructed provided the camera calibration parameters are known. Perspective projection can be modeled using the equation shown in (1) where (u, v) are the pixel coordinates in the image, (x, y, z) are the coordinates of the point in the camera coordinate frame, o_x and o_y are offsets that deal with shifting the coordinates from top left corner of the camera to its Principle Point (the point where the optical axis pierces the sensor), and (f_x, f_y) are the effective focal lengths of the camera in pixels in the x and y directions, respectively.

$$(u, v) = \left(f_x \times \frac{x}{z} + o_x, f_y \times \frac{y}{z} + o_y\right) \tag{1}$$

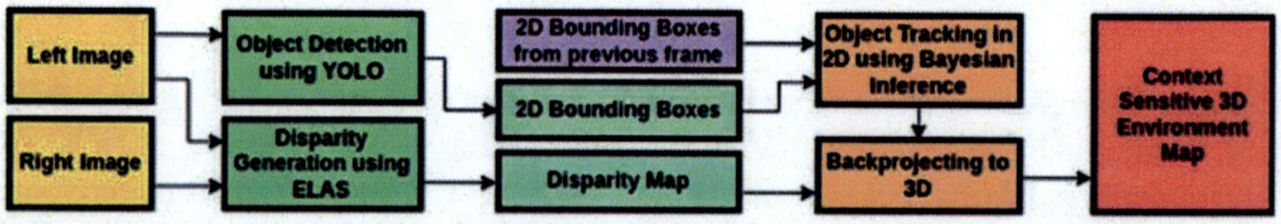

Fig. 2 Stereo input from a calibrated camera pair is used to compute a disparity map using our accelerated version of ELAS. At the same time, objects in the left image are detected using YOLOv4-Tiny and Bayesian inference is used to track them in 2D. Disparity map and camera calibration data are then used to project the image along with the bounding boxes (drawn around the objects detected) from 2D to 3D

Using (1), we can write the 2D to 3D transformation as shown in (2). It is worth noting that the projection from 3D to 2D yields a single point, whereas the reverse, for a single camera, yields a ray in 3D. Hence, we add another camera as shown in (3), where b is the baseline length between the two cameras, (u_l, v_l) is the point in the left camera's image and (u_r, v_r) is the point in the right camera's image.

$$(x, y) = \left(\frac{z \times (u - o_x)}{f_x}, \frac{z \times (v - o_y)}{f_y}\right), z > 0 \tag{2}$$

$$\begin{aligned} (u_l, v_l) &= \left(f_x \times \frac{x}{z} + o_x, f_y \times \frac{y}{z} + o_y\right) \\ (u_r, v_r) &= \left(f_x \times \frac{x - b}{z} + o_x, f_y \times \frac{y}{z} + o_y\right) \end{aligned} \tag{3}$$

We then solve for the point of intersection of the rays from each camera to find the location of the point in 3D as shown in (4). The denominator term $(u_l - u_r)$ is known as the disparity $(d(x, y))$, and the 3D depth (z) of a point is inversely proportional to its disparity. This disparity is computed by observing a pattern in the left image and scanning along the epipolar line to find a match in the right image and calculating the distance between them. We have parallelized Support Match Search, Delaunay Triangulation, Disparity Plane Computation, Disparity Grid Creation, Matching, and Adaptive Mean Filtering in the ELAS algorithm.

$$(x, y, z) = \left(\frac{b(u_l - o_x)}{(u_l - u_r)}, \frac{bf_x(v_l - o_y)}{f_y(u_l - u_r)}, \frac{bf_x}{(u_l - u_r)}\right) \tag{4}$$

3.2 3D Object Tracking Using Bayesian Inference

As humans, when we move through our environment, our actions not only depend on what we see currently but are also affected by our understanding of the environment and the objects in it (our prior knowledge) and also the recent history of the state of the environment (a measure of likelihood). Keeping track of objects from one frame to the next allows us to predict their positions in subsequent frames which could then be used for planning ahead.

We perform object detection on the left input image for multiple classes using YOLOv4-Tiny. These objects are detected in 2D space. We transform these 2D bounding boxes into 3D using the calculated disparity map of the frame and the camera calibration parameters. The position vector of the object in 3D, (X_c, Y_c, Z_c), is determined by averaging the position vectors of all the n points contained within the 2D bounding box as described by (5) where (x_i, y_i) is within the 2D bounding box.

$$\begin{bmatrix} X_c \\ Y_c \\ Z_c \end{bmatrix} = \frac{\sum_i^n \begin{bmatrix} x_i \\ y_i \\ z_i \end{bmatrix}}{n} \tag{5}$$

We employ Bayesian inference to track objects between multiple frames. According to Bayes Theorem, for two events θ and r,

$$P(r|\theta) = \frac{P(\theta|r) \times P(r)}{\int P(\theta|r') \times P(r')\mathrm{d}r'}. \tag{6}$$

Here, $P(\theta|r)$ is the Likelihood, $P(r)$ is the Prior, and $\int P(\theta|r') \times P(r')\mathrm{d}r'$ is the Normalizing Constant. We can think of r as the change in position vector (direction of motion) of an object and θ as its recent history of the direction of motion. Then, the above equation can be used to answer the question: "What is the probability of the object moving in the direction r given its recent history of motion θ?".

By performing object tracking in 2D and projecting the predicted position onto 3D, we save computational complexity by eliminating the need to perform tracking in the third dimension. This is possible as the depth data gets encoded in the depth map and can be extracted from there. Note that any further mentions of r will be referring to the change in the 2D position vector of the object being tracked. Any mean or standard deviation applied on r will hence be the mean or standard deviation of the change in position vector. We have chosen the Prior and Likelihood to follow Normal distributions as this seems to accurately model random object movement. Hence,

$$\text{Prior} = \frac{\exp(-(r - \mu_P)^2/2\sigma_P^2)}{\sigma_P\sqrt{2\pi}}, \tag{7}$$

$$\text{Likelihood} = \frac{\exp(-(r - \mu_L)^2/2\sigma_L^2)}{\sigma_L\sqrt{2\pi}}. \tag{8}$$

The probability, $P(r|\theta)$ of the object moving in the direction r, given the prior knowledge that r behaves as a normally distributed variable defined by $\{\mu_P, \sigma_P\}$, its recent history (Likelihood) behaving as a normally distributed variable defined by $\{\mu_L, \sigma_L\}$, and assuming the normalizing constant to be N, is given by the equation,

$$P(r|\theta) = \frac{\text{Prior} \times \text{Likelihood}}{N}. \tag{9}$$

Maximizing (9), we obtain:

$$r = \frac{\mu_L.\sigma_P^2 + \mu_P.\sigma_L^2}{\sigma_P^2 + \sigma_L^2}. \tag{10}$$

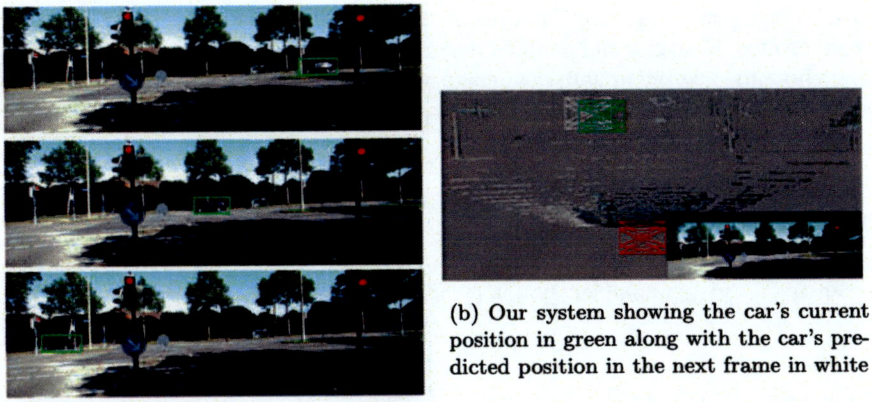

(a) Car moving through an intersection from right to left

(b) Our system showing the car's current position in green along with the car's predicted position in the next frame in white

Fig. 3 Bayesian inference prediction

The above equation holds if we were to make reasonable assumptions of what the Prior distribution is. We could also obtain a simplified version of the predictor by taking the limit of the prior's mean to be zero and its standard deviation to be infinity. In such a scenario, the mean of the recent history of position vectors gives us the direction in which the object is most likely to move in. With this knowledge, when a set of frames in the sequence are considered, the path traced out by the objects in the scene can be computed, and hence, we can predict the positions of the objects in subsequent frames.

Figure 3b shows one case where the system predicts the position of a black car as it passes through the intersection. The car (marked with a green box) starts on the right of the frame and travels toward the left. Given the context of the three images in Fig. 3a, it becomes clear that the object is traveling from right to left. On each frame of this scene, the system is predicting the vehicle position (marked as a white box) in the subsequent frame with less than 15 pixels of error. The position of the camera setup is marked as a red box.

4 Dataset

The KITTI dataset, made by equipping a vehicle with two pairs of color and grayscale video cameras, having a total of four cameras, has been used for evaluating our work. In total, this dataset has 6 h of diverse scenarios, capturing real-world traffic situations that range from freeways over urban and rural areas with many dynamic and static objects. Accurate ground truth is provided by a Velodyne laser scanner and a GPS localization system. The data is calibrated, timestamped, and synchronized and rectified. Both raw and rectified image sets are available to use from the dataset. We chose to experiment with raw image sequences [16].

5 Experimental Evaluation

All experimental evaluation was done on our laptop with an Intel i7-10750H @ 2.60GHz × 12 and Nvidia GeForce GTX 1650 Ti (Mobile).

5.1 Disparity Generation

As seen in Table 1, our version of ELAS shows a 43.56% improvement over the stock version. When subsampling is enabled, only every second pixel is evaluated, which is generally adequate in robotics applications. This would still produce a much denser point cloud compared to a sparse LiDAR. With this option turned on, our version shows a 30.67% improvement over stock ELAS. Processing times per frame were measured on image pairs (of size (375, 1242) each) from multiple sequences of the KITTI raw dataset and were averaged to obtain the results shown in 1. Additionally, our accelerated version of ELAS inherits the original ELAS implementation's accuracy. In this table, D1-all is the percentage of stereo disparity outliers in the frame, averaged over all ground truth pixels (a metric defined by the KITTI dataset). This table also shows how our implementation of ELAS compares to learned approaches like LEAStereo [11], Lac+GwcNet [12], and UPFNet [14] in terms of Runtime and accuracy. Though ELAS is outperformed in accuracy, it more than compensates with speed, which is crucial for small-scale robotics applications running on edge devices.

Table 1 Comparing the performance of different depth perception algorithms on a pair of 375 × 1242 images

Algorithm	Processing time (ms)	D1-all^{+} (%)
Lac+GwcNet	650.00	1.77
LEAStereo	300.00	1.65
UPFNet	250.00	1.62
Stock ELAS	169.34 (67.00[a])	9.72
Parallelized ELAS (Ours)	95.56 (46.45[a])	9.72

[a] With subsampling

[b] These results are obtained after comparing the predictions to ground truth LiDAR data from the KITTI dataset

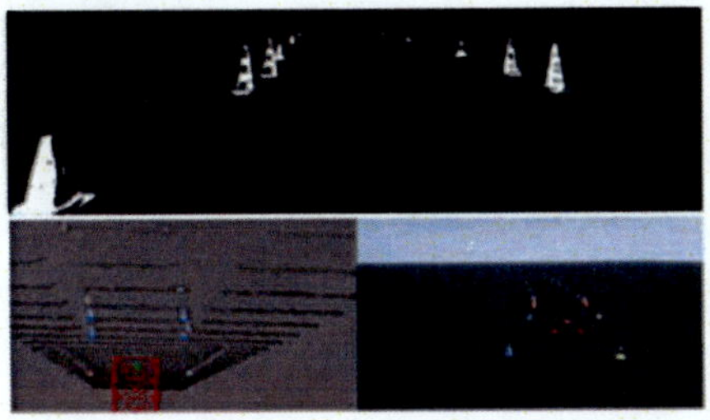

(a) Color masking is used to segment cones.

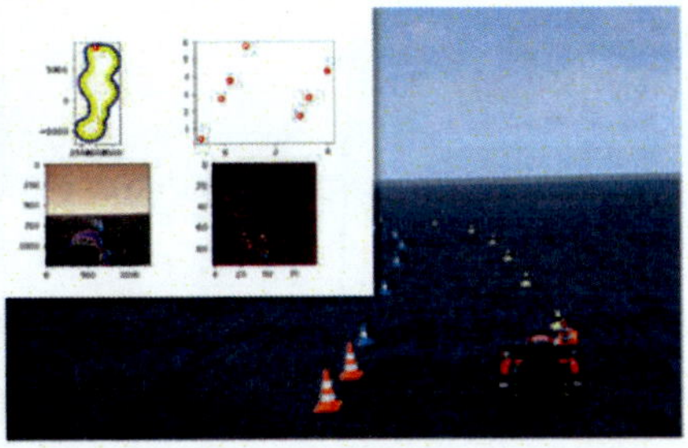

(b) A Top Down projection of the point-cloud generated (Bird's eye view) is fed as input to the DQN agent

(c) The DQN agent was trained in three stages. In the first stage, it was punished for exiting the track or colliding with a cone. In the second stage, the reward function was modified to also encourage the agent to maintain higher speeds throughout the track. Finally, in the third stage, lap times were also accounted for while calculating the reward. This was done to prevent the agent from receiving too many punishments in the initial stages and teach it multiple concepts gradually.

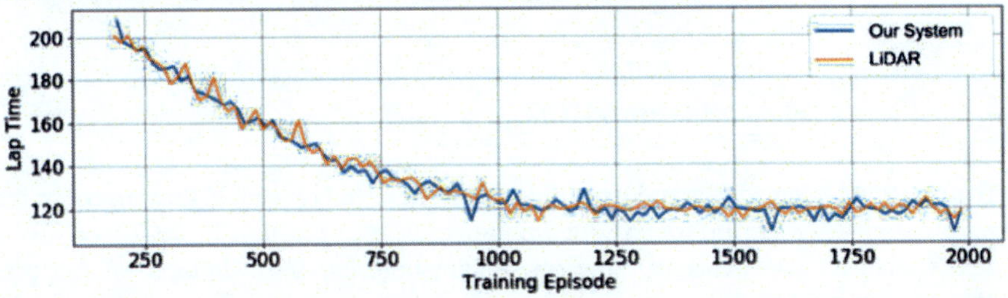

(d) Comparing the performance of the DQN agent (Lap time) when using the BEV from our Vision Based Depth Perception vs Lidar as input

Fig. 4 DQN agent training in FSDS

5.2 *Formula Student Driverless Simulator*

With the objective of demonstrating our system's utility, we trained a DQN agent to drive a car within the Formula Student Driverless Simulator based on Microsoft's Airsim [17]. As shown in Figs. 4a and 4b, we set up a track with cones and calibrated a stereo camera pair (placed on the car's nose cone) using a checkerboard pattern in

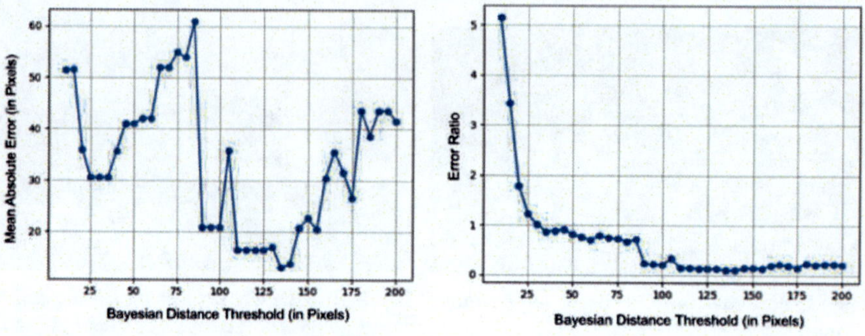

(a) Mean Absolute Error vs Threshold (b) Error Ratio vs Threshold

Fig. 5 Finding the optimum threshold value

the simulator to obtain the distortion, rectification, and projection matrices for the same. The DQN agent was fed the top down projection (Bird's eye view) of the 3D point cloud generated by our vision system as input and was trained for 2000 episodes in stages as shown in Fig. 4c. It converged at around 1200 episodes. We then trained another DQN agent in a similar manner with LiDAR replacing our vision-based depth perception system and compared their performance as shown in Fig. 4d. From this graph, it is evident that our vision-based system shows performance comparable if not slightly better than that of LiDAR, in terms of number of episodes to converge and Lap Time after convergence.

5.3 Bayesian Prediction Error

To predict the upcoming x and y positions of all detected objects in the current frame, we employ Bayesian inference. As we know the true position of the object in the next frame, we can measure the error. A threshold has been defined for mapping objects across multiple frames and if an object in the previous frame is beyond this distance in pixels from the object in the current frame, our system considers them to be separate entities. Based on our observations of the mean error and error ratio (calculated as the ratio of mean error to the distance threshold), we set the threshold to 135 pixels to obtain a mean position prediction error of around 13 pixels as shown in Fig. 5a. Note that this is specific to the KITTI dataset and the threshold will change depending on the use case. We plot error ratio vs the aforementioned threshold as shown in Fig. 5b and use the elbow method to find the optimum threshold value.

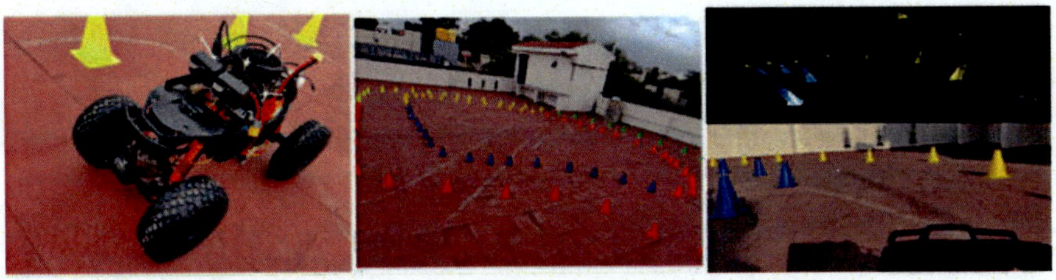

Fig. 6 Our drive-by-wire car equipped with a stereo camera system and the test track it was driven on. Real-time rendering of the 3D environment projection is taking place 50 Hz

6 Real-World Testing

After obtaining promising results in simulation, we constructed a small drive-by-wire car and mounted a stereo camera system comprising of two 640×480 VGA cameras as shown in Fig. 6, to test our system in the real world. The camera pair was calibrated using "stereoCameraCalibrator", part of the Computer Vision toolbox in Matlab [18]. The car was driven around the track and the video stream from the stereo pair was fed in as input to our system to produce 3D projections of the environment 50 Hz. Such a signal can be fed into SLAM systems to produce 3D scene mapping that is superior when compared to pure monocular SLAM. This is done in real time all while not needing any LiDAR units. Our system is thus able to provide a rich signal with low latency which is ideal for real-time applications.

7 Conclusion and Future Work

Though our system performs well in the simulation, it is still prone to calibration errors that arise in the real world where camera setups have intrinsic and extrinsic errors (neither distortion free nor perfectly geometric). As of today, extremely accurate results (rivaling that of LiDARs) cannot be achieved in real time in the real world purely from vision. However, cost of the hardware is greatly reduced with some trade-off in accuracy. Hence, after extensive experimentation with vision-based depth perception and LiDAR, a conclusion can be drawn that, if the use case is for unidirectional short-range applications, such as small robots traveling at moderate speeds, a good stereo system can even eradicate the need for expensive Time of Flight (ToF) sensors like LiDAR and hence aid toward much cost-cutting. Techniques like Bayesian inference to predict object position in subsequent frames are effective and have real-world applications. Using a more effective block-matching approach or a learned approach to disparity map generation can improve prediction results.

References

1. Aleotti, F., Tosi, F., Poggi, M., Mattoccia, S.: Generative adversarial networks for unsupervised monocular depth prediction. In: Proceedings of the European Conference on Computer Vision (ECCV) Workshops, p. 0 (2018)
2. OpenMP Architecture Review Board: OpenMP application program interface version 3.0 (Nov 2015), https://www.openmp.org/wp-content/uploads/openmp-4.5.pdf
3. Vingelmann, P., Fitzek, F.H.: Cuda, release: 10.2.89. NVIDIA (2020), https://developer.nvidia.com/cuda-toolkit
4. Group, K.: The OpenGL® graphics system: A specification. https://www.khronos.org/registry/OpenGL/specs/gl/glspec46.core.pdf (Oct 2019)
5. Wang, Y., Chao, W.L., Garg, D., Hariharan, B., Campbell, M., Weinberger, K.Q.: Pseudo-lidar from visual depth estimation: Bridging the gap in 3d object detection for autonomous driving. In: Proceedings of the IEEE/CVF Conference on Computer Vision and Pattern Recognition, pp. 8445–8453 (2019)
6. Garg, R., Bg, V.K., Carneiro, G., Reid, I.: Unsupervised CNN for single view depth estimation: Geometry to the rescue. In: European conference on computer vision, pp. 740–756. Springer, Berlin (2016)
7. Chakravarty, P., Narayanan, P., Roussel, T.: Gen-slam: Generative modeling for monocular simultaneous localization and mapping. In: 2019 International Conference on Robotics and Automation (ICRA), pp. 147–153. IEEE (2019)
8. Wang, R., Pizer, S.M., Frahm, J.M.: Recurrent neural network for (un-) supervised learning of monocular video visual odometry and depth. In: Proceedings of the IEEE/CVF Conference on Computer Vision and Pattern Recognition, pp. 5555–5564 (2019)
9. Aleotti, F., Tosi, F., Poggi, M., Mattoccia, S.: Generative adversarial networks for unsupervised monocular depth prediction. In: Proceedings of the European Conference on Computer Vision (ECCV) Workshops, p. 0 (2018)
10. Zhao, C., Sun, Q., Zhang, C., Tang, Y., Qian, F.: Monocular depth estimation based on deep learning: an overview. Sci. China Technol. Sci. 1–16 (2020)
11. Cheng, X., Zhong, Y., Harandi, M., Dai, Y., Chang, X., Li, H., Drummond, T., Ge, Z.: Hierarchical neural architecture search for deep stereo matching. Adv. Neural Inf. Proc. Syst. **33** (2020)
12. Liu, B., Yu, H., Long, Y.: Local similarity pattern and cost self-reassembling for deep stereo matching networks (2021). 10.48550/ARXIV.2112.01011, https://arxiv.org/abs/2112.01011
13. Mao, Y., Liu, Z., Li, W., Dai, Y., Wang, Q., Kim, Y.T., Lee, H.S.: Uasnet: Uncertainty adaptive sampling network for deep stereo matching. In: 2021 IEEE/CVF International Conference on Computer Vision (ICCV), pp. 6291–6299 (2021). 10.1109/ICCV48922.2021.00625
14. Wu, Z., Wu, X., Zhang, X., Wang, S., Ju, L.: Semantic stereo matching with pyramid cost volumes. In: 2019 IEEE/CVF International Conference on Computer Vision (ICCV), pp. 7483–7492 (2019). 10.1109/ICCV.2019.00758
15. Bochkovskiy, A., Wang, C.Y., Liao, H.Y.M.: Yolov4: Optimal speed and accuracy of object detection. arXiv preprint arXiv:2004.10934 (2020)
16. Geiger, A., Lenz, P., Stiller, C., Urtasun, R.: Vision meets robotics: The kitti dataset. Int. J. Rob. Res. **32**(11), 1231–1237 (2013)
17. Shah, S., Dey, D., Lovett, C., Kapoor, A.: Airsim: High-fidelity visual and physical simulation for autonomous vehicles. In: Field and service robotics. pp. 621–635. Springer, Berlin (2018)
18. The Mathworks, Inc., Natick, Massachusetts: MATLAB version 9.11.0.1837725 (R2021b) (2021)

Development of an Automated Algorithm to Quantify Optic Nerve Diameter Using Ultrasound Measures: Implications for Optic Neuropathies

Vishal Gupta, Maninder Singh, Rajeev Gupta, Basant Kumar, and Deepak Agarwal

Abstract The paper presents a novel computational and image processing algorithm for automatic measurement of optic nerve diameter (OND) from B-scan ultrasound images acquired in a traumatic cohort. The OND is an important diagnostic parameter for the detection and therapeutic planning of several diseases and trauma cases. The automated measurement of OND may provide reliable information for predicting intracranial pressure (ICP) in traumatic patients. In the proposed method, the automatic measurement of the OND involves pre-processing an ultrasound image, followed by retinal globe detection, optic nerve localization, optic nerve tip detection and measurement of OND. The proposed algorithm measures the OND in the perpendicular direction to the optic nerve axis, considering the orientation of the optic nerve in an automated framework. The developed algorithm automatically calculates the OND. The study includes Twenty-four traumatic individuals with optic nerve pathologies whose B-scan ultrasound images of the optic nerve were manually obtained in the axial plane. The accuracy of the automatic measurement has been quantified by comparing it with manually measured OND by medical experts. A low Percent Root Mean Square Difference (PRD) value of 14.92% between automated and manually measured OND is found as an accuracy measure for the automatic measurement of OND. The proposed algorithm automatically and accurately determines the optic nerve diameter from an eye ultrasound image.

Keywords Optic nerve diameter · Intracranial pressure · Ultrasound

---

V. Gupta
Centre for Development of Telematics, Telecom Technology Centre of Government of India, New Delhi, India

M. Singh (✉) · R. Gupta · B. Kumar
Electronics and Communication Engineering Department, Motilal Nehru National Institute of Technology Allahabad, Prayagraj, India
e-mail: manibiet04@gmail.com

D. Agarwal
JPNATC, All India Institute of Medical Sciences, New Delhi, India

M. Tistarelli et al. (eds.), *Computer Vision and Machine Intelligence*, Lecture Notes in Networks and Systems 586,
https://doi.org/10.1007/978-981-19-7867-8_23

1 Introduction

Optic nerve sonography with the B-scan technique has been standardized and implemented mainly for the assessment of ocular disorders [1–3]. The optic nerve and its sheath diameters have been investigated in healthy and impaired subjects using A-scan and B-scan ultrasonography, CT scan, and MRI [4–9]. The variation in the normal range of OND reported in various studies may be due to different ethnicities and races, different applied methodologies, populations, and sizes. The manual measurement of the OND is highly dependent on the skill and experience levels of the neuroradiologist and it may lead to false-positive diagnostics in certain situations. For this, an ultrasound image was chosen, and with the help of a caliper instrument (using lines and a marker), the neuroradiologist determined the OND manually. During the manual OND measurement, the neurologist marked a horizontal line at 3 mm below the optic nerve to find OND, without considering the orientation of the optic nerve axis from vertical; this demands accurate measurement of OND. Since measuring these diagnostic parameters needs experienced and a trained medical expert/ radiologist, precise measurement of these parameters is an open research problem. Few authors have reported the research studies related to OND measurement through ultrasonography. Chen et al. [10] reported the OND and ONSD for healthy Chinese adults. They correlated OND with ONSD with a median OND/ONSD ratio of 0.63 (range from 0.59 to 0.67), independent of gender, age, height, weight, and transverse eyeball diameter. Wolf et al. [11] compared OND measurements to an MRI sequence in 33 German populations and concluded the MRI was more precise than ultrasound. The change of correlation of OND with ONSD will occur, and therefore the OND/ONSD ratio might be helpful in detecting intracranial hypertension [12].

This study presents a novel computational and image processing algorithm for automatically measuring OND from ultrasound images acquired in a traumatic cohort. The main contributions of the presented work are mentioned as follows:

a. The automated algorithm is developed for the OND measurement using B-scan ultrasound images of traumatic patients. The algorithm is tested and compared with the manual measurement performed by the radiologists.
b. The proposed algorithm calculates the OND in the perpendicular direction to the optic nerve axis, considering the orientation of the optic nerve and thus, providing a more clinically accurate OND measurement.
c. The algorithm automatically measures OND that can be effectively used in clinical settings by embedding the algorithm into the ocular ultrasound machines to achieve accurate, repeatable, and operator-independent results.
d. Computing the Percent root mean Difference (PRD) between algorithm generated and the manually measured OND values, the accuracy of the proposed algorithm is validated.

The automated measurement of OND can substantially improve the accuracy of the measurement leading to diagnostic conclusions. To the best of our knowledge, no research work has been reported on automated measurement of OND from an eye ultrasound image. Thus, this study proposes a novel ultrasound image post-processing algorithm for the automatic measurement of OND. The rest of the paper is organized as follows. In Sect. 2, the method is explained in detail, which includes the image processing steps needed for the measurement of the OND. Section III describes the experimental results for the proposed method. Finally, Sections IV and V provide discussion and overall conclusion, respectively.

2 Methods and Materials

The eye ultrasound image dataset of 24 adult patients with traumatic brain injury (TBI) was acquired in DICOM format from Trauma Center of All India Institute of Medical Sciences (AIIMS), India. Patients included adults of ages ranging between 20 and 55 years. Our proposed method developed a novel digital image processing and intelligent computing algorithm for the automated measurement of OND from an eye ultrasound image. Obtaining diagnostic information from ultrasound images is a challenging task due to its blurry nature and presence of speckle noise [13]. The proposed algorithm for automated measurement of OND eliminated the manual dependence to reduce such limitations. The acquired images were first converted into lossless BMP format. The proposed OND measurement algorithm was designed and implemented in Python using the OpenCV library (Open-Source Computer Vision) and several other supporting libraries, as shown in Fig. 1. The algorithm is broadly divided into three functional steps: (Step-1) Image pre-processing and retina globe detection, (Step-2) Optic nerve localization and segmentation, and (Step-3) OND calculation. Detailed operations of each functional block diagram are explained in the following sections:

2.1 Image Pre-processing and Retina Globe Detection

In this step, the algorithm is first performed to read the original image of the B-scan ultrasound, as shown in Fig. 2a, followed by pre-processing and filtering operations to remove the unwanted part and associated noise in the target ultrasound image

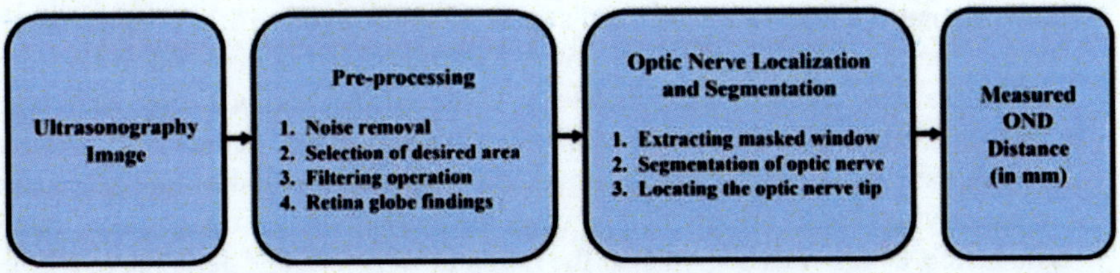

Fig. 1 Block diagram of proposed automated measurement of OND

(a) (b) (c)

Fig. 2 **a** Original image, **b** pre-processed, and **c** thresholded binarized image by filtering

[14–17]. Apply erosion operation and median filtering with a kernel size of 5 × 5 to smooth the ultrasound image and remove speckle noise. Figure 2b illustrates the pre-processing of an ultrasound image.

It is crucial to automatically identify the retina structure and separate it from the optic nerve region in the original image. The unwanted black portion other than the retina globe captures the retinal area; a white border is created along the boundary of the image and then thresholding the ultrasound image to yield a complete black retina region as depicted in Fig. 2c. Thus, the steps in this algorithm detected the retina globe in the image by ellipse fitting in the largest contour present in the target image, as shown in Fig. 3.

2.2 Optic Nerve Localization and Segmentation

In this step, using the ellipsoidal contour of the retinal globe, the algorithm used ellipse enlargement, masking and binary thresholding methods to extract an approximate window that would indeed contain a part of the optic nerve. The image window containing the optic nerve was obtained by finding the largest black contour in a masked window in the neighborhood of the bottom part of an ellipse fitted in the retina globe detected in previous step. The optic nerve localization and segmentation steps are explained in detail as follows:

1. The multiple black ellipses obtained have a centroid, minor axis, and major axis and orientation. Their start and end angles range from 60 to 120 degrees (angle taken clockwise from horizontal). Vary the minor and major axis in the loop (130 iterations taken here) to create a whole ellipse region to get an image. This region acts as a window to mask the original image so that only specific probabilistic localization areas are covered. The extracted window image is added to the original image arithmetically to get the masked image in Fig. 4 illustrated, and hence, a particular region near the retina is obtained. The optic nerve will lie somewhere within this elliptical window.

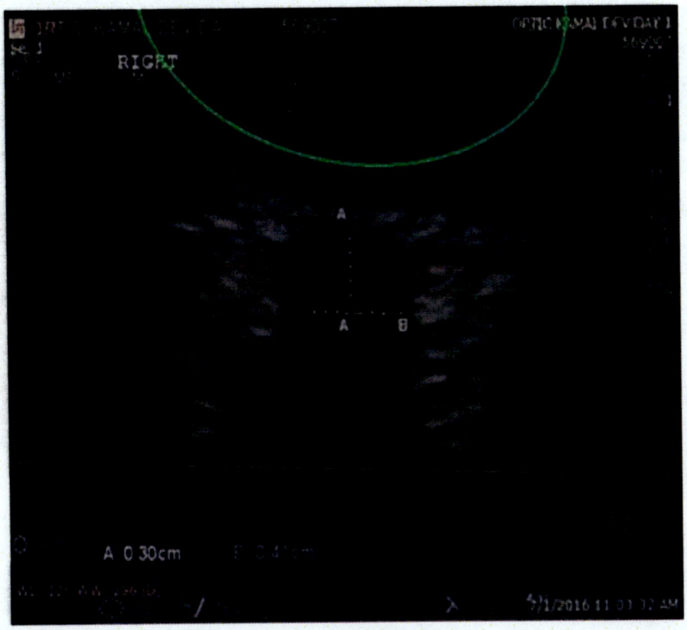

Fig. 3 Retina globe detection: fitted ellipse in retinal part

Fig. 4 Window image added to get masked image

2. Apply binary thresholding (Otsu's binarization), which automatically finds the optimum threshold value based on peaks in the histogram. The filtering methods and morphological operations (opening and closing to remove a small group of pixels and fill up holes, respectively) performed to get a binary image are observed in Fig. 5.
3. Find all contours in the binary image and find the area-wise largest contour. Determine moments of contour and hence calculate centroid to get a rough estimate of the location of the optic nerve. The contour plot with its centroid and rectangle to be extracted is demonstrated in Fig. 6.

Fig. 5 Binarized image after thresholding and filtering

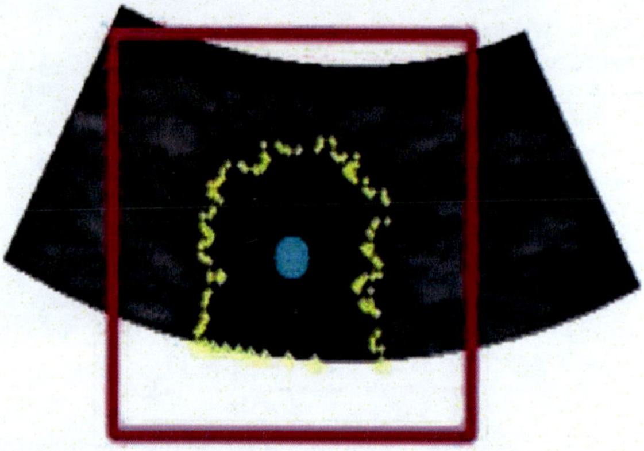

Fig. 6 Masked image plotted with the largest contour with its centroid

2.3 Optic Nerve Diameter Measurement

The optic nerve diameter window extracted in step 2 above was processed further to find the OND measurement. After a series of processing and filtering stages, the optic nerve was automatically segmented, and a contour was fitted inside the optic nerve. Again, a tip of the optic nerve was located as refer in Fig. 7 and using mathematical equations, OND was determined at a distance of 3 mm from the tip of the optic nerve.

Fig. 7 Grayscale image with contour and line to find two points p_1 and p_2 with contour

The coordinate of the point $(c_{x,\text{dia}}, c_{y,\text{dia}})$ from where perpendicular is to be drawn as nerve diameter (N_D) pixels below optic nerve tip coordinates is given as:

$$\left(c_{x,\text{dia}}, c_{y,\text{dia}}\right) = c_{x,\text{tip}} + N_D\cos(\beta), c_{y,\text{tip}} + N_D\sin(\beta) \tag{1}$$

where $(c_{x,\text{tip}}, c_{y,\text{tip}})$ is the tip of the optic nerve.

The equation for the optic nerve diameter axis at 3 mm below the optic nerve tip is given as:

$$z = \left(-\frac{1}{m}\right)\left(x + c_{x,\text{dia}}\right) + c_{y,\text{dia}} \tag{2}$$

where $(c_{x,\text{dia}}, c_{y,\text{dia}})$ is the coordinate of the point where perpendicular is to be drawn, m is defined as the slope of the line, and x is the independent variable.

In next step, the raster scanning of the grayscale image as shown in Fig. 7 is performed to find pixels with intensity 255, which are the intersection points for the line within the contour. In case of more than two points, take a threshold d_{th} so if two points are d distance apart where $d < d_{\text{th}}$, then merge them to a single point. And in the case where $d > d_{\text{th}}$ ($d_{\text{th}} = 20$ in our case), take the farthest two points. Consider the two intersection points such as p_1 and p_2, taking x coordinates as negative to make them positive. The two points are plotted with a line joining them on an image, which are end points of OND at required distance and calculate Euclidean distance

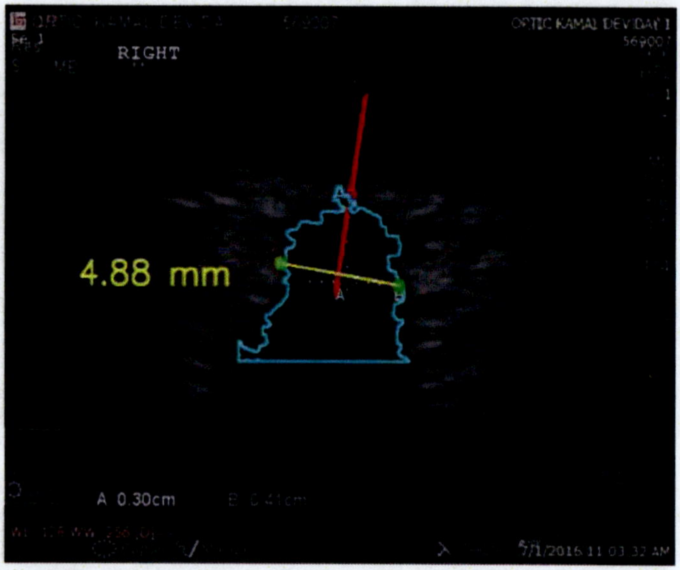

Fig. 8 Final result: green circle shows optic nerve extreme diameter points, and the yellow line shows OND. The result is marked on the left side in the mm scale. OND marked by a neurologist is in the bottom center as B

between the points p_1 and p_2 which simply gives required OND in pixels. Final OND in millimeter-scale can be obtained by dividing OND in pixels with mm to pixels conversion factor, which was 23.47 in our case. The final result is shown in Fig. 8.

3 Result

The data were successfully processed through the proposed algorithm as shown in Fig. 9 where the algorithm is applied to multiple subjects. The developed algorithm determined the optic nerve axis and its orientation, tip of the optic nerve, optic nerve diameter axis at 3 mm below the optic nerve tip along the optic nerve axis, endpoints of OND along the diameter axis, and finally, the optic nerve diameter as shown in Fig. 8. The Mean OND, as predicted by the algorithm, was 5.00 mm.

The results of OND determined by the neuroradiologist were considered gold standards and compared with the algorithm predictions. The accuracy of the algorithm in predicting the OND was reported. Percent Root Mean Square Difference (PRD) was chosen as the performance parameter to indicate the accuracy of the OND measurement using the proposed automated algorithm compared to the manual measurement performed by the neuroradiologist. PRD of manual and automatically measured OND was calculated as:

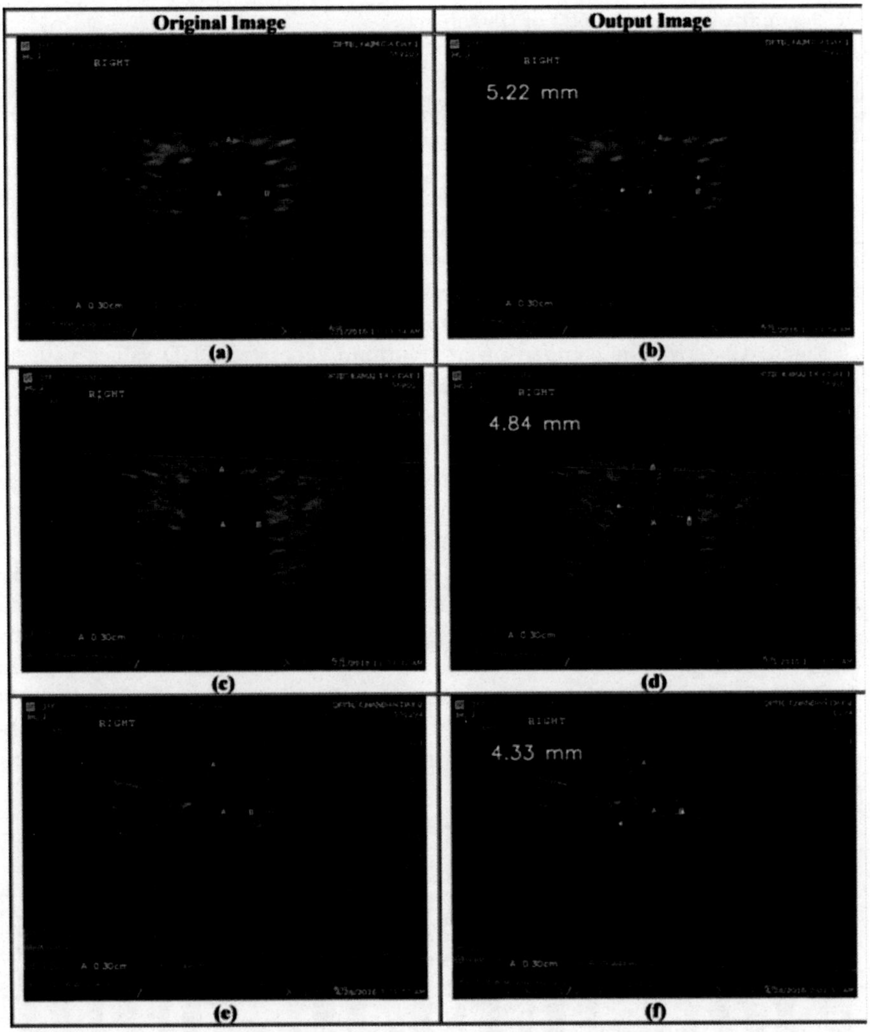

Fig. 9 The result of the algorithm when applied to multiple subjects where images (**a, c, e**) show raw image and images (**b, d, f**) shows output image indicating contour for the optic nerve, line from the retina to the optic nerve, optic nerve tip and the line connecting two dots at ends of OND represents required diameter. Automatically measured OND in mm is shown in the top left corner of each output image

$$\mathrm{PRD} = \sqrt{\frac{\sum_{i=0}^{N}(u_i - v_i)^2}{\sum_{i=0}^{N} v_i^2}} \times 100 \tag{3}$$

where u_i = Automated measured OND of ith subject,

v_i manually measured OND of ith subject,
N Number of subjects.

Furthermore, the proposed OND results for qualitative comparison, a bar graph of each measure (Dataset A = Result of OND by the proposed automated algorithm and Dataset B = Result of OND by conventional manual marking by neuroradiologist) was created. There was no significant difference between the predicted and neuroradiologist determined OND values suggesting that the paired means were similar. The algorithm over-predicted the OND in 15 subjects and under-predicted it in the remaining subjects. The maximum and minimum difference in prediction was 1.8 mm and 0.1 mm with mean absolute difference of 0.63 mm.

Figure 10 shows the comparison of algorithm generated OND values with the manual measurement done by the radiologists. A total of 24 patients were admitted to the All India Institute of Medical Sciences (AIIMS) trauma center, New Delhi. The mean and standard deviation of the obtained OND values have been compared with other reported works. The automated generated OND values reported the mean and standard deviation of 4.99 ± 0.76 mm and the manually generated OND values reported 4.88 ± 1.02 mm. Geeraerts et al. [19] reported the OND measurement for the sample size of 37, its mean and standard deviation was 4.53 ± 0.33 mm, whereas Lochner et al. [20] obtained the value 3.08 ± 0.38 mm for the sample size of 20. This implies that a large sample size is needed for further validation of the OND measurement. Further, the correlation is obtained for the OND measurement, validating the findings of automated and manual measurement. The correlation coefficient of the automated measured OND and manually obtained by the radiologist is found to be 0.71. The PRD value for the manual and automatically measured OND was 14.92%, considered satisfactory.

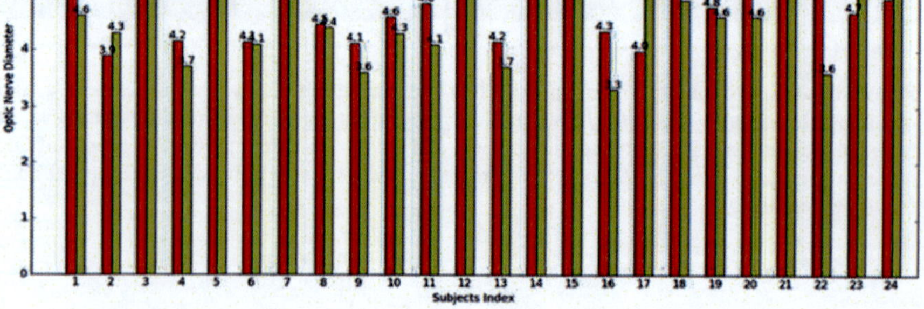

Fig. 10 Comparison of the manual and automatic OND values for 24 patients

4 Discussion

This study presented a novel computational and image processing algorithm for automatically measuring OND from eye ultrasound images acquired in a traumatic cohort. This study proposes an accurate technique to measure OND. It can be effectively used in clinical settings by embedding the algorithm into the ocular ultrasound machines for achieving accurate, repeatable, and operator-independent results. Manual measurement of the OND is highly dependent on the skill and experience levels of the neuroradiologist, and it may lead to false-positive diagnostics in certain situations. It was observed during this study that while measuring OND, neuroradiologists do not consider the optic nerve orientation and always draw a vertical line from the optic nerve tip without considering the optic nerve axis. Hence, the optic nerve diameter is measured in the horizontal direction. The proposed algorithm calculated the OND in the perpendicular direction to the optic nerve axis, considering the orientation of the optic nerve and thus, providing a more clinically accurate OND measurement. However, in hindsight, a detailed study on the effect of considering the correct optic nerve axis while determining the OND is warranted.

Literature reports conflicting data on the relationship of OND with ICP. Geeraerts and colleagues [18] reported that OND (2.65 $\pm$ 0.28 mm) determined using T2-weighted MRI acquisition in 38 patients with TBI was not significantly different than healthy cohort. Whereas the same group [19] reported a positive correlation between OND and ICP in a study on TBI population ($n = 37$) using ocular ultrasound with OND measurements of 4.53 $\pm$ 0.33 mm. Another study reported OND of 5.63 $\pm$ 0.69 mm in pediatric population with raised ICP which was significantly different than a controlled cohort. Our results of OND measurement by neuroradiologist were 4.88 $\pm$ 1.02 mm which were similar to those previously reported. While we could not relate the OND measurements with ICP, the mean value and ranges (3.3 mm–6.8 mm) corroborated well with the literature [19]. Compared to our results, the mean OND values in 20 healthy volunteers with no eye disorders and with normal ICP are reported to be 3.08 $\pm$ 0.38 mm [20]. While ONSD is reported to have a stronger correlation with ICP as well as changes in ICP [19–21], it can only be used to monitor ICP as the baseline ONSD value that relates to a healthy ICP value may not be established a priori [22–24]. However, ONSD is also regarded as a reliable measure [19] over OND due to difficulties differentiating the optic nerve from its sheaths. No significant difference found between the gold standard and predicted results confirmed the algorithm's accuracy. Such information may prove crucial to the neuroradiologist in confirming their diagnosis and correcting any errors in the manual evaluation. While the cohort used in this study had a TBI, OND measures also play an essential role in detecting optic nerve atrophy in patients with multiple sclerosis [25].

5 Conclusion

This study successfully implemented the concept of computer vision and intelligent computing to acquire a suitable input image from a given video sample of eye ultrasound to obtain an accurate OND measure in patients with TBI. This study had certain limitations. There were no details available for the level of TBI in the patients, and thus we could not categorize and analyze the results in depth. Further, there was no information available for ICP in this cohort, limiting our ability to determine the correlation between OND and ICP. While the ONSD measure is a better indicator of ICP, our future research work will focus on developing an automated algorithm for the measurement of ONSD.

References

1. del Saz-Saucedo, P., et al.: Sonographic assessment of the optic nerve sheath diameter in the diagnosis of idiopathic intracranial hypertension. J. Neurol. Sci. **361**, 122–127 (2016)
2. De La Hoz Polo, M., et al.: Ocular ultrasonography focused on the posterior eye segment: what radiologists should know. Insights Imaging **7**(3), 351–364 (2016)
3. Sahoo, ,S.S., Agrawal, D.: Correlation of optic nerve sheath diameter with intracranial pressure monitoring in patients with severe traumatic brain injury. Indian J. Neurotrauma **10**, 9–12 (2013)
4. Li, J., Wan, C.: Non-invasive detection of intracranial pressure related to the optic nerve. Quant. Imaging Med. Surg. **11**(6), 2823–2836 (2021). https://doi.org/10.21037/qims-20-1188
5. Kishk, N.A., et al.: Optic nerve sonographic examination to predict raised intracranial pressure in idiopathic intracranial hypertension: the cut-off points. Neuroradiology J. **31**(5), 490–495 (2018)
6. Nguyen, B.N., et al.: Ultra-high field magnetic resonance imaging of the retrobulbar optic nerve, subarachnoid space, and optic nerve sheath in emmetropic and myopic eyes. Transl. Vis. Sci. Technol. **10**(2), 8 (2021). https://doi.org/10.1167/tvst.10.2.8
7. Kimberly, H.H., Noble, V.E.: Using MRI of the optic nerve sheath to detect elevated intracranial pressure. Crit Care **12**, 181 (2008)
8. Kalantari, H., Jaiswal, R., Bruck, I., et al.: Correlation of optic nerve sheet diameter measurements by computed tomography and magnetic resonance imaging. Am. J. Emerg. Med. **31**, 1595–1597 (2013)
9. Gupta, S., Pachisia, A.: Ultrasound-measured optic nerve sheath diameter correlates well with cerebrospinal fluid pressure. Neurol. India **67**(3), 772 (2019)
10. Chen, H.,Ding, G.S., Zhao, Y.C., Yu, R.G., Zho, J.X.: Ultrasound measurement of optic nerve diameter and optic nerve sheath diameter in healthy Chinese adults. BMC Neurology **15**, 106 (2015)
11. Lagreze, W.A., Lazzaro, A., Weigel, M., Hansen, H.-C., Hennig, J., Bley, T.A.: Morphometry of the retrobulbar human optic nerve: comparison between conventional sonography and ultrafast magnetic resonance sequences. Invest. Ophthalmol. Vis. Sci. **48**(5), 1913–1917 (2007)
12. Kilker, B.A., Holst, J.M., Hoffmann, B.: Bedside ocular ultrasound in the emergency department. Eur J Emerg Med. **21**(4), 246–253 (2014)
13. Mateo, J.L., Fernández-Caballero, A.: Finding out general tendencies in speckle noise reduction in ultrasound images. Expert Syst. Appl. **36**(4), 7786–7797 (2009)
14. Michailovich, O.V., Tannenbaum, A.: Despeckling of medical ultrasound images. IEEE Trans. Ultrason. Ferroelectr. Freq. Control **53**(1), 64–78 (2006)
15. Burckhardt, C.B.: Speckle in ultrasound B-mode scans. IEEE Trans. Sonics Ultrason. **25**(1), 1–6 (1978)

16. Tasnim, T., Shuvo, M.M., Hasan, S.: Study of speckle noise reduction from ultrasound B-mode images using different filtering techniques. In 2017 4th International Conference on Advances in Electrical Engineering (ICAEE), pp. 229–234. IEEE (2017)
17. Joel, T., Sivakumar, R.: An extensive review on Despeckling of medical ultrasound images using various transformation techniques. Appl. Acoust. **138**, 18–27 (2018)
18. Geeraerts, T., et al.: Use of T2-weighted magnetic resonance imaging of the optic nerve sheath to detect raised intracranial pressure. Critical Care (London, England) **12**(5), R114 (2008). https://doi.org/10.1186/cc7006
19. Geeraerts, T., Merceron, S., Benhamou, D., Vigué, B., Duranteau, J.: Non-invasive assessment of intracranial pressure using ocular sonography in neurocritical care patients. Intensive Care Med. **34**, 2062–2067 (2008). https://doi.org/10.1007/s00134-008-1149-x
20. Lochner, P., Coppo, L., Cantello, R., et al.: Intra- and interobserver reliability of transorbital sonographic assessment of the optic nerve sheath diameter and optic nerve diameter in healthy adults. J. Ultrasound **19**, 41–45 (2016). https://doi.org/10.1007/s40477-014-0144-z
21. Pansell, J., Bell, M., Rudberg, P., Friman, O., Cooray, C.: Optic nerve sheath diameter measurement by ultrasound: Evaluation of a standardized protocol. J. Neuroimaging **32**(1), 104–110 (2022)
22. Youm, J.Y., Lee, J.H., Park, H.S.: Comparison of transorbital ultrasound measurements to predict intracranial pressure in brain-injured patients requiring external ventricular drainage. J. Neurosurg. **23**, 1–7 (2021)
23. Nag, D.S., Sahu, S., Swain, A., Kant, S.: Intracranial pressure monitoring: gold standard and recent innovations. World J. Clin. Cases **7**, 1535–1553 (2019)
24. Du, J., Deng, Y., Li, H., Qiao, S., Yu, M., Xu, Q., et al.: Ratio of optic nerve sheath diameter to eyeball transverse diameter by ultrasound can predict intracranial hypertension in traumatic brain injury patients: a prospective study. Neurocrit. Care **32**, 478–485 (2020)
25. Titlic, M., Erceg, I., Kovacevic, T., et al.: The correlation of changes of the optic nerve diameter in the acute retrobulbar neuritis with the brain changes in multiple sclerosis. Coll. Antropol. **29**, 633–636 (2005)

APFNet: Attention Pyramidal Fusion Network for Semantic Segmentation

Krishna Chaitanya Jabu, Mrinmoy Ghorai, and Y. Raja Vara Prasad

Abstract This paper proposes a novel method for multi-spectral semantic segmentation of urban scenes based on an Attention Pyramidal Fusion Network (APFNet) using RGB colour and infrared (IR) images. APFNet incorporates a attention spatial pyramid pooling network with attention module for fusion of features extracted by computing correlation between RGB and IR images. This method enhances feature representation and makes use of complementary properties from multi-spectral images. We have used three fusion techniques such as summation, concatenation, and an extension to concatenation for the proposed fusion network. The proposed network is compared with other state-of-the-art networks and tested on RGB-IR dataset. The experimental results show that the proposed APFNet improves multi-spectral semantic segmentation outcomes with high accuracy in classification and localisation.

Keywords Multi-spectral · Convolutional neural networks · Fusion · Semantic segmentation

1 Introduction

Semantic segmentation algorithms enable computers to analyse images at the pixel level and classify the region of interest in the image based on the predefined set of classes. Autonomous driving and path planning medical image diagnosing are some of the important application of semantic segmentation technique.

K. C. Jabu (✉) · M. Ghorai · Y. R. V. Prasad
Department of Electronics and Communication Engineering, Indian Institute of Information Technology, Sri City, Chittoor, Andhra Pradesh, India
e-mail: krishnachaitanya.j18@iiits.in

M. Ghorai
e-mail: mrinmoy.ghorai@iiits.in

Y. R. V. Prasad
e-mail: yrv.prasad@iiits.in

M. Tistarelli et al. (eds.), *Computer Vision and Machine Intelligence*,
Lecture Notes in Networks and Systems 586,
https://doi.org/10.1007/978-981-19-7867-8_24

In recent years, several semantic segmentation algorithms based on deep convolutional neural networks have been proposed. Some of the methods use visible images to train the semantic segmentation network. Visible images which are captured using standard visible cameras perform consistently during bright lighting scenarios; however, their performance is depreciated during poor lighting conditions. To address this issue, IR images are also utilised which are acquired with thermal imaging cameras record the contours of objects based on their heat distributions, which are unaffected by lighting conditions.

Multi-spectral semantic segmentation technique involves pictorial data interpretation based on the features extracted from both RGB and IR images. For combining the features collected from RGB and IR images, existing multi-spectral semantic segmentation models employed a summing or concatenation operation in the networks.

In this paper, an attention pyramidal fusion network has been proposed for performing RGB-IR multi-spectral semantic segmentation using three different fusion approaches. The proposed fusion techniques incorporate CBAM [14] attention block and an attention atrous spatial pyramid pooling network (AASPP) to segment objects robustly. The proposed AASPP conducts convolutions at several dilation rates that helps in collecting image features at multiple resolutions and attains attention to enhance the feature representation.

2 Literature Review

Recently, convolutional neural network (CNN) has been playing an important role in solving computer vision tasks like semantic segmentation. In addition, an encoder-decoder structure based on fully convolutional network (FCN) [3] has been introduced for semantic segmentation through SegNet [4]. To accelerate the performance of encoder-decoder, UNet [5] builds skip connections from encoder to decoder which prevents network from degradation. It helps the network to hold more information from the original images.

RefineNet [7] introduced Multi-resolution Fusion which enables the acquisition of image details across several resolutions, as well as chained residual pooling, which enables the efficient capture of image background information. Deeplab [3] adopted atrous convolutions which helps to acquire more field of view during convolution and hence enhance semantic segmentation performance. All of these algorithms are effective to analyse RGB images in properly illuminated lighting conditions. But the same level of precision cannot be maintained when processing images obtained in low-light environments.

To address this problem, MFNet [8] and RTFNet [9] introduced multi-spectral semantic segmentation. These networks use multi-spectral data from IR and RGB images to improve semantic segmentation performance, particularly when processing images are captured in dark lighting conditions. The multi-spectral features extracted from the infrared and RGB data are integrated using channel-wise concatenation in

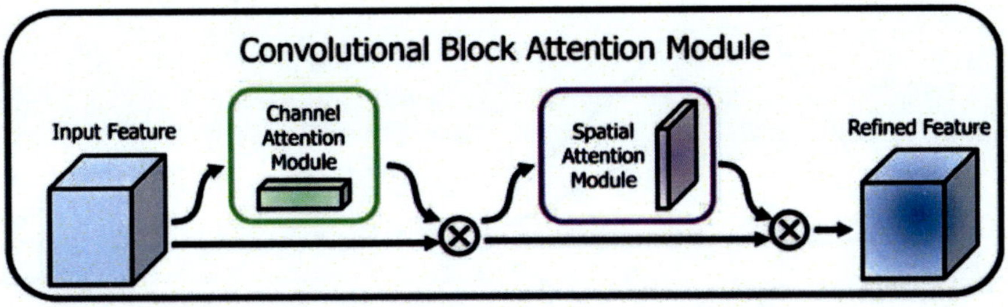

Fig. 1 Convolutional block attention module (CBAM). This diagram is taken from the paper [14]

MFNet and channel-wise summation in RTFNet. Similarly, FuseNet [10] is a multi-spectral semantic segmentation architecture that is intended to conduct semantic segmentation under indoor situations by combining RGB and depth information. FuseNet incorporates VGG-16 [2] as the encoder of the network, and fusion is performed by summation operation.

Siamese convolutional neural network [13] proposed an approach for fusing IR and RGB data, intends to calculate weights for IR and RGB pixels at their corresponding positions on the image. The fusion is done by performing weighted sum of IR and RGB pixel values. This technique takes cross-spectrum correlation into account, but excludes contextual relationships when evaluating the RGB or IR pixel weight in feature space.

In natural language processing, the attention mechanism outperformed problems based on encoder-decoder. Subsequent versions of this technology, or adaptations thereof, were used in a number of other applications, including speech processing and computer vision [14–18]. Non-local neural networks [15] employed a self-attention technique to conduct video classification by processing RGB images. Furthermore, in DANet [16], developed two self-attention modules to improve the spatial and channel representations of feature maps.

AFNet [12] proposed a co-attention mechanism for fusing multi-spectral features from IR and RGB images. AFNet incorporates an attention fusion module that allows for the spatial acquisition of weighted correlations across different spectral feature maps.

The goal of this paper is to present a new APFNet architecture for semantic segmentation tasks by leveraging the advantages of deep learning on fusing IR and RGB data.The proposed architectures mainly aim to effectively utilise atrous convolutions and channel attention modules where atrous convolutions enable to comprehend the images from diverse field of view and channel attention map exploits the inter channel relationship of RGB and IR features efficiently. This paper presents three fusion techniques which are based on summation and concatenation operations. Fusion is accomplished in the proposed architectures using an attention atrous spatial pooling pyramids AASPP as shown in Fig. 2 along with attention block called as CBAM [14].

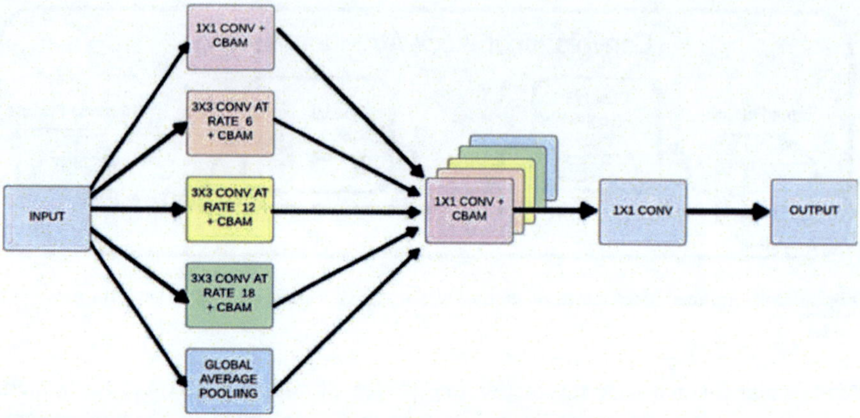

Fig. 2 Attention atrous spatial pyramidal pooling (AASPP) network

The CBAM [14] is incorporated into the architecture for attaining channel-specific attention. The proposed AASPP module is based on architecture described in [6, 11] additionally having attention blocks included into it helps for adjusting the convolutional field of view which is obtained by performing atrous convolutions as well attains channel-specific attention. The suggested architecture is similar to that of encoder-decoder architecture where encoders calculate feature space and decoder is developed to decrypt the feature space into a semantic segmented image. The key contributions of the proposed work can be summarised as follows.

- Propose an attention pyramidal fusion network for semantic segmentation by extracting and fusing features from IR and RGB images based on ResNet encoder model.
- Propose to incorporate two modules one for achieving channel-wise attention (CBAM) and other (AASPP). Parallel atrous convolutions performed within AASPP module helps to effectively capture multi-scale information and the channel attention module, which elevates the channel level IR and RGB features in our architecture to enhance semantic segmentation performance.
- Analyse different fusion architectures including proposed concatenation fusion architecture.
- Show experimental results and comparison with state-of-the-art approaches as well as validate the proposed model under different light conditions of images.

3 The Proposed Network

The primary goal of the proposed approach is to maximise the performance of semantic segmentation by using the potentials of both infrared and RGB images. In this paper, a novel approach of multi-spectral semantic segmentation using CBAM [14] and AASPP network is proposed in order to improve the result of segmentation of urban scenes. The performance of the proposed model has been analysed using several fusion approaches.

3.1 Encoder Network

In the proposed architecture, two parallel encoders are designed to extract information from RGB and IR images simultaneously. These two encoder modules are build on ResNet-152 [1] architecture, and they are structurally similar except to the number of input channels in the first layer. The encoder with RGB image as input contains three input channels, and the encoder with IR image contains one input channel. To prevent a significant loss of spatial information in feature maps, we eliminate average pooling and fully connected layers in the ResNet architecture [1]. This also contributes to size reduction of proposed model. The proposed ResNet architecture starts with an initial block that is composed of three layers successively: a convolutional layer, a batch normalisation layer, and a ReLU activation layer. The model is trained using ResNet-152 block. The ResNet block is followed with an attention atrous spatial pooling pyramid AASPP which is based on architecture from [6, 11] along with CBAM [14] attention blocks incorporated into its first 4 branches of the pyramid as shown in Fig. 2.

The AASPP module consists of atrous convolutional layer which helps in adjusting the convolution effective field of view. Let y represent the output signal, w represent the filter, and ar represent the atrous rate. For each location i on output y, atrous convolution performed on the input feature map x is defined as:

$$y[i] = \sum_{j} x[i + ar.j].w[j]. \tag{1}$$

As shown in the preceding Eq. 1, the field of vision adjusted by a parameter ar called atrous rate which is an effective metric for improving the field of view filters without increasing the computational complexity. Standard convolution is a specific instance of atrous convolution when $ar = 1$. The AASPP module shown in Fig. 2 is composed of five branches, where four of those branches contain an atrous convolutions along with CBAM[14] channel attention block as shown in the Fig. 1 and last branch performs global average pooling for the input of AASPP. Each of the five branches are followed by CBAM attention module.

CBAM module as shown in Fig. 1 aids in achieving channel-specific attention. Channel attention is computed by squeezing the spatial dimension of the input feature map as shown in Eq. 2 from [14].

$$M_c(F) = \sigma(MLP(AvgPool(F)) + MLP(MaxPool(F))). \tag{2}$$

In the above equation, M_c denotes the required channel attention map and MLP denotes multi layer perceptron. The average and max pooling from Eq. 2 help to aggregate the spatial features generating 2 spatial context descriptors. Based on the type of fusion, the number of input channels to the AASPP module varies between 2048 (for channel-wise summation-based fusion) and 4096 (channel-wise concatenation-based fusion).

3.2 *Decoder Network*

The decoder network of the proposed APFNet is as indicated in [11] contains an up-sampler. The decoder receives the fused (concatenation or summation) low level features from the early layers of ResNet along with output of AASPP as input. The extracted low level features are processed by 1X1 convolutions in order to reduce number of channels which is later concatenated with output of AASPP. The concatenated output is further processed using 3X3 convolutions before being sent to an up-sampler. The up-sampler generates segmented image.

4 Fusion Architectures

As stated above, the performance of the semantic segmentation model is evaluated using multiple fusion architectures like summation, concatenation, and gated fusion. The results are analysed for all the implemented architectures during generic day and night conditions.

4.1 *Fusion Architecture 1 (Summation)*

This architecture as shown in Fig. 3 involves a summation operator for combining IR and RGB features. The output from the intermediate layers of the ResNet from IR

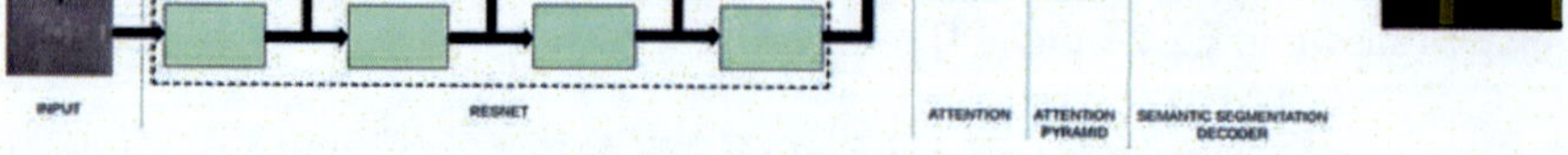

Fig. 3 Summation-based fusion architecture

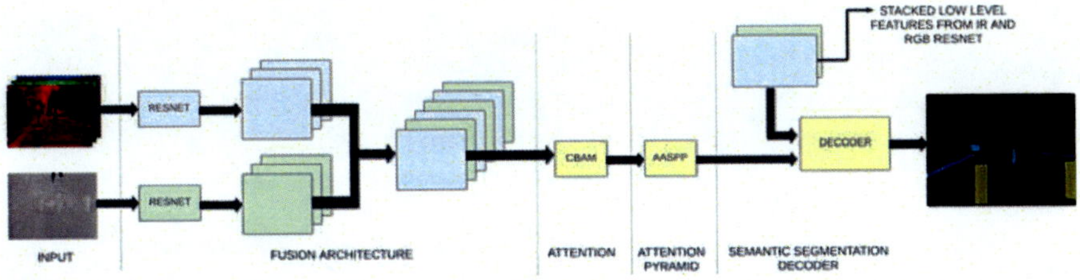

Fig. 4 Concatenation-based fusion architecture

and RGB networks are summed up and processed further to the next RGB ResNet block. The output obtained from the summation at the final ResNet block is given as input to the CBAM module followed by a AASPP module which is further followed by the decoder network for extracting segmented images.

4.2 Fusion architecture 2 (concatenation)

This architecture as shown in Fig. 4 involves a concatenation operator with an CBAM module for combining IR and RGB features. The output of ResNet from IR and RGB networks are stacked up channel-wise and processed further to a CBAM attention module. The output obtained from the CBAM block is given as input to the AASPP module which is further followed by the decoder network for extracting segmented images.

4.3 Fusion architecture 3 (concatenation at AASPP)

The ResNet layer output of RGB and IR networks as shown in Fig. 4 is individually fed into two CBAM modules later concatenated at the pyramidal level as shown in Fig. 5. The concatenated output is further processed using 1×1 convolutions of the AASPP which is further followed by the decoder network.

5 Experimental Results and Observations

In this section, we analyse our proposed APFNet architectures and compare it to state-of-the-art networks using extensive tests on a publicly available dataset.

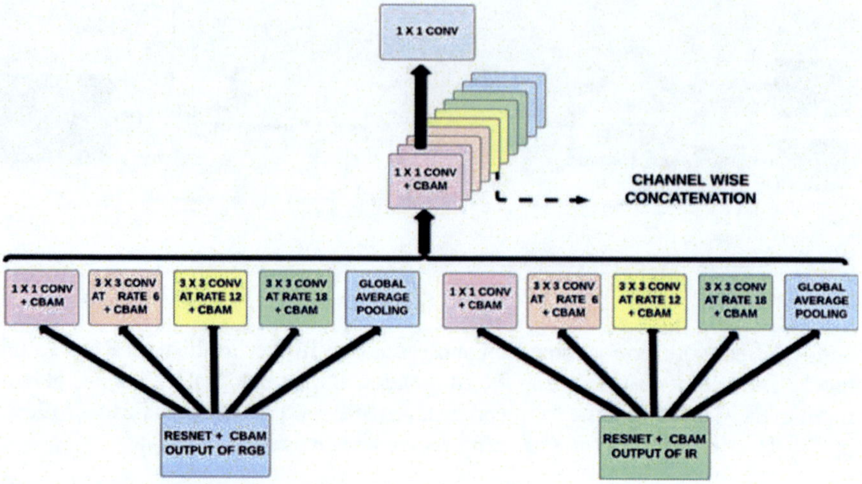

Fig. 5 Concatenation at AASPP

5.1 *Dataset*

The RGB-IR dataset contains 1569 4-channel RGB-IR urban scenario images that are captured by a IngRec R500 camera, of the four channels first three belong to RGB while the fourth channel is IR image. The dataset collection contains 820 images that are taken in the day light and the other 749 that are taken at low-light conditions. The resolution of the images in the dataset is 480 × 640. There are nine semantic segmentation classes identified in the ground truth images, namely car, person, bike, curve, car stop, colour cone, bump, guardrail, and background. The dataset is split into training dataset, testing dataset, and validation dataset in the ratio of 2:1:1 (784:393:392) as described in [8].

5.2 *Experimental Setup*

Our suggested APFNet is implemented using PyTorch 1.10. Our APFNet was trained on a Google Colab using a Tesla P100 GPU. Due to the fact that GPU memory is restricted to 16 GB, we adapt the batch sizes for various networks. APFNet is trained using the pre-trained ResNet weights given by PyTorch, with the exception of the first convolutional layer of ResNet. Because we are using a thermal encoder, we only have one channel of input data, but ResNet is built for three.

For training, we use the Stochastic Gradient Descent (SGD) optimisation technique. The momentum and weight decay rates are, respectively, set at 0.9 and 0.0005. The starting learning rate is set to 0.01, and the model is trained with a batch size of 4. To progressively lower the learning rate, we use the polynomial decay learning rate scheduler approach.

5.3 Evaluation Metrics

Mean Accuracy (mAcc) and Mean Intersection over Union (MIoU) are the 2 quantitative evaluation metrics chosen for evaluating the semantic segmentation model. First, accuracy or recall and intersection over union for each labelled class is calculated. Finally mean of these metrics are calculated which is called as Mean Accuracy (mAcc) and Mean Intersection over Union (MIoU).

$$mAcc = \frac{1}{N}\sum_{i=1}^{N}\frac{TP_i}{TP_i + FN_i}, \tag{3}$$

$$mIoU = \frac{1}{N}\sum_{i=1}^{N}\frac{TP_i}{TP_i + FP_i + FN_i}. \tag{4}$$

From the above Eqs. 3 and 4 TP, TN are defined as true positives and true negatives and FP and FN are defined as false positives and false negatives. Here, N denotes number of classes in the dataset which is 9. MIoU is defined as the mean of ratio of true positives to the summation of true positives and false negatives for all the N classes, and similarly, mAcc is defined as the mean of ratio of true positives to the summation of true positives, false positives, and false negatives for all the N classes.

5.4 Overall Experimental Results

For the verification of the proposed APFNet-based fusion architectures, the results are compared with other state-of-the-art networks like MFNet, RTFNet, and AFNet which are trained and tested on the dataset indicated in [8]. mIoU and mAcc are the evaluation metrics chosen for comparing the networks. Table 1 presents the network's quantitative comparative findings.

Due to the fact that the majority of pixels in the dataset are unlabeled, the assessment results for the unlabeled class are comparable across various networks. They include less information and are omitted from the table. As shown in Table 1, our APFNet with fusion architecture 2 produce the best results across all the state-of-the-art networks in terms of mAcc and mIoU metrics. Fusion architecture 3 gave good results in terms of mAcc and performing comparatively in terms of mIoU with other state-of-the-art approaches. Fusion architecture 1 gave higher mAcc and mIoU than

Table 1 Quantitative comparisons of MFNet, RTFNet, AFNet, and multiple fusion architectures of the proposed APFNet over eight classes, with Acc, IoU along with consolidated metrics mAcc and mIoU %. Bold text indicates the highest values for both the metrics

Model name	Car		Person		Bike		Curve		Car stop		Guardrail		Colour cone		Bump		mAcc	mIoU
	mAcc	mIoU	mAcc	mIoU	mAcc	mIoU	mAcc	mIoU	mAcc	mIoU	mAcc	mIoU	mAcc	mIoU	mAcc	mIoU		
MFNet	72.9	60.9	60.2	53.4	54.4	43.1	26.3	22.9	11.9	9.43	0.0	0.0	20.3	18.8	25.2	23.5	41.1	36.5
RTFNet	91.1	84.3	76.4	66.1	67.4	55.8	60.1	42.4	38.7	30.1	1.1	0.6	4.1	3.1	14.5	11.4	50.3	43.5
AFNet	91.2	86.0	76.3	67.4	72.8	**62.0**	49.8	43.0	35.3	28.9	**24.5**	**4.6**	50.1	44.9	61.0	**56.6**	62.2	54.6
Fusion architecture 1	93.4	86.3	**78.5**	68.5	66.0	55.4	48.5	36.5	28.61	34.0	10.4	1.6	48.5	41.9	**70.3**	48.7	61.0	51.6
Fusion architecture 2	**94.5**	85.4	**78.5**	**70.30**	79.6	60.9	**67.5**	**45.0**	56.9	39.4	0.7	0.4	52.8	44.5	68.8	49.5	66.4	**54.8**
Fusion architecture 3	92.9	**87.5**	75.7	68.0	**81.9**	56.0	58.3	42.6	**68.3**	**40.7**	0.0	0.0	**56.5**	**46.9**	66.3	45.3	**66.5**	53.9

MFNet and RTFNet but performed lower than AFNet. This illustrates the superiority of fusion architectures 2 and 3 over other state-of-the-art implementations. By comparing the APFNet versions, we can find that APFNet with fusion architecture 2 outperforms the others substantially. There are some Acc and IoU values in the table which are close to 0, most notably in the Guardrail class.

As seen in the dataset article [8], the dataset's classes are very imbalanced. Because the Guardrail class contains the smallest percentage of the pixels, there are extremely little training data for it. We feel that the models are not sufficiently trained on this class owing to a lack of training data, which results in the 0.0 outcomes during the test. Additionally, the test dataset contains 393 photos, but only 4 images have the Guardrail class. Thus, we believe that the test dataset's very low pixel count for the Guardrail class is another cause for the 0.0 findings.

5.5 *Experimental Analysis of Under Different Illuminations*

To further validate the APFNet in various lighting circumstances, comparisons are made utilising all daylight and night images, as demonstrated in Table 2. When compared with state-of-the-art studies, our fusion architectures 2 and 3 approach produces higher mAcc and MIoU values in low-light situations. Additionally, for APFNet Fusion architectures 2 and 3, the metrics of images taken at night are greater than those taken during the day. This implies that the proposed RGB-IR APFNet Fusion architectures 2 and 3 is advantageous for low lighting scenarios, as the fusion of IR image features may provide rich details and considerably enhance the final segmentation results in low-light conditions.

Table 2 Daytime and nighttime outcomes are compared with state-of-the-art implementations for all fusion types. Bold text indicates the highest values for both the metrics

Model name	Daytime		Nighttime	
	mAcc	mIoU	mAcc	mIoU
MFNet	42.6	36.1	41.4	36.8
RTFNet	49.3	41.7	47.4	41.6
AFNet	54.5	**48.1**	60.2	53.8
Fusion architecture 1	64.6	45.4	58.2	51.6
Fusion architecture 2	**70.2**	46.9	64.9	55.7
Fusion architecture 3	68.3	45.6	**65.0**	**55.9**

Table 3 Adaptation study with and without CBAM

Model name	Without CBAM		With CBAM	
	mAcc	mIoU	mAcc	mIoU
Fusion architecture 1	60.7	51.7	61.0	51.6
Fusion architecture 2	60.9	52.9	66.4	54.8
Fusion architecture 3	62.4	53.3	66.5	53.9

Table 4 Adaptation study varying number of atrous convolutions

Number of branches	Dilation rates	mAcc	MIoU
3	[1,6]	63.0	53.1
5	[1, 6, 12, 18]	66.4	54.8
7	[1, 6, 12, 18, 24, 30]	64.0	54.6

5.6 Adaptation Study

Adaptation study is performed to verify the importance of CBAM and how number of branches in AASPP pyramid effect semantic segmentation outcomes.

Significance of CBAM This analysis is performed by removing the attention module between ResNet and AASPP for implemented fusion architectures. From Table 3, we can infer that the semantic segmentation outcomes are better for fusion architectures 2 and 3 when attention module between backbone(ResNet) and AASPP is involved in the architecture. The performance of fusion architecture 1 decreased for the metric mAcc and increased for the metric mIoU with removal of CBAM module between backbone(ResNet) and AASPP from the architecture.

Number of Atrous Convolutions in AASPP For performing this study, fusion architecture 2 is chosen and the number of atrous convolutions in the AASPP architecture is varied from 2, 4, and 6. From Table 4, we can observe that increasing or decreasing the number of atrous convolutions from 4 is reducing models performance. Fusion architecture 2 gave best results when the model has 4 atrous convolutions at dilation rates [1, 6, 12, 18].

6 Conclusions and Future Directions

The research proposes three fusion strategies using attention pyramidal network (AASPP) for performing multi-spectral semantic segmentation using RGB and IR image inputs. The standard ResNet encoders are modified by adding CBAM modules at the end to improve feature extraction. The proposed APFNet utilises AASPP module in order to optimise the fusion of multi-spectral signals. The multi-spectral properties are enhanced using the generated attentions to enhance the RGB-IR fusion results.

Quantitative assessments of the proposed fusion architectures demonstrate that the fusion architecture 2 outperforms state-of-the-art networks such as FuseNet, MFNet, RTFNet, and AFNet in terms of both classification and localisation accuracy while fusion architecture 3 outperforms state-of-the-art networks in terms of classification accuracy. Additional evaluations reveal the APFNet-based architectures has superior performance for segmentation tasks using images obtained at low-light conditions. To summarise, using the proposed fusion architectures 2 and 3 improved the results of RGB-IR multi-spectral semantic segmentation.

References

1. He, K., Zhang, X., Ren, S., Sun, J.: Deep residual learning for image recognition. In: 2016 IEEE Conference on Computer Vision and Pattern Recognition (CVPR). pp 770–778 (2016)
2. Liu, S., Deng, W.: Very deep convolutional neural network based image classification using small training sample size. In: 2015 3rd IAPR Asian Conference on Pattern Recognition (ACPR). pp 730–734 (2015)
3. Shelhamer, E., Long, J., Darrell, T.: Fully convolutional networks for semantic segmentation. IEEE Trans. Pattern Anal. Mach. Intell. **39**, 640–651 (2017). https://doi.org/10.1109/TPAMI.2016.2572683
4. Badrinarayanan V, Kendall A, Cipolla R (2017) SegNet: A Deep Convolutional Encoder-Decoder Architecture for Image Segmentation. IEEE Trans. Pattern Anal. Mach. Intell. **39**, 2481–2495. https://doi.org/10.1109/TPAMI.2016.2644615
5. Ronneberger, O., Fischer, P., Brox, T.: U-Net: convolutional networks for biomedical image segmentation. In: Navab, N., Hornegger, J., Wells, W.M., Frangi, A.F. (eds.) Medical Image Computing and Computer-Assisted Intervention—MICCAI 2015, pp. 234–241. Springer International Publishing, Cham (2015)
6. Chen, L.-C., Papandreou, G., Kokkinos, I., Murphy, K., Yuille, A.: DeepLab: semantic image segmentation with deep convolutional nets, atrous convolution, and fully connected CRFs. IEEE Trans. Pattern Anal. Mach. Intell. https://doi.org/10.1109/TPAMI.2017.2699184
7. Lin, G., Milan, A., Shen, C., Reid, I.: RefineNet: multi-path refinement networks for high-resolution semantic segmentation. In: 2017 IEEE Conference on Computer Vision and Pattern Recognition (CVPR), pp. 5168–5177. IEEE Computer Society, Los Alamitos, CA, USA (2017)
8. Ha, Q., Watanabe, K., Karasawa, T., Ushiku, Y., Harada, T.: MFNet: towards real-time semantic segmentation for autonomous vehicles with multi-spectral scenes. In: 2017 IEEE/RSJ International Conference on Intelligent Robots and Systems (IROS), pp 5108–5115 (2017)
9. Sun, Y., Zuo, W., Liu, M.: RTFNet: RGB-thermal fusion network for semantic segmentation of urban scenes. IEEE Rob. Autom. Lett. **4**, 2576–2583 (2019). https://doi.org/10.1109/LRA.2019.2904733
10. Hazirbaz, C., Ma, L., Domokos, C., Cremers, D.: FuseNet: incorporating depth into semantic segmentation via fusion-based CNN architecture (2016)
11. Chen, L.-C., Zhu, Y., Papandreou, G., Schroff, F., Adam, H.: Encoder-decoder with atrous separable convolution for semantic image segmentation. In: Ferrari, V., Hebert, M., Sminchisescu, C., Weiss, Y. (eds.) Computer Vision ECCV 2018., pp. 833–851. Springer International Publishing, Cham (2018)

12. Xu, J., Lu, K., Wang, H.: Attention fusion network for multi-spectral semantic segmentation. Pattern Recognition Lett. **146**, 179–184 (2021). https://doi.org/10.1016/j.patrec.2021.03.015
13. Piao, J., Chen, Y., Shin, H.: A new deep learning based multi-spectral image fusion method. Entropy **21**, 570. https://doi.org/10.3390/e21060570
14. Woo, S., Park, J., Lee, J.-Y., Kweon, I.S.: CBAM: Convolutional block attention module. In: Ferrari, V., Hebert, M., Sminchisescu, C., Weiss, Y. (eds.) Computer Vision—ECCV 2018, pp. 3–19. Springer International Publishing, Cham (2018)
15. Wang, X., Girshick, R., Gupta, A., He, K.: Non-local neural networks. In: 2018 IEEE/CVF Conference on Computer Vision and Pattern Recognition. pp. 7794–7803 (2018)
16. Fu, J., Liu, J., Tian, H., Li, Y., Bao, Y., Fang, Z., Lu, H.: Dual attention network for scene segmentation. In: 2019 IEEE/CVF Conference on Computer Vision and Pattern Recognition (CVPR). pp. 3141–3149 (2019)
17. Mishra, S.S., Mandal, B., Puhan, N.B.: Multi-level dual-attention based CNN for macular optical coherence tomography classification. IEEE Signal Process. Lett. **26**, 1793–1797 (2019). https://doi.org/10.1109/LSP.2019.2949388
18. Bastidas, A.A., Tang, H.: Channel attention networks. In: Proceedings of the IEEE/CVF Conference on Computer Vision and Pattern Recognition (CVPR) Workshops (2019)

Unsupervised Virtual Drift Detection Method in Streaming Environment

Supriya Agrahari and Anil Kumar Singh

Abstract Real-time applications generate an enormous amount of data that can potentially change data distribution. The underline change in data distribution concerning time causes concept drift. The learning model of the data stream encounters concept drift problems while predicting the patterns. It leads to deterioration in the learning model's performance. Additional challenges of high-dimensional data create memory and time requirements. The proposed work develops an unsupervised concept drift detection method to detect virtual drift in non-stationary data. The K-means clustering algorithm is applied to the relevant features to find the stream's virtual drift. The proposed work reduces the complexity by detecting the drifts using the *k* highest score features suitable with high-dimensional data. Here, we analyze the data stream's virtual drift by considering the changes in data distribution of recent and current window data instances.

Keywords Data stream mining · Concept drift · Clustering · Learning model · Adaptive model

1 Introduction

The data stream is a continuous flow of data instances. The sequence of data instances is produced from various applications, such as cyber security data, industrial production data, weather forecasting data, and human daily activity data [1]. Vast volumes of data characterize the data streams originated at high frequencies. The data stream mining learning model performs predictive analysis of data samples. But due to the dynamic behavior of the data stream, accurate prediction becomes difficult for a single learning model because the training samples of data instances are insufficient to define the complexity of problem space and degrade the learning model's accuracy.

S. Agrahari (✉) · A. K. Singh
Motilal Nehru National Institute of Technology Allahabad,Prayagraj, India
e-mail: supriyagrahari@gmail.com

A. K. Singh
e-mail: ak@mnnit.ac.in

M. Tistarelli et al. (eds.), *Computer Vision and Machine Intelligence*,
Lecture Notes in Networks and Systems 586,
https://doi.org/10.1007/978-981-19-7867-8_25

In the streaming environment, the data distribution changes are considered concept drift, negatively impacting the learning model's performance. So, there is a requirement to get through concept drift while prediction is performed for the data stream. The drift detector is generally coupled with a prediction (or learning) model [2]. The detector raises a signal for the prediction model whenever the changes in the environment are detected. After that, the prediction model adapts as per the knowledge present in the recent data instances and discards the current prediction model to cope with the stream's drifts.

Data stream considers the sequence of data examples (or instance) that arrives online at varying speed. There is a possibility of changes in features or class definitions compared with past knowledge (i.e., concept drifts) [3]. In practice, there are limited, deferred, or even no labels present in the streaming data. It happens because the label for the incoming data instances cannot be quickly obtained for reasons, such as the higher labeling cost in real-world applications. It limits the many concept drift detectors, such as distribution-based, error-rate-based, and statistical-based methods [4].

Several drift detectors need labeled data to observe the model's performance, which helps to identify whether the model is obsolete. However, the labeled data in a streaming environment is not available in several applications, and providing a label is a cost-ineffective process. In such cases, supervised learning is no longer efficient. Hence, the unsupervised method offers a more practical approach in the streaming environment. On the other hand, the high-dimensional data creates high computational complexity, so the relevant feature selection is a way to provide an efficient process of model building.

We propose an unsupervised drift detection method that detects the drifts based on significant differences between the recent and current window data distribution. Drift detection is performed using the relevant features identified by the chi-square test. Dimensionality reduction of the feature space directly results in a faster computation time [5]. K-means clustering is also performed on the data stream because it is the most popular and simplest clustering algorithm. The significant contributions of the paper are as follows:

- We propose an unsupervised drift detection method in the streaming environment.
- The proposed work performs drift detection efficiently with no labeling requirement.
- The experiment performs with k highest score features. So, it minimizes the computational complexity of the learning model.

The paper is organized as follows: Section 2 discusses several existing drift detection methods. Section 3 gives a detailed description of the proposed work with their pseudo-code and workflow diagram. Section 4 contains the experimental evaluation, case study and results. Section 5 describes the conclusion and future work.

2 Related Work

In this section, we briefly discuss previous research related to concept drift. There are two categories of concept drift detection methods: supervised and unsupervised. The supervised approach assumes that the true labels (or target values) of incoming data, instances are available after prediction. So, they generally use the error or prediction accuracy of the learning model as the main input to their detection technique, whereas the unsupervised approaches do not require labels in their techniques. In this section, we emphasize unsupervised concept drift detection approaches.

de Mello et al. [6] focuses on Statistical Learning Theory to provide a theoretical framework for drift detection. It develops a plover algorithm that detects the drift using statistical measures such as mean, variance, kurtosis, and skewness. It utilizes power spectrum analysis to measure drift detection's data frequencies and amplitudes.

SOINN+ [7] is self-organizing incremental neural network with a forgetting mechanism. From the incoming data instances, it learns a topology-preserving mapping to a network structure. It demonstrates clusters of arbitrary shapes in streams of noisy data. Souza et al. [3] present an Image-Based Drift Detector (IBDD) for high-dimensional and high-speed data streams. It detects the drifts based on pixel differences. Huang et al. [5] present a new unsupervised feature selection technique for the handling of high-dimensional data.

OCDD is unsupervised One-Class Drift Detector [8]. It uses a sliding window and a one-class classifier to identify the change in concepts. The drift is detected when the ratio of false predictions is higher than a threshold. Pinto et al. [9] present a SAMM, an automatic model monitoring system. It is a time- and space-efficient unsupervised streaming method. It generates alarm reports along with a summary of the events. In addition, it provides information about important features to explain the concept drift detection scenario.

DDM [10], ADWIN [11], ECDD [12], SEED [13], SEQDRIFT2 [14], STEPD [15], and WSTDIW [16] are compared with the proposed approach to compare performance in terms of classification accuracy.

3 Proposed Work

In the streaming environment, the learning model requires processing data instances as fast as the data becomes available because there is limited memory to process it. In this regard, an online algorithm can process data sequentially or form a window for computation to work well with limited memory and time.

In the proposed work, the data stream is defined as D_s=[x_i, x_{i+1}, ..., x_{i+n}, ...], and D_s is d dimensional data matrix. Each x represents feature vector at different timestamp. The change in data distribution (P) between different timestamp t_l and t_m, i.e., $P_{t_l} \neq P_{t_m}$ is considered as concept drift. In the paper, the virtual drift detection is perfomed by identified the change in feature vector distribution over time. In such case, the boundaries of data distribution of data instances remain same.

3.1 Proposed Drift Detection Method

This section illustrates the working of the proposed drift detection method. The method is model-independent and unsupervised. The general workflow diagram of the proposed work is shown in Fig. 1. The pseudo-code of the proposed detector is described in Algorithms 1 and 2.

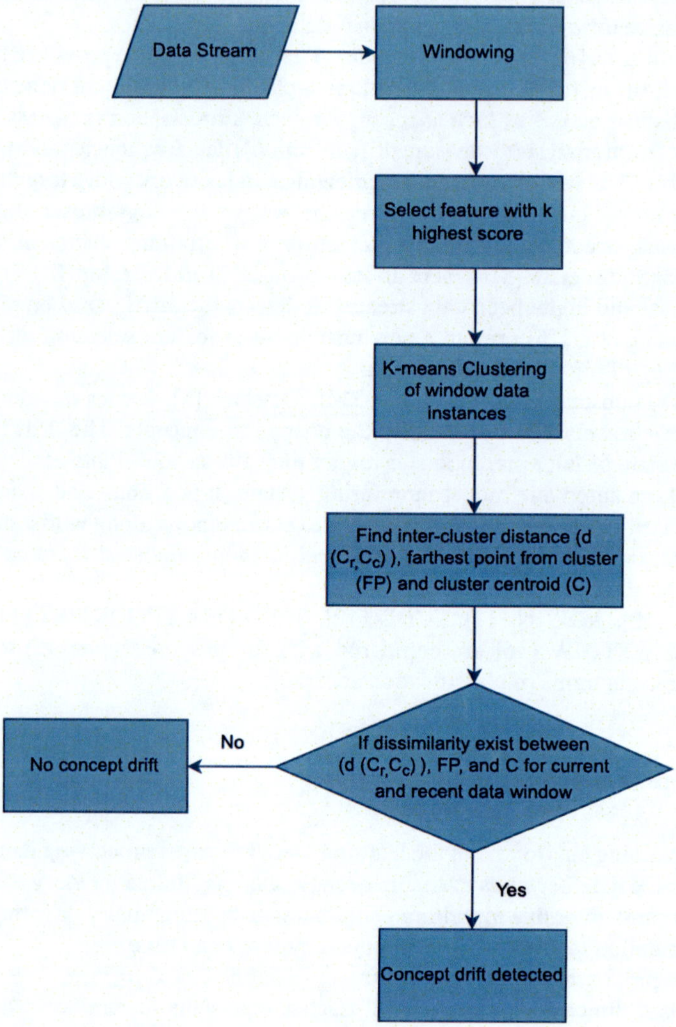

Fig. 1 General workflow diagram of proposed work

The proposed method detects the drifts using two data windows where w_c is the current window. When new data instances exist, the previous w_c becomes a recent window w_r to accommodate the current window with newly available data.

We utilize the chi-square test [17] as the statistical analysis to find the association or difference between recent and current window data instances. The chi-square is a popular feature selection method. It evaluates data features separately in terms of the classes. It is necessary to discretize the range of continuous-valued features into intervals. The chi-squared test compares the obtained values of a class's frequency due to the split to the expected frequency of the class. Let N_{ij} be the number of Ci class samples in the jth interval among the N examples, and M_{Ij} be the number of samples in the jth interval. $E_{ij} = M_{Ij}\ |C_i|/N$ is the anticipated frequency of N_{ij}. A particular data stream's chi-squared statistic is thus defined as Eq. 1.

$$\chi^2 = \sum_{i=1}^{C}\sum_{j=1}^{I} \frac{(N_{ij} - E_{ij})}{E_i j} \tag{1}$$

The number of intervals is denoted by I. The higher the obtained value, the more useful the relevant features are. In this way, the best feature is extracted from the window data based on k highest score. The selection of features eliminates the less important information and reduces the method's complexity.

Algorithm 1: Windowing of data stream

Data: *Data Stream*: D_s, *Current Window*: w_c, *Recent Window*: w_r.
Result: *Window of data instances.*

```
Initialize current window size;
while stream has data instance do
    if w_c ≠ Full then
        Add data instances into the current window;
    else
        DriftDetector(w_c)
```

K-means clustering algorithm utilizes the high score features of the current window. K-means clustering is an unsupervised learning technique, and it is used to split the unlabeled data into non-identical groups. The random selection of data points as a cluster center is performed, and further, the distances between the centroids and the data points are calculated. It assigns each data instance to its nearest centroid. We store these cluster centroids for further evaluation as defined in Eqs. 2 and 3. For the new incoming data instances, a new cluster center is selected.

$$C_r = \{C_{r_i}, C_{r_{i+1}}, \ldots, C_{r_{i+n}}\} \tag{2}$$

$$C_c = \{C_{c_i}, C_{c_{i+1}}, \ldots, C_{c_{i+n}}\} \tag{3}$$

Algorithm 2: Drift detection method

Data: *Data Stream*: D_s; *Current Window*: w_c; *Recent Window*: w_r; *Temporary Variable* : l; *Cluster Center*: C_i; *Array*: $C_r, C_c, FP, d(C_r, C_c)$.
Result: *Drift detection.*

Function `DriftDetector(`w_c`):`
  if l==0 **then**
    Select features according to the k highest scores using the Chi-Square test;
    Apply K-means clustering on selected features data instances;
    Find center of each cluster(C_{r_i}) and store it in C_r;
    Calculate squared distance to cluster center to identify data point furthest from the centers and store result in FP_i;
  else
    Select features according to the k highest scores using Chi-Square test;
    Apply K-means clustering on selected festures data instances;
    Find center of each cluster (C_{c_i}) and store it in C_c;
    Calculate squared distance to cluster center to identify data point furthest from the centers and store result in FP_{i+1};
    Calculate the euclidean distance between the cluster centers and store outcome in $d_i(C_r, C_c)$;
    if $C_{r_i} \neq C_{c_i} \vee FP_i \neq FP_{i+1} \vee d_i(C_r, C_c) \neq d_{i+1}(C_r, C_c)$ **then**
      Return True;
    else
      Return False;

The centroid distance $d(C_r, C_c)$ between the centroid of recent window (C_r) and the centroid of the current window (C_c) is evaluated with the help of Euclidean distance, as shown in Eq. 4.

$$d(C_r, C_c) = \sqrt{(C_{r_i} - C_{c_i})^2 + (C_{r_j} - C_{c_j})^2} \tag{4}$$

Figure 2 demonstrates the cluster centroid and farthest point (FP) of a cluster, respectively. The virtual drift is detected when their boundary remains the same, but the cluster data distribution varies over time. The proposed work signals the virtual drift when one of the three conditions is satisfied, as shown below.

$$f(x) = \begin{cases} C_{r_i} \neq C_{c_i} \vee FP_i \neq FP_{i+1} \vee d_i(C_r, C_c) \neq d_{i+1}(C_r, C_c), & \text{return True} \\ \text{False}, & \text{otherwise} \end{cases}$$

The above condition shows that concept drift is detected if there is a change in centroid or farthest point or intercluster distance; otherwise, no drift is flagged. When the drift is signaled, the learning model rebuilds with the new incoming data instances to overcome concept drift in the data stream.

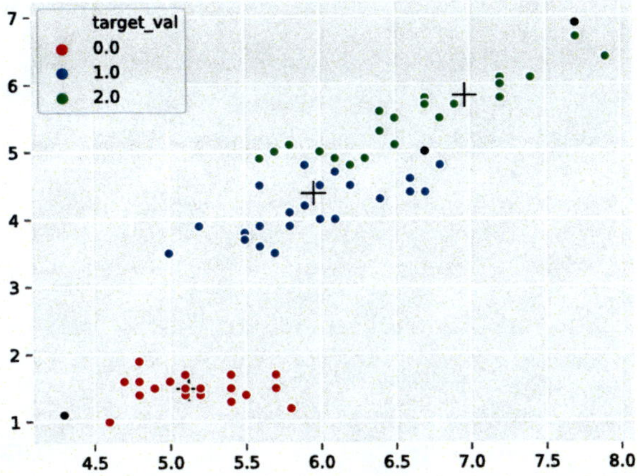

Fig. 2 Demostration of cluster centroid (+) and farthest point of each cluster centroid (•)

4 Evaluation and Results

This section discusses the experimental setup, case study, experimental datasets, and experimental results and analyses. The experiment is implemented in Python using libraries Scikit-learn and Scikit-multiflow. The online method does not contain a built-in drift adaptation mechanism, so the proposed work can be applied there. Here, Interleaved Test-Then-Train [18] approach is utilized for evaluation purposes. It depicts that the learning model prediction is performed when the new data instances arrive, and further, an update is done in the model.

Here, the sliding window mechanism is used. The window shrinks whenever the drift is flagged; otherwise, it expands. For evaluation purposes, the window size is taken as 50. When the drift is detected, the size of the window is reduced by half of its current size.

4.1 Case Study on Iris Dataset

The iris dataset includes three iris species with 150 number of instances and four attributes. There is no missing value presents in the dataset. The dataset is characterized as multivariate and the atrribute characteristics are real.

It has some properties about flowers. The two flower species are not linearly separable from each other. At the same time, one of them is linearly separable from the other two. Figure 3 demonstrates the data window clustering at different timestamps. There are three timestamps data windows in which the centroid and distribution of clusters are more similar. So, there is no drift found in these data windows.

Figure 4 exhibits that the cluster with data points in red color drifted from timestamp t_p to t_q. It suggests that the data distribution of a cluster changes concerning time.

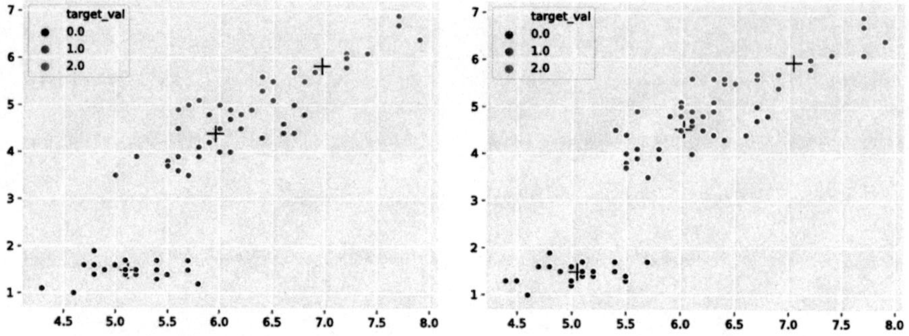

(a) Clustering of current window data at t_l time stamp

(b) Clustering of current window data t_n time stamp

Fig. 3 Data window clustering at different timestamp (no drift scenario)

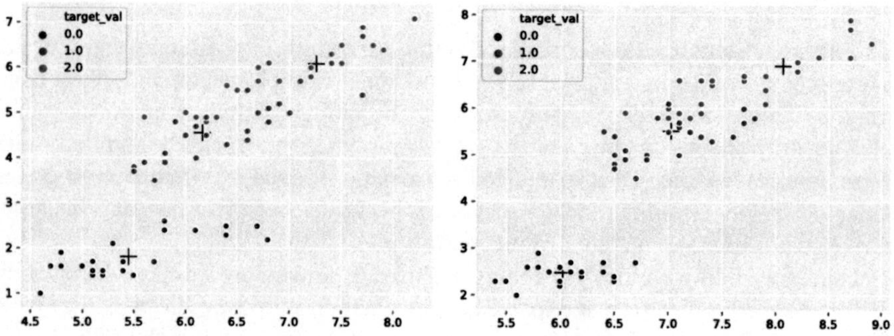

(a) Clustering of current window data at timestamp t_p

(b) Clustering of current window data at timestamp t_q

Fig. 4 Data window clustering at different timestamp (drift scenario)

4.2 Datasets

Synthetic datasets

- **LED dataset**: It works by predicting the digits that appear on the seven-segment LED display. There is 10 noise in this multivariate dataset. There are 24 attributes in total, all of which are categorical data. The qualities are represented in the form of 0 or 1 on the LED display. It indicates whether or not the reciprocal light is turned on. The ten percent noise indicates that each attribute vector has a ten percent chance of being reversed. A drift is defined as a change in the value of a characteristic.

- **SINE dataset**: In the dataset, there are two contexts: Sine1, where $y_i = sin(x_i)$, and Sine2, where $y_i = 0.5 + 0.3 \times sin(3\pi x_i)$. Reversing the context as mentioned earlier condition detects concept drift.
- **Agrawal dataset**: The information in the dataset pertains to people who are interested in taking out a loan. They are divided into two groups: group A and group B. The data collection includes age, salary, education level, house value, zip code, and other variables. There are ten functions in all, but only five are used to construct the dataset. The attribute value can be both numeric and nominal. The notion drift occurs both quickly and gradually in this case.

Real-time datasets

- **Airlines dataset**: There are two target values in the dataset. It determines if a flight is delayed. The analysis is based on factors such as flight, destination airport, time, weekdays, and length.
- **Spam Assassin dataset**: Based on e-mail communications, the data collection comprises 500 attributes. The values of all characteristics are binary. It shows whether a word appears in the e-mail. Does there appear to be a progressive change in spam texts over time?
- **Forest cover dataset**: The dataset includes 30 × 30 m cells in Region 2 of the US Forest Service (USFS). There are 54 qualities, 44 of which are binary values and 10 of which are numerical values. It depicts many characteristics such as height, vegetation appearances, disappearances, and so on. It's a normalized set of data.
- **Usenets dataset:** Usenets is a dataset that combines usenet1 and usenet2 to create a new dataset. It's a compilation of twenty different news organizations. The user labels the communications in the order of their interest. In both datasets, there are 99 properties.

4.3 Experimental Results and Analyses

Synthetic and real-time datasets are used in the experiment. In the synthetic dataset, abrupt and gradual drift contains datasets are written as Abr and Grad, respectively. In addition, the number of data instances is also mentioned with the particular dataset. The suggested method is compared with existing methods that use the Hoeffiding Tree classifier. At the end of the data stream, the mean accuracy of each window is utilized to calculate classification accuracy (or average mean accuracy). The mean accuracy is calculated by the ratio of the number of correct predictions to the total number of predictions of each window.

In terms of classification accuracy, the suggested technique using the Hoeffding base classifier behaves as follows (see Table 1). Sine (Grad-20K), Airlines and Usenets dataset exhibit a decrease in classification accuracy. At the same time, Agrawal (Abr-20K) dataset shows a marginal decrease in classification accuracy. In addition to this, Agrawal (Abr-50K), Agrawal (Abr-100K), Agrawal (Abr-50K),

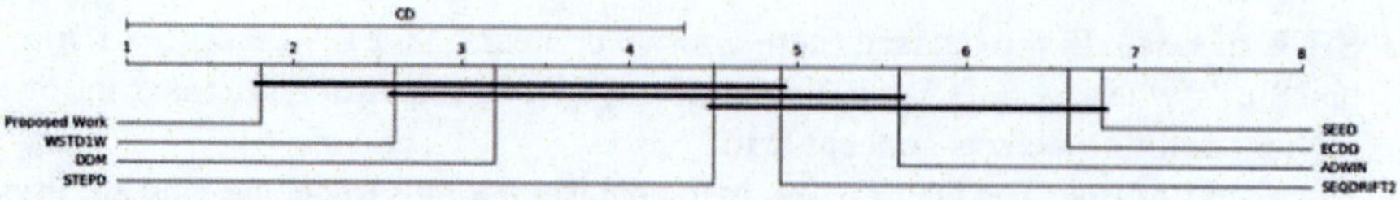

Fig. 5 Critical distance (CD) diagram based on classification accuracy of methods with HT classifier

Agrawal (Grad-20K), Agrawal (Grad-50K), Agrawal (Grad-100K), Forest Cover, and Spam Assassin manifest a significant increase in classification accuracy.

We use the *Friedman* test with *Nemenyi-post-hoc* analysis (Demšar) to validate the statistical significance of the performance of the proposed method and the compared methods utilizing NB and HT classifier. The null hypothesis $H0$ states that equivalent methods have the same rank. The *Friedman* test is based on this assumption. We compare eight strategies using ten datasets in this test. Each method is ranked according to its performance in terms of classification accuracy (see Table 1).

As mentioned by Demšar, a *Nemenyi-post-hoc* analysis is performed, and the Critical Difference (CD) is calculated. The proposed technique apparently outperforms ADWIN, ECDD, and SEED substantially (see Fig. 5).

5 Conclusion

In the streaming environment, the learning model has the ability to obtain new information. It updates information by applying the forgetting mechanism and rebuilding the learning model using further information. Several drift detection algorithms assume that the label of data instances is available after the learning model's prediction. But in the real-time scenario, it is not feasible. The paper proposes an unsupervised drift detection method to detect virtual drift in non-stationary data. It minimizes the complexity of data by selecting the k-high score features of data samples. So, it works efficiently with high-dimensional streaming data.

The future direction of the proposed work is that it can be applied to various application domains. Future work will add the outlier detection technique with the proposed drift detection method. In addition, distinguishing between noise and concept drift is an open challenge.

Table 1 Comparison of classification accuracy between proposed method and existing methods using HT classifier

Dataset	ADWIN	DDM	ECDD	SEED	SEQDRIFT2	STEPD	WSTD1W	Proposed work
LED (Grad-20K)	63.09	70.73	67.59	55.60	60.97	65.64	69.06	**82.40**
Sine (Grad-20K)	85.54	86.82	85.22	85.49	86.69	86.01	86.79	83.04
Agrawal (Abr-20K))	64.50	65.21	64.27	64.64	64.63	65.27	65.76	65.29
Agrawal (Abr-50K)	65.88	70.01	65.40	65.39	67.29	66.47	69.89	**88.24**
Agrawal (Abr-100K)	66.63	73.09	66.96	65.71	68.70	67.06	70.52	**88.29**
Agrawal (Grad-20K)	63.62	65.27	63.26	63.84	63.37	64.26	64.98	**88.24**
Agrawal (Grad-50K)	65.48	69.20	65.69	65.04	66.83	66.02	68.58	**88.49**
Agrawal (Grad-100K)	66.35	73.48	66.43	65.47	68.49	66.89	70.22	**88.40**
Airlines	66.70	65.35	63.66	**66.71**	66.60	65.73	**66.71**	58.47
Forest cover	67.73	67.14	67.39	67.32	67.68	67.62	68.18	**70.73**
Spam assassin	91.87	89.34	88.39	90.90	89.70	91.42	91.80	**91.89**
Usenets	68.41	71.01	**72.75**	68.65	66.31	71.95	71.58	63.07
Average rank	5.6	3.2	6.6	6.8	4.9	4.5	2.6	1.8

Bold significes the highest accuracy of method with respect to particular dataset

References

1. Agrahari, S., Singh, A.K.: Concept drift detection in data stream mining: A literature review. J. King Saud Univer. Comput. Inf. Sci. (2021). ISSN 1319-1578. https://doi.org/10.1016/j.jksuci.2021.11.006. URL https://www.sciencedirect.com/science/article/pii/S1319157821003062
2. Agrahari, Supriya, Singh, Anil Kumar: Disposition-based concept drift detection and adaptation in data stream. Arab. J. Sci. Eng. **47**(8), 10605–10621 (2022). https://doi.org/10.1007/s13369-022-06653-4
3. Souza, V., Parmezan, A.R.S., Chowdhury, F.A., Mueen, A.: Efficient unsupervised drift detector for fast and high-dimensional data streams. Knowl. Inf. Syst. **63**(6), 1497–1527 (2021)
4. Xuan, Junyu, Jie, Lu., Zhang, Guangquan: Bayesian nonparametric unsupervised concept drift detection for data stream mining. ACM Trans. Intell. Syst. Technol. (TIST) **12**(1), 1–22 (2020)
5. Huang, H., Yoo, S., Kasiviswanathan, S.P.: Unsupervised feature selection on data streams. In: Proceedings of the 24th ACM International on Conference on Information and Knowledge Management, pp. 1031–1040 (2015)
6. de Mello, R.F., Vaz, Y., Grossi, C.H., Bifet, A.: On learning guarantees to unsupervised concept drift detection on data streams. Expert Syst. Appl. **117**, 90–102 (2019)
7. Wiwatcharakoses, C., Berrar, D.: Soinn+, a self-organizing incremental neural network for unsupervised learning from noisy data streams. Expert Syst. Appl. **143**, 113069 (2020)
8. Gözüaçık, Ömer., Can, Fazli: Concept learning using one-class classifiers for implicit drift detection in evolving data streams. Artif. Intell. Rev. **54**(5), 3725–3747 (2021)
9. Pinto, F., Sampaio, M.O.P., Bizarro, P.: Automatic model monitoring for data streams. *arXiv preprint* arXiv:1908.04240 (2019)
10. Gama, J., Medas, P., Castillo, G., Rodrigues, P.: Learning with drift detection. In: Brazilian Symposium on Artificial Intelligence, pp. 286–295. Springer, Berlin (2004)
11. Bifet, Albert: Adaptive learning and mining for data streams and frequent patterns. ACM SIGKDD Explor. Newsl. **11**(1), 55–56 (2009)
12. Ross, G.J., Adams, N.M., Tasoulis, D.K., Hand, D.J.: Exponentially weighted moving average charts for detecting concept drift. Pattern Recogn. Lett. **33**(2), 191–198 (2012)
13. Huang, D.T.J., Koh, Y.S., Dobbie, G., Pears, R.: Detecting volatility shift in data streams. In: 2014 IEEE International Conference on Data Mining, pp. 863–868 (2014). https://doi.org/10.1109/ICDM.2014.50
14. Pears, R., Sakthithasan, S., Koh, Y.S.: Detecting concept change in dynamic data streams. Mach. Learn. **97**(3), 259–293 (2014)
15. Nishida, K., Yamauchi, K.: Detecting concept drift using statistical testing. In: Corruble, V., Takeda, M., Suzuki, E. (eds.) Discovery Science, pp. 264–269, Springer, Berlin (2007). ISBN 978-3-540-75488-6
16. de Barros, R.S.M., Hidalgo, J.I.G., de Lima Cabral, D.R.: Wilcoxon rank sum test drift detector. Neurocomputing **275**, 1954–1963 (2018)
17. Franke, T.M., Ho, T., Christie, C.A.: The chi-square test: Often used and more often misinterpreted. Am. J. Eval. **33**(3), 448–458 (2012)
18. Gama, João., Žliobaitė, Indrė, Bifet, Albert, Pechenizkiy, Mykola, Bouchachia, Abdelhamid: A survey on concept drift adaptation. ACM Comput. Surv. (CSUR) **46**(4), 1–37 (2014)
19. Demšar, Janez: Statistical comparisons of classifiers over multiple data sets. J. Mach. Learn. Res. **7**, 1–30 (2006)

Balanced Sampling-Based Active Learning for Object Detection

Sridhatta Jayaram Aithal, Shyam Prasad Adhikari, Mrinmoy Ghorai, and Hemant Misra

Abstract Active learning focuses on building competitive models using less data by utilizing intelligent sampling, thereby reducing the effort and cost associated with manual data annotation. In this paper an active learning method based on a balanced sampling of images with high and low confidence object detection scores for training an object detector is proposed. Images with higher object prediction scores are sampled using the uncertainty measure proposed by Yu et al. which utilizes detection, classification and distribution statistics. Though this method encourages balanced distribution in sampling, a deeper look into the sampled distribution reveals that the under-represented classes in the initial labeled pool remain skewed throughout the subsequent active learning cycles. To mitigate this problem, in each active learning cycle, we propose to sample an equal proportion of images with high and low confidence object prediction scores from the model trained in the last cycle, where the low confidence prediction sample selection is based on the model's prediction scores. Experiments conducted on the UEC Food 100 dataset show that the proposed method performs better than the baseline random sampling, CALD and low confidence prediction sampling method by +4.7, +8.7, and +3.1 mean average precision (mAP), respectively. Moreover, consistently superior performance of the proposed method is also demonstrated on the PASCAL VOC'07 and PASCAL VOC'12 datasets.

Work done as intern at Swiggy.

S. J. Aithal (✉) · M. Ghorai
IIIT Sricity, Sathyavedu 517646, India
e-mail: sridhattajayaram.a@iiits.in

M. Ghorai
e-mail: mrinmoy.ghorai@iiits.in

S. P. Adhikari · H. Misra
Applied Research Swiggy, Bangalore 560103, India
e-mail: shyam.pa@swiggy.in

H. Misra
e-mail: hemant.misra@swiggy.in

M. Tistarelli et al. (eds.), *Computer Vision and Machine Intelligence*,
Lecture Notes in Networks and Systems 586,
https://doi.org/10.1007/978-981-19-7867-8_26

Keywords Active learning · Intelligent sampling · Data annotation · Object detection · Balanced sampling · Balanced distribution

1 Introduction

Active learning is the process of building competitive models with fewer annotated data with human in the loop as shown in Fig. 1. It is an active area of research in deep learning. It's importance is more in case of training object detection models. The image annotation process is more time consuming due to the need for annotating bounding boxes around the object for localization along with the class labels. Active learning can help alleviate the need for annotating large amount of data and hence reduce the amount of manual labor and time.

Active learning is well studied in the area of computer vision, specifically for image classification [1–3], where the classification uncertainty of the model is used for selecting informative samples for training. Active learning has also been explored for object detection tasks [2, 4–7], where the predominant method for sampling informative images from a unlabeled pool is based on image uncertainties computed using a model pre-trained on a smaller subset of labeled data. However, there are differences in the way the uncertainties are computed.

In this paper we propose active learning for object detection as shown in Fig. 2 in which half of the informative images in a budget cycle are sampled using the consistency-based method (CALD: Consistency-based Active Learning method for object Detection) [7], whereas the rest of the images are sampled from images with low model prediction scores. CALD assists in sampling informative images from well-represented classes, whereas selecting images from the low prediction score regime encourages sampling from under-represented classes. The proposed method thus benefits from this diversity in sampling. We show that the proposed method achieves competitive detection performance in fewer active learning cycles than the baseline random sampling and CALD.

Rest of the paper is organized as follows: Sect. 2 summarizes the related work whereas the method and experiments are described in Sects. 3 and 4, respectively, followed by conclusion in Sect. 5.

2 Related Works

Most of the works in the literature on active learning for object detection focus on uncertainty scoring mechanisms. The sampling of most informative samples is done, so that the best performing model can be trained within a specified annotation budget. The self-supervised sample mining [8] method is based on an approach where the region proposals are pasted on the cross validation samples to measure their classification prediction consistency, and determine whether to annotate the sample in a

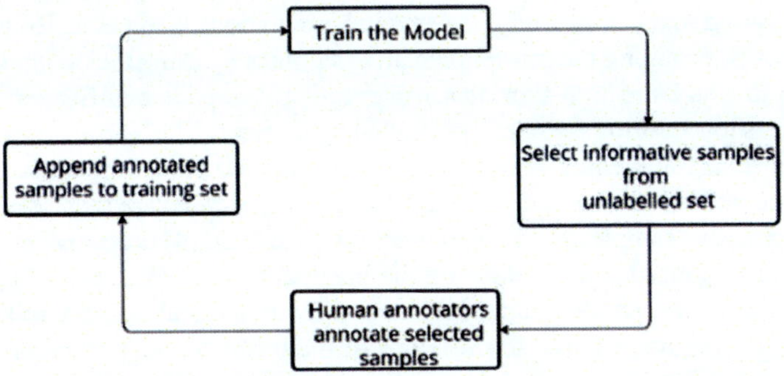

Fig. 1 Active learning

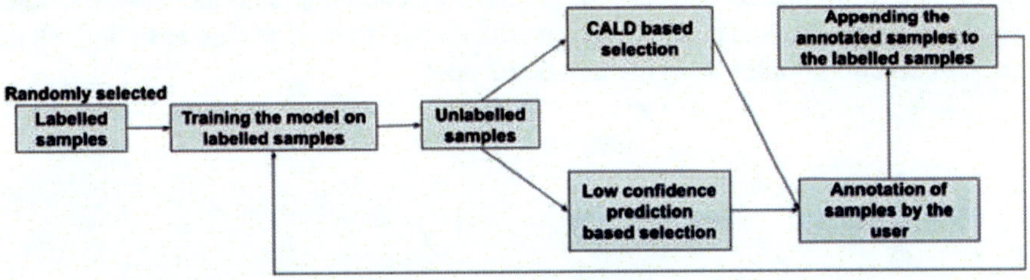

Fig. 2 Block diagram of the proposed active learning method

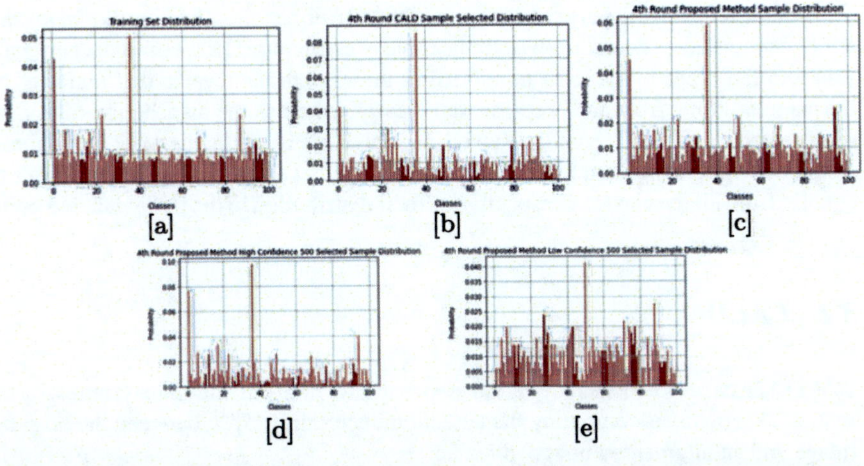

Fig. 3 Distribution of **a** UEC Food 100 training set, **b** Samples selected with CALD, **c** Samples selected using the proposed method, **d** Samples with higher prediction scores selected from the proposed method, **e** Samples with lower prediction scores selected from the proposed method

self-supervised manner or send it for manual annotation. Wang et al. [6] proposed a method for determining the most informative samples by using the confidence-based sampling in which the high-scoring samples are processed in a self-supervised manner whereas low-scoring samples are labeled by humans. The learning loss approach [2] uses a separate learning module known as loss prediction module to predict the loss for the unlabeled samples; the samples with highest loss are then selected as the most informative samples for manual labeling. Kao et al. [4] proposed to use a combination of two metrics for measuring the informativeness of an object hypothesis; namely localization stability and localization tightness. Localization stability is computed by passing noise corrupted versions of the original image through the model and measuring the variations of the predicted object location, whereas localization tightness is based on the overlapping ratio between the region proposal and the final prediction in 2-stage models. CALD [7] uses the combination of consistency-based scoring mechanism and the distribution with respect to labeled classes to find out the most uncertain samples from the unlabeled pool.

3 Method

3.1 Problem Definition

Most of the recent works in active learning focus on either classification-based uncertainty estimation or detection-based uncertainty estimation. However a deeper analysis into the types of samples mined using these methods reveals that majority of the samples come from well-represented classes, whereas the distribution of under-represented classes still remains skewed, as shown in Fig. 3b In this work we propose a method wherein the sample diversity is maintained as shown in Fig. 3c by sampling equally from all the classes irrespective of their distribution in the initial labeled pool.

3.2 CALD

CALD [7] uses a consistency-based approach to find out the uncertainties in an image. This involves computing the detection uncertainty, S_1^{IoU}, between the original image and an augmented image given as:

$$S_1^{IoU} = \mathrm{argmax}_{I_i'} IoU(I_i', I) \tag{1}$$

where, I is the original image, I' is an augmented version of I generated by adding color jitter, Gaussian noise, grid distortion, ISO noise, etc. on the original image, and IoU is the bounding box overlap between the predicted object location in I and I'. The next step is to compute classification score uncertainty, S_2^{Dist}, which is calculated as the divergence between the predicted class-wise probabilities and given as:

$$S_2^{\text{Dist}} = (1 - JS(I_{\text{prob}}, I'_{\text{prob}})) \tag{2}$$

where $JS(I_{\text{prob}}, I'_{\text{prob}})$, is the Jensen-Shannon (JS) divergence between the class-wise probabilities of the original unlabeled pool of images, I_{prob}, and their corresponding augmented versions, I'_{prob}. JS divergence between two discrete probability distributions P and Q can be computed as:

$$JS(P||Q) = \frac{1}{2}\left(\sum_{x \in X} P(x) \log\left(\frac{P(x)}{Q(x)}\right)\right) - \frac{1}{2}\left(\sum_{x \in X} P(x) \log\left(\frac{Q(x)}{P(x)}\right)\right) \tag{3}$$

Next, the distribution uncertainty score, S_3^{Dist}, is given as :

$$S_3^{\text{Dist}} = (1 - JS(I'_{\text{prob}}, I_{\text{prob}}^{\text{lab}})) + (1 - JS(I_{\text{prob}}, I_{\text{prob}}^{\text{lab}})) \tag{4}$$

where $I_{\text{prob}}^{\text{lab}}$, represents the class-wise probabilities obtained from the already labeled pool of images. Finally, all of the uncertainty scores computed in Eqs. (1), (2) and (4) are summed up,

$$S_{\text{high\_conf}} = S_1^{\text{IoU}} + S_2^{\text{Dist}} + S_3^{\text{Dist}} \tag{5}$$

to get the final uncertainty score of the image, $S_{\text{high\_conf}}$. The images are then sampled in ascending order of, $S_{\text{high\_conf}}$, till the budget criteria is satisfied.

3.3 Proposed Method

A deeper look into the distribution of samples selected using CALD reveals that the under-represented classes in the initial labeled pool remain skewed throughout the subsequent active learning cycles, as shown in Fig. 3b The classes which are well-represented in the initial random sampling, continue to remain so, whereas under-represented classes remain in minority. To have a balanced distribution of samples from all the classes, we propose to sample images where the model has low prediction scores in addition to sampling using CALD, as shown in Fig. 2 Images with low object detection scores are sampled based on,

$$S_{\text{low\_conf}} = max(I_{\text{prob}}) \tag{6}$$

where $S_{\text{low\_conf}}$, is the maximum of all the predicted class-probabilities, I_{prob}, across all the predicted bounding boxes in an image, I. The selection of samples is based on, $S_{\text{low\_conf}}$, whereby all the samples with, $S_{\text{low\_conf}} < threshold$, are probable candidates for selection. The samples are selected in ascending order of $S_{\text{low\_conf}}$ until the budget criteria is satisfied. The threshold is decided empirically in every cycle so as to satisfy the budget needs. From Fig. 3 we see that sampling, 50% using CALD, Fig. 3d, and sampling the rest 50% from low confidence prediction samples, Fig. 3e, encourages balanced sampling of classes, Fig. 3c, and the sampled distribution is similar to the overall data distribution. In Sect. 4, we show that the diversity induced by the proposed method leads to training competitive models within a less budget cycle.

4 Experiments

4.1 Dataset Used

The dataset used in experiments are Pascal VOC 2007 [9], Pascal VOC 2012 [10] and UEC Food 100 dataset [11]. The Pascal VOC 2007 has 20 annotated classes with 5011 training and 4952 validation samples. Similarly, Pascal VOC 2012 has 20 annotated classes with 5717 training and 5823 validation samples. The UEC Food 100 dataset has 100 classes with 9466 training and 1274 validation samples. Annotation budget for Pascal VOC 2007 and VOC 2012 is 500 samples per cycle, whereas 1000 samples per cycle is the budget for UEC Food 100.

4.2 Metrics Used

The metrics for the experiments conducted on UEC Food 100 dataset are mean average precision (mAP) and mean average recall (mAR) and mean F1 (mF1) on validation dataset at IoU threshold =0.5:0.05:0.95, whereas mAP at IoU threshold = 0.5 is the metric for experiments on Pascal VOC 2007 and VOC 2012 for better comparison with existing methods.

4.3 Model

We have considered Faster RCNN [12], a two stage detector, with Resnet-50 fpn [13] backbone pre-trained on coco dataset for all of our experiments. In each cycle, the models were trained with a learning rate of 0.0003 and decay of 0.0005. Cosine Annealing Warm Restart [14] was used with batch size = 1. In each cycle, UEC Food 100 was trained for 40 epochs, whereas VOC-07 and VOC-12 were trained for 20 epochs using stochastic gradient descent with momentum. While training, the data was augmented using random horizontal flip with a probability of 0.5. The weight factor in the Eq. (2) as discussed in paper [7] is not considered. The ablation study of the weight factor in paper [7] indicates that even when the weight factor is not considered there is no significant change in the performance.

Table 1 Active learning result on UEC Food 100 dataset after 4 cycles

Active learning methods	mAP @IoU=0.5:0.05:0.95	mAR @IoU=0.5:0.05:0.95	mF1 @IoU=0.5:0.05:0.95
Proposed method	**46.5**	**62.8**	**53.4**
Random Passive Learning	41.8	58.5	48
CALD [7]	37.8	57.1	45
Low confidence	43.4	60.5	50
75% CALD and 25% Low confidence	45	61.4	51.9
25% CALD and 75% Low confidence	45.4	62.2	52

The definition for the bold in table: Proposed Method- Balanced Sampling

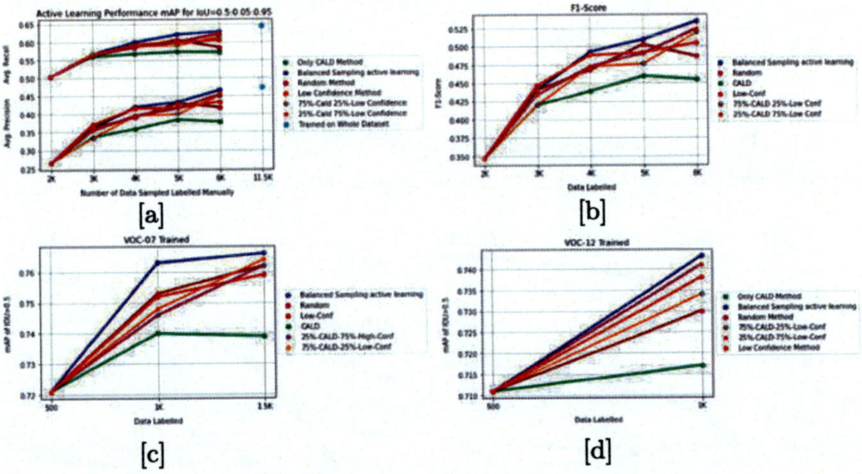

Fig. 4 **a** UEC Food 100 active learning results, **b** F1 Score results of active learning on UEC Food 100, **c** Pascal VOC-07 results, **d** Pascal VOC-12 results

4.4 Results

The results of the proposed method compared to the baseline random passive sampling, CALD and low confidence prediction samples on the UEC Food 100 data are presented in Table 1, and the comparative results of these methods during each active learning cycle is presented in Fig. 4a and b. Moreover, results of different sampling proportions (25:75, 50:50, 75:25) between CALD and low confidence sampling are also presented.

From Table 1, and Fig. 4a and b we see that sampling equally (50:50) from CALD and low-scoring samples results in the best performing model within less active learning cycles. We consider this as our best performing model. Moreover, we also

Table 2 Active learning result on VOC-07 dataset after 2 cycles

Active learning methods	mAP @IoU=0.5
Proposed method	**76.6**
Random passive learning	76.2
CALD [7]	73.9
Low confidence	75.9
75% CALD and 25% Low confidence	76.4
25% CALD and 75% Low confidence	76.2

The definition for the bold in table: Proposed Method- Balanced Sampling

Table 3 Active learning result on VOC-12 dataset after 1 cycle

Active learning methods	mAP @IoU=0.5
Proposed method	**74.3**
Random passive learning	73
CALD [7]	71.7
Low confidence	73.8
75% CALD and 25% Low confidence	73.4
25% CALD and 75% Low confidence	74.1

The definition for the bold in table: Proposed Method- Balanced Sampling

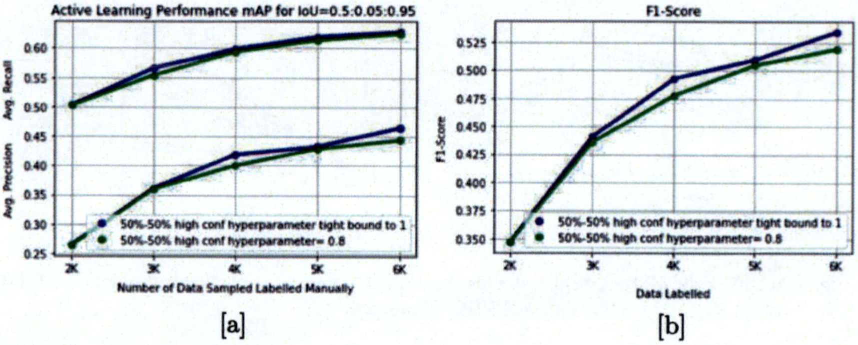

Fig. 5 Comparison of using different thresholds for sampling high-scoring images for UEC Food 100 dataset **a** Average precision and recall curves, **b** F1 Score

see that our best model performs better than random passive sampling, CALD and low confidence samples by +4.7/+8.7/+1.3 mAP, respectively. Also compared to the model trained on the whole training dataset, the proposed method leads to a model with similar performance using 50% less training data. Similarly, comparisons of the proposed method presented in Tables 2 and 3, and Fig. 4c and d for the VOC-07 and VOC-12 dataset, respectively, show that the proposed method consistently performs better than other methods.

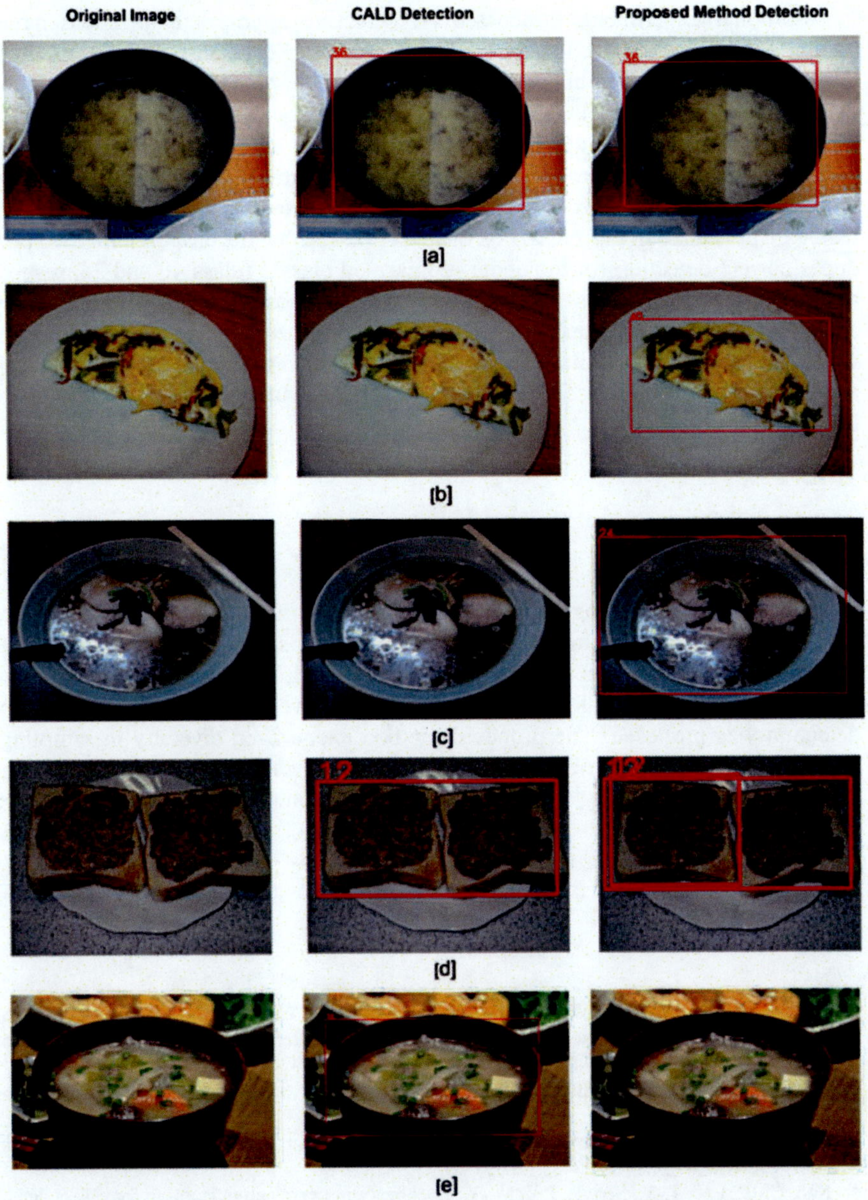

Fig. 6 Visual comparison of detection results of CALD and the proposed method on UEC food 100 dataset (**a**) for well-represented class (class 36), (**b**) and (**c**)for under-represented classes (class 40 and 24, respectively), (**d**) overlapping detection, (**e**) missed detection

Additional experiments conducted for selecting appropriate threshold hyper parameters for CALD used in the proposed method, Fig. 5, shows that setting the threshold for detection confidence close to 1 results in better performance than the default value of 0.8. This leads us to believe that balanced sampling from extremes of the high-scoring and low-scoring samples results in better performing models.

Few detection results are presented in Fig. 6, where the results from the proposed method are compared with CALD. The well represented class (class 36) detection result are presented in Fig. 6a. In Fig. 6b and c, we see that the proposed method produces correct detection for the under-represented classes (class 40 and 24, respectively), while CALD misses these detections. This can be attributed to the way sampling is done in the proposed method. CALD focuses on selecting samples from well-represented class, while the proposed method seeks to sample equally from well and under-represented classes. Some missed and overlapping detection results are presented in Fig. 6d and e.

5 Conclusion

While most of the existing works on active learning have focused on coming up with novel methods for computing uncertainty measures, not many of them focus on the underlying distribution of the sampled data, which leads to sub-optimal model training. In this work a balanced sampling-based active learning method for object detection was proposed. The proposed method encouraged diversity in sampling, whereby the sampled distribution was similar to the overall data distribution. Maintaining this class diversity was shown to produce competitive models within less annotation budget.

References

1. Bengar, J.Z., van de Weijer, J., Fuentes, L.L., Raducanu, B.: Class-balanced active learning for image classification. In: Proceedings of the IEEE/CVF Winter Conference on Applications of Computer Vision, pp. 1536–1545 (2022)
2. Yoo, D., Kweon, I.S.: Learning loss for active learning. In Proceedings of the IEEE/CVF Conference on Computer Vision and Pattern Recognition, pp. 93–102 (2019)
3. Sener, O., Savarese, S.: Active learning for convolutional neural networks: a core-set approach. arXiv preprint arXiv:1708.00489 (2017)
4. Kao, C.-C., Lee, T.-Y., Sen, P., Liu, M.-Y.: Localization-aware active learning for object detection. In: Asian Conference on Computer Vision, pp. 506–522. Springer, Berlin (2018)
5. Haussmann, E., Fenzi, M., Chitta, K., Ivanecky, J., Xu, H., Roy, D., Mittel, A., Koumchatzky, N., Farabet, C., Alvarez, J.M.: Scalable active learning for object detection. IEEE Intell. Veh. Symp. (iv). IEEE **2020**, 1430–1435 (2020)
6. Wang, K, Lin, Liang, Yan, Xiaopeng, Chen, Ziliang, Zhang, Dongyu, Zhang, Lei: Cost-effective object detection: active sample mining with switchable selection criteria. IEEE Trans. Neural Netw. Learn. Syst. **30**(3), 834–850 (2018)

7. Yu, W., Zhu, S., Yang, T., Chen, C., Liu, M.: Consistency-based active learning for object detection. arXiv preprint arXiv:2103.10374 (2021)
8. Wang, K., Yan, X., Zhang, D., Zhang, L., Lin, L.: Towards human-machine cooperation: Self-supervised sample mining for object detection. In: Proceedings of the IEEE Conference on Computer Vision and Pattern Recognition, pp. 1605–1613 (2018)
9. Everingham, M., Van Gool, L., Williams, C.K.I., Winn, J., Zisser- man, A.: The PASCAL visual object classes challenge 2007 (VOC2007) results (2007). http://www.pascal-network.org/challenges/VOC/voc2007/workshop/index.html
10. Everingham, M., Van Gool, L., Williams, C.K.I., Winn, J., Zisser-man, A.: The PASCAL Visual object classes challenge 2012 (VOC2012) results (2012). http://www.pascal-network.org/challenges/VOC/voc2012/workshop/index.html
11. Matsuda, Y., Hoashi, H., Yanai, K.: Recognition of multiple-food images by detecting candidate regions. In: Proceeding of the IEEE International Conference on Multi- media and Expo (ICME) (2012)
12. Ren, S., He, K., Girshick, R., Sun, J.: Faster r-cnn: Towards real-time object detection with region proposal networks. Adv. Neural Inf. Process. Syst. **28** (2015)
13. Lin, T.-Y., Dollar, P., Girshick, R., He, K., Hariharan, B., Belongie, S.: Feature pyramid networks for object detection. In: Proceedings of the IEEE conference on computer vision and pattern recognition, pp. 2117–2125 (2017)
14. Loshchilov, I., Hutter, F.: Sgdr: Stochastic gradient descent with warm restarts. arXiv preprint arXiv:1608.03983 (2016)

Multi-scale Contrastive Learning for Image Colorization

Ketan Lambat and Mrinmoy Ghorai

Abstract This paper proposes a multi-scale contrastive learning technique for image colorization. The image colorization task aims to find a map between the source gray image and the predicted color image. Multi-scale contrastive learning for unpaired image colorization has been proposed to transform the gray patches in the input image into the color patches in the output image. The contrastive learning method uses input and output patches and maximizes their mutual information to infer an efficient mapping between the two domains. We propose a multi-scale approach for contrastive learning where the contrastive loss is determined from different resolutions of the source image to improve the color quality. We further illustrate the effectiveness of the proposed approach through experimental outcomes and comparisons with cutting-edge strategies in image colorization tasks.

Keywords Colorization · Contrastive learning · GAN

1 Introduction

Image colorization deals with mapping binary or grayscale or infrared images into color images. It requires a semantic understanding of the scene and knowledge of color representation in the world. It has many applications like colorization of historical images [1], black and white cartoons [2], nightvision infrared images [3], remote sensing images [4], old archival videos [5], and many more.

Most of the current approaches for image colorization use deep convolutional neural networks as its foundation. However, we can classify these techniques into two groups, one category consists of fully automatic methods and the other category is semi-automatic where user intervention is required. Fully automatic image colorization [6, 7] is the main goal of this paper. Though these methods can generate

K. Lambat (✉) · M. Ghorai
Indian Institute of Information Technology, Sri City, Chittoor, India
e-mail: ketanvijay.l18@iiits.in

M. Ghorai
e-mail: mrinmoy.ghorai@iiits.in

M. Tistarelli et al. (eds.), *Computer Vision and Machine Intelligence*,
Lecture Notes in Networks and Systems 586,
https://doi.org/10.1007/978-981-19-7867-8_27

color images without user intervention, the result of image colorization is not always satisfactory.

This paper is focused on deep learning architecture with a contrastive learning mechanism. In the contrastive learning technique, features of a dataset are learnt by teaching the model which data points are similar or different, without providing labels. Thus, the model knows which features of a dataset are similar and which ones are different, and by combining contrastive learning with generative techniques, desired results can be obtained.

Contrastive learning is frequently employed in image-to-image translation, which translates a picture from one domain to an image from another [8, 9]. Some examples of image-to-image conversion tasks are style transfer, image in-painting, image colorization, etc. Contrastive learning is an effective unsupervised machine learning [10, 11] tool for visual representation. When translating one image into another, this is primarily done using the Generative Adversarial Networks (GAN) along with encoders and decoders. In this paper, we have trained GAN to colorize grayscale images using multi-scale contrastive learning. We propose multi-scale contrastive learning to extract the feature map in different scales and combine scale-based contrastive losses to get the final loss. Each object can appear different in different scales and varied features can be extracted when an image is considered in different resolutions. Taking this into consideration, we have obtained the contrastive loss for the images in different resolutions which helps to capture various features of the images in different levels of detail.

2 Related Work

The goal of image colorization, which is an image-to-image translation process, is to create a new image from an input image. In this section, we discuss some of the GAN models such as usual GAN, Pix2Pix GAN, CycleGAN, and basic contrastive learning.

2.1 Generative Adversarial Network

The outputs of Generative Adversarial Networks (GANs) are excellent in image generation, style transfer, image editing, medical image synthesis, and representation learning among others [12]. Image-to-image translation is also primarily done using the Generative Adversarial Networks (GAN). It is a generative model and backbone of various cutting-edge image-to-image translation methods. Generative modeling is an unsupervised learning task in machine learning where the patterns or similar features in an image are detected automatically by the model to generate new examples as output which are similar to the items in the original dataset.

The secret to GAN's effectiveness is the adoption of an adversarial loss, which forces the generated images to be visually identical to the target images. Since producing such images is a common task for computer graphics, this loss is particularly useful in these situations. An adversarial loss is used to learn the mapping, making it unable to distinguish between translated and target domain images [12].

A GAN is made up of two networks, a discriminator D and a generator G. The generator network takes in an input which can be some random noise as well to generate a sample based on the initial weights. This generated sample is the input for the discriminator network along with sample from the target image. The job of discriminator network is to classify whether the generated sample is a real target image or a fake one. This classification trains the generator network to generate samples which are indistinguishable from the target images in a way that the discriminator network can be deceived. If the discriminator network labels the generated sample as a real image, we can say that the model has converged. Thus, the generator and the discriminator operate similar to a min-max game algorithm, where the two networks are each other's adversary and hence the name Generative Adversarial Networks.

2.2 *Pix2Pix*

The Pix2Pix model [13] was a major breakthrough for the image-to-image translation tasks. It is a general-purpose GAN model also called as the conditional GAN, where the output image is generated based on a conditional input image. This model gave very promising results for applications not limited to style transfer like labels to real image, sketch to portrait, aerial to map, and background removal.

GAN is a generative model that learns a mapping $G : z \rightarrow y$, i.e., from random noise vector z to the output image y [14]. Whereas, Pix2Pix or conditional GANs $G : \{x, z\} \rightarrow y$, learn a mapping from input image x and random noise vector z, to y [13].

This model however requires a paired dataset where each image in the input domain needs to have a complementary image in the output domain. Collecting such a dataset can itself be a time-consuming and almost impossible task.

2.3 *CycleGAN*

The CycleGAN technique [12] removed the requirement of paired images required for training a network. Here, the image-to-image translation model is trained in an unsupervised manner using unpaired examples, i.e., there is no mapping from the input to the output domain in the training set. This model opened a lot of opportunities for further work in this field and can be used for style transfer in paintings, season translation in images, converting objects from one to another, etc.

This model works by using two GANs in parallel, where one translates images from domain $A \rightarrow B$ and the other from domain $B \rightarrow A$. While cycle consistency losses are used to prevent the learned mappings G and F from contradicting one another, adversarial losses are used to align the distribution of generated images with the distribution of data in the target domain [12]. One drawback of this model was that the model size was huge and training time was considerably high owning to two GANs used for one iteration.

2.4 Contrastive Learning

Next, we look at the Contrastive Unpaired Translation (CUT) model [8], which could achieve better results than cycleGAN while using one GAN model itself thus also reducing the training time. Here, we take multiple patches from the input and target images and two corresponding patches are encouraged to map to a similar point in learnt feature space in contrast to the other non-corresponding negative patches.

The key principle is that, in comparison with other random patches, a created output patch should look more like its matching input patch. The mutual information between the respective input and output patches is maximized using a multilayer, patchwise contrastive loss, which in an unpaired setting allows for one-sided translation. Since the negative patches were extracted from the original image alone, this method can work even with the dataset having one image for input and output domain each.

Here, images from the input domain, $\chi \subset \mathbb{R}^{H \times W \times C}$ are translated to look close to an image in the output domain, $\gamma \subset \mathbb{R}^{H \times W \times 3}$ using a dataset of unpaired images $X = \{x \in \chi\}$, $Y = \{y \in \gamma\}$. In this method, the mappings are learnt in one direction only, and thus, the use of inverse auxiliary generators and discriminators is avoided. A Noise Contrastive Estimation framework [10] is used to enhance the mutual information between input and output using a Patchwise Contrastive Loss (PatchNCE Loss), where the likelihood of choosing a positive sample over a negative sample is estimated as the cross-entropy loss.

3 Proposed Multi-scale Contrastive Learning Technique

The proposed approach is based on the contrastive learning technique mentioned in the CUT Model [8]. To create a meaningful mapping between the input and output domain, this technique maximizes the mutual information between the input and output patches. In the base CUT Model [8], the contrastive loss is obtained using a single resolution of the image, whereas we have used different resolutions of the image to calculate the patchwise contrastive loss. Different resolutions of an image contain different types of color information. A larger patch helps to get the general color theme of an image while the smaller patches help to obtain the

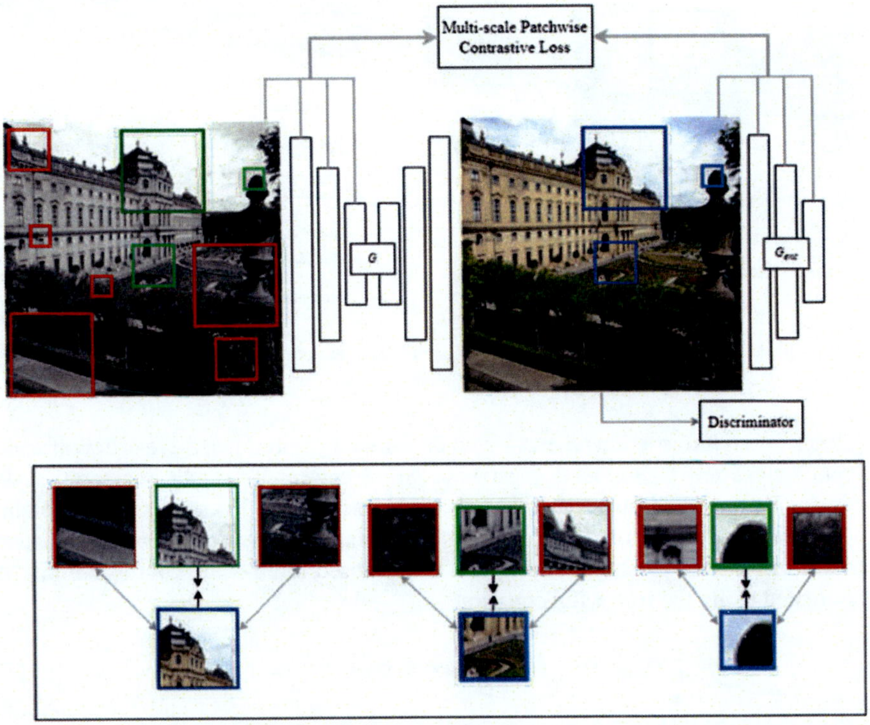

Fig. 1 Illustration of **multi-scale contrastive learning for image colorization** a patch in the generated output should be similar to a patch in input that corresponds to it, while dissimilar to the other negative patches. Here, the green patches match the blue ones, while the red patches are the negative contrastive samples

textural information. Thus, using this idea a multi-scale contrastive learning method is developed, where different resolutions or scales of the input image are taken with normalized weights to calculate the contrastive loss attempting to improve the output images' color quality.

In Fig. 1, we see that multiple patches of different scales have been picked in the input grayscale image. Similarly, a few patches which correspond to the patches in the source image have been identified in the destination domain image. The green patches match the blue ones, and the contrastive learning model tries to generate an image such that these two patches are brought closer. At the same time, the negative patches shown in red color are not identical to the blue patches and are in contrast to them, and thus, the model also makes efforts to increase the loss between them. This learning technique can also be trained on single images as the contrastive representation and the patches are formulated within the same image.

To get the multi-scale contrastive loss, we first generate three different resolutions of the input image, represented by L_{256}, L_{128}, and L_{64}. Different features of an image

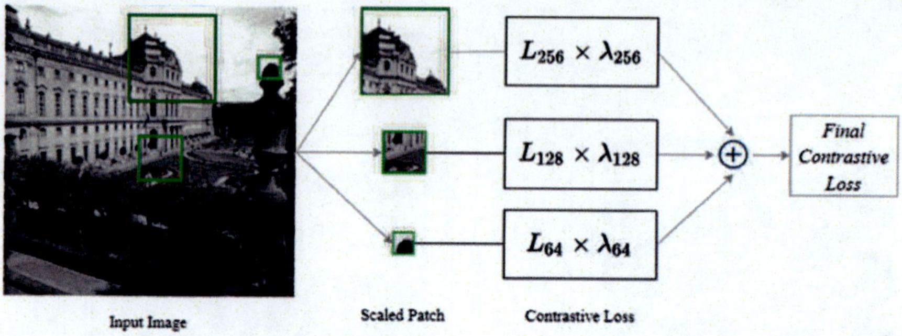

Fig. 2 Illustration showing the calculation of the multi-scale contrastive loss

give different information when viewed in different resolutions, these different scale input images help the model to learn these varied features. To get the final multi-scale contrastive loss, the features or loss of each scale is multiplied by a normalized weight. Then, the contrastive loss is determined by summing up the normalized contrastive loss of each scale as shown in Fig. 2. The proposed multi-scale contrastive loss is computed using the following equation.

$$\text{total\_nce\_loss} = L_{256} \times \lambda_{256} + L_{128} \times \lambda_{128} + L_{64} \times \lambda_{64}. \quad (1)$$

where L_a is the loss of a $a \times a$ sample and λ_a is the weight for L_a.

4 Experimental Setup

In this section, we show the outcomes of the suggested approach and compare them to certain state-of-the-art methods.

4.1 Models Used

We train the CycleGAN [12] and CUT Model [8] using their default settings on the below-mentioned datasets and obtain the FID scores for these methods.

To work on our proposed multi-scale contrastive learning technique, we use the default settings of CUT model [8] with a multi-scale approach. The model by default uses 256 × 256 scale patches only for calculating the PatchNCE loss. As a part of our experiment, we use additional 128 × 128 and 64 × 64 scaled patches along with some normalized weights. We arrived at these weights after trying out multiple combinations on a smaller dataset. Based on the weights for $\lambda_{256}, \lambda_{128}, and \lambda_{64}$, further training was done to obtain the results.

4.2 Datasets Used

To check the Image colorization results, two publicly available datasets have been used, Cityscapes dataset [15] and Imagenet dataset [16].

Cityscapes Dataset The Cityscapes dataset [15] contains images of German city streets, with 2975 training and 500 validation images. Each image is of 256 × 256 resolution. The original dataset only includes RGB photos and semantic label images. The grayscale images required for the image colorization are obtained with the use of the OpenCV library [17], from the respective RGB pictures. Thus, the obtained grayscale images and the provided color images form a part of our final dataset.

ImageNet Dataset ImageNet [16] is a massive collection of annotated pictures used in computer vision research. We have used the ImageNet Large Scale Visual Recognition Challenge 2012 (ILSVRC2012) [18] subset and further taken 1000 training and 500 validation photos for performing this experiment in the available GPU resources. The images provided are the colored RGB images, and for creating a dataset for image colorization training, we used the OpenCV library [17] to convert the RGB images to their equivalent single channel images. Grayscale photos are used as the input domain while color images are used as the target domain in the final dataset.

4.3 Metrics Used

The score of Fréchet Inception Distance (FID) is used to assess the outputs using the default setting of *pytorch-fid* [19] implementation. A lower FID score is better as it conveys that the curves of the images are closer.

Colorization is a multi-modal problem, where a given object can have multiple plausible colors. On re-colorizing an RGB image, a single random sample represents another plausible colorization. It is unlikely to have the same colors as the ground-truth RGB image. So while evaluating the image colorization model, it is advised to use distribution-level metrics such as the FID and NOT per-image metrics such as LPIPS or SSIM [20].

The Fréchet distance, or *d*, is a mathematical measure of how similar two curves are that takes into account the positions and arrangements of the curves' points. Consider a person walking with a dog tied with a leash (see Fig. 3). The minimum length of leash required such that both can move through curves in the same direction without needing any additional slack is described as the Fréchet distance.

In a deep network space, the FID score calculates the divergence between real and artificial pictures and estimates their distribution. The FID analyzes one of the deeper layers of the Inception v3, a convolutional neural network's mean and standard deviation instead of simply comparing images pixel by pixel [21–23].

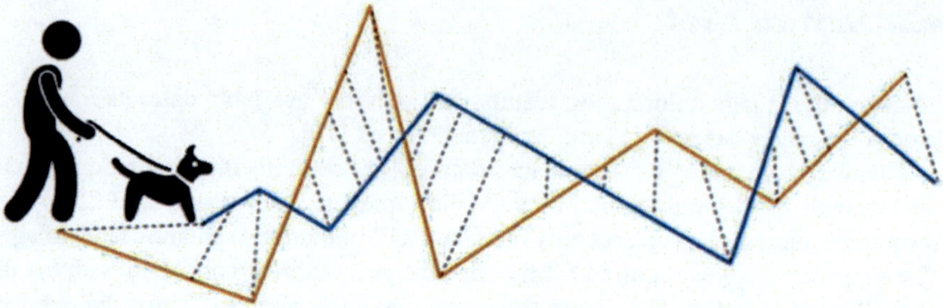

Fig. 3 Fréchet distance is the minimum leash length needed to walk the curves from the beginning to the end [21]

5 Experimental Results and Analysis

We carry out experiments using the mentioned experiment setup. We train the cycleGAN [12] and CUT [8] models and get the output for the Cityscapes [15] and ImageNet [16] dataset using their default settings. The results of these are mentioned in Table 1, and the generated images are present in col 3 and 4 of Figs. 4 and 5.

For the multi-scale contrastive loss, an important step is to find out the weights to be multiplied by the contrastive loss for each scale. We use a subset of the dataset to find out which combination of normalized weights gives the best output for our evaluation metric, the FID score. We find that $\lambda_{64} = 0.25, \lambda_{128} = 0.25, and \lambda_{256} = 0.50$ are the best suitable weights for $L_{64}, L_{128}, and L_{256}$, respectively.

Using these best-normalized weights in Eqn 1, we train the proposed multi-scale contrastive learning model on the datasets. The FID scores for the generated images are mentioned in Table 1, and the generated images are present in col 5 of Figs. 4 and 5.

In all the 3 models, cycleGAN [12], CUT Model [8], and the proposed multi-scale contrastive learning approach, the main structure of the image has been successfully generated, with no artifacts present during image restoration. The boundaries of the images are clearly defined and match that of the source image and the real-world image. Even in regions containing textures like grass, animal fur, and skin, there are no blurry spots or mismatching pixels.

Regarding the color of the generated image, we see a decent enough color restoration. Some random color patches can be seen in the output of the cycleGAN model for the ImageNet dataset (col 3, Fig. 5), which can be seen as a failure there. Some portions of the image are still gray while some are very brightly colored with an incorrect color (e.g., blue color for arm). The CUT Model output (col 4, Fig. 5) does not have any such abnormal colors but some parts of the image remain uncolored. For the cityscapes dataset, image colorization has been done satisfactorily.

The output of our proposed approach is visually superior to other approaches. Although there is no perfect color restoration, we do not observe any random color regions and the entire image is colorized as well. Our approach gives better results

Fig. 4 Qualitative results of cityscapes dataset. col 1: input grayscale image col 2: ground-truth image col 3: image generated using cycle GAN model col 4: image generated using CUT Model col 5: image generated using proposed multi-scale contrastive learning approach

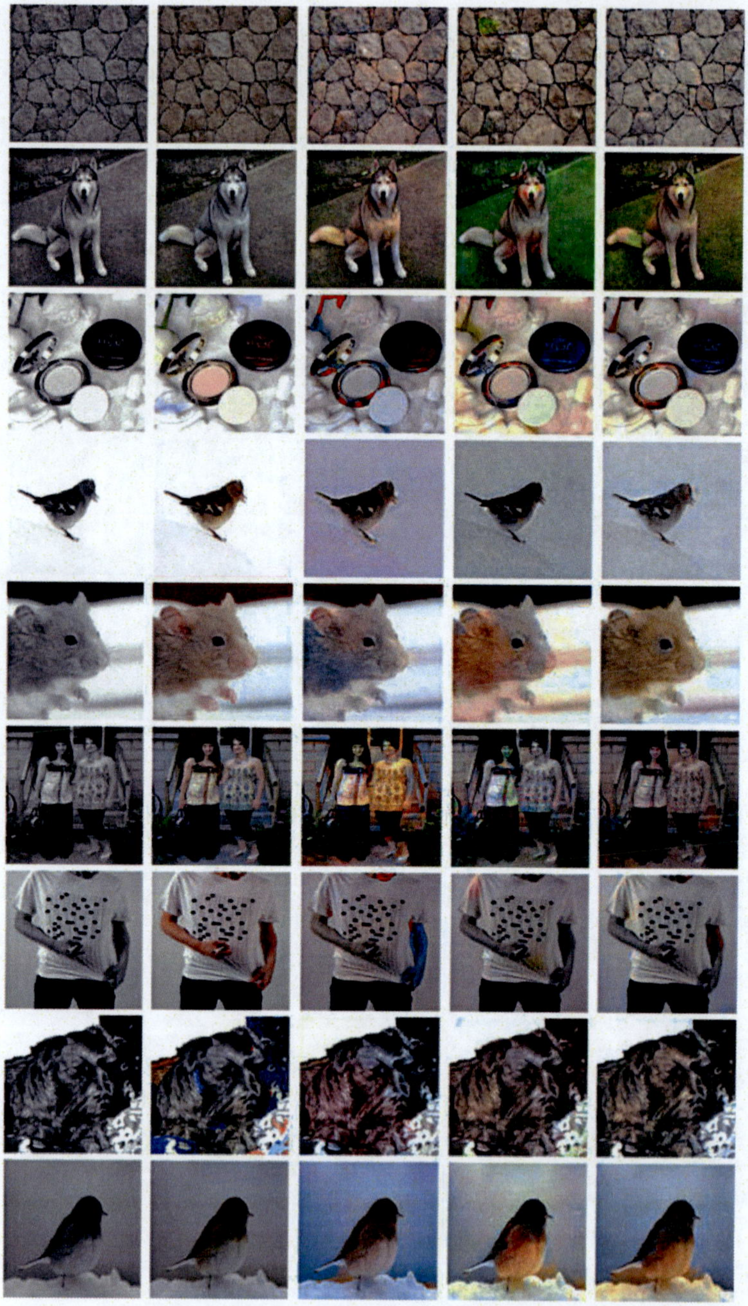

Fig. 5 Qualitative results of ImageNet dataset. col 1: input grayscale image col 2: ground-truth image col 3: image generated using cycle GAN model col 4: image generated using CUT Model col 5: image generated using proposed multi-scale contrastive learning approach

Table 1 FID Scores for the generated images

Dataset/model	CycleGAN	CUT model	Multi-scale contrastive learning
Cityscapes	22.9655	29.4273	**29.1117**
ImageNet	53.8637	63.6680	**61.0883**

in terms of FID as we see lower FID scores than the default CUT Model [8]. Our approach does not outperform CycleGAN [12] quantitatively, based on FID scores, but overall, the qualitative outcomes are better. The results of the experiment are available in Table 1 in bold font.

6 Conclusion

We propose multi-scale contrastive learning for unpaired image colorization problem by increasing their shared knowledge between the gray input and the color output. We capture feature in multiple scales to preserve color based on the texture information in patches. In addition, we find contrastive loss in different scales and combine them to get final contrastive loss. In most circumstances, this provides us with better results than other state-of-the-art procedures. Further, we can improve the image colorization by considering gradient of an image to better distinguish color based on smooth and sharp region in the image.

References

1. Joshi, M.R., Nkenyereye, L., Joshi, G.P., Islam, S.M., Abdullah-Al-Wadud, M., Shrestha, S.: Auto-colorization of historical images using deep convolutional neural networks. Mathematics **8**(12), 2258 (2020)
2. Sýkora, D., Buriánek, J., Žára, J.: Colorization of black-and-white cartoons. Image Vis. Comput. **23**(9), 767–82 (2005)
3. Toet, A.: Colorizing single band intensified night vision images. Displays **26**(1), 15–21 (2005)
4. Gravey, M., Rasera, L.G., Mariethoz, G.: Analogue-based colorization of remote sensing images using textural information. ISPRS J. Photogrammetry Remote Sens. **1**(147), 242–54 (2019)
5. Geshwind, D.M.: Method for colorizing black and white footage. US Patent 4,606,625 (1986)
6. Zhang, R., Isola, P., Efros, A.A., Colorful image colorization. In: European Conference on Computer Vision, vol. 8, pp. 649–666. Springer, Cham (2016)
7. Su, J.W., Chu, H.K., Huang, J.B.: Instance aware image colorization. In: Proceedings of the IEEE/CVF Conference on Computer Vision and Pattern Recognition, pp. 7968–7977 (2020)
8. Park, T., Efros, A.A., Zhang, R., Zhu, J.Y.: Contrastive learning for unpaired image-to-image translation. In: European Conference on Computer Vision. pp. 319-345. Springer (2020)

9. Han, J., Shoeiby, M., Petersson, L., Armin, M.A.: Dual contrastive learning for unsupervised image-to-image translation. In: Proceedings of the IEEE/CVF Conference on Computer Vision and Pattern Recognition, pp. 746–755 (2021)
10. Oord, A.v.d., Li, Y., Vinyals, O.: Representation learning with contrastive predictive coding. arXiv preprint arXiv:1807.03748 (2018)
11. Chen T, Kornblith S, Norouzi M, Hinton G. A simple framework for contrastive learning of visual representations. In: International Conference on Machine Learning, pp. 1597-1607. PMLR (2020)
12. Zhu, J.Y., Park, T., Isola, P., Efros, A.A.: Unpaired image-to-image translation using cycle-consistent adversarial networks. In: IEEE International Conference on Computer Vision (ICCV) (2017)
13. Isola, P., Zhu, J.Y., Zhou, T., Efros, A.A.: Image-to-image translation with condi- tional adversarial networks. In: Proceedings of the IEEE Conference on Computer Vision and Pattern Recognition. pp. 1125–1134 (2017)
14. Goodfellow, I., Pouget-Abadie, J., Mirza, M., Xu, B., Warde-Farley, D., Ozair, S., Courville, A., Bengio, Y.: Generative adversarial nets. In: NIPS (2014)
15. Cordts, M., Omran, M., Ramos, S., Rehfeld, T., Enzweiler, M., Benenson, R., Franke, U., Roth, S., Schiele, B.: The cityscapes dataset for semantic urban scene understanding. In: Proceedings of the IEEE Conference on Computer Vision and Pattern Recognition (CVPR) (2016)
16. Deng, J., Dong, W., Socher, R., Li, L.-J., Li, K., Fei-Fei, L.: Imagenet: a large-scale hierarchical image database. In: 2009 IEEE Conference on Computer Vision and Pattern Recognition, pp. 248–255 (2009)
17. Bradski, G.: The OpenCV library. Dr. Dobb's J. Softw. Tools (2000)
18. Russakovsky, O.*, Deng, J.*, Su, H., Krause, J., Satheesh, S., Ma, S., Huang, Z., Karpathy, A., Khosla, A., Bernstein, M., Berg, A.C., Fei-Fei, L.: (* = equal contribution) ImageNet large scale visual recognition challenge. IJCV (2015)
19. https://github.com/mseitzer/pytorch-fid
20. Kumar, M., Weissenborn, D., Kalchbrenner, N.: Colorization transformer. In: ICLR (2021)
21. https://www.coursera.org/lecture/build-better-generative-adversarial-networks-gans/frechet-inception-distance-fid-LY8WK
22. https://en.wikipedia.org/wiki/Frechet_distance
23. https://www.oreilly.com/library/view/generative-adversarial-networks/9781789136678/9bf2e543-8251-409e-a811-77e55d0dc021.xhtml

Human Activity Recognition Using CTAL Model

Mrinal Bisoi, Bunil Kumar Balabantaray, and Soumen Moulik

Abstract In recent years, computer vision-related applications and research works are rapidly increasing daily. There are so many kinds of computer vision tasks like object detection, robotics, gait analysis, medical diagnosis, crime detection, and many more. Along with these Human Activity Recognition or HAR is also one kind. The challenges of HAR are as follows: First, it is computationally expensive and a time-consuming process, and second, it is a video data means it has spatial and temporal features, so it is hard to achieve good performance. To overcome these issues, we have proposed "a hybrid CNN model(Convolutional Triplet Attention LSTM (CTAL) model)" for activity recognition on the UCF50 dataset. In this paper, we will discuss about the used data set, proposed model, experiments, and results.

Keywords Activity recognition · Action recognition · Computer vision · Video classification · UCF50 · Deep learning (DL) · Convolutional neural network (CNN) · Long short-term memory (LSTM) · Triplet attention

1 Introduction

In recent times, advancement in Deep Learning (DL) and Machine Learning(ML) has led to their widespread use in image processing tasks. Image classification, image segmentation, object detection, etc., have benefited a lot due to DL-based algorithms. But the computer vision especially video classification tasks are quite complex and computationally expensive. Human activity recognition is also one of the popular video-based computer vision task. We get motivation for the human activity recognition because of wide variety of applications like gaming, human-robot interaction,

M. Bisoi (✉) · B. K. Balabantaray · S. Moulik
Department of Computer Science, National Institute of Technology Meghalaya, Shillong, India
e-mail: t20cs003@nitm.ac.in

B. K. Balabantaray
e-mail: bunil@nitm.ac.in

S. Moulik
e-mail: smoulik@nitm.ac.in

M. Tistarelli et al. (eds.), *Computer Vision and Machine Intelligence*,
Lecture Notes in Networks and Systems 586,
https://doi.org/10.1007/978-981-19-7867-8_28

rehabilitation, sports, health monitoring, video surveillance, and robotics. Convolutional neural network (CNN) is conventionally used for optical flow features and to extract deep RGB features. Recently, 3D CNN has been widely used by researchers for activity recognition [1–4]. But in [22] shows that 2D CNN outperforms than 3D CNN in video classification task.

In this paper, we propose a hybrid model to perform activity recognition on the UCF50 database. As it has more classes than the number of classes of many other datasets, camera motion, various camera angles, and dynamic clattered backgrounds make the activity recognition more challenging.

This paper is organized into seven sections: Sect. 1 deals with an introduction and basic discussions about human activity recognition, Sect. 2 deals with related work on human activity recognition, Sect. 3 deals with the used database and our model structure, Sect. 4 deals with the experimental environment and the experiment, Sect. 5 deals with the result and the comparison, Sect. 6 deals with the result discussion and comparison to the existing state-of-the-art (SoTA) method, and Sect. 7 consists of our conclusion and future scope.

2 Related Work

To solve the activity recognition-related challenges, various kinds of approaches are proposed. In vrskova et al. [4], 3D CNN was used to classify human activity, and 2D CNN with fully connected layers was used for video motion classification in the paper [5] but the no of batch size and epochs are low in both the papers.

In [6], obaidi et al. used Neuromorphic Visual Sensing (NVS) for local and global feature extraction and then KNN and quantum-enhanced SVM was used separately for human action recognition.

Jalal et al. [7] proposed pseudo-2D stick model for feature extraction, and then the k-ary tree hashing algorithm was used over-optimized feature vectors for event classification.

To extract silhouette images and background subtraction Ramya et al. [8] used correlation coefficient-based frame difference, and then from silhouette distance, transform-based features and entropy features were given as input to a neural network for action recognition.

For action recognition, Xu et al. [9] proposed semi-supervised discriminant multi-manifold analysis (SDMM) using least square loss function along with spectral projected gradient (SPG) and Karush–Kuhn–Tucker (KKT) condition.

In [10], Compact Descriptors for visual search (CDVS) was used for feature extraction, and then, a human detector detects the person. Both features were sent to CDVS trajectories, and finally, standard SVM was used for action recognition.

To reduce the dimension of feature space, He et al. [11] used fuzzy vector quantization method of human action recognition and got membership vectors, and finally, an Extreme Learning Machine (ELM) was used with a single hidden layer to classify membership vectors, and on that basis, action recognition is done.

Banerjee et al. [12] used unsupervised self-tuned spectral clustering technique to temporarily cluster video frames into consistent snippets, then coupled snippet features with two-level video features, and then used CNN to extract global frame features and a few local motion features. Finally, multi-class SVM was used for action classification.

For action recognition, Zhang et al. [13] used two layers of CNN to extract information from different levels and added a linear dynamic system (LDS) to capture temporal structures of actions.

In [14], the input data was divided into different clusters; further, these data were divided into coherent and non-coherent groups. Both groups were learned using K-singular value decomposition dictionary learning, and then, the orthogonal projection-based selection was used to get an optimal dictionary and combined both. Finally, that dictionary was updated using limited Broyden-Fletcher-Goldfarb-Shanno optimization to classify the action recognition videos.

Uddin et al. [15] introduced adaptive local motion descriptor (ALMD) which was used for local textures, then spark machine learning library random forest was used for action recognition.

In [16], first dense sampling was done in each spatial scale, then Spatio-temporal Pyramid (STP) was used for embedding structure information, and a multi-channel was used to combine different features (e.g., trajectory shape, HOF, HOG, MBH). Finally, bag-of-features representation and SVM classifier were used to classify action recognition.

3 Database and Model Structure

For the human activity recognition, in the first step, we extract the data frames from the human activity videos. In the second step, we propose a CNN-based hybrid model CTAL which is the combination of CNN [17], triplet attention [18], and bidirectional LSTM [19]. We use triplet attention because at negligible computational overhead it can provide significant performance gain and cross-dimension interaction. Finally, we train and evaluate the model on UCF50 dataset.

3.1 UCF50 Dataset

UCF50 [20] is one of the realistic but challenging dataset because of camera motion, various camera angles, object appearances, and dynamic background. It contains 6676 human activity videos of 50 action categories. These videos are grouped into 25 groups based on some common features like similar background, same person, similar viewpoints, and so on.

This dataset is the extended version of UCF-11 [21]. It has 50 action categories collected from YouTube. Some of the categories are Basketball, BenchPress, Billiards,

Fig. 1 Some categories of the dataset

BreastStroke, Diving, Fencing, HighJump, HorseRace, HorseRiding, JumpRope, Kayaking, Mixing, PlayingGuitar, PlayingPiano, PoleVault, RopeClimbing, Rowing, TrampolineJumping, WalkingWithDog, YoYo, etc. For a better understanding of the dataset, Fig. 1 is given with some of the categories of human activity videos.

3.2 Model Architecture

Data Preprocessing Video means changing many frames per second time. Data preprocessing is very important for making data in a format that can be given to the model for training. Therefore, at first, we extract random 25 data frames of height and width of 64 pixels from each video and create a different database.

As neural networks process inputs using small weight values, hence input with larger values can disrupt or slow down the training process. Therefore, in the next stage, image normalization is done to map each pixel value from the range [0,1] to [0,255].

Before giving the newly created data frames as input to the model, we split the newly created database in which 80% of the data is used for training and 20% of the data is used for testing and validation.

Model Structure At first, we implement the triplet attention module [18], and the time distributed layer is used in Keras to make it compatible with our model. Then, we use the functional API of TensorFlow to build our model.

In our model, there are six convolutional triplet attention blocks, three biLSTM layers, and four fully connected dense layers. Finally, for output classification, we use one dense layer with the softmax classifier.

The extracted data frames are used as input to the first convolutional triplet attention block. After that 5 more blocks are there, each block contains a time distributed conv2D layer with tanh activation because of less number of parameters and independence of local variations, followed by a time distributed max-pooling layer to get the abstract form of the representation, and batch normalization is used to standardize the input to next layer for each mini-batch, and lastly, a triplet attention layer is added for cross-dimension interaction which removes the information bottleneck. The filter sizes of the convolutional layer are as follows: 16, 32, 64, 128, 256, and 512. In the last 4 blocks, we add the dropout layer also with the value 0.2 to prevent the network from over-fitting.

Then, we flatten the output result and for enabling additional training and send it through three bidirectional LSTM layers with cell sizes 512, 256, and 128. Next, four fully connected dense layers are added of size 128, 64, 32, and 32 with tanh activation. Finally, for classification, we use one dense layer with a softmax classifier. For a better understanding of model structure, Fig. 2 is given.

4 Environment and Experiment

4.1 Experimental Environment

TensorFlow framework is used to develop our model architecture. We have used Quadro RTX 6000 for training and inferencing.

4.2 Experiment

Before starting the model compilation and training, we use the early stopping function on the validity loss with patience 20. Then, we do not fix the learning rate, instead of that we use reduce LR on plateau function to get the best accuracy from the model.

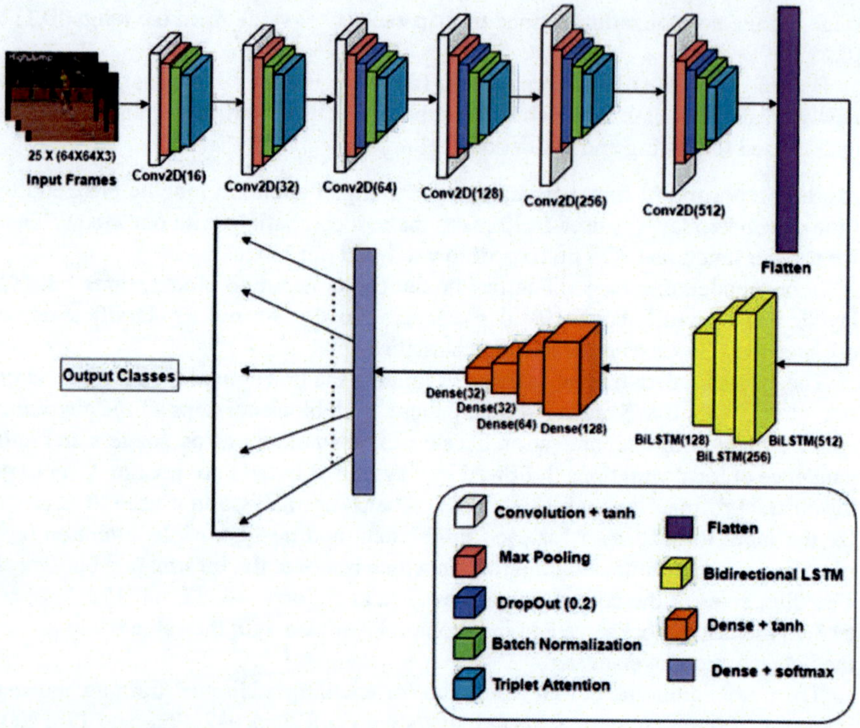

Fig. 2 Architecture of CTAL model

For the compilation of the model, we use adam optimizer to provide more efficient weights to the model, and categorical cross-entropy as loss function, and for evaluating our model, we use accuracy, precision, and recall as performance metrics.

Finally, we train our model using 256 batch size and it runs for 180 epochs and we use shuffle true to reduce the variance and make the model less overfit.

Besides that, to make a comparison we use 3D CNN instead of 2D CNN in our model, and in another experiment, we use a normal convolutional block without a modified triplet attention block. In the next section, the result obtained through these experiments is compared in Table 1.

5 Result and Comparison

After training our main model with multiple repetitions, when we check the validation of the model, we get the mean with confidence intervals of validation accuracy (91.85 $\pm$ 0.15)%, mean with confidence intervals of validation precision (95.20 $\pm$ 0.16)%

Table 1 Comparison between our experimental models

Our experimental model	Total parameters (in Million)	Total floating point operations (m)	Mean accuracy (%)
CTAL without triplet attention block	3.228	6.45	89.75
CTAL with 3D CNN block	6.371	3.23	91.62
CTAL model	3.229	3.23	**91.85**

The highest accuracy is marked in bold

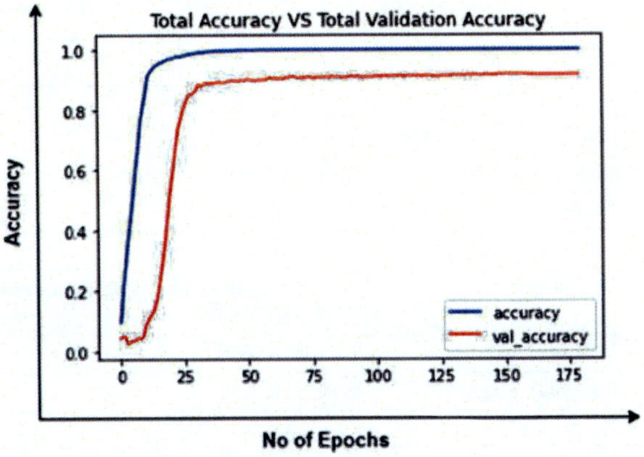

Fig. 3 ROC curves of training accuracy versus validation accuracy

and mean with confidence intervals of validation recall (89.94 ± 0.19)% for the mean validation loss of 0.4102.

After testing our model with multiple repetitions, we get (91.41± 0.21)% mean testing accuracy, and the mean testing loss is 0.4137. ROC curves of the max result evaluation and the loss evaluation are given in Figs. 3 and 4.

And we compare our model's mean with confidence interval accuracy to the accuracy of the other model in Table 2.

6 Result Discussion and Comparison to SoTA Method

In this section, first, we will discuss the result of our three experimental models. In one experimental model, we remove the triplet attention block from our main CTAL model and found that in that model there are 3.228 Million total parameters which are almost the same as our main CTAL model but the total floating point operations are almost double compared to our main CTAL model along with that the mean accuracy of the human activity recognition is 2.10% less compare to our main CTAL model.

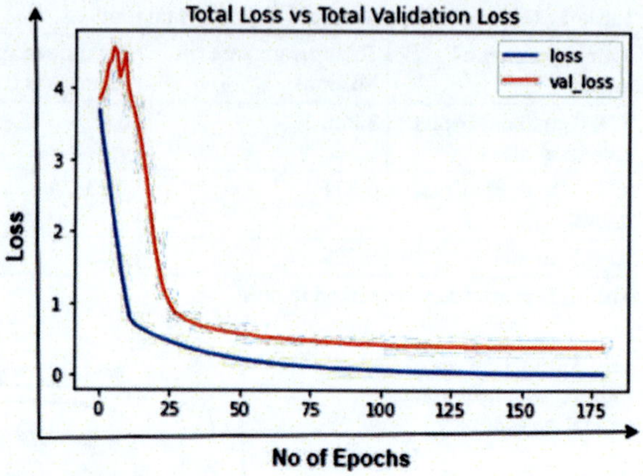

Fig. 4 ROC curves of training loss versus validation loss

Table 2 Comparison with other model result

Author	Used dataset	Accuracy (%)
Vrskova et al. [4]	UCF50	82.2
Obaidi et al. [6]	Emulator UCF50, Rerecorded UCF50	E UCF50: KNN as classifier:- 69.45
Jalal et al. [7]	UCF50	90.48
Ramya et al. [8]	UCF50	80
Xu et al. [9]	UCF50	89.84
Dasari et al. [10]	UCF50	54.8
CTAL model (our)	UCF50	**(91.85 ± 0.15)**

The highest accuracy is marked in bold

In another experimental model, instead of 2D CNN we use 3D CNN layer and found that the number of total parameters is 6.371 Million which is almost double compared of our main CTAL model, so training the model also took more time along with that the mean accuracy of the human activity recognition is 0.23% less compare to our main CTAL model though the total floating point operations are same.

Next as a state-of-the-art method, Singh et al. in [23] used hand-crafted features, and on the discriminative sparse dictionary of these features, sparse codes are computed along with that differential motion descriptor is used to track the motion in the videos, and Support Vector Machine (SVM) is used as a classifier to recognize the human activity and achieved 97.50% accuracy on UCF50 dataset. The main demerit of the method is that it used hand-crafted features so the computation time is so high (100 s per video) but in our model, the computation time is very low (11 s per video) compared to the state-of-the-art method and our model is lightweight too. The comparison between our CTAL model and the state-of-the-art model is given in Table 3.

Table 3 Comparison with state-of-the-art model

Author	Year	Merit	Demerit	Accuracy (%)
Singh et al. [23]	2022	Highest accuracy till now	High computational time (100 s/video)	97.50
CTAL model (our)	–	Low computational time (11 s/video), and number of parameters are also low	Can not achieve the highest accuracy	(91.85 ± 0.15)

7 Conclusion

In this paper, we present a hybrid model ("convolutional triplet attention lstm or CTAL model") in which we use CNN to extract deep RGB frames, then use triplet attention to gain significant performance and cross-dimension interaction, and use bidirectional LSTM for extra data flow features. On the UCF50 dataset using our model, we achieve 91.85% mean accuracy which is a better result than many other papers.

In the near future, this work will be extended with respect to adversarial learning and apply other methods also to achieve a better result for human activity recognition.

References

1. Ouyang, Xi., Shuangjie, Xu., Zhang, Chaoyun, Zhou, Pan, Yang, Yang, Liu, Guanghui, Li, Xuelong: A 3D-CNN and LSTM based multi-task learning architecture for action recognition. IEEE Access **7**, 40757–40770 (2019)
2. Hu, Zheng-ping, Zhang, Rui-xue, Qiu, Yue, Zhao, Meng-yao, Sun, Zhe: 3D convolutional networks with multi-layer-pooling selection fusion for video classification. Multimedia Tools Appl. **80**(24), 33179–33192 (2021)
3. Boualia, S.N., Amara, N.E.: 3D CNN for human action recognition. In: 2021 18th International Multi-Conference on Systems, Signals & Devices (SSD), pp. 276–282. IEEE (2021)
4. Vrskova, Roberta, Hudec, Robert, Kamencay, Patrik, Sykora, Peter: Human activity classification using the 3DCNN architecture. Appl. Sci. **12**(2), 931 (2022)
5. Luo, Y., Yang, B.: Video motions classification based on CNN. In: 2021 IEEE International Conference on Computer Science, Artificial Intelligence and Electronic Engineering (CSAIEE), pp. 335–338. IEEE (2021)
6. Al-Obaidi, Salah, Al-Khafaji, Hiba, Abhayaratne, Charith: Making sense of neuromorphic event data for human action recognition. IEEE Access **9**, 82686–82700 (2021)
7. Jalal, Ahmad, Akhtar, Israr, Kim, Kibum: Human posture estimation and sustainable events classification via pseudo-2D stick model and K-ary tree hashing. Sustainability **12**(23), 9814 (2020)

8. Ramya, P., Rajeswari, R.: Human action recognition using distance transform and entropy based features. Multimedia Tools Appl. **80**(6), 8147–8173 (2021)
9. Xu, Zengmin, Ruimin, Hu., Chen, Jun, Chen, Chen, Jiang, Junjun, Li, Jiaofen, Li, Hongyang: Semisupervised discriminant multimanifold analysis for action recognition. IEEE Trans. Neural Netw. Learn. Syst. **30**(10), 2951–2962 (2019)
10. Dasari, R., Chen, C.W.: Mpeg cdvs feature trajectories for action recognition in videos. In: 2018 IEEE Conference on Multimedia Information Processing and Retrieval (MIPR), pp. 301–304. IEEE (2018)
11. He, W., Liu, B., Xiao, Y.: Multi-view action recognition method based on regularized extreme learning machine. In: 2017 IEEE International Conference on Computational Science and Engineering (CSE) and IEEE International Conference on Embedded and Ubiquitous Computing (EUC), vol. 1, pp. 854–857. IEEE (2017)
12. Banerjee, B., Murino, V.: Efficient pooling of image based CNN features for action recognition in videos. In: 2017 IEEE International Conference on Acoustics, Speech and Signal Processing (ICASSP), pp. 2637–2641. IEEE (2017)
13. Zhang, L., Feng, Y., Xiang, X., Zhen, X.: Realistic human action recognition: when cnns meet lds. In: 2017 IEEE International Conference on Acoustics, Speech and Signal Processing (ICASSP), pp. 1622–1626. IEEE (2017)
14. Wilson, S., Krishna Mohan, C.: Coherent and noncoherent dictionaries for action recognition. IEEE Signal Process. Lett. **24**(5), 698–702 (2017)
15. Uddin, M.A., Joolee, J.B., Alam, A., Lee, Y.K.: Human action recognition using adaptive local motion descriptor in spark. IEEE Access **5**, 21157–21167 (2017)
16. Wang, Heng, Kläser, Alexander, Schmid, Cordelia, Liu, Cheng-Lin.: Dense trajectories and motion boundary descriptors for action recognition. Int. J. Comput. Vis. **103**(1), 60–79 (2013)
17. Shin, H.C., et al.: Deep convolutional neural networks for computer-aided detection: CNN architectures, dataset characteristics and transfer learning. IEEE Trans. Med. Imaging **35**(5), 1285–1298 (2016)
18. Misra, D., Nalamada, T., Arasanipalai, A.U., Hou, Q.: Rotate to attend: convolutional triplet attention module. In: Proceedings of the IEEE/CVF Winter Conference on Applications of Computer Vision, pp. 3139–3148 (2021)
19. Graves, Alex, Schmidhuber, Jürgen.: Framewise phoneme classification with bidirectional LSTM and other neural network architectures. Neural Networks **18**(5–6), 602–610 (2005)
20. Reddy, Kishore K., Shah, Mubarak: Recognizing 50 human action categories of web videos. Mach. Vis. Appl. **24**(5), 971–981 (2013)
21. Liu, J., Luo, J., Shah, M.: Recognizing realistic actions from videos "in the wild". In: 2009 IEEE Conference on Computer Vision and Pattern Recognition, pp. 1996–2003. IEEE (2009)
22. Kurmanji, M., Ghaderi, F.: A comparison of 2D and 3D convolutional neural networks for hand gesture recognition from RGB-D data. In: 2019 27th Iranian Conference on Electrical Engineering (ICEE). IEEE (2019)
23. Singh, K., et al.: A sparse coded composite descriptor for human activity recognition. Expert Syst. **39**(1), e12805 (2022)

Deep Learning Sequence Models for Forecasting COVID-19 Spread and Vaccinations

Srirupa Guha and Ashwini Kodipalli

Abstract Time series forecasting constitutes an important aspect of traditional machine learning prediction problems. It includes analyzing historic time series data to make future predictions and thus enabling strategic decision-making. It has several practical applications throughout various industries including weather, finance, engineering, economy, healthcare, environment, business, retail, social studies, etc. On the verge of increasing COVID-19 cases globally, there is a need to predict the variables like daily cases, positive tests, vaccinations taken, etc. using state-of-the-art techniques for monitoring the spread of the disease. Understanding underlying patterns in these variables and predicting them will help monitor the spread closely and also give insights into the future. Sequence models have demonstrated excellent abilities in capturing long-term dependencies and hence can be used for this problem. In this paper, we present two recurrent neural network-based approaches to predict the daily confirmed COVID-19 cases, daily total positive tests and total individuals vaccinated using LSTM and GRU. Our proposed approaches achieve a mean absolute percentage error of less than 1.9% on the COVID-19 cases in India time series dataset. The novelty in our research lies in the long-term prediction of daily confirmed cases with a MAPE of less than 1.9% for a relatively long forecast horizon of 165 days.

Keywords COVID-19 cases time series forecasting · Recurrent neural networks for COVID-19 cases prediction · Prediction of COVID-19 total vaccinations · Long-Short-Term-Memory and Gated-Recurrent-Units for COVID-19 cases prediction

---

S. Guha
Indian Institute of Science Bangalore, Bangalore, India

A. Kodipalli (✉)
Department of Artificial Intelligence and Data Science, Global Academy of Technology, Bangalore, India
e-mail: dr.ashwini.k@gat.ac.in

M. Tistarelli et al. (eds.), *Computer Vision and Machine Intelligence*,
Lecture Notes in Networks and Systems 586,
https://doi.org/10.1007/978-981-19-7867-8_29

1 Introduction

A time series constitutes a set of observations recorded for consecutive time points. Earlier, statistical forecasting techniques and regression-based techniques like Moving Average, Simple and Damped Exponential Smoothing, Holt–Winters, ARIMA, SARIMA, SARIMAX, UCM were used. These traditional statistical methods perform well in business cases where the datasets have a countable and finite number of predictive variables and are highly explainable. On the other hand, in business cases with massive amounts of data, machine learning forecasting methods like LSTM, Random Forest, K-nearest neighbors regression, CART, Support Vector Regression, etc. exhibit higher accuracy and performance rates, but they are not as easy to interpret. The deep learning sequence models are predominantly recurrent neural networks like LSTMs and GRUs which have proven to be very efficient in learning the temporal dependence from the data. LSTMs are able to learn the context without requiring it to be pre-specified. GRUs are improved versions of the standard recurrent neural network where the vanishing gradient problem of a standard RNN is solved through the use of update and reset gates.

Coronavirus disease 2019 (COVID-19), caused by severe acute respiratory syndrome coronavirus 2 (SARS-CoV-2), has been rapidly spreading worldwide since 2019, leading to an ongoing pandemic. This has highly impacted the people's livelihoods and economy across the globe. In this novel research work, we propose recurrent neural network-based approaches to forecast daily COVID-19 cases, the daily positive COVID-19 tests and total vaccinations taken in India in 2021, based on historic data.

2 Literature Survey

Parul et al. [1] demonstrated with recurrent neural network (RNN)-based LSTM variants in the study which includes bi-directional, convolutional and deep LSTM models and observed prediction errors with less than 3%. Sourabh et al. [2] considered the confirmed COVID-19 cases in India and USA, and a comparative study was done. Another variant of RNN considered was stacked LSTM in addition to the previous variants mentioned in this literature survey. It was found that convolutional LSTM was better in terms of accuracy and efficiency. Autoregressive Integrated Moving Averages (ARIMA), Prophet approaches are considered in the work by Jayanthi et al. [3] and found that stacked LSTM yielded error of less than 2%. Liao et al. [4] proposed a new model for prediction of COVID-19 outbreak which is called as SIRVD by using deep learning. The model is a combination of LSTM and other time series prediction models. Shiu et al. [5] used New Zealand data to forecast the COVID-19 pandemic spread. A comparative study of the various results was done based on the prediction of outbreak of the pandemic in other countries. Vikas et al. [6] applied simple average, moving average, Holt's linear trend method, Naïve method,

Holt–Winters method and ARIMA model for the analysis. They found that Naïve-based method came up with better results than other implemented techniques. LSTM is found out to be the best technique for time series forecasting of the COVID-19 pandemic spread and has proven lesser errors compared to any other machine learning techniques as further demonstrated by Mohan et al. [7] and Chimmula et al. [8]. Gulf Co-operation council data are considered by Ghany et al. [9] for the prediction of COVID-19 pandemic spread.

2.1 *Research Gap and Motivation*

The forecasting models proposed in the existing literature include both classical machine learning techniques and deep learning architectures. It was observed that in order to increase the accuracy of the predictions as well as focus on reliable long-term forecasts, deep learning sequence model-based approaches are more suitable. However, for increased accuracy, the complexity of the architecture is likely to increase and this introduces the vanishing gradient problem, thereby slowing down the learning. The existing literature does not have advanced deep learning architectures simplified for predicting daily COVID-19 cases, daily positive tests and total individuals vaccinated in India with an accuracy of less than 2% by using a simple sequence model-based architecture. This paper aims to bridge this gap by proposing simplified LSTM and GRU architectures which predict daily confirmed COVID-19 cases with a state-of-the-art MAPE of less than 2% on test dataset. Due to long-term nature of the forecasts, this paper has limited the forecasting models to only deep learning-based sequence models. The novelty of this work is summarized as below: The architectures proposed in this paper consist of three or less hidden layers thereby reducing the complexity and ensuring steady learning. At the same time, these models are able to predict the daily confirmed cases with a MAPE of less than 2% on the test dataset with a relatively long forecast horizon of 165 days. Table 1 shows a comparison of the proposed approach with the results of best performing models of the existing work on predicting the variables number of confirmed cases and positive cases:

3 Proposed Approach, Details of Implementation and Methodology

Long Short-Term Memory (LSTM) The state-of-the-art RNN architecture LSTM has time and again proved to be a good fit for modeling time series data because of its ability to capture long-term dependences while solving the vanishing gradient problems commonly observed in the vanilla RNNs. Figure 1 shows the sequence of steps generated by unrolling a cell of the LSTM network, and Fig. 2 shows a vanilla

Table 1 Comparison of proposed approach with existing work

Reference	Variable	Algorithm	Data source	Forecast horizon	Metric
Proposed work	Confirmed cases	LSTM and GRU	Kaggle	165 days	1.9%
Parul et al. [1]	Positive reported cases	LSTM variants	Ministry of Health and Family Welfare (Government of India)	6 days	MAPE: 5.05%
Sourabh et al. [2]	Confirmed cases	LSTM variants	The Ministry of Health and Family Welfare, Government of India and Centers for Disease Control and Prevention, U.S Department of Health and Human Services	30 days	MAPE: 2.17%
Jayanthi et al. [3]	Confirmed global cases	ARIMA, LSTM Variants and Prophet	Center for Systems Science and Engineering (CSSE) at Johns Hopkins University, USA	60 days	MAPE: 0.2%
Liao et al. [4]	Infected cases	Time-dependent SIRVD epidemic model	Johns Hopkins University System Science and Engineering Center (CSSE) and the website - Our World In Data	7 days	MAPE: 5.07%
Vikas et al. [6]	Cases	Classical machine learning methods: ARIMA, moving average, HWES, etc.	WHO "Data WHO Coronavirus Covid-19 cases and deaths-WHO-COVID-19-global-data"	30 days	RMSE: 110.09
Mohan et al. [7]	Daily confirmed and cumulative confirmed cases	Hybrid statistical machine learning models	The website of Covid19india	180 days	MAPE: 0.06%
Chimmula et al. [8]	Cases	LSTM	Johns Hopkins University and Canadian Health authority	30 days	Accuracy: 92.67%
Ghany et al. [9]	Confirmed cases	LSTM	Johns Hopkins dataset (JHU CSSE, 2020)	60 days	MAPE: 44.66

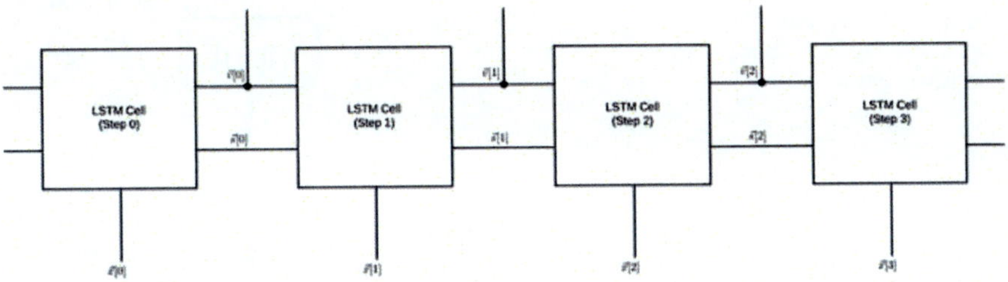

Fig. 1 Sequence of steps generated by unrolling a cell of the LSTM network (displaying four steps for illustration)

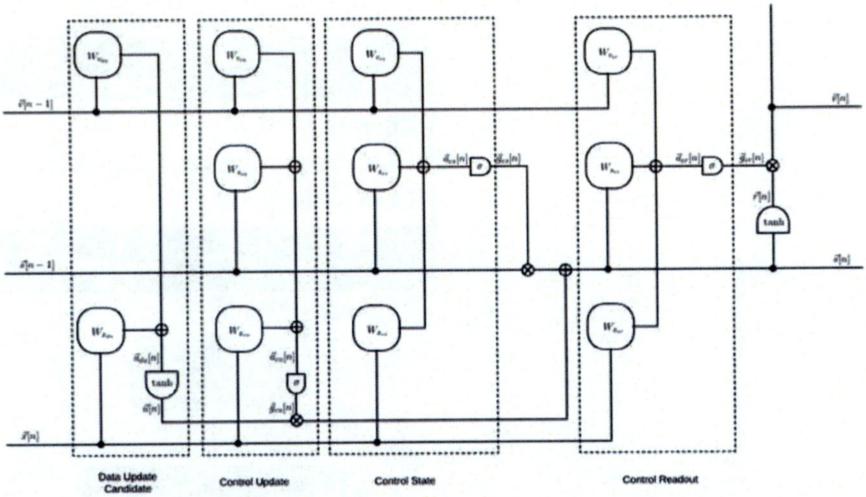

Fig. 2 Vanilla LSTM network cell

LSTM network cell. (Source: Fundamentals of recurrent neural network (RNN) and Long Short-Term Network (LSTM) by Alex Sherstinsky, https://arxiv.org/pdf/1808.03314.pdf) For prediction of total confirmed cases, the LSTM architecture implemented comprises a single LSTM layer with 200 units, with a dropout of 0.1, followed by 3 dense layers. Figure 3 shows the LSTM architecture proposed in this paper for prediction of daily confirmed COVID-19 cases in India. The following are additional specifications: Optimizer: Adam, Loss function: Binary Cross Entropy, Validation Metrics: MAPE, RMSE. The LSTM network is trained with the following hyperparameter settings: Epochs = 400, Batch Size = 5, Validation Split = 0.2, Verbose = 2

For prediction of daily positive tests, the LSTM architecture implemented comprises a single LSTM layer with 400 units, with a dropout of 0.2, followed by a dense layer with 20 units. The other hyperparameter settings remain the same as that for daily confirmed cases.

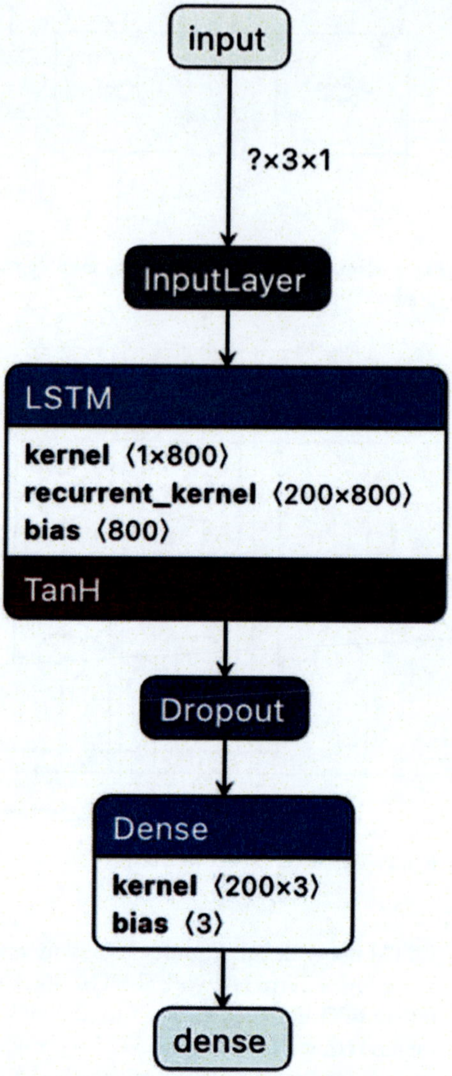

Fig. 3 LSTM architecture for forecasting daily total confirmed cases

For prediction of daily total individuals vaccinated, the LSTM architecture with same hyperparameter settings was used to forecast as used for predicting the above two variables. The LSTM architecture implemented comprises a single LSTM layer with 200 units, with a dropout of 0.2, followed by three dense layers.

Gated Recurrent Unit (GRU) GRUs are simpler than LSTMs, consisting of reset and update gates and also solve the vanishing gradient problem. GRUs can sometimes outperform LSTMs especially when the dataset size is relatively small. Figure 4

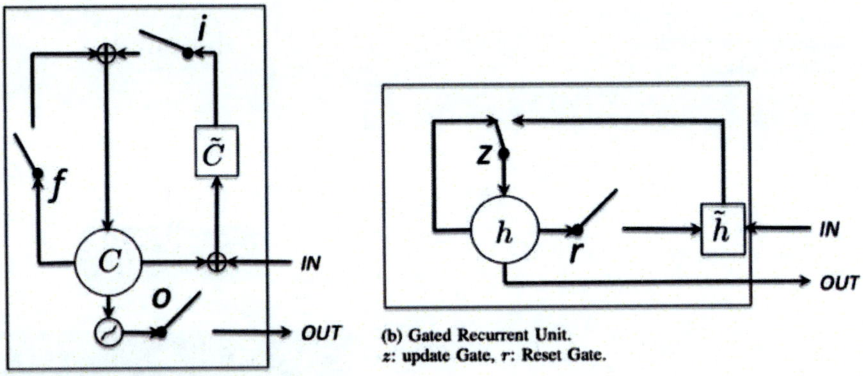

(a) Long Short Term Memory Unit.
i: Input Gate, f: Forget Gate, o: Output Gate.

(b) Gated Recurrent Unit.
z: update Gate, r: Reset Gate.

Fig. 4 Long short-term memory unit and gated recurrent unit

shows a comparison between the LSTM and GRU cells (Source: Gated Recurrent Unit (GRU) for Emotion Classification from Noisy Speech by Rajib Rana, Julien Epps, Raja Jurdak, Xue Li, Roland Goecke, Margot Breretonk and Jeffrey Soar, https://arxiv.org/pdf/1612.07778.pdf). For prediction of daily total confirmed cases, the GRU architecture implemented comprises a single GRU layer with 100 units, followed by a dense layer of 512 units and ReLU activation function, with a dropout of 0.2, followed by a dense layer with a single unit with linear activation function. Figure 5 shows the GRU architecture proposed in this paper for prediction of daily confirmed COVID-19 cases in India. The following are additional specifications: Optimizer: Adam, Loss Function: Binary Cross Entropy, Validation Metrics: MAPE, RMSE. The GRU network is trained with the following hyperparameter settings: Epochs = 400, Batch Size = 5, Validation Split = 0.2, Verbose = 2.

For prediction of daily positive tests, the GRU architecture implemented comprises a single GRU layer with 600 units, followed by a dense layer of 512 units, with a dropout of 0.2, followed by a dense layer with 20 units and the linear activation function. The other hyperparameter settings remain the same.

For prediction of daily total individuals vaccinated, the same hyperparameter settings were used for the GRU implementation as for the above two variables. The GRU architecture implemented for the prediction of daily total individuals vaccinated comprises a single GRU layer with 400 units, followed by a dense layer of 512 units with ReLU activation function, with a dropout of 0.2, followed by a dense layer with 40 units and the linear activation function.

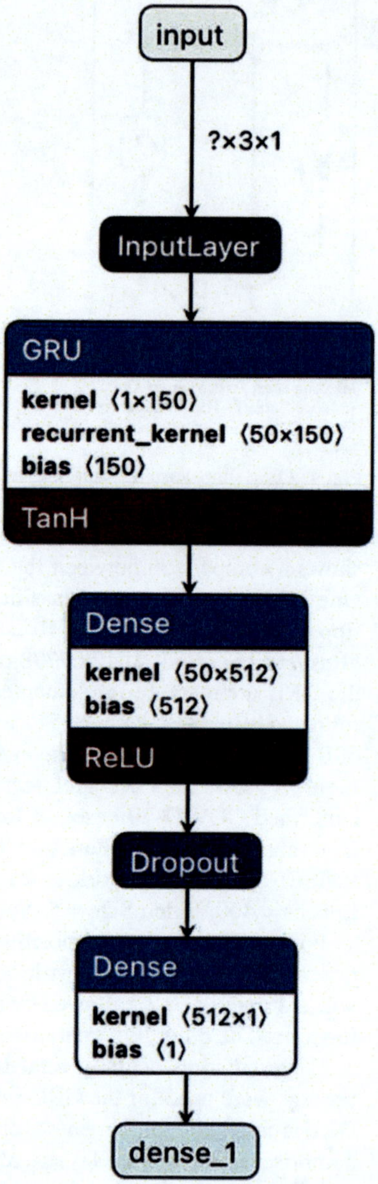

Fig. 5 GRU architecture for forecasting daily total confirmed cases

4 Model Evaluation Metrics

1. Mean Absolute Percentage Error (MAPE) Mean absolute percentage error (MAPE) is used for evaluating predictive performance of the architectures. The formula is as follows:

$$\text{MAPE} = \frac{100\%}{n} \sum_{t=1}^{n} \left| \frac{A_t - F_t}{A_t} \right|$$

where A_t is the actual value and F_t is the forecast value. The absolute value of the difference between actual and forecasted values is divided by the actual value and the ratio is thus obtained for every time point t. The summation of these ratios for all t values from 1 to n when divided by the number of time points n and taken as a percentage gives the MAPE.

2. Root Mean Squared Error (RMSE) The root mean squared error (RMSE) is another popular metric which is used in this paper as a forecasting model evaluation metric. It computes the squares of differences between the actual and forecasted values for T time points, sums these values and divides by the number of forecast time points, T to obtain the mean squared error (MSE). The square root of MSE gives the RMSE metric. The formula is as follows:

$$\text{RMSE} = \sqrt{\frac{\sum_{t=1}^{T} (\hat{y}_t - y_t)^2}{T}}.$$

where $\hat{y}_t$ denotes the predicted value for time t, y_t, denotes the actual value of the forecasted variable, and T denotes the total number of time points or observations for which forecasts are generated and evaluated.

5 Dataset Details

1. Case Time Series dataset: This dataset contains 642 records of the following relevant variables for monitoring COVID-19 across 28 states and 8 union territories of India recorded daily from January 30, 2020 to October 31, 2021: daily total confirmed cases and daily positive tests

2. India Vaccines dataset: This dataset contains 154 records of vaccination doses taken for COVID-19 across 28 states and 8 union territories recorded daily from January 15, 2021 to June 17, 2021, in India for the vaccines Covaxin, Oxford/AstraZeneca:

The dataset is obtained from Kaggle:
https://www.kaggle.com/sudalairajkumar/covid19-in-india

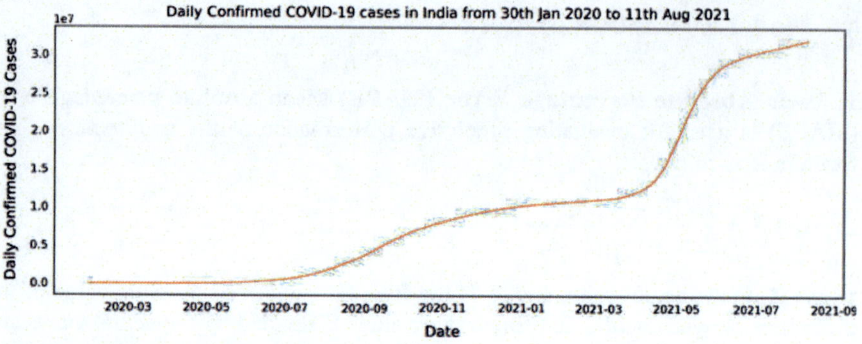

Fig. 6 Daily confirmed COVID-19 cases in India

Table 2 Lagged series

t	(*t*-1)	(*t*-2)	(*t*-3)
Series	1st lagged series	2nd lagged series	3rd lagged series

Table 3 Daily confirmed cases dataset split

Dataset size	Train size	Validation size	Test size
557	314	78	165

5.1 Daily Total Confirmed Cases

The dataset comprises 557 records of total number of daily confirmed COVID-19 cases in India from January 31, 2020 to October 31, 2021, after dropping the records with missing values, as shown in Fig. 6. The daily confirmed total cases data shows a peak in Sep 2021. We added three additional features, representing the 1st, 2nd and 3rd lagged series of the daily confirmed cases, as shown in Table 2. The train, validation and test split of the dataset are as shown in Table 3:

5.2 Daily Positive Tests

The dataset comprises 494 records of total number of daily positive COVID-19 tests in India from April 2020 to July 2021 after dropping the records with missing values. (Fig. 7). The train, validation and test split of the dataset are as shown in Table 4: The daily total positive tests data shows a peak between July 2020 and Oct 2020.

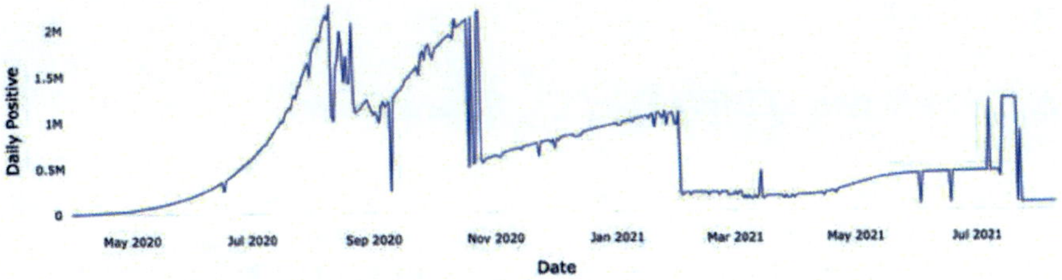

Fig. 7 Daily positive COVID-19 tests in India

Table 4 Daily positive tests dataset split

Dataset size	Train size	Validation size	Test size
494	278	69	147

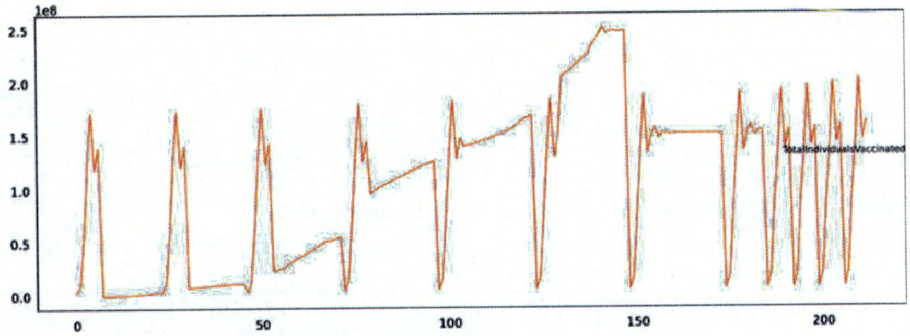

Fig. 8 Daily total individuals vaccinated in India after rolling mean imputation

Table 5 Total individuals vaccinated dataset split

Dataset size	Train size	Validation size	Test size
210	120	29	61

5.3 *Total Individuals Vaccinated*

The dataset comprises 210 records of total number of total individuals vaccinated in India from Jan 2021 to Dec 2021 after dropping the rows with missing values. The dataset being highly volatile with zero values in between, a rolling mean imputation with window 3 was performed to smoothen the dataset. Figure 8 shows the plot of the daily total individuals vaccinated after the rolling mean imputation. The train, validation and test split of the dataset are as shown in Table 5: The daily total individuals vaccinated data shows the highest peak between June 2021 and July 2021.

6 Results

6.1 Prediction of Daily Total Confirmed Cases

LSTM Results After 400 epochs of training, the following are the train, validation and test performances of the LSTM model for prediction of daily total confirmed cases as shown in Table 6. The results (Figs. 9, 10 and 11) show that the training MAPE and loss fluctuates while the validation MAPE and loss oscillates around a constant value. The model exhibits a MAPE of 0.019 or 1.9% and RMSE of 9.476 on the test dataset.

GRU Results After 400 epochs of training, the following are the train, validation and test performances of the GRU model for prediction of daily total confirmed cases are shown in Table 7. The results (Figs. 12, 13, and 14) show that the train and validation MAPE and loss for GRU has higher fluctuations as epochs progress than LSTM before converging. The GRU is able to attain a MAPE of 0.014 or 1.4% and RMSE of 7.987 on test dataset. Thus it is observed that the GRU architecture exhibits slightly better performance than LSTM for prediction of daily total confirmed COVID-19 cases.

Table 6 LSTM model performance for daily total confirmed cases

Train MAPE	Train loss	Validation MAPE	Validation loss	Test MAPE	Test RMSE
38.7005	6.3708e−05	0.6252	1.8621e−05	0.019 (or 1.9%)	9.476

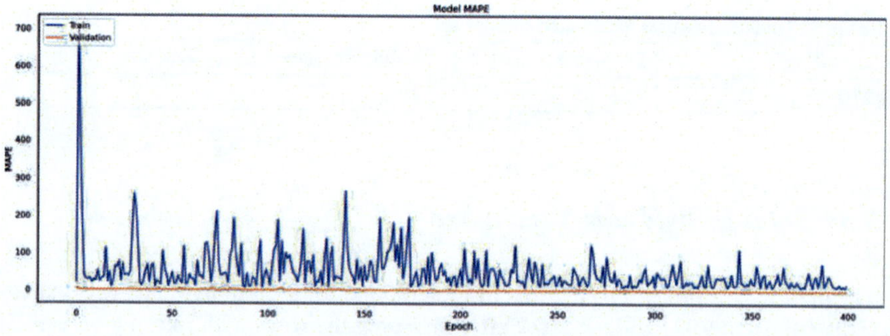

Fig. 9 Plot of LSTM train MAPE and validation MAPE for daily total confirmed cases with number of epochs

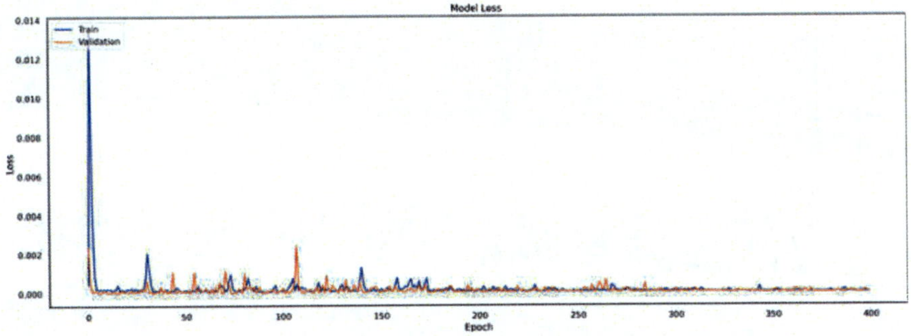

Fig. 10 Plot of LSTM train loss and validation loss for daily total confirmed cases with number of epochs

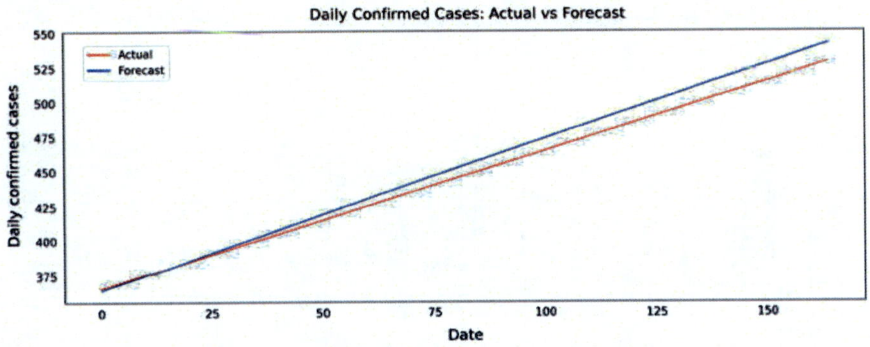

Fig. 11 Plot of LSTM actual and forecasted daily total confirmed cases

Table 7 GRU performance for daily total confirmed cases

Train MAPE	Train loss	Validation MAPE	Validation loss	Test MAPE	Test RMSE
52.2943	2.3555e−04	1.2221	5.8993e−05	0.014 (or 1.4%)	7.987

6.2 Prediction of Daily Positive Tests

LSTM Results After 400 epochs of training, the LSTM results for prediction of daily positive tests are shown in Table 8. Figure 15 shows the variation of LSTM train and validation MAPE with training epochs, and Fig. 16 shows a similar plot for train and validation loss. Figure 17 shows the comparison of actual and LSTM predictions on the test set.

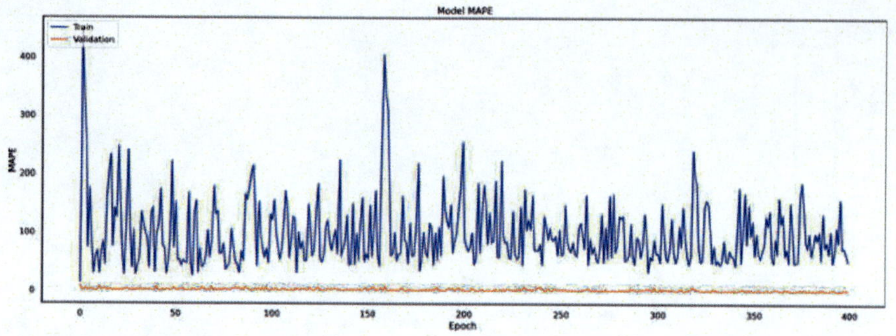

Fig. 12 Plot of GRU train MAPE and validation MAPE for daily total confirmed cases with number of epochs

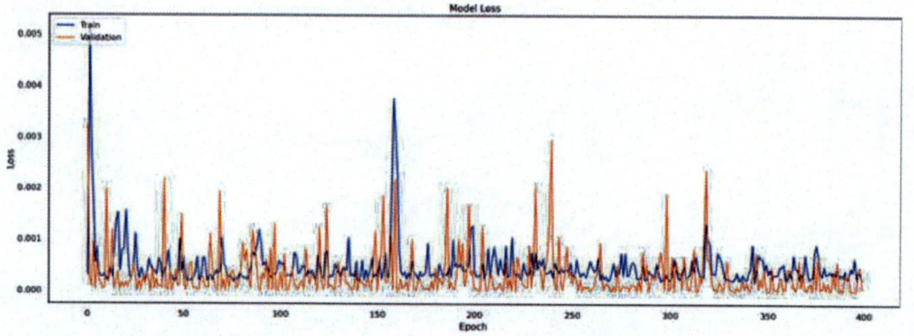

Fig. 13 Plot of GRU train loss and validation loss for daily total confirmed cases with number of epochs

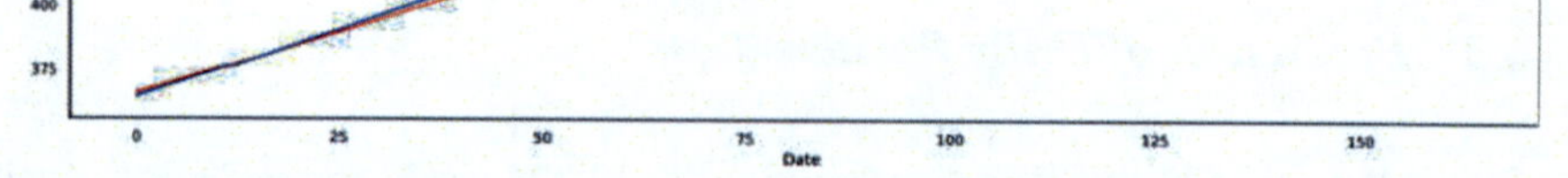

Fig. 14 Plot of GRU actual and forecasted for daily total confirmed cases

Table 8 LSTM performance for daily positive tests

Train MAPE	Train loss	Validation MAPE	Validation loss	Test MAPE	Test RMSE
37.2580	0.0052	18.5765	0.0093	0.226 or 22.6%	52.468

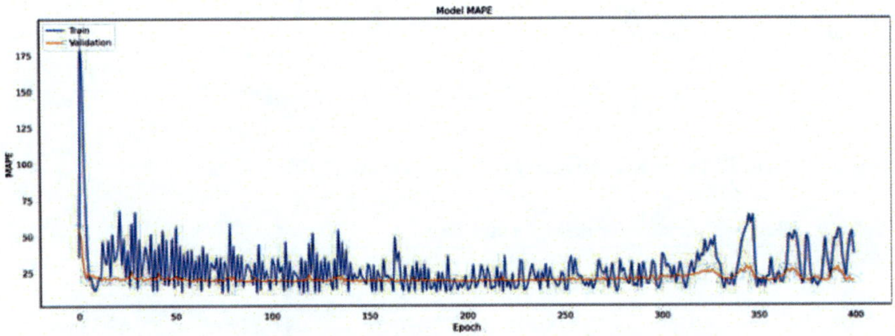

Fig. 15 Plot of LSTM train MAPE and validation MAPE with number of epochs for daily positive tests

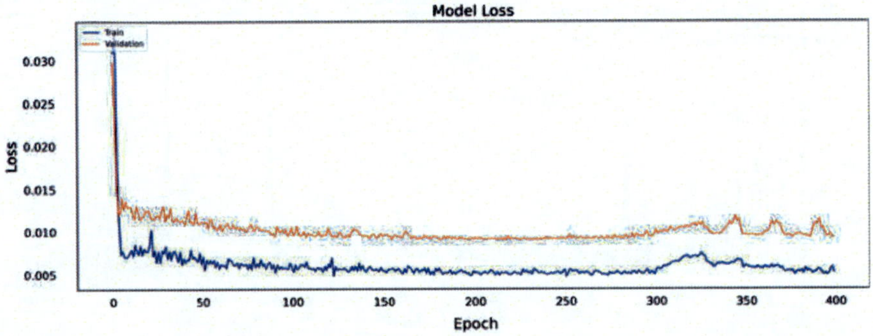

Fig. 16 Plot of LSTM train loss and validation loss with number of epochs for daily positive tests

GRU Results After 400 epochs of training, the GRU results for prediction of daily positive tests are shown in Table 9: Fig. 18 shows the variation of GRU train and validation MAPE with training epochs, and Fig. 19 shows a similar plot for train and validation loss. Figure 20 shows the comparison of actual and GRU predictions on the test set.

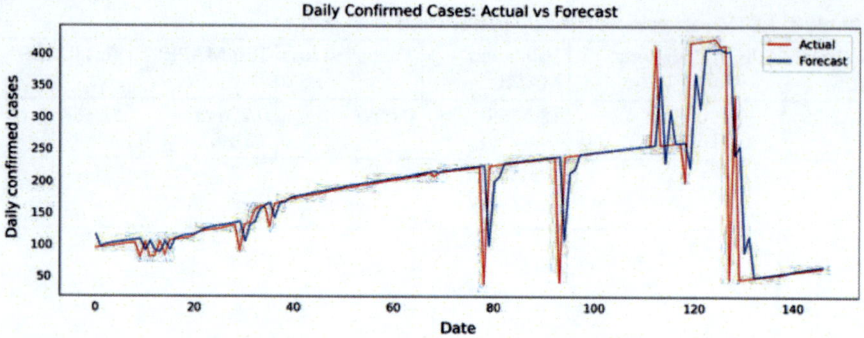

Fig. 17 Plot of LSTM actual and forecasted daily positive COVID-19 tests

Table 9 GRU performance for daily positive tests

Train MAPE	Train loss	Validation MAPE	Validation loss	Test MAPE	Test RMSE
73.2404	0.0056	27.0296	0.0157	0.319 or 31.9%	60.585

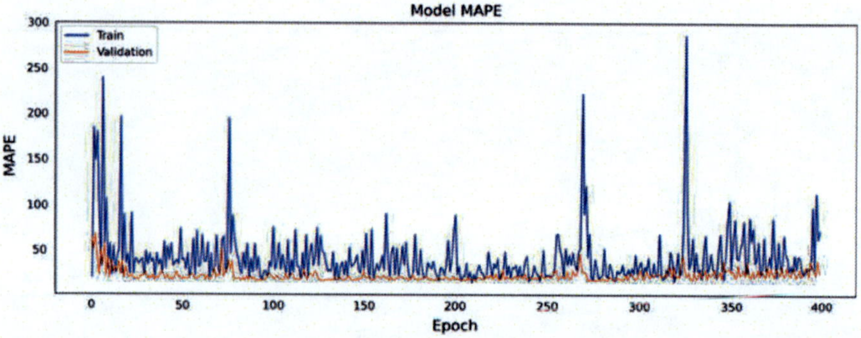

Fig. 18 Plot of GRU train MAPE and validation MAPE with number of epochs for daily positive COVID-19 tests

6.3 Prediction of Total Individuals Vaccinated

LSTM Results After 400 epochs of training, the LSTM results for prediction of daily total individuals vaccinated are as shown in Table 10. Figure 21 shows the variation of LSTM train and validation MAPE with training epochs, and Fig. 22 shows a similar plot for train and validation loss. Figure 23 shows the comparison of actual and LSTM predictions on the test set.

GRU Results After 400 epochs of training, the GRU results for prediction of daily total individuals vaccinated are shown in Table 11. Figure 24 shows the variation of

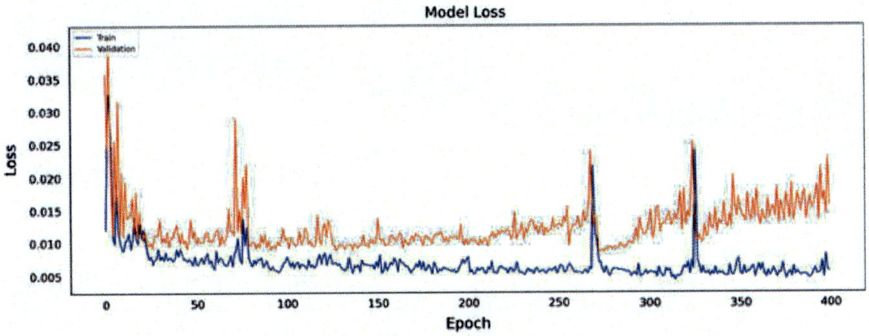

Fig. 19 Plot of GRU train loss and validation loss with number of epochs for daily positive COVID-19 tests

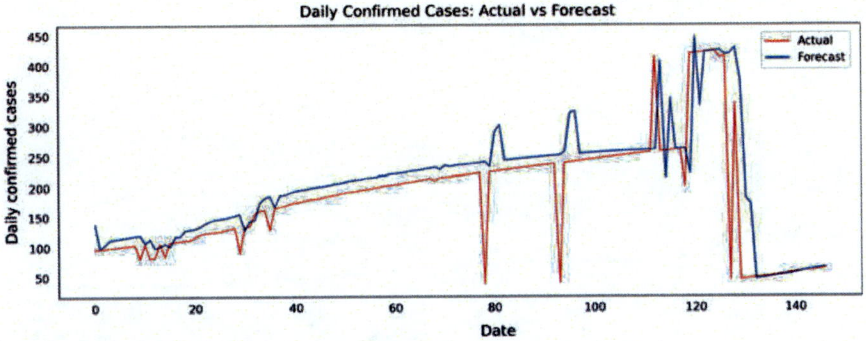

Fig. 20 Plot of GRU actual and forecasted daily positive COVID-19 tests for daily positive COVID-19 tests

Table 10 LSTM performance for total individuals vaccinated

Train MAPE	Train loss	Validation MAPE	Validation loss	Test MAPE	Test RMSE
2598910	0.0076	81.9045	0.1158	0.610 or 60.1%	53.170

GRU train and validation MAPE with training epochs and Fig. 25 shows a similar plot for train and validation loss. Figure 26 shows the comparison of actual and GRU predictions on the test set.

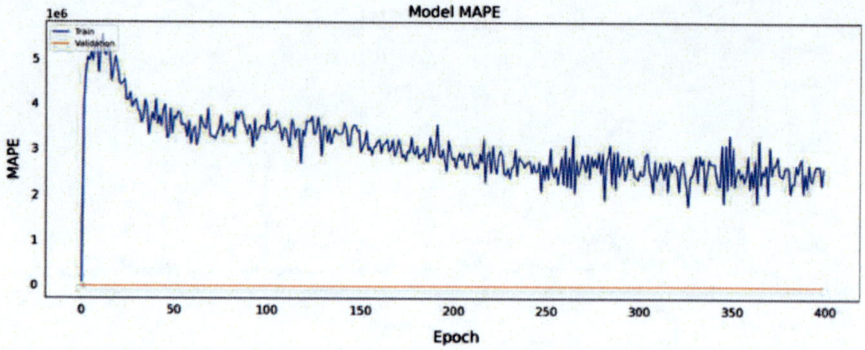

Fig. 21 Plot of LSTM train MAPE and validation MAPE with number of epochs for daily total individuals vaccinated

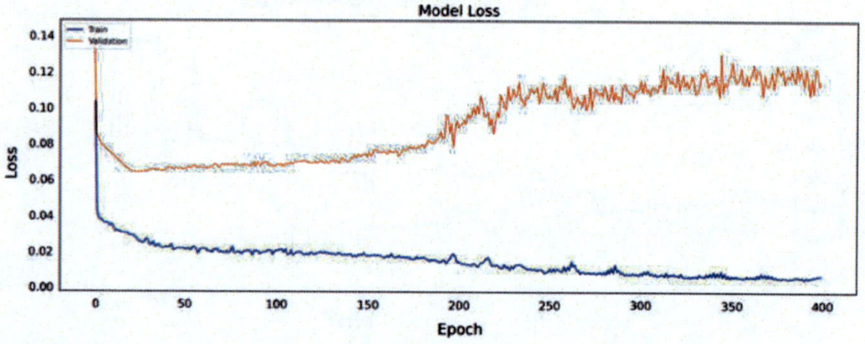

Fig. 22 Plot of LSTM train loss and validation loss with number of epochs for daily total individuals vaccinated

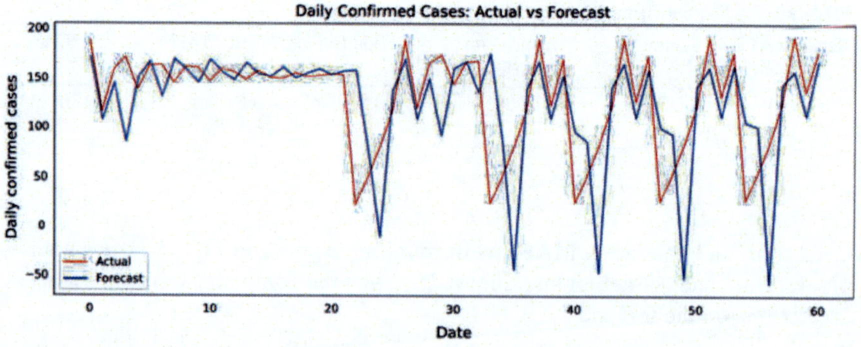

Fig. 23 Plot of LSTM actual and forecasted for daily total individuals vaccinated

Table 11 GRU performance for total individuals vaccinated

Train MAPE	Train loss	Validation MAPE	Validation loss	Test MAPE	Test RMSE
2346084	0.0081	70.1029	0.0689	0.581 or 58.1%	44.971

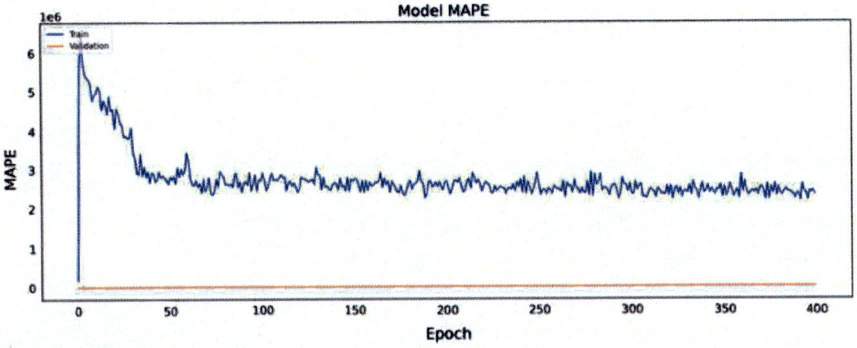

Fig. 24 Plot of GRU train MAPE and validation MAPE with number of epochs for daily total individuals vaccinated

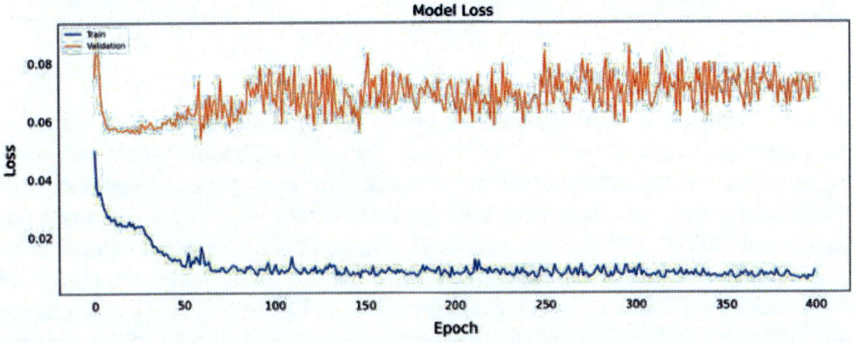

Fig. 25 Plot of GRU train loss and validation loss with number of epochs for daily total individuals vaccinated

7 Discussion

For daily confirmed cases, both LSTM and GRU are able to fit the training data well and generalize well on both the validation and test datasets. Out of the two architectures, GRU performs better on all three datasets—train, validation and test. Thus, GRU is the better architecture for prediction of daily total confirmed cases. For daily positive tests, owing to fluctuations in the data, the performance degrades for both the architectures as compared to daily total confirmed cases. LSTM performs

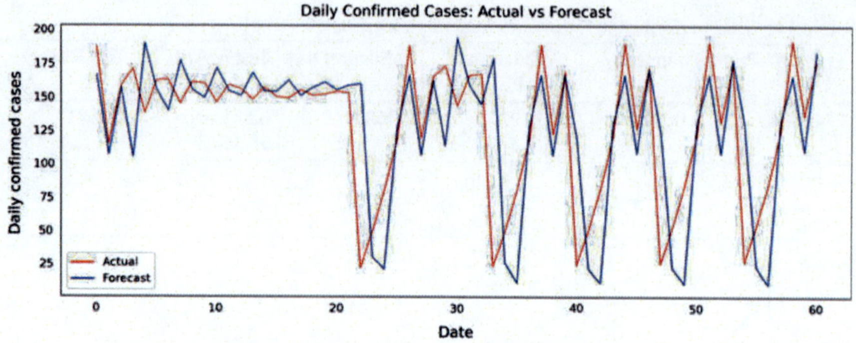

Fig. 26 Plot of GRU actual and forecasted total individuals vaccinated

Table 12 LSTM and GRU performance comparison for all variables

Model	Data	Metric	Variable		
			Confirmed cases	Positive tests	Individuals vaccinated
LSTM	Test	MAPE	0.02	0.2	0.61
		RMSE	9.5	52.5	53.2
GRU	Test	MAPE	0.01	0.3	0.6
		RMSE	7.98	60.6	44.97

slightly better on the test dataset and hence LSTM is the better architecture for prediction of daily positive COVID-19 tests. For total individuals Vaccinated, owing to imbalance in the data and still fluctuations after rolling mean imputations, the performance degrades further for both the architectures as compared to daily total confirmed COVID-19 cases and daily total positive COVID-19 tests. Both the LSTM and GRU architectures have comparable performances for the daily total individuals vaccinated data. Table 12 shows the comparison of LSTM and GRU performances for different variables. (Note: Metrics are rounded up to 1 decimal place wherever applicable)

The proposed simplified sequence models outperform the existing deep learning architectures by exhibiting a MAPE of less than 2% on the test dataset for daily COVID-19 cases and daily positive tests in India. Hence, these architectures can be used to predict these variables with state-of-the-art accuracy.

8 Conclusion

The deep learning recurrent neural networks efficiently capture the time series patterns, both short term and long term and are ideal for prediction in both cases.

LSTM and GRU, the standard RNN architectures, were implemented for prediction of COVID-19 daily total confirmed cases, total positive tests and total individuals vaccinated. While daily total confirmed cases shows an upward trend over time, total daily positive tests and total daily individuals vaccinated show seasonal patterns with fluctuations. Daily total confirmed cases were predicted with an accuracy of more than 98% by both the architectures. For daily positive tests and total individuals vaccinated, collection of more sample and feature pre-processing will help increase prediction accuracy for both the RNN architectures.

References

1. Arora, P., Kumar, H., Panigrahi, B.K.: Prediction and analysis of COVID-19 positive cases using deep learning models: a descriptive case study of India. Chaos, Solitons & Fractals **139**, 110017 (2020)
2. Shastri, S., Singh, K., Kumar, S., Kour, P., Mansotra, V.: Time series forecasting of Covid-19 using deep learning models: India-USA comparative case study. Chaos, Solitons & Fractals **140**, 110227 (2020)
3. Devaraj, J., Elavarasan, R.M., Pugazhendhi, R., Shafiullah, G.M., Ganesan, S., Jeysree, A.K., Khan, I.A., Hossain, E.: Forecasting of COVID-19 cases using deep learning models: is it reliable and practically significant? Results Phys. **21**, 103817 (2021)
4. Liao, Z., Lan, P., Fan, X., Kelly, B., Innes, A., Liao, Z.: SIRVD-DL: a COVID-19 deep learning prediction model based on time-dependent SIRVD. Comput. Biol. Med. **138**, 104868 (2021)
5. Kumar, Shiu, Sharma, Ronesh, Tsunoda, Tatsuhiko, Kumarevel, Thirumananseri, Sharma, Alok: Forecasting the spread of COVID-19 using LSTM network. BMC Bioinform. **22**(6), 1–9 (2021)
6. Chaurasia, V., Pal, S. (2020) Application of machine learning time series analysis for prediction COVID-19 pandemic. Res. Biomed. Eng. 1–13 (2020)
7. Mohan, S., Solanki, A.K., Taluja, H.K., Singh, A.: Predicting the impact of the third wave of COVID-19 in India using hybrid statistical machine learning models: A time series forecasting and sentiment analysis approach. Comput. Biol. Med. **144**, 105354 (2022)
8. Chimmula, V.K.R., Zhang, L.: Time series forecasting of COVID-19 transmission in Canada using LSTM networks. Chaos, Solitons & Fractals **135**, 109864 (2020)
9. Ghany, K.K.A., Zawbaa, H.M., Sabri, H.M.: COVID-19 prediction using LSTM algorithm: GCC case study. Inform. Med. Unlocked **23**, 100566 (2021)

Yoga Pose Rectification Using Mediapipe and Catboost Classifier

Richa Makhijani, Shubham Sagar, Koppula Bhanu Prakash Reddy, Sonu Kumar Mourya, Jadhav Sai Krishna, and Manthan Milind Kulkarni

Abstract Yoga will be unfruitful if the individual performing it does not have proper posture. Attending yoga classes and getting training sessions can be expensive and time-consuming, and most existing yoga-based applications are just focused on pose classification and posture correction for a small number of yoga poses, and most of them do not provide output in the form of audio. The dataset containing 13 (398 images) postures is analyzed with OpenCV, and the key points of the pose are retrieved using MediaPipe. These retrieved key points are used for calculating angles from the pose, and together the key points and angles are used to train machine learning models for distinguishing yoga postures. The improperly positioned body part was identified using the average values of each key point and computed angles. We were able to achieve 98.9% accuracy using CatBoost classifier. We have executed this model on 82 practical images and obtained 93.9% accuracy on yoga classification and 78.05% accuracy for identifying the improperly positioned body part.

Keywords Yoga · Yoga posture · OpenCV · MediaPipe · Machine learning · Catboost classifier

1 Introduction

Our busy daily routines cause long-term health issues such as diabetes, cardiac arrest, and obesity because we do not put much attention on our mental and physical health. Around 39% of the world's population is overweight. Exercise helps us lose weight, but it may require equipment and physical strength to perform. Yoga, on the other hand, consists of lightweight movements that can improve physical and psychological wellness. For elderly people, yoga is a better choice for improving their fitness. Visiting the yoga training centers or taking online yoga courses and getting coached by a personal trainer are neither cheap nor accessible to everyone, and finding an

R. Makhijani (✉) · S. Sagar · K. B. P. Reddy · S. K. Mourya · J. S. Krishna · M. M. Kulkarni
Department of Computer Science and Engineering, Indian Institute of Information Technology Nagpur, Nagpur, Maharashtra, India
e-mail: richa.makhijani@iiitn.ac.in

M. Tistarelli et al. (eds.), *Computer Vision and Machine Intelligence*, Lecture Notes in Networks and Systems 586,
https://doi.org/10.1007/978-981-19-7867-8_30

affordable yoga session that fits into one's schedule can be difficult. Pre-recorded yoga tutorials are not as effective as yoga practiced at a training center.

In this research, we proposed an approach that includes identifying a wider range of yoga poses as well as poorly positioned body parts when compared to previous studies. We developed a program that can identify a pose and deliver feedback for an improperly positioned body part using computer vision and machine learning techniques. We used a data set containing 13 yoga postures, each with 32 images. The MediaPipe library and openCV are used to extract key points from the images. The angles of joints in the image are calculated using the key points. The key points and angles are used to train the appropriate machine learning model, and we found that the CatBoost classifier does have the maximum accuracy.

2 Related Works

Human actions, such as yoga poses, can be detected in a variety of ways. The suggested model in [1] compares the posture of a user's live feed with an instructor's recorded footage. Using OpenPose and a PC camera, the system first detects a pose. The difference between an instructor's and a user's body angles is then calculated. If it exceeds a certain threshold, the approach recommends that that bodily component be corrected. Only three postures were evaluated: opening arms, one leg standing for a child, and one leg standing for an adult girl. In this paper, OpenPose was proposed as a human estimation model that requires a lot of memory. As a result, this technique is unsuitable for real-time analysis.

For posture analysis, [2] creates a Real-Time Infinity Yoga Tutor application. There are two key modules in the system: (1) the module for pose estimation (OpenPose) and (2) the module for detecting poses (Deep learning model). Based on its knowledge, Infinity Yoga Tutor is a yoga posture detection and correction application that employs a mobile-based approach to correct inappropriate yoga postures in people who practice yoga based on their knowledge and by viewing yoga videos or utilizing yoga apps. Although there are several tools for detecting yoga postures, there are few systems for correcting improperly performed yoga postures. It is also a very expensive operation when compared to a mobile phone camera. As a result, it is not ideal for a yoga posture detecting system.

This research [3] offers a yoga assistant mobile application based on the recognition of human key points and the calculation of angles at such key points. To accommodate diverse yoga poses, an improved score calculation technique is proposed. More humanized designs, such as voice service, have been included. This model is only used for three postures. Here, for identifying yoga poses, the improved score is used, which does not perform better on diverse yoga poses. Here, there is a limit to how many outputs they can provide because they only show four depending on angle deviations but not the wrong body component.

A four-stage system is proposed in this paper [4]: which includes a photographing device (input image), a model of pose detection (MediaPipe), comparison with reference data and geometric analysis, and feedback and postural correction (Display or Audio message). In this study, five important yoga postures were trained and tested: Virabhadrasana, Utkatasana, Vrikshasana, Phalakasana, and Trikonasana. The same body angles between two lines drawn from MediaPipe are used to compare poses. When executing this system on a computer, the performance is consistently 30 frames per second. They have limitations, such as not considering the depth distance from the camera, whereas we have addressed the z component/depth distance as well.

This paper [5] suggests an approach that consists of two modules: preprocessing and native (real-time) applications. To obtain 17 fundamental body points, the video frames are fed into the Posenet Class. The distance between each body component is determined using the Euclidean distance, and the angles are calculated for comparison using the cosine law. They suggested using a multi-staged CNN to forecast the erroneous body component, but we employed machine learning techniques instead.

This paper [6] makes use of a big dataset that includes over 5500 (self-generated) images of 10 various yoga positions. The body's important points are collected using tf-pose in this article. The angles of joints in the human body are computed and used as features in machine learning models. Using a random forest classifier, this dataset is evaluated on several machine learning classification models and obtains a 99% accuracy. The most significant flaw we discovered is that this paper is only used to detect yoga postures, not improper body parts, and it is not done in real time.

In this paper [7], the body's essential points are gathered using OpenPose. Key points in both 2D and 3D are extracted here and used for further research. The model provided here is a simple neural network sequential model with two hidden layers and one output layer. Finally, we can observe that with 2D key points, the accuracy is 85.8%, and with 3D key points, it is 100%. Only, 3 asanas are discussed in this text. We can see that their sample size is quite tiny, so they are achieving high accuracy, which is not good for the model, and it is also biased toward the available data.

3 Methodology

The proposed methodology consists of four stages (see Fig. 1). They are,

1. Preprocessing part of the system.
2. Keypoint extraction and angle calculation.
3. Training models for yoga identification.
4. Identifying improperly positioned body part.

3.1 Pre-processing Part of the System

We have collected the dataset from this site [8]. This dataset contains 107 asanas, but due to the data size of each asana and the discrete nature of asanas, we only

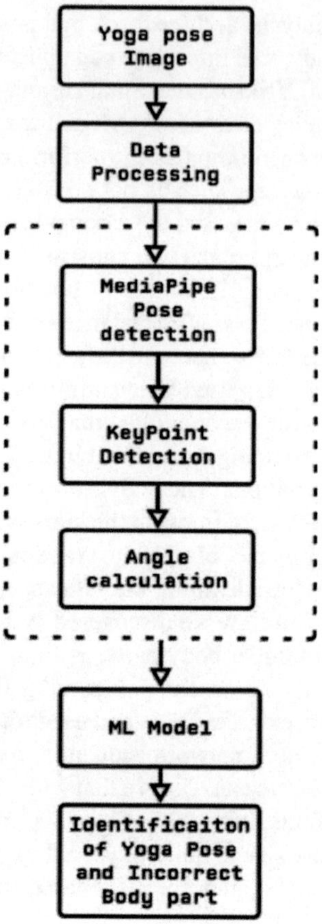

Fig. 1 Methodology

considered 13 of them for analysis. To make the data seem clearer, all of the images are oriented to the right. Finally, we have a dataset of 398 photos, with 25 to 35 images in each asana after they have been cleaned, which will be used for further analysis.

3.2 *Key Point Extraction and Angle Calculation*

After preprocessing the data, the data is passed to mediapipe library where the 33 key points are extracted. Figure 2 illustrates these crucial key points. These key points are made up of the point's *x*, *y*, and *z* coordinates. The angles at certain key points are then calculated using the *x* and *y* coordinates. As indicated in Fig. 3, there are 13

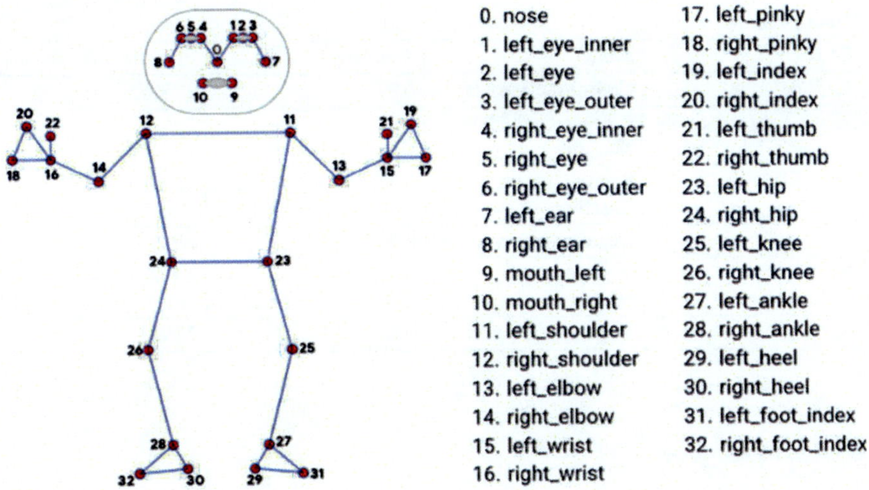

Fig. 2 Key points [9]

Angle	Title	Right		
		Key point 1	Key point 2	Key point 3
RU1	Right Elbow	16	14	12
RU2	Right Shoulder	14	12	24
RU3	Right Abdomen	12	24	26
RL1	Right Knee	24	26	28
RL2	Right Foot	26	28	32

Angle	Title	Left		
		Key point 1	Key point 2	Key point 3
LU1	Left Elbow	11	13	15
LU2	Left Shoulder	13	11	23
LU3	Left Abdomen	11	23	25
LL1	Left Knee	23	25	27
LL2	Left Foot	25	27	31

Fig. 3 Angles notation

key points at which angles have been computed. The 2-argument arctangent(atan2(y, x)) formula is used to determine the angle between that point (x, y) and the X+ axis in relation to the origin. Finally, we have put all of the key points and angles into a data frame, along with a column for discrete asanas. This data frame is then used to do model analysis.

Table 1 Model analysis

Sr. No.	Model	Accuracy	F1 score
1.	CatBoost classifier	0.99	0.98985
2.	K nearest neighbors classifier	0.97	0.97052
3.	Random forest classifier	0.97	0.9698
4.	Decision tree classifier	0.96	0.9597
5.	Support vector machine	0.9	0.902
6.	Logistic regression	0.89	0.8913

3.2.1 Mediapipe

MediaPipe is a machine learning pipeline architecture for processing time-series data such as video and audio. This cross-platform framework is compatible with PCs and servers, as well as android, iOS, and embedded devices like the Raspberry Pi and Jetson Nano. It is Google's open-source cross-platform framework for building perception pipelines. It is widely used for dataset preparation pipelines for ML training and ML inference pipelines.

3.3 Training Models for Yoga Identification

This data frame is split into two datasets (train and test) while keeping the image distribution in each discrete asana consistent. There are 298 and 100 images in the train and test datasets, respectively. We ran these datasets through several classification machine learning models and reviewed the results. The results of several models are depicted in Table 1. According to the results, the CatBoost classifier [10] produces better results, with a 99% accuracy and a 98.9 f1-score. After training, every one of these models is saved using the pickle library, which can be used to identify discrete asanas.

3.4 Identifying Improper Body Part

We calculate the average of these values of discrete asanas from the data frame including key points and angles and save in a json file for easier access to the data. Now, if a new image is submitted to this model, we can determine the accuracy of each asana by extracting key points and angles. We maintain an accuracy criterion of 85%, over which the asana is performed properly, and below this and larger than 40, we claim that the asana is performed improperly. We calculate the deviation between the mean values recorded in json and the new image's key points and angles. While doing that asana, the label with the highest deviation value is detected as the improper body part.

4 Type of System

We have proposed two types of systems (see Fig. 4). They are as follows:

1. Static System
2. Dynamic/Real-Time System.

4.1 Static System

Static images are saved in a test folder, after which the key points and angles are retrieved and provided to the model for asana classification. After identifying the asana, the highest deviation value is calculated using key points and angles, yielding the improperly positioned body part. All of these findings are saved in a csv file, where the original wrong body part and the calculated improperly positioned body part are compared.

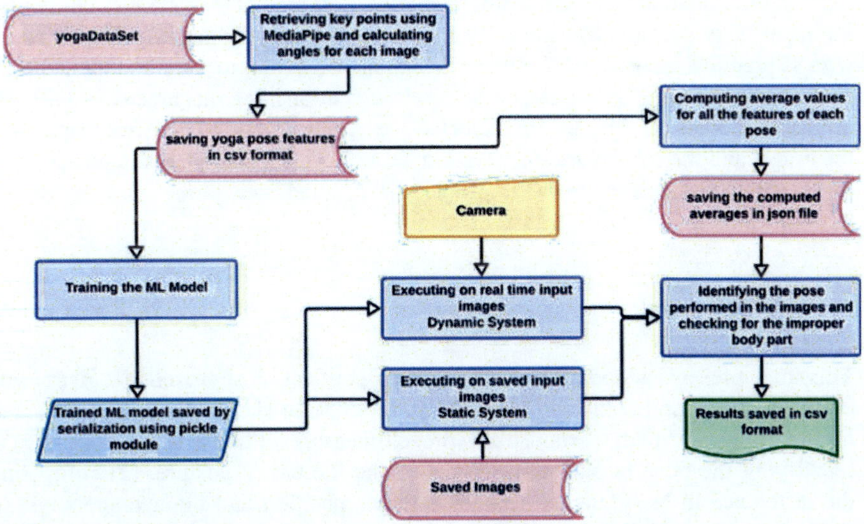

Fig. 4 Design of system

Table 2 Yoga pose detection analysis

Data	Correct pose detected	Wrong pose detected	Total
Practical data (Self-generated)	77	5	82

Table 3 Yoga improper body part analysis

Data	Identified Improper Body Part Correctly	Identified Improper Body Part Incorrectly	Total
Practical data (Self-generated)	64	18	82

4.2 Dynamic/Real-Time System

Key points and angles are gathered and delivered to the model for asana classification using real-time camera input. The maximum deviation value is determined using key points and angles after identifying the asana, producing the inaccurate body part. For a better user experience, the output is displayed as well as delivered as audio.

5 Results

On the dataset (398 images), we trained and evaluated various machine learning models, and the CatBoost classifier with the highest accuracy of 99.89% was chosen for improper body part identification. We have also used a new dataset (self-generated) with 82 practical images in different environment conditions to test the model. Table 2 shows that after using the model, the asana classification has an accuracy of 93.9024% on the new practical dataset. We detected the improper body part after classifying the asana, and the results are displayed in Table 3. This figure yields in an accuracy of 78.0488% for identifying the improperly positioned body part.

6 Conclusion

This study presents a MediaPipe-based asana identification approach that gives users verbal guidance and feedback to help them rectify non-standard asana poses based on their improper body parts. This approach measures a user's asana pose by: (1) identifying the pose or skeleton, (2) determine the user's posture, (3) computing the difference in body angles between a user's position and the average angle of that pose, and (4) designating the improper part of the user by the place where the highest difference is identified as the improperly positioned body part of the user.

The effectiveness of the approach is demonstrated by the achieved results, which provide an accuracy of 78.1% for the practical dataset, which contains images of people of diverse ages, genders, and physical appearances performing different yoga poses in various environments.

7 Future Work

Only, 13 yoga asanas are currently identified using the proposed approach. Because there are so many yoga asanas, developing a posture estimate model that properly detects all of them has been a major problem. The MediaPipe library will have a difficult time correctly identifying overlaps between persons or bodily parts. While performing asanas, there is also scope for improvement in identifying improper body parts. This approach may also be utilized to create a full-fledged application that will aid in the proper practice of asanas.

References

1. Thar, M.C., Winn, K.Z.N., Funabiki, N.: A proposal of yoga pose assessment method using pose detection for self-learning. In: 2019 International Conference on Advanced Information Technologies (ICAIT) (pp. 137–142). IEEE (2019). 10.1109/AITC.2019.8920892
2. Rishan, F., De Silva, B., Alawathugoda, S., Nijabdeen, S., Rupasinghe, L., Liyanapathirana, C.: Infinity yoga tutor: Yoga posture detection and correction system. In: 2020 5th International Conference on Information Technology Research (ICITR) (pp. 1–6). IEEE (2020). 10.1109/ICITR51448.2020.9310832
3. Huang, R., Wang, J., Lou, H., Lu, H., Wang, B.: Miss yoga: a yoga assistant mobile application based on keypoint detection. In: 2020 Digital Image Computing: Techniques and Applications (DICTA) (pp. 1–3). IEEE (2020). 10.1109/DICTA51227.2020.9363384
4. Anilkumar, A., KT, A., Sajan, S., KA, S.: Pose estimated yoga monitoring system. Available at SSRN **3882498** (2021). https://doi.org/10.2139/ssrn.3882498
5. Chiddarwar, G.G., Ranjane, A., Chindhe, M., Deodhar, R., Gangamwar, P.: AI-based yoga pose estimation for android application. Int. J. Inn. Sci. Res. Tech. **5**, 1070–1073 (2020). 10.38124/IJISRT20SEP704
6. Agrawal, Y., Shah, Y., Sharma, A.: Implementation of machine learning technique for identification of yoga poses. In: 2020 IEEE 9th International Conference on Communication Systems and Network Technologies (CSNT) (pp. 40–43). IEEE (2020). 10.1109/CSNT48778.2020.9115758
7. Narayanan, S.S., Misra, D.K., Arora, K., Rai, H.: Yoga pose detection using deep learning techniques. In: Proceedings of the International Conference on Innovative Computing and Communication (ICICC) (2021). 10.2139/ssrn.3842656
8. Kaggle: Yoga pose image classification dataset. https://www.kaggle.com/shrutisaxena/yoga-pose-image-classification-dataset. Last Accessed: 14 Dec 2021
9. Mediapipe https://google.github.io/mediapipe/solutions/pose.html. Last Accessed: 17 Jan 2022
10. Catboost: https://wwww.catboost.ai/en/docs/. Last Accessed: 22 Apr 2022

A Machine Learning Approach for $PM_{2.5}$ Estimation for the Capital City of New Delhi Using Multispectral LANDSAT-8 Satellite Observations

Pavan Sai Santhosh Ejurothu, Subhojit Mandal, and Mainak Thakur

Abstract $PM_{2.5}$, a dangerous air pollutant, mapping on a city level scale, plays a crucial role in the development of sustainable policies toward balanced ecology and a pollution-free society. Recently, multispectral and hyperspectral satellite imagery promise a high capability toward detecting the places with soaring atmospheric pollution and aerosol information. The multispectral imagery uses the ambient surface reflectance from the surface of the earth in the visible spectrum bands. LANDSAT-8 satellite provides multispectral observations over the surface of the earth with 30m resolution. We develop various machine learning models for $PM_{2.5}$ estimation for one of the most highly polluted cities in the world, Delhi, the captial city of India using LANDSAT-8 observations and ground-level $PM_{2.5}$ data. A feasible multispectral-based $PM_{2.5}$ estimation model is established in this study, which promises high-resolution $PM_{2.5}$ mapping from LANDSAT-8 imagery with an acceptable level of accuracy.

Keywords $PM_{2.5}$, Machine learning, Landsat 8

1 Introduction

The urban population are suffering from various health-related issues and environmental problems worldwide. Air pollution is one of the threatening health hazards to urban masses. Particulate matters $PM_{2.5}$ (size less than 2.5 μm) is one of those pollutants which severely affects human health and has multiplicative negative impacts on the overall life cycle. It can penetrate through the human cells and cause various

P. S. S. Ejurothu (✉) · S. Mandal · M. Thakur
Indian Institute of Information Technology, Sri City, India
e-mail: pavansaisantosh.e18@iiits.in

S. Mandal
e-mail: subhojit.m@iiits.in

M. Thakur
e-mail: mainak.t@iiits.in

M. Tistarelli et al. (eds.), *Computer Vision and Machine Intelligence*,
Lecture Notes in Networks and Systems 586,
https://doi.org/10.1007/978-981-19-7867-8_31

human health-related ailments. Furthermore, not limited to human health, these particles also impact cloud dynamics and atmospheric radiation balance, which induces a shift in other meteorological variables. In the recent decades, remote sensing methods are proven to be very useful for developing a city-based air pollution monitoring framework and providing insights to the corresponding city administration to design sustainable policies for a healthy urban environment. With the increased launch of satellites, broader high-resolution coverage, and continuous evolution of spectral imaging systems, satellite observations promise a high potential in developing spectra-derived $PM_{2.5}$ estimation methods for $PM_{2.5}$ monitoring framework. A band represents the segment of the electromagnetic spectrum. This spectrum is exerted by spectral sensors on board the satellite. The satellite imagery instruments measure the surface reflectance of a particular band of the electromagnetic spectrum. The vertical aerosol profile can be derived via an empirical model from the surface reflectance information of various spectrum bands. Usually, the fine spectral resolution of the spectral signals of particular bands has proven to be helpful in retrieving high-definition and clear atmospheric composition.

The ground-level aerosol information usually differs from the satellite-derived vertical column loading of aerosols. Due to this, the ground monitoring stations observations are used to build a ground-level aerosol/particulate matter conversion model. Nevertheless, the relationship between satellite-derived aerosol optical depth (AOD) and ground-level aerosol/particulate matter concentration is still not understood effectively and robustly up to this date. Various factors such as terrain information, cloud cover, local atmospheric conditions, and the onboard satellite instrument errors affect these satellite imagery-based conversion models. Generally, these kinds of conversion models do not work uniformly over all the geographic locations. Therefore, researchers develop these conversion models based on observations for a particular region to make the model perform robustly and capture local geographical information in the modeling.

Notably, hyperspectral satellite imageries can also be used as an alternative option, but minimal data availability is a barrier to using such data for ground-level $PM_{2.5}$ estimation. The moderate resolution imaging spectroradiometer (MODIS), MISR, OMI, VIIRS, and CALIPSO also provide atmospheric aerosol information. However, due to the coarse spatial resolution of the data products and limited data availability over the region, employing these instrument data products for city-based air pollution monitoring becomes challenging. On the other hand, LANDSAT-8 satellite [1] onboard operational land imager (OLI) and thermal infrared sensor (TIRS) instruments are capable of capturing the high-resolution multispectral image, which is used for retrieving various geoscience-related variables as well as $PM_{2.5}$. Machine learning algorithms have proven to be successful in computer vision, medical imaging, information technology, drug development, climate modeling, etc., for their inherent capability to extract underlying complex features. In this work, we developed various machine learning models using LANDSAT-8 imagery for ground-level $PM_{2.5}$ concentration estimation, and those captures the complex relationships among the LANDSAT 8 provided 8 spectral band data (total 11 bands) and ground-level $PM_{2.5}$ concentrations. Being one of the most polluted cities in the world, our region of interest is selected as New Delhi, National Capitial Region (NCR) of India. With significant accuracy, our proposed model can be deployed for ground-level $PM_{2.5}$ estimation using LANDSAT-8 imagery.

2 Related Works

Several studies utilized LANDSAT-8 imagery for obtaining surface temperature characteristics, heat and moisture dynamics for agricultural interests, forestry mapping, geological mineral mapping, etc. Though it is successful in these respective areas of geoscience-related problems, very limited studies have been done in the field of air pollution mapping using multispectral observations. Very few works investigate the multispectral data and ground $PM_{2.5}$ concentration conversion model in the context of a highly polluted city like New Delhi using machine learning algorithms. Alvarez-Mendoza et al. [2] proposed an empirical PM_{10} estimation model using the surface reflectance bands of LANDSAT-7 ETM+, LANDSAT-8 OLI/TIRS, and AQUA-Terra/MODIS sensor observations and ground-level observations from 9 monitoring stations in Quito, the capital of Ecuador. Mishra et al. [3] developed a multispectral empirical $PM_{2.5}$ estimation model using multiple linear regression model (MLR) for 10 monitoring stations in Delhi with 7 LANDSAT-8 scenes. The linear regression algorithm has been used in this study. Very few data samples were used in this work for the multispectral to $PM_{2.5}$ data conversion, which may not be sufficient for capturing the conversion effect due to different weather conditions and atmospheric energy dynamics. Also this model does not study in detail the full potential of the multispectral imagery for ground-level $PM_{2.5}$ estimation. MLR algorithm is unable to capture the non-linear relationship between the multispectral band features and ground $PM_{2.5}$ concentrations. On the contrary, the machine learning models have shown potential in unveiling complex relationships between the input features and target variables. Among the machine learning models, multiple linear regression (Yunping et al. [4]), random forest, support vector machine (Nengcheng et al. [5]), Gaussian process regression, XGBoost, and artificial neural network models have been used for ground $PM_{2.5}$ estimation from various space-borne satellite observations (e.g., MODIS). Xue et al. [6] developed a spatio-temporally weighted random forest model to estimate ground-level $PM_{2.5}$ concentrations from MODIS AOD data. Although, these works explored different ML models for estimating ground-level particulate matter concentrations, and very few studies have considered LANDSAT-8 observations as an input option while using ML models.

Delhi, the capital of India, faces a number of sustainability-related concerns as a result of its urbanization efforts, heavy traffic, air pollutant diffusion and transportation-related problems, and harsh weather conditions. These issues make Delhi a significant study region for air pollution monitoring than any other city in the world. The contribution of this study can be listed as follows:

1. Development of a point-wise estimation model using multispectral satellite imagery to predict ground-level $PM_{2.5}$ concentration for the city of Delhi.
2. Use of state-of-the-art machine learning models on the spectral reflectance data for high-resolution (30 m) $PM_{2.5}$ concentration estimation for the city of Delhi.

3 LANDSAT-8 Data Description

Landsat 8 Level-1 product cannot be used in the study due to the presence of calibration error. Also, it is not processed through a procedure called absolute radiometric calibration, which is essential for extracting ground-level aerosol information. Instead, the Level-2 product undergoes absolute radiometric calibration, atmospheric correction, and terrain correction from the Level-1 product. The Level-1 data product comes with 11 multispectral bands. The corresponding wavelength and resolution of each of the bands are shown in Table 1. For Level-2 data, bands no. 8 and 9 are not present. The band no. 10 product is resampled to 30 m resolution for scientific studies. The Level-2 data products are downloaded from USGS Earth Resources Observation and Science (EROS) Website. Notably, bands no. 10 and 11 are thermal bands captured by thermal infrared sensor (TIRS). Band 1 provides information about coastal aerosol, and bands no. 2 (blue), 3 (green), and 4 (red) also provide crucial information about atmospheric aerosol. A LANDSAT-8 scene over Delhi is shown in Fig. 1, also the corresponding spectral bands are shown.

Table 1 Band information of LANDSAT-8 satellite images

Bands	Wavelength (micrometers)	Resolution (m)
Band 1—coastal aerosol	430–450	30
Band 2—blue	450–510	30
Band 3—green	530–590	30
Band 4—red	640–670	30
Band 5—near infrared (NIR)	850–880	30
Band 6—shortwave infrared (SWIR) 1	1570–1650	30
Band 7—shortwave infrared (SWIR) 2	2110–2290	30
Band 8—panchromatic	500–680	15
Band 9—cirrus	1360–1380	30
Band 10—thermal infrared (TIRS) 1	10600–11190	100
Band 11—thermal infrared (TIRS) 2	11500–12510	100

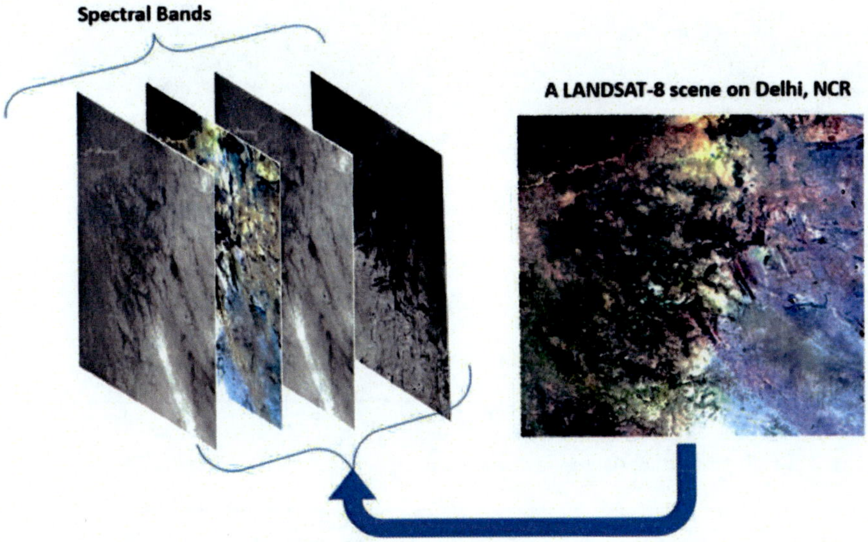

Fig. 1 Different spectral band of a LANDSAT-8 scene over Delhi on 14th October, 2020

4 Study Area

Broader New Delhi (the National Capital Region of India) is one of the most populated cities of India, which covers a 1484 km^2 area. Industrial emissions (12%) and traffic-related pollution around (67%) (Rizwan et al. [7]) are a few of the primary sources of air pollution in this city. Central Pollution Control Board (CPCB) controlled 40 ground monitoring stations measure ambient air pollutant concentrations in a 15 minutes interval. The locations of these monitoring stations are shown in Fig. 2.

5 Problem Formulation and Methodology

The band-wise surface reflectance values of LANDSAT-8 Collection-2 Level-2 Science Product (LC08_L2SP) are used for the $PM_{2.5}$ estimation model as input in this study. Generally, bands 1, 2, 3, and 4 are essential bands for extracting ground-level $PM_{2.5}$ concentrations. However, due to cloud-related noise, preprocessing-based noise, and other atmospheric correction-based processing noise, we considered all other bands as features for the model development. The $PM_{2.5}$ estimation regression problem can be formulated as follows:

$$PM_{2.5} = f(B_1, B_2, B_3, \ldots, B_7, B_{10}) \tag{1}$$

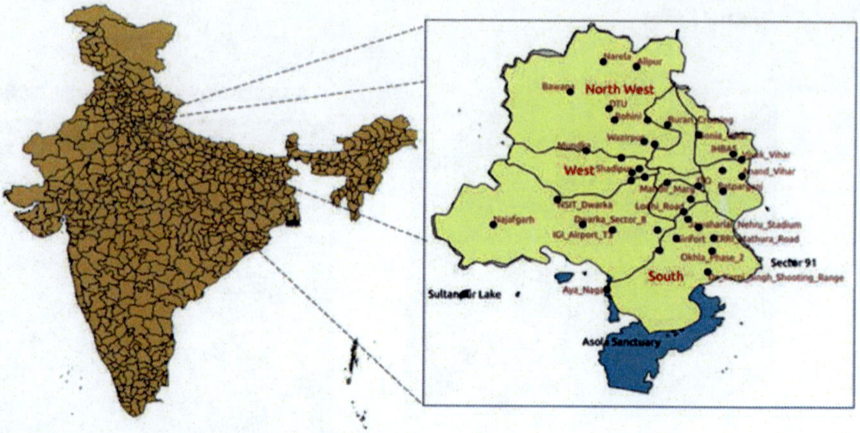

Fig. 2 CPCB monitoring stations in Delhi

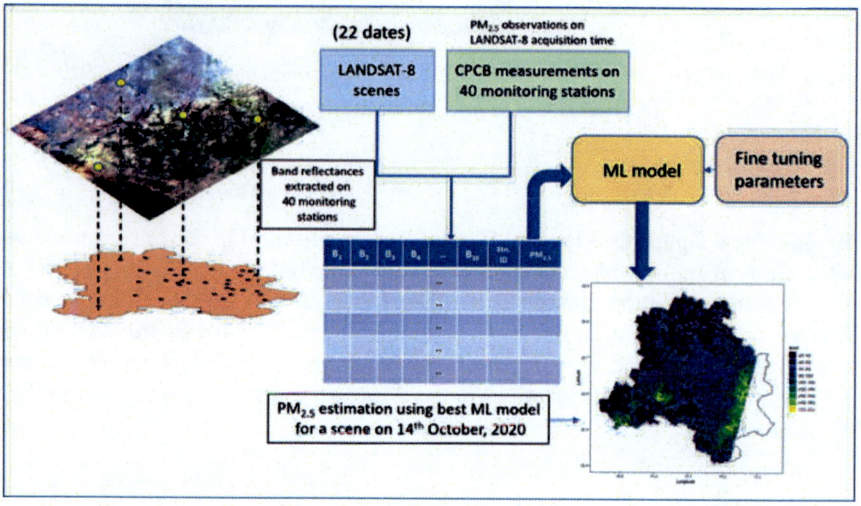

Fig. 3 Methodology

where B_i is the reflectance of ith band. The function f captures the complex relationship between the band reflectance and ground monitoring stations' $PM_{2.5}$ concentration while training using the machine learning (ML) models. The modeling methodology is shown in Fig. 3. The LANDSAT-8 satellite passes over the study region once every 16 days between 10:45 AM and 11 AM Indian Standard Time (IST). Also, the CPCB monitoring stations' $PM_{2.5}$ observations are collected on the corresponding LANDSAT-8 acquisition dates and during satellite passing time over Delhi. For 40 monitoring stations for each date, the LANDSAT-8 band features are extracted and matched with the corresponding $PM_{2.5}$ concentration. A dataset is created with band reflectance features and $PM_{2.5}$ concentrations for each date and each monitoring station location. The dates in which the band features as well as $PM_{2.5}$ concentrations are unavailable, they are removed. The dataset has been created with a total of 808 samples after cleaning. In the next step, several machine learning models were employed for modeling $PM_{2.5}$ concentration estimation. Well-known models such as random forest, XGBoost, weighted random forest (Liu et al. [8]), ensemble trees, and SVM models were used to develop the modeling framework. The fivefold cross-validation method was chosen to compare the accuracy matrices. The following error metrics were computed to compare the model performances: coefficient of determination R^2, root mean square error (RMSE) [9], and mean absolute error (MAE) [10].

The best model is employed on a random day which is out of the training dataset dates. The model is deployed to estimate $PM_{2.5}$ concentration on 14th October 2020.

6 Results and Discussion

Our study extracted 22 dates from the LC08_L2SP data acquisition date-time stamp. These data are acquired within the date 1st January 2020 to 1st June 2020 (shown in Table 2), covering all the CPCB monitoring stations in Delhi. Among all the bands, first four band's density histogram is shown in Fig. 4. The correlation between the bands and ground-level $PM_{2.5}$ concentrations is shown in Fig. 5. $PM_{2.5}$ has shown to have higher correlations with band no. 1, 2, 3, 4, and 10.

Accuracy measures in Table 3 show the fivefold cross-validation on 80% of the dataset. It is observed that the random forest model provides better accuracy than all other models. The random forest model shows superiority over-weighted random forest model. Though Gaussian process regression provides smooth estimates but it captures the underlying Gaussianity from the training dataset, which is not much reactive to the cloud-related noise or instrument noise. The top 4 performing models' (random forest, weighted random forest, Gaussian process regression, and XGBoost) performances on the 20% test data are shown in the form of the actual versus predicted graph in Fig. 6. Among other models, the XGBoost model shows a significantly good

Table 2 The LANDSAT-8 acquisition dates used in this study

Sl. No.	Date	Sl. No.	Date	Sl. No.	Date	Sl. No.	Date
1	09-01-2020	7	26-02-2020	13	14-04-2020	18	23-05-2020
2	16-01-2020	8	04-03-2020	14	21-04-2020	19	01-06-2020
3	25-01-2020	9	13-03-2020	15	30-04-2020	20	08-06-2020
4	01-02-2020	10	20-03-2020	16	07-05-2020	21	17-06-2020
5	10-02-2020	11	29-03-2020	17	16-05-2020	22	24-06-2020
6	17-02-2020	12	05-04-2020				

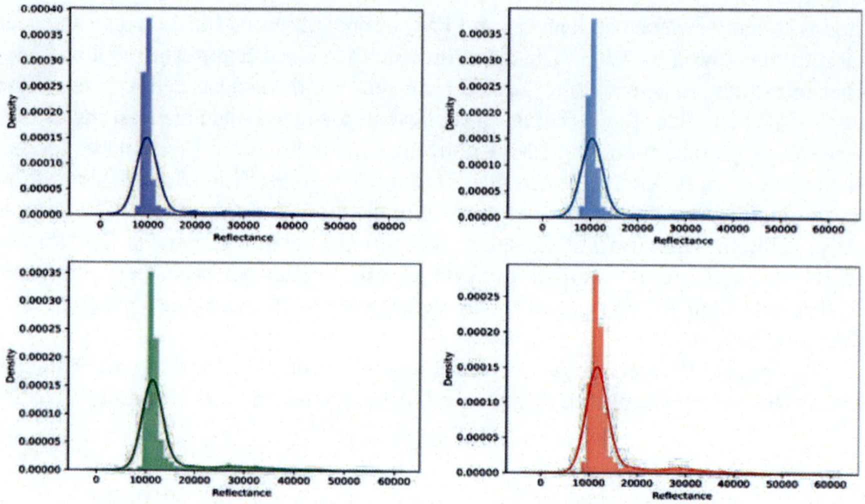

Fig. 4 Histogram of band ID 1, 2, 3, and 4

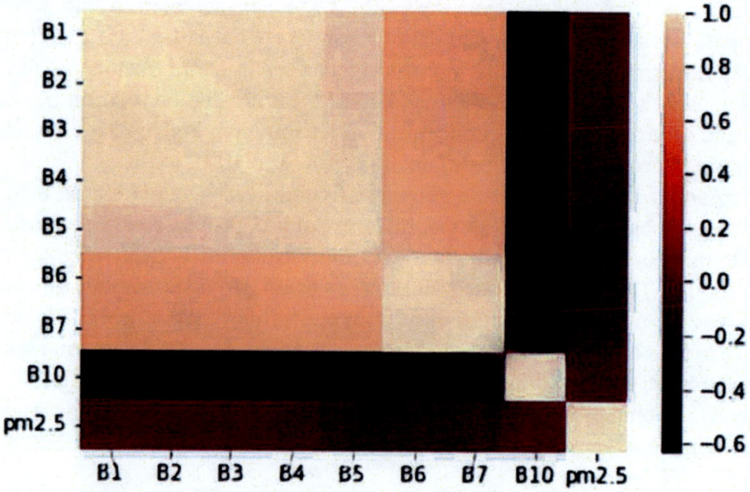

Fig. 5 Band correlation

performance with respect to ensemble models or support vector machine models. Notably, Mishra et al. [3] used multiple linear regression model and found R^2 value 0.93 on the test data for a dataset extracted from 7 LANDSAT-8 images; where a very less number of imageries were considered in the study. We developed the models on a considerably larger dataset. Also, the use of multiple linear regression underestimates for higher $PM_{2.5}$ values and overestimates for lower $PM_{2.5}$ values. The MLR model performance on our dataset is shown in Table 3. It can be observed

Table 3 Fivefold cross-validation results for different models

Sl. No.	Algorithm	RMSE (in $\mu g/m^3$)	MAE (in $\mu g/m^3$)	R^2
1	Random forest	44.12	32.24	0.54
2	Weighted random forest	44.90	33.29	0.52
3	Gaussian process regression	47.07	33.99	0.50
4	XGBoost	46.81	34.04	0.49
5	Ensemble trees	50.57	36.11	0.43
6	SVM	60.15	41.19	0.19
7	Multiple linear regression	61.31	45.08	0.16

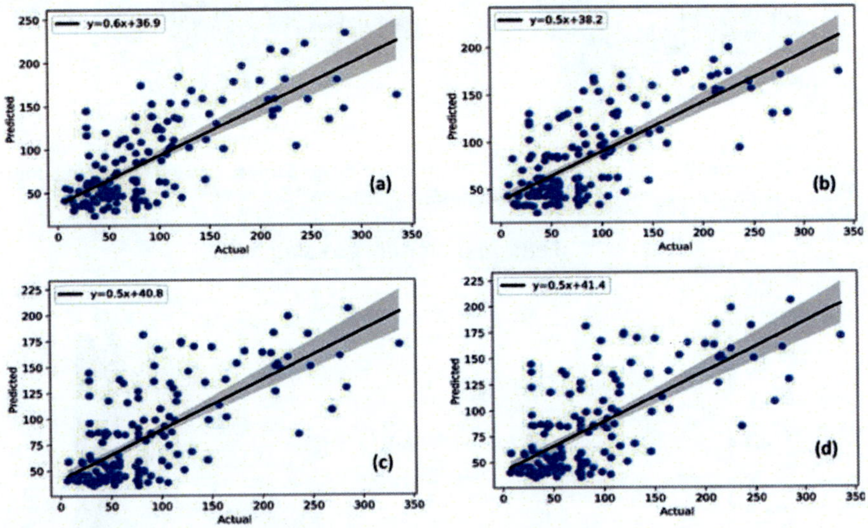

Fig. 6 Actual versus predicted plot for **a** random forest **b** weighted random forest **c** Gaussian process regression **d** XGBoost model

that the machine learning models show superiority over MLR model. The different ML model estimated $PM_{2.5}$ map on 14th October 2020 for Delhi from LANDSAT-8 scene is shown in Fig. 7. We can observe that the top 4 performing model provides almost similar estimates. The feature importance in the model is computed using the mean decrease impurity (MDI) [11] method. The feature importance is shown in Fig. 8. The B_{10} band feature has high importance score in the modeling.

Several studies (e.g.: Karlson et al. [12], Topouzelis et al. [13]) have pointed out the utility of random forest (RF) model for classification problems related to various remote sensing applications. Mellor et al. [14] found random forest model to be insensitive to noisy and imbalanced training label. It provides reliable predictions using an ensemble of multiple decision trees. As a result, it can make use of the best

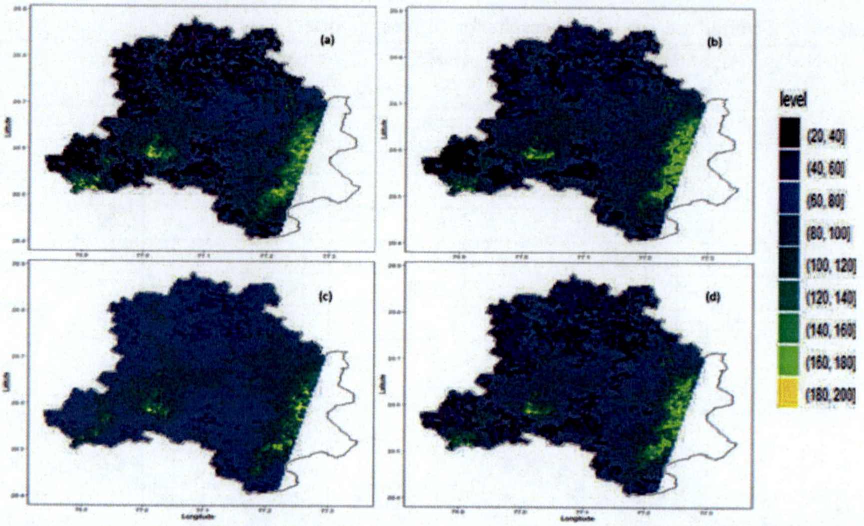

Fig. 7 $PM_{2.5}$ prediction for 14th October, 2020 using **a** random forest, **b** weighted random forest, **c** Gaussian process regression, **d** XGBoost models

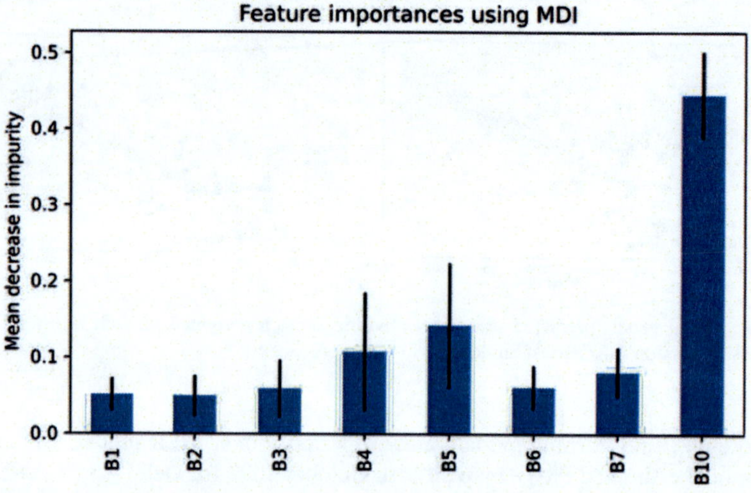

Fig. 8 Feature importance using MDI of the random forest model

decision trees' power and can detect any little spectral difference caused by haze from air pollution. These could be reasons why the random forest model performs better than other machine learning models for our dataset.

This methodology uses the concept of atmospheric radiative transfer processes (Zhang et al. [15]) for atmospheric aerosol to estimate $PM_{2.5}$ concentration for the city of Delhi. The existing aerosol in the atmosphere causes the variation of surface reflectance of the optical signals emitted from multispectral satellite sensors. We consider 22 LANDSAT-8 observations over the span of 6 months of time as the inclusion of further observations becomes quite expensive process. Notably, the methodology includes 22 LANDSAT-8 images, allowing the $PM_{2.5}$ conversion model to learn the complex interplay between multispectral data and $PM_{2.5}$ under a variety of meteorological circumstances. Further, the modeling can be extended over a large dataset; extracted from the LANDSAT-8 imageries for multiple years to achieve the robustness of the multispectral reflectance and ground-level $PM_{2.5}$ relationship.

7 Conclusion and Future Work

In this study, a point-based satellite acquired multispectral reflectance values to $PM_{2.5}$ conversion model is developed for the capital city of Delhi using a bunch of important machine learning models. Compared with related literature studies, these models perform on the dataset with satisfactory accuracy. The random forest model promises superiority with respect to other machine learning models for $PM_{2.5}$ estimation from multispectral LANDSAT-8 observations. The fivefold cross-validation of random forest model obtains a prediction accuracy of $R^2 = 0.54$, RMSE $= 32.24\ \mu g/m^3$, and MAE $= 44.12\ \mu g/m^3$. The accuracy can be increased by denoising the band features of multispectral data, which inherits noises from cloud masking algorithms and other preprocessing techniques. Furthermore, one can enlarge the dataset using more image acquisitions over the ground monitoring stations.

References

1. USGS, Department of the Interior U.S. Geological Survey, Landsat 8 (L8) Data Users Handbook (2019, November). Available online at https://www.usgs.gov/media/files/landsat-8-data-users-handbook. https://doi.org/10.5066/P9OGBGM6
2. Alvarez-Mendoza, C.I., Teodoro, A.C., Torres, N., Vivanco, V.: Assessment of remote sensing data to model PM10 estimation in cities with a low number of air quality stations: a case of study in Quito, Ecuador. Environments **6**(7), 85 (2019)
3. Mishra, R.K., Agarwal, A., Shukla, A.: Predicting ground-level PM2.5 concentration over Delhi using Landsat 8 satellite data. Int. J. Remote Sens. **42**(3), 827–838 (2021)
4. Chen, Y., Han, W., Chen, S., Tong, L.: Estimating ground-level PM2.5 concentration using Landsat 8 in Chengdu, China. In *Remote Sensing of the Atmosphere, Clouds, and Precipitation V* (Vol. 9259). SPIE (2014)
5. Chen, N., Yang, M., Du, W., Huang, M.: PM2.5 estimation and spatial-temporal pattern analysis based on the modified support vector regression model and the 1 km resolution MAIAC AOD in Hubei, China. ISPRS Int. J. Geo-Inf. **10**(1), 31 (2021)
6. Xue, W., et al.: Inferring near-surface PM2.5 concentrations from the VIIRS deep blue aerosol product in China: a spatiotemporally weighted random forest model. Remote Sens. **13**(3), 505 (2021)

7. Rizwan, S.A., Nongkynrih, B., Gupta, S.K.: Air pollution in Delhi: its magnitude and effects on health. Indian J. Community Med. **38**(1), 4 (2013)
8. Liu, Z.-S., Siu, W.-C., Huang, J.-J.: Image super-resolution via weighted random forest. In: 2017 IEEE International Conference on Industrial Technology (ICIT). IEEE (2017)
9. Bali, V., Kumar, A., Gangwar, S.: Deep learning based wind speed forecasting-a review. In: 2019 9th International Conference on Cloud Computing, Data Science & Engineering (Confluence). IEEE (2019)
10. Senthil, K.P.: Improved prediction of wind speed using machine learning. EAI Endorsed Trans. Energy Web **6**(23) (2019)
11. Scornet, E. Trees, forests, and impurity-based variable importance. arXiv preprint arXiv:2001.04295 (2020)
12. Karlson, M., Ostwald, M., Reese, H., Sanou, J., Tankoano, B., Mattsson, E.: Mapping tree canopy cover and aboveground biomass in Sudano-Sahelian woodlands using Landsat 8 and random forest. Remote Sens. **7**(8), 10017–10041 (2015)
13. Topouzelis, K., Psyllos, A.: Oil spill feature selection and classification using decision tree forest on SAR image data. ISPRS J. Photogrammetry Remote Sens. **68**, 135–143 (2012)
14. Mellor, A., Boukir, S., Haywood, A., Jones, S.: Exploring issues of training data imbalance and mislabelling on random forest performance for large area land cover classification using the ensemble margin. ISPRS J. Photogrammetry Remote Sens. **105**, 155–168 (2015)
15. Zhang, Y., et al.: Satellite remote sensing of atmospheric particulate matter mass concentration: advances, challenges, and perspectives. Fundam. Res. **1**(3), 240–258 (2021)

Motion Prior-Based Dual Markov Decision Processes for Multi-airplane Tracking

Ruijing Yang, Xiang Zhang, Guoqiang Wang, and Honggang Wu

Abstract Multi-airplane tracking (MAT) is the foundation of airport video surveillance. Besides common problems in tracking such as occlusion, this task is further complicated by the specific challenges in airport. For example, due to the low-compact shape of the airplane, its appearance will change drastically when turning, frequently resulting in failures in data association. In this paper, airplane motion prior-based dual Markov decision processes (DMDPs) are proposed for MAT. In airport surface, the airplane has only two direction modes such as straight and turn and two velocity modes such as acceleration and uniform speed. Such motion prior can be used to cope with the challenges in MAT. Firstly, we create a motion state MDP (msMDP) where each target has three motion patterns: straight/constant velocity (S/CV), straight/constant acceleration (S/CA), and curve/constant velocity (C/CV). This is consistent with the motion of airplane. Secondly, after the computation of affinity matrix based on msMDP, another tracking state MDP (tsMDP) which manages the lifetime of airplane is use to assist the data association step within the DMDPs framework. Finally, the effectiveness of the presented algorithm is verified based on the AGVS-T1(a new change detection dataset for airport ground video surveillance benchmark).

Keywords Multi-airplane tracking · Airport surface surveillance · Markov decision processes

---

R. Yang · X. Zhang (✉)
University of Electronic Science and Technology of China, Chengdu 611731, Sichuan, China
e-mail: uestchero@uestc.edu.cn

Yangtze Delta Region Institute (Quzhou), University of Electronic Science and Technology of China, Quzhou 324000, Zhejiang, China

G. Wang · H. Wu
The Second Research Institute of Civil Aviation Administration of China, Chengdu 610041, Sichuan, China

M. Tistarelli et al. (eds.), *Computer Vision and Machine Intelligence*,
Lecture Notes in Networks and Systems 586,
https://doi.org/10.1007/978-981-19-7867-8_32

1 Introduction

With the vigorous development of civil aviation transportation, the airport surface becomes more and more busy and crowded. Multi-airplane tracking (MAT) is fundamental and crucial in airport surface surveillance, which aims to locate all airplanes and maintain their identities in consecutive frames. The most widely used strategy in MAT is tracking-by-detection. Candidate airplane regions are generated in each frame with object detector, and then, they are linked to form object trajectories with data association.

Although multiple object tracking (MOT) has achieved a great success in real applications, e.g., pose estimation [1], there is still a lack of effective solutions for airplane tracking. Besides common problems in tracking, there are some special issues in MAT and among them, the low-compact shape of airplane brings great challenges. The airplane is a rigid target with thin strips and many protrusions. Such low-compact object looks very different when the motion pattern changes, e.g., from straight ahead to turning. A single motion model is difficult to deal with airplane tracking. Therefore, the key to solving this problem is to design a strategy to manage multiple motion states. We notice that the airplane can only speed uniformly on the curves. In other words, the motion patterns of the airplane can be predicted. This discovery inspires us to make use of motion prior of airplane for better MAT.

We follow the tracking-by-detection framework and propose airplane motion prior-based dual Markov decision processes (DMDPs) for MAT. Firstly, a motion state MDP (msMDP) is developed, where each airplane has three motion patterns: straight/constant velocity (S/CV), straight/constant acceleration (S/CA), and curve/constant velocity (C/CV). This is consistent with the motion prior of the airplane. Secondly, after the computation of affinity matrix based on msMDP and object detection, another tracking state MDP (tsMDP) which manages the lifetime of airplane is used to assist the data association step within the DMDPs framework. The tsMDP has four patterns, such as tentative, tracked, lost, and deleted. Finally, comparison experiments are conducted on the AGVS-T1 benchmarks [2], and the results clearly demonstrate that our tracker is effective for MAT.

2 Related Work

MOT has been researched for many years and also has been applied in video surveillance. In this section, after a brief review of MAT, an important issue related to tracking, motion modeling, is also discussed.

2.1 MAT

Nowadays, the video surveillance systems may be omnipresent. Video surveillance system has a large amount of video content, which makes real-time management and search of multiple video data challenging. Kumar et al. [3–5] consider capturing the access dependency between multiple video views and apply the target tracking algorithm to extract low activity frames, which reduces the unimportant video content and achieves real-time management and access to video content.

With the vigorous development of civil aviation transportation, video surveillance for airport surface has attracted much attention [6], and MAT naturally becomes an important task. Unlike pedestrian tracking, tracking multiple airplanes have its own challenges. For example, the appearance of airplane does not change when it is moving on the straight ways but changes drastically when turning. Therefore, the MOT algorithms in fundamental research generally are not suitable for MAT, and new algorithms should be developed to cope with the specific challenges in airport surface. For example, Mian et al. [7] propose a modified KLT tracking algorithm by the use of a feature clustering criterion under the KLT framework [8], which not only improves the tracking accuracy of airplane, but also can be used in complex environment .

2.2 Motion Modeling

The premise of MAT is to establish an appropriate motion model to describe the motion state of the targets. In this respect, Singer et al. [9] propose a single model algorithm to track the targets. In practice, each airplane on the airport surface has multiple maneuvers such as turning and accelerating, so that a single motion model is difficult to handle MAT. To this end, multiple models (MM) can be used to better represent the motion state of the airplane at different positions on the airport scene. The first generation MM method was pioneered by Magill [10], who proposed an adaptive method to estimate the random sampling process described by the initial unknown parameter vector. Its characteristic is that each element filter and all other element filters are independent. In a further method, Li et al. [11, 12] consider the switching between models and proposed a variable structure multiple model (VSMM) algorithm, which is developing rapidly and is becoming the state-of-the-art of MM estimation. In this work, we make use of the specific prior information in airport surface, e.g., airplane motion prior and surface structure prior, to build multiple motion models for effective MAT.

3 Proposed Method

The flowchart of the proposed method is shown in Fig. 1. Our method follows the tracking-by-detection framework, and there are two main components, such as the motion state MDP (msMDP) and the tracking state MDP (tsMDP).

The msMDP contains three sub-states to manage the motion-prior-based motion modeling for each airplane. Then, the motion prediction is combined with object detection to compute the affinity matrix. After the Hungarian algorithm solves the optimal assignment, the final tracking state of all targets can be updated with the tsMDP, where four sub-states are included to manage the lifetime of each airplane. Next, we firstly introduce the problem formulation in our method and then describe the details of each component.

3.1 Problem Formulation

3.1.1 Markov Decision Process Formulation

The MDP consists of the tuple $(\mathcal{S}, \mathcal{A}, T(\bullet, \bullet), R(\bullet, \bullet))$:

- The target state $S_i \in \mathcal{S}$ describes the state of each target.
- The action $A_i \in \mathcal{A}$ is a description of each target behavior.
- The Markov state transition function $T(\bullet, \bullet) : \mathcal{S} \times \mathcal{A} \rightarrow \mathcal{S}$ describes the impact of each action in current state.
- The reward function $R(\bullet, \bullet) : \mathcal{S} \times \mathcal{A} \rightarrow R$ defines the reward feedback after the action.

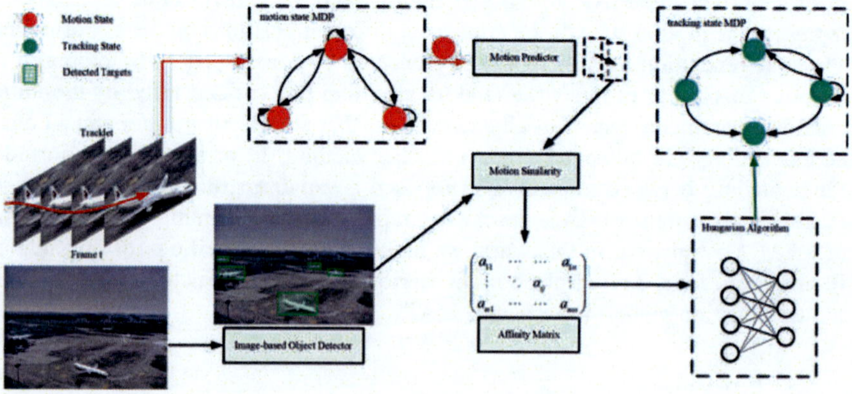

Fig. 1 Flowchart of DMDPs

In this paper, we use two Markov decision process to complete the tracking task, which can be formulated as follows:

$$S = S^M \cup S^T \tag{1}$$

where S^M is the motion state and S^T is the tracking state.

3.2 msMDP

Next, we introduce msMDP in detail from the target state, action and transfer function, and reward function.

3.2.1 Target State

As shown in Fig. 2, the motion state in msMDP is divided into three sub-states: straight/constant velocity, straight/constant acceleration, and curve/constant velocity, which can be formulated as follows:

$$S^M = S^M_{S/CA} \cup S^M_{S/CV} \cup S^M_{C/CV} \tag{2}$$

Figure 2 also shows the transition relationship between the three sub-states. Each state can maintain the existing state, and some states can switch to each other. When the airplane is accelerating in the straight line, the state can be transformed from "straight/constant velocity" to "straight/constant acceleration". When the airplane enters the curve from the straight, the state can be transformed from "straight/constant velocity" to "curve/constant velocity". The opposite also exists. In addition, because the inertia of the airplane is too large, it must maintain a constant speed before and

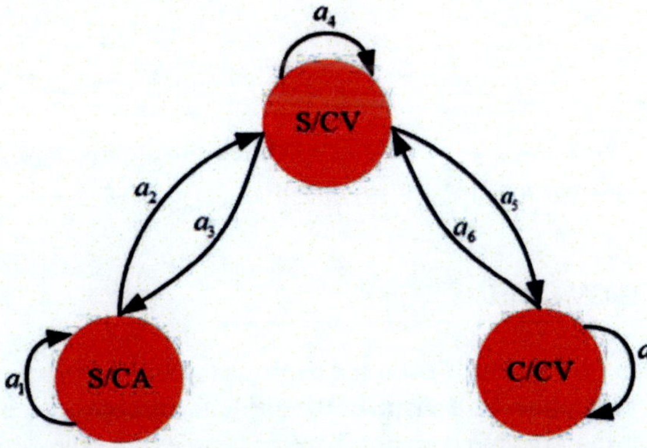

Fig. 2 msMDP in our framework

after turning, so there is no state transition between “straight/constant acceleration” and “curve/constant velocity”.

3.2.2 Actions and Transition Function

Seven possible transitions are designed between the states of each target. Given the current state and an action, we can transition to a new state for the target, which can be formulated as follows:

$$\mu_t^{S_j^M} = T(\mu_{t-1}^{S_i^M}, a_{S_i^M, S_j^M}) \tag{3}$$

where $\mu_t^{S_j^M}$ indicates that the target’s tracking state at time t is S_j^M, $\mu_{t-1}^{S_i^M}$ indicates that the target’s tracking state at time $t-1$ is S_i^M, $a_{S_i^M, S_j^M}$ denotes that executing action on a S_i^M state target transfer into S_j^M state, and $T(\bullet, \bullet)$ is the Markov state transition function.

3.2.3 Reward Function

Similar to the tracking state MDP, we solve the reward function to obtain the maximum reward value. Next, we introduce the decisions designed to S/CA sub-state, C/CV sub-state, and S/CV sub-state.

S/CA sub-state

In the S/CA sub-state, the decision is to transfer to the S/CV state or keep S/CA state. We define the state of each target as $\mathrm{x} = [x, y, s, r, \dot{x}, \dot{y}, \dot{s}]^T$, x, and y represent the horizontal and vertical pixel location of the center of the target, while the scale s and r represent the scale (area) and the aspect ratio of the target bounding box, respectively. We calculate the reward function as follows:

$$R_{S/\mathrm{CA}}(\mu_{t-1}^{S_i^M}, a_{S_i^M, S_j^M}) = \begin{cases} +\theta_{S/\mathrm{CA}}, \dot{x} \geq \dot{x}_{\mathrm{thd}} || \dot{y} \geq \dot{y}_{\mathrm{thd}} \\ -\theta_{S/\mathrm{CA}}, \dot{x} < \dot{x}_{\mathrm{thd}} \& \dot{y} < \dot{y}_{\mathrm{thd}} \end{cases} \tag{4}$$

where $(\dot{x}_{\mathrm{thd}}, \dot{y}_{\mathrm{thd}})$ are the specified thresholds. $\theta_{S/\mathrm{CA}} = +1$ if $a_{S_i^M, S_j^M} = a_{S_{S/\mathrm{CA}}^M, S_{S/\mathrm{CA}}^M}$, and $\theta_{S/\mathrm{CA}} = -1$ if $a_{S_i^M, S_j^M} = a_{S_{S/\mathrm{CA}}^M, S_{S/CV}^M}$ in Fig. 2.

C/CV sub-state

In the C/CV sub-state, the decision is to transfer to the S/CV state or keep C/CV state. Due to the rigid shape of the airplane, its appearance will dynamically change when it in curve movement. Therefore, we extract the appearance feature of the airplane by deep neural network. In this work, we design the reward function in a C/CV state with appearance feature representation:

$$R_{C/CV}(\mu_{t-1}^{S_i^M}, a_{S_i^M, S_j^M}) = \begin{cases} +\theta_{C/CV}, E_{\text{dis}}(F_t, F_{t-\tau_{\text{app}}}) > d_{\text{thd}} \\ -\theta_{C/CV}, E_{\text{dis}}(F_t, F_{t-\tau_{\text{app}}}) \le d_{\text{thd}} \end{cases} \quad (5)$$

where d_{thd} is a specified threshold, $E_{\text{dis}}(\bullet, \bullet)$ is the cosine distance, F_t is detection object feature, $F_{t-\tau_{\text{app}}}$ is the appearance feature of the $t - \tau_{\text{app}}$ frame on the tracklet, and τ_{app} is the time interval between the two frames used to compute $E_{\text{dis}}(\bullet, \bullet)$. $\theta_{C/CV} = +1$ if $a_{S_i^M, S_j^M} = a_{S_{C/CV}^M, S_{S/CV}^M}$, and $\theta_{C/CV} = -1$ if $a_{S_i^M, S_j^M} = a_{S_{C/CV}^M, S_{C/CV}^M}$ in Fig. 2.

S/CV sub-state

In the S/CV sub-state, the msMDP needs to decide whether to keep the target as S/CV, convert it to a S/CA state or C/CV state. Here, the solution of the reward function is similar to the reverse process of $R_{S/\text{CA}}$ and $R_{C/CV}$.

3.2.4 Motion Modeling

Since the msMDP includes three sub-states, there are three motion models to correspond to each sub-state. For S/CV and S/CA, Kalman filter and variable parameter Kalman filter are separately utilized to complete state estimation. Due to the rigid shape of the airplane, its appearance will dynamically change when the airplane is turning. To solve this problem, we perform motion prediction using SiameseRPN, which is a single object tracker based on template matching. It is an improvement of SiamFC network. Its feature extraction network is the same as SiamFC [13], but the difference is that it introduces the regional recommendation network (RPN) in the field of target detection. Therefore, it is suitable for tracking when the appearance of the target changes dynamically. Figure 3 shows the network structure of SiameseRPN single target tracker.

3.3 tsMDP

Next, we describe tsMDP briefly. This module is mainly followed by reference [14]. As shown in Fig. 4, the tracking state of an airplane is divided into four sub-states in tsMDP: tentative, tracked, lost, and deleted, which can be formulated as follows:

$$S^T = S_{\text{tentative}}^T \cup S_{\text{tracked}}^T \cup S_{\text{lost}}^T \cup S_{\text{deleted}}^T \quad (6)$$

The transition relationship has the four sub-states. The "tentative" is the initial state of any trajectory. Once a target is recognized by the pretrained object detector, set it to "tentative" state. When "tentative" state target has been matched for several

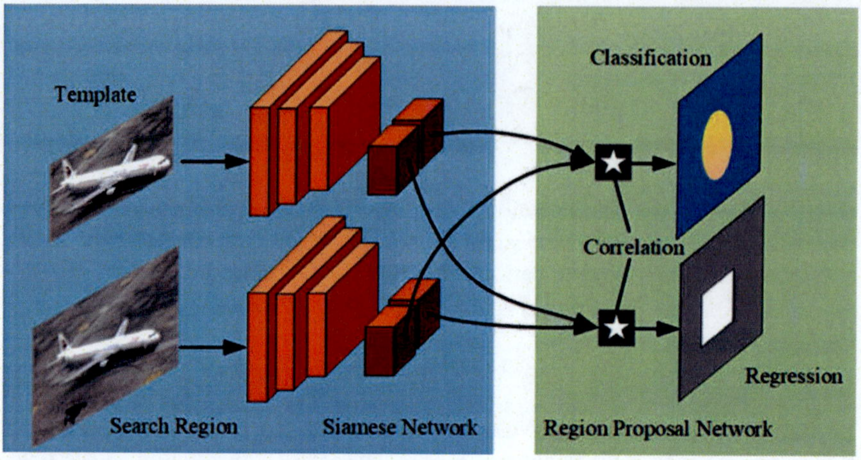

Fig. 3 Network architecture of the SiameseRPN

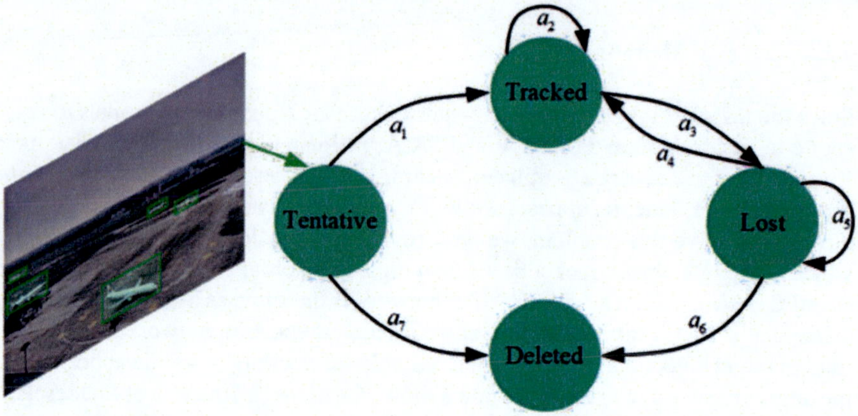

Fig. 4 tsMDP in our framework

consecutive frames, set it as the "tracked" state, if it is not matched, set it as the "deleted" state. A "tracked" target can keep "tracked" state, or convert to "lost" state due to occlusion or out of view. A "lost" target also can keep "lost" state, or convert back to "tracked" state if the target comes back into view, or convert to "deleted" if the target disappears for the preset time. Finally, the "deleted" is the final state of any trajectory.

4 Experiments

We complete extensive experiments on the AGVS-T1 dataset to evaluate the proposed multi-airplane tracker. In this section, we first introduce the implementation details of our method and the benchmark used for experiments and then compare the presented method with state-of-the-art algorithms.

4.1 Implementation Details

4.1.1 Object Detection and SimaeseRPN Tracker

Since our multi-airplane tracker is a tracking-by-detection framework, we use Faster R-CNN [15] and SDP [16] as our detector. Faster R-CNN uses ResNet101 [17] for feature extraction. SDP investigates two new strategies : scale-dependent pooling and layerwise cascaded rejection classifiers. In our experiment, we first train Faster R-CNN, SDP, and SimaeseRPN tracker on ImageNet [18] and then fine-tune the model on the AGVS datasets.

4.1.2 Data Association

Our data association is based on the Hungarian algorithm of affinity matrix. Firstly, the previous tracklet position is predicted by the msMDP. Then, we adopt the IoU distance to calculate the similarity between the predicted position and the current detection to form an affinity matrix. Finally, the Hungarian algorithm is used for data association according to the affinity matrix and update the tracklets.

4.2 Dataset

The AGVS-T1 benchmark dataset [2] is specially designed for industrial airport video surveillance. AGVS-T1 is a large-scale dataset, which includes 25 long videos (S1–S25) with a total of about 100,000 frames and accurate tracking ground truth. Besides common challenges in tracking such as occlusion and appearance changes, AGVS-T1 also contain some real challenges faced by airport surface tracking, e.g., haze, camouflage, shadow, and multi-scale airplanes. AGVS-T1 can be used to evaluate the performance of existing tracking algorithms in real airport environment.

4.3 Benchmark Evaluation

We evaluate the performance of our tracker on the AGVS-T1 benchmark, and some tracking results are shown in Fig. 5. Six state-of-the-art algorithms with public codes are chosen for comparison: Sort [19], CMOT [20], DeepSort [21], MDP [14], MHT [22], and MOTDT [23], and the comparison results are shown in Tables 1 and 2. Note that we use two object detectors, Faster R-CNN [15] and SDP [16], in combination with the trackers for evaluation. As shown, our method outperforms many previous state-of-the-art trackers in AGVS-T1 benchmark. Based on faster R-CNN detector, we improve 1.2% in MOTA, 0.7% in MOTP, 0.3% in IDF1, 1.4% in ML, 43 in ID Sw. compared with the second best published tracker. And based on SDP detector, we improve 1.3% in MOTA, 0.9% in MOTP, 0.8% in MT, 1.3% in ML, 44 in ID Sw. compared with the second best published tracker. All in all, the above performance improvement can show that our tracker is more robust than other trackers in dealing with the problems of target occlusion and shape rigidity.

In addition, we find that the performance of all comparison algorithms on AGVS-T1 is relatively better than that on other tracking datasets, e.g., MOT series. This may be because AGVS-T1 is relatively simple and contains only a single scene and a single object type, and the object motion speed is quite slow. In addition, there is less data for some challenge types in AGVS-T1, such as occlusion and various illumination change. Therefore, this dataset needs to be improved.

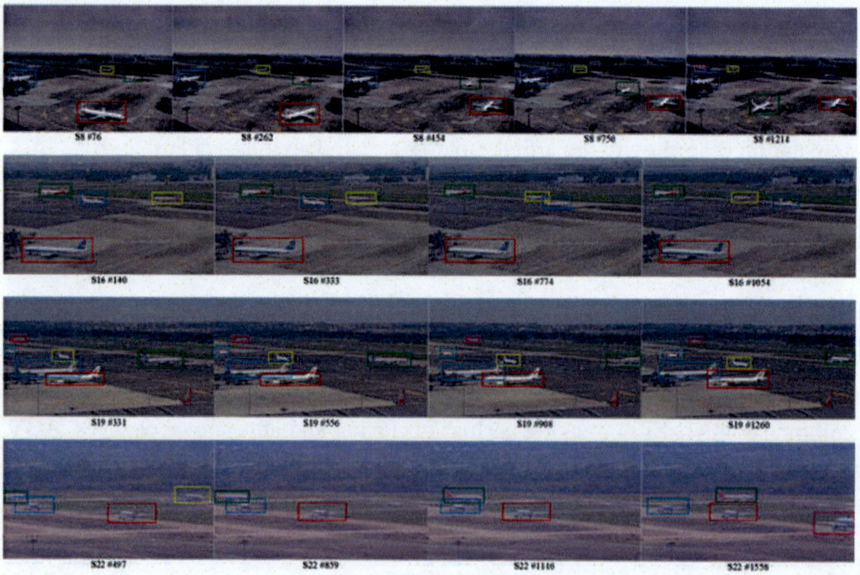

Fig. 5 Some tracking results by our method on AGVS-T1 benchmark. The same identity is labeled by box with the same color

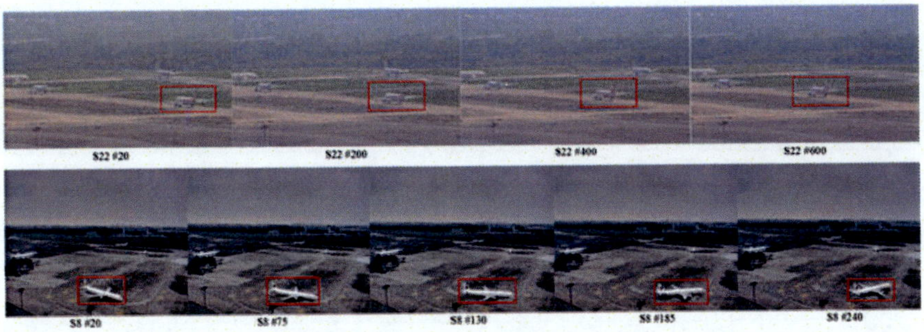

Fig. 6 Two examples of Kalman filter and siamese tracker on the AGVS-T dataset. Top row: the results of Kalman filter in straight motion. Bottom row: the results of SiamesePRN tracker in turning motion. Almost all airplanes are tracked correctly by our method

Table 1 Tracking performance on AGVS-T1 dataset

Detector	Tracker	MOTA (%)↑	MOTP (%)↑	IDF1 (%)↑	MT (%)↑	ML (%)↓	ID Sw.↓
Faster R-CNN [15]	SORT [19]	77.9	81.5	80.4	65.6	17.5	404
Faster R-CNN [15]	CMOT [20]	77.6	81.3	80.1	65.0	17.8	410
Faster R-CNN [15]	DeepSORT [21]	79.8	84.8	83.8	67.2	16.3	376
Faster R-CNN [15]	MDP [14]	78.6	83.3	82.6	66.1	18.0	356
Faster R-CNN [15]	MHT [22]	80.7	85.6	85.1	67.9	16.1	312
Faster R-CNN [15]	MOTDT [23]	81.7	88.2	85.8	**69.0**	14.9	314
Faster R-CNN [15]	**Ours**	**82.9**	**88.9**	**86.1**	68.6	**13.5**	**271**

Best scores are marked in bold

Table 2 Tracking performance on AGVS-T1 dataset

Detector	Tracker	MOTA (%)↑	MOTP (%)↑	IDF1 (%)↑	MT (%)↑	ML (%)↓	ID Sw.↓
SDP [16]	SORT [19]	78.4	83.3	81.5	66.1	16.8	391
SDP [16]	CMOT [20]	78.0	83.1	81.2	65.8	17.3	403
SDP [16]	DeepSORT [21]	80.3	85.3	84.6	68.0	15.6	358
SDP [16]	MDP [14]	79.3	83.5	83.3	66.8	17.4	341
SDP [16]	MHT [22]	81.3	86.2	86.0	68.8	15.1	301
SDP [16]	MOTDT [23]	82.5	88.5	**86.3**	69.4	14.3	306
SDP [16]	**Ours**	**83.8**	**89.4**	86.2	**70.2**	**13.0**	**262**

Best scores are marked in bold

Table 3 Ablation of DMDPs on the effectiveness of tsMDP on the AGVS-T1 datasets

Method	MOTA (%)	MOTP (%)	ML (%)	ID Sw.
DMDPs w/o a4	77.5	81.3	17.8	90
DMDPs	**81.6**	**86.1**	**12.8**	**68**

Best scores are marked in bold

Table 4 Ablation of DMDPs on the effectiveness of msMDP on the AGVS-T datasets

Method	MOTA (%)↑	MOTP (%)↑	MT (%)↑	ML (%)↓
B1	78.7 %	83.7	71.8	13.8
B2	75.4	80.9	66.7	16.6
B3	81.2	85.8	70.8	**12.5**
B4	**81.6**	**86.1**	**73.6**	12.8

Best scores are marked in bold

4.4 Ablation Study

In this work, the main purpose of our approach is to address target occlusion by tsMDP and solve the shape rigidity problems by msMDP in airplane tracking. To provide a transparent demonstration of the improvement of each component, we present the ablation study on the partial AGVS-T1 dataset, and all of experimental verification results were carried out under fair conditions.

4.4.1 tsMDP

In this part, the MDP of the tracking state successfully manages the lifetime of an airplane. To analyze the influence of the tracking state MDP in our tracker's performance, we conduct an experiment by disabling a component at one time and then examining the performance drop in terms of the MOTA score on the AGVS-T1 dataset.

We disable action a4 in lost state (Fig. 4). In this case, the target in the lost state cannot return to the tracking state. We can see a large amount of performance loss in Table 3. This can successfully prove the effectiveness of tsMDP.

4.4.2 msMDP

The msMDP can effectively deal with the tracking problem caused by the rigid shape of the airplane. In order to verify the contribution of four modules in our algorithm, we establish four baseline methods by using one module at a time. Each baseline method is described as follows: B1: We only use Kalman filter motion model; B2: We only use variable parameter Kalman filter motion model; B3: We only use SimaeseRPN tracker motion model; B4: The proposed method.

As shown in Table 4, the proposed method B4 has improved overall performance compared to B1, B2, and B3. Especially, the improvement in MOTA and MT can clearly show the effectiveness of the motion state MDP in handling the shape rigidity of the airplanes. Some visual examples of the Kalman filter and SiameseRPN tracker on the AGVS-T1 dataset are shown in Fig. 6.

5 Conclusion

In this paper, we have proposed a dual Markov decision processes (DMDPs) for multi-airplane tracking (MAT) problem. The core idea of our method was to make full use of the motion prior information in airports scene, that was, the motion patterns of the airplane could be predicted. Based on such prior information, we formulated the airplane motion state as decision making in Markov decision processes (MDP), where the movement of an airplane was modeled in a MDP with three sub-states (S/CV, S/CA, and C/CV). Furthermore, another MDP which modeled the lifetime of an airplane with four sub-states (tentative, tracked, lost, and deleted) was adopted to assist the first MDP under the DMDPs framework. Experimental results on both benchmark and in real scene demonstrated that our method was effective in airport surface surveillance, we also find that there are other prior knowledge related to the moving airplane, such as the ADS-B data including the geodetic coordinates of each airplane. Therefore, the future research direction is to exploit more prior information for better MAT.

Acknowledgements This work was supported by National Science Foundation of China (U19A2052, U1733111), National Key R&D Program of China (2021YFB1600500), Chengdu Science and Technology Project (2021-JB00-00025-GX), Key R&D Program of Sichuan Province (2020YFG0478), the Project of Quzhou Municipal Government (2021D012).

References

1. Bao, Q., Liu, W., Cheng, Y., Zhou, B., Mei, T.: Pose-guided tracking-by-detection: robust multi-person pose tracking. IEEE Trans. Multimedia **23**, 161–175 (2021)
2. Agvs. [online]. www.agvs-caac.com. Accessed May 2019
3. Kumar, K., Shrimankar, D.D.: F-des: fast and deep event summarization. IEEE Trans. Multimedia **20**(2), 323–334 (2018)
4. Kumar, K., Shrimankar, D.D.: Deep event learning boost-up approach: delta. Multimedia Tools Appl. **77**(20), 26635–26655 (2018)
5. Kumar, K., Shrimankar, D.D., Singh, N.: Event bagging: a novel event summarization approach in multiview surveillance videos. In: 2017 International Conference on Innovations in Electronics, Signal Processing and Communication (IESC), pp. 106–111 (2017)
6. Besada, J., Garcia, J., Portillo, J., Molina, J., Varona, A., Gonzalez, G.: Airport surface surveillance based on video images. IEEE Trans. Aerosp. Electron. Syst. **41**(3), 1075–1082 (2005)
7. Mian, A.S.: Realtime visual tracking of aircrafts. In: 2008 Digital Image Computing: Techniques and Applications, pp. 351–356 (2008)

8. Kalal, Z., Mikolajczyk, K., Matas, J.: Tracking-learning-detection. IEEE Trans. Pattern Anal. Mach. Intell. **34**(7), 1409–1422 (2012)
9. Singer, R.A., Stein, J.J.: An optimal tracking filter for processing sensor data of imprecisely determined origin in surveillance systems. In: IEEE Conference on Decision and Control, pp. 171–175 (1971)
10. Magill, D.: Optimal adaptive estimation of sampled stochastic processes. IEEE Trans. Autom. Control **10**(4), 434–439 (1965)
11. Rong Li, X.: Multiple-model estimation with variable structure. ii. model-set adaptation. IEEE Trans. Autom. Control **45**(11), 2047–2060 (2000)
12. Li, X.P., Youmin Zhang, Xiaorong Zhi: Multiple-model estimation with variable structure: model-group switching algorithm. In: IEEE Conference on Decision and Control, vol. 4, pp. 3114–3119 (1997)
13. Cen, M., Jung, C.: Fully convolutional siamese fusion networks for object tracking. In: 2018 25th IEEE International Conference on Image Processing (ICIP), pp. 3718–3722 (2018)
14. Xiang, Y., Alahi, A., Savarese, S.: Learning to track: Online multi-object tracking by decision making. In: IEEE International Conference on Computer Vision, pp. 4705–4713 (2015, December)
15. Ren, S., He, K., Girshick, R.B., Sun, J.: Faster R-CNN: towards real-time object detection with region proposal networks. IEEE Trans. Pattern Anal. Mach. Intell. **39**(6), 1137–1149 (2017)
16. Yang, F., Choi, W., Lin, Y.: Exploit all the layers: fast and accurate cnn object detector with scale dependent pooling and cascaded rejection classifiers. In: IEEE Conference on Computer Vision and Pattern Recognition, pp. 2129–2137 (2016)
17. He, K., Zhang, X., Ren, S., Sun, J.: Deep residual learning for image recognition. In: IEEE Conference on Computer Vision and Pattern Recognition, pp. 770–778 (2016, June)
18. Deng, J., Dong, W., Socher, R., Li, L., Kai Li, Fei-Fei, Li: Imagenet: a large-scale hierarchical image database. In: IEEE Conference on Computer Vision and Pattern Recognition, pp. 248–255 (2009)
19. Bewley, A., Ge, Z., Ott, L., Ramos, F., Upcroft, B.: Simple online and realtime tracking. In: IEEE International Conference on Image Processing, pp. 3464–3468 (2016)
20. Bae, S., Yoon, K.: Robust online multi-object tracking based on tracklet confidence and online discriminative appearance learning. In: IEEE Conference on Computer Vision and Pattern Recognition, pp. 1218–1225 (2014)
21. Wojke, N., Bewley, A., Paulus, D.: Simple online and realtime tracking with a deep association metric. In: IEEE International Conference on Image Processing, pp. 3645–3649 (2017, September)
22. Kim, C., Li, F., Ciptadi, A., Rehg, J.M.: Multiple hypothesis tracking revisited. In: IEEE International Conference on Computer Vision, pp. 4696–4704 (2015)
23. Chen, L., Ai, H., Zhuang, Z., Shang, C.: Real-time multiple people tracking with deeply learned candidate selection and person re-identification. In: IEEE International Conference on Multimedia and Expo, pp. 1–6 (2018, July)

URL Classification on Extracted Feature Using Deep Learning

Vishal Kumar Sahoo, Vinayak Singh, Mahendra Kumar Gourisaria, and Anuja Kumar Acharya

Abstract The widespread adoption of the World Wide Web (WWW) has brought about a monumental transition toward e-commerce, online banking, and social media. This popularity has presented attackers with newer opportunities to scam the unsuspecting—malicious URLs are among the most common forms of attack. These URLs host unsolicited content and perpetrate cybercrimes. Hence classifying a malicious URL from a benign URL is crucial to enable a secure browsing experience. Blacklists have traditionally been used to classify URLs, however, blacklists are not exhaustive and do not perform well against unknown URLs. This necessitates the use of machine learning/deep learning as they improve the generality of the solution. In this paper, we employ a novel feature extraction algorithm using 'urllib.parse', 'tld', and 're' libraries to extract static and dynamic lexical features from the URL text. IPv4 and IPv6 address groups and the use of shortening services are detected and used as features. Static features like https/http protocols used show a high correlation with the target variable. Various machine learning and deep learning algorithms were implemented and evaluated for the binary classification of URLs. Experimentation and evaluation were based on 450,176 unique URLs where MLP and Conv1D gave the best overall results with 99.73% and 99.72% accuracies and F1 Scores of 0.9981 and 0.9983, respectively.

Keywords Malicious URL detection · 1-Dimensional convolutional neural network (Conv1D) · Deep Learning · Machine learning · Multi-layer perceptron (MLP)

V. K. Sahoo · V. Singh · M. K. Gourisaria (✉) · A. K. Acharya
School of Computer Engineering, KIIT Deemed to be University, Bhubaneswar, Odisha 751024, India

V. K. Sahoo
e-mail: v.vk650@gmail.com

M. Tistarelli et al. (eds.), *Computer Vision and Machine Intelligence*,
Lecture Notes in Networks and Systems 586,
https://doi.org/10.1007/978-981-19-7867-8_33

1 Introduction

Resources on the Internet are located using the Uniform Resource Locator (URL). A malicious URL is a link devised to facilitate scam attacks and frauds. Clicking on such URLs can download malware that compromises a machine's security and the data stored within it or expose users to phishing attacks. With the rapid expansion of the Internet, there has been a rise in such URL-based cyber-attacks making malicious URL attacks the most popular form [1] of cyber-attack. There are two main components of an URL [2]. The protocol being used is denoted by the protocol identifier; The IP address or a domain name to locate resources is specified by the resource name. It implies the existence of a structure within URLs that attackers exploit to deceive users into executing malicious codes, redirecting users to unwanted malicious websites or downloading malware. While the following attack techniques use malicious URLs to attack their targets [3–5]. Phishing, Drive-by Download, and Spam. Phishing attacks refer to types of attacks that trick users into revealing sensitive information by posing as a genuine URL. Drive-by-Download attacks take place when malware is unintentionally downloaded on a system. Finally, spams are unsought for messages to promote or phishing.

The increase in malicious URL [1] attacks in recent years have necessitated research in the area to help detect and prevent malicious URLs. Blacklists and rule-based detection are the basis of most URL detection techniques [6]. While this form of detection achieves high accuracy, with the variety of attack types and new ways to attack users and the vast magnitude of scenarios in which such attacks can take it is hard to build a general solution and that's why blacklists do not perform well against newer URLs. These limitations have led researchers to employ machine learning techniques to classify a URL as malicious or benign [7–10] in search of a robust general solution. Machine learning and deep learning have contributed a lot in various fields like the diagnosis of tumors [11], diabetes classification [12], retinal disease detection [13], and support for digital agricultural systems [14], algorithms like Naïve Bayes and Support Vector Machine have been used to study twitter sentiments [15] and inferring chronic kidney failure [16]. In a Machine Learning approach, URLs are collected, out of which, features are extracted that can be interpreted mathematically by the model and learn a prediction function that can classify an unknown URL as malicious or benign.

In this paper, we use lexical features extracted from the URL and use machine learning algorithms like Logistic Regression, Random forest, K-Nearest Neighbor (KNN), XGBoost, Gaussian NB, Decision tree, Multi-Layer Perceptron(MLP) and a deep learning technique like 1-dimensional convolutional neural network(Conv1D). The structure of this paper is as follows. Section 2 reviews some recent works in the literature on malicious URL detection. Data preparation in Sect. 3. Section 4 describes the algorithms used. Experimental results are provided in Sect. 5. The paper concludes in Sect. 6.

2 Related Work

This section summarizes some of the recent work in the field of malicious URL classification.

2.1 Malicious URL Classification Based on Signature

Researchers have applied signature-based URL classification for a long time. Here a blacklist of known malicious URLs is established. Sun et al. [17] 2016 built AutoBLG. It automatically generates blacklists. Adding prefilters ensures that the blacklist is always valid. When an URL is accessed, a database query is generated against the blacklist. If the URL exists in the database, a warning is generated. Although this way of classification works well for known URLs, it quickly fails to classify unknown URLs, and it is tedious to update a blacklist.

Prakash et al. in 2006 [6] proposed a predictive blacklisting technique called PhishNet. It is a predictive blacklisting technique that detects phishing attacks. It exploits the fact the attacks change the structure of the URL when launching a phishing attack. PhishNet collected 1.55 million children and 6000 parent URL in 23 days. It was concluded that similar phishing URLs were 90% similar to each other. PhishNet is 80 times faster than Google's API on average. Like any blacklisting approach, this approach also suffers from the fact that blacklisting techniques simply do not perform well against unknown URLs that haven't previously been blacklisted.

A preliminary evaluation was presented in 2008 by Sinha et al. [18], where they evaluate popular blacklists with more than 7000 hosts from an academic site. It was concluded that these backlists have significant false negatives and false positives.

2.2 Machine Learning-Based URL Classification

URL datasets have trained on all three types of machine learning methods to achieve high accuracy even with unknown URLs. Supervised, Unsupervised, and Semi-Supervised methods have been used to classify URLs.

Ma et al. [19] in 2009 used statistical attributes from URLs and compared the proposed classification model with Gaussian NB, Logistic Regression models, and Support Vector Machine experiments. This approach gained an accuracy of 97%. Although high accuracies were achieved, this approach is not scalable.

Sahoo et al. in their study [2] in 2017 evaluated many models such as a URL dataset Naïve Bayes, Decision Trees, SVM, Logistic Regression, Random Forest, and Online Learning. It was studied that heavily imbalanced data can produce high results even if labels every URL is benign. In our paper, we properly balance the data to prevent false high accuracies. While Vundavalli et al. [20] in 2020 used

Logistic Regression, Neural Network, and Naïve Bayes on parameters like URL and domain identity to detect phishing websites in their paper. Naïve Bayes got the highest accuracy of 91%.

Aydin et al. [21] in 2020 presented methods for phishing URL detection. They extracted 133 separate features and then used Gain Ratio and ReliefF algorithms for feature selection. Naïve Bayes, Sequential Minimal Optimization, and J48 classifiers were used. They obtained modest results of 97.18% on J48 (Gain Ratio) and 98.47% on J48 (ReliefF). Bharadwaj et al. [22] in 2022 used GloVe to represent words as global vectors by considering global word-word co-occurrence statistics. They used SVM and MLP with statistical features and GloVe and obtained the highest accuracy of 89% with MLP.

In our paper, we extract lexical features including Character, Semantic groups, and Host-based groups. In Sect. 5 the accuracy of using said features will be given. Table 1. Summarizes all the related works cited in a tabular form.

3 Data Preparation

This section gives a detailed explanation of the dataset used. The feature extraction methods. The pre-processing steps that were taken to make the dataset training ready.

3.1 Dataset Used

The dataset used was published on Kaggle [23] in 2019 by Siddharth Kumar. The acquired dataset was collected from various sources such as PhisTank a database of phishing URLs and Malware Domains Blacklist which is a database of URLs that infect the system with malware. The dataset contains 450,176 unique URLs. 77% of said URLs are benign and 23% malicious. The dataset has 3 column attributes: 'index', 'url', 'label', 'result' (Fig. 1).

3.2 Feature Extraction

This dataset does not include any trainable features. Hence, we use the novel feature extraction algorithm to extract features from the given URLs in the dataset. The algorithm uses lamdas and iterates through all the URLs to extract the following features were extracted from the URL. Lexical features used:

Hostname Length: uses "urlparse" from the urllib.parse library to extract the hostname string from the URL using the "urlparse(i).netloc" where "i" is the current URL in the iteration. The string is then passed to the len() function to get the length of the string.

Table 1 Tabular representation of related work

Ref. No.	Author name	Techniques used	Description	Shortcomings
[2]	Sahoo et al. (2017)	Gaussian Naïve Bayes, Decision Trees, SVM, Logistic Regression, Random Forest, Online Learning	Evaluates many models on static features that were extracted from the URL	Does not explore dimensionality reduction techniques like PCA
[6]	Prakash et al. (2010)	PhishNet	Proposes a system to enumerate known phishing sites to discover new phishing URLs. Match dissected URLs against a blacklist using a matching algorithm. Fast	Can't discover newer URLs with different structures in the past. Low accuracy against unknown URLs
[16]	Sun et al. (2016)	AutoBGL	Automatically adds malicious URLs to a blacklist. Obtains high accuracy against blacklisted URLS	Very low accuracy against unknown URLs that have not been blacklisted
[18]	Ma et al. (2009)	Gaussian Naïve Bayes, Logistic Regression, Support Vector Machine	Extracted statistical features from URLs to use for training machine learning	The approach is not scalable. Did not compare with Trees-based models. Small dataset of only 30,000 URLs
[19]	Vundavalli et al. (2020)	Gaussian Naïve Bayes, Decision Trees, SVM, Logistic Regression, Random Forest, Online Learning	Explores URL detection using features like domain identity	No Deep Learning or Tree-based algorithm explored
[20]	Aydin et al. (2020)	Gaussian Naïve Bayes, Sequential Minimal Optimization, J48	Presents phishing URL detection methods using 36/58 extracted features	High feature space dimension. Modest accuracy of 97–98%
[21]	Bharadwaj et al. (2022)	Support Vector Machine, Artificial Neural Network	Used a new feature GloVe that represents words as a global vector. Reduced error by 63.33%	Very low accuracies

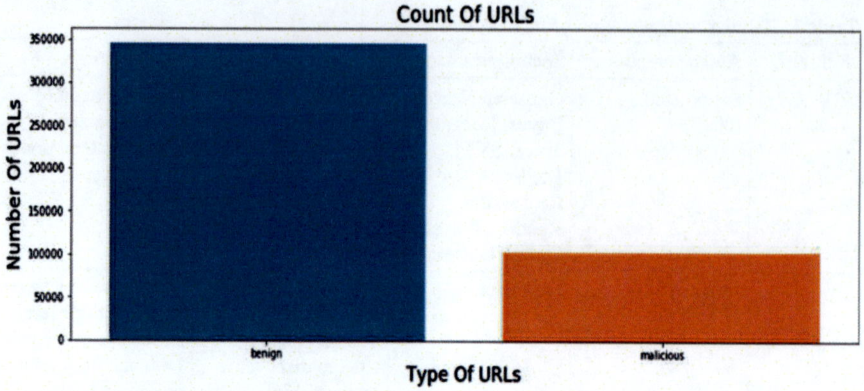

Fig. 1 Class distribution

Url Length: calculated by passing the URL string to len().
Path Length: Output of "urlparse(i).path" is passed to the len() function to calculate the length of the path of the URL.
First Directory Length: Output of "urlparse(i).path" is split using "urlpath.split('/')[1]", whose length is calculated using len().
'%' Count: "i.count('%')" where "i" is the URL string.
Top-Level Domain Length: Its length is calculated by counting the number of characters in the Top-Level Domain obtained the "get_tld" function from the "tld" library.
'@' Count: "i.count('@')" where "i" is the URL string.
'-' Count: "i.count('-')" where "i" is the URL string.
'?' Count: "i.count('?')" where "i" is the URL string.
'.' Count: "i.count('.')" where "i" is the URL string.
'=' Count: "i.count(' = ')" where "i" is the URL string.
Number of Directories: Count the number of "/" in the path of URL string.
'http' Count: "i.count('http')" where "i" is the URL string.
'www' Count: "i.count('www')" where "i" is the URL string.
Digits Count: Count of numeric characters from the URL string.
Letters Count: Count of alphabetical characters from the URL string.
Binary Features: Use of Shortening URL: Searches from a dictionary for shortening services using "search(dict, i)" from "re" library where "i" is the URL string and "dict" is the dictionary.
Use of IP or not: Searches from a dictionary of IP groups (IPv4 and IPv6) for IP groups using "search(dict, i)" from "re" library where "i" is the URL string and "dict" is the dictionary.

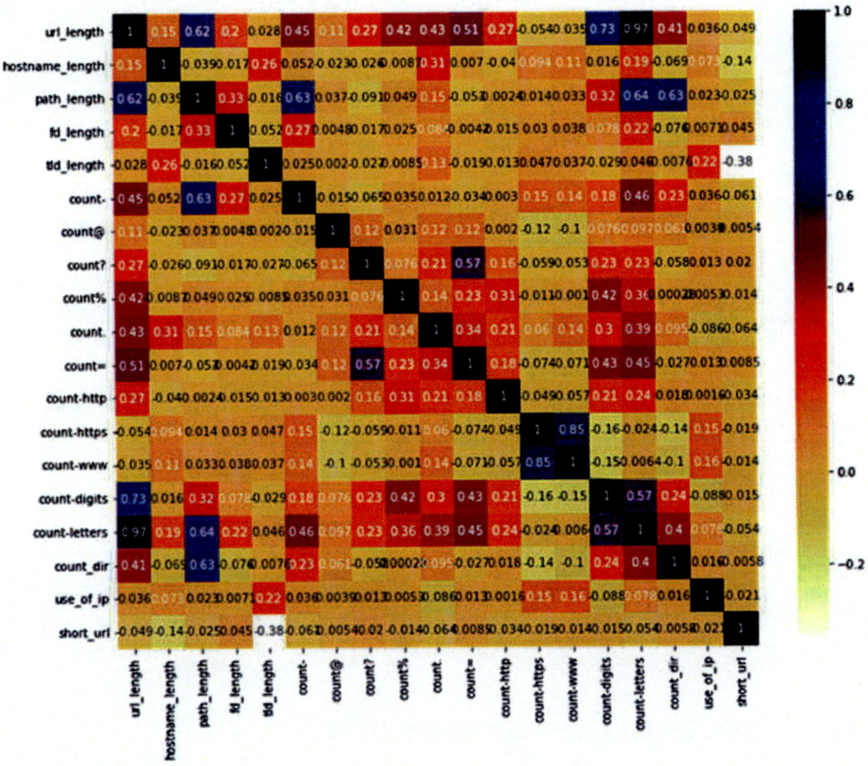

Fig. 2 Correlation matrix

3.3 Data Exploration

An essential pre-processing step. In this sub-section, we explore the dataset and perform feature selection. There were no missing values found on this dataset and there were no features that had zero variance. The feature called 'count-letters' had a high correlation of 0.97 with 'url-length' as seen in Fig. 2. It is higher than the threshold of 0.85. Hence we dropped the 'count-letters' column.

3.4 Over-Sampling

After splitting the data for training and testing, we oversample the training set to balance the unbalanced data to eliminate any bias toward benign URLs which would hamper learning. The 'malicious' URL class was randomly oversampled to match 75% of benign URLs in terms of quantity. SMOTE technique was not used in this study due to its generating random values nature based on the k values. This is

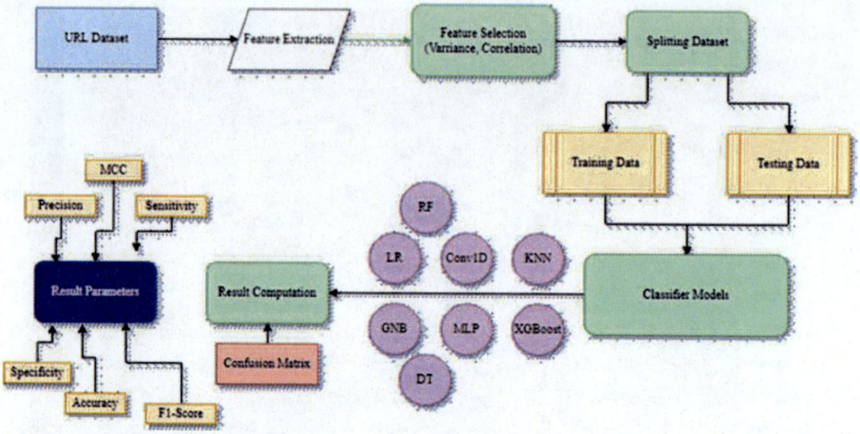

Fig. 3 Workflow for building and evaluating the models

undesirable because it may produce undesirable noise in the training data that may not be reflective of the real-world URLs. Figure 3 Illustrates the workflow used in this study.

3.5 Train-Test Split and Hardware Used

The dataset was split in an 80:20 ratio with a random state of 1. 80% of the data was used for training the models and 20% for testing. All the models were run on Jupyter Notebook environment with Python 3, Scikit-Learn, and Tensorflow 2.8 on a system running on Windows 11 with Intel i7 10th Generation and 16 GB RAM at 3200 MHz.

4 Technology Used

This paper employs many machine learning algorithms such as Decision tree, Logistic Regression, SVM, KNN, XGBoost, Random forest, Gaussian NB Multi-Layer Perceptron (MLP) and a deep learning technique like 1-dimensional convolutional neural network (Conv1D). These techniques have been used to predict liver disease [24], medical disease analysis [25] and for the early prediction of heart disease [26]. Table 2 briefly describes all the models used in this paper.

Table 2 Tabular illustration of classifier models used and their training parameters

Classifier	Description
Logistic regression (LR)	Utilizes a logistic function used for classifying categorical variables. They are statistical models. The sigmoid function implemented in linear regression is used in logistic regression. {Parameters: default}
XGBoost	Boosting technique based on random decision tree ensemble algorithm. Residual error updates after every tree learns from the preceding tree. Can handle sparse and also can implement regularization which prevents over-fitting. {Parameters: default}
Decision tree (DT)	Based on a flowchart-like tree structure. Each node is a test case, branches are the possible outcomes of the node, and terminates into a leaf (label). Breaks complex problems into simpler problems by utilizing a multilevel approach. {Parameters: criterion = 'entropy', random_state = 0, max_depth = None}
Random Forest (RF)	A form of bagging, non-parametric ensemble algorithm. It is effective at increasing test accuracy while also reducing over-fitting. {Parameters: default}
K-nearest neighbor (KNN)	Classification points are calculated using Euclidian distances to find the best K values for a given dataset. KNNs are very time-consuming. Follows a non-parametric method. {Parameters: n_neighbors = 9, metric = minkowski, p = 2}
Multi-layer perceptron (MPL)	Multi-Layer Perceptron (MLP) also called Artificial Neural Network (ANN) are fully interconnected neural networks where all the neurons are interconnected with each other. All the interconnected layers use an activation function, usually ReLU. Output layer activated using a sigmoid function in case of binary classification. Used in applications of face recognition [27], speech recognition [28], etc.
Gaussian naïve bayes (GNB)	Follows a Gaussian Normal system and the Bayes Theorem where we find the probability of one event occurring given that another already happened. {Parameters: default}

(continued)

Table 2 (continued)

Classifier	Description
1-Dimensional convolutional neural network (Conv1D)	The model consists of three Conv1D layers each paired with 1-dimensional max pooling with no padding. The output of the model is flattened and processed by 3 fully connected perceptron neural networks. Various real-world applications such as stock price prediction on daily stocks [29], pathfinding for mobile robots in unstructured environments [30] and automatically tagging music [31] and Mycobacterium tuberculosis detection [32]

5 Experimental Results

In this section, experimental results are discussed in detail. All the machine learning models, boosting techniques and deep learning models are evaluated against standard scoring metrics for a classification problem and the best mode is chosen based on the standard evaluation metrics. Metrics used are Sensitivity, Specificity, Precision, Matthews Correlation Coefficient (MCC), Negative Predicted Value (NPV), False Discovery Rate (FDR), False Negative Rate (FNR), F1 Score, and Accuracy.

Keras Tuner was used to go through all the permutations of hidden layers that can be used and found the best MLP architecture. In this architecture Tensorflow implicitly defines the initial input layer of 18 input neurons corresponding to the feature space. The input layer is followed by 6 fully connected hidden layers with 512, 256, 128, 32, 16 and 2 neurons, respectively. All the hidden layers use the ReLU activation function and the kernel initializer used is "he_uniform". Finally, there is an output layer of 1neuron activated by the sigmoid activation function for binary classification. The model was trained over 10 epochs over a batch size of 64 and a validation split of 0.2 was used. Binary Cross-Entropy loss function and Adamax optimizer were employed.

The Conv1D architecture has 3 1d-convolutional layers of units 128, 64 and 64, respectively, followed by a MaxPool1D layer of pool size 2 and stride 2. The architecture is then flattened and linked with 4 fully connected layers with 3471, 847, 847 and 521 neurons all activated with ReLU. Each fully connected layer undergoes Batch Normalization and Dropout at a rate of 0.5 to prevent over-fitting. Output layers contain 1 neuron with sigmoid activation. Figure 4a and b show Loss versus Epoch of MLP and Conv1D while training.

Table 3 presents the results of all the classifiers. All the metrics were calculated using the results from the confusion matrix generated upon the test set.

MLP achieves an accuracy of 99.73% accuracy because MLP can learn complex and non-linear relations feasibly. Random Forest and Conv1D get an accuracy score of 99.72% on the test set. On the other hand, Gaussian Naïve Bayes has the lowest accuracy of 99.18% among all the models tested, although Gaussian NB had the

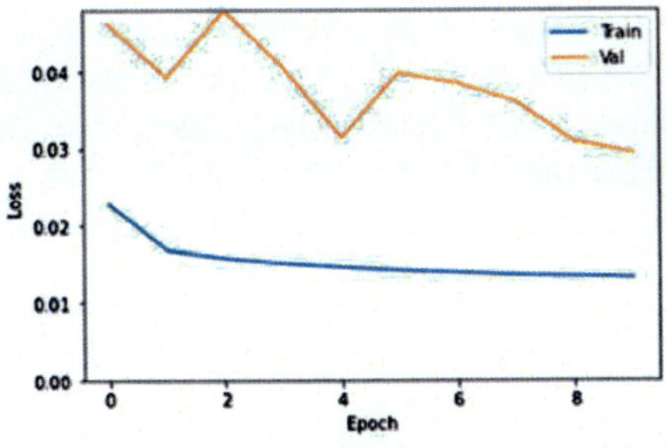

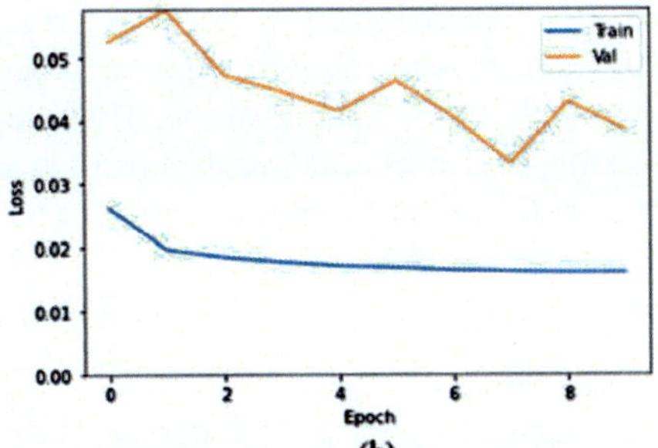

Fig. 4 **a** Loss versus Epoch during training (MLP) **b** Loss versus Epoch during training (Conv1D)

Table 3 Model performance table

Model	Accuracy	Precision	Sensitivity	Specificity	F1 Score	MCC	CM (2 × 2)	
LR	0.9966	0.9978	0.9978	0.9929	0.9978	0.9906	68,939	149
							153	20,795
Conv1D	0.9972	0.9988	0.9975	0.9962	0.9981	0.992	69,004	84
							171	20,777
KNN	0.9966	0.9979	0.9977	0.993	0.9925	0.9905	68,942	146
							158	20,790
DT	0.9961	0.9975	0.9975	0.9916	0.9975	0.9892	68,912	176
							171	20,777
RF	0.9972	0.9985	0.9979	0.9949	0.9982	0.9922	68,981	107
							149	20,799
GNB	0.9918	0.9913	0.998	0.972	0.9947	0.9774	68,488	600
							134	20,814
XGBoost	0.9974	0.9987	0.9979	0.9957	0.9983	0.9927	68,999	89
							144	20,804
MLP	0.9973	0.999	0.9975	0.9967	0.9983	0.9925	69,020	68
							173	20,775

highest Sensitivity score of 0.998 followed by XGBoost and Random Forest with a score of 0.9979. MLP takes the highest scores in Specificity and Precision of 0.9967 and 0.999, respectively. Gaussian NB has the lowest Specificity score (0.972) and the lowest Precision score of 0.9913. MLP and XGBoost have the highest F1 Score and MCC Score of 0.9983 and 0.9927 (0.9925 for MLP), respectively. Gaussian NB scores the lowest in FNR of 0.002 and the highest NPV score of 0.9936. Finally, MLP has the lowest FDR score of 0.001 followed by Conv1D. From Table 3. it can be inferred that Conv1D is a close competitor of MLP. A confusion matrix (CM) is a table that is used to define the performance of a classification algorithm. A confusion matrix visualizes and summarizes the performance of a classification algorithm. All the metrics were derived from the given CM (2X2). FN, TP, TN, and FP stand for False Negative, True Positive, True Negative, and False Positive, respectively. On testing the feature extraction algorithm and the classifiers on a different URL dataset. Conv1D had the highest accuracy of 92.47% followed by Logistic Regression with an accuracy of 90.31%.

6 Conclusion and Future Work

In this paper, lexical features were extracted from URLs and used to train various models. The empirical results show us that MLP is the best overall performing model across all metrics. This is true because of MLP's ability to learn complex and non-linear relations easily. The techniques described in this paper can be implemented on IT security systems. Decision Tree was the worst classifier model among all the models based on the standard metric results tested in this paper.

For future work, we would develop a self-feature-extracting neural network model that could provide an end-to-end solution with high accuracy and implement recurrent neural networks like LSTM and gated recurrent units to possibly obtain higher accuracies with low training times and computational cost. These models do not account for the new font-based URL attacks. Further incorporation of font-based URLs in training data could help in the detection of this type of attack.

References

1. Internet Security Threat Report (ISTR) 2019–Symantec.: https://www.symantec.com/content/dam/symantec/docs/reports/istr-24-2019-en.pdf. Last Accessed 17 Mar 2022
2. Sahoo, D., Liu, C., Hoi, S.C.: Malicious URL detection using machine learning: a survey (2017). arXiv preprint arXiv:1701.07179
3. Khonji, M., Iraqi, Y., Jones, A.: Phishing detection: a literature survey. IEEE Commun. Surv. Tutorials **15**(4), 2091–2121 (2013)
4. Cova, M., Kruegel, C., Vigna, G.: Detection and analysis of drive-by-download attacks and malicious JavaScript code. In Proceedings of the 19th International Conference on World Wide Web, pp. 281–290. (2010)
5. Heartfield, R., Loukas, G.: A taxonomy of attacks and a survey of defence mechanisms for semantic social engineering attacks. ACM Comput. Surv. (CSUR) **48**(3), 1–39 (2015)
6. Prakash, P., Kumar, M., Kompella, R.R., Gupta, M.: Phishnet: predictive blacklisting to detect phishing attacks. In: 2010 Proceedings IEEE INFOCOM, pp. 1–5. IEEE (2010)
7. Garera, S., Provos, N., Chew, M., Rubin, A.D.: A framework for detection and measurement of phishing attacks. In: Proceedings of the 2007 ACM Workshop on Recurring Malcode, pp. 1–8. (2007)
8. Khonji, M., Jones, A., Iraqi, Y.: A study of feature subset evaluators and feature subset searching methods for phishing classification. In: Proceedings of the 8th Annual Collaboration, Electronic Messaging, Anti-Abuse and Spam Conference, pp. 135–144. (2011)

9. Kuyama, M., Kakizaki, Y., Sasaki, R.: Method for detecting a malicious domain by using whois and dns features. In: The Third International Conference on Digital Security and Forensics (DigitalSec2016), vol. 74 (2016)
10. Ma, J., Saul, L.K., Savage, S., Voelker, G.M.: Learning to detect malicious urls. ACM Trans. Intell. Syst. Technol. (TIST) **2**(3), 1–24 (2011)
11. Singh, V., Gourisaria, M.K., Harshvardhan, G.M., Rautaray, S.S., Pandey, M., Sahni, M., ... Espinoza-Audelo, L.F.: Diagnosis of intracranial tumors via the selective CNN data modeling technique. Appl. Sci. **12**(6), 2900 (2022)
12. Das, H., Naik, B., Behera, H.S.: Classification of diabetes mellitus disease (DMD): a data mining (DM) approach. In: Progress in Computing, Analytics and Networking, pp. 539–549. Springer, Singapore (2018)
13. Sarah, S., Singh, V., Gourisaria, M.K., Singh, P.K.: Retinal disease detection using CNN through optical coherence tomography images. In 2021 5th International Conference on Information Systems and Computer Networks (ISCON), pp. 1–7. IEEE (2021)
14. Panigrahi, K.P., Sahoo, A.K., Das, H.: A cnn approach for corn leaves disease detection to support digital agricultural system. In: 2020 4th International Conference on Trends in Electronics and Informatics (ICOEI), vol. 48184, pp. 678–683. IEEE (2020)
15. Chandra, S., Gourisaria, M.K., Harshvardhan, G.M., Rautaray, S.S., Pandey, M., Mohanty, S.N.: Semantic analysis of sentiments through web-mined twitter corpus. In CEUR Workshop Proceedings, vol. 2786, pp. 122–135. (2021)
16. Pramanik, R., Khare, S., Gourisaria, M.K.: Inferring the occurrence of chronic kidney failure: a data mining solution. In: Gupta, D., Khanna, A., Kansal, V., Fortino, G., Hassanien, A.E. (eds.) Proceedings of Second Doctoral Symposium on Computational Intelligence. Advances in Intelligent Systems and Computing, vol. 1374, Springer, Singapore (2022)
17. Sun, B., Akiyama, M., Yagi, T., Hatada, M., Mori, T.: Automating URL blacklist generation with similarity search approach. IEICE Trans. Inf. Syst. **99**(4), 873–882 (2016)
18. Sinha, S., Bailey, M., Jahanian, F.: Shades of grey: on the effectiveness of reputation-based "blacklists". In: 2008 3rd International Conference on Malicious and Unwanted Software (MALWARE), pp. 57–64. IEEE (2008)
19. Ma, J., Saul, L.K., Savage, S., Voelker, G.M.: Beyond blacklists: learning to detect malicious web sites from suspicious URLs. In: Proceedings of the 15th ACM SIGKDD International Conference on Knowledge Discovery and Data Mining, pp. 1245–1254. (2009)
20. Vundavalli, V., Barsha, F., Masum, M., Shahriar, H., Haddad, H.: Malicious URL detection using supervised machine learning techniques. In: 13th International Conference on Security of Information and Networks, pp. 1–6. (2020)
21. Aydin, M., Butun, I., Bicakci, K., Baykal, N.: Using attribute-based feature selection approaches and machine learning algorithms for detecting fraudulent website URLs. In: 2020 10th Annual Computing and Communication Workshop and Conference (CCWC), pp. 0774–0779. IEEE (2020)
22. Bharadwaj, R., Bhatia, A., Chhibbar, L. D., Tiwari, K., Agrawal, A.: Is this url safe: detection of malicious urls using global vector for word representation. In: 2022 International Conference on Information Networking (ICOIN), pp. 486–491. IEEE (2022)
23. https://www.kaggle.com/datasets/siddharthkumar25/malicious-and-benign-urls. Last Accessed 3 Mar 2022
24. Singh, V., Gourisaria, M.K., Das, H.: Performance analysis of machine learning algorithms for prediction of liver disease. In: 2021 IEEE 4th International Conference on Computing, Power and Communication Technologies (GUCON), pp. 1–7. IEEE (2021)
25. Das, H., Naik, B., Behera, H.S.: Medical disease analysis using neuro-fuzzy with feature extraction model for classification. Inform. Med. Unlocked **18**, 100288 (2020)
26. Sarah, S., Gourisaria, M.K., Khare, S., Das, H.: Heart disease prediction using core machine learning techniques—a comparative study. In: Advances in Data and Information Sciences, pp. 247–260. Springer, Singapore (2022)
27. Magesh Kumar, C., Thiyagarajan, R., Natarajan, S.P., Arulselvi, S., Sainarayanan, G.: Gabor features and LDA based face recognition with ANN classifier. In: 2011 International Conference on Emerging Trends in Electrical and Computer Technology, pp. 831–836. IEEE (2011)

28. Wijoyo, S., Wijoyo, S.: Speech recognition using linear predictive coding and artificial neural network for controlling the movement of a mobile robot. In: Proceedings of the 2011 International Conference on Information and Electronics Engineering (ICIEE 2011), Bangkok, Thailand, pp. 28–29. (2011)
29. Jain, S., Gupta, R., Moghe, A.A.: Stock price prediction on daily stock data using deep neural networks. In: 2018 International Conference on Advanced Computation and Telecommunication (ICACAT), pp. 1–13. IEEE (2018)
30. Visca, M., Bouton, A., Powell, R., Gao, Y., Fallah, S.: Conv1D energy-aware path planner for mobile robots in unstructured environments. In: 2021 IEEE International Conference on Robotics and Automation (ICRA), pp. 2279–2285. IEEE (2021)
31. Kim, T., Lee, J., Nam, J.: Sample-level CNN architectures for music auto-tagging using raw waveforms. In: 2018 IEEE International Conference on Acoustics, Speech and Signal Processing (ICASSP), pp. 366–370. IEEE (2018)
32. Singh, V., Gourisaria, M.K., Harshvardhan, G.M., Singh, V.: Mycobacterium tuberculosis detection using CNN ranking approach. In: Gandhi, T.K., Konar, D., Sen, B., Sharma, K. (eds.) Advanced Computational Paradigms and Hybrid Intelligent Computing. Advances in Intelligent Systems and Computing, vol. 1373. Springer, Singapore (2022)

Semi-supervised Semantic Segmentation for Effusion Cytology Images

Shajahan Aboobacker, Deepu Vijayasenan, S. Sumam David, Pooja K. Suresh, and Saraswathy Sreeram

Abstract Cytopathologists analyse images captured at different magnifications to detect the malignancies in effusions. They identify the malignant cell clusters from the lower magnification, and the identified area is zoomed in to study cell level details in high magnification. The automatic segmentation of low magnification images saves scanning time and storage requirements. This work predicts the malignancy in the effusion cytology images at low magnification levels such as 10× and 4×. However, the biggest challenge is the difficulty in annotating the low magnification images, especially the 4× data. We extend a semi-supervised learning (SSL) semantic model to train unlabelled 4× data with the labelled 10× data. The benign F-score on the predictions of 4× data using the SSL model is improved 15% compared with the predictions of 4× data on the semantic 10× model.

Keywords Effusion cytology · Deep neural network · Semantic segmentation · Semi-supervised learning

1 Introduction

Cytopathologists examine the effusion samples visually through a microscope to detect the malignant cells. Automation of malignancy detection in effusion cytology images helps to minimize the manual work. Most of the literature addresses detection of malignancy in high magnification images such as 40× [2, 16, 18, 19]. However, cytopathologists usually observe the sample through the microscope at different objective magnifications. One advantage of images captured at a lower magnification is that they cover a larger area in a single frame than images taken at higher magnification [6, 13]. They also examine the morphological and textural behaviour

S. Aboobacker · D. Vijayasenan (✉) · S. S. David
National Institute of Technology Karnataka, Surathkal, Karnataka 575025, India
e-mail: deepu.senan@gmail.com

P. K. Suresh · S. Sreeram
Kasturba Medical College Mangalore, Manipal Academy of Higher Education, Manipal, Karnataka 575001, India

M. Tistarelli et al. (eds.), *Computer Vision and Machine Intelligence*,
Lecture Notes in Networks and Systems 586,
https://doi.org/10.1007/978-981-19-7867-8_34

of the clusters of cells at the lower magnification [3, 5, 14]. Cytopathologists typically identify the malignant regions from the lower magnification. The identified area is zoomed in to study the cell level details at higher magnification [6, 13]. Similarly, in automatic segmentation, one could identify the potentially malignant region in low resolution. This information can be used to perform a higher magnification scan of the malignant area alone to save scanning time and memory.

However, automatic segmentation at lower magnification is challenging because of missing and/or blurring features such as nuclei and texture. Therefore, reliable annotations are not possible at a low magnification level. There are techniques for identifying the region of interest in low magnification images [21]. In contrast, this method utilizes whole slide images (WSIs) that are publicly accessible and labelled at a higher magnification. The WSI, which is already labelled, is used to generate lower magnification images. However, labelling a new dataset at the lowest magnification is challenging.

This work predicts the malignancy of effusion cytology images at lower magnification levels (10× and 4×) rather than at a higher magnification level such as 40×. Sample images are shown in Fig. 1. It can be seen that individual cells are easily distinguishable in 40× data, while in the low magnification images, only clusters of cells can be observed. We have five classes: benign, malignant, cytoplasm around the cells, individual cells that we name as the isolated class, and the background. Each pixel in the image is classified into one of these classes. This type of classification is called semantic segmentation. We were able to train a conventional semantic system for 10× images. But, it was not possible for 4× images because labelling low-resolution images itself was impossible. We extend a semi-supervised model (SSL) for the semantic segmentation of 4× data. The SSL model includes the unlabelled 4× along with the labelled 10× data for the training.

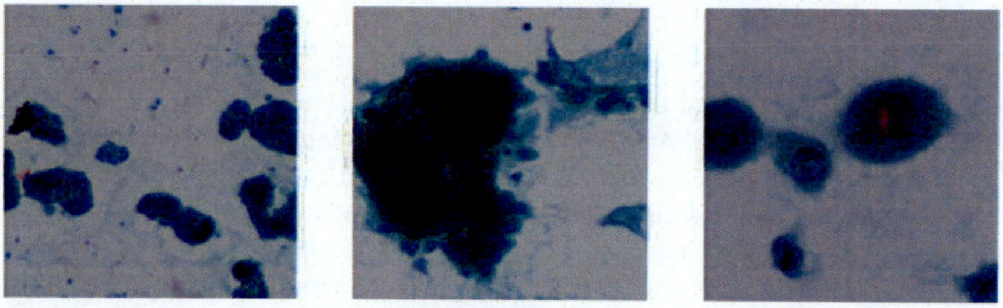

(a) 4X (b) 10X (c) 40X

Fig. 1 Effusion cytology images at different magnifications

2 Semi-supervised Learning

Semi-supervised learning (SSL) utilizes the unlabelled data as additional information to the labelled data or vice versa [17]. Even if the unlabelled data are not in abundance, it can provide additional information which helps training the labelled data to improve classification performance [17].

The basic principles of SSL are based on assumptions like data being smooth in a low-dimensional manifold and sparse on class boundaries. The smoothness assumption assumes any two input points that are close by should have the same labels. The manifold assumption tells data points on the same low-dimensional manifold to have the same label. The low-density assumption is also related to the smoothness assumption, where the decision boundary should pass through the low-density region, hence minimum entropy [17].

One common approach to semi-supervised learning is to add a loss term for the unlabelled data. This loss term, based on the above assumptions, employs entropy minimization and consistency regularization. This results in better generalization of the trained model [4].

In self-training methods such as pseudo-label [9] and noisy student [20], pseudo-labels are generated by predicting a batch of unlabelled images and selecting the maximum confidence class. The model is trained for both labelled and unlabelled images with a cross-entropy loss. Consistency regularization is employed in temporal ensembling [11], pi-Model [11], virtual adversarial training [10] and mean teacher [15]. Random augmentations on the images or perturbations such as dropout on the model are performed. Then, a mean squared difference is computed between the original and perturbed unlabelled data predictions and a cross-entropy loss for the labelled data. In the hybrid method, such as MixMatch [4] and FixMatch [12], pseudo-labels are generated, along with consistency regularization.

In this work, we adapt the MixMatch algorithm [4] for a semantic segmentation problem. The MixMatch algorithm for classification performs SSL as follows: K different random augmentations are performed for each of the unlabelled images. Predictions are computed for each image for all augmentations and are averaged. A temperature sharpening, as shown in equation (1), is applied to this averaged image to get the pseudo-label.

$$\text{Sharpen}(\tilde{y}_{u_i}, T) = \tilde{y}_{u_i}^{\frac{1}{T}} \Big/ \sum_{j=1}^{C} \tilde{y}_{u_j}^{\frac{1}{T}} \tag{1}$$

where $\tilde{y}_{u_i}$ is the prediction value for each class. C is the no. of classes and T is the temperature.

The total loss is calculated as

$$\mathcal{L}_{\text{Total}} = \mathcal{L}_{\text{CE}} + \lambda\, \mathcal{L}_{\text{MSE}} \tag{2}$$

Cross-entropy loss $\mathcal{L}_{\mathrm{CE}}$ is used for labelled images, and the mean squared error $\mathcal{L}_{\mathrm{MSE}}$ is used for the unlabelled images. This algorithm outperforms all other methods in the classification of the CIFAR and SVHN data sets [4]. However, MixMatch addresses only classification problems. We extend it to perform semantic segmentation.

3 SSL for Semantic Segmentation

Semantic segmentation is the process of classifying each pixel of an image into different object classes. When the input images are augmented, corresponding label images are also spatially altered. Therefore, averaging the predicted output for MixMatch will lead to incorrect results. We introduce reverse augmentation of the predicted label to address this issue. This ensures that all the predicted labels are spatially aligned.

The algorithm performs multiple random augmentations on the data and creates pseudo-labels. These pseudo-labels are used to perform SSL with a MobileUNet model using the loss function in equation (5). The two individual steps of SSL are detailed below.

3.1 Pseudo-Label Generation

Pseudo-labels for the unlabelled data are generated from different augmented versions of the same image. These augmentations are randomly selected from the given set of augmentations, such as horizontal flip, vertical flip, height-width shift, zoom and crop. Each unlabelled image is augmented K times. Each of these images is passed through the model to predict a label image. This label image is spatially altered (e.g. shifted, zoomed in/out, ...) with respect to the input image as seen in Fig. 2. These label images cannot be averaged as they do not have spatial correspondence. Hence, we have to reverse the augmented images back to their normal size and position before averaging the predicted outputs. We compute parameters required for the reverse augmentations from the parameters that are used for performing the individual random augmentations.

The reverse augmented predicted images are then averaged to obtain a single image. This image is sharpened to reduce the entropy of label distribution. The output of the sharpening function in equation (1) is used as the pseudo-label. Boundary pixels are affected during the generation of pseudo-labels. The images are padded with zeros around the boundaries to address the boundary problem. The padding helps to restore the pixels extended outside the boundary portion during the augmentation, such as shifting and zooming out. The zeros padded at each boundary side are calculated based on the shifting and zooming parameters. The predictions of the padded portion are classified to one of the available classes as it does not belong to any of the classes.

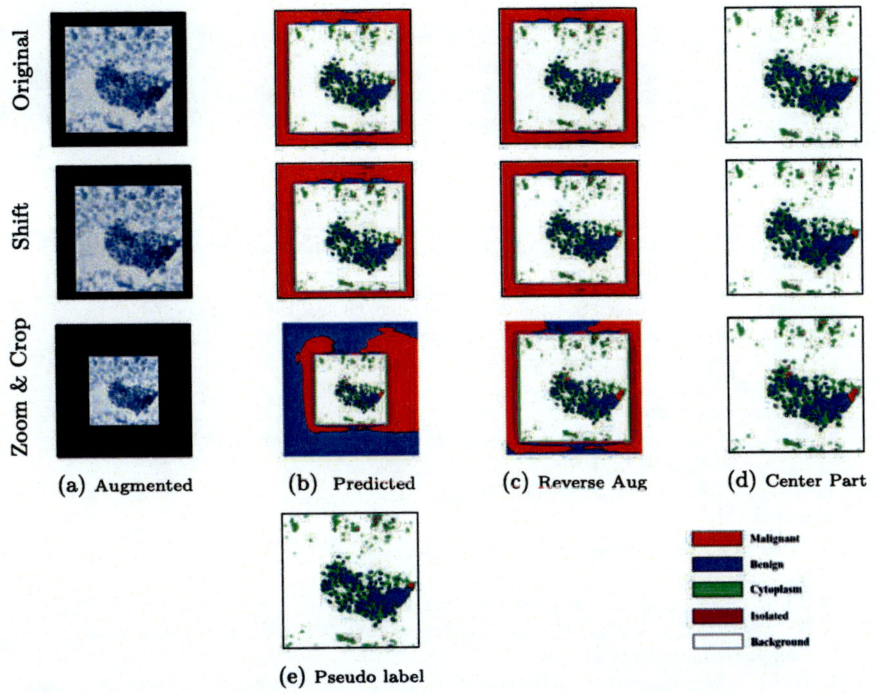

Fig. 2 Different types of augmentation and corresponding reverse augmentation for a sample image

However, the padded portion is removed before generating the pseudo-label, and only, the centre part is used for calculating the loss. Hence, the augmentation step is free from any errors.

3.2 Network Training

During the training process, the model is trained with both labelled and unlabelled data. The labelled data (x_l) are trained using pixel-wise cross-entropy as a loss function with the one-hot encoded label (y_l). Pseudo-labels (y_u) are created for the unlabelled data (x_u) as explained in Sect. 3.1. This pseudo-label is used for an unlabelled batch in training with pixel-wise mean squared error as a loss function. The combined loss function is the weighted sum of this mean square error and cross-entropy for labelled data. It is defined as:

$$\mathcal{L}_s = \frac{1}{B} \sum_{p \in B} H(y_l, P_{model}(\hat{y}_l | \hat{x}_l; \theta) \tag{3}$$

$$\mathcal{L}_u = \frac{1}{KB} \sum_{p \in KB} \|(y_u - P_{model}(\hat{y}_{u,k}|\hat{x}_{u,k};\theta)\|_2^2 \tag{4}$$

$$\mathcal{L} = \mathcal{L}_s + \lambda_u \, \mathcal{L}_u \tag{5}$$

Algorithm 1: Proposed method for semi-supervised semantic segmentation of effusion cytology images

---

Input: Labelled Images (x_l) and their labels (y_l), Unlabelled images (x_u), Number of Augmentations (K), Augmentation parameters for labelled data (ϕ_l), Augmentation parameters for unlabelled data($\phi_{u,k}$), Sharpening Temperature (T)

Output: $P_{model}(y|x;\theta)$

for $k = 1$ **to** K **do**

  $\hat{x}_{u,k} = Augment(x_u, \phi_{u,k})$ K Random Augmentations of unlabelled image

  $y_{u,k} = P_{model}(y|\hat{x}_{u,k};\theta)$

  $\tilde{y}_{u,k} = ReverseAugment(y_{u,k}, \phi_{u,k})$

end for

$\tilde{y}_u = \frac{1}{K}\sum_k \tilde{y}_{u,k}$ Averaging the predictions

$y_u = \text{Sharpen}(\tilde{y}_u, T)$ Pseudo Label

$X_L = (x_l, y_l)$ Labelled image and its label

$X_U = (\hat{x}_{u,k}, \hat{y}_u)$ $k \in (1, 2, ...K)$ K augmentations of an unlabelled image and its augmented pseudo label

Update the parameters of the model $\mathcal{L}_s$ and $\mathcal{L}_u$ are supervised and unsupervised loss respectively

P_{model} *by* $\mathcal{L} = \mathcal{L}_s + \lambda_u \, \mathcal{L}_u$ λ_u is the weightage of unsupervised loss

---

whereas $\mathcal{L}_s$ is the supervised loss, $\mathcal{L}_u$ is the unsupervised loss, and the hyperparameter λ_u is the weightage of unsupervised loss in the combined loss function. B is the batch size, K is the number of augmentations in the unlabelled batch, and p is every pixel in the batch. $H(y_l, \hat{y}_l)$ is the cross-entropy between y_l, $\hat{y}_l$ distributions.

The proposed method is provided in Algorithm 1.

4 Experiments and Results

To evaluate the extended model, we compare the results of SSL with the baseline supervised model. This section explains the dataset, results of the baseline model and the semi-supervised model.

4.1 Data Set

The data set consists of 345 effusion cytology images of 30 patients with the size 1920 × 1440 pixels, of which 212 are at 40× magnification, 83 are at 10× magni-

Table 1 Number of sub-images in training, validation and testing

Dataset	Training	Validation	Testing
10×	4955	131	192
4× resampled	7319	677	419

fication, and the remaining 50 are at 4× magnification. Images contain clusters of benign and malignant cells and isolated cells. We annotated the 10× images in a semi-automatic fashion. Initially, using the help of 40× data and its label, a Gaussian mixture model (GMM) is created and predicted for 10× images with multiple numbers of components and manually corrected the label according to the understanding of the cluster. Subsequently, the annotations are refined and validated by a cytopathologist. The 4× data are not annotated as it was challenging to differentiate the isolated cells and cluster of cells. The cytoplasm labelling is also challenging at this magnification level.

The 10× and 4× data are randomly partitioned into training (80), validation (26), and test (27) sets. The images are cropped into sub-images to handle the memory management in the computation process. The sub-images which have less than 5% foreground pixels are ignored. The number of sub-images in each set is detailed in Table 1.

We consider 10× data as labelled and 4× data as unlabelled. The 4× data are upsampled to 10× magnification. Here, we have two data sets, the original 10× data and the 10× upsampled from 4× data, which we named as 4× resampled.

4.2 Metrics

The pixel-wise classification performance is evaluated using F-score, as defined in equation (8), the harmonic mean of precision and recall. Reporting the overall F-score is irrelevant as the number of background pixels is very high compared to the foreground pixels. Therefore, we calculated the F-score without considering the background class pixels and named it foreground F-score. The benign and malignant F-scores are also important; hence, we calculate the F-scores of benign and malignant separately. We also find the average of benign and malignant F-scores.

$$\text{Precision} = \frac{\text{TP}}{\text{TP} + \text{FP}} \tag{6}$$

$$\text{Recall} = \frac{\text{TP}}{\text{TP} + \text{FN}} \tag{7}$$

$$F - \text{score} = 2 \times \frac{\text{Precision} \times \text{Recall}}{\text{Precision} + \text{Recall}} \tag{8}$$

True positive (TP) tells the number of pixels correctly classified; false positive (FP) is negative pixels that are falsely classified as positive, and false negative (FN) is positive pixels that are falsely classified as negative.

4.3 Baseline

We implemented the proposed algorithm using the Keras framework. We used MobileUNet as our baseline model [7]. MobileUNet is an extended version of UNet, which is proven to work with very few training images [7]. MobileUNet uses separable convolution over regular convolution, which minimizes the complexity cost and model parameters. MobileUNet has shown excellent results in the semantic segmentation of cytology images [1].

The baseline model is trained only on 10× data. The 4× resampled images are predicted using this baseline model, and the results obtained are shown in Table 2. We trained another baseline model, namely ResUNet++ [8] and compared the validation results. MobileUNet performed better in terms of malignant, average and foreground F-scores. Hence, we continue with MobileUNet as our baseline model.

4.4 Semi-supervised

We trained the SSL model with labelled 10× data and unlabelled 4× resampled data. Each image from the 4× resampled data is randomly augmented from the distribution shown in Table 3.

Table 2 Results of 4× resampled validation images on the baseline models

Baseline Model	No. of Images	Malignant F-score	Benign F-score	Average F-score	Foreground F-score
MobileUNet	677	0.88	0.52	0.7	0.77
ResUNet++	677	0.72	0.52	0.62	0.75

Table 3 Distribution of random augmentations

Augmentations	Range	Probability
Horizontal Flip	–	0.5
Vertical Flip	–	0.5
Height width shift	0 to 5 pixels	0.5
Zoom	80% to 120%	0.2

Table 4 Foreground F-score of 4× resampled data for different λ and K values

λ	K		
	2	**3**	**4**
0.3	0.76	0.75	0.77
1	0.76	0.77	0.77
3	0.76	0.76	0.78
10	**0.79**	0.74	0.76

Table 5 Results of 4× resampled test images on the baseline and extended SSL model

Model	No. of images	Malignant F-score	Benign F-score	Average F-score	Foreground F-score
Baseline	419	0.87	0.53	0.7	0.75
Extended SSL model	419	0.87	0.61	0.74	0.77

The training is started after initializing the weights from the 10× model. There are two hyperparameters to be tuned; the weightage of unlabelled loss (λ) defined in the Eq. (5) and no. of random augmentations for unlabelled data (K). We initially trained the SSL for 20 epochs with different combinations of λ and K to tune these hyperparameters. RMSProp is used as the optimizer with a learning rate of 10^{-4} and a decay rate of 10^{-4}. The results are shown in Table 4. We have chosen λ as 10 and K as 2 based on these results.

The extended semi-supervised model is trained with 10× and 4× resampled data with the chosen values for hyperparameters λ and K for 200 epochs. This trained model is evaluated for the test data as shown in Table 5. The foreground F-score has been improved from 0.75 to 0.77. The benign F-score is improved from 0.53 to 0.61.

Figure 3 shows the original ground truth and the predictions from the baseline and the proposed methods. In the baseline, it can be seen that a lot of benign pixels are getting misclassified as malignant pixels. This results in a low F-score for benign pixels. In the proposed method, more benign pixels are getting classified correctly. This is shown an improvement of 15% in the F-score of benign pixels for the test set.

4.5 Classification

The aim of this work is to identify the malignant areas in the low magnification image to scan those areas in high magnification. Hence, we classify each sub-image into a benign or a malignant image. Any image with malignant pixels greater than a predefined threshold is considered malignant. The sub-images that are classified as malignant will be further scanned at high magnification. The threshold is chosen such

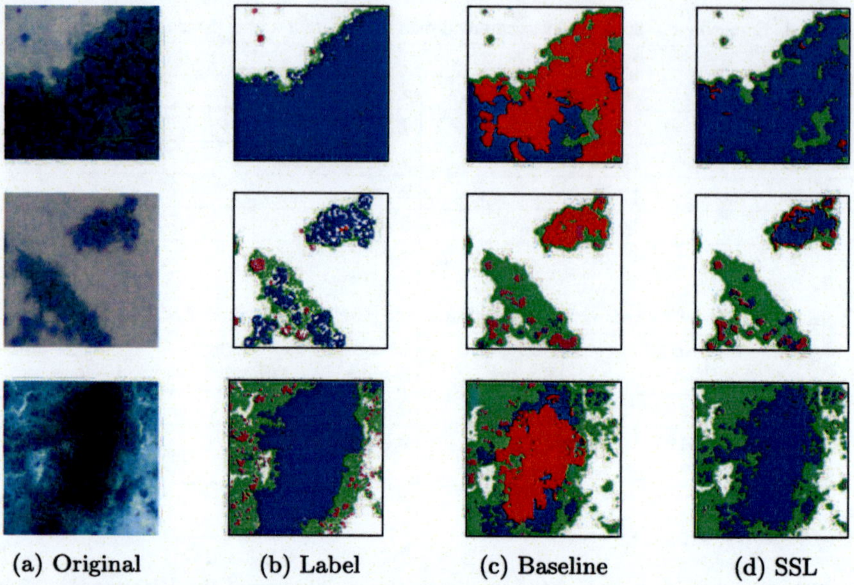

Fig. 3 Predictions of a few samples on the baseline model and the extended SSL model

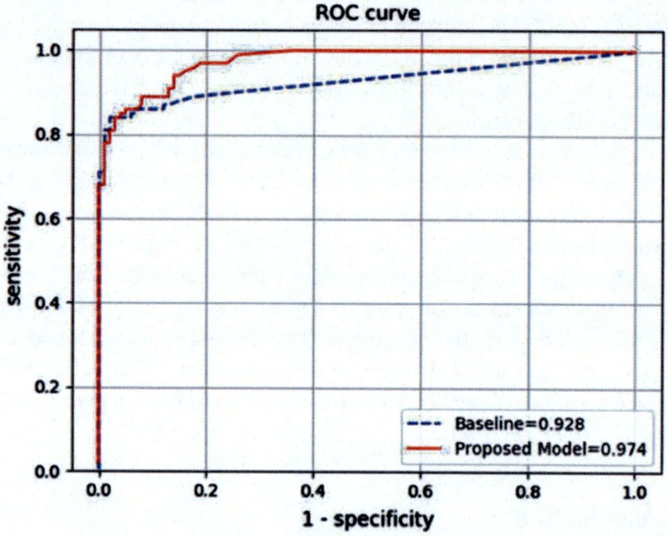

Fig. 4 Receiver operating characteristics (ROCs) curve with AUC

Table 6 The proportion of area excluded from scanning at a higher magnification for various sensitivity values in the baseline and the proposed model

Models Sensitivity	1.0	0.95	0.9	0.85
Baseline	0	0	0	65.39
Proposed	41.05	55.37	58.71	66.83

that sensitivity should be as close as possible to 100%, such that not even a single malignant image is missed. Figure 4 shows the receiver operating characteristics (ROCs) curves for the baseline and the proposed models.

We chose the operating point from the ROC at different sensitivity values. The proportion of area excluded from scanning at a higher magnification is shown in Table 6 for these operating points.

The entire 419 sub-images should be scanned at higher magnification while using the baseline model. But, only 247 sub-images have to be scanned while using the proposed semi-supervised model at sensitivity value one. Therefore, the proposed model avoids scanning 41% of the low magnification images, thereby saving the scanning and processing time.

5 Conclusion

Automatic segmentation of low magnification images helps to save scanning time and storage requirements. This work predicted the malignancy of effusion cytology images at lower magnification levels 10× and 4×. The 10× images can be labelled but not the 4× images. Hence, we proposed a semi-supervised semantic segmentation method to train the unlabelled 4× data with the labelled 10× data. The F-score of benign pixels is improved by 15% compared to the baseline model without affecting the malignant F-score. The sub-area that needs to be scanned at higher magnification in the whole slide is reduced by around 41%. We have manually upsampled the 4× data using a bicubic filter to use in the SSL model. In the future, we will explore SSL models based on adversarial training.

References

1. Aboobacker, S., Vijayasenan, D., David, S.S., Suresh, P.K., Sreeram, S.: A deep learning model for the automatic detection of malignancy in effusion cytology. In: 2020 IEEE International Conference on Signal Processing, Communications and Computing (ICSPCC), pp. 1–5
2. Barwad, A., Dey, P., Susheilia, S.: Artificial neural network in diagnosis of metastatic carcinoma in effusion cytology. Cytometry Part B Clin. Cytometry **82**(2), 107–111 (2012)

3. Belsare, A., Mushrif, M.: Histopathological image analysis using image processing techniques: an overview. Signal Image Process. **3**(4), 23 (2012)
4. Berthelot, D., Carlini, N., Goodfellow, I., Papernot, N., Oliver, A., Raffel, C.A.: Mixmatch: a holistic approach to semi-supervised learning. Adv. Neural Inf. Process. Syst. **32** (2019)
5. Gurcan, M.N., Boucheron, L.E., Can, A., Madabhushi, A., Rajpoot, N.M., Yener, B.: Histopathological image analysis: a review. IEEE Rev. Biomed. Eng. **2**, 147–171 (2009)
6. Higgins, C.: Applications and challenges of digital pathology and whole slide imaging. Biotech. Histochem. **90**(5), 341–347 (2015)
7. Howard, A.G., Zhu, M., Chen, B., Kalenichenko, D., Wang, W., Weyand, T., Andreetto, M., Adam, H.: Mobilenets: Efficient convolutional neural networks for mobile vision applications. arXiv preprint arXiv:1704.04861 (2017)
8. Jha, D., Smedsrud, P.H., Riegler, M.A., Johansen, D., De Lange, T., Halvorsen, P., Johansen, H.D.: Resunet++: an advanced architecture for medical image segmentation. In: 2019 IEEE International Symposium on Multimedia (ISM), pp. 225–2255. IEEE (2019)
9. Lee, D.H., et al.: Pseudo-label: The simple and efficient semi-supervised learning method for deep neural networks. In: Workshop Challenges Representation Learn., ICML. vol. 3, p. 896 (2013)
10. Miyato, T., Maeda, S.i., Koyama, M., Ishii, S.: Virtual adversarial training: a regularization method for supervised and semi-supervised learning. IEEE Trans. Pattern Anal. Mach. Intell. **41**(8), 1979–1993 (2018)
11. Samuli, L., Timo, A.: Temporal ensembling for semi-supervised learning. In: Proceedings of International Conference on Learning Representations (ICLR), vol. 4, p. 6 (2017)
12. Sohn, K., Berthelot, D., Carlini, N., Zhang, Z., Zhang, H., Raffel, C.A., Cubuk, E.D., Kurakin, A., Li, C.L.: Fixmatch: simplifying semi-supervised learning with consistency and confidence. Adv. Neural Inf. Process. Syst. **33**, 596–608 (2020)
13. Spanhol, F.A., Oliveira, L.S., Petitjean, C., Heutte, L.: A dataset for breast cancer histopathological image classification. IEEE Trans. Biomed. Eng. **63**(7), 1455–1462 (2015)
14. Ta, V.T., Lezoray, O., Elmoataz, A., Schüpp, S.: Graph-based tools for microscopic cellular image segmentation. Pattern Recognit. **42**(6), 1113–1125 (2009)
15. Tarvainen, A., Valpola, H.: Mean teachers are better role models: weight-averaged consistency targets improve semi-supervised deep learning results. Adv. Neural Inf. Process. Syst. **30** (2017)
16. Teramoto, A., Yamada, A., Kiriyama, Y., Tsukamoto, T., Yan, K., Zhang, L., Imaizumi, K., Saito, K., Fujita, H.: Automated classification of benign and malignant cells from lung cytological images using deep convolutional neural network. Inform. Med. Unlocked **16**, 100205 (2019)
17. Van Engelen, J.E., Hoos, H.H.: A survey on semi-supervised learning. Mach. Learn. **109**(2), 373–440 (2020)
18. Win, K., Choomchuay, S., Hamamoto, K., Raveesunthornkiat, M.: Detection and classification of overlapping cell nuclei in cytology effusion images using a double-strategy random forest. Appl. Sci. **8**(9), 1608 (2018)
19. Win, K.Y., Choomchuay, S., Hamamoto, K., Raveesunthornkiat, M., Rangsirattanakul, L., Pongsawat, S.: Computer aided diagnosis system for detection of cancer cells on cytological pleural effusion images. BioMed. Res. Int. (2018)
20. Xie, Q., Luong, M.T., Hovy, E., Le, Q.V.: Self-training with noisy student improves imagenet classification. In: Proceedings of IEEE/CVF Conference on Computer Vision and Pattern Recognition, pp. 10687–10698 (2020)
21. Zeiser, F.A., da Costa, C.A., de Oliveira Ramos, G., Bohn, H.C., Santos, I., Roehe, A.V.: Deepbatch: a hybrid deep learning model for interpretable diagnosis of breast cancer in whole-slide images. Exp. Syst. Appl. **185**, 115586 (2021)

Image Augmentation Strategies to Train GANs with Limited Data

Sidharth Lanka, Gaurang Velingkar, Rakshita Varadarajan, and M. Anand Kumar

Abstract Training modern generative adversarial networks (GANs) to produce high-quality images requires massive datasets, which are challenging to obtain in many real-world scenarios, like healthcare. Training GANs on a limited dataset overfits the discriminator on the data to the extent that it cannot correctly distinguish between real and fake images. This paper proposes an augmentation mechanism to improve the dataset's size, quality, and diversity using a set of different augmentations, namely flipping of images, rotations, shear, affine transformations, translations, and a combination of these to form some hybrid augmentation. Fretchet distance has been used as the evaluation metric to analyze the performance of different augmentations on the dataset. It is observed that as the number of augmentations increase, the quality of generated images improves, and the Fretchet distance reduces. The proposed augmentations successfully improve the quality of generated images by the GAN when trained with limited data.

Keywords GANs · Data augmentation · Fretchet distance

S. Lanka (✉) · G. Velingkar · R. Varadarajan · M. Anand Kumar
Department of Information Technology, National Institute of Technology Karnataka, Surathkal, Mangalore, India
e-mail: sidharthlanka.191it251@nitk.edu.in

G. Velingkar
e-mail: gaurang.191it113@nitk.edu.in

R. Varadarajan
e-mail: rakshitajps.191it140@nitk.edu.in

M. Anand Kumar
e-mail: m_anandkumar@nitk.edu.in

M. Tistarelli et al. (eds.), *Computer Vision and Machine Intelligence*,
Lecture Notes in Networks and Systems 586,
https://doi.org/10.1007/978-981-19-7867-8_35

1 Introduction

With the advent of technology and the rise of artificial intelligence, research in machine learning methods like deep learning, reinforcement learning, and GANs has grown exponentially. Every day, new technologies are invented and introduced. Deep learning is gaining popularity among scholars. Neural networks have found their way into every machine learning activity currently being conducted [1]. Generative adversarial networks (GANs) are a subset of neural networks lately gaining much traction. In machine learning, generative modeling is an unsupervised learning task that automatically finds and learns regularities or patterns in particular data. The model may be used to create many novel outputs that might have been obtained from the original dataset.

A GAN is a generative model that employs two neural networks, such as the generator and the discriminator. The generator creates bogus pictures and sends them to the discriminator, differentiating between actual and bogus images. Both are trained concurrently until the generator successfully tricks the discriminator, indicating that it produces realistic pictures. GANs have various real-world applications, from health to photo editing, photo design, and video prediction. One of the causes for the dramatic increase in research on the topic is the seemingly limitless availability of photographs on the Internet. There are also several high-quality image-only datasets available online. The difficulty emerges when the same is applied to real-world settings. Modern GANs need around 10^5 images to generate good quality images. In fields such as medicine, where an extensive dataset is difficult to obtain, it is necessary to use existing resources efficiently. This is a significant impediment to the use of GANs in this field.

Other barriers to gathering big enough image datasets for a particular purpose include privacy, copyright difficulties, image quality, the time required, and cost. This impedes research in critical areas such as medicine. Thus, the capacity to attain high accuracy with a small dataset is a huge accomplishment. As a result, it is vital to develop GANs capable of producing high-quality pictures even with a limited dataset. With a short dataset, the discriminator overfits the dataset and provides no valuable input to the generator to help it improve. Deep learning models employ augmentations to solve the problem of overfitting. GANs trained on a more extensive dataset outperform those trained on a small dataset. If not managed appropriately, a poorly augmented dataset might result in GANs that create images with noise even though the dataset is noise-free. This 'leaking' of augmentations is undesirable in any GAN and should be addressed when the model is being built.

A summary of all the research conducted has been described in Sect. 2. This paper proposes several augmentations to the CIFAR-10 dataset, described in Sect. 3.2, which has been chosen to replicate a limited dataset scenario. A small subset of the dataset consisting of images of only birds was used, which had only 5000 images in total. A GAN model trained on this dataset produced underwhelming results, as expected. A list of augmentations applied has been listed under Sect. 3.3. With augmentations applied to the dataset, the quality of generated images improved significantly, as expressed in Sect. 4 further below in this paper.

2 Literature Survey

Current research involves finding better augmentation strategies to get up-to-the-standard results even when the available data is limited. While it is preferred to have large datasets rather than increase the dataset through such augmentation plus generative methods, these methods are being developed for instances where the amount of data available is less due to privacy and other reasons.

Models have been built and proposed for this purpose and analyzed for performance. These models include traditional augmentation such as rotation, zooming, cropping, scaling, translating, and advanced augmentations such as StyleTransfer. Further, dynamic augmentation methods have been proposed which can fine-tune themselves depending upon the adversarial detections by the GAN models. A summary of all the research done earlier is shown in Table 1.

Affine transformations are widely used as data augmentation methods for training deep learning models involving images [2]. While majority of the research focuses on such methods individually, a combination of different techniques has great potential to improve augmentation results which is being explored in this paper.

3 Methodology

The project first applies augmentations to the CIFAR-10 Birds Dataset to expand it. The GAN model is trained on this expanded dataset with an overall size of 55000. The Fretchet inception distances are plotted along with the generator and discriminator losses after training. A grid of generated images is plotted too. The architecture of the models involved can be seen in Fig. 1.

3.1 Architecture

See Fig. 1 and Table 2.

3.2 Dataset

CIFAR-10 dataset [7] is chosen for use in this project. It is a commonly used dataset for computer vision applications. It contains 60,000 colored images of size (32 × 32), which belong to 10 different classes, namely airplanes, cars, birds, cats, deer, dogs, frogs, horses, ships, and trucks. The ten classes have 6000 images each in the dataset. This dataset is chosen because it has lower resolution images and can thus be trained faster, making it convenient to experiment with different kinds of augmentations and evaluate their performance.

Table 1 Summary of literature survey

Ref No.	Authors	Study description	Conclusions/results
[3]	Tero Karras, Miika Aittala, Janne Hellsten, Samuli Laine, Jaakko Lehtinen and Timo Aila	This paper proposes an adaptive discriminator augmentation mechanism to stabilize the training significantly with restricted data.	The proposed model worked well on multiple training datasets with only a few thousand images. The FID recorded for limited data dataset CIFAR-10 also improved
[2]	A. Mikołajczyk and M. Grochowski	This paper not only performed comparisons between many data augmentation methods in classifying images (such as traditional scaling and zooming) but also GANs and StyleTransfer. It also proposed a novel method of augmentation based on StyleTransfer	Traditional methods proved to be most convenient but were vulnerable to adversarial attacks. The StyleTransfer augmentation outperformed all
[4]	X. Zhang, Z. Wang, D. Liu, Q. Lin and Q. Ling	This paper proposes a deep adversarial augmentation mechanism to stabilize the training significantly with restricted amount of data	The proposed model worked well on three real-world limited datasets. The model also proved to work for the task of detecting objects
[5]	Ö. Ö. Karadağ and Ö. Erdaş Çiçek	This paper analyzes the classifier's performance on images augmented by three methods: Gaussian blurring, dropout of regions, and GAN	The results showed that while GAN outperformed the other two augmentation strategies in terms of the efficiency of the classifier and it is heavily dependent on the size of the dataset used
[6]	T. A. Korzhebin and A. D. Egorov	This paper compared four traditional image augmentation methods along with transfer learning strategies and was applied to five different classification models	The results showed that the larger the model, the fewer the differences caused by different applied transfer strategies

The dataset is loaded with the help of a *dataloader*. Since it is required to improve the working of a GAN with a limited dataset, only images belonging to the *Birds* class are used for training. Thus, 5000 images are selected from the birds class as the limited dataset for augmentations and generative modeling.

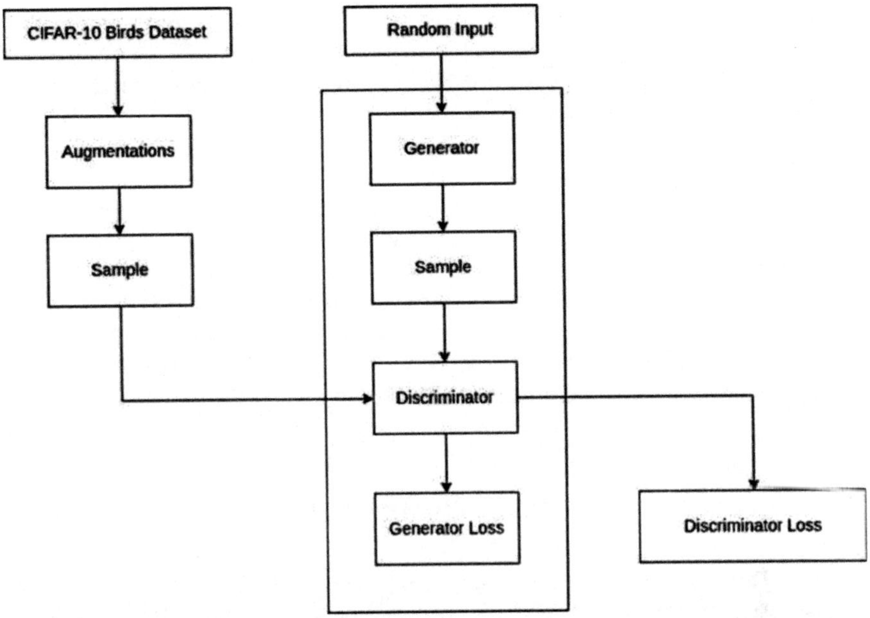

Fig. 1 Architecture of the models involved

3.3 Data Augmentation

Various primary data augmentations were initially applied to the images, including image inversion, rotation, cropping, flipping, and more. After analyzing the performance of these data augmentations on the images, those improving the results to the most considerable extent were selected and combined to form hybrid functions with expectations of improving the model outcomes even further. The final chosen augmentations functions produce images that increase the dataset size while also decreasing the FID value, thus indicating a better quality of images in the extended dataset and better performance by the GAN model on using these images together.

The ten data augmentation functions finally being applied to the images to get the extended dataset are as follows:

1. Flip: The image is flipped along both the axes.
2. Shear: The image is randomly stretched in one of the axial planes between degree values of 5 and 30.
3. Rotation: The image is rotated by a random degree value between 5 and 30.
4. Translation: The image is shifted along the axes by a random pixel value between 0 and 5.
5. Affine transformation: An affine transformation is applied with a degree value between 0 and 30.

6. Brightness: The image's brightness is changed by increasing or decreasing it by a factor of any value up until 2.
7. Contrast: The image's contrast is changed by increasing or decreasing it by a factor of any value up until 2.
8. Hybrid augmentation 1: The image is flipped randomly along both axes, and then, Gaussian noise is added to the image.
9. Hybrid augmentation 2: The image is flipped randomly along both axes, then rotated randomly by a degree value between 0 and 30. Then, the image is shifted along the axes by a random pixel value between 0 and 5. Finally, an affine transformation is applied with a degree value between 0 and 30.
10. Hybrid augmentation 3: The image is flipped randomly along both axes, then rotated randomly by a degree value between 0 and 30. The image is then shifted along the axes by a random pixel value between 0 and 5.

3.4 Generative Adversarial Networks Model

GANs [8] are a unique approach to training a generative model that frames the task as a supervised learning problem [9] with two sub-models: the generator model, which we train to produce new instances, and the discriminator model, which attempts to categorize examples as real or false (generated). In an adversarial zero-sum game, the two models are trained until the discriminator model is fooled about half the time, suggesting that the generator model generates convincing instances.

The generator takes on the role of a forger, attempting to create lifelike images from random noise (in our example). Meanwhile, the discriminator acts as an evaluator, attempting to distinguish real from fake images. The generator tries to trick the discriminator as much as possible by getting the images (for example) closer to reality in each stage, causing the discriminator to categorize them as genuine. By identifying its images as fake, the discriminator directs the generator to produce more realistic images. This min-max game is continued until a Nash equilibrium is reached, which is defined as a situation where both players in a game know the equilibrium strategy of the opponent. Neither of them can gain anything by changing their strategy. This is the ideal place to stop training the GAN. In this paper, both the generator and the discriminator have been built using deep convolutional GANs (DCGANs) [10, 11], a variant of the regular vanilla GAN where deep convolutional layers are used in both generator and discriminator instead of using dense layers.

3.4.1 Generator

The seeds and some hyperparameters are initialized before the training starts. The generator starts by taking in random noise from the latent space of dimension 100 here and then performs the mapping to an image of size $(3 \times 64 \times 64)$. The DC generator has five convolutional layers, where a 2D transposed convolutional operator is used over the input planes, and this layer is also known as a *fractionally strided convolution*. Along with the stride, these layers are used for upsampling or producing the images.

Batch normalization is applied to all convolutional transpose layers. Batch normalization stabilizes learning by adjusting the input to the mean and variance of each unit to zero. This assists in resolving training challenges caused by insufficient initialization and the gradient flow in deeper models. This was critical in getting deep generators to learn because it prevented the generator from collapsing all samples into a single point, a common type of failure in GANs. However, sample oscillation and model instability occurred when batch normalization was applied to all layers simultaneously. As a result, neither the generator output layer nor the discriminator input layer received batch normalization.

The generator uses the ReLU activation function in all layers except the output layer. The *tanh* function is used in the output layer, as adopting a bounded activation for the output layer allows the model to learn to saturate and cover the training distribution's color space more quickly.

3.4.2 Discriminator

The discriminator is an image classification model whose primary function is to distinguish fake images from actual images. The dataset is fed to the discriminator as the actual images while building the dataset from scratch. The discriminator uses five Convolution 2D layers. In the discriminator, all pooling layers are replaced by strided convolutions.

All layers except the first and final layer apply batch normalization to stabilize the learning process, like in the case of the generator.

Except for the output layer, all the other layers use the leaky ReLU activation function [12] as it is useful with helping gradients flow easily through the architecture. The final output layer in the discriminator uses the sigmoid activation function since the model's output must be a true or false value. The sigmoid activation function works best to serve this purpose.

Both the generator and discriminator were trained for 50 epochs, using Adam optimizers [13] and binary cross entropy loss [14] on both the models. The same learning rate is used for both the discriminator and the generator, and the training is carried out from scratch without using a pre-trained model.

3.4.3 Model Evaluation

The quality of the generated images from the generator after training the GAN and the performance of the model are measured using an evaluation metric known as Frechet inception distance (FID) [15]. The purpose of the FID score was to compare the statistics of a collection of synthetic images to the statistics of a collection of real photos from the target domain to evaluate synthetic images. The Inception v3

Table 2 Fretchet distance for different learning rates

Generator LR	Discriminator LR	Mean FID
0.0005	0.0005	261.147
0.0001	0.0004	247.124
0.00005	0.0001	273.167

model is used for calculating the FID score. The coding layer of the model is used to capture computer vision-specific properties of an input image. For a collection of real and synthetic images, these activations are computed. The calculation for FID is divided into three main parts:

1. The Inception network to extract 2048-dimensional activations from the pool3 layer for real and generated samples, respectively.
2. By computing the mean and covariance of the pictures, the activations are summarized as a multivariate Gaussian. The activations across the collection of real and generated images are then calculated.
3. Finally, Wasserstein-2 distance is calculated for the mean and covariance of real and generated images.

A lower FID suggests greater image quality; a larger score indicates a lower image quality, and the connection may be linear. When systematic distortions, such as the addition of random noise and blur, are performed, lower FID scores correspond with higher quality outputs.

4 Results and Analysis

The discriminator and generator learning rate hyperparameters are tuned in such a way to achieve the best Fretchet distance for a given number of epochs. After experimenting with several learning rates over the non-augmented dataset, the models which give the lowest Mean F.D are the learning rates used to experiment on the final augmented dataset. A summary of the experiments conducted is shown in Table 2.

In general, when it comes to the loss functions of GANs, both the generator and discriminator loss are very non-intuitive. Thus, both the losses do not follow a specific pattern but are a sequence of periodic up and down movements for both the discriminator and the generator. However, the generator loss, when quantified, will mostly be higher than the discriminator loss. As the generator and discriminator are competing, the improvement of one of them automatically corresponds to the increase in loss on the other model. This happens until the other model learns better about the newly received loss, increasing the loss of its competitor by improving its performance.

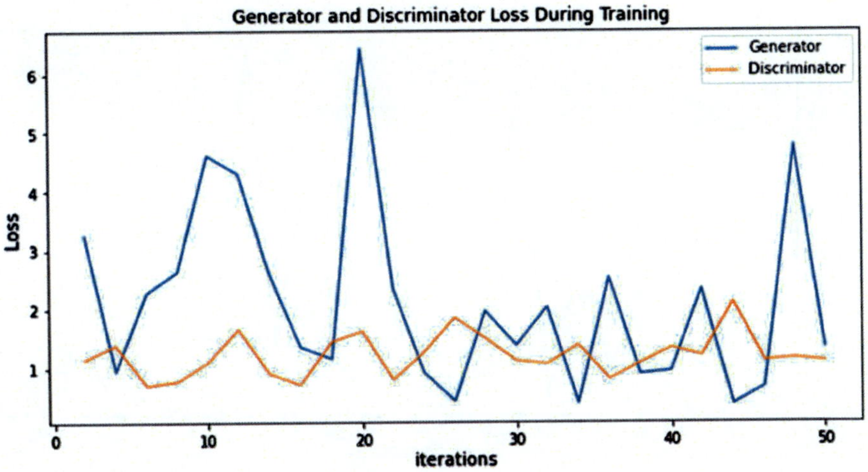

Fig. 2 Generator versus discriminator losses without augmentations

Figures 2 and 3 show the generator versus discriminator losses without and with augmentation, respectively. On observing the loss graphs with and without augmentation, we see that the generator loss fluctuates between extraordinarily high and low values after very few epochs in training when augmentations have been applied. This level of fluctuation is not observed on the graph without augmentation. This is because the discriminator is fed by the augmented dataset, on which it trains better and faster for the same number of epochs and has much more evident improvements in each epoch. Thus, in contrast, the generator must improve to counter the discriminator's loss, and thus, we notice that the generator loss significantly falls and rises periodically. This shows that the generator is learning and improving faster according to its loss graph trends when the augmented dataset is used compared to when the augmented dataset is not.

The Fretchet distances obtained by the cumulative augmentations were compared with those when no augmentations were applied. It can be seen from Fig. 4 that as the number of augmentations increases, the Fretchet distance keeps getting better, implying that the quality of images is increasing. A small overlap between the Fretchet distance plots when three and six augmentations are applied shows that the change is not very significant even though there are improvements compared to the model when no augmentations are applied. Table 3 provides further readings to support this observation. However, when more augmentations are applied, including the hybrid augmentations, a significant improvement can be seen as there is no overlap between the plots. It can be inferred that the hybrid augmentations amplify the effect of applying individual augmentations to produce significantly better results than those with individual augmentations. A more diverse and extensive dataset is produced, which helps generate good quality images.

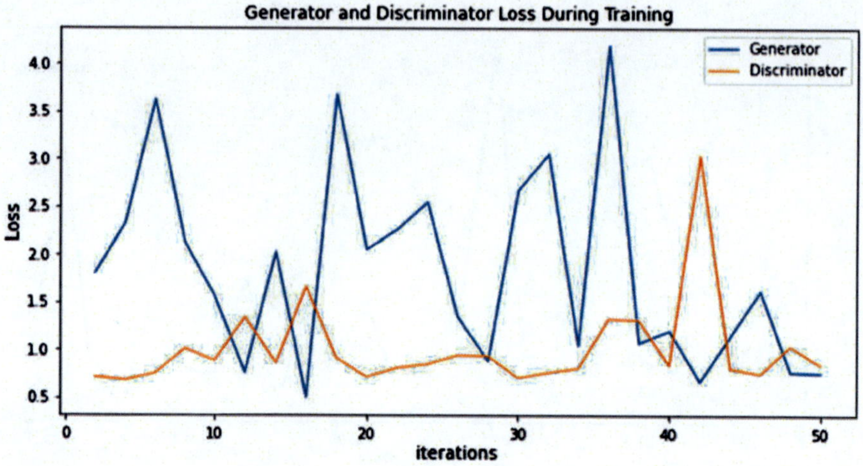

Fig. 3 Generator versus discriminator losses with augmentations

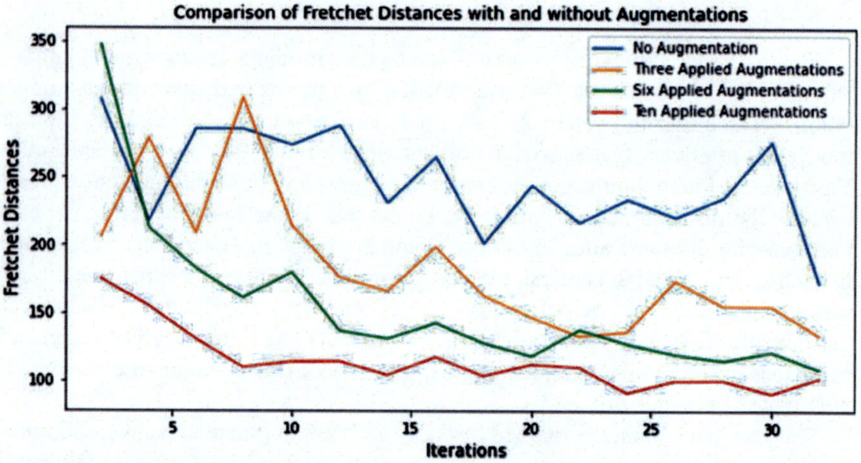

Fig. 4 Comparison of Fretchet distances with and without augmentations

Table 3 Summary of Fretchet distances during training

Augmentations	Max. FID	Min. FID	Mean FID
No augmentations	307.511	172.224	247.124
Three augmentations	308.959	133.108	184.740
Six augmentations	347.223	108.573	154.264
Ten augmentations	175.324	91.103	114.610

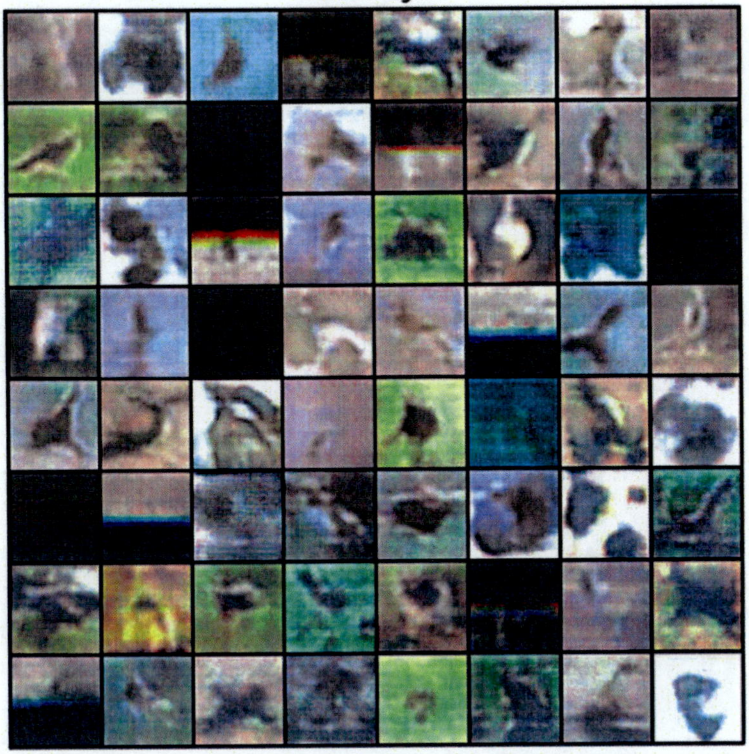

Fig. 5 Images generated by the GAN when trained on the partial augmented dataset

Table 3 provides a cumulative analysis of the applied augmentations. A set of ten augmentations in total were defined and applied progressively on the dataset. The first set of observations with augmentations corresponds to flip, rotation, and shear augmentations applied to the dataset. It can be seen that FID reduces, implying an improved quality of generated images. Next, three more augmentations were added changes in brightness, contrast, and horizontal and vertical translations. This further improved results. The final set of ten augmentations includes all of the augmentations mentioned previously in the *Data Augmentation* section of *Methodology*. It can be seen that this significantly improves results due to the presence of hybrid augmentations that introduces a higher degree of diversity in the dataset.

Figure 5 displays a list of generated images when only six augmentations were applied to the dataset. It can be seen that there is a lot of noise in the form of dark images generated. Figure 6 shows an improvement as the fully augmented dataset generates lesser noise.

The results of this paper complement the findings and research currently being conducted in the field. While majority of the strategies being used for augmentation

Fig. 6 Images generated by the GAN when trained on the fully augmented dataset

are simple transformations, a combination of the best performing transformations has great potential [2] to produce desired results, as we are exploring in this paper. Experiments with the CIFAR-10 dataset have proven to be useful in the past [4]. With a lot of research being conducted on the same, we have extended the idea to provide scope for further research and comparisons to be conducted. The results are highly dependent on the size of the dataset used [5], and as we have used a subset of CIFAR-10, it is incongruous to compare our results quantitatively with the others where the entire dataset has been used. A generic algorithm means that we could not apply transfer learning strategies which have shown immense potential and success [6] in the data augmentation field.

5 Conclusion and Future Work

This paper proves that the quality of generated images by the GAN model significantly improves upon using several augmentations to prevent discriminator overfitting and enlarge a limited dataset, as we see the Fretchet distance decreases with the increase in the selected augmentations, thus giving better quality of results. This model developed in this paper has successfully zeroed in on a systematic set of augmentations that give a limited dataset diversity while ensuring that 'leaking' to the GAN model does not occur. Thus, the systematic procedure followed during these augmentations can be used on higher resolution datasets and more complex GAN models to obtain more promising results in future. By testing the augmentations out on multiple datasets and training the model for an extended period, more observations can be made to support the developed method and obtained results on training limited datasets for generative adversarial networks (GANs). Such experiments can be carried out in future to obtain better results.

References

1. Barrachina, D.G.E, Boldizsar, A., Zoldy, M., Torok, A.: Can neural network solve everything? Case study of contradiction in logistic processes with neural network optimisation. In: 2019 Modern Safety Technologies in Transportation (MOSATT), pp. 21–24 (2019)
2. Mikołajczyk, A., Grochowski, M.: Data augmentation for improving deep learning in image classification problem. In: 2018 International Interdisciplinary PhD Workshop (IIPhDW), pp. 117–122 (2018)
3. Karras, T., Aittala, M., Hellsten, J., Laine, S., Lehtinen, J., Aila, T.: Training generative adversarial networks with limited data. In Proceedings of the 34th International Conference on Neural Information Processing Systems, NIPS'20, Red Hook, NY, USA. Curran Associates Inc. (2020)
4. Zhang, X., Wang, Z., Liu, D., Ling, Q.: Dada: deep adversarial data augmentation for extremely low data regime classification, pp. 2807–2811 (2019)
5. Karadağ, Ö.Ö., Çiçek, Ö.E.: Experimental assessment of the performance of data augmentation with generative adversarial networks in the image classification problem. In: 2019 Innovations in Intelligent Systems and Applications Conference (ASYU), pp. 1–4 (2019)
6. Korzhebin. T.A., Egorov, A.D.: Comparison of combinations of data augmentation methods and transfer learning strategies in image classification used in convolution deep neural networks. In: 2021 IEEE Conference of Russian Young Researchers in Electrical and Electronic Engineering (ElConRus), pp. 479–482 (2021)
7. Ho-Phuoc, T.: CIFAR10 to compare visual recognition performance between deep neural (2018, November)
8. Goodfellow, I., Pouget-Abadie, J., Mirza, M., Xu, B., Warde-Farley, D., Ozair, S., Courville, A., Bengio, Y.: Generative adversarial networks. Adv. Neural Inf. Process. Syst. **3** (2014)
9. Odena, A.: Semi-supervised learning with generative adversarial networks (2016, June)
10. Radford, A., Metz, L., Chintala, S.: Unsupervised representation learning with deep convolutional generative adversarial networks (2015, November)
11. Chapagain, A.: Dcgan–image generation (2019, February)
12. Xu, B., Wang, N., Chen, T., Li, M.: Empirical evaluation of rectified activations in convolutional network (2015, May)
13. Kingma, D., Adam, J.B.: A method for stochastic optimization. In: International Conference on Learning Representations (2014, December)
14. Ruby, Usha, Yendapalli, Vamsidhar: Binary cross entropy with deep learning technique for image classification. Int. J. Adv. Trends Comput. Sci. Eng. **9**, 10 (2020)
15. Nunn, E., Khadivi, P., Samavi, S.: Compound Frechet inception distance for quality assessment of gan created images (2021, June)

Role of Deep Learning in Tumor Malignancy Identification and Classification

Chandni, Monika Sachdeva, and Alok Kumar Singh Kushwaha

Abstract Tumor is a life-threatening disease that is characterized as an abnormal lump of tissue formed in any part of the human body. Malignant tumors have tendency to become worse and invade other parts of body. Timely diagnosis can surely improve duration and chances of survival. The use of machine learning techniques is surging in the field of oncology as it can significantly reduce the surgeon's workload and make a better prognosis of patient conditions. However, "Deep learning" is a specific class of machine learning algorithms that uses Neural Networks (NN) and is able to achieve higher performance in medical diagnosis and detection. This study reviews deep learning-based techniques in the detection and categorization of tumor and summarize the key findings.

Keywords Machine learning · Deep learning · Convolution neural network · Tumor classification · Benign tumor · Malignant tumor

1 Introduction

A tumor is an unusual growth of tissue in any part of the body. This unusual tissue growth in one place and that cannot spread is called a benign tumor. If such growth of tissue is unbridled and can spread to neighboring tissues; then it is cancerous, also called malignant tumor or cancer. Tumor can be in different sites of human body like brain, skin, lungs, bone, cervical, etc. Figure 1 depicts tumor in the brain using Medical Resonance Imaging (MRI). Due to the severity of the disease, there are several diagnostic tools and techniques to improve the survival rate and revamp

Chandni (✉) · M. Sachdeva
Department of CSE, I.K. Gujral Punjab Technical University, Kapurthala, India
e-mail: chandnikathuria5@gmail.com

M. Sachdeva
e-mail: monika@ptu.ac.in

A. K. S. Kushwaha
Department of CSE, Guru Ghasidas Vishwavidyalaya, Bilaspur, India

M. Tistarelli et al. (eds.), *Computer Vision and Machine Intelligence*,
Lecture Notes in Networks and Systems 586,
https://doi.org/10.1007/978-981-19-7867-8_36

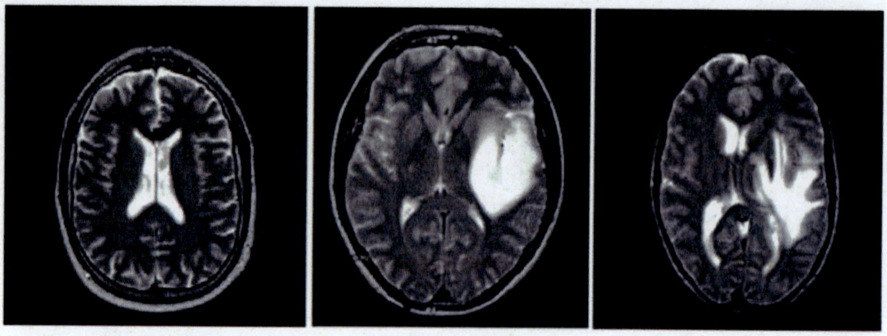

a. Healthy brain b. Benign tumor c. Malignant tumor

Fig. 1 MRI visualization of healthy brain (**a**) and brain having tumor (**b**, **c**)

the life standard of patient. It takes a lot of effort to visually identify and categorize tumors from medical images and such examination is also subjected to human fallacy. Artificial Intelligence (AI) has made eminent progress in medical and healthcare [1, 2] in past few years. It encompasses various mathematical techniques to support automated reasoning and decision making. Hence AI algorithms are also effective in detecting and classifying tumors. Machine learning (ML) is sub-field of AI that can learn pattern and relations in high-dimensional data to support predictions and decision making [3]. It enables machine to learn from data that may be labeled or unlabeled. Accordingly ML techniques are primarily categorized as supervised and unsupervised.

AI today is dominated by deep learning, which is a form of supervised ML technique. It is influenced by the ways that the human brain processes information and uses Artificial Neural Networks (ANN). These networks do not need set of rules created by humans; instead, they use a vast amount of data to map the input to a set of labels [4]. DL designs [5] can be "Supervised" such as Recurrent Neural Networks (RNN), Gated Recurrent Units (GRU), Long Short-Term Memory (LSTM), and Convolutional Neural Networks or "Unsupervised" such as Boltzmann Machines, Auto-encoders and Generative Adversarial Networks (GAN). CNN is one the most powerful and suitable DL design for processing of spatial data such as images.

Effectiveness of classical ML techniques relies in their data representation method, i.e., learning features from raw data. Hence feature engineering has been important research aspect in last years for the application of ML in healthcare. There are various methods like Harris Corner Detection, Histogram of Oriented gradients (HOG), Scale-Invariant Feature Transform (SIFT), Speeded-Up Robust Features (SURF), Bag of Words (BoW), etc. With deep learning this can be done in an automated way. Deep learning methods allow hierarchical non-linear feature learning by using set of hidden layers, where initial layers extracts or learns the low-level features and higher layers extract the high-level features.

Many imaging techniques provide medical data for cancer detection and classification in the form of images of the body part that is suspected of abnormal cell growth. These techniques commonly include Ultrasound, CT scan (Computed Tomography), X-Ray, MRI, etc. ML and DL techniques can thus learn patterns or features of tumor or cancer from medical image datasets to classify them as benign or malignant or specific type of malignancy depending on the site of cancer. Thus, these techniques serve as promising tool for automatic identification and classification of tumor malignancy at early stages, which can overcome challenges that prevail in manual examination. This study presents a review of deep learning techniques applied to biomedical datasets for tumor identification and classification. It also identifies various challenges that must be addressed in future research in this area.

2 Literature Review

ML and DL methods are being extensively applied to oncology [6, 7]. Feature extraction and classification are two principal steps in tumor malignancy detection and classification. Conventional ML approach implements these steps separately whereas DL offers end to end training of network with automated feature learning and classification. The study of literature has revealed that there are three different approaches in applying ML and DL methods: (i) Pure ML approach uses ML methods for both steps. (ii) Pure DL approach uses DL for both steps. (iii) Hybrid approach uses blend of ML and DL. This section qualitatively describes various studies from the literature. Pioneer works from the current literature are summarized in Table 1. It presents the technique used for preprocessing, approach used for feature extraction, classification, dataset along with a comparative view of performance.

Every year, hundreds of women lose their lives to diseases like breast cancer. AI has been used to quickly and effectively detect and classify breast tumor or cancer. Detection model proposed in [8] proves superiority of CNN over Support Vector Machine (SVM) and Random Forest (RF) classifier. 145 features are extracted with prior segmentation using K-means clustering, boundary region removal, regions ranking, and region growing method. After feature extraction, classification is done using SVM, RF, and CNN. Modified U-Net segmentation with Inception V3, ResNet50, DenseNet121, MobileNetV2, VGG16 [9], also provided good improvement in classification results. Besides the classification using full image, techniques have been suggested where model is trained using labeled dataset with Region of Interest (ROI) information. The full image classification model is then initialized with the patch classifier's parameters. Authors in [11] suggested an approach that introduced mass detection and patch extraction based on the feature matching of different regions using Maximally Stable Extremal Regions (MSER) [12]. Extracted patches are then classified with modified AlexNet and GoogleNet. Patch extraction reduced volume of data to be processed and improved accuracy as well. ResHist model [13] with 152 layers based on ResNet50 architecture, having 13 residual blocks also provides comparable performance for breast cancer classification. Model is also good at handling vanishing gradient problem. Majority of work on breast cancer

Table 1 Comparative analysis of recent studies on cancer classification

References	Organ	Technique	Dataset and reported accuracy
[8]	Breast	Pre-processing using speckle noise removal, image normalization, and image enhancement. 145 features classified using SVM, RF, and CNN	**151 tumor lesions from cooperating hospital** SVM: 80% RF: 77.78% CNN: 85.42%
[9]	Breast	Segmentation of tumor with U-Net architecture and DenseNet121, Inception V3, ResNet50, VGG16, MobileNet V2 for classification. Inception V3 offered best results	**DDSM** [34] Inception V3 + MSU : 98.87% **MIAS** [35] Inception V3 + MSU: 96.87% **CBIS-DDSM** [36] Inception V3 + MSU: 94.18%
[10]	Breast	CNN architecture of VGG16 and ResSNet-50 and hybrid of both is used initially for patch classification using ROI annotations. Patch classifier weights are the used for initialization of whole image classifier. Four model averaging provided best results	**CBIS-DDSM** [36] AUC-0.91, sensitivity-86.1%, specificity-80% **INbreast** [37] AUC-0.98, sensitivity-86.7%, specificity-96.1%
[11]	Breast	Patch extraction and classification using AlexNet and GoogleNet separately	**AlexNet, GoogleNet accuracy respectively** **DDSM** [34] 100%, 98.46% **INbreast** [37] 100%, 88.24% **MIAS** [35] 98.53%, 91.58% **Data from Egypt National Cancer Institute** 97.89%, 88.24%
[13]	Breast	Residual network of 152 layers with 13 residual blocks based on ResNet50 architecture	**BreaKHis** [38] 92.5%
[15]	Brain	At phase-1 two channel CNN is used to classify healthy and glioma samples. At phase-2 RCNN is used to classify Glioma samples into meningioma and pituitary class	**Dataset from two leading hospitals and Kaggle** Phase-1: 98.2% Phase-2: 100%
[16]	Brain	GoogleNet with transfer learning is used for feature extraction and classification with SVM, KNN, Dense layers of CNN. SVM outperformed among all.	**Figshare** [39] Deep transfer learned model: 92.3 ± 0.7% Deep features+SVM features: 97.8 ± 0.2% Deep features+KNN: 98.0 ± 0.4%
[17]	Brain	Hybrid CNN-SVM. Input modality is threshold segmented brain MRIs	**BRATS2015** [40] 98.495%

(continued)

Table 1 (continued)

References	Organ	Technique	Dataset and reported accuracy
[18]	Brain	Three deep learning models: MobileNetV2 ,VGG19, InceptionV3 are investigated for classification of healthy and tumor images from brain X-ray dataset	**Brain MRI dataset from kaggle** [41] MobileNetV2: 92% VGG19: 88.22% InceptionV3: 91%
[19]	Brain	Glioma, meningioma, and pituitary classification with eighteen layer deep CNN. Performance is evaluated on cropped, uncropped and segmented dataset	**Figshare** [39] Cropped images: 98.93% Uncropped images: 99% Segmented images: 97.6%
[20]	Brain	CNN for segmenting pre-processed images. GoogleNet transfer learned model for feature extraction and classification	Segmentation, classification accuracy **BRATS2018** [42] 96.5%, 96.4% **BRATS2019** [43] 97.5%, 97:3% **BRATS2020** [44] 98%, 98.7%
[21]	Brain	Fuzzy deformable fusion model for segmentation + LBP feature extraction, statistical features+ Binary classification of brain MRIs using CNN	**BRATS** [45] 95.3% **SimBRATS** 96.3%
[24]	Skin	Deep CNN used for binary classification of skin lesions. Pre-processing of dataset is done to reduce noise artifacts. Model outperformed standard CNN models like AlexNet, ResNet, VGG16, DenseNet, MobileNet	**HAM10000**[46] 93.16%
[25]	Skin	Features (like color, shape, texture, etc.) extracted and classified using CNN and SMTP loss function. Method outperformed CNN-based classification	**Melanoma dataset** [47] For Stage 1, 2, 3: 96% For Stage 1, 2: 92%
[26]	Skin	Transfer learning is performed using pre-trained ImageNet weights and fine tuning of EfficientNets variants B0–B7 is performed for multiclass skin cancer classification into seven classes	**HAM10000** [46] 87.9%

classification has been done using ultrasound images, there is need to explore other modalities like CT, MRI to improvise classification performance.

Benign brain tumors such as meningioma, pituitary tumors, and astrocytomas have slow progression rate as compared to malignant brain tumors like oligodendrogliomas, astrocytomas [14]. They are also graded as I, II, III, and IV by World Health Organization, depending upon rate of growth of cancerous cells. Fully automated CNN model with two channels is investigated in [15] for classification into glioma and healthy class. The same model is then applied as the feature extractor of RCNN to detect tumor regions and classify them into meningioma and pituitary class. Features extracted using deep network can also be classified by applying ML classifiers. Similar model suggested in [16] implements transfer learning with GoogleNet to extract features from brain MRIs. Classification is the done using SVM and K-Nearest Neighbor (KNN). Hybrid CNN-SVM implemented in [17] for binary classification, also use threshold segmentation. It thus improves performance with additional cost of prior segmentation of tumor region. Three pre-trained CNN variants- MobileNetV2, VGG19, and InceptionV3 are used in [18] to classify brain X-ray images with transfer learning from ImageNet. MobileNetV2 has offered good results on small dataset of brain X-ray images. Using cropped, uncropped, and segmented brain MRIs, an investigation was made for glioma, meningioma, and pituitary tumor classification. Uncropped lesions exhibited the highest accuracy [19], this denotes the contribution of background information to the classification results. A five stage pipeline with preprocessing, skull stripping, tumor segmentation using CNN, post processing, and classification is suggested in [20]. The model has offered better results in comparison with other models suggested in the literature for BRATS dataset. Method suggested in [21] blends Local Binary Pattern (LBP) features with statistical features, followed by classification using CNN for better results.

Skin cancer is also widespread and very difficult to detect because of similarity with various other skin diseases [22, 23]. Melanoma is the most prevalent type of skin cancer. Deep CNN [24] initialized with ImageNet weights is implemented for binary classification into benign and malignant skin cancer. Suggested CNN architecture provides better results than standard models like AlexNet, VGGNet. Melanoma stage classification system proposed in [25] uses CNN with loss function based on similarity measure for classification of 81 features extracted from skin images. The proposed approach outperforms SVM classifier. Effective scaling of the network across different dimensions can help the model to learn more fine features from images. EfficientNets implement this uniform scaling idea. Experiments with different variants of EfficientNets for multiclass classification of skin cancer have proved improvement in accuracy and efficiency of model achieved with scaling. A study on one of the malignant forms of bone tumor called "osteosarcoma" is presented in [27]. The performance of six deep networks InceptionV3 and NASNet-Large, VGG16, VGG19, ResNet50, DenseNet201 was evaluated where VGG19 model achieved the highest accuracy in both binary and multiclass classification.

To improve the generalizability of DL-based solution to tumor classification, approach for detecting multiple lung diseases-COVID-19, pneumonia, and lung cancer from chest X-ray and CT images is suggested [28]. Model combines data collected from various open sources and investigated four architectures: VGG19+CNN, ResNet152V2, ResNet152V2 with Gated Recurrent Unit (GRU), and ResNet152V2 with Bi-GRU. The VGG19+CNN architecture outperformed the other three proposed models with 98.05% accuracy. Dense CNN model DenseNet121 has also illustrated good performance for malignancy detection from of lung X-ray images [29]. GAN is used to synthesize data for ResNet50-based lung cancer classification model [30], reported accuracy is 98.91%. This signifies that GAN can be used to address the issue of imbalance dataset.

3 Discussion

Deep learning techniques have received a lot of attention in the research papers mentioned in the literature. Among various DL designs, CNN performs most well when used to classify tumors. This is because of its influential feature learning potential. RNN, LSTM on other hand are more suitable for processing sequential or temporal data. Further there are different variants of CNN used for tumor classification, each mitigating certain issues. Like residual neural networks alleviate vanishing gradient problem. Inception architecture makes efficient utilization of computing resource that makes it suitable for processing large medical datasets. Classical ML techniques like decision tree, Naïve Bayes, KNN, SVM, etc., have also been used in last years for the automated detection and diagnosis of cancer [31–33]. However comparative results mentioned in articles have marked notable improvement in classification performance with the use of DL techniques over classical ML techniques. The use of hybrid methods, combining ML with DL is also promising to improve results. Pre-processing of images and augmentation of data has also contributed to improve performance with DL methods. Various techniques in literature have also implemented segmentation of tumor region before classification that is an additional overhead. Without segmentation of Region of Interest (ROI) also, state-of-art results have been achieved.

Undoubtedly, AI is playing a big role in clinical decision making and has become transformational force in tumor diagnosis. Still there are different challenges and issues that need focus. DL methods are expensive in terms of cost and time. Due to inter and intra class similarity among various types of tumor, misclassification occurs that leads to false diagnosis. Hence misclassification rate must be reduced. Almost all prediction models have been verified on just one cancerous spot. In order to boost the models' generalizability, they must be evaluated on other cancer locations as well. Effectiveness of DL lies in sufficient training of networks using enough and well-annotated medical images to learn patterns and features of tumor or cancer. But there are relatively few public datasets. Performance evaluation also becomes difficult as many authors have worked with non-public datasets. Large number of un-annotated medical images can also be obtained from clinical setups and unsupervised learning techniques can be used to label them. This will help to increase availability of training data for supervised learning models based on CNN for classification. Methods need to be more robust in a way that slight variations in images must not affect their classification performance.

4 Conclusion

According to the study of recent literature, deep learning techniques have a great deal of promise to speed up the tumor diagnosis. Comparing the observed accuracy to conventional techniques, there is a remarkable performance improvement. There are various CNN designs with greater precision as the number of hidden layers increase. Hence, to learn a varied range of features and boost classification accuracy, advanced CNN architectures must be investigated. One of the drawbacks of deep learning models is their hidden computational nature. This makes it difficult to apply them in clinical settings as they lack transparency. Some authors have used non-public datasets for training and did not fully disclose training process. Hence future works should address these issues. Challenges like resource efficiency, management of large scale datasets should also be addressed in order to make ML/DL-based systems more practical.

References

1. Manne, R., Kantheti, S.C.: Application of artificial intelligence in healthcare: chances and challenges. Curr. J. Appl. Sci. Technol. **40**(6), 78–89 (2021)
2. Chen, M., Decary, M.: Artificial intelligence in healthcare: an essential guide for health leaders. Healthc. Manage. Forum **33**(1), 10–18 (2020)
3. Sarker, I.H.: Machine learning: algorithms, real-world applications and Research Directions. SN Comp. Sci. **2**(3), (2021)
4. Wu, H., Liu, Q., Liu, X.: A review on deep learning approaches to image classification and object segmentation. Comp. Mater. Continua. **60**(2), 575–597 (2019)
5. Alzubaidi, L., Zhang, J., Humaidi, A.J., Al-Dujaili, A., Duan, Y., Al-Shamma, O., Santamaría, J., Fadhel, M.A., Al-Amidie, M., Farhan, L.: Review of deep learning: concepts, CNN architectures, challenges, applications, future directions. J. Big Data **8**(1), (2021)
6. He, Y., Zhao, H., Wong, S.T.C.: Deep learning powers cancer diagnosis in digital pathology. Comput. Med. Imaging Graph. **88**, 101820 (2021)
7. Kourou, K., Exarchos, K.P., Papaloukas, C., Sakaloglou, P., Exarchos, T., Fotiadis, D.I.: Applied machine learning in cancer research: a systematic review for patient diagnosis, classification and prognosis. Comput. Struct. Biotechnol. J. **19**, 5546–5555 (2021)
8. Chang, Y.W., Chen, Y.R., Ko, C.C., Lin, W.Y., Lin, K.P.: A novel computer-aided-diagnosis system for breast ultrasound images based on BI-RADS categories. Appl. Sci. **10**(5), 1830 (2020)
9. Salama, W.M., Aly, M.H.: Deep learning in mammography images segmentation and classification: automated CNN approach. Alexandria Eng. J. **60**(5), 4701–4709 (2021)
10. Shen, L., Margolies, L.R., Rothstein, J.H., Fluder, E., McBride, R., Sieh, W.: Deep learning to improve breast cancer detection on screening mammography. Sci. Rep. **9**(1), (2019)
11. Hassan, S.A., Sayed, M.S., Abdalla, M.I., Rashwan, M.A.: Breast cancer masses classification using deep convolutional neural networks and transfer learning. Multimed. Tools Appl. **79**(41–42), 30735–30768 (2020)
12. Hassan, S.A., Sayed, M.S., Abdalla, M.I., Rashwan, M.A.: Breast Detection of breast cancer mass using MSER detector and features matching. Multimed. Tools Appl. **78**(14), 20239–20262 (2019)
13. Gour, M., Jain, S., Sunil Kumar, T.: Residual learning based CNN for breast cancer histopathological image classification. Int. J. Imaging Syst. Technol. **30**(3), 621–635 (2020)

14. Amin, J., Sharif, M., Haldorai, A., Yasmin, M., Nayak, R.S.: Brain tumor detection and classification using Machine Learning: a comprehensive survey. Compl. Intell. Syst. (2021)
15. Kesav, N., Jibukumar, M.G.: Efficient and low complex architecture for detection and classification of brain tumor using RCNN with two channel CNN. J. King Saud Univ. Comp. Inform. Sci. (2021)
16. Deepak, S., Ameer, P.M.: Brain tumor classification using deep CNN features via transfer learning. Comput. Biol. Med. **111**, 103345 (2019)
17. Khairandish, M.O., Sharma, M., Jain, V., Chatterjee, J.M., Jhanjhi, N.Z.: A hybrid CNN-SVM threshold segmentation approach for tumor detection and classification of MRI Brain Images. IRBM, (2021)
18. Tazin, T., Sarker, S., Gupta, P., Ayaz, F.I., Islam, S., Monirujjaman Khan, M., Bourouis, S., Idris, S.A., Alshazly, H.: A robust and novel approach for brain tumor classification using convolutional neural network. Comput. Intell. Neurosci. **2021**, 1–11 (2021)
19. Alqudah, A.M.: Brain tumor classification using deep learning technique—a comparison between cropped, uncropped, and segmented lesion images with different sizes. Int. J. Adv. Trends Comp. Sci. Eng. **8**(6), 3684–3691 (2019)
20. Gull, S., Akbar, S., Khan, H.U.: Automated detection of brain tumor through magnetic resonance images using convolutional neural network. Biomed. Res. Int. **2021**, 1–14 (2021)
21. Kumar, S., Mankame, D.P.: Optimization driven deep convolution neural network for brain tumor classification. Biocybern. Biomed. Eng. **40**(3), 1190–1204 (2020)
22. Adegun, A., Viriri, S.: Deep learning techniques for skin lesion analysis and melanoma cancer detection: a survey of state-of-the-art. Artif. Intell. Rev. **54**(2), 811–841 (2021)
23. Kareem, O.S., Abdulazeez, A.M., Zeebaree, D.Q.: Skin lesions classification using deep learning techniques. Asian J. Res. Comp. Sci. **9**(1), 1–22 (2021)
24. Ali, M.S., Miah, M.S., Haque, J., Rahman, M.M., Islam, M.K.: An enhanced technique of skin cancer classification using deep convolutional neural network with transfer learning models. Mach. Learn. Appl. **5**, 100036 (2021)
25. Patil, R. Bellary, S.: Machine learning approach in melanoma cancer stage detection. J. King Saud Univ. Comp. Inform. Sci., 1319–1578 (2020)
26. Ali, K., Shaikh, Z.A., Khan, A.A., Laghari, A.A.: Multiclass skin cancer classification using efficientnets—a first step towards preventing skin cancer. Neurosci. Informat. **2**(4), 100034 (2022)
27. Anisuzzaman, D.M., Barzekar, H., Tong, L., Luo, J., Yu, Z.: A deep learning study on osteosarcoma detection from histological images. Biomed. Signal Process. Control **69**, 102931 (2021)
28. Ibrahim, D.M., Elshennawy, N.M., Sarhan, A.M.: Deep-chest: multi-classification deep learning model for diagnosing COVID-19, pneumonia, and lung cancer chest diseases. Comput. Biol. Med. **132**, 104348 (2021)
29. Ausawalaithong, W., Thirach, A., Marukatat, S., Wilaiprasitporn, T.: Automatic lung cancer prediction from chest X-ray images using the deep learning approach. In: 2018 11th Biomedical Engineering International Conference (BMEiCON), pp. 1–5. IEEE, Chiang Mai, Thailand (2018)
30. Salama, W.M., Shokry, A., Aly, M.H.: A generalized framework for lung Cancer classification based on deep generative models. Multimed. Tools Appl., 1–18 (2022)
31. Huang, Q., Chen, Y., Liu, L., Tao, D., Li, X.: On combining biclustering mining and AdaBoost for breast tumor classification. IEEE Trans. Knowl. Data Eng. **32**(4), 728–738 (2019)
32. Anji Reddy, V. and Soni, B.: Breast cancer identification and diagnosis techniques. In: Machine Learning for Intelligent Decision Science. Algorithms for Intelligent Systems. Springer, Singapore, pp. 49–70 (2020). https://doi.org/10.1007/978-981-15-3689-2_3
33. Vidya, M., Karki, M.V.: Skin cancer detection using machine learning techniques. In: 2020 IEEE International Conference on Electronics, Computing and Communication Technologies, pp. 1–5, IEEE.
34. USF Digital Mammography Home Page, http://www.eng.usf.edu/cvprg/mammography/database.html. Last Accessed 8 Dec 2021

35. Mammographic Image Analysis Society (mias) database v1.21, Apollo Home, https://www.repository.cam.ac.uk/handle/1810/250394. Last Accessed 11 Oct 2021
36. CBIS-DDSM—The Cancer Imaging Archive (TCIA) Public Access—Cancer Imaging Archive Wiki, https://wiki.cancerimagingarchive.net/display/Public/CBIS-DDSM. Last Accessed 22 Jan 2022
37. InBreast, Kaggle, https://www.kaggle.com/martholi/inbreast. Last Accessed 1 Dec 2021
38. Breast cancer histopathological database (BreakHis), https://web.inf.ufpr.br/vri/databases/breast-cancer-histopathological-database-breakhis/. Last Accessed 22 Nov 2021
39. Brain tumor dataset, figshare, https://figshare.com/articles/dataset/brain_tumor_dataset/1512427. Last Accessed 16 Dec 2021
40. Challenges "BRATS2015, BRATS—SICAS Medical Image Repository, https://www.smir.ch/BRATS/Start2015. Last Accessed 11 Jan 2022
41. Brain Tumor Dataset, Kaggle, https://www.kaggle.com/preetviradiya/brian-tumor-dataset/code. Last Accessed 27 Feb 2022
42. Brats-2018, Kaggle, https://www.kaggle.com/sanglequang/brats2018. Last Accessed 16 Feb 2022
43. Brats2019_1, Kaggle, https://www.kaggle.com/anassbenfares/brats2019-1. Last Accessed 16 Feb 2022
44. BRATS2020 dataset (training + validation), Kaggle, https://www.kaggle.com/awsaf49/brats20-dataset-training-validation/code. Last Accessed 16 Feb 2022
45. Menze Bjoern, H., et al.: The multimodal brain tumor image segmentation benchmark (BRATS). IEEE Trans. Med. Imaging **34**(10), 1993–2023 (2015)
46. The HAM10000 dataset, a large collection of multi-source dermatoscopic images of common pigmented skin lesions, Harvard Dataverse, https://dataverse.harvard.edu/dataset.xhtml?persistentId=doi%3A10.7910%2FDVN%2FDBW86T. Last Accessed 11 Jan 2022
47. Machine learning methods for binary and multiclass classification of melanoma thickness from dermoscopic images, https://www.uco.es/grupos/ayrna/ieeetmi2015. Last Accessed 6 Jan 2022

Local DCT-Based Deep Learning Architecture for Image Forgery Detection

B. H. Shekar, Wincy Abraham, and Bharathi Pilar

Abstract A convolutional neural network model efficient in forgery detection in images regardless of the type of forgery is proposed. The AC coefficients in the block DCT of the entire image are analysed for the suspected forgery operation. The feature vector is extracted from the non-overlapping blocks of size 8×8 of the image. It consists of the standard deviation and non-zero counts of the block DCT coefficients of the image and its cropped version. The image is first converted to YCbCr colour space. The feature vector is extracted for all three channels. We then supply this feature vector as input to the deep neural network for detection. We have trained the DNN using CASIAv1 and CASIAv2 datasets separately and tested them. The train test ratio used is 80:20 for experimentation. Experimentation results on standard datasets, namely CASIA v1 and CASIA v2, reveal the efficiency of the proposed approach. A comparison with some of the existing approaches shows the proposed approach's performance in terms of detection accuracy.

Keywords Image forgery · Discrete cosine transform · Deep learning · CNN

1 Introduction

Image forgery has now become a common phenomenon due to the availability of a large number of image capturing and manipulation tools today. Image forgery can be done by repositioning objects within the same image, by copying and placing objects in a different region in the same image(copy-move) as shown in Fig. 1 or by placing objects from other images(splicing)as shown in Fig. 2. In this scenario, it becomes difficult for the common man to decide whether what has been seen is real or fake. The authenticity of images is questionable and needs to be verified before they can

B. H. Shekar · W. Abraham (✉)
Department of Computer Science, Mangalore University, Konaje, India
e-mail: wincya@gmail.com

B. Pilar
Department of Computer Science, University College, Mangalore, India

M. Tistarelli et al. (eds.), *Computer Vision and Machine Intelligence*,
Lecture Notes in Networks and Systems 586,
https://doi.org/10.1007/978-981-19-7867-8_37

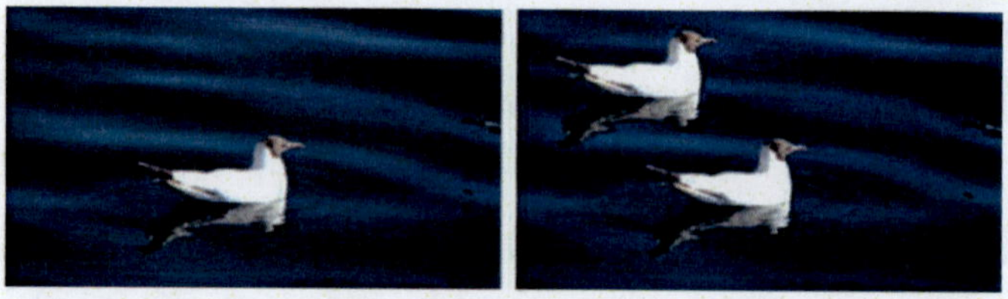

Fig. 1 A sample unaltered and copy-move image (left to right) from CVIP copy-move forgery dataset

Fig. 2 A sample unaltered (first two) and spiced image from CASIA V1.0 dataset

be used for various purposes like as evidence in a court of law or before publishing it through the various media. The lack of robust tools for image forgery detection makes it difficult or impossible to check the authenticity.

Like there are many methods and tools available for performing forgery in digital content, several methods exist to hide the manipulation done on the content. The extent of anti-forensics is such that, even highly efficient algorithms fail under some circumstances to detect forgery. So, while the detection methods are being developed, anti-forensics must be taken care of. Machine learning algorithms for forgery detection need a large dataset, the lack of which leads to poor results. Though many forgery detection methods have been developed, many of them lack robustness concerning the different types of forgery in the same image. Other than anti-forensics, some distortions which occur in the image due to processing done on the image by various social media websites or compression media, make detection difficult. Hence, the research in this area of forgery detection is very challenging and has attracted interest from researchers for the past several years.

2 Review of Literature

Here, we present an overview of the various methods that exist in the state-of-the-art literature for forgery detection. Gani et. al. [1] propose a robust technique for copy-move forgery detection based on sign information of local DCT and cellular automata. K-d tree-based nearest neighbour matching method is used to find the duplicate areas in the image. Cao et al. [2] propose a method in which DCT is performed on the fixed-size blocks of the image. To reduce the size of the feature vector, only, four coefficients are extracted from each block. Though it performs well on multiple copy-move forgery, noise, and blur, there are problems with false matches. In the technique proposed by He et al. [3] for splicing detection, the transition probability matrices in the DCT domain produce the Markov features which are expanded to obtain the correlation between the DCT coefficients across the blocks and in between the blocks. Additional features in the DWT domain are constructed to obtain the dependencies amongst wavelet coefficients across the scale, position, and orientation. SVM-RFE is used for feature selection to reduce computational costs. Li et al. [4] propose a method to perform image splicing detection using Markov features in Quaternion discrete cosine transform (QDCT) domain. Classification is performed by SVM after extracting QDCT coefficients from within the blocks and between the blocks.

The above techniques detect either copy-move or splicing but not both. But, there exists some methods which can detect image forgery irrespective of the type of forgery. Dong et.al. [5] proposed photo forensics from JPEG dimples that makes use of the artefact introduced in JPEG compression. The artefact introduced by various camera models differs from each other, and this becomes helpful in forgery detection by analyzing the associated correlation energy of different blocks. But, the method fails when the image is operated using software that does not produce dimples. Muhammad et al. [6] proposed a method using steerable pyramid transform (SPT) and local binary pattern (LBP). They highlighted the significant role of chrominance channels in forgery detection. The method proposed by Alahmadi et al. [7] can perform both copy-move and splicing forgery detection for an image. It performs image texture analysis using local binary pattern (LBP) and discrete cosine transform (DCT). The LBP code of each block of the forged image is transformed into the DCT domain, and the standard deviation of these DCT block coefficients is computed. However, the technique does not consider post-processing operations performed on the image. Vidyadharan et al. [8] make use of various image texture representations like binary statistical image features, LBP, binary Gabor pattern and local phase quantization to form the feature vector. Random forest classifier performs classification and claims to have significant accuracy.

Prakash et al. [9] proposed another technique for the detection of copy-move as well as splicing forgery. Features from different colour spaces are extracted using the method based on Markov random process. However, the used dataset was not thoroughly tested for the various types of forgeries. The method proposed by Saleh et al. [10] performs image forgery detection using an SVM classifier. It uses multi-scale Weber local descriptors (WLDs), and features are extracted from the chrominance components of the image. Dua et al. [11] proposed a method wherein the YCbCr conversion of the image is done. The local DCT features are extracted from each Y, Cb and Cr channel. The changes that occur after forgery operation in the statistical properties of the AC coefficients and the number of non-zero coefficients provide a clue as to whether the image is forged or not. The features are extracted from the image and the image after cropping. The feature vector which is extracted is used by the SVM classifier for classification. Features corresponding to all three channels are extracted, and the final feature vector is created by concatenating the extracted feature vectors from all three channels.

'Digital image forgery detection using deep learning approach' [12] proposes an algorithm for splicing detection using VGG-16 architecture. Image patches are supplied as input to the model and obtain the classification result. The authors claim 97.8% accuracy for the CASIA V2 dataset.

Ali et al. [13] propose another deep learning method that detects both splicing and copy-move forgeries. The variations in image compression are modelled using CNN architecture. The difference between the original and decompressed images is used as the feature vector. A validation accuracy of 92.23 % is obtained.

In the method proposed by Sudiatmika et al. [14], error level analysis on each image is used as the prominent feature for forgery detection using a deep learning model. Validation accuracy of 88.46% is obtained.

Jiangqun et al. [15] use a deep convolutional neural network (CNN) with a multi-semantic CRF-based attention model for image forgery detection and localization. Boundary transition artefacts which arise due to forgery are captured using the conditional random field-based attention model. Average precision of 0.610 for CASIA1 and 0.699 for CASIA2 are obtained. Many of these methods rely on various transforms of the image instead of dealing with it in the spatial domain. Discrete Wavelet transform (DWT), discrete cosine transform (DCT) and singular value decomposition (SVD) are some of these. An extensively used transform is DCT, as it offers promising results in terms of computational complexity and the level of classification accuracy. At the same time, the use of deep learning models for image forgery detection can be seen in the literature which offers high detection accuracy. Hence, we have been motivated to use DCT features with deep learning architecture for forgery detection. The method by Dua et al. [11] stands as the basis for our work. We extract the features as proposed by Dua et al. [11] and give them to DNN for classification which yields better results for the CASIA v1 dataset.

Section 3 describes the proposed forgery detection algorithm in detail. In Sect. 4, the experimental results of the proposed approach are presented, and Sect. 5 is the conclusion of the paper.

3 Proposed Method

The proposed approach based on local DCT features is presented here. Tampering in the image causes variation in the statistical measures of the block DCT coefficients (BDCT) which can be used as an indication of the forgery. The dominant features are obtained from the image as explained in the paper by Dua et al. [11]. Initially, the image is converted to the YCbCr colour space, and then, all three channels are used for further processing. Each input image channel is first divided into non-overlapping

8×8 blocks, and the DCT of each block is computed and stored column-wise in a matrix containing 64 rows. The number of columns in this matrix will be equal to the number of 8×8 blocks in the image.

Next, the image is cropped by deleting the first four rows and four columns, and the above process is repeated. From the two matrices created which contain the DCT coefficients of all blocks column-wise, the standard deviation of the values in each row is found except for the row corresponding to the DC coefficient. The number of non-zero coefficients is also found in each row except the first row. This results in a 63×2 sized array containing the standard deviation and the number of non-zero DCT coefficients of the 63 rows in the matrix. Thus, there are two such arrays corresponding to the image and the cropped one. Next, these arrays are combined, flattened, and treated as the feature vector. Figure 3 shows the flow diagram of the proposed image forgery detector based on DCT domain features. These features are fed into a deep neural network model for detection.

We have used a deep neural network model using CNN for the classification of images. Convolutional neural networks are found to perform well for various image processing tasks due to the presence of various types of layers in them. It does feature extraction and then classification on its own. We have relied more on

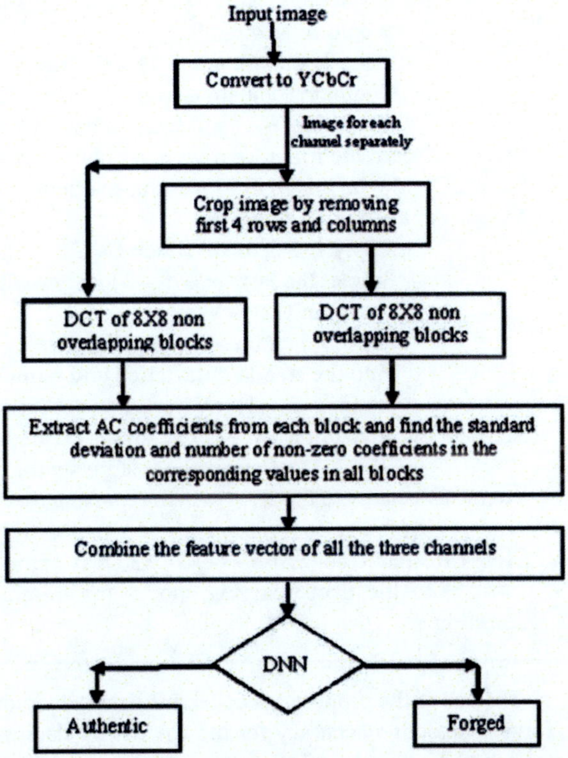

Fig. 3 Design flow of the local DCT feature extraction for forgery detection

the classification capability of the CNN model rather than the feature extraction, as indicated by the preprocessing done on the input before supplying it to the model. It is because of the understanding that standard deviation and the number of non-zero DCT coefficients in the image and the image after cropping is a discriminative feature for forgery detection. Further processing on the extracted features is done by the CNN model which enhances its discriminating capability. Thus, CNN is used here as a discriminator.

4 Algorithm

Steps involved in feature extraction and forgery detection using DNN

Step1: Convert the image to YCbCr colour space.
Step2: For each colour sub-image do
Divide the image in to N non-overlapping m x n blocks
Step3: For each m x n block do
Apply DCT to get Block DCT
Extract the last $(m * n - 1)$ AC coefficients from Block DCT as a column vector V
Place V in a $(m * n - 1)$ xN array
Compute standard deviation and number of ones of each row of the $(m * n - 1)$ xN array to get $(m * n - 1)$ x2 array
Step4: Discard the first four rows and columns of the image
Step5: Divide the image in to $N1$ non-overlapping m xn blocks
Step6: For each m xn block do
Apply DCT to get Block DCT
Extract the last $(m * n - 1)$ AC coefficients from Block DCT as a column vector V
Place V in a $(m * n - 1)$ x$N1$ array
Find the standard deviation and number of ones of each row of the $(m * n - 1)$ x$N1$ array to get $(m * n - 1)$ x2 array.
Step7: Concatenate $(m * n - 1)$ x2 arrays created in steps 3 and 6 to get $(m * n - 1)$ x4 array, the feature vector
Step8: Concatenate the array which represent the feature vector for the subimage to form $m * n - 1$ x4 x3 array
Step9: Reshape the $(m * n - 1)$ x4 x3 array appropriately, give it as input to the deep learning model for training and prediction for forgery detection

---

The use of the proposed deep learning model yields considerable improvement in forgery detection accuracy for the CASIA v1 dataset. Here, we describe the architecture of the deep learning system which very well performs the classification of the

Table 1 Summary of the DNN architecture

Layer	Filter size	No. of neurons	Activation
Conv1	7×7	42	Relu
Conv2	5×5	32	Relu
Conv3	3×3	16	Relu
Conv4	3×3	8	Relu
Conv5	3×3	1	Relu
FullyConnected1	–	40	Relu
FullyConnected2	–	1	Sigmoid

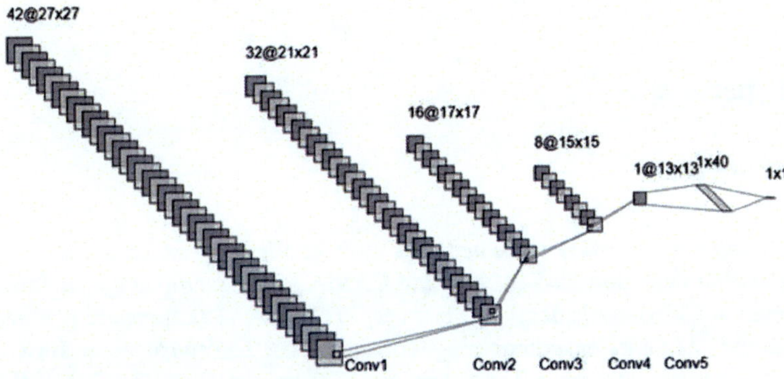

Fig. 4 Architecture of the DNN for forgery detection

image. To build a forgery detection classifier, we adopt a CNN architecture with 5 convolution layers Conv1 to Conv5 and two fully connected layers. One-dimensional feature vector of size 729 of each image is reshaped to (27, 27, 1) and fed to Conv1 as input. Table 1 shows the summary of the architecture.

Conv1 uses a filter of size 7×7; Conv2 uses a filter of size 5×5, and the convolutional layers Conv3 to Conv5 use a filter of size 3×3. The number of filters used in layers Conv1 to Conv5 is 42,32,16,8, and 1 respectively. Since local features are extracted and become part of the computation of standard deviation and number of ones as a feature vector for the image, no further exploitation of local features is impossible and unnecessary too.

Also, since the image is already pre-processed and features extracted, initial convolutional layers themselves are deep enough to extract more abstract features out of the input data. Use of the pooling layers and batch normalization is avoided. All the convolutional layers use 'relu' activation. The first fully connected layer uses 40 neurons and 'relu' activation while the second fully connected layer uses one neuron with 'sigmoid' activation which performs the classification. Figure 4 shows the architecture of the DNN for forgery detection. We have used a ratio of 80:20 for splitting the dataset for training and testing, respectively.

Table 2 Forgery detection accuracy for CASIA v1 and CASIA v2 datasets

Method	CASIA V1 (%)	CASIA V2 (%)
Proposed approach	**95.08**	**97.50**
Vidyadharan et al. [8]	94.13	97.03
Muhammad et al. [6]	94.89	97.33
Saleh et al. [10]	94.19	96.61
Dua et al. [11]	93.2	98.3
Ali et al. [13]	–	92.23
Sudiatmika et al. [14]	88.46	–
Kusnetsov et al. [12]	–	97.8

Bold for the result of proposed approach

5 Experiments

5.1 *Dataset*

The proposed approach is evaluated on a more challenging and realistic dataset for manipulation detection CASIA v1.0 and CASIA v2.0 developed by the Institute of Automation Chinese Academy of Sciences. In CASIA v1.0, there are 800 authentic images and 921 tampered colour images of size 384×256 and all are in JPEG format without any post-processing. Thus, there are 1721 images altogether. In CASIA v2.0, there are 7491 authentic and 5123 forged multiple-sized colour images with sizes varying from 240×160 to 900×600 pixels. Several post-processing operations are applied to the images across edges. In addition, the images are available in JPEG format with different quality factors and also in uncompressed form.

5.2 *Performance Analysis*

The proposed system is capable of classifying the images as authentic or forged regardless of the type of forgery. Accuracy is used as a performance evaluation metric. Accuracy is the percentage of correct predictions for the test data. In the experiment, first, the features are extracted from the whole CASIA v1 and CASIA v2 datasets. Then, the feature vectors are fed to the DNN separately, and the performance metrics are calculated. Adam optimizer with a learning rate of 0.001 is used. Initial decay rates beta1 = 0.9 and beta2 = 0.999 are used. 0.0001 is set as an exponential decay parameter with an epsilon value of 1e-08. The DNN is trained using a batch size of 10. Table 2 shows the accuracy of the proposed method compared to the state-of-the-art methods for the CASIA v1 and CASIA v2 datasets.

6 Conclusion

In this paper, we presented a deep neural network model which is efficient in image forgery detection regardless of the type of forgery. The changes in the statistical features of the block DCT coefficients are captured as feature vectors. The extracted features are used for classification by the DNN model. We have trained the DNN using CASIA v1 and CASIA v2 datasets separately and tested them. Experimentation results on standard datasets, namely CASIA v1 and CASIA v2, reveal the efficiency of the approach proposed by us. Comparison with some of the existing approaches demonstrates the proposed approach's performance in terms of detection accuracy.

References

1. Gani, G., Qadir, F.: A robust copy-move forgery detection technique based on discrete cosine transform and cellular automata. J. Inf. Secur. Appl. (2020). [Elsevier]
2. Cao, Y., Gao, T., Fan, L., Yang, Q.: A robust detection algorithm for copy-move forgery in digital images. Forensic Sci. Int. (2012). [Elsevier]
3. He, Z., Lu, W., Sun, W., Huang, J.: Digital image splicing detection based on Markov features in DCT and DWT domain. Pattern Recogn. (2012). [Elsevier]
4. Li, C., Ma, Q., Xiao, L., Li, M., Zhang, A.: Image splicing detection based on Markov features in QDCT domain. Neurocomputing (2017). [Elsevier]
5. Dong, J., Wang, W., Tan, T.: Casia image tampering detection evaluation database. In: 2013 IEEE China Summit and International Conference on Signal and Information Processing, IEEE (2013)
6. Muhammad, G., Al-Hammadi, M.H., Hussain, M., Bebis, G.: Images forgery detection using steerable pyramid transform and local binary pattern. Mach. Vis. Appl. (2014). [Springer]
7. Amani, A., Hussain, M., Hatim, A., Muhammad, G., Bebis, G., Mathkour, H.: Passive detection of image forgery using DCT and local binary pattern. Signal Image Video Process. (2017). [Springer]
8. Vidyadharan, D.S., Thampi, S.M.: Digital image forgery detection using compact multi-texture representation. J. Intell. Fuzzy Syst. (2017). [IOS Press]
9. Prakash, C.S., Kumar, A., Maheshkar, S., Maheshkar, V.: An integrated method of copy-move and splicing for image forgery detection. Multimedia Tools Appl. (2018). [Springer]
10. Saleh, S.Q., Hussain, M., Muhammad, G., Bebis, G.: Evaluation of image forgery detection using multi-scale weber local descriptors. In: International Symposium on Visual Computing, Springer (2013)
11. Dua, S., Singh, J., Harish, P.: Image forgery detection based on statistical features of block DCT coefficients. Procedia Comput. Sci. (2020). [Elsevier]
12. Kuznetsov, A.: Digital image forgery detection using deep learning approach. J. Phys. Conf. Ser. **1368**, 032028 (2019)
13. Ali, S.S., Ganapathi, I.I., Vu, N.-S., Ali, S.D., Saxena, N., Werghi, N.: Image forgery detection using deep learning by recompressing images. Electronics (2022)
14. Sudiatmika, I.B.K., Rahman, F., Trisno, T.: Image forgery detection using error level analysis and deep learning. TELKOMNIKA Telecommun. Comput. Electron, Control (2018)
15. Rao, Y., Ni, J., Xiea, H.: Multi-semantic CRF-based attention model for image forgery detection and localization. Signal Process. (2021). [Elsevier]

Active Domain-Invariant Self-localization Using Ego-Centric and World-Centric Maps

Kanya Kurauchi, Kanji Tanaka, Ryogo Yamamoto, and Mitsuki Yoshida

Abstract The training of a next-best-view (NBV) planner for visual place recognition (VPR) is a fundamentally important task in autonomous robot navigation, for which a typical approach is the use of visual experiences that are collected in the target domain as training data. However, the collection of a wide variety of visual experiences in everyday navigation is costly and prohibitive for real-time robotic applications. We address this issue by employing a novel *domain-invariant* NBV planner. A standard VPR subsystem based on a convolutional neural network (CNN) is assumed to be available, and its domain-invariant state recognition ability is proposed to be transferred to train the domain-invariant NBV planner. Specifically, we divide the visual cues that are available from the CNN model into two types: the output layer cue (OLC) and intermediate layer cue (ILC). The OLC is available at the output layer of the CNN model and aims to estimate the state of the robot (e.g., the robot viewpoint) with respect to the world-centric view coordinate system. The ILC is available within the middle layers of the CNN model as a high-level description of the visual content (e.g., a saliency image) with respect to the ego-centric view. In our framework, the ILC and OLC are mapped to a state vector and subsequently used to train a multiview NBV planner via deep reinforcement learning. Experiments using the public NCLT dataset validate the effectiveness of the proposed method.

Keywords Visual robot place recognition · Domain-invariant next-best-view planner · Transferring convnet features

K. Kurauchi · K. Tanaka (✉) · R. Yamamoto · M. Yoshida
University of Fukui, 3-9-1 bunkyo, Fukui, Japan
e-mail: tnkknj@u-fukui.ac.jp

R. Yamamoto
e-mail: mf220362@g.u-fukui.ac.jp

M. Yoshida
e-mail: mf210369@g.u-fukui.ac.jp

M. Tistarelli et al. (eds.), *Computer Vision and Machine Intelligence*,
Lecture Notes in Networks and Systems 586,
https://doi.org/10.1007/978-981-19-7867-8_38

1 Introduction

The training of a next-best-view (NBV) planner for visual place recognition (VPR) is fundamentally important for autonomous robot navigation. VPR is typically formulated as a passive single-view image classification problem in visual robot navigation [1], with the aim of classifying a view image into one of the predefined place classes. However, such passive formulation is ill-posed, and its VPR performance is significantly dependent on the presence or absence of landmark-like objects in the view image [2]. To address this ill-posedness, several studies have extended the passive VPR task to an active multiview VPR with NBV planning [3]. Active VPR is formulated as a state-to-action mapping problem that aims to maximize the VPR performance while suppressing the expected costs of action and sensing at future viewpoints. The standard approach involves training the NBV planner using the visual experience that is collected in the target domain as training data [4–6]. However, this requires the collection of a large, diverse set of visual experiences for everyday navigation, which is expensive and prohibitive for real-time robotic applications.

In this study, as opposed to the common path of training from visual experiences, we exploit a deep convolutional neural network (CNN) as a source of training data (Fig. 1). It has been demonstrated that a CNN classifier that is employed by a standard single-view VPR module [1] can provide domain-invariant visual cues [7], such as activations that are invoked by a view image. This motivated us to reuse the available CNN classifier as the source of domain-invariant visual cues. To this end, we reformulate the NBV problem to use domain-invariant visual cues instead of a raw view image as the visual input to the NBV planner. Thus, synthetic visual input is used as the training data to train an NBV planner via reinforcement learning, in the spirit of the recent paradigm of simulation-based training [8]. Because the already available CNN model is reused for supervision, it is not necessary to collect training data in the target domain.

Specifically, we divide the visual cues that are available from the CNN model into two types: the output layer cue (OLC) and intermediate layer cue (ILC). The OLC is available at the output layer of the CNN model as an estimate of the robot state (e.g., a viewpoint estimate) with respect to the world-centric view [9]. The OLC is useful for an NBV planner to learn the optimal state-specific motion planning on a prebuilt environment map. The ILC is available within the intermediate layers of the CNN model as a high-level description of visual contents (e.g., a saliency image) with respect to the ego-centric view [10]. The ILC is useful for an NBV planner to learn the optimal scene-specific behavior given a view image.

We subsequently exploit the OLC and ILC within a practical framework for active VPR. The ILC is implemented as a saliency image, which can be obtained within the intermediate layers of the CNN model using a saliency imaging technique [10]. The dimension reduction from the saliency map to a state vector is trained using a lightweight proxy NBV task. The OLC is implemented as a place-specific probabilistic distribution vector (PDV), which is obtained at the output layer of the CNN

(a) ILC in the ego-centric view (b) OLC in the world-centric view

Fig. 1 We divide the visual cues that are available from the CNN model into two types such as the OLC and ILC (visualized by the heat maps) and fuse the OLC and ILC into a new state vector to reformulate the NBV planning as a domain-invariant task

model. A sequential robot self-localization framework [11] is employed to integrate a sequence of place-specific PDVs from multiple viewpoints during an active VPR task. Thereafter, the OLC and ILC are fused into a compact state vector via unsupervised information fusion [12]. An NBV planner is trained as state-to-action mapping using deep Q-learning with delayed rewards [13] with the experience replay strategy. Experiments using the publicly available NCLT dataset validate the effectiveness of the proposed framework.

2 Related Work

Active VPR plays a particularly important role in scenarios in which robots navigate featureless environments. Typical examples of such scenarios include indoor VPR tasks in featureless passageways and classrooms and outdoor VPR tasks in featureless road environments [3]. Other important scenarios include active SLAM tasks, in which an alternative VPR task known as active loop closing is considered to recognize revisits using an incomplete environmental map that is being built and often featureless [14].

Motion planning based on deep reinforcement learning is being studied recently, mainly in the application domain of local planning. In [4], a map-less navigation framework with a multiview perception module based on an attention mechanism is presented to filter out redundant information caused by multi-camera sensing. In [5], a new target-driven visual navigation framework is presented to enable rapid adaptation to new sensor configurations or target objects with a few shots. While these existing studies do not address the use of global maps, we are interested in active self-localization, and the use of global maps in visual local planning.

Although most existing active VPR methods are non-deep, in recent years, attempts have been made to boost active VPR using deep learning. In [15], an active VPR model known as the "active neural localizer" was trained using neural networks. This model consists of a Bayes filtering component for VPR, perceptual component for estimating the observation likelihood, and structured component for expressing beliefs. It can be trained from data end to end and achieves high VPR accuracy while minimizing the cost of action and sensing during multiview VPR tasks. An efficient active SLAM framework was presented in [16]. Although the use of learning techniques for exploration is well motivated, the end-to-end training of NBV planners is costly and prohibitive. To address this computational load, a neural SLAM module, global policy module, and local policy module were combined in a hierarchical framework. It was possible to reduce the search space during learning significantly without sacrificing the performance by means of this hierarchical module configuration.

Whereas these existing works were aimed at the efficient domain adaptation of an NBV planner, we attempt to realize a domain-invariant NBV planner that does not require adaptation to a new domain. This approach is motivated by the recent paradigm of long-term VPR, in which robots are required to be domain invariant rather than domain adaptive. However, these long-term VPR frameworks have primarily been studied for passive VPR applications [1]. Our research extends this invariant VPR framework from passive to active. We reformulate the NBV planning task as a simulation-based training problem in which the already available passive VPR module is reused as a teacher model; thus, no overhead of training costs per domain is required. Specifically, the focus of our research is on the reuse of the CNN and knowledge transfer from VPR to NBV, which has not been explored in the previous studies.

Using the combination of ego-centric and world-centric maps as input to an action planner has been very popular in the field of mobile robotics for over 20 years [16]. However, many of these approaches have assumed that the relationship between the ego-centric and world-centric coordinate systems is given or precisely measured (e.g., w/ LRF scanner). This does not apply to the general "lost robot" problem, including the VPR applications of this paper. Our approach is motivated by the availability of OIC and ILC available from the recent deep learning-based VPR model, which are proposed to be represented by the world-centric and ego-centric coordinate systems in this study.

3 Approach

Our goal is to extend a typical passive single-view VPR task to active multiview VPR. Single-view VPR is a passive image classification task that aims to predict the place class $c \in C$ from a view image s for a predefined set C of places. Multiview VPR aims to make a prediction from a view sequence rather than from a single view. Naturally, the VPR performance is strongly dependent on the view sequence. Hence, active

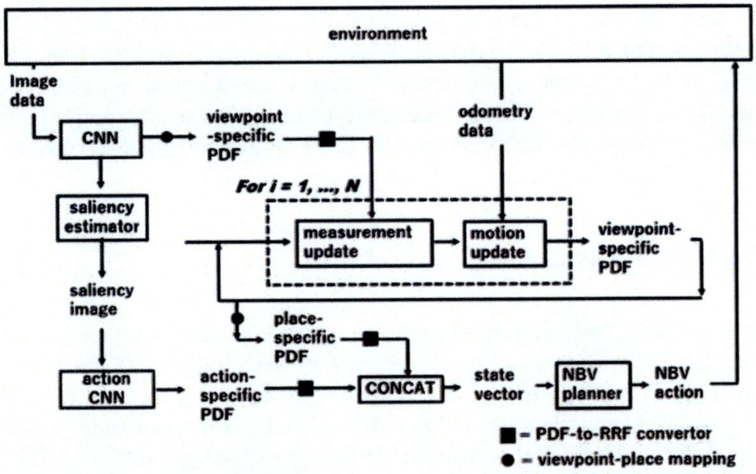

Fig. 2 Active VPR framework. The NBV planner is trained by transferring two types of ConvNet features: OLC ("viewpoint-specific PDV") and ILC ("saliency image"), from the already available CNN model

VPR with viewpoint control plays an important role in multiview VPR. The training of the NBV planner is formulated as a machine learning task with delayed rewards. That is, in the training stage, the reward for the success or failure of the multiview VPR task is delayed until the final viewpoint in each training episode. Then, in the test stage, at each step $t \in [0, T]$, the NBV action a_t is planned incrementally based on the action-sensing sequence $(a_1, s_1) \cdots (a_{t-1}, s_{t-1})$, and the VPR performance at the final viewpoint $t = T$ in the test episodes is expected to be maximized.

Figure 2 presents the active VPR framework. In our approach, a CNN model is assumed to be pretrained as a visual place classifier (Sect. 3.1), and the aim is for its domain-invariant state recognition ability to be transferred to the NBV planner. The CNN model provides two domain-invariant cues: the OLC (Sect. 3.2) and ILC (Sect. 3.3). These cues are subsequently fused into a single-state vector to be transferred to the NBV planner. As no supervision is available in autonomous robotics applications, an unsupervised fusion method is adopted (Sect. 3.4). Subsequently, an NBV planner in the form of state-to-action mapping is trained via deep Q-learning with delayed rewards (Sect. 3.5). Each of these steps is described in detail below.

3.1 VPR Model

The CNN model is trained as a visual place classifier to classify a given view image into one of the $|C|$ predefined place classes via a standard protocol of self-supervised learning [9]. The number of training epochs is set to 300,000. Prior to training, the training view sequence is partitioned into travel distances of 100 m. Accordingly, the view images along the training view sequence are divided into $|C|$ class-specific sets of training images with successive time stamps. For supervision, each training image is annotated with a pseudo-ground-truth viewpoint, which can be reconstructed from the training view sequence using a structure-from-motion technique [17]. Notably, this process is fully self-supervised and does not require human intervention.

3.2 CNN Output-Layer Cue

A multiview VPR task aims to integrate a sequence of ego-motion and perception measurements incrementally into an estimate of the *viewpoint* of the robot in the form of a viewpoint-specific PDV (Fig. 1b). This task is formulated as sequential robot self-localization via a Bayes filter [18]. It consists of two distinctive modules, namely motion updates and perception updates, as illustrated in Fig. 2. The inputs to these modules are the odometry and visual measurements at each viewpoint. The output from the Bayes filter is the belief of the most recent viewpoint in the form of a viewpoint-specific PDV. The Markov localization algorithm [18] is employed to implement the Bayes filter. The state space is defined as a one-dimensional space that represents the travel distance along the training view sequence and a spatial resolution of 1 m is used.

The viewpoint-specific PDV that is maintained by the Bayes filter subsystem can be mutually converted into a place-specific PDV that is maintained by the VPR system. The conversion from the viewpoint-specific PDV to the place-specific PDV is defined as the operation of marginalization for each place class. The conversion from the place-specific PDV to the viewpoint-specific PDV is defined as an operation of normalization with the size of the place-specific set of viewpoints.

The $|C|$-dim place-specific PDV vector represents the knowledge to be transferred from the VPR to the NBV modules. It is also used as the final output of the multiview VPR task at the final viewpoint of a training/test episode, which is subsequently used to compute the reward in the training stage or to evaluate the performance in the test stage.

3.3 CNN Intermediate Layer Cue

The aim of the saliency imaging model is to summarize where the highly complex CNN "looks" in an image for evidence for its predictions (Fig. 1a). It is trained via the saliency imaging technique in [10] during the VPR training process. Once it is trained, the model predicts a grayscale saliency image from the intermediate signal of the CNN. This saliency image provides pixel-wise intensity, indicating the part of the input image that is the most responsible for the decision of the classifier. Notably, the model is domain invariant; that is, it is trained only once in the training domain, and no further retraining is required.

A saliency image is too high dimensional to be used directly as input into an NBV planner. Its dimensionality is proportional to the number of image pixels and has an exponential impact on the cost of simulating the possible action-sensing sequences for multiview VPR planning.

To address this issue, a dimension reduction module is trained using a proxy NBV task. The only difference between this proxy NBV task and the original NBV task is the length of the training/test episode. That is, in the proxy task, only the episodes that consist of a single action (i.e., $T = 1$), which aims to classify a view image into an *optimal* NBV action, are considered. The optimal action for a viewpoint in a training episode is defined as the action that provides the highest VPR performance among action set A at that viewpoint.

The main benefit of the use of such a proxy single-view task is that it enables the reformulation of the NBV task as an image classification task instead of the multiview NBV planning task. That is, as opposed to the original task of multiview active VPR, reinforcement learning is not required to train such an action CNN.

Once it has been trained, the action CNN classifier can be viewed as a method for dimension reduction, which maps an input saliency image to a compact $|A|$-dim action-specific PDV. Any model architecture can be used for this dimension reduction, and in our implementation, a CNN classifier model with a standard training protocol is used.

3.4 Reciprocal Rank Transfer

A discriminative feature known as the reciprocal rank feature (RRF) is used as the input for the VPR-to-NBV knowledge transfer. In the field of multimodal information retrieval [23], the RRF is a discriminative feature that can model the output of a ranking function with unknown characteristics (e.g., a retrieval engine), and it can be used for cross-engine information fusion from multiple retrieval engines. As VPR is an instance of a ranking function, the RRF representation can be applied to an arbitrary VPR model. It should be noted that it is a ranking-based feature, and thus, it can make use of the excellent ranking ability of the CNN model.

Specifically, an RRF vector is processed in the following procedure. First, it is computed by sorting the elements in a given PDV vector in the descending order. Then, the two types of PDV cues, the OLC and ILC, are obtained in the form of $|C|$-dim and $|A|$-dim RRF vectors. Then, they are concatenated into a $(|A| + |C|)$-dim vector.

3.5 Training NBV Planner

The NBV planner is trained using a deep Q-learning network (DQN) [19]. A DQN is an extension of Q-learning [13] that addresses the computational intractability of standard table-like value functions for high-dimensional state/action spaces. The basic concept of the DQN is the use of a deep neural network (instead of a table) to approximate the value function.

The RRF vector is used as the state vector for the DQN. The action set consists of 30 discrete action sets consisting of 1, 2, ..., and 30 m forward movements along the training viewpoint trajectory. The number of training episodes is set to 300,000. The experience replay technique is used to stabilize the DQN training process. The implementation of the DQN follows the original work in [19].

The reward is determined based on whether the prediction of the multiview VPR in the training episode matches the correct answer. First, a sequence of odometry and image data in a training episode is integrated into an estimate of the viewpoint-specific PDV using the Bayes filter. The viewpoint-specific PDV is subsequently mapped onto a place-specific PDV. Thereafter, the place-specific PDV is translated into the top-1 place class ID. If this top-1 prediction result matches the ground-truth place class, a positive reward of +1 is assigned; otherwise, a negative reward of −1 is assigned.

4 Experiments

4.1 Settings

The public NCLT dataset [20] was used in the experiments. This dataset contains empirical data that were obtained by operating a Segway robot at the University of Michigan North Campus at different times of the day and in different seasons.

Images from the front-facing camera of the onboard sensor LadyBug3 were used as the main modality. Moreover, GPS data from an RTK-GPS were used to reconstruct the ground truth and to simulate the movements of the robot during the training and testing episodes. The images exhibited various appearances, including snow cover and falling leaves, as illustrated in Fig. 3.

Notably, in our scenario of a domain-invariant NBV planner, the model was trained only once in the training domain, and no further retraining for new test domains could be conducted. Five sessions, namely "2012/1/8," "2012/1/15," "2012/3/25," "2012/8/20," and "2012/9/28," were used as the test domains. The number of test episodes was set to 5000. One session, "2012/5/26," which had one of the largest area coverages and longest travel distances (6.3 km), was used as the training domain.

The number of actions per episode was set to $T = 3$. That is, the estimate at the 3rd viewpoint of the Bayes filter, which integrated measurements from the 0th $\cdots$ 3rd viewpoints, was used as the final output of our active VPR system.

The dataset was preprocessed as follows: The viewpoint trajectories of the robot were discretized using a spatial resolution of 1 m in terms of the travel distance. When multiple view images belonged to the same discretized viewpoint, the image with the youngest time stamp was used. The starting point of each episode was randomly

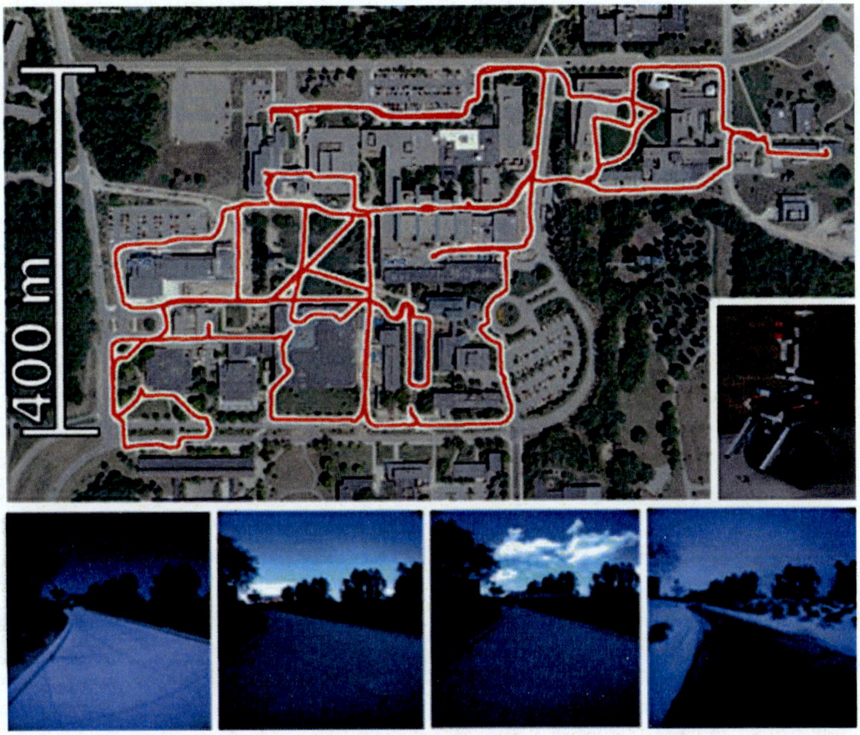

Fig. 3 Experimental environment. Top: entire trajectories and Segway vehicle robot. Bottom: views from on board front-facing camera in different seasons

selected. For an episode in which the starting viewpoint was very close to the unseen area in the workspace, the robot was often forced out by an action to the unseen area, and such a test episode was simply discarded and replaced with a newly sampled episode.

The proposed method was compared with four comparative methods: the single-view method, random method, OLC-only method, and ILC-only method. A passive single-view VPR scenario was assumed in the single-view method. This method could be viewed as a baseline for verifying the significance of multiview methods. The random method was a Naive multiview method in which the action at each viewpoint in each episode was randomly sampled from action set A. This method could be viewed as a baseline for verifying the significance of the NBV planning. The OLC-only method was an ablation of the proposed method in which the NBV planner used the OLC as a $|C|$-dim state vector. The ILC-only method was another ablation of the proposed method in which the NBV planner used the ILC as an $|A|$-dim state vector.

The mean reciprocal rank (MRR) [21] was used as the performance index. The MRR is commonly used to evaluate information retrieval algorithms, where a larger

MRR indicates better performance. The maximum value of MRR = 1 when the ground-truth class can be answered first for all queries (i.e., test episodes). The minimum value of MRR = 0 when the correct answer does not exist within the shortlist for all queries.

4.2 Results

The proposed method and four comparative methods were evaluated. We were particularly interested in investigating the robustness of the viewpoint planning against changes in the environments and the contribution of the viewpoint planning to improving the recognition performance.

Figure 4 depicts examples of view sequences along the planned viewpoint trajectories for the proposed and random methods. The results of these two methods exhibited a clear contrast. For example A, the proposed NBV method yielded a good viewpoint at which landmark-like objects such as tubular buildings and streets were observed, whereas the random method yielded featureless scenes. For example B, major strategic forward movements were planned by the proposed method to avoid featureless scenes, whereas short travel distance movements by which the robot remained within featureless areas were yielded by the random method. For example C, the proposed method was successful in finally moving to the front of the characteristic building, whereas the random method provided many observations that did not include landmark-like objects.

Table 1 presents the performance results. It can be observed that the performance of the single-view method deteriorated significantly in the test on "2012/1/15." This is because the overall appearance of the scene was very different from that in the training domain, mainly owing to snow cover. A comparison of the random and

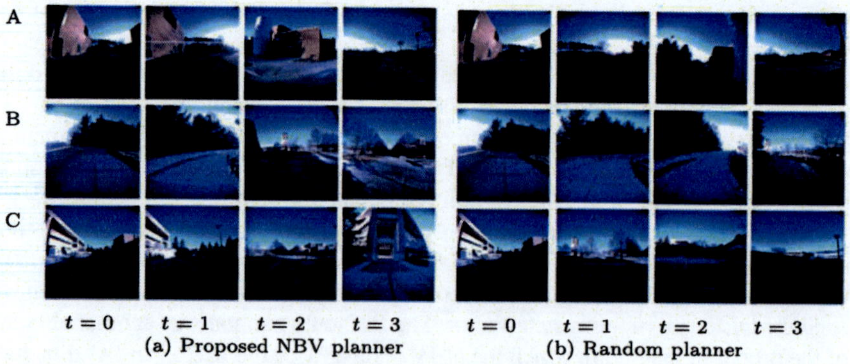

Fig. 4 Example view images at the 0th⋯3rd NBVs planned by the proposed method (**a**) and by the random method (**b**) are depicted for three different starting viewpoints

Table 1 Performance results

Test session	2012/1/8	2012/1/15	2012/3/25	2012/8/20	2012/9/28
Single-view [1]	0.441	0.293	0.414	0.345	0.365
Random	0.547	0.413	0.538	0.457	0.542
OLC-only [22]	0.619	0.471	0.579	0.497	0.567
ILC-only	0.625	0.457	0.585	0.494	0.608
Proposed	**0.647**	**0.493**	**0.596**	**0.518**	**0.623**

Bold indicates the method with the best score

single-view methods revealed that the former outperformed the latter in all test data, according to which the significance of the multiview VPR was confirmed. The ILC-only method outperformed the random method in all test data, indicating that the ILC is effective for active VPR. A comparison of the ILC-only method and OLC-only method demonstrated that the superiority or inferiority of these two methods is significantly dependent on the type of test scene. As expected, the OLC was relatively ineffective from the early viewpoints of the multiview VPR task, as inferences from the measurement history could not be leveraged. However, the ILC was relatively ineffective when the input scene was not discriminative because the saliency image was less reliable. For both methods, buildings often functioned as effective landmarks, especially in outdoor environments. There are two main reasons. First, the excellent object recognition capabilities of CNN allowed the landmarks to distinguish from each other if the entire building was visible in the field image at sufficient resolution. Second, when the distance to the building was too close or too far, the building land arc could be observed properly by moving the robot forward by the travel distance learned in the training domain. In the early stages of each episode, OLC was often ineffective. In fact, its effectiveness was comparable to that of the random method. The main reason is that the OLC integrates past inference results in the episode and is not sufficiently informative in the early stages. It can be concluded that the OLC and ILC have complementary roles with different advantages and disadvantages. Importantly, the proposed method, which combines the advantages of the two methods, outperformed the individual methods in all test domains.

5 Conclusions

A new training method for NBV planners in VPR has been presented. In the proposed approach, simulation-based training is realized by employing the available CNN model as a teacher model. The domain-invariant state recognition capability of the CNN model is transferred, and the NBV planner is trained to be domain invariant. Furthermore, two types of independent visual cues have been proposed for extraction from this CNN model. Experiments using the public NCLT dataset demonstrated the effectiveness of the proposed method.

References

1. Masone, C., Caputo, B.: A survey on deep visual place recognition. IEEE Access **9**, 19516–19547 (2021)
2. Berton, G., Masone, C., Paolicelli, V., Caputo, B.: Viewpoint invariant dense matching for visual geolocalization. In: 2021 IEEE/CVF International Conference on Computer Vision, ICCV 2021, Montreal, QC, Canada, October 10–17, 2021, pp. 12149–12158. IEEE (2021)
3. Khalvati, K., Mackworth, A.K.: Active robot localization with macro actions. In: 2012 IEEE/RSJ International Conference on Intelligent Robots and Systems, IROS 2012, Vilamoura, Algarve, Portugal, October 7–12, 2012, pp. 187–193. IEEE (2012)
4. Huang, X., Chen, W., Zhang, W., Song, R., Cheng, J., Li, Y.: Autonomous multi-view navigation via deep reinforcement learning. In: 2021 IEEE International Conference on Robotics and Automation (ICRA), pp. 13798–13804. IEEE (2021)
5. Luo, Q., Sorokin, M., Ha, S.: A few shot adaptation of visual navigation skills to new observations using meta-learning. In: 2021 IEEE International Conference on Robotics and Automation (ICRA), pp. 13231–13237. IEEE (2021)
6. Kretzschmar, H., Markus, S., Christoph, S., Burgard, W.: Socially compliant mobile robot navigation via inverse reinforcement learning. Int. J. Robotics Res. **35**(11), 1289–1307 (2016)
7. Hausler, S., Garg, S., Xu, M., Milford, M., Fischer, T.: Patch-netvlad: multi-scale fusion of locally-global descriptors for place recognition. In: IEEE Conference on Computer Vision and Pattern Recognition, CVPR 2021, virtual, June 19–25, 2021, pp. 14141–14152. Computer Vision Foundation/IEEE (2021)
8. Zhao, W., Queralta, J.P., Westerlund, T.: Sim-to-real transfer in deep reinforcement learning for robotics: a survey. In: 2020 IEEE Symposium Series on Computational Intelligence, SSCI 2020, Canberra, Australia, December 1–4, 2020, pp. 737–744. IEEE (2020)
9. Krizhevsky, A., Sutskever, I., Hinton, G.E.: Imagenet classification with deep convolutional neural networks. Commun. ACM **60**(6), 84–90 (2017)
10. Fong, R.C., Vedaldi, A.: Interpretable explanations of black boxes by meaningful perturbation. In: IEEE International Conference on Computer Vision, ICCV 2017, Venice, Italy, October 22–29, 2017, pp. 3449–3457. IEEE Computer Society (2017)
11. Burgard, W., Fox, D., Thrun, S.: Markov localization for mobile robots in dynamic environments. CoRR. abs/1106.0222 (2011)
12. Cormack, G.V., Clarke, C.L.A., Büttcher, S.: Reciprocal rank fusion outperforms condorcet and individual rank learning methods. In: Allan, J., Aslam, J.A., Sanderson, M., Zhai, C.X., Zobel, J. (eds.) Proceedings of the 32nd Annual International ACM SIGIR Conference on Research and Development in Information Retrieval, SIGIR 2009, Boston, MA, USA, July 19–23, 2009, pp. 758–759. ACM (2009)
13. Chevtchenko, S.F., Ludermir, T.B.: Learning from sparse and delayed rewards with a multilayer spiking neural network. In: 2020 International Joint Conference on Neural Networks, IJCNN 2020, Glasgow, United Kingdom, July 19-24, 2020, pp. 1–8. IEEE (2020)
14. Lee, E.M., Choi, J., Lim, H., Myung, H.: REAL: rapid exploration with active loop-closing toward large-scale 3d mapping using UAVs. In: IEEE/RSJ International Conference on Intelligent Robots and Systems, IROS 2021, Prague, Czech Republic, September 27–October 1, 2021, pp. 4194–4198. IEEE (2021)
15. Chaplot, D.S., Parisotto, E., Salakhutdinov, R.: Active neural localization. arXiv preprint arXiv:1801.08214 (2018)
16. Chaplot, D.S., Gandhi, D., Gupta, S., Gupta, A., Salakhutdinov, R.: Learning to explore using active neural SLAM. In: 8th International Conference on Learning Representations, ICLR 2020, Addis Ababa, Ethiopia, April 26–30, 2020. OpenReview.net (2020)
17. Schonberger, J.L., Frahm, J.-M.: Structure-from-motion revisited. In: Proceedings of the IEEE conference on computer vision and pattern recognition, pp. 4104–4113 (2016)
18. Thrun, S., Burgard, W., Fox, D.: Probabilistic Robotics (Intelligent Robotics and Autonomous Agents). MIT Press (2005)

19. Mnih, V., Kavukcuoglu, K., Silver, D., Rusu, A.A., Veness, J., Bellemare, M.G., Graves, A., Riedmiller, M.A., Fidjeland, A., Ostrovski, G., Petersen, S., Beattie, C., Sadik, A., Antonoglou, I., King, H., Kumaran, D., Wierstra, D., Legg, S., Hassabis, D.: Human-level control through deep reinforcement learning. Nature **518**(7540), 529–533 (2015)
20. Carlevaris-Bianco, N., Ushani, A.K., Eustice, R.M.: University of Michigan north campus long-term vision and lidar dataset. Int. J. Robot. Res. **35**(9), 1023–1035 (2016)
21. Liu, L., Özsu, M. T.: Encyclopedia of database systems **6**, Springer, (2009)
22. Kanya, K., Kanji, T.: Deep next-best-view planner for cross-season visual route classification. In: 2020 25th International Conference on Pattern Recognition (ICPR), pp. 497–502. IEEE (2021)
23. Mourão, A., Martins, F., Magalhaes, J.: Multimodal medical information retrieval with unsupervised rank fusion. Comput. Med. Imaging Graph. **39**, 35–45 (2015). https://doi.org/10.1016/j.compmedimag.2014.05.006

Video Anomaly Detection for Pedestrian Surveillance

Divakar Yadav, Arti Jain, Saumya Asati, and Arun Kumar Yadav

Abstract With the increase in video surveillance technology, modern human beings have more viable options to enhance safety, security, and monitoring. Automatic video surveillance is an option that provides remote monitoring with little human effort and is a computer vision task. There is no end to the applications of automatic video surveillance such as traffic monitoring, theft detection, fight detection. These are important in various places like industrial, residential and official buildings, roads, and many more. The key objective of the present study is to monitor the pedestrian streets and to provide safety and security by identifying anomalous events. However, tracking an anomalous event in itself is a tricky task because of changes in the definition of an anomaly in different scenarios. In this research, high-level features are used to enhance anomaly detection performance using an auto-encoder model. The features are derived from the pre-trained models, and the contextual properties are derived from the extracted features. The datasets used for anomaly detection on the pedestrian streets are UCSD Pedestrian Street Peds1 and Peds2. The performance is evaluated on the Receiver Operating Characteristic (ROC) curve, Area under Curve (AUC), Precision-Recall curve, Average Precision, and Equal Error Rate (EER) value.

Keywords Auto-encoder · Equal error rate · Mean squared error (MSE) · Contextual anomaly · Video surveillance

D. Yadav (✉) · S. Asati · A. K. Yadav
Computer Science and Engineering, NIT, Hamirpur, Himachal Pradesh 177005, India
e-mail: divakaryadav@ignou.ac.in

A. Jain
Computer Science and Engineering, Jaypee Institute of Information Technology, Noida, Uttar Pradesh 201304, India

M. Tistarelli et al. (eds.), *Computer Vision and Machine Intelligence*,
Lecture Notes in Networks and Systems 586,
https://doi.org/10.1007/978-981-19-7867-8_39

1 Introduction

With advancements in hi-tech technology, the use of surveillance cameras is increasing day by day, and so are the problems with their effective usage too. One such major challenge is to a lookout at all these surveillance screens simultaneously. Also, even if surveillance cameras have caught irregularities in the video data, they are not handled within time. Henceforth, real-time anomaly detection technology is a must that is capable of automatically recognizing irregularities [1, 2]. Several researchers have discussed numerous anomaly detection challenges, one among them is the lack of a clear description of anomalies themselves. Generally, the anomalies are considered outliers within the normal distribution of the training samples. And so, most researchers rather than taking it as a classification problem, state it as an outlier detection problem [3]. The applications of anomaly detection include–fraud prevention, preventive maintenance, error detection, and system monitoring. Across numerous industries like finance, security, energy, medical, information technology, social media, e-commerce, etc. revealing anomalies in real-time streaming data has major practical applications [4].

In addition, surveillance cameras are used to monitor and manage traffic widely [5–7]. Compared with other sensor systems, these can provide more valuable and plentiful information, such as ground induction loops and bridge sensors. In video scenes, identifying the position of lanes is a critical basis for the extraction of traffic parameters in lanes and the semantic investigation of vehicular performance, and also traffic events [8]. Generally, specifying the lane position manually is the most straightforward technique during system installation. However, at present, this mode is rarely used because of the error in systems running and the method's inflexibility. Specifically, the method is not appropriate for Pan-Tilt-Zoom (PTZ) cameras, which are generally used in video-based traffic observing systems nowadays for their flexibility [9–11]. It is allowed to adjust parameters online in the case of a PTZ camera, and when these constraints alter, the position of lanes must be redetected in video images. Moreover, the consequent processes like extraction of traffic parameters in lanes, investigation of traffic events, and behavior of vehicles are disturbed. The accurate and speedy extraction directly affects the interrupted time of the consequent operations [12].

As the moving objects have merely been identified and tracked chiefly in the existing video surveillance systems which lack detection and recognition of objects' behavior in the surveillance scene [13]. The key objective of monitoring the scene is first to detect and then investigate the infrequent event or person's unusual behavior in actual life. Such tasks are handled manually in a long video sequence that is neither be applied nor be effective. Moreover, the video surveillance arrangement has previously lost its novel intention for keenly intervening and nearly turned out to be a tool for offering video evidence later. The intelligent identification of unusual behavior can also save much-storing space and avoid the employees finding and gathering enormous evidence after the unlawful actions [14]. Nowadays, because of the reducing costs of video cameras, an enormous number of surveillance cameras have been mounted. Industrial applications of intelligent video surveillance are extensively used to reduce the human resources to analyze large-scale video data [15, 16]. For intelligent surveillance, various essential tools like object tracking, privacy security, pedestrian recognition, gait investigation, vehicle template detection, video summarization, face and iris recognition, and crowd calculating are established [17]. As a favorable direction, predicting future frames promotes an enormous growth for unsupervised learning without detailed data annotation but is needed for broadly studied overseen learning. It also helps in several kinds of applications, such as intention estimate in robotic, Unmanned Aerial Vehicles (UAVs), and self-governing driving systems [18–21]. Commonly, the construction of a foretelling model can solve the problem of future frame estimation, and the next frame(s) can be predicted by leveraging the content of prior and existing frames. The predictive models are deliberated as unsupervised deep learning frameworks. There are countless models to train them, by taking frames foreseen in a video as ground truths. Predictive models are generally based on auto-encoders, recurring neural networks or generative adversarial nets, which can be skilled without outside supervision [22].

On the other hand, facing the multifaceted and dynamic scenes, the diversity in appearance and motion representation remains in great trouble for effective prediction [23]. Therefore, surveillance cameras have broad applications in residential, industrial, and official buildings for security, safety, and efficiency. Studies show that abnormal behavior detection or object movement identification still cannot be made more efficiently. Instead, it is carried out manually at many places, which increases the need for a workforce. An exponential increase can be seen in the availability of streaming, time-series data across every industry. Mainly, it is driven by the upsurge in the Internet of Things (IoT) [24] and associated real-time data sources. At present, there is a vast number of applications with sensors that produce significant and uninterruptedly changing data. Thus, video anomaly detection requires more research work to be carried out. Therefore, the present study is carried out to investigate further and to improve the video anomaly detection approach for pedestrian surveillance. The research contributions are as follows:-

- To propose a deep learning-based auto-encoder model designed for video anomaly detection for pedestrian surveillance.
- To experiment with a publically available dataset intended for video anomaly detection for pedestrian surveillance.
- To analyze results using evaluation metrics aimed at video anomaly detection for pedestrian surveillance.

The article is organized in the following ways. Section 2 mentions the related work. Section 3 elaborates the proposed methodology which includes datasets, preprocessing, feature additions, training, and testing. Section 4 discusses the result analysis in terms of Average under Curve (AUC), Average Precision (AP), and Equal Error Rate (ERR). Section 5 concludes the paper.

2 Related Work

There is wide use of surveillance cameras at generic places like banks, shops, houses, traffic signals, offices, etc. And so, arises numerous video anomaly detection methods that focus on camera video streams utilizing large-scale training data. In an investigation, lane detection is carried out through a video-based transportation monitoring system. The three main steps of the study are–the abstraction of the road region, the adjustment of the dynamic camera, and the setting of the three virtual detecting lines. Experimental outcomes have shown that the position of the lane center can be identified efficiently through the proposed method. Also, the safety of traffic surveillance systems can be improved significantly [12]. Another anomaly video detection study has described a method based on spatiotemporal constraints to recognize these abnormal behaviors. Further, a multi-target tracking algorithm based on the intersection area is used to attain precise spatiotemporal parameters. The validation by comparison with experimental results has illustrated that the strange running behavior in surveillance videos can be effectively detected by combining these two algorithms [14].

In another study, a unique video event detection system is analyzed which both spatial and temporal contexts are considered. To distinguish the video, firstly, the Spatio-temporal video separation is performed, and then a new descriptor based on region, called motion context is suggested to express the nine information about both the motion and the appearance of the Spatio-temporal section. Also, compact random projections are adopted to increase the rapidity of the examination process. The validation by experimental and state-of-the-art methods has shown advantages of the algorithm [17]. Some state-of-the-art techniques are used for video anomaly detection by proposing the Numenta Anomaly Benchmark (NAB) approach, which sets a benchmark on a dataset, performance evaluation, and code library, setting it as an important tool for anomaly detection algorithms [25]. NAB dataset consists of 58 data files, with 1000–22,000 data instances for a total of 365,551 data points, all labeled by hand. The performance metric that is developed for this approach is called the NAB score.

Detection of the anomaly using deep multiple instances learning ranking model to identify anomaly score, to sort out anomalous clips from normal clips has a disadvantage of failure in identifying normal group activities and reducing the accuracy [2]. However, the significant contribution of the authors is the dataset comprising 13 different crime activities collected altogether with annotations, and this is the largest available video surveillance dataset to date. The performance metric used is AUC, with 75.41% with proposed constraints and 74.44% without proposed constraints. Sparse coding-based anomaly detection is used to take input training data and learn the dictionary from normal events, only for video anomaly detection. In testing, if the event cannot be formed with the atoms from the learned dictionary that event is considered unusual, and a false alarm rate is used for anomaly performance evaluation [26]. The performance metrics used are AUC and EER, 86.1 and 22.0%, respectively, for Anomaly Net. Compared with other state-of-art methods the UMN dataset gives comparable results with lesser EER values.

Hough forest is used to learn mapping in Hough space for encoding space and time action class distribution [27]. The approach is applied to three different datasets, and the accuracy is achieved at around 90%, but the approaches compared are not deep learning methods. However, lesser accuracy is also observed in the case of complex videos, abrupt changing of frames, and zooming in or out of frame.

Object-centric methods [28] are also used to detect anomalies in videos using SSD and auto-encoders for feature extraction of both motion and appearance, then applying CNN [29, 30] for classification, making it a multiclass classification. The snippets of abnormal events are identified with weakly supervised learning. Anomaly detection on various data helps identify any unusual event that might contradict the regular behavior pattern of data. This strategy is applicable in video anomaly detection using contextual features of data to gain more embedded features of videos.

From the above-discussed literature, it is clear that there are three types of anomalies: point anomaly, contextual anomaly, and collective anomaly. In the point anomaly, data point values lie far away from entire data points. In the contextual anomaly, data point values significantly deviate from the other data points in the same context. And, in the collective anomaly, a collective set of data points deviate from all other data points. In addition, it is clear that contextual anomaly plays a vital role in a video stream and is mined in our approach to increase the abnormality interpretability.

3 Proposed Methodology

The proposed approach is mainly focused on the crowd or pedestrian behavior on the streets. Pre-trained models are chosen for object detection, object classification, and background segmentation. The anomaly detection method is developed using extracted features from the pre-trained model and adding more features. The steps followed in the proposed methodology are depicted in Fig. 1 and are discussed as stated below.

3.1 Data Collection

The dataset that is used for the abnormality detection on the pedestrian streets is taken from the UCSD anomaly detection dataset [31]. This dataset is prepared from pedestrian streets where the cameras are mounted at a height and recorded crowds, pedestrians walking down the pathway with varying crowd density. Anomalies included bikers, skaters, vehicles, carts, and pedestrians walking on grass, or across the pathway. Dataset is divided into two clips–Peds1 having 34 training and 36 testing video samples, and Peds2 having 16 training and 12 testing video samples.

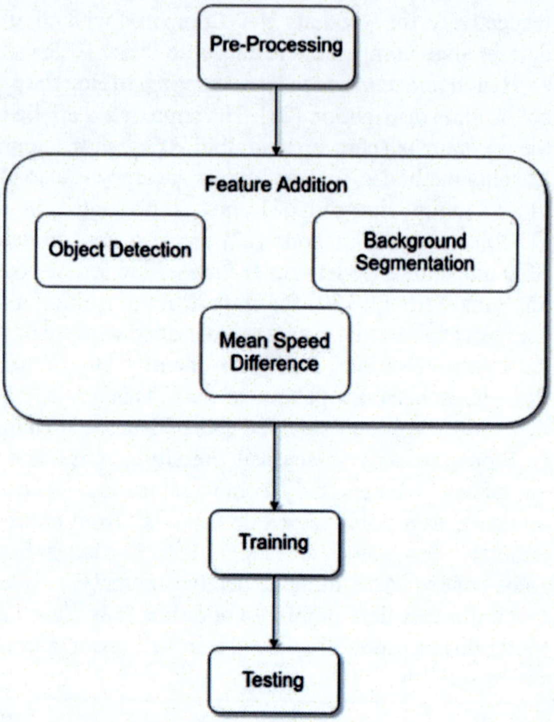

Fig. 1 Flow diagram of proposed methodology

3.2 *Pre-processing*

For pre-processing of the dataset, first, the videos are sliced to frames of 200 each, and the CSV file of Peds1 and Peds2 datasets contains 92 columns with all the details of frames with speed, the velocity of pedestrians, location of pedestrians in frames, etc. are used for speed adjustment of the pedestrian concerning a frame. The speed column is updated using the distance formula (Eq. 1) for the velocity of x and y as velocity_x and velocity_y, respectively.

$$\text{Speed} = \left(\text{velocity}_x^2 + \text{velocity}_y^2\right)^{0.5} \tag{1}$$

The resolution for the Peds1 dataset is 158 × 238, and for Peds2 is 240 × 360. And so, the pixel resolution for the Peds1 dataset is resized to 1.5 times, making its resolution close to that of Peds2.

3.3 Feature Addition

There are various possible causes for abnormality, such as motion, appearance, and location. All of these are considered for abnormal event identification. Panoptic Feature Pyramid Networks (PFPN) is used for background segmentation. This PFPN model is then run on the Detectron2 platform, a Facebook AI Research Software. The model is pre-trained on the COCO dataset [32]. The matrix output of background segmentation is not used directly in our abnormality detection model, but a contextual feature extraction method is used to convert the output to a scalar.

The Joint Detection and Embedding model (JDE) is used for Pedestrian detection and tracking. JDE simultaneously outputs the predicted location and appearance embedding of targets in a single forward pass. It uses Feature Pyramid Network (FPN) architecture. The joint objective is a weighted linear sum of losses from every scale and component. An automatic learning scheme is adopted for loss weights by using the concept of task-independent uncertainty. ResNet 101(R101) implemented on the Detectron2 platform is used for the appearance feature. R101 is a 101 layers deep Convolutional Neural Network (CNN) trained on the COCO dataset. The output of this includes a vector with 80 object categories. This vector is directly used for the abnormality detection model as input. For the mean speed feature, the mean speed is calculated for the whole frame from the detected speeds of pedestrians within each frame. Deviation of a pedestrian from this mean speed indicates the presence of some anomaly either the pedestrian is on some vehicle or is running. The mean speed of all pedestrians present in a frame is termed frame speed. Contextual features also play a significant role in maximizing the performance of abnormality detection in videos.

3.4 Training

Convolutional Auto Encoder (CAE) is used for abnormality detection of behavior patterns in pedestrians over the pedestrian street for training the model. This CAE is encompassed with a denoising filter making it a Denoising Auto Encoder (DAE). For DAE, noise is added to the training dataset, which serves as an input for the CAE. The architecture of CAE is simple for which input passes through an input layer, and then it is encoded at the encoder. The encoded input is passed through a decoder which decodes it and reconstructs the data to be as close as possible to the original input. The obtained data is compared to the original data, if the reconstruction error is lesser then the sample is considered normal, else if the reconstruction error is substantial then the sample is abnormal.

To construct deep DAE, three fully connected hidden layers are added. The number of units in the input is the same as in the output, determined by the input feature space. The layer node numbers that provided the best experimentation results are 50, 30, and 50, respectively. Compressed input features are stored in the middle code layer (30). The activation function of a node defines the node's output for a given set of inputs.

So the activation function used for the training is the sigmoid function because it outputs the values in the range from 0 to 1 since the output of the proposed model is a probability of whether the segment is an abnormal or normal instance in a video.

Batch normalization is used to stabilize the learning process of machines and reduce the number of training epochs required to train the deep neural networks. Due to fewer memory requirements, computational efficiency, suitability for large data sets, and appropriateness for noise gradients, Adam optimizer is used for optimization. This method uses mean squared error (MSE) as a reconstruction error to identify anomalies. Training epochs required are 33 with a batch size of 120.

3.5 Testing

The features compiled through the auto-encoder model are the crucial highlight of this model. The feature extractor model is fed with test data, and the Mean Squared Error (MSE) is calculated (Eq. 2).

$$\text{MSE} = \frac{1}{n}(\text{Actual value} - \text{Observed value})^2 \tag{2}$$

where n is the number of data points.

The test data is already been pre-processed along with the training data during the pre-processing phase. The high-level features that are obtained from the auto-encoder model are also fed to the classifier model through the MSE prediction.

4 Results

For the proposed model, the first loss curve is plotted to verify the model's worth. The epochs used are 33, and the batch size is 120. For these epochs and the batch size, the model loss curve is plotted and is shown in Fig. 2.

The training data loss decreases, and test data loss also decreases with the increased iterations. The curve for test data is slightly above the training curve. Therefore, it can be concluded that the auto-encoder model slightly overfits a learning model because the loss of data decreases with an increase in epochs. The metrics used to compare the model's performance are Receiver Operating Characteristic (ROC) curve and Precision-Recall (PR) curve. ROC curve plots the graph between True Positive Rate (TPR) and False Positive Rate (FPR) at all thresholds. TPR is also called Recall. The False Negative Rate (FNR) indicates the Miss Rate or False Alarm.

The Area under Curve (AUC) is an area under the ROC curve that depicts the aggregate performance for all possible thresholds. High precision is related to low FPR, which means the model generates more accurate results. The high recall is related to FNR, which means the model is returning all positive results majorly.

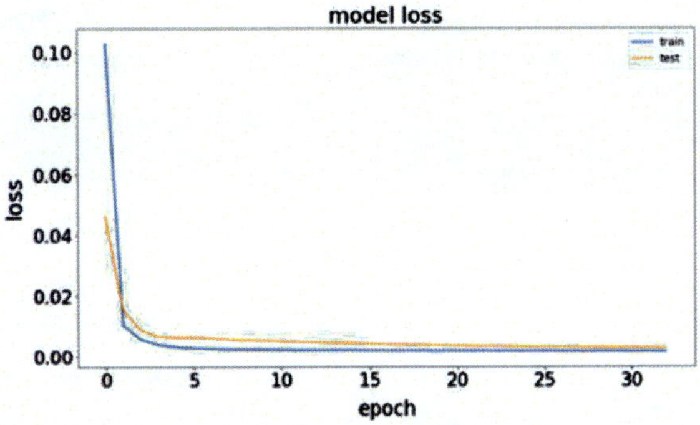

Fig. 2 Auto-encoder model loss curve

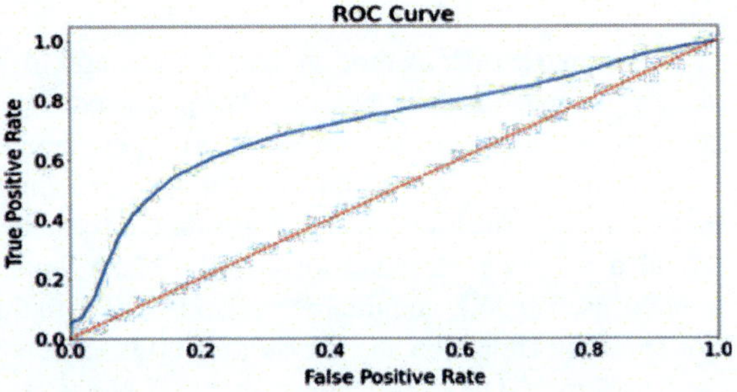

Fig. 3 ROC curve for model performance metrics

Therefore, the ideal model has high precision and high recall. The Equal Error Rate (EER) determines the threshold for which the False Acceptance Rate (FAR) is equal to the False Rejection Rate (FRR). The lower EER implies higher accuracy of the model. The ROC curve for the performance evaluation is shown in Fig. 3. Here, FPR is represented on the x-axis, and TPR is represented on the y-axis. Also, the red line indicates the half area of the curve, i.e., the curve over the red line has an AUC of more than 50% and below the red line has an AUC of lesser than 50%.

From Fig. 4, it is observed that the Precision-Recall curve is high, indicating good model performance. The numerical values of all metrics are shown below:

- The AUC value obtained is 71.74%.
- The Average Precision (AP) value is 91.63%.

The EER value is 0.32, and the threshold value is 0.13.

5 Conclusion and Future Scope

This research is carried out to contribute toward better automatic surveillance. The study focuses on pedestrian behavior, walking down the pedestrian street, and identification of anomalous events. The possible causes for anomaly event identification

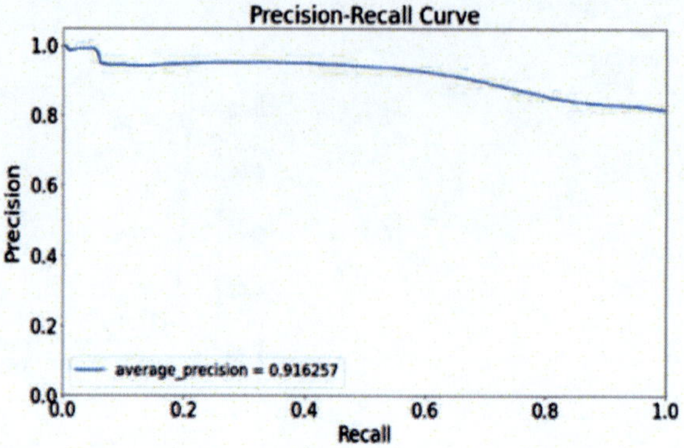

Fig. 4 Precision-Recall curve for model performance metrics

are recognized such as motion, appearance, and location to enhance the safety and security of persons down the lane. The proposed methodology works with contextual features of videos to increase the overall performance of the model. It is observed that the features compiled through the auto-encoder model are the crucial highlight of this model. The high-level features that are obtained from the auto-encoder model are also fed to the classifier through the Mean Square Error (MSE) prediction. The epochs used are 33, and the batch size is 120 with the Adam optimizer.

The Joint Detection and Embedding (JDE) model is used for pedestrian detection and tracking. JDE uses Feature Pyramid Networks for background segmentation which is run on the Detectron2 platform. The model is pre-trained with ResNet 101 having 101 layered deep Convolutional Neural Networks on the COCO dataset. An automatic learning scheme is adopted for loss weights by using the concept of task-independent uncertainty. For the mean speed feature, the mean speed is calculated for the whole frame from the detected speeds of pedestrians within each frame. Deviation of a pedestrian from this mean speed indicates the presence of some anomaly, i.e., either the pedestrian is on some vehicle or is running. Convolutional Auto Encoder (CAE) is used for abnormality detection of behavior patterns in pedestrians over the pedestrian street for the model. This CAE is encompassed with a denoising filter making it a Denoising Auto Encoder (DAE). The layer node numbers that provided the best experimentation results are 50, 30, and 50, respectively, using the sigmoid function. The output of the proposed model is the probability of whether the segment is an abnormal or normal instance in a video.

There is a decrease in the training and test data loss with an increase in epochs. The curves- Receiver Operating Characteristic (ROC) and Precision-Recall (PR) are plotted as performance metrics to evaluate the system performance. ROC plots between True Positive Rate (TPR) and False Positive Rate (FPR) at all thresholds. TPR is also called Recall. False Negative Rate (FNR) indicates the Miss Rate or False Alarm. High precision is related to low FPR, and high recall is related to FNR. Equal Error Rate (EER) determines the threshold for which False Acceptance Rate (FAR) is equal to False Rejection Rate (FRR). The lower EER implies higher accuracy of the model. And, Area under Curve (AUC) is an area under the ROC curve that depicts the aggregate performance for all possible thresholds. AUC value is obtained as 71.74%, Average Precision (AP) value is 91.63%, EER value is 0.32, and the threshold value is 0.13, respectively.

In the future, this work can be extended further to cover other video anomaly datasets, including store monitoring, subway surveillance, highway road traffic detection, and so on. A better model for background segmentation and pedestrian tracking can also be created to enhance the features level. Some high-level features such as action recognition can also be added to improve the system performance.

References

1. Xu, K., Sun, T., Jiang, X.: Video anomaly detection and localization based on an adaptive intra-frame classification network. IEEE Trans. Multimed. **22**(2), 394–406 (2019). https://doi.org/10.1109/TMM.2019.2929931
2. Nagrath, P., Dwivedi, S., Negi, R., & Singh, N. Real-Time Anomaly Detection Surveillance System. In: Proceedings of Data Analytics and Management, pp. 665–678 (2022). https://doi.org/10.1007/978-981-16-6289-8_54
3. Franklin, R. J., Dabbagol, V.: Anomaly detection in videos for video surveillance applications using neural networks. In: 2020 Fourth International Conference on Inventive Systems and Control (ICISC), pp. 632–637 (2020). https://doi.org/10.1109/ICISC47916.2020.9171212
4. Ahmad, S., Purdy, S.: Real-time anomaly detection for streaming analytics (2016). arXiv preprint arXiv:1607.02480. https://doi.org/10.48550/arXiv.1607.02480
5. Mehboob, F., Abbas, M., Rauf, A., Khan, S.A., Jiang, R.: Video surveillance-based intelligent traffic management in smart cities. In: Intelligent Video Surveillance, p. 19 (2019).
6. Parkyns, D.J., Bozzo, M.: CCTV Camera sharing for improved traffic monitoring. In: IET Road Transport Information and Control Conference and the ITS United Kingdom Members' Conference (RTIC 2008), Manchester, UK (2008). https://doi.org/10.1049/ic.2008.0771
7. Baran, R., Rusc, T., Fornalski, P.: A smart camera for the surveillance of vehicles in intelligent transportation systems. Multimed. Tools Appl. **75**(17), 10471–10493 (2016). https://doi.org/10.1007/s11042-015-3151-y
8. Nee, J., Hallenbeck, M.E., Briglia, P.: Surveillance Options for Monitoring Arterial Traffic Conditions (No. WA-RD 510.1). Washington State Department of Transportation (2001).
9. Komagal, E., Yogameena, B.: Foreground segmentation with PTZ camera: a survey. Multimed. Tools Appl. **77**(17), 22489–22542 (2018). https://doi.org/10.1007/s11042-018-6104-4
10. Bimbo, A.D., Dini, F., Pernici, F., Grifoni, A.: Pan-Tilt-Zoom Camera Networks, pp. 189–211 (2009).
11. de Carvalho, G.H., Thomaz, L.A., da Silva, A.F., da Silva, E.A., Netto, S.L.: Anomaly detection with a moving camera using multiscale video analysis. Multidimens. Syst. Sign. Process. **30**(1), 311–342 (2019). https://doi.org/10.1007/s11045-018-0558-4
12. Ren, J., Chen, Y., Xin, L., Shi, J.: Lane detection in video-based intelligent transportation monitoring via fast extracting and clustering of vehicle motion trajectories. Math. Probl. Eng. **2014**(156296), 1–12 (2014). https://doi.org/10.1155/2014/156296
13. Paul, M., Haque, S.M., Chakraborty, S.: Human detection in surveillance videos and its applications—a review. EURASIP J. Adv. Sign. Process. **2013**(176), 1–16 (2013). https://doi.org/10.1186/1687-6180-2013-176
14. Zhu, Y.Y., Zhu, Y.Y., Zhen-Kun, W., Chen, W.S., Huang, Q.: Detection and recognition of abnormal running behavior in surveillance video. Math. Probl. Eng. **2012**(296407), 1–14 (2012). https://doi.org/10.1155/2012/296407

15. Tu, N.A., Wong, K.S., Demirci, M.F., Lee, Y.K.: Toward efficient and intelligent video analytics with visual privacy protection for large-scale surveillance. J. Supercomput. **77**(12), 14374–14404 (2021). https://doi.org/10.1007/s11227-021-03865-7
16. Zhang, G., Xu, B., Liu, E., Xu, L., Zheng, L.: Task placement for crowd recognition in edge-cloud based urban intelligent video systems. Clust. Comput. **25**(1), 249–262 (2022). https://doi.org/10.1007/s10586-021-03392-3
17. Cong, Y., Yuan, J., Tang, Y.: Video anomaly search in crowded scenes via spatio-temporal motion context. IEEE Trans. Inf. Forensics Secur. **8**(10), 1590–1599 (2013). https://doi.org/10.1109/TIFS.2013.2272243
18. Rudenko, A., Palmieri, L., Herman, M., Kitani, K.M., Gavrila, D.M., Arras, K.O.: Human motion trajectory prediction: a survey. Int. J. Robot. Res. **39**(8), 895–935 (2020). https://doi.org/10.1177/0278364920917446
19. Zunino, A., Cavazza, J., Volpi, R., Morerio, P., Cavallo, A., Becchio, C., Murino, V.: Predicting intentions from motion: the subject-adversarial adaptation approach. Int. J. Comp. Vis. **128**(1), 220–239 (2020). https://doi.org/10.1007/s11263-019-01234-9
20. Stocco, A., Weiss, M., Calzana, M., Tonella, P.: Misbehavior prediction for autonomous driving systems. In: Proceedings of the ACM/IEEE 42nd International Conference on Software Engineering, pp. 359–371 (2020). https://doi.org/10.1145/3377811.3380353
21. Hildmann, H., Kovacs, E.: Using unmanned aerial vehicles (UAVs) as mobile sensing platforms (MSPs) for disaster response civil security and public safety. Drones **3**(3), 59 (2019). https://doi.org/10.3390/drones3030059
22. Mogren, O.: C-RNN-GAN: Continuous Recurrent Neural Networks with Adversarial Training (2016). arXiv preprint arXiv:1611.09904. https://doi.org/10.48550/arXiv.1611.09904
23. Li, S., Fang, J., Xu, H., Xue, J.: Video frame prediction by deep multi-branch mask network. IEEE Trans. Circuits Syst. Video Technol. **31**(4), 1283–1295 (2020). https://doi.org/10.1109/TCSVT.2020.2984783
24. Kushwah, R., Batra, P.K., Jain, A.: Internet of things architectural elements, challenges and future directions. In: 2020 6th International Conference on Signal Processing and Communication (ICSC), pp. 1–5 (2020). https://doi.org/10.1109/ICSC48311.2020.9182773
25. Lavin, A., Ahmad, S.: Evaluating Real-Time Anomaly Detection Algorithms—The Numenta Anomaly Benchmark. In: 2015 IEEE 14th International Conference on Machine Learning and Applications (ICMLA), Miami, FL, USA, pp. 38–44 (2015). https://doi.org/10.1109/ICMLA.2015.141
26. Zhou, J.T., Du, J., Zhu, H., Peng, X., Liu, Y., Goh, R.S.M.: Anomalynet: an anomaly detection network for video surveillance. IEEE Trans. Inf. Forensics Secur. **14**(10), 2537–2550 (2019). https://doi.org/10.1109/TIFS.2019.2900907
27. Serrano, I., Deniz, O., Espinosa-Aranda, J.L., Bueno, G.: Fight recognition in video using hough forests and 2D convolutional neural network. IEEE Trans. Image Process. **27**(10), 4787–4797 (2018). https://doi.org/10.1109/TIP.2018.2845742
28. Ionescu, R.T., Khan, F.S., Georgescu, M.I., Shao, L.: Object-centric auto-encoders and dummy anomalies for abnormal event detection in video. In: Proceedings of the IEEE/CVF Conference on Computer Vision and Pattern Recognition, pp. 7842–7851 (2019).
29. Yadav, A.K., Jain, A., Lara, J.L.M., Yadav, D.: Retinal blood vessel segmentation using convolutional neural networks. In: Proceedings of the 13th International Joint Conference on Knowledge Discovery, Knowledge Engineering and Knowledge Management (IC3K 2021), Vol. 1: KDIR, pp. 292–298 (2021). https://doi.org/10.5220/0010719500003064
30. Siddiqui, F., Gupta, S., Dubey, S., Murtuza, S., Jain, A.: Classification and diagnosis of invasive ductal carcinoma using deep learning. In: 2020 10th International Conference on Cloud Computing, Data Science and Engineering (Confluence), pp. 242–247 (2020). https://doi.org/10.1109/Confluence47617.2020.9058077
31. UCSD Anomaly Detection Dataset. Accessed Jan 2022. http://www.svcl.ucsd.edu/projects/anomaly/dataset.html
32. Lin, T.Y., Maire, M., Belongie, S., Hays, J., Perona, P., Ramanan, D., Dollár, P., Zitnick, C.L.: Microsoft Coco: common objects in context. In: European Conference on Computer Vision, pp. 740–755 (2014). https://doi.org/10.1007/978-3-319-10602-1_48

Cough Sound Analysis for the Evidence of Covid-19

Nicholas Rasmussen, Daniel L. Elliott, Muntasir Mamun, and KC Santosh

Abstract Efficient techniques for Covid-19 screening tests are crucial to preventing infection rates. Although symptoms present differently in different sociodemographic groups, cough is still ubiquitously presented as one of the primary symptoms in severe and non-severe infections alike. Audio/speech processing is no exception in health care. In this paper, we implemented a convolutional neural network (CNN) algorithm to analyze 121 clinically verified cough audio files from the volunteer group 'Virufy' for Covid-19 screening. Using a single relatively small CNN with a large, fully connected dense layer trained on melspectrograms alone, we achieved 0.933 test accuracy and an AUC of 0.967 on a small dataset. Our results are competitive with state-of-the-art results with both small and large datasets.

Keywords Cough sound · Covid-19 · CNN

1 Introduction

Covid-19 has scourged the planet since it was declared a pandemic by the World Health Organization on January 30, 2020. There are 523,786,368 total cases and 6,279,667 deaths worldwide as of May 24, 2022[1]. Furthermore, there are approximately 11,752,673 people currently fighting this infection. Despite what humans have done to combat this virus, the disruption to humanity's normal function has been significant. The greatest hope for ending this pandemic is vaccines, of which 3,327,841,570 have been distributed globally[1]. However, are there other ways that we can help to combat this virus and end the pandemic quickly? The answer to

[1] WHO coronavirus (covid-19) dashboard, 2021. https://covid19.who.int/ last accessed on 5/24/22.

N. Rasmussen (✉) · D. L. Elliott · M. Mamun · KC Santosh
2AI: Applied AI Research Lab—Computer Science, University of South Dakota, Vermillion, SD 57069, USA
e-mail: nicholas.rasmussen@coyotes.usd.edu

KC Santosh
e-mail: santosh.kc@usd.edu

M. Tistarelli et al. (eds.), *Computer Vision and Machine Intelligence*,
Lecture Notes in Networks and Systems 586,
https://doi.org/10.1007/978-981-19-7867-8_40

that question may lie in the screening techniques employed to help people decide whether to obtain a laboratory test and quarantine to avoid infecting others. Screening techniques, especially cough sounds, can be deployed at a low cost in resource-constrained regions to help people who do not have access to vaccines. This deployment is possible because even state-of-the-art microphones sell for less than 80 U.S. Dollars, which can then be used to screen thousands of people. Furthermore, smartphone technology presents easy access to people worldwide since each contains a quality microphone.

Cough presentation rate (CPR) is the percentage of Covid-19 patients with cough as a symptom and is high in those infected by Covid-19. In Covid-19 infections, cough is ubiquitous in nearly all socio-demographic groups as one of the most common symptoms regardless of severity. Other factors, such as age, also have no bearing on the presentation of cough symptoms. In some cases, studies found that a sample of patients presented cough as a symptom 90% of the time [1]. These statements are further elaborated upon in Sect. 2. These facts highlight that Covid-19 is a respiratory disease, and as such, it affects the lungs in ways that produce discernible features that are recognized by AI-guided tools.

Cough can be used to help diagnose and predict many diseases, such as asthma, COPD, and even stroke. Recently, researchers have leveraged AI-guided tools to help search for specific markers of diseases. For example, in Rudraraju et al. [2], the authors found that AI-guided tools can help classify a patient's lung problems as obstructive or restrictive. Also, in Han et al. [3], the authors used cough and other valuable data to estimate the severity level of illnesses. As such, AI-guided tools have an apparent possibility to get leveraged for screening cough sounds for Covid-19 infections.

There is a statistical difference between clinically validated and non-clinically validated datasets, which is expounded upon in Sect. 2. We found that clinically validated datasets used to train AI-guided tools consistently achieved better performance than those without clinical validation. This study tested this further using a small dataset of 121 cough samples with clinical validation to see if this continues to hold.

The remainder of this paper is as follows. Section 2 presents related works utilizing AI-guided tools for screening Covid-19 infections using cough sound. Section 3 describes the dataset and CNN architecture. Section 4 presents the results. We also provide a modicum of discussion in Sect. 5 Finally, Sect. 6 concludes our paper.

2 Related Works

AI has contributed a lot in analyzing speech/audio signal for the evidence of possible abnormalities [4–6]. Covid-19 is no exception. Several studies have been conducted on the general symptomology of the Covid-19 disease to help characterize the disease and its effects on humans. Throughout our review, multiple delimiting factors separated the studied patients into groups. These delimiting factors are socio-demographics, the severity level of the illness, and age. In this text, we use cough presentation rate (CPR) to show the number of people who present a cough in the entire study. Two of the studies reviewed showed differences in the overall presented symptoms of different ethnicity; however, cough remained one of the most common symptoms at high rates regardless [7, 8]. Next, in Lapostolle et al. [1], a socio-demographical study of french people, the CPR was 90% of the 1,487 patients observed. The severity level of the illness is also a non-factor in the commonality of the cough symptom. Vaughan et al. [9] split patients into different groups based on outpatient, inpatient, or emergency care services. This study found that cough was still one of the most common, if not the most common, symptoms among these different groups, with an average CPR of 86% across the entire study. Also, the fact that cough is one of the most common symptoms among non-severe infections is essential because this group generalizes well to the population at large since most of the people infected by Covid-19 have non-severe infections [10]. Age was the last factor that helped to separate patients into groups. Based on the study with the largest number of samples in our review at 3986 patients, age was a non-critical factor in determining if the cough was presented as a common symptom [11]. Based on the studies we reviewed, cough is one of the most common symptoms in Covid-19 infections regardless of any observed factors.

A key delimiting factor in our review of AI-guided tools is whether the datasets used to train the tool had their clinical information verified. Repeatedly, we found that studies with datasets that are clinically verified had better performance metrics than those studies that did not have their datasets clinically verified. For example, the study with the highest performance in classifying Covid-19 from non-Covid-19 coughs was Laguarta et al. [12], with a 97.1% accuracy and a sample size of 5320 coughs. Furthermore, Hassan et al. [13] performance with a 97% accuracy. Next, Andreu-Perez et al. [14] had a sensitivity of 96%. Lastly, Wei et al. [15] had 76% accuracy, and Imran et al. had 93%. Although these accuracies are lower than their counterparts, both studies attempted a multiclass classifier for different diseases, including Covid-19, and found a 98.7% and 94% specificity in detecting Covid-19, respectively. The other studies without clinical verification of their datasets did worse than those with clinical validation. Although three studies performed admirably with Mouawad et al. [16] at 91% accuracy, Pahar et al. [17] at 95% accuracy, and Anupam et al. 96.9% [18] accuracy, either these studies sensitivity was lower, or questions remain concerning their data selection techniques. The remaining non-clinically verified studies did not perform as well with Bansal et al. [19] at 70.58% accuracy, Shimon et al. [20] at 74% accuracy, Mohammed et al. [21] at 77% accuracy, Vrindavanam et al. [22] at 83.9% accuracy, and Dash et al. [23] at 85.7% accuracy. From these studies, we can see that clinical validation of their datasets helped performance in the majority of cases. Based on this fact, we decided to confirm this phenomenon with our experiment using a small clinically validated dataset.

For a far more comprehensive review on the state-of-the-art methods being used as of early 2022 for detecting Covid-19 using cough sounds, please refer to Santosh et al. [24].

3 Method

In this section, we present our experimental methodology. We begin by describing the dataset we collected. Afterward, we detail how we split the data between testing, training, and validation sets. Lastly, we explain general CNN architecture and the proposed architecture used in our study.

3.1 Dataset Collection

The dataset for the current experiment was collected from an online repository. It consists of 121 segmented cough samples from 16 patients. Among them, 48 cough recordings are Covid-19 positive, and 73 are Covid-19 negative. This dataset was labeled with a positive or negative Covid-19 status based on PCR testing result. Other information such as patient demographics including age, gender, and medical history are contained in the metadata. This clinical data is collected in the hospital under the supervision of physicians. Informed patient consent is also gathered and anonymity preserved. Prepossessing included creating segmented coughs based on identifying periods of relative silence and splitting the recording using zero padding to fit all of the segments into a standard array. The audio segments which were not coughs or had extreme background noise were removed. The collected audio records were in MP3 format. Table 1 is provided to show patients' data that have generated cough samples.

3.2 Feature Extraction and Selection

Using the dataset of 121 cough samples for analysis, we imported the audio processing library 'librosa' [25] in our implementation for speech analysis. It helps extract the audio features at the frame and segment level, covering frequency-based, structural, statistical, and temporal attributes into actionable data. Although other literature reviewed in Sect. 2 used multiple different types of data such as melspectrograms, mel-frequency cepstral coefficient, regular spectrograms, and even raw cough audio to feed into multiple different AI models. Melspectrograms were chosen based on the efficient use of melspectrograms in highest accuracy papers in the reviewed literature. For better understanding, audio's melspectrograms for both Covid-19 positive and negative are shown in Fig. 1.

Table 1 Information of the 16 patients in the dataset

Covid-19 test	Age	Gender	Medical history	Smoker	Patient reported symptoms
Negative	53	Male	None	Yes	None
Positive	50	Male	Congestive heart failure	No	Shortness of breath.
Negative	43	Male	None	No	Sore throat
Positive	65	Male	Asthma or chronic lung disease.	No	Shortness of breath, New or worsening cough
Positive	40	Female	None	No	Sore throat, loss of taste, loss of smell
Negative	66	Female	Diabetes with complications	No	None
Negative	20	Female	None	No	None
Negative	17	Female	None	No	Shortness of breath, sore throat, body aches
Negative	47	Male	None.	No	New or worsening cough.
Positive	53	Male	None.	Yes	Fever, breath shortness, worsening cough, taste loss
Positive	24	Female	None.	No	None
Positive	51	Male	Diabetes with complications.	No	Fever new or worsening cough, sore throat
Negative	53	Male	None.	No	None
Positive	31	Male	None.	No	Shortness of breath, new or worsening cough
Negative	37	Male	None.	No	None
Negative	24	Female	None.	No	New or worsening cough

3.3 *CNN Architecture*

3.3.1 Basics of CNNs

Convolutional neural networks (CNNs) are a type of neural network that can extract features from input and, after that, perform classification. It is mainly composed of 3 types of layers known as convolution, pooling, and dense layers. The convolution layer is responsible for deriving features from input and is composed of filters that convolute over the input to compute structural aspects in every pass. This layer is generally followed by a pooling layer that downsamples the input based on either maximizing the structural characteristics or averages them. This layer reduces computational overhead and can enhance the structural features that matter most. A CNN can have multiple convolutions and pooling layers in multifarious sequences. At times, a CNN might not have a pooling layer at all. The dense layer then performs

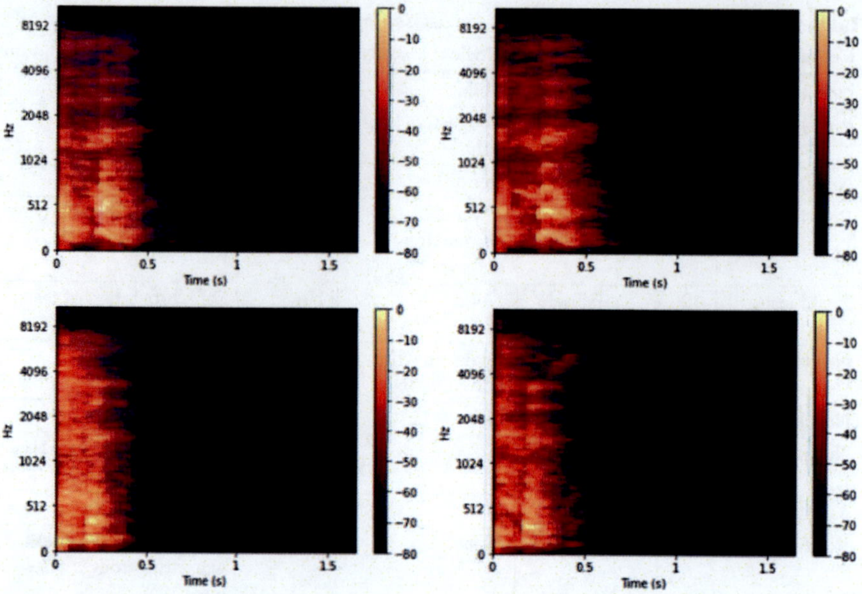

Fig. 1 Melspectrogram samples: Covid-19 positive (first row) and Covid-19 negative (second row)

the classification after flattening the characteristics from the convolutional and pooling layers into one dimension. The dense layer is also known as the fully connected layer. A network might have single or multiple dense layers with a varying number of neurons. However, the output of the final dense layer generally corresponds to the number of classes the network tries to classify the image as. In the case of our experiment, this corresponded to a single binary classification of zero or one.

3.3.2 Proposed CNN

In our study, the input was initially passed to two convolution layers, the first layer with the largest kernel size of 7 by 7 at 32 filters, the second layer with the second largest kernel size of 5 by 5 at 64 filters. Combining these two layers at the top of the network gives the network 52,864 nodes on the entire image before passing it to a subsequent pooling layer. After this, initial first layer combo, the network does a 2 by 2 max-pooling operation. Then, the number of 3 by 3 convolutional layer and 2 by 2 max-pooling layer repeats can vary based on how large the image is. In this instance, only, two repeats were used with 256 filters each. The final convolutional layer uses a 2 by 2 kernel with double the amount of filters (512). Finally, the output from the last convolution layer is flattened and passed to a 256 dimensional dense layer. The thought for using a smaller dense layer of 256 is to compact the large

Table 2 Hyper-parameters used in different layers of the proposed CNN

Layer	Strides	Window size	Filters	Parameters	Output
Convolution 1	1 × 1	7 × 7	32	1,600	(122 × 65 × 32)
Convolution 2	1 × 1	5 × 5	64	51,264	(118 × 61 × 64)
Max Pooling	2 × 2	2 × 2	N/A	N/A	(59 × 30 × 64)
Convolution 3	1 × 1	3 × 3	256	147,712	(57 × 28 × 256)
Max Pooling	2 × 2	2 × 2	N/A	N/A	(28 × 14 × 256)
Convolution 4	1 × 1	3 × 3	256	590,080	(26 × 12 × 256)
Max Pooling	2 × 2	2 × 2	N/A	N/A	(13 × 6 × 256)
Convolution 5	1 × 1	2 × 2	512	524,800	(12 × 5 × 512)
Flatten	N/A	N/A	N/A	N/A	(30,720)
Dense 1	N/A	N/A	256	7,864,576	(256)
Dense 2	N/A	N/A	1	257	0 or 1
Total	N/A		N/A	9,180,289	N/A

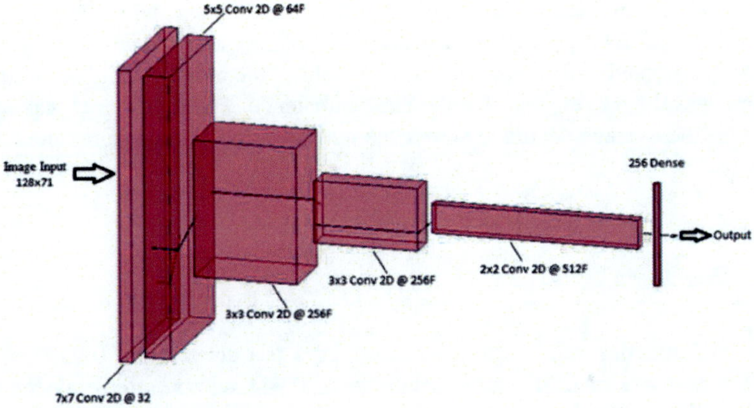

Fig. 2 Graphical representation of the model

amount of information being given from the final convolution layer and make sense of it in a smaller space before being passed to the final classification layer, which classifies the image as a zero or one. All layers use a ReLU activation function and a dropout of 0.2, while the last layer uses a sigmoid activation with no dropout for the final output. Table 2 has been provided to show the above layout of the proposed CNN architecture.

The model was trained with a learning rate of .01, 150 epochs, and a batch size consisting of the entire training set (66). A binary cross-entropy-based loss function was used along with an adamax optimizer. The details of the number of generated parameters for the different layers are presented in Table 2, and Fig. 2 provides a graphical representation.

4 Experiments

In what follows, we have compiled the results of our experiment. Overall, we achieved a high accuracy rating. This result is in line with other experiments that used clinically verified cough data.

4.1 *Validation*

The dataset was split into testing and training sets prior to utilizing k-fold cross-validation to explore different ANN architectures. This procedure was done to compensate for the limited number of samples while preventing bias toward the dataset.

The hold-out test set was composed of 30 samples (approximately 25% of the data) selected by random batch sampling. Twelve samples were Covid-19 positive, and 18 samples were Covid-19 negative. The remaining 91 samples (36 positives and 55 negatives) were used for training and validation. Of these, 66 samples were used for training and 25 samples for validation. The training and validation samples were randomly sampled at the beginning of each fold of the training/validation loop, and nothing was done to account for the class imbalance. The best model was selected based on the average accuracy across the ten folds and tested against the hold-out testing set.

4.2 *Our Results*

After implementing our method, we achieved a test accuracy of 0.9333, AUC of 0.967, sensitivity of 0.916, and specificity of 0.944 without adjusting the output threshold. Thus, although the dataset is relatively small, we have shown that clinically verified cough datasets can be leveraged to generate highly accurate results for Covid-19 screening using only cough sounds. This fact stresses that if a Covid-19 screening application gets built, it may use clinically verified cough data to generate a machine learning model that is effective in early detection of the Covid-19 disease.

4.3 *Dropout*

After completing the main experiment, following the methods based on the highest validation accuracy, we also ran the same model against the test set with different dropout probabilities for training. As Kovacs et al. [26] have shown, generally, adding dropout increases the overall performance of a CNN model in speech acoustic datasets, especially if they are noisy. Since coughing is akin to human speech and is

Table 3 Covid-19 screening performance based on cough sounds: accuracy (ACC, in %), area under the curve (AUC), sensitivity (SEN), and specificity (SPEC) for different levels of dropout

Dropout rate	Performance			
	ACC	AUC	SEN	SPEC
0	0.80	0.87	0.583	0.944
0.05	0.767	.803	0.50	0.944
0.1	0.867	0.935	0.75	0.944
0.125	0.90	0.986	0.833	0.944
0.15	.867	0.944	0.75	0.944
0.2	0.933	0.968	0.916	0.944
0.25	0.73	0.875	0.417	0.944
0.3	0.833	0.907	0.667	0.944
0.5	0.70	0.690	0.583	0.778

inherently noisy because of the nature of a cough, we tested if dropout would increase overall accuracy. Based on the results in Table 3, we can see that the optimal dropout for this model and dataset lies between .1 and .2 dropout. Although .2 dropout produced the best validation results and scored the highest test accuracy without tuning the output threshold of the model, we can see that .125 dropout produced the highest AUC of all the tests. This fact suggests that we may be able to tune the output threshold of the model for better results. Also, in general, we can see that dropout helped identify covid-19 as it increased the sensitivity of the model and produced far better results than no dropout. However, we can also see a cut-off point of approximately .3 dropout; afterward, results seem to degrade. Other dropout types, such as spacial/channel dropout, could have been explored and may have increased overall results. However, they were not used at the time of this experiment.

4.4 Comparative Analysis

In Table 4, we have compiled a list of studies performed with only clinically verified cough samples to compare our model. These studies show the sample sizes of the studies along with performance results such as accuracy (ACC), an area under the curve (AUC), sensitivity (SEN), and specificity (SPEC) when available. As seen from the table, our results align with the expectations from other clinically verified cough datasets. Therefore, if producers of machine learning models for Covid-19 cough screening can collect datasets with clinically verified data, they will produce a highly accurate machine that will save lives.

Table 4 Covid-19 screening performance based on cough sounds: accuracy (ACC, in %), area under the curve (AUC), sensitivity (SEN), and specificity (SPEC) on 'laboratory-confirmed datasets'

Authors (year)	Cough samples	Performance			
		ACC	AUC	SEN	SPEC
Wei et al. [15]	1283	0.76	–	0.99	0.95
Imran et al. [27]	543	0.93	–	0.94	0.91
Laguarta et al. [12]	5320	0.971	0.97	0.985	0.942
Hassan et al. [13]	80	0.97	0.974	0.964	–
Andreu-Perez et al. [14]	8380	–	0.988	0.964	0.962
Best Proposed (2022)	121	0.933	0.967	0.916	0.944

5 Discussion

In this section, we discuss some points of interest in the research that we have done. As follows, this section will discuss the improvement dropout had on this model. Also, we will discuss some possible improvements to the model that could generate higher accuracy.

Dropout regularization is generally accepted in the signal processing community as improving the performance and generalization of speech acoustic models [26]. We have shown that dropout regularization also improves the performance of a cough-based Covid-19 detector by a significant margin, as much as 13 points of accuracy. We can say with little doubt that speech acoustic model techniques can also generalize to cough acoustic model techniques. Furthermore, different dropout types could further improve this model's performance, such as spacial/channel dropout. This type of dropout could be applied to the first two layers of the model and should, in theory, improve the results.

Threshold tuning is commonly used in the industry to help the performance of many different model types for various imbalanced classification problems [28]. Since Covid-19 detection is inherently an imbalanced classification problem, strong arguments can be made for tuning the output threshold of the model instead of using the standard 0.5 cut-off point. Based on our dropout testing, we can see that our model achieves a 98.6% AUC with 0.125 dropout. Theoretically, we could use output threshold tuning to get this model to outperform our "best" model on this dataset and provide the second-best results compared to the other 'laboratory-confirmed datasets' in our comparative analysis.

6 Conclusion and Future Work

In this paper, we have proposed a cough sound-based model to detect Covid-19 positive cases. We have evaluated our model on a clinically verified 'Virufy' dataset and achieved an accuracy of 93.33% which is competitive with other state-of-the-art models with clinically verified datasets. Our model opens the possibility of developing an early-stage screening tool based on cough sound. However, as our data sample size is minimal for a test of this nature, we are collecting more data for future works to improve our model's robustness and increase its generalizability to more datasets.

References

1. Lapostolle, F., Schneider, E., Vianu, I., Dollet, G., Roche, B., Berdah, J., Michel, J., Goix, L., Chanzy, E., Petrovic, T., et al.: Clinical features of 1487 covid-19 patients with outpatient management in the greater paris: the covid-call study. Internal Emerg. Med. **15**(5), 813–817 (2020)
2. Rudraraju, G., Palreddy, S.D., Mamidgi, B., Sripada, N.R., Sai, Y.P., Vodnala, N. K., Haranath, S.P.: Cough sound analysis and objective correlation with spirometry and clinical diagnosis. Inf. Med. Unlocked **19** (2020)
3. Han, J., Qian, K., Song, M., Yang, Z., Ren, Z., Liu, S., Liu, J., Zheng, H., Ji, W., Tomoya, K., et al.: An early study on intelligent analysis of speech under covid-19: Severity, sleep quality, fatigue, and anxiety. Interspeech 2020 (2020)
4. Santosh, KC.: Speech processing in healthcare: can we integrate? Intell. Speech Signal Process. 1–4 (2019)
5. Mukherjee, H., Sreerama, P., Dhar, A., Obaidullah, S.K.M., Roy, K., Mahmud, M., Santosh, KC.: Automatic lung health screening using respiratory sounds. J. Med. Syst. **45**(2) (2021)
6. Mukherjee, H., Salam, H., Santosh, KC.: Lung health analysis: adventitious respiratory sound classification using filterbank energies. Int. J. Pattern Recogn. Artif. Intell. **35**(14), 2157008 (2021)
7. Gayam, V., Chobufo, M.D., Merghani, M.A., Lamichhane, S., Garlapati, P.R., Adler, M.K.: Clinical characteristics and predictors of mortality in African-Americans with Covid-19 from an inner-city community teaching hospital in New York. J. Med. Virol. **93**(2), 812–819 (2020)
8. Weng, C.-H., Saal, A., Butt, W.W.W., Chan, P.A.: Characteristics and clinical outcomes of Covid-19 in Hispanic/Latino patients in a community setting: a retrospective cohort study. J. Med. Virol. **93**(1), 115–117 (2020)
9. Vaughan, L., Veruttipong, D., Shaw, J.G., Levy, N., Edwards, L., Winget, M.: Relationship of socio-demographics, comorbidities, symptoms and healthcare access with early Covid-19 presentation and disease severity. BMC Infect. Dis. **21**(1), 1–10 (2021)
10. Coronavirus, 2021. Last accessed on 18 July 2021
11. Macedo, M.F.C., Pinheiro, I. M., Carvalho, C.J.L., Hilda, C.J.R., FragaH.C., Isaac, P.C. , Montes, S.S., O.A.C., Alves, L.A., Saba, H., Márcio, L. V., Araújo, M.L, et al.: Correlation between hospitalized patients' demographics, symptoms, comorbidities, and Covid-19 pandemic in Bahia, Brazil. PLOS ONE **15**(12), 1–15 (2020)
12. Laguarta, J., Hueto, F., Subirana, B.: Covid-19 artificial intelligence diagnosis using only cough recordings. IEEE Open J. Eng. Med. Biol. **1**, 275–281 (2020)
13. Hassan, A., Shahin, I., Alsabek, M.B.: Covid-19 detection system using recurrent neural networks. In: 2020 International Conference on Communications, Computing, Cybersecurity, and Informatics (CCCI) (2020). https://doi.org/10.1109/ccci49893.2020.9256562

14. Andreu-Perez, J., Perez-Espinosa, H., Timonet, E., Kiani, M., Giron-Perez, M.I., Benitez-Trinidad, A.B., Jarchi, D., Rosales, A., Gkatzoulis, N., Reyes-Galaviz, O.F., Torres, A., Reyes-Garcia, C.A., Ali, Z., Rivas. F.: A generic deep learning based cough analysis system from clinically validated samples for point-of-need Covid-19 test and severity levels. IEEE Trans. Serv. Comput. 1–1 (2021)
15. Wei, W., Wang, J., Ma, J., Cheng, N., Xiao, J.: A real-time robot-based auxiliary system for risk evaluation of Covid-19 infection. Interspeech 2020 (2020)
16. Mouawad, P., Dubnov, T., Dubnov, S.: Robust detection of Covid-19 in cough sounds. SN Comput. Sci. **2**(1) (2021)
17. Pahar, M., Klopper, M., Warren, R., Niesler, T.: Covid-19 cough classification using machine learning and global smartphone recordings. Comput. Biol. Med. **135** (2021)
18. Anupam, A., Mohan, N.J., Sahoo, S., Chakraborty, S.: Preliminary diagnosis of Covid-19 based on cough sounds using machine learning algorithms. In: 2021 5th International Conference on Intelligent Computing and Control Systems (ICICCS) (2021). https://doi.org/10.1109/iciccs51141.2021.9432324
19. Bansal, V., Pahwa, G., Kannan, N.: Cough classification for Covid-19 based on audio mfcc features using convolutional neural networks. In: 2020 IEEE International Conference on Computing, Power and Communication Technologies (GUCON) (2020). https://doi.org/10.1109/gucon48875.2020.9231094
20. Shimon, C., Shafat, G., Dangoor, I., Ben-Shitrit, Asher: Artificial intelligence enabled preliminary diagnosis for Covid-19 from voice cues and questionnaires. J. Acoust. Soc. Am. **149**(2), 1120–1124 (2021)
21. Mohammed, E.A., Keyhani, M., Sanati-Nezhad, A., Hossein Hejazi, and Behrouz H. Far. An ensemble learning approach to digital corona virus preliminary screening from cough sounds. Sci. Rep. **11**(1) (2021)
22. Vrindavanam, J., Srinath, R., Shankar, H.H., Nagesh, G.: Machine learning based Covid-19 cough classification models—a comparative analysis. In: 2021 5th International Conference on Computing Methodologies and Communication (ICCMC) (2021)
23. Dash, T.K., Mishra, S., Panda, G., Satapathy, S.C.: Detection of Covid-19 from speech signal using bio-inspired based cepstral features. Pattern Recogn. **117** (2021)
24. Santosh, KC., Rasmussen, N., Mamun, M., Aryal, S.: A systematic review on cough sound analysis for Covid-19 diagnosis and screening: is my cough sound Covid-19? PeerJ Comput. Sci. **8**, 1–16 (2022)
25. Lella, K.K., Pja, A.: Automatic diagnosis of covid-19 disease using deep convolutional neural network with multi-feature channel from respiratory sound data: cough, voice, and breath. Alexandria Eng. J. **61**, 1319–1334 (2021). https://doi.org/10.1016/j.aej.2021.06.024
26. Kovács, G., Tóth, L., Van Compernolle, D., Ganapathy, S.: Increasing the robustness of cnn acoustic models using autoregressive moving average spectrogram features and channel dropout. Pattern Recog. Let. **100**, 44–50 (2017)
27. Imran, A., Posokhova, I., Qureshi, H.N., Masood, U., Riaz,M.S., Ali, K., John, C.N., Hussain, M.D.I., Nabeel, M.: Ai4covid-19: Ai enabled preliminary diagnosis for covid-19 from cough samples via an app. Inf. Med. Unlocked **20**, 100378 (2020)
28. Brownlee, J.: A gentle introduction to threshold-moving for imbalanced classification. Mach. Learn Mastery (2021). https://machinelearningmastery.com/threshold-moving-for-imbalanced-classification/

Keypoint-Based Detection and Region Growing-Based Localization of Copy-Move Forgery in Digital Images

Akash Kalluvilayil Venugopalan and G. Gopakumar

Abstract The issues caused due to image manipulations are common these days. Furthermore, it causes severe troubles in news broadcasting, social media, and digital media forensics. Mainly, the types of image manipulations are divided into four; they are splice forgery, copy-move, morphing, and retouching. Among them, copy-move forgery is one of the most challenging manipulations to detect since it does not change image characteristics while performing the copy-move forgery operation. In this paper, we propose a copy-move forgery detection and localization scheme that detect forgery regions in the image even though the forgery image undergoes translation, scaling, and rotation attacks. The scheme uses the SIFT algorithm for keypoints extraction from the forgery image, DBSCAN to cluster these keypoints, and Hu's invariant moments are used to identify similarity between two suspicious regions in the image. Lastly, a region growing is performed around these detected regions to localize copy-move forgery regions. The scheme has experimented with a CoMoFoD dataset which is publicly available, and the result shows that the proposed scheme outperforms the state-of-the-art non-deep learning-based copy-move forgery techniques in terms of recall, FNR, F1-score, and also in computational time.

Keywords Copy-move forgery detection (CMFD) · SIFT algorithm · DBSCAN clustering · Hu's invariant moments · Region growing

1 Introduction

Image manipulation is a common issue that we can see in social media, news broadcasting, and many other domains. Different computer applications are used for image manipulations, providing a comfortable environment for the user to make any changes

A. Kalluvilayil Venugopalan (✉) · G. Gopakumar
Department of Computer Science and Engineering, Amrita Vishwa Vidyapeetham, Amritapuri, India
e-mail: amenp2ari20001@am.students.amrita.edu

G. Gopakumar
e-mail: gopakumarg@am.amrita.edu

M. Tistarelli et al. (eds.), *Computer Vision and Machine Intelligence*,
Lecture Notes in Networks and Systems 586,
https://doi.org/10.1007/978-981-19-7867-8_41

to the original/authentic image. Therefore, image tampering is immensely popular and can be performed easily. The image manipulations are either done to make the picture more attractive to the audience or hide some content in the original image. Sometimes, it is complicated to identify these manipulations [1] with our naked eyes. As we know, technology is advancing day by day and even in the field of image manipulation. As a result, image manipulation tools are getting developed to such a level that the editor can easily create manipulations to digital images so that an average person with a naked eye cannot figure out forgery regions in an image. Some standard tools used for manipulating images are Adobe Photoshop, GIMP (GNU Image Manipulation Program), Corel PaintShop, etc. Image manipulation can cause severe issues in many domains that deal with digital images, such as forensics, fake news creation, and social media. To submit an image to a court for a case, verifying that the image has not undergone any tampering is mandatory. Therefore, we should be able to detect the forgery region in an image as accurately as possible and decide the authenticity of an image. If the tampered image is given to the court for a particular case, the court may end up in a false conclusion about the case. Due to this reason, it is relevant for the forensic department to prove whether a particular image has undergone any tampering or not.

The forgery image detection can be categorized into two types: active [2] and passive approaches [3]. The active approach requires pre-processing on the image, such as adding a signature or watermark to the image and is often done during image creation. And, if the signature or watermark extracted from the forgery image matches the original image, then we can conclude that the image is forged. The passive or blind approach is more difficult compared to the active approach as we do not consider any signature or watermark in the image. The passive approach includes image pixel-based techniques, format-based techniques, etc.

Image manipulation can also be classified into several categories: copy-move forgery, splice forgery, retouching, and morphing [3]. Copy-move forgery detection is one of the challenging tasks to detect as the copied source is placed in the same image resulting in similar image properties for both source and target regions. This contrasts with the image splice forgery, where [4] the source and target are from two different images. The splice forgery means that the portion of the first image is copied and added to a second image creating a forged image. The forgery region can undergo image transformations such as scaling, translation, rotation, and flip in both copy-move and copy-splice forgery. Here, copy-move and splicing require passive approaches to detect manipulations. Our research explored different copy-move forgery detection techniques on available datasets intending to incorporate novel ideas for improving the performance of the copy-move forgery detection technique and came up with a copy-move forgery detection and localization technique that can detect forgery with higher TPR, F1-score, and lesser FNR. Also, the technique only requires less computational time.

The structure of the paper is organized as follows. In Sect. 2, we have discussed about the related works, Sect. 3 discusses the implementation of proposed scheme, Sect. 4 contains results that we obtained from the purposed scheme, and Sect. 5 is the conclusion.

2 Related Works

Many researchers came up with different approaches for detecting copy-move forgery [5, 6] in the image. This has been one of the major topics for a few decades. Some of the techniques can only detect forgery, while others can detect and localize regions of copy-move forgery in the given input image. Block-based technique for copy-move forgery detection and localization can give better results, but the algorithm used for block-based detection is [7] computationally exhaustive since it uses a greedy search for detection. Whereas detecting and localizing copy-move forgery using keypoints can work much faster than the block-based method, but we should compromise on the model performance. Many authors came up with different approaches to improve the results of keypoint-based detection techniques. Some of the authors used scale invariant feature transform (SIFT) to detect keypoints and then used the nearest neighbor method to match these keypoints that are contributing to the copy-move forgery regions [8]. And, since the deep learning is emerging field in these century, many researchers also tried to develop CMFD techniques using DL-based methods [9]. However, most of these techniques, even though they detect forgery regions, may fail to detect copy-move forgery if the image undergoes any image transformations, and another limitation of most of the techniques is that it requires more amount of memory as well as may take a longer time for computation. Therefore, in our research, we developed an unsupervised machine learning-based technique that has lesser computation time and can also detect and localize copy-move forgery even though the image has undergone attacks such as scaling, rotation, translation, and noise adding.

3 Working and Implementation of the Proposed Approach

We have developed a technique to detect and localize copy-move forgery regions in the given copy-move forgery image. The technique is mainly robust to translation, scaling, and rotation attacks. The flow diagram of implemented copy-move forgery detection (CMFD) technique is shown in Fig. 1. The primary components of the architecture are scale invariant feature transform (SIFT), keypoint clustering using the DBSCAN algorithm, distance calculation of blocks using Hu's invariant moments, and a region growing algorithm. Below we have given a detailed step-by-step procedure of the proposed scheme.

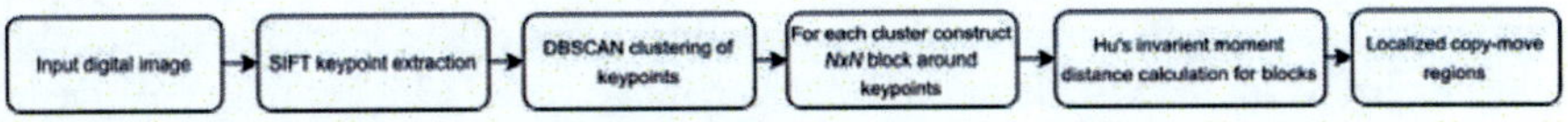

Fig. 1 Basic flow diagram of proposed approach

3.1 SIFT Keypoint Extraction and DBSCAN Clustering

In the initial step, given input image (X), the proposed scheme extracts SIFT keypoint from the input image(X). Here, if two regions of the image undergo copy-move forgery, the keypoints extracted from these regions will be similar in their properties and different from the authentic regions of the image (X). More precisely, we can say that if two pixels are undergone copy-move forgery in an image, then the keypoints corresponding to these two regions will show the same properties. And, then, if we try to find a similarity or distance between these two keypoints, we will find that both keypoints are similar. In that case, we can simply apply a clustering algorithm to group keypoints extracted from copy-move regions and authentic regions into two different clusters. However, we cannot assume that image(X) will only have an authentic region and exactly one copy-move forgery region. There can be multiple forgeries in an image. Hence, using a clustering algorithm like the K-means, which requires an exact number of clusters (k) as the hyperparameter will not be good. Therefore, it is better to use a clustering algorithm that does not require the number of clusters 'k' as its hyper-parameter. A clustering algorithm called DBSCAN clustering only takes two hyperparameters for clustering, and they are *epsilon*(ϵ) and *minPoints* [10]. Moreover, the algorithm will give all the possible clusters(k) within the data or keypoints. Using DBSCAN algorithm, it is possible to cluster the keypoints from multiple copy-move forgery regions. We should also notice that if multiple forgery regions exist in the image(X), the keypoints from these regions might form different clusters. In our implementation, we make use of this concept to detect forgery. Once keypoints are extracted using SIFT algorithm [11], we pass the keypoints to DBSCAN algorithm to cluster the keypoints by looking at the keypoint density in the feature space. Ideally, keypoints from authentic regions form one cluster, and the remaining clusters might be formed due to the presence of keypoints from the copy-move forgery regions in the image(X).

3.2 Building N × N Blocks Around Extracted Keypoints

After SIFT keypoints extraction and clustering, we take each cluster from the clusters (k) and construct an $N \times N$ block by keeping each keypoint in this cluster at the center of the block. This process will be applied to all the clusters generated by the previous step. By applying this strategy, we are capturing an $N \times N$ portion of the image from where we have extracted keypoints. Since we have grouped keypoints into different clusters, the cluster(s) that has keypoints from copy-move forgery region(s) and keypoints from its corresponding authentic regions can be localized separately in the image (X) for the visualization. However, this gives us a higher false positive rate (FPR) and false negative rate (FNR) for the task of copy-move forgery localization. FPR is high because we have localized some of the authentic regions which does not participate in copy-move along with the copy-move forgery regions. And as we know SIFT algorithm can also miss some relevant keypoints that contribute to copy-move forgery regions in the image (X). Therefore, localizing using these keypoints after block construction will introduce some false negatives (FN). The implemented architecture tackles this problem in the next two steps by calculating Hu's moment distance between each block and region growing method.

3.3 *Calculating the Distance Between Each Pair of Blocks Within the Cluster*

This part of the algorithm reduces false positives(FP) in the previously detected blocks. In this step, the distance between each pair of blocks within every 'k' cluster is calculated to ensure that the pair of blocks are from the copy-move forgery region of the image (X). The distance is calculated with the help of Hu's invariant moments [12].

$$\left.\sum_{k=0}^{p}\sum_{m=0}^{q}\sum_{n=0}^{r} H(b_k^n, b_k^m)\right\} \text{for all } m \neq n \tag{1}$$

where

$H(b_k^n, b_k^m)$ is the Hu's invariant moments between blocks b_k^n and b_k^m

$$H_i = \sum_{i=0}^{6} -sign(H_i)log|H_i| \tag{2}$$

$$D(b_i, b_j) = \sum_{M=0}^{6} \frac{H_M^{b_i} - H_M^{b_j}}{H_M^{b_i}} * 10^3 \tag{3}$$

Equation 2 takes log transform for all the 7 moments, and Eq. 3 takes distance between ith and jth blocks.

Hu's invariant moments will give a seven-element vector as the output when an input image is passed, which will be invariant to translation, rotation, scaling, and image reflection. Here, as shown in Eq. 1, we calculate Hu's invariant moments for each pair of blocks for every 'k' number of clusters developed from earlier steps. Where $(k_1, k_2, ...k_p)$ is the total number of clusters derived from DBSCAN clustering, and (b_k^m, b_k^n) is the blocks in the clusters where 'm' and 'n' are the index number of the block in the cluster. After determining the Hu's seven-element vector for every block within this cluster, then making use of these seven-element vectors calculate the distance between each of them. The distance between each pair of blocks (b_i, b_j) is calculated using Eq. 3.

$$\delta \leq D(b_i, b_j) \tag{4}$$

In the Eq. 3, $H_m^{b_i}$ is the Hu's invariant moment vector of block 'b_i', 'M' represents moment position in the seven-element vector, and this is also followed in the case of block 'b_j'. The equation will give a scalar value which is the distance between two pairs of blocks. Here, if two pair of the block is from the copy-move forgery region and its corresponding source region, then the distance among them will be approximately close to 0, and if the distance is higher than a predetermined threshold (δ), then those pair of blocks (b_i, b_j) have not participated in the copy-move forgery process. In our experiments, we evaluated the approach with different thresholds 'δ' and came up with an optimal δ-value that gave better results in our experiments. And any pair of the block that gives a distance less than 'δ', those blocks are considered as the copy-move forgery regions (i.e., as shown in Eq. 4). To reduce the computation complexity, the technique takes each cluster from 'k' number of clusters and compares the similarity distance for each pair of blocks within that cluster only. This will reduce the computational time to a great extent. And by this approach, we can ensure that the blocks are from a copy-move forgery region or not, which will help us reduce FPs.

3.4 Region Growing on the Previously Detected Blocks

Even after reducing false positives present in the detected block, the number of true positives(TP) will probably be lesser in the detected block list. This is expected because we use a keypoint-based detection method to detect forgery regions. Since we are using SIFT algorithm, the algorithm will fail to extract a major portion of keypoints from the copy-move forgery regions and their source regions. This issue is faced in every keypoint-based copy-move forgery detection technique. Our proposed method uses a region growing-based strategy to tackle this problem. The proposed region growing technique takes each detected block after eliminating FP's using the previously discussed method. The developed region growing algorithm takes each block and finds the nearest non-overlapping $N \times N$ blocks. To perform this, the region growing algorithm takes the previously detected pair of blocks b_i and b_j from the cluster. Here, b_i is the block from the copy-move forgery region, and b_j is the block from its source region. Now, by keeping b_i and b_j at the center find the top, bottom, left, and right $N \times N$ block of both b_i and b_j blocks as shown in Fig. 2. Let us consider that T_i, B_i, R_i, and L_i are the nearest $N \times N$ blocks of block b_i, and T_j, B_j, R_j, and L_j are the nearest $N \times N$ blocks of b_j. We again calculate their distance using Hu's invariant moments to find whether these blocks are from copy-move forgery. This means the distance from T_i and T_j will be computed, and if the distance is less than the threshold 'δ', we consider these blocks are from a copy-move forgery region, else we reject these blocks and conclude it is from the authentic region. This distance calculation is performed on B_i and B_j, R_i and R_j, and L_i and L_j also. Later, all the newly detected blocks will again undergo a region growing process to further localize the copy-move forgery regions. This allows us to increase pixel-

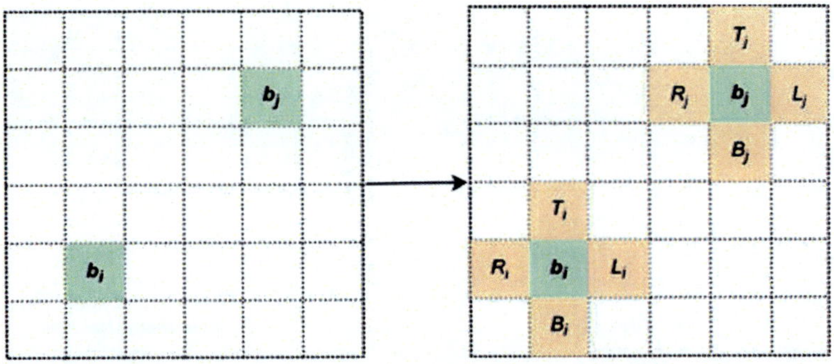

Fig. 2 Region growing performed around $N \times N$ block b_i and b_j. Where T_i, B_i, R_i, L_i, T_j, B_j, R_j, and L_j are the $N \times N$ blocks derived from previously detected blocks using region growing

level localization results and increase TPR. Finally, the proposed scheme localizes the copy-move forgery region in a binary image with all these detected blocks.

4 Experiment Results

We have experimented the proposed architecture of copy-move forgery detection with a public dataset called *CoMoFoD* [13]. All the experiments are conducted on a 2-Core 2.20 GHz Intel(R) Xeon(R) CPU with 12 GB RAM, and the programming language used is Python 3.7. We have also conducted different analyzes on the algorithm, and all the analysis is discussed in this section. The detection recall, FNR, accuracy, precision, F1-score, and ROC-AUC are calculated to determine how well the scheme can distinguish between copy-move forgery and authentic images. Pixel-level localization results are also computed in the experiment that we have conducted. A detailed description of our analyzes and experiment is given in this section.

In this experiment, the most challenging task is setting up the right value of thresholds that can detect and localize copy-move forgery efficiently. Several experiments are conducted to obtain an optimal hyperparameter for DBSCAN clustering and Hu's invariant moment distance calculation. *epsilon*(ϵ) and *minPoints* of DBSCAN are set to 60 and 2, respectively. And we have collected 100 samples from the CoMoFoD dataset to obtain an optimal value for 'δ'. These collected samples are tested with various δ-values such as 0.5, 1, 2, 3, 4, and 5 and then calculated corresponding δ-value's recall and FNR to determine which 'δ' has maximum recall and minimum FPR and FNR. The recall plot is given in Fig. 3a. All three metrics were identified as saturated after $\delta = 2$. And till $\delta = 2$, the three metrics rapidly increase their results. Therefore, we have picked a reasonable optimal threshold of $\delta = 2.5$ for conducting further experiments.

After setting the desired hyperparameter and threshold, we tested the proposed scheme with *CoMoFoD* dataset. Each image from test samples is first given to SIFT

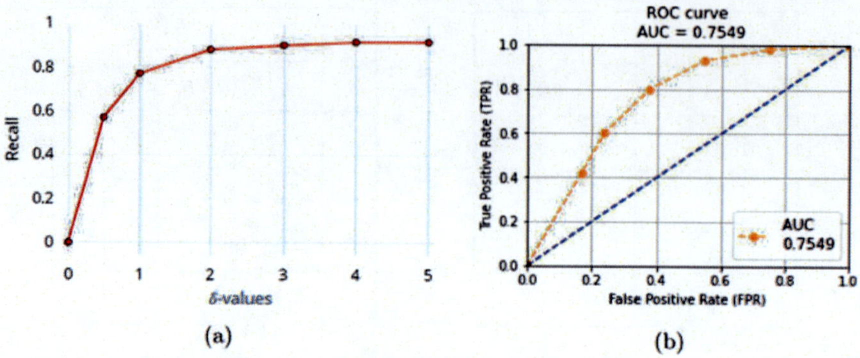

Fig. 3 In above figure, **a** shows recall of proposed approach by varying 'δ', and **b** is the ROC achieved by setting 'δ' to 2.5 for CoMoFoD dataset

algorithm to extract keypoints from the image and then clustered these keypoints using the DBSCAN clustering algorithm. In the next step, each keypoints from kth cluster are taken, and an $N \times N$ block is constructed by keeping this keypoint at the center of the block. In the experiment, the block size used is 4×4. However, trying with larger or smaller block sizes is also possible, but this can affect pixel-level localization results. This means that if the block size is increased above a specific limit, then pixel-level localization TPR can be improved, but this can also increase false-positive blocks in localization, affecting the overall performance of the proposed scheme. Once the blocks are constructed, the distance between each pair of blocks within the cluster is calculated, and if any pair of blocks have a distance less than $\delta = 2.5$, then those blocks are considered as copy-move forgery blocks. This process is done for all the 'k' clusters obtained after DBSCAN clustering. Finally, a region growing is applied to previously detected blocks by placing those blocks at the center of the growing algorithm, as we have mentioned in Sect. 3.4. For the region growing with Hu's invariant moment distance calculation, we have used $\delta = 2.5$, the same threshold used to detect initial blocks of forgery. After region growing, the proposed scheme gives a binary image of the copy-move forgery region(s) of the forgery image. The localized outputs by the proposed scheme under different attack are shown in Fig. 4.

The obtained results from the proposed scheme are summarized below. The detection recall obtained by testing with CoMoFoD is 91.52%, and the false negative rate (FNR) is 10.27%. A detailed performance metrics table is given in Table 1. A comparison between the proposed technique's results and some of the famous research on copy-move forgery detection is summarized in Table 2. The best precision, recall, and F1-score obtained from various methods are highlighted in bold in Table 2. It should be noted that the proposed approach outperforms other copy-move forgery detection methods in terms of recall and F1-score. The significance of bold in Table 2 is that the proposed approach achieved the highest recall and F1-score, as indicated by the bolded values in the last row of the table. The proposed method of CMFD outper-

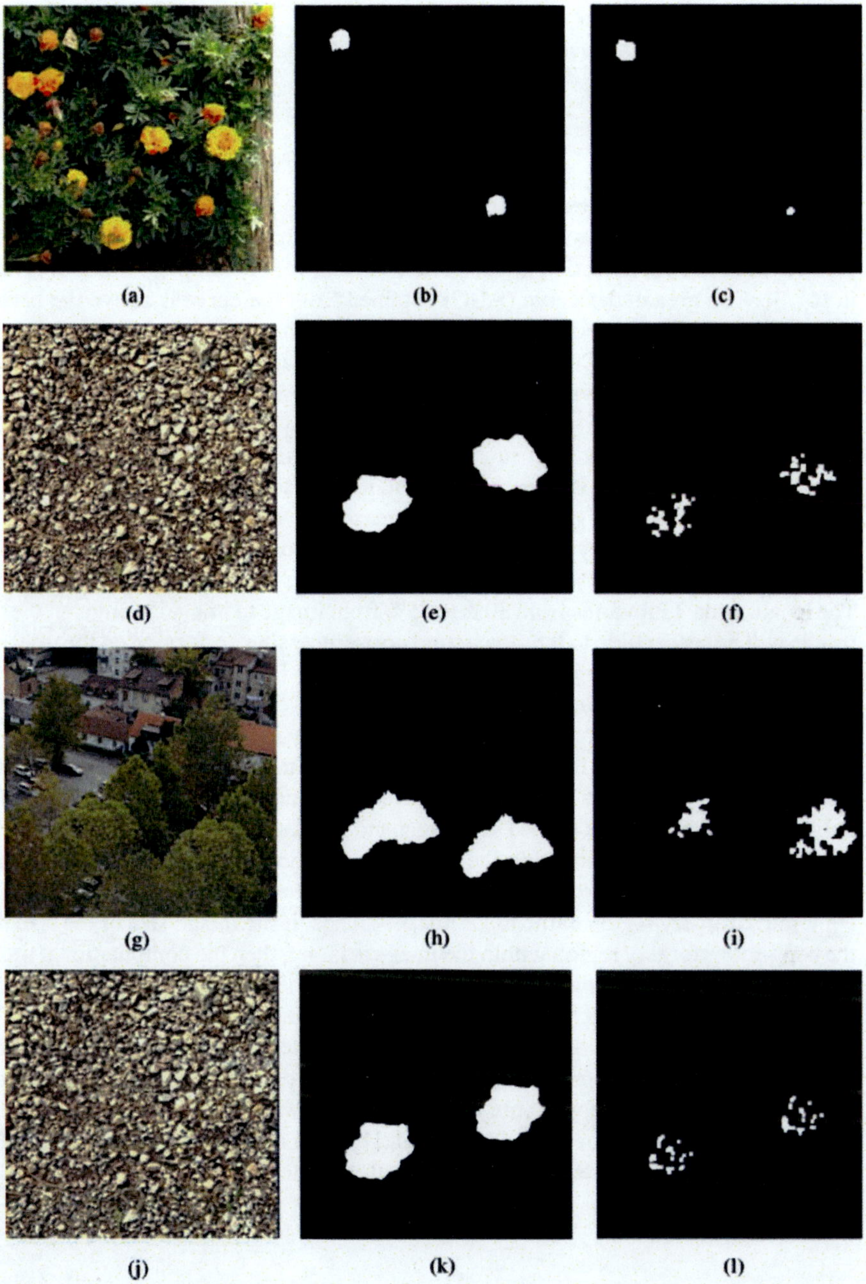

(a) (b) (c) (d) (e) (f) (g) (h) (i) (j) (k) (l)

Fig. 4 In the above figure, **a–c** is copy-move forgery image with a translation attack, **d–f** is an example of a rotation attack, **g–i** shows scaling attack, and **j–i** is an image with noise adding. The left column's image is copy-moved, and its ground truth is shown in the middle column. The output of the proposed approach for each attack is shown in the right column

Table 1 Performance metric of detection with threshold $\delta = 2.5$

Recall	Accuracy	Precision	$F1$-score
0.9152	0.7024	0.7433	0.8203

forms other techniques in terms of recall and F1-score, and the approach even shows nearly good performance when compared with the deep learning-based approaches [9, 14]. The ROC for detection is also plotted for CoMoFoD samples, which is given in Fig. 3b. The area under curve (AUC) obtained from the curve is above the baseline by 75.49%. The proposed scheme gives satisfactory results in terms of recall, FNR, F1-score, and ROC-AUC for the CoMoFoD dataset. Since SIFT and Hu's invariant moments are invariant to translation, scaling, rotation, and illumination, the proposed method can also detect and localize copy-move forgery regions even though the image goes through such attacks. The technique can detect copy-move forgery regions from the images which contain Gaussian noise, but detection and localization recall reduce gradually with increase in σ or noise. The average pixel-level localization accuracy obtained with the CoMoFoD dataset for the scheme is 91.62%, precision = 82.51%, and FNR = 0.99%, which is also a satisfactory result. The localization TPR varies from 30% to 55% from image to image. Moreover, these results will be susceptible to the attacks and post-processing performed on the image.

As discussed in earlier sections, many copy-move forgery detection techniques require a longer time to detect and localize copy-move forgery. Block-based techniques can even take hours to detect and localize copy-move forgery regions in the image, whereas keypoint-based techniques generally take an average time of 5–10 min to detect and localize. In our analysis, we have found that the proposed scheme only takes an average time of 9.41 s to detect and localize copy-move forgery, which is a better result when comparing the computational time with many existing techniques (Fig. 5). But this can vary with the size of the image and the size of the copy-move forgery region in the input image. That is, if the image size or the size of the copy-move forgery region within the image is larger, then the computational time will also increase proportionally with these sizes. When these sizes are increased, the number of regions that should be captured by the region growing algorithm will also increase, increasing computational time. Also, note that the proposed scheme is tested with a 2-core Intel processor with 12 GB RAM due to the system limitations, and the proposed scheme was able to detect forgery within a short period of time, whereas techniques proposed by Chien et al. [15] and other techniques such as EB [16] and ECEB [17] are tested by Chien et al. on Intel i7-6700 CPU with 32 GB RAM [15].

Table 2 Comparison of the proposed scheme's detection results with existing copy-move forgery detection techniques on the CoMoFoD dataset

	Precision	Recall	F1-score
Chien [15]	0.7019	0.8461	0.7672
Yue wu [9]	**0.8532**	0.7875	0.8009
Jian [18]	0.5446	0.8504	0.6639
Amerini [8]	0.7	0.875	0.7777
Yaqi Liu [14]	0.5927	0.8220	0.6318
Ours	0.7433	**0.9152**	**0.8203**

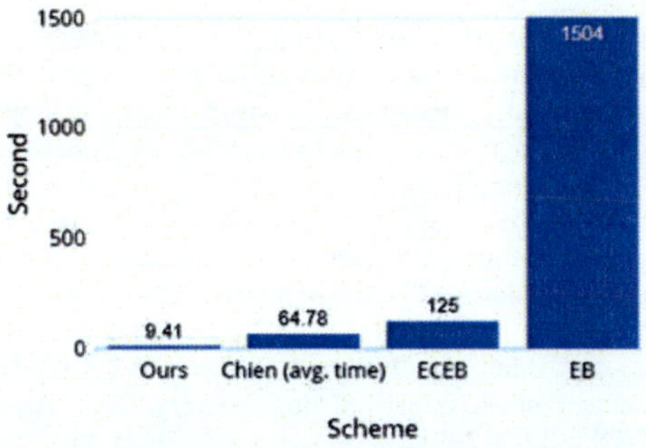

Fig. 5 Computational time required for the proposed method, Chien [15], expanding block algorithm(EB) [16], and enhanced cluster expanding block algorithm (ECEB) [17]. (Note that the proposed scheme is tested on a 2-core Intel processor with 12 GB RAM, and it only took an average time of 9.41sec to detect and localize forgery, whereas the ECEB, EB, and scheme proposed by Chien et al. are tested on Intel i7-6700 CPU with 32 GB RAM [15].)

5 Conclusion

The proposed copy-move forgery detection method uses scale invariant feature transform(SIFT), DBSCAN clustering, and Hu's invariant moment's distance calculation, and an region growing is used to improve localization results. The experiment with *CoMoFoD* dataset shows the proposed scheme can also detect forgery regions even though the copy-move forgery region undergoes attacks such as translation, scaling, and rotation. Also, the experiment results show that the proposed scheme outperforms existing techniques in terms of detection recall, FNR, and F1-score. And another significant achievement of the proposed scheme is that it can detect and localize the forgery region within a short period of time, where most of the existing techniques require higher computational time for detection and localization of copy-move forgery regions.

References

1. Hrudya, P., Nair, L.S., Adithya, S.M., Unni, S.M.,Poornachandran, P.: Digital image forgery detection on artificially blurred images. In: 2013 International Conference on Emerging Trends in Communication, Control, Signal Processing and Computing Applications (C2SPCA), pp. 1–5 (2013). 10.1109/C2SPCA.2013.6749392
2. Menon, S.S., Mary Saana, N.J., Deepa, G.: Image forgery detection using hash functions, vol. 8 (2019)
3. Abidin, A.B.Z., Majid, H.B.A., Samah, A.B.A., Hashim, H.B.: Copy-move image forgery detection using deep learning methods: a review, vol. 2019 (2019, December). 10.1109/ICRIIS48246.2019.9073569
4. Gopal, D., G, G.: A deep learning approach to image splicing using depth map*. ICADCML (2022)
5. Lu, S., Hu, X., Wang, C., Chen, L., Han, S., Han, Y.: Copy-move image forgery detection based on evolving circular domains coverage. Multimedia Tools Appl. 1–26 (2022)
6. Nair, G.S., Gitanjali Nambiar, C., Rajith, N., Nanda, K., Nair, J.J.: Copy-move forgery detection using beblid features and dct. In: Innovations in Computational Intelligence and Computer Vision, pp. 409–417. Springer, Berlin (2022)
7. Narayanan, S.S., Gopakumar, G.: Recursive block based keypoint matching for copy move image forgery detection. In: 2020 11th International Conference on Computing, Communication and Networking Technologies (ICCCNT), pp. 1–6 (2020). 10.1109/ICCCNT49239.2020.9225658
8. Amerini, I., Ballan, L., Caldelli, R., Bimbo, A.D., Serra, G.: A sift-based forensic method for copy-move attack detection and transformation recovery. IEEE Trans. Inf. Forensics Secur. **6** (2011). 10.1109/TIFS.2011.2129512
9. Wu, Y., Abd-Almageed, W., Natarajan, P.: Busternet: detecting copy-move image forgery with source/target localization, vol. 11210. LNCS (2018)
10. Ester, M., Kriegel, H.P., Sander, J., Xu, X., et al.: A density-based algorithm for discovering clusters in large spatial databases with noise. In: kdd, vol. 96, pp. 226–231 (1996)
11. Lowe, D.G.: Distinctive image features from scale-invariant keypoints. Int. J. Comput. Vision **60**(2), 91–110 (2004)
12. Hu, M.K.: Visual pattern recognition by moment invariants. IRE Trans. Inf. Theor. **8** (1962). 10.1109/TIT.1962.1057692
13. Tian, X., Zhou, G., Xu, M.: Image copy-move forgery detection algorithm based on orb and novel similarity metric. IET Image Process. **14** (2020)
14. Liu, Y., Guan, Q., Zhao, X.: Copy-move forgery detection based on convolutional kernel network (2017). 10.48550/ARXIV.1707.01221, https://arxiv.org/abs/1707.01221
15. Chen, C.C., Lu, W.Y., Chou, C.H.: Rotational copy-move forgery detection using sift and region growing strategies. Multimedia Tools Appl. **78** (2019)
16. Lynch, G., Shih, F.Y., Liao, H.Y.M.: An efficient expanding block algorithm for image copy-move forgery detection. Inf. Sci. **239** (2013)
17. Chen, C.C., Wang, H., Lin, C.S.: An efficiency enhanced cluster expanding block algorithm for copy-move forgery detection. Multimedia Tools Appl. **76** (2017)
18. Li, J., Li, X., Yang, B., Sun, X.: Segmentation-based image copy-move forgery detection scheme. IEEE Trans. Inf. Forensics Secur. **10** (2015)

Auxiliary Label Embedding for Multi-label Learning with Missing Labels

Sanjay Kumar and Reshma Rastogi

Abstract Label correlation has been exploited for multi-label learning in different ways. Existing approaches presume that label correlation information is available as a prior, but for multi-label datasets having incomplete labels, the assumption is violated. In this paper, we propose an approach for multi-label classification when label details are incomplete by learning auxiliary label matrix from the observed labels, and generating an embedding from learnt label correlations preserving the correlation structure in model coefficients. The approach recovers missing labels and simultaneously guides the construction of model coefficients from the learnt label correlations. Empirical results on multi-label datasets from diverse domains such as image & music substantiate the correlation embedding approach for missing label scenario. The proposed approach performs favorably over four popular multi-label learning techniques using five multi-label evaluation metrics.

Keywords Multi-label · Missing labels · Auxiliary labels · Label correlation embedding

1 Introduction

Real-world classification problems often associate a data instance with multiple classes and are inherently multi-label. Such problems are prevalent in diverse domains such as prediction of protein functions [23], multi-topic document categorization [20], and image annotation [1]. Each of the associated labels often convey a unique semantics related with the data instance. Some common solution approaches either change the problem into a binary classification problem or adapt some existing classification method for multi-label learning [19]. In their simplest form, these

S. Kumar (✉) · R. Rastogi
South Asian University, Chanakyapuri, New Delhi, Delhi 110021, India
e-mail: sanjayksau@gmail.com

R. Rastogi
e-mail: reshma.khemchandani@sau.ac.in

M. Tistarelli et al. (eds.), *Computer Vision and Machine Intelligence*,
Lecture Notes in Networks and Systems 586,
https://doi.org/10.1007/978-981-19-7867-8_42

approaches consider each label independently and ignore any existing label correlation. For datasets with correlated labels, it can be beneficial to incorporate label correlations for effective model training.

A common occurrence in multi-label problems is the lack of datasets with complete label information. Capturing label information is expensive and error-prone, and often the ambiguity in label assignment results in partially labeled training data [16]. Multi-label learning with labels missing is an arduous problem and relatively less researched. Many multi-label methods presume availability of complete label information and do not account for missing labels. Missing labels add to the complexity as information useful for learning such as label correlations are no longer available directly.

In absence of full label information, label correlations are less reliable and may exhibit inconsistent inter-label relationships. Recovering missing labels as part of model building improves the quality of label correlation details, and the recovered label correlation structure can be used to guide label vector prediction for new data instances. Based on the potential benefit of recovered labels and hence label correlations, in this paper, we develop a multi-label technique relying on learning auxiliary labels and creating an embedding from learnt label correlations when part of labels are missing.

Related work on multi-label learning is covered in Sect. 2. The algorithm for model proposed is discussed in Sect. 3. Empirical results and related discussion are in Sect. 4, and lastly, Sect. 5 covers paper conclusion.

2 Related Work

In multi-label learning, we commonly presume a complete label assignment for the training data. This assumption breaks down frequently especially for large label spaces, as generally, the labels are only partially observed [21]. Developing a multi-label model by ignoring the unobserved labels will result in a degraded classifier. Several approaches have been proposed in the past for handling unobserved labels. Popular techniques for multi-label learning with incomplete labels can be categorized as: pre-processing methods, transductive methods, and synchronized methods. Pre-processing methods try to approximate the missing labels first and then train the model for new instances [21, 22]. Transductive methods attempt multi-label learning in a transductive setting, i.e., consider missing label recovery as their primary task [6]. Synchronized methods consider the issue of missing labels by reconstructing missing labels and training the model together. Most recent multi-label learning methods [9, 15, 28] tackle classifier training and missing label recovery at the same time.

Class labels are often correlated in multi-label datasets, and exploiting label correlations can supply vital information for building a multi-label classifier [8]. Multi-label learning methods exploiting label correlations are grouped as 1st order and 2nd order or high-order based on the scope of label correlations. 1st order approaches [1, 26] presume labels to be independent, 2nd order approaches [7, 11] capture pairwise label correlations, and higher order label correlation approaches[17] capture correlations across all labels. Some algorithms capture not only positive but negative label correlations as well for model building [4].

Manifold learning is also a popular multi-label learning paradigm [2, 10, 25]. Some methods utilize label manifold and the feature space structure to restore the missing labels and build the classification model [18]. Another assumption in multi-label learning is label matrix low rank, and several multi-label learning methods explore a low-dimensional subspace for labels [3, 12, 14]. Low-dimensional label space embedding along with exploring label correlations in the original label space has also been used for multi-label text classification problems [13]. Approaches such as [9] model label correlations by constructing an embedding matrix from label correlations and utilize it as model coefficients.

Multi-label approaches when labels are missing utilize label correlations but do not directly explore label correlation relationship with model coefficients via matrix embedding. Our model is motivated from this and employs label correlations and label embedding technique for multi-label classification. In the proposed method, we utilize the relationship that label correlations exhibit to recover missing labels and to simultaneously guide the construction of regression coefficients from the learnt label correlations. The following section discuses the proposed method in detail.

3 Problem Formulation and Algorithm Proposed

Consider a training dataset $\{(x_k, y_k)\}_{k=1}^{n}$ for multi-label classification having n training samples with k-th data instance $x_k \in \mathbb{R}^d$ and associated class label vector $y_k \in \{0, 1\}^l$ where l stands for the total number of class labels. Let $X = \{x_1, x_2, \ldots, x_n\} \in \mathbb{R}^{n\times d}$ stands for the data matrix with d as feature dimension. $Y = \{y_1, y_2, \ldots, y_n\} \in \{0, 1\}^{n\times l}$ represents the label matrix for X where 1 indicates a label is present, 0 indicates the label is either not applicable or unobserved. Given a dataset $\{X, Y\}$, the multi-label learning aims to model a classifier for predicting label vector $y \in \{0, 1\}^l$ for a previously unseen instance $x \in \mathbb{R}^d$. Using the symbols discussed, we will prepare a model for multi-label learning when label details are incomplete.

We start with least squared loss for empirical risk minimization given by:

$$\min_{W} \frac{1}{2}\|XW - Y\|_F^2 + \mathcal{R}(W) \tag{1}$$

where $\|.\|_F$ is the Frobenius norm, $W \in \mathbb{R}^{d\times l}$ is the model coefficient matrix, and $\mathcal{R}(W)$ is the regularizer for the coefficient matrix W.

3.1 Learnt Label Correlations Embedding

We aim to learn the label correlations and exploit it for missing label recovery. By learning the label correlations containing inter-label relationships, we can augment the incomplete label matrix with missing label information and strengthen the prediction capability of the model. Let $R \in \mathbb{R}^{l \times l}$ be the correlation matrix for labels where the uv-th entry in matrix denotes the correlation for the u-th and v-th labels. Incorporating the label correlations for supplementing the label matrix Y, the minimization problem can be expressed as follows:

$$\min_{W,R} \frac{1}{2}\|XW - YR\|_F^2 + \frac{\lambda_1}{2}\|YR - Y\|_F^2 + \mathcal{R}(W, R) \tag{2}$$

where $\mathcal{R}(W, C)$ contains the regularizer for the coefficient matrix W and learnt label correlations R. Note that YR above will represent the auxiliary label matrix containing full label information. Second term above in optimization ensures that the observed relevant label information is maintained.

We can further approximate model coefficients from the learnt label correlations R by decomposing it into an embedding comprised of the model's regression coefficients [9, 27]. Incorporating label correlation decomposition, the minimization problem can be updated to:

$$\begin{aligned}\min_{W,R} \frac{1}{2}\|XW - YR\|_F^2 &+ \frac{\lambda_1}{2}\|YR - Y\|_F^2 \\ &+ \frac{\lambda_2}{2}\|R - \alpha W^T W\|_F^2 + \mathcal{R}(W, R)\end{aligned} \tag{3}$$

where $\alpha \geq 0$ balances the learnt label correlations and its decomposition to model coefficients. By default, we set $\alpha = 1$. Note that R represents the auxiliary correlations that will be learnt as part of the optimization problem.

For regularization, we use Frobenius norm. In particular, $\|W\|_F^2$ is a good choice for selecting the quality features specific to the labels. Replacing the regularizer $\mathcal{R}(W, R)$ with Frobenius norm for W and R, we rewrite the minimization problem as follows:

$$\begin{aligned}\min_{W,R} \frac{1}{2}\|XW - YR\|_F^2 &+ \frac{\lambda_1}{2}\|YR - Y\|_F^2 \\ &+ \frac{\lambda_2}{2}\|R - \alpha W^T W\|_F^2 + \frac{\lambda_3}{2}\|W\|_F^2 + \frac{\lambda_4}{2}\|R\|_F^2\end{aligned} \tag{4}$$

where λs denotes the tradeoff parameters.

3.2 Instance Similarity

Similar instances as per the smoothness assumption have higher odds of similar looking label vector. To let the instance's similarity steer label predictions, we can include the assumption in the optimization formulation by defining similarity $p_{uv} = exp^{-\|x_u - x_v\|_2^2}$ where x_u and x_v represent the uth and vth instances. For similar instances x_u and x_v, the predicted label vectors, say, f_u and f_v, have good odds to be alike. The regularizer for the smoothness can be prepared as follows:

$$\begin{aligned} R_{\text{inst}} &= \frac{1}{2} \sum_{u,v=1}^{n} p_{uv} \| f_u - f_v \|_2^2 \\ &= Tr((XW)^T L_I XW) \end{aligned}$$

where XW denotes the prediction matrix for training instances in X, and $L_I \in \mathbb{R}^{n \times n}$ denotes the graph Laplacian for instance similarity.

Incorporating the regularizer R_{inst} for instance similarity with tradeoff parameter λ_5, the final minimization problem becomes as follows:

$$\begin{aligned} \min_{W,R} \frac{1}{2} \|XW - YR\|_F^2 + \frac{\lambda_1}{2} \|YR - Y\|_F^2 + \frac{\lambda_2}{2} \|R - \alpha W^T W\|_F^2 \\ + \frac{\lambda_3}{2} \|W\|_F^2 + \frac{\lambda_4}{2} \|R\|_F^2 + \frac{\lambda_5}{2} tr((XW)^T L_I XW) \end{aligned} \quad (5)$$

3.3 Optimization

The minimization problem in Eq. 5 consists of all quadratic terms and hence convex and smooth. The problem is composed of two unknowns: the model coefficient matrix W and the learnt label correlation matrix R. We represent the final objective function using $Q(\Theta)$ where $\Theta = \{W, R\}$ is the set of function parameters. The objective function can be optimized by alternative minimization strategy by iteratively minimizing over W and R using gradient descent.

Updating W : Fixing R, gradient of $Q(\Theta)$ w.r.t W is given by:

$$\begin{aligned} \nabla_W Q(\Theta) =& X^T (XW - YR) + 2\lambda_2(-\alpha W)(R - \alpha W^T W) \\ &+ \lambda_3 W + \lambda_5 X^T L_I XW \end{aligned} \quad (6)$$

Updating R: Fixing W, gradient of $Q(\Theta)$ w.r.t R is given by:

$$\begin{aligned} \nabla_R Q(\Theta) =& - Y^T (XW - YR) + \lambda_1 Y^T (YR - Y) \\ &+ \lambda_2 (R - \alpha W^T W) + \lambda_4 R \end{aligned} \quad (7)$$

Using the aforementioned mathematical derivations, we formalize the optimization steps based on gradient descent in Algorithm 1.

Algorithm 1 ALEML Optimization Algorithm

Input: Training instance matrix $X \in \mathbb{R}^{n \times d}$, class labels $Y \in \{0, 1\}^{n \times l}$, α, regularization parameters $\lambda_1, \lambda_2, \lambda_3, \lambda_4$, and λ_5, learning rate η_1 and η_2

Output: W, R

Initialization: $W_0 \in \mathbb{R}^{d \times l}$, *and* $R_0 \in \mathbb{R}^{l \times l}$ *randomly, t=0.*

REPEAT until convergence

1: Compute $\nabla_W Q(W_t, R_t)$ according to equation (6).

2: Update W as

$$W_{t+1} = W_t - \eta_1 \nabla_W Q(W_t, R_t)$$

3: Compute $\nabla_R Q(W_{t+1}, R_t)$ according to equation (7).

4: Update R as

$$R_{t+1} = R_t - \eta_2 \nabla_R Q(W_{t+1}, R_t)$$

5: $W_t = W_{t+1}$; $R_t = R_{t+1}$; $t = t + 1$

RETURN W, R

---

4 Experiments

4.1 Datasets

Four multi-label datasets[1] are used for comparison and performance measurement with other multi-label algorithms. The datasets represent different domains, label counts, and label cardinality. Dataset specific attributes like label and instance count and label cardinality are illustrated in Table 1.

4.2 Multi-label Metrics

Proposed model's performance is assessed using five multi-label metrics [24]. Let $M = (x_k, Y_k)_{k=1}^{p}$ be the test set where x_k and Y_k represent the kth instance and label vector with q classes, respectively. $h(x_k)$ represents predicted labels and $f(x_k, y)$ gives the model's confidence score in y.

---

[1] http://www.uco.es/kdis/mllresources/.

Table 1 Multi-label datasets' attributes

Dataset	Features	Instances	Labels	LCard	Domain
Medical	1449	978	45	1.25	Text
cal500	68	502	174	26.04	Music
Image	294	2000	5	1.24	Image
Genbase	1185	662	27	1.25	Biology

Hamming loss evaluates the misclassification rate for an instance-label pair.

$$HL = \frac{1}{p}\sum_{k=1}^{p}\frac{1}{q}|h(x_k)\Delta Y_k| \tag{8}$$

Average precision calculates the average portion of present labels ranked more than a specific label.

$$AP = \frac{1}{p}\sum_{k=1}^{p}\frac{1}{|Y_k|}\sum_{l\in Y_k}\frac{|l_k|}{rank_f(x_k, l)} \tag{9}$$

where $l_k = \{l'|rank_{f(x_k,l')} \leq rank_{f(x_k,l)}, l' \in Y_k\}$.

Ranking loss is the portion of negative labels ranked more than positive labels.

$$RL = \frac{1}{p}\sum_{k=1}^{p}\frac{|\{(y', y'')|f(x_k, y') \leq l_k\}|}{|Y_k||\bar{Y}_k|} \tag{10}$$

where $l_k = f(x_k, y'')|(y', y'') \in Y_k \times \bar{Y}_k)$.

Coverage is the average number of additional labels included to cover all relevant labels.

$$\text{Cov} = \frac{1}{n}\sum_{i=1}^{n}[[\max rank(x_i, j) - 1|j \in Y_i^{+}]]$$

AUC is mean positive instances portion of labels ranked more than negative ones.

$$\text{AUC} = \frac{1}{q}\sum_{k=1}^{q}\frac{|\{(x, x')|f(x, y_k) \geq f(x', y_k), (x, x') \in I_k \times \bar{I}_k\}|}{|I_k||\bar{I}_k|} \tag{11}$$

where I_k and $\bar{I}_k$ stand for relevant and irrelevant data instances of the kth class label.

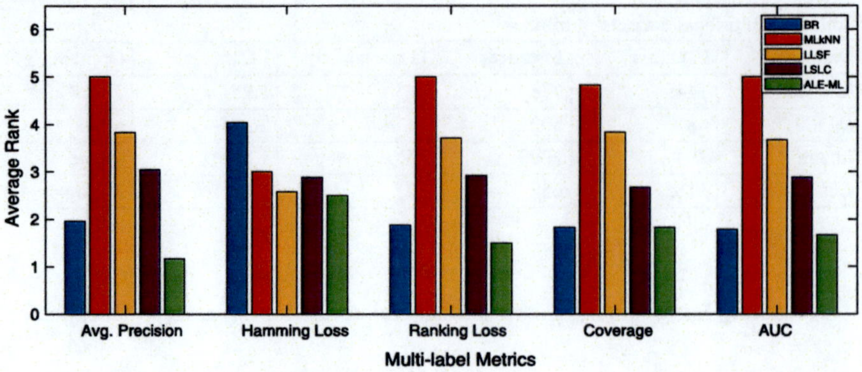

Fig. 1 Consolidated rank of compared algorithms for each evaluation metrics. A lower value denotes a better rank

4.3 Baselines

For baselining model performance, we match it up with four well-known multi-label classification algorithms. Algorithms' details are as follows:

- Binary relevance (BR) views a multi-label problem as a composition of several binary classification problems, one for each label. We use least squared linear regression to build a binary classifier per label for this comparison.
- ML-kNN [26] is a multi-label approach based on k nearest neighbors. k values are explored in $\{5, 7, \ldots, 11\}$.
- LLSF [7] learns label-specific features for multi-label learning. The hyperparameters are searched in the range $\{2^{-10}, 2^{-9}, \cdots, 2^{10}\}$.
- LSLC [4] learns label-specific features and incorporates positive and negative correlations among labels for multi-label learning. Hyperparameter values are explored in $\{2^{-10}, 2^{-9}, \cdots, 2^{1}\}$.
- Auxiliary label embedding for multi-label learning with missing labels (ALEML) is the algorithm proposed in this paper. It learns label correlations and embeds it as model coefficients. α is set as 1, and other hyperparameters' search range is $\{10^{-7}, 10^{-6}, \cdots, 10^{2}\}$.

4.4 Empirical Results and Discussion

For performance comparison, datasets are partitioned using 5-fold cross validation. For creating training data with missing labels, labels are randomly dropped to achieve missing label rate of 0.1, 0.3 and 0.5. Tables 2, 3, and 4 show the results achieved at each missing label rate on each of the four datasets for five multi-label metrics. $\uparrow(\downarrow)$ indicates a higher (lower) metric value is better. Best performance in each row is

Table 2 Missing rate 0.1

Datasets	Metrics	BR	MLKNN	LLSF	LSLC	ALE-ML
cal500	AP↑	0.5004	0.2595	0.4550	0.5003	**0.5019**
	HL↓	0.1400	0.1384	0.1443	0.1388	**0.1381**
	RL↓	**0.1797**	0.8285	0.2332	**0.1797**	**0.1797**
	Cov↓	0.7475	0.8716	0.8696	**0.7467**	0.7494
	AUC↑	0.8167	0.6883	0.7648	0.8167	**0.8168**
Genbase	AP↑	**0.9933**	0.9475	0.9909	**0.9933**	0.9927
	HL↓	0.0052	0.0060	**0.0042**	0.0050	0.0051
	RL↓	0.0053	0.0642	0.0053	0.0054	**0.0034**
	Cov↓	0.0183	0.0613	0.0181	0.0185	0.0149
	AUC↑	0.9840	0.9372	0.9848	0.9838	**0.9873**
Medical	AP↑	0.9009	0.6146	0.6386	0.8996	**0.9070**
	HL↓	0.0126	0.0169	0.0446	**0.0124**	0.0125
	RL↓	0.0163	0.4359	0.1424	0.0172	**0.0144**
	Cov↓	0.0276	0.3063	0.1699	0.0290	**0.0247**
	AUC↑	0.9779	0.7274	0.8483	0.9767	**0.9804**
Image	AP↑	0.7771	0.6768	0.7558	0.7747	**0.7777**
	HL↓	0.2189	**0.1819**	0.1981	0.2076	0.2074
	RL↓	**0.1823**	0.5947	0.2091	0.1835	0.1834
	Cov↓	0.2019	0.2989	0.2212	**0.2018**	0.2040
	AUC↑	**0.7750**	0.6468	0.7496	0.7742	0.7731

highlighted. The results show that the proposed method performs favorably overall in comparison with other evaluated methods for each missing label rate.

Figure 1 shows metric wise average rank for participating algorithms. ALEML is the best performing algorithm over the datasets for all five metrics considered for this evaluation. The performance of proposed method is better for average precision when compared with other multi-label metrics. ML-*k*NN falls behind other label correlations-based algorithms on each evaluation metric on average in this comparison .

Similar to LSLC and LLSF, ALEML also utilizes label correlations in the model building, especially for supplementary label correlations useful for missing label recovery. The embedding approach preserves the label correlation information in the model coefficients and improves the model prediction capability. Results highlight that utilizing the learnt label correlations for constructing model coefficients embedding improves the multi-label learning capability of the model when labels are missing.

Table 3 Missing rate 0.3

Datasets	Metrics	BR	MLKNN	LLSF	LSLC	ALEML
cal500	AP↑	0.4989	0.1785	0.4370	0.4987	**0.4996**
	HL↓	0.1453	0.144	**0.1438**	0.1448	0.145
	RL↓	**0.1809**	0.9397	0.2469	0.1810	0.1815
	Cov↓	0.7511	0.8720	0.8884	**0.7510**	0.7530
	AUC↑	**0.8155**	0.6546	0.7510	**0.8155**	0.8151
Genbase	AP↑	0.9815	0.6501	0.9635	0.9807	**0.9827**
	HL↓	0.0145	0.0229	**0.0135**	0.0145	0.0145
	RL↓	0.0128	0.3786	0.0240	0.0137	**0.0100**
	Cov↓	0.0281	0.1197	0.0485	0.0302	**0.0232**
	AUC↑	0.9745	0.8437	0.9443	0.9723	**0.9789**
Medical	AP↑	0.8926	0.4966	0.5806	0.8912	**0.8958**
	HL↓	0.0178	0.0197	0.0483	0.0177	**0.0176**
	RL↓	0.0238	0.5858	0.1804	0.0244	**0.0218**
	Cov↓	0.0372	0.3784	0.2221	0.0381	**0.0343**
	AUC↑	0.9694	0.6545	0.8016	0.9687	**0.9718**
Image	AP↑	0.7677	0.6089	0.7404	0.7628	**0.7724**
	HL↓	0.2305	**0.2008**	0.211	0.2235	0.2234
	RL↓	**0.1907**	0.7228	0.2245	0.1932	0.1926
	Cov↓	**0.2105**	0.3346	0.2349	0.2113	0.2134
	AUC↑	**0.7639**	0.5898	0.7310	0.7608	0.7620

We employ Nemenyi test to analyze if performance difference is significant among the matched up algorithms [5]. Nemenyi critical difference formula is $q_\beta\sqrt{\frac{t(t+1)}{6M}}$. t, the algorithm count is set to 5 and M, and the total data points are set to 12 (4×3). For significance level $\beta = 0.5$, q_β equals 2.728, and critical difference is 1.7609. Figure 2 shows the algorithms' performance on five multi-label metrics based on Nemenyi test. Algorithms which are not connected by horizontal line differ by more than one critical difference unit and differ significantly. As shown in the figure, ALEML performs markedly well than algorithms on the extreme left and competitively with rest of the others.

Table 4 Missing rate 0.5

Datasets	Metrics	BR	MLKNN	LLSF	LSLC	ALEML
cal500	AP↑	0.4971	0.1266	0.4113	0.4970	**0.4978**
	HL↓	0.1498	0.149	**0.1474**	0.1498	0.1497
	RL↓	**0.1820**	0.9942	0.2695	**0.1820**	0.1827
	Cov↓	0.7509	0.8721	0.9093	**0.7493**	0.7537
	AUC↑	**0.8143**	0.6304	0.7284	**0.8143**	0.8137
Genbase	AP↑	0.9748	0.2478	0.9368	0.9744	**0.9765**
	HL↓	0.0195	0.0389	**0.0178**	0.0189	0.0192
	RL↓	0.0179	0.8025	0.0446	0.0189	**0.0146**
	Cov↓	0.0340	0.1919	0.0767	0.0360	**0.0282**
	AUC↑	0.9683	0.7821	0.9162	0.9662	**0.9742**
Medical	AP↑	0.8756	0.2501	0.4946	0.8750	**0.8784**
	HL↓	0.0234	0.0248	0.0523	0.0233	**0.0232**
	RL↓	0.0279	0.8533	0.2438	0.0285	**0.0277**
	Cov↓	**0.0407**	0.4602	0.2840	0.0417	0.0409
	AUC↑	**0.9667**	0.5712	0.7414	0.9659	0.9660
Image	AP↑	0.6935	0.5547	0.6806	0.6772	**0.7361**
	HL↓	0.2315	0.213	**0.2191**	0.2239	0.2238
	RL↓	0.2626	0.7950	0.2858	0.2943	**0.2216**
	Cov↓	0.2616	0.3358	0.2815	0.2852	**0.2369**
	AUC↑	0.6982	0.5877	0.6776	0.6710	**0.7309**

5 Conclusion

In this paper, we present a learnt label correlation embedding-based approach when part of label information is unobserved in multi-label datasets. Auxiliary label matrix is learnt from observed labels for missing label reconstruction, and the structure of the learnt label correlations is preserved in the model coefficients through generated embedding. The proposed approach provides a unified framework for completing missing labels and predicting labels for new data instances. Experiments and comparison with four multi-label learning methods using five multi-label metrics validate the proposed learnt correlation embedding approach. Future work involves exploring the model coefficient construction by combining learnt label correlations embedding and features similarity embedding in a unified framework.

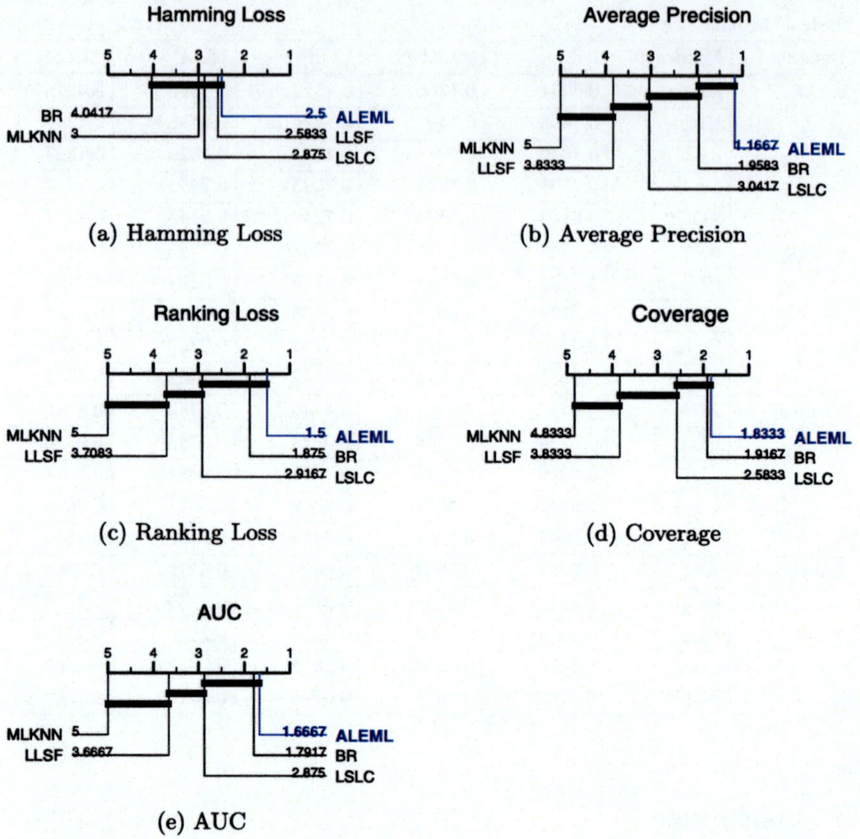

Fig. 2 Statistical comparison of competing algorithms with critical difference of 1.76 at significance level 0.05 using Nemenyi test. Significantly different algorithms are not connected through same horizontal lines

References

1. Boutell, M.R., Luo, J., Shen, X., Brown, C.M.: Learning multi-label scene classification. Pattern Recogn. **37**(9), 1757–1771 (2004)
2. Cai, Z., Zhu, W.: Multi-label feature selection via feature manifold learning and sparsity regularization. Int. J. Mach. Learn. Cybern. **9**(8), 1321–1334 (2018)
3. Chen, Y.N., Lin, H.T.: Feature-aware label space dimension reduction for multi-label classification. Adv. Neural Inf. Process. Syst. **25** (2012)
4. Cheng, Z., Zeng, Z.: Joint label-specific features and label correlation for multi-label learning with missing label. Appl. Intell. **50**(11), 4029–4049 (2020)
5. Demšar, J.: Statistical comparisons of classifiers over multiple data sets. J. Mach. Learn. Res. **7**, 1–30 (2006)
6. Goldberg, A., Recht, B., Xu, J., Nowak, R., Zhu, J.: Transduction with matrix completion: three birds with one stone. Adv. Neural Inf. Process. Syst. **23** (2010)

7. Huang, J., Li, G., Huang, Q., Wu, X.: Learning label specific features for multi-label classification. In: 2015 IEEE International Conference on Data Mining, pp. 181–190. IEEE (2015)
8. Huang, J., Qin, F., Zheng, X., Cheng, Z., Yuan, Z., Zhang, W., Huang, Q.: Improving multi-label classification with missing labels by learning label-specific features. Inf. Sci. **492**, 124–146 (2019)
9. Huang, J., Xu, Q., Qu, X., Lin, Y., Zheng, X.: Improving multi-label learning by correlation embedding. Appl. Sci. **11**(24), 12145 (2021)
10. Huang, R., Jiang, W., Sun, G.: Manifold-based constraint Laplacian score for multi-label feature selection. Pattern Recogn. Lett. **112**, 346–352 (2018)
11. Kumar, S., Rastogi, R.: Low rank label subspace transformation for multi-label learning with missing labels. Inf. Sci. **596**, 53–72 (2022)
12. Lin, Z., Ding, G., Hu, M., Wang, J.: Multi-label classification via feature-aware implicit label space encoding. In: International Conference on Machine Learning, pp. 325–333. PMLR (2014)
13. Liu, H., Chen, G., Li, P., Zhao, P., Wu, X.: Multi-label text classification via joint learning from label embedding and label correlation. Neurocomputing **460**, 385–398 (2021)
14. Liu, W., Wang, H., Shen, X., Tsang, I.: The emerging trends of multi-label learning. IEEE Trans. Pattern Anal. Mach. Intell. (2021)
15. Ma, Z., Chen, S.: Expand globally, shrink locally: discriminant multi-label learning with missing labels. Pattern Recogn. **111**, 107675 (2021)
16. Rastogi, R., Kumar, S.: Discriminatory label-specific weights for multi-label learning with missing labels. Neural Process. Lett. (2022)
17. Read, J., Pfahringer, B., Holmes, G., Frank, E.: Classifier chains for multi-label classification. Mach. Learn. **85**(3), 333 (2011)
18. Tan, A., Ji, X., Liang, J., Tao, Y., Wu, W.Z., Pedrycz, W.: Weak multi-label learning with missing labels via instance granular discrimination. Inf. Sci. **594**, 200–216 (2022)
19. Tsoumakas, G., Katakis, I.: Multi-label classification: an overview. Int. J. Data Warehousing Min. **3**(3) (2006)
20. Tsoumakas, G., Katakis, I., Vlahavas, I.: Mining multi-label data. In: Data Mining and Knowledge Discovery Handbook, pp. 667–685. Springer,Berlin (2009)
21. Wu, B., Liu, Z., Wang, S., Hu, B.G., Ji, Q.: Multi-label learning with missing labels. In: 2014 22nd International Conference on Pattern Recognition, pp. 1964–1968. IEEE (2014)
22. Wu, B., Lyu, S., Ghanem, B.: Ml-mg: Multi-label learning with missing labels using a mixed graph. In: Proceedings of the IEEE International Conference on Computer Vision, pp. 4157–4165 (2015)
23. Wu, J.S., Huang, S.J., Zhou, Z.H.: Genome-wide protein function prediction through multi-instance multi-label learning. IEEE/ACM Trans. Comput. Biol. Bioinf. **11**(5), 891–902 (2014)
24. Wu, X.Z., Zhou, Z.H.: A unified view of multi-label performance measures. In: International Conference on Machine Learning, pp. 3780–3788. PMLR (2017)
25. Zhang, J., Luo, Z., Li, C., Zhou, C., Li, S.: Manifold regularized discriminative feature selection for multi-label learning. Pattern Recogn. **95**, 136–150 (2019)
26. Zhang, M.L., Zhou, Z.H.: Ml-knn: a lazy learning approach to multi-label learning. Pattern Recogn. **40**(7), 2038–2048 (2007)
27. Zhang, Y., Yeung, D.Y.: A regularization approach to learning task relationships in multitask learning. ACM Trans. Knowl. Disc. Data (TKDD) **8**(3), 1–31 (2014)
28. Zhu, Y., Kwok, J.T., Zhou, Z.H.: Multi-label learning with global and local label correlation. IEEE Trans. Knowl. Data Eng. **30**(6), 1081–1094 (2017)

Semi-supervised Semantic Segmentation of Effusion Cytology Images Using Adversarial Training

Mayank Rajpurohit, Shajahan Aboobacker, Deepu Vijayasenan, S. Sumam David, Pooja K. Suresh, and Saraswathy Sreeram

Abstract In pleural effusion, an excessive amount of fluid gets accumulated inside the pleural cavity along with signs of inflammation, infections, malignancies, etc. Usually, a manual cytological test is performed to detect and diagnose pleural effusion. The deep learning solutions for effusion cytology include a fully supervised model trained on effusion cytology images with the help of output maps. The low-resolution cytology images are harder to label and require the supervision of an expert, the labeling process time-consuming and expensive. Therefore, we have tried to use some portion of data without any labels for training our models using the proposed semi-supervised training methodology. In this paper, we proposed an adversarial network-based semi-supervised image segmentation approach to automate effusion cytology. The semi-supervised methodology with U-Net as the generator shows nearly 12% of absolute improvement in the f-score of benign class, 8% improvement in the f-score of malignant class, and 5% improvement in mIoU score as compared to a fully supervised U-Net model. With ResUNet++ as a generator, a similar improvement in the f-score of 1% for benign class, 8% for the malignant class, and 1% in the mIoU score is observed as compared to a fully supervised ResUNet++ model.

Keywords Image segmentation · Effusion cytology · Semi-supervised learning

1 Introduction

Pleural effusion is a medical condition in which an excessive amount of fluid gets accumulated inside the pleural cavity. This excessive accumulation of fluid is secondary to signs of inflammations, infections, malignancies, etc. Usually, a manual cytological test is performed to detect and diagnose pleural effusion. A sufficient

M. Rajpurohit · S. Aboobacker · D. Vijayasenan (✉) · S. Sumam David
National Institute of Technology Karnataka, Surathkal, Karnataka 575025, India
e-mail: deepu.senan@gmail.com

P. K. Suresh · S. Sreeram
Kasturba Medical College Mangalore, Manipal Academy of Higher Education, Manipal, Karnataka 575001, India

M. Tistarelli et al. (eds.), *Computer Vision and Machine Intelligence*,
Lecture Notes in Networks and Systems 586,
https://doi.org/10.1007/978-981-19-7867-8_43

amount of stained and centrifuged fluid sample is taken on a glass slide for examination. The cytopathologists examine the slides visually using a microscope to detect malignant cells. This complete process is called effusion cytology. The examination involves images of different resolutions ranging from 4 to 40×. The cytologist first examines low-resolution images (usually 4 and 10×) to detect regions of interest (cluster level malignancies). Those regions which are found malignant at the cluster level are then examined at a higher resolution (40×) to detect malignant cells (cell level malignancies) [1, 2]. This process is time-consuming as scanning of complete images is done at different resolutions and then analyzed. There are a few deep learning approaches that have been recently used to automate this process of effusion cytology [3–5]. But nearly all of them are fully supervised, which means for every image segmentation, label map is required. The segmentation maps are prepared manually either by the cytopathologists or under their supervision. Hence, the preparation of labels for all the images is both an expensive and time-consuming process. The work attempts to segment the areas of the image available in the lowest resolution, i.e., 4×. This could reduce the time and effort required for the effusion cytology process. Since the 4× images are hard to label, we have used this data without any labels along with 10× data with labels in a semi-supervised manner. Semi-supervised learning techniques involve training a model with a small amount of labeled data along with unlabeled data. Deep learning models trained on these techniques perform very similarly to the models trained with fully supervised techniques. The use of a semi-supervised training methodology could reduce the time, cost, and effort required to prepare the labels for image data. Therefore, an adversarial network-based semi-supervised image segmentation approach for effusion cytology is proposed.

Semi-supervised learning generally utilizes entropy minimization, consistency regularization, or a hybrid of both techniques to train a model. Pseudo label [6] and noisy student [7] are examples of entropy minimization techniques. Temporal ensembling [8], mean teacher [9], and virtual adversarial training [10] are few examples of consistency regularization techniques. There are a few hybrid semi-supervised techniques available that utilize both entropy minimization and consistency regularization such as FixMatch [11] and MixMatch [12]. Cui et al. [13] proposed a robust loss-based semi-supervised training to enhance the model robustness toward noise in pseudo-labels for brain tumor segmentation task. Hung et al. [14] proposed an adversarial network for semantic segmentation. This network consists of a DeepLabV3 [15] network as the generator to generate the semantic maps. A fully convolutional network (FCN) serves as a discriminator which differentiates between the generated probability map distribution and the ground truth distribution. The generator is trained to generate semantic map till the discriminator finds it hard to classify whether a pixel is from the ground truth distribution or it is from the semantically generated label distribution. The proposed discriminator improves the semantic segmentation performance of the generator when the adversarial loss is coupled with categorical cross-entropy loss. This method uses unlabeled images to enhance the performance [14].

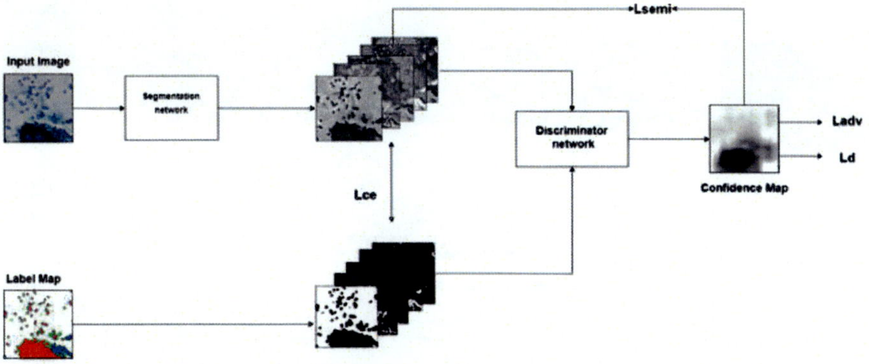

Fig. 1 Overview of proposed methodology

2 Adversarial Network-Based Semi-supervised Image Segmentation Methodology

An overview of the proposed training methodology is given in Fig. 1. It consists of two networks, a generator network that carries out the semantic segmentation task and generates semantic maps. Given an input image of shape H × W × 3 the generator gives an output of shape H × W × C which is the class probability map where H and W are the height and width of the image, and C is the number of classes. The discriminator network is an FCN that is fed by class probability maps of shape H × W × C either from ground truth or generated distribution. The output of the discriminator is a confidence map of the shape H × W × 1. Every pixel of this confidence map gives the probability of the sample pixel whether it belongs to the ground truth distribution or from the generated distribution [16].

The generator model is trained on cross-entropy loss L_{ce} between the original mask and predicted mask, adversarial loss L_{adv} the adversarial loss, and L_{semi} semi-supervised loss calculated from confidence maps. The discriminator is trained on L_D to optimize the generator predictions. This methodology can support training with both labeled and unlabeled images. When a labeled image is fed to the generator, it generates a corresponding semantic map, and L_{ce} categorical cross-entropy loss is calculated. The discriminator is fed by both generated and original maps, and binary cross-entropy loss L_D is calculated. The discriminator is trained by backpropagating on loss L_D. An adversarial loss term is calculated by combining L_{ce} and L_D as per $L_{adv} = L_D$ for both predicted and original masks as input to the discriminator. When this loss is used for training the discriminator, it is termed as L_D, and when this loss is to be combined with L_{ce}, this is termed as L_{adv}, and generator is trained by backpropagating on this combined loss of

$$L_{seg} = L_{ce} + k_{adv} \times L_{adv} \tag{1}$$

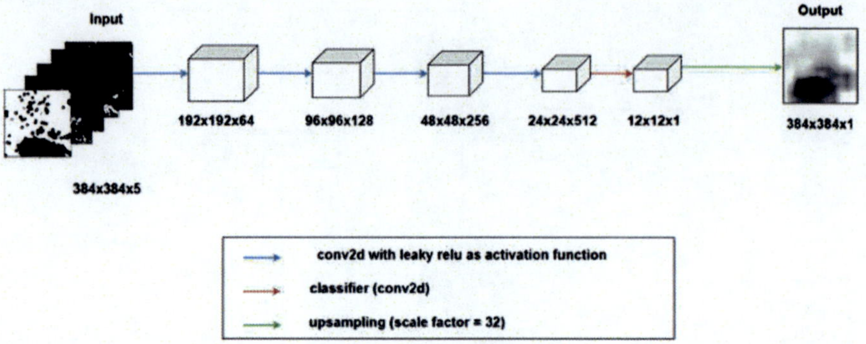

Fig. 2 Fully convolutional network (FCN)-based discriminator

For unlabeled images, categorical cross-entropy loss cannot be calculated. Hence, the generated map is fed to the discriminator directly, and only high confidence output labels from the confidence map are taken to calculate a binary cross-entropy loss term L_{semi} which is combined with L_{adv} and is then used for training of the generator.

$$L_{\text{seg}} = L_{\text{ce}} + k_{\text{adv}} \times L_{\text{adv}} + k_{\text{semi}} \times L_{\text{semi}} \tag{2}$$

where L_{ce}, L_{adv}, and L_{semi} are multi-categorical cross-entropy loss, adversarial loss term and semi-supervised loss term respectively and k_{adv} and k_{semi} are hyperparameters to weight the loss terms.

2.1 Network Architecture

Adversarial networks generally consist of two models: generator and discriminator. The work of a generator network is to predict the output. The discriminator tries to discriminate whether the input fed to it belongs to the ground truth distribution or from the generated label distribution [16]. Two different types of generators are used for the experiments, viz., standard U-Net [17] and ResUNet++ [18]. It is proven that U-Net works with very few training images and yields more precise segmentation for medical images. On the other hand, ResUNet++ usually trains faster and yields good performance on medical image data with fewer parameters. Hence, these two models are chosen as the generators. The last layer of the models gives the output class probability maps as H × W × C.

The discriminator network is a fully convolutional network similar to FCN used in Hung et al. [14]. It consists of 5 convolutional layers with a kernel shape of 4×4. The number of channels of the output layers is 64, 128, 256, 512, and 1, respectively, as shown in Fig. 2. Every convolutional layer is followed by a ReLU layer. The one channel output confidence map generated is then upsampled to the input image shape.

2.2 Loss Function

The input image is denoted by X_m of shape H × W × 3, and the generator network is denoted by $G(.)$ which takes an image as input and gives output $G(X_m)$ of the shape H × W × C. The discriminator network is abbreviated as $D(.)$, and it takes either one-hot encoded ground truth or the output semantic probability map of shape H × W × C as input and gives the confidence map of shape H × W × 1 as output. The one-hot encoded ground truth is denoted by Y_m. The generator is trained with a combined multi-task loss term as given in Eq. (2). The categorical cross-entropy loss between the ground truth Y_m and the output probability map $G(X_m)$ is termed as L_{ce} given in Eq. (3).

Consider training with labeled data, an image X_m is input to generator $G(.)$, giving $G(X_m)$ as the output semantic map. One-hot encoded ground truth is Y_m, and categorical cross-entropy loss between them is given by:

$$L_{ce} = -\sum_{h,w}\sum_{c \in C} Y_m^{(h,w,c)} \log G(X_m)^{(h,w,c)} \tag{3}$$

The adversarial loss for fully convolutional network $D(.)$ is given in Eqs. (4) and (5). With this loss, the segmentation network is trained to mislead the discriminator. Equation (4) is the adversarial loss for the predicted mask fed to the discriminator, whereas Eq. (5) gives adversarial loss for the original mask fed to the discriminator.

$$L_{adv} = -\sum_{h,w} \log\left(D(G(X_m))^{(h,w)}\right) \tag{4}$$

$$L_{adv} = -\sum_{h,w} \log\left(D(Y_m)^{(h,w)}\right) \tag{5}$$

In the case of unlabeled data, labels are not available, and the training is done in a semi-supervised manner. The loss term L_{ce} cannot be calculated as ground truth is not available. The L_{adv} can be determined because it only requires the output of the discriminator. In addition to these loss terms, the discriminator is trained with a self-supervised framework. The discriminator-generated confidence map $D(G(X_m))$ can be taken to choose the high confidence intervals, whose distribution is close to the distribution of ground truth. A high confidence map can be generated by applying a threshold to the generated confidence map. The self-supervised and one-hot encoded ground truth $\hat{Y}_m$ are element-wise set with $\hat{Y}_m^{(H,W,C^*)} = 1$ if $c^* = \text{argmax}_c G(X_m)^{(H,W,C)}$. The corresponding semi-supervised loss term is then defined as given in Eq. (6).

$$L_{semi} = -\sum_{h,w}\sum_{c \in C} I(D(G(X_m))^{(h,w)} > T_{semi}).\hat{Y}_m^{(h,w,c)} \log(G(X_m)^{(h,w,c)}), \tag{6}$$

where T_{semi} is the threshold and $I(.)$ is the indicator function. Training is done by backpropagating on L_{seg} which is given in Eq. (2).

3 Experiments and Result

We conducted two sets of experiments in this study. In the first set, models were trained using different amounts of 10× training data. These experiments are conducted to show that the proposed semi-supervised adversarial network-based methodology works. In the second experiment, we used complete 10× labeled data and 4× unlabeled data for training models.

3.1 Dataset

This research work is carried out jointly with the Department of Pathology, KMC Mangalore. The usage of data for research purposes is permitted and approved by the KMC Mangalore Institutional Ethics Committee (IEC KMC MLR 01-19/34; 16 Jan 2019). There is a total of 345 pleural effusion cytology images of size 1920 × 1440 distributed among 30 patients, detailed in Table 1. The masks for images are prepared with the help of an experienced cytologist. Masks consist of 5 different colors map to a particular class, viz., red for malignant, blue for benign, magenta for inflammatory cells, green for cytoplasm, and white serves as background. This image size is very large; hence, patches of size 384 × 384 × 3 are created from the images. The dataset is divided into training, validation, and test sets. The split is independent of the patient. The training, validation, and test split details of the dataset before and after creating patches are shown in Table 2. An example sub-image and mask are shown in Fig. 3. The reason for creating patches is that it helps us with memory management while training the models, and it helps in introducing more effective data augmentation.

Table 1 Image distribution

Magnification	Benign	Malignant	Total
40×	127	85	212
10×	38	45	83
4×	17	33	50

Table 2 Training, validation, and test split of the original images and patches

Magnification	Train		Validation		Test	
	Images	Patches	Images	Patches	Images	Patches
10×	49	4955	17	131	17	192
4×	30	7319	10	677	10	519

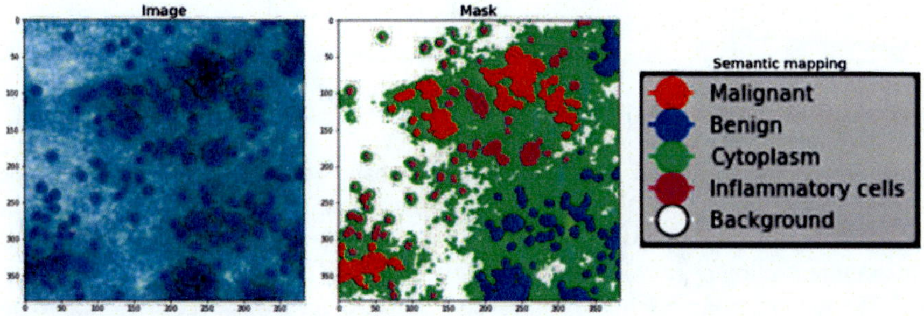

Fig. 3 Effusion cytology image patch of the size 384 × 384 × 3 and corresponding semantic mapping

3.2 *Experiment 1*

In the first experiment, we use three different sets of a different number of 10× image data to train three models:

1. A fully supervised semantic segmentation model which is U-Net with categorical cross-entropy loss L_{ce} as the loss term.
2. A fully supervised adversarial network with U-Net as the generator and FCN as the discriminator and the loss term used for generator is Eq. (1).
3. A semi-supervised adversarial network with some amount of labeled data and the rest images was taken unlabeled. The generator and discriminator are U-Net and FCN, respectively. The combined loss term used for training the generator is Eq. (2).

These three models were trained in three different sets of different amounts of image data used for training, i.e., with 25% training images, 50% training images, and 100% training images. For example, if an experiment is conducted with 25% training data, the U-Net and fully supervised adversarial network are trained with 25% of the training images in a fully supervised manner. While the semi-supervised adversarial network is trained with 25% labeled images and the rest 75% unlabeled images. Hence, the first training scheme in experiment 1 is that we trained three models using 25% of the training images, i.e., 1239 images out of the total 4955 training images of 10× magnification. Data augmentation as mentioned previously is used in every experiment. k_{semi} is taken as 0.3. All the models were trained for 150 epochs, Adam optimizer with a max learning rate of 10^{-3} and β_1, β_2 as 0.9,

Table 3 Standard evaluation on 10× test data, models trained with 25, 50, and 100% labeled data

Model	25% labeled data			50% labeled data			100% labeled data		
	F-Ben	F-Mal	mIoU	F-Ben	F-Mal	mIoU	F-Ben	F-Mal	mIoU
L_{ce}	0.57	0.83	0.68	0.66	0.85	0.73	0.67	0.84	0.73
$L_{ce} + L_{adv}$	0.59	0.83	0.69	0.67	0.86	0.73	0.66	0.84	0.73
$L_{ce} + L_{adv} + L_{semi}$	**0.62**	**0.87**	**0.73**	**0.72**	**0.89**	**0.75**	–	–	–

0.99, respectively. For all the experiments, we have used data augmentations such as random horizontal and vertical flips, crop and pad, shift, scale, and rotate. The standard evaluation involves f-score calculations for benign, malignant classes, and mean intersection over union score(mIoU score) refer the Table 3.

The second training scheme employed in experiment 1 was exactly similar to the first scheme in all aspects, except for the fact that this time 50% of training data is used for training the models. Training data of 2478 labeled images were taken to train the first two models. An additional 2477 unlabeled images were taken for training the semi-supervised model for the 50% data scheme. In the third set when 100% labeled data are taken, the adversarial network becomes equivalent to a semi-supervised adversarial network; hence, only two models could be trained under this scheme with all 4955 labeled training images. Table 3 shows the evaluations on test data of 10× magnification.

These results show that when the adversarial loss term is added, the performance improves. When a semi-supervised loss term is added, it better improves, and when more data are used to train the models, respective performances further improves. Also, the semi-supervised model trained with 50% labeled and rest 50% unlabeled data performs better than U-Net trained on 100% training data.

3.3 Experiment 2

The results of experiment 1 make it clear that the semi-supervised training strategy works well. Experiment 2 is conducted with complete 10× training data as labeled data and 4× training data as unlabeled data to train models in a semi-supervised manner such that performance of the model on 4× test data(lowest resolution effusion cytology data) improves. Also, the benign images can be screened out in the lowest resolution images as precisely as possible so that fewer images have to be further examined under higher resolution to save time. Four models were trained in experiment 2 which are as follows:

Table 4 Standard evaluation for U-Net on 10× and 4× test data

Training scheme	10× test			4× test		
	F-Ben	F-Mal	mIoU	F-Ben	F-Mal	mIoU
L_{ce}	0.67	0.84	0.73	0.52	0.83	0.57
$L_{ce} + L_{adv} + L_{semi}$	**0.79**	**0.95**	**0.80**	**0.64**	**0.91**	**0.62**

Table 5 Standard evaluation for ResUNet++ on 10× and 4× test data

Training scheme	10× test			4× test		
	F-Ben	F-Mal	mIoU	F-Ben	F-Mal	mIoU
L_{ce}	0.72	0.89	0.74	0.59	0.79	0.57
$L_{ce} + L_{adv} + L_{semi}$	**0.75**	**0.90**	**0.76**	**0.60**	**0.87**	**0.58**

1. A fully supervised U-Net trained on 4955 labeled images of 10× magnification with categorical cross-entropy loss L_{ce} as the loss term.
2. A fully supervised ResUNet++ trained on 4955 labeled images of 10× magnification with categorical cross-entropy loss L_{ce} as the loss term.
3. A semi-supervised adversarial network with 10× labeled data of 4955 images and 4× unlabeled data of 7319 images. The generator and discriminator are U-Net and FCN respectively. The combined loss term used for the training generator is given in Eq. (2).
4. A semi-supervised adversarial network with 10× labeled data of 4955 images and 4× unlabeled data of 7319 images. The generator and discriminator are ResUNet++ and FCN, respectively. The combined loss term used for training generator is given in Eq. (2).

The models having U-Net and ResUNet++ as generators are evaluated on 10× and 4× testing data, respectively, as shown in Tables 4 and 5, respectively. Figures 4 and 5 show predictions of models on test images of 10× magnification. Figures 6 and 7 show predictions of models on the test images of 4× magnification.

Results of experiment 2 also show that the models when trained under proposed semi-supervised methodology with added unlabeled data improve the performance. This strategy works well for two different resolutions of data also. Since the addition of 4× (different resolution than 10×), unlabeled data improved the performance of the model on not only the 4× test data but also on the 10× test data.

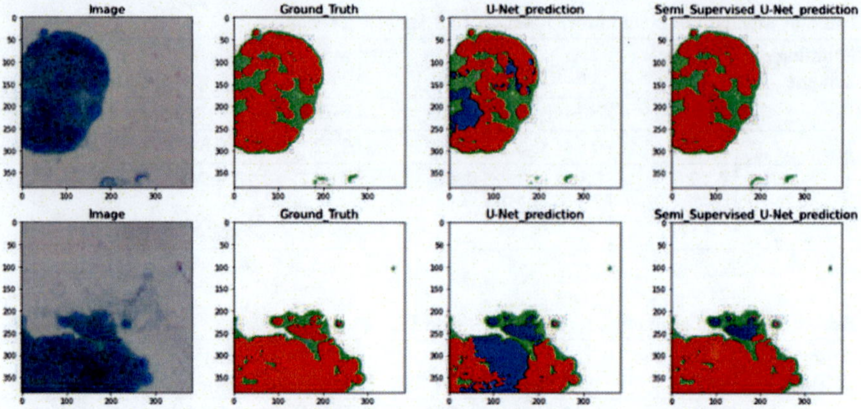

Fig. 4 Experiment 2: predictions on 10× magnification test images for U-Net and corresponding semi-supervised trained U-Net generator

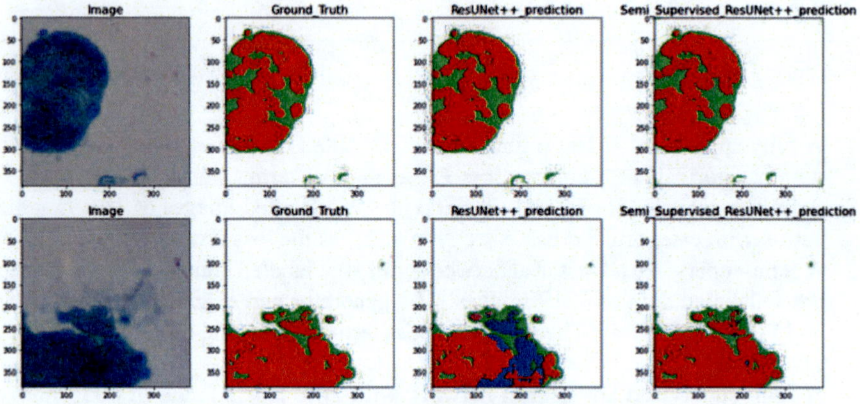

Fig. 5 Experiment 2: predictions on 10× magnification test images for ResUNet++ and corresponding semi-supervised trained ResUNet++ generator

4 Conclusion

The proposed adversarial network-based methodology for semi-supervised image segmentation has been evaluated by performing two experiments. In the first experiment, when models were trained with 25% labeled data, the f-score of benign class improved by nearly 5%, the f-score of malignant class improved by 4%, and the mIoU score improved by 5% as compared to U-Net model. Similarly for models trained with 50% labeled data, the f-score of benign class improved by 5%, for malignant class it improved by 4%, and the mIoU score improved by 2% as compared to U-Net model. This experiment concludes that the proposed semi-supervised methodology works well in the test environment.

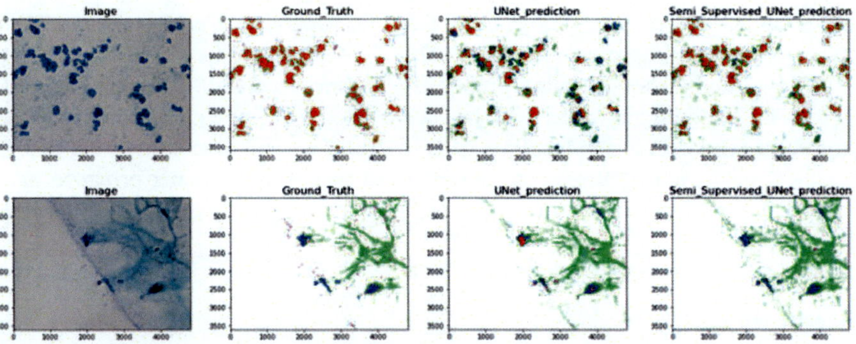

Fig. 6 Experiment 2: predictions on 4× magnification test images for U-Net and corresponding semi-supervised trained U-Net generator

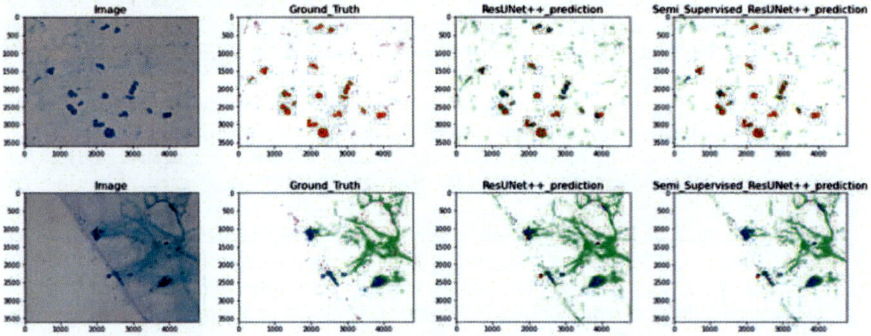

Fig. 7 Experiment 2: predictions on 4× magnification test images for ResUNet++ and corresponding semi-supervised trained ResUNet++ generator

For experiment 2, if we consider 10× test data, the U-Net generator trained in an adversarial semi-supervised manner shows improvements in the f-score of the benign class by 12%, the f-score of the malignant class by 11%, and the mIoU score by 7% as compared to the fully supervised U-Net model. The ResUNet++ generator trained with semi-supervised adversarial methodology shows the improvements in the f-score for the benign class by 3%, the f-score of the malignant class by 1%, and the mIoU score improved by 2% as compared to the fully supervised ResUNet++ model.

Considering 4× test data, when the U-Net generator is trained in the adversarial semi-supervised manner, the f-score of the benign class improved by 12%, the f-score of the malignant class improved by 8%, and the mIoU score improved by 5% as compared to the fully supervised U-Net model. Similarly, when the ResUNet++ generator is trained in the adversarial semi-supervised manner, the f-score of benign class improved by 1%, the f-score for malignant class improved by 8%, and the mIoU score improved by 1% as compared to the fully supervised ResUNet++ model. All these percentage improvements are absolute.

These results show that the proposed adversarial network-based methodology works well and improves the performance of models using unlabeled data. Hence, for the scenarios in which labeling data are expensive and time-consuming (usually medical data), this strategy can be employed to improve the performance of the model.

References

1. Ehya, H.: Effusion cytology. Clin. Lab. Med. **11**, 443–467 (1991)
2. Lepus, C., Vivero, M.: Updates in effusion cytology. Surg. Pathol. Clin. **11**, 523–544 (2018)
3. Aboobacker, S., Vijayasenan, D., David, S., Suresh, P., Sreeram, S.: A deep learning model for the automatic detection of malignancy in effusion cytology. In: 2020 IEEE International Conference On Signal Processing, Communications And Computing (ICSPCC), pp. 1–5 (2020)
4. Win, K., Choomchuay, S., Hamamoto, K., Raveesunthornkiat, M.: Artificial neural network based nuclei segmentation on cytology pleural effusion images. In: 2017 International Conference On Intelligent Informatics And Biomedical Sciences (ICIIBMS), pp. 245–249 (2017)
5. Win, K., Choomchuay, S., Hamamoto, K., Raveesunthornkiat, M.: Comparative study on automated cell nuclei segmentation methods for cytology pleural effusion images. J. Healthc. Eng. **2018** (2018)
6. Seibold, C., Reiß, S., Kleesiek, J., Stiefelhagen, R.: Reference-guided Pseudo-Label Generation for Medical Semantic Segmentation (2021). ArXiv Preprint ArXiv:2112.00735
7. Xie, Q., Luong, M., Hovy, E., Le, Q.: Self-training with noisy student improves imagenet classification. In: Proceedings of the IEEE/CVF Conference on Computer Vision and Pattern Recognition, pp. 10687–10698 (2020)
8. Laine, S., Aila, T.: Temporal Ensembling for Semi-supervised Learning (2016). ArXiv Preprint ArXiv:1610.02242
9. Tarvainen, A., Valpola, H.: Mean teachers are better role models: Weight-averaged consistency targets improve semi-supervised deep learning results. Adv. Neural Inform. Proces. Syst. **30** (2017)
10. Miyato, T., Maeda, S., Koyama, M., Ishii, S.: Virtual adversarial training: a regularization method for supervised and semi-supervised learning. IEEE Trans. Pattern Anal. Mach. Intel. **41**, 1979–1993 (2018)
11. Sohn, K., Berthelot, D., Carlini, N., Zhang, Z., Zhang, H., Raffel, C., Cubuk, E., Kurakin, A., Li, C.: Fixmatch: simplifying semi-supervised learning with consistency and confidence. Adv. Neural Inform. Proces. Syst. **33**, 596–608 (2020)
12. Berthelot, D., Carlini, N., Goodfellow, I., Papernot, N., Oliver, A., Raffel, C.: Mixmatch: a holistic approach to semi-supervised learning. Adv. Neural Inform. Proces. Syst. **32** (2019)
13. Cui, W., Akrami, H., Joshi, A., Leahy, R.: Semi-supervised Learning Using Robust Loss (2022). ArXiv Preprint ArXiv:2203.01524
14. Hung, W., Tsai, Y., Liou, Y., Lin, Y., Yang, M.: Adversarial learning for semi-supervised semantic segmentation. In: Proceedings of the British Machine Vision Conference (BMVC) (2018)
15. Chen, L., Papandreou, G., Kokkinos, I., Murphy, K., Yuille, A.: Deeplab: Semantic image segmentation with deep convolutional nets, atrous convolution, and fully connected crfs. IEEE Trans. Pattern Anal. Mach. Intel. **40**, 834–848 (2017)
16. Goodfellow, I., Pouget-Abadie, J., Mirza, M., Xu, B., Warde-Farley, D., Ozair, S., Courville, A., Bengio, Y.: Generative adversarial nets. Adv. Neural Inform. Proces. Syst. **27** (2014)

17. Ronneberger, O., Fischer, P., Brox, T.: U-net: convolutional networks for biomedical image segmentation. In: International Conference on Medical Image Computing and Computer-Assisted Intervention, pp. 234–241 (2015)
18. Jha, D., Smedsrud, P., Riegler, M., Johansen, D., De Lange, T., Halvorsen, P., Johansen, H.: Resunet++: an advanced architecture for medical image segmentation. In: 2019 IEEE International Symposium on Multimedia (ISM), pp. 225–2255 (2019)

RNCE: A New Image Segmentation Approach

Vikash Kumar, Asfak Ali, and Sheli Sinha Chaudhuri

Abstract Semantic image segmentation based on deep learning is gaining popularity because it is giving promising results in medical image analysis, automated land categorization, remote sensing, and other computer vision applications. Many algorithms have been designed in recent years, yet there is scope for further improvement in computer vision research. We have proposed a unique ensemble method called Ranking and Nonhierarchical Comparison Ensemble (RNCE) for semantic segmentation of landcover images based on the Ranking and Nonhierarchical Comparison methodology. Our approach has been tested on pretrained models showing improved accuracy and mean IoU with respect to the existing method. The code is available at: https://github.com/vekash2021/RNCE.git.

Keywords Remote sensing · Pixel classification · Image segmentation · Landcover classification

1 Introduction

Computer vision is the area in which machines learn how to extract information from visual data and understand the surrounding at various levels of detail and abstraction. It has a wide range of applications such as object detection, tracking, semantic segmentation, instance segmentation, etc. This paper deals with semantic segmentation. The objective of semantic segmentation is to classify each pixel in an image into one of several predefined categories. It is one of the techniques used in remote sensing imaging applications that aims to classify each pixel in a landcover image. The availability of small space-grade equipment has led to an increase in the number of small satellites which will be used for a wide range of remote sensing applications. The reduction in production costs for satellites has resulted in democratized access to the area. Satellite imaging (a type of remotely sensed data) has grown in popularity and demand in recent years, with imagery that was previously available to a few research

V. Kumar (✉) · A. Ali · S. S. Chaudhuri
Electronics and Telecommunication Engineering, Jadavpur, Kolkata, West Bengal 700032, India
e-mail: vikashkr2110@gmail.com

M. Tistarelli et al. (eds.), *Computer Vision and Machine Intelligence*,
Lecture Notes in Networks and Systems 586,
https://doi.org/10.1007/978-981-19-7867-8_44

teams becoming more widely available. Furthermore, commercial satellite imagery has evolved from pixel push to content delivery as a result of new market growth; this makes us more understanding of satellite image features. The openly to-be-had satellite image dataset attracts interest from researchers making better models for image segmentation.

With the rise of machine learning and deep learning techniques, a convolutional neural network (CNN) is showing desired results in the field of computer vision and is being widely utilized in the automatic classification of objects, regions, etc. The CNN is highly successful in extracting useful features from images leading to several research work applying a deep convolutional network for the classification of various types of images replacing traditional image semantic segmentation methods. The classic method of image segmentation relies on clustering technologies like the Random Walks and the Markov model, to name a few. The problem with the traditional image segmentation algorithm is a mathematical approach that is susceptible to noise and requires a lot of interaction between humans and computers for accurate segmentation that takes time and effort. Earlier researchers utilized textures and spectral features of remote sensing images to distinguish them, but the details of the information thus obtained were not sufficient for complete and accurate segmentation. The demand for automatic information retrieval and processing arises when the amount of data to be processed increases.

The deep learning-based algorithm solved the problem of the traditional way of image segmentation. But predicting the category label for every image pixel is very challenging for an individual model. Every model does not give a satisfactory result on a particular dataset. We proposed a Ranking and Nonhierarchical Comparison Ensemble (RNCE) method for enhancing prediction performance in this research, with the goal of constructing an effective model by combining different models. The advantage of the Ranking and Nonhierarchical Comparison (RNC) [1] method is that it can reduce the number of comparisons between multiple entities so that consistency is automatically maintained via the determination of priorities. The motive for this initiative is that land monitoring is a difficult undertaking for the government since it requires a large amount of staff and time. Using the proposed method, it is easy to segment the structures of roads, forests, water bodies, open areas, and buildings in urban planning, which will help very much in efficient urban land development and planning.

The remaining sections of this paper are divided as follows: Sect. 2 provides an overview of related work on Semantic Segmentation of Satellite Images. Section 3 presents the methodology of this work, Sect. 4 discusses the outcomes, while Sect. 5 analyses the findings and suggests scope for further research.

2 Literature Survey

In the last several years, there has been a lot of work on semantic segmentation on satellite images. In the last several years, there has been a lot of work on semantic segmentation on satellite images. Li et al. [2] proposed a model for building footprint generation by new boundary regularization networks in an end-to-end generative adversarial network (GAN) of satellite images. Bao et al. [3] proposed a model for semantic segmentation using improved deeplabV3+ for which mobilenetv3 is used for feature extraction as well as pyramid structure to expand the receptive field. Zeng et al. [4] suggested a method of post-processing of semantic segmentation using K-means clustering for improving the accuracy of model. Heryadi et al. [5] presented a study of various pretrained ResNets versions as feature extractors in deeplab model to improve the segmentation of satellite images. Baghbaderani et al. [6] proposed spectrum separation techniques to obtain effective spectral data representations for semantic segmentation of satellite images. They showed that appropriately extraction of characteristics from images and using them as input to a deep learning network can increase the efficiency of landcover classifications. Zhang et al. [7] implemented a two-stage practical training technique for a building damage evaluation model based on the Mask R-CNN. The training of ResNet 101 as a building feature extractor in the Mask R-CNN is the first stage. They use a model learned stages one to generate a more deep learning architecture that can distinguish buildings with varying degrees of damage from satellite images in stage two. Vakalopoulou et al. [8] had proposed a solution based on support vector machines (SVMs). The SVM model has been employed for the classification of the training datasets. The resulting map gives a score for each pixel inside the class, the buildings were extracted using the Markov Random Field (MRF)-based model which improved the detection of building in the map. Hansen and Salamon [9] suggested that grouping together networks with comparable functions and configurations improved the individual network's prediction performance. Zou et al. [10] developed a rapid ensemble learning method that combines deep convolutional networks and random forests to shorten learning time while working with minimal training data.

To classify the region of interest, most of the techniques in the existing literature rely on a single deep learning model. We can't rely on a single deep learning model to predict the desired result because satellite images are complicated and difficult to segment due to the similarity in the texture of different regions. A combination of more than one model is required to predict the desired result, which is called an ensemble method. There is no such work that has been performed in the existing literature, which is completely focused on integrating the model to predict pixel-wise semantic segmentation of remote sensing images.

3 Proposed Architecture

In this section, we have discussed the proposed work, which is divided into three parts, data acquisition, model selection, loss function, and training.

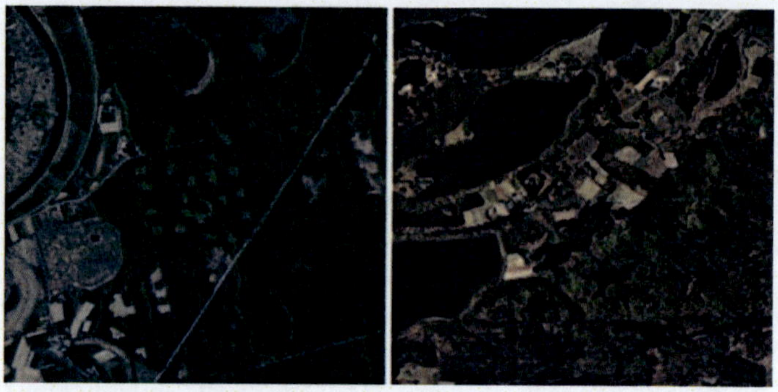

Fig. 1 Sample images from LandCover datasets

3.1 Data Acquisition

For semantic segmentation of satellite images, we used the LandCover.ai dataset, which is an open-source resource. The LandCover.ai [11] (Land Cover from Aerial Imagery) collection is used to map houses, forests, lakes, and roads from aerial images, which were captured with three spectral bands of RGB and a spatial resolution of 25 or 50 cm per pixel. They are from various years and flights (2015–2018). We chose 41 orthophoto tiles from various nations throughout the globe to ensure that the dataset is as diverse as possible (as shown in Fig. 1). Each tile is approximately 5 km^2. There are 33 images with a 25 cm resolution (about 9000 × 9500 pixels) and eight images with a 50 cm resolution (approximately 4200 × 4700 pixels).

3.2 Model Selection

We have selected different pretrained models as an encoder for the ensemble technique such as VGG (Visual Geometry Group), ResNet, Inception-ResNet-v2, EfficientNet, and MobileNet. The VGG [12] architecture is a multilayer deep CNN architecture. With VGG-16 or VGG-19, the term "deep" refers to the number of layers, which are sixteen and nineteen convolutional layers, respectively. Groundbreaking object recognition models are built on the VGG architecture. VGGnet effectively exceeds the baseline of many tasks and datasets after ImageNet. In Residual Networks(ResNets) [13] learns residual functions concerning the input layers, instead of learning unreferenced functions. Instead of every stacked layer fitting into a desired underlying mapping, residual nets allow these layers to fit into a residual mapping. They build networks by stacking residual blocks on top of each other; for example, a ResNet-50 contains 50 layers made up of these blocks. The convolutional neural network InceptionResNet-v2 was trained on over one million images from the ImageNet database. The network contains 164 layers deep which help in classifying images into 1000 object categories. As an end result, the network learns rich function representations for a huge variety of images. EfficientNet [14] is based on a CNN structure and scaling approach that uniformly scales all depth/width/resolution dimensions with the usage of a compound coefficient. Unlike the traditional method, which scales these elements freely, the EfficientNet scaling approach uses a set of consistent scaling coefficients to scale network width, depth, and resolution uniformly. MobileNet [15] increases mobile models state-of-the-art performance on a variety of tasks and benchmarks, as well as across a range of model sizes. It uses depth-wise separable convolutions. It employs separable convolutions that can be separated in depth. When compared to networks with regular convolutions of the same depth in the nets, the number of parameters is dramatically reduced. As a result, lightweight deep neural networks are created.

3.3 Proposed Model

Figure 2 depicts the flow diagram of the proposed model for landcover segmentation. In our framework, we used distinct pretrained models as encoders for feature extraction, each of which has several layers, such as ResNet50, which has 50 deep layers. Inception Resnet-v2 contains 162 deep layers. The expansion path on the decoder side is made up of four blocks. A deconvolution layer with stride 2 is present in each block, and Concatenation with the corresponding cropped feature map from the contracting path is done. We use skip connections to get more precise locations at every stage of the decoder by concatenating the output of the transposed convolution layers with the feature maps from the encoder at the same level. Two 3×3 convolution layers are employed, as well as the ReLU activation function (with batch normalization). On the same landcover dataset, the model was trained with distinct encoders and decoders. The model is ensembled after training to predict the segmented region in the landcover images. The ensemble approach is a methodology for integrating numerous models to increase prediction performance by aggregating the predictions of multiple independent models. The ensemble model can be produced by mixing numerous modelling techniques or by utilizing various training datasets. It is mostly used to decrease generalization mistakes in deep learning algorithms. The ensemble strategy of n-segmentation models is proposed in this research to improve the accuracy performance of single models, where n denotes the number of single models merged at a time. The decision-making process for allocating weight to each model is not simple, nevertheless, when comparing alternatives with basic qualities and a small number of comparisons, priority may be allocated quickly. However, in the event of several qualities on a large number of comparisons, simplify the attributes to make a judgement by omitting part of them or exchanging cognitive effort for decision accuracy, thus reducing decision-making accuracy. As a result, we use the RNC approach to allocate weights to our model. The hierarchy structure is used to

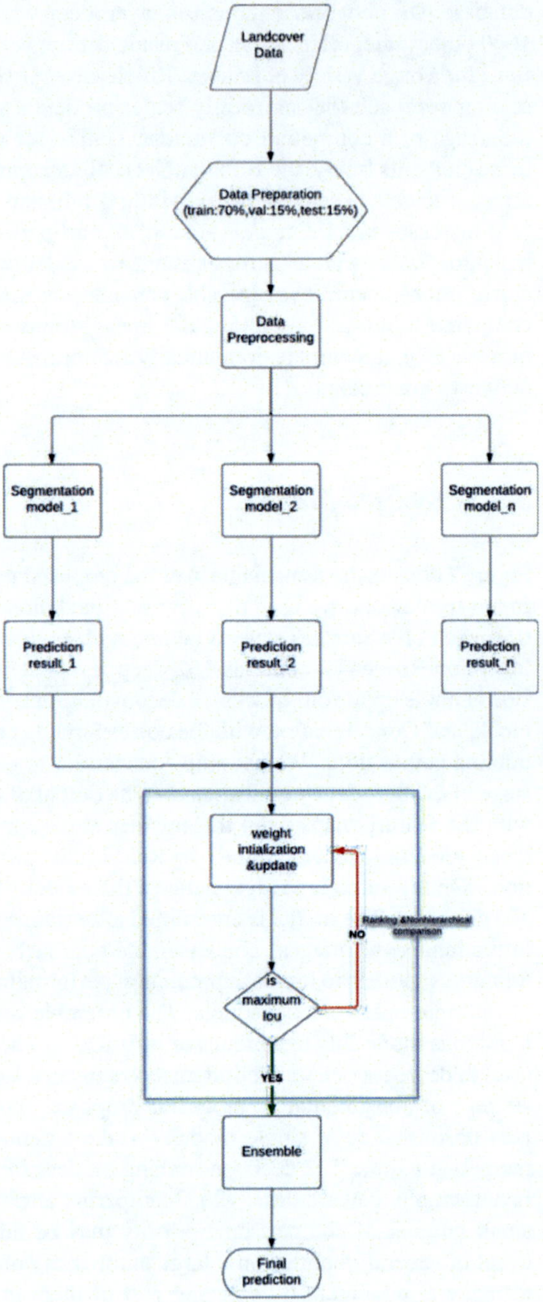

Fig. 2 Flow diagram for proposed model

define the priority of many characteristics in this manner. It decides the priority by utilizing characteristics and entities to form a hierarchy. Second, priorities for entities within each category are established. Third, by assigning a priority to things from distinct groups that have the same priority. Setting a priority among entities with adjoined priorities is the fourth step. The weight of each model is allocated once the priority of all submissions has been decided.

The model was trained with different encoders and decoders on the same landcover dataset. After the training, all individual model is evaluated using accuracy and MeanIoU. In this paper, we have used five different trained models for ensemble to improve the performance of the prediction result. For better comparison of model performance, We have made a combination of three one at a time, after that initialize the weights to each one of them then it is going through the loop for checking the condition of maximum mean IoU. If mean IoU is not maximum, then weights are updated using ranking and nonhierarchy comparison to each one of the models using four steps. During the loop, it is going through all possible combination of weights for finding the maximum mean IoU. Loop is running until to get maximum mean IoU; once the maximum means IoU comes to the loop, condition breaks and gives the final prediction results.

3.4 Loss Function and Training

Focal loss is used to deal with the problem of the class imbalance trouble. A modulation term is applied to the cross-entropy loss function, making it efficient and smooth to learn for tough examples. It is a dynamically scaled cross-entropy loss, in which the scaling element decreases as confidence within the proper class grows. The formula by which focal loss is calculated is given below.

$$\mathrm{FL}(P_t) = -\alpha_t((1 - P_t)^{\gamma}) \log(P_t) \tag{1}$$

This α balances out the amount of examples in each class. This α is either determined through cross-validation or is inversely proportional to the number of instances for a certain class. The γ phrase is used to concentrate on the difficult instances. The likelihood of proper categorization defines the hard and easy cases. If it is low, then we are dealing with a hard sample. The modulating factor remains near 1 in this situation and is unaffected. The modulating factor vanishes to 0 when it is high. The simple samples are given less weight as a result of this.

During the training of the model, we are running the model for 80 epochs in which we set the rate of learning such that if the performance of the model does not improve, the learning rate is automatically reduced to improve the model performance and also set some limit conditions in which if model performance does not improve to some range of predefined epoch, training stops, and model is saved.

4 Experimental Results and Comparisons

The experimental result of the proposed method has been shown in Table 1. This has been implemented on Python 3.7.13 using the Python library. All experiments were performed on Intel(R) Xeon(R) CPU @ 2.30 GHz and Tesla T4 GPU with 13GB of RAM. Qualitative and quantitative approaches are used to test the performance of the proposed model. The result is divided into two parts; in the first part, all the results are compared in accuracy and mean IoU, and in the second part, results are compared with visual prediction results.

4.1 Accuracy

The quantity of correct predictions provided by a model is measured by accuracy. For multiclass classification, the accuracy value goes from 0 to 1, with 1 being the highest level of accuracy. The proportion of match between the actual and anticipated classes determines the value.

Table 1 Comparison of accuracy and mean IoU between different models

Method	Accuracy	MeanIoU
UNET	41.26	23.08
DeepLabv3+ OS 16	91.39	81.81
DeepLabv3+ OS 8	93.01	83.43
VGG19	89.48	79.9
Inception-ResNet-v2	92.73	83.15
ResNet-50	91.32	81.74
EfficientNet-b0	90.29	80.71
MobileNetV2	89.69	80.11
VGG19+ResNet-50+Inception-ResNet-v2	93.14	83.56
ResNet-50+Inception-ResNet-v2+ EfficientNet-b0	93.16	83.58
Inception-ResNet-v2+ EfficientNet-b0+ MobileNetV2	93.34	83.76
VGG19+MobileNetV2 +Inception-ResNet-v2	92.46	82.88
Inception-ResNet-v2+ EfficientNet-b0+VGG19	91.34	81.76
ResNet-50+Inception-ResNet-v2+ MobileNetV2	93.21	83.63
VGG19+MobileNetV2+ ResNet-50	91.76	82.18

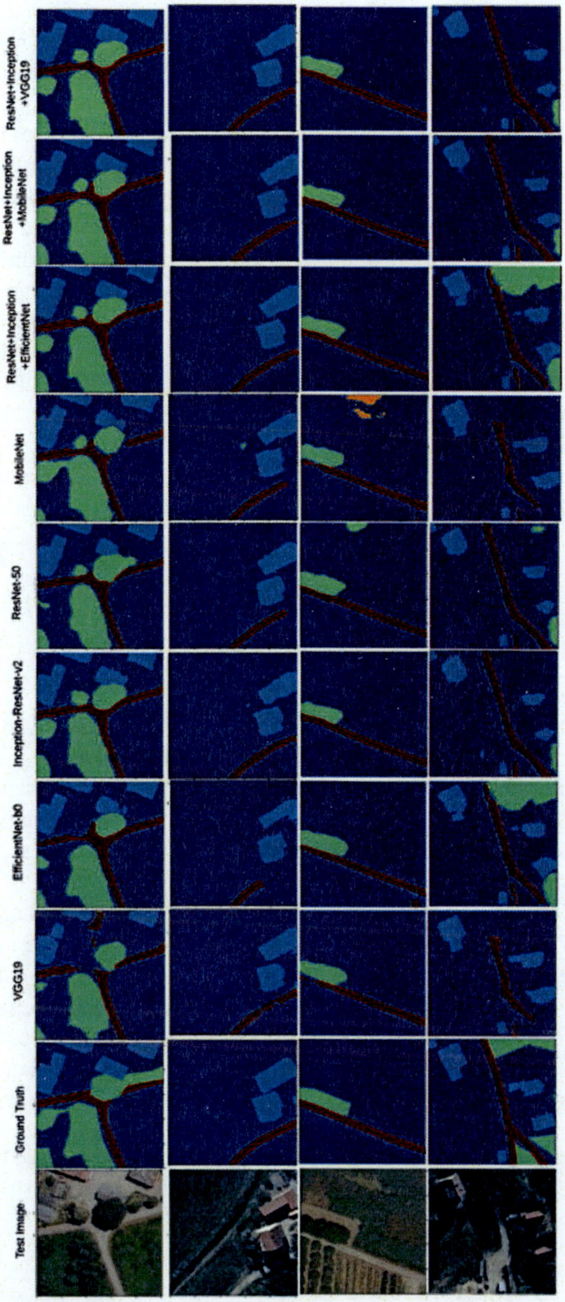

Fig. 3 Prediction result of different models

4.2 Mean Intersection-Over-Union

Performance was measured using Intersection-over-Union (IoU). The IoU is a statistic for determining a classifier's accuracy. To evaluate semantic image segmentation, the measure mean Intersection-over-Union is used. It first computes the IoU value for each semantic class and then averages the values over all classes.

The landcover dataset was first transformed to greyscale. The resultant data and annotated masks are transferred to the deep learning model for training. The segmentation results of these models were evaluated on accuracy and mean IoU. The comparison between the state-of-the-art model and the proposed model is shown in Table 1. The table is divided into three columns, where one column represents the method name; the second one is accuracy, and the third one is mean IoU. The table shows that, virtually under all scenarios, the ensemble technique outperforms the individual method in terms of assessment measures. The ensemble technique has a higher accuracy and mean IoU. As shown in Table 1, the ensemble technique of Inception-ResNet-v2+ EfficientNet-b0+ MobileNetV2 provides maximum accuracy and maximum mean IoU than all other methods (Fig. 3). Table 1 displays the qualitative results, and it is evident from this table that ensemble-based approaches produce better results.

5 Conclusion

The requirement for land mapping is a necessity nowadays because of changes in environmental conditions. The traditional approach is not enough to meet the requirement. As a result, the research focuses on a methodology for segmenting landcover areas gathered from satellite images into separate land mapping regions. It offered a method for automatically mapping buildings, woods, open areas, water bodies, and roads using semantic segmentation of landcover optical images. The proposed method outperforms the recent state-of-the-art method, which was applied by other authors using DeepLabv3+ OS 16 and DeepLabv3+ OS 8 on the same landcover dataset. Our proposed method combines Inception-ResNet-v2+ EfficientNet-b0+ MobileNetV2 which gives better accuracy and mean IoU. Even though the suggested model might be improved, the positive findings obtained lead us to believe that the model could be used for land monitoring tasks such as landscape change detection, urban planning, etc.

Acknowledgements This work has been carried out in the Digital Control and Image Processing Lab, ETCE Department, Jadavpur University.

References

1. Song, B., Kang, S.: A method of assigning weights using a ranking and nonhierarchy comparison. Adv. Decis. Sci. **2016**, 1–9 (2016). https://doi.org/10.1155/2016/8963214
2. Li, Q., Zorzi, S., Shi, Y., Fraundorfer, F., Zhu, X.X.: End-to-end semantic segmentation and boundary regularization of buildings from satellite imagery. In: IEEE International Geoscience and Remote Sensing Symposium IGARSS **2021**, 2508–2511 (2021). https://doi.org/10.1109/IGARSS47720.2021.9555147
3. Bao, Y., Zheng, Y.: Based on the improved Deeplabv3+ remote sensing image semantic segmentation algorithm. In: 2021 4th International Conference on Advanced Electronic Materials, Computers and Software Engineering (AEMCSE), pp. 717–720 (2021). https://doi.org/10.1109/AEMCSE51986.2021.00148
4. Zeng, X., Chen, I., Liu, P.: Improve semantic segmentation of remote sensing images with K-mean pixel clustering: A semantic segmentation post-processing method based on k-means clustering. In: 2021 IEEE International Conference on Computer Science, Artificial Intelligence and Electronic Engineering (CSAIEE), pp. 231–235 (2021). https://doi.org/10.1109/CSAIEE54046.2021.9543336
5. Heryadi, Y., Soeparno, H., Irwansyah, E., Miranda, E., Hashimoto, K.: The Effect of Resnet Model as Feature Extractor Network to Performance of DeepLabV3 Model for Semantic Satellite Image Segmentation (2021). https://doi.org/10.1109/AGERS51788.2020.9452768
6. Baghbaderani, R.K., Qi, H.: Incorporating spectral unmixing in satellite imagery semantic segmentation. IEEE International Conference on Image Processing (ICIP) **2019**, 2449–2453 (2019). https://doi.org/10.1109/ICIP.2019.8803372
7. Zhao, F., Zhang, C.: Building damage evaluation from satellite imagery using deep learning. In: 2020 IEEE 21st International Conference on Information Reuse and Integration for Data Science (IRI), pp. 82–89 (2020). https://doi.org/10.1109/IRI49571.2020.00020
8. Vakalopoulou, M., Karantzalos, K., Komodakis, N., Paragios, N.: Building detection in very high resolution multispectral data with deep learning features. In: IEEE International Geoscience and Remote Sensing Symposium (IGARSS) **2015**, 1873–1876 (2015). https://doi.org/10.1109/IGARSS.2015.7326158
9. Hansen, L.K., Salamon, P.: Neural network ensembles. IEEE Trans. Pattern Anal. Mach. Intel. **12**(10), 993–1001 (1990). https://doi.org/10.1109/34.58871
10. Zuo, Y., Drummond, T.: Fast Residual Forests: Rapid Ensemble Learning for Semantic Segmentation. CoRL (2017)
11. Boguszewski, A., Batorski, D., Ziemba-Jankowska, N., Zambrzycka, A., Dziedzic, T.: LandCover.ai: Dataset for Automatic Mapping of Buildings, Woodlands and Water from Aerial Imagery (2020)
12. Simonyan, K., Zisserman, A.: Very Deep Convolutional Networks for Large-Scale Image Recognition. arXiv 1409.1556 (2014)
13. He, K., Zhang, X., Ren, S., & Sun, J. (2016). Deep residual learning for image recognition. 770–778. https://doi.org/10.1109/CVPR.2016.90
14. Tan, M., Le, Q. V.: EfficientNet: Rethinking Model Scaling for Convolutional Neural Networks. ArXiv abs/1905.11946 (2019)
15. Sandler, M., Howard, A., Zhu, M., Zhmoginov, A., Chen, L.-C.: MobileNetV2: inverted residuals and linear bottlenecks. 4510–4520 (2018). https://doi.org/10.1109/CVPR.2018.00474

Cross-Media Topic Detection: Approaches, Challenges, and Applications

Seema Rani and **Mukesh Kumar**

Abstract Developing technologies and new social media platforms (Facebook, YouTube, Flickr, and Twitter) are transforming how people communicate with each other and scrutinize information. Different social media users prefer different mode of information for expressing their views. For instance, Twitter users prefer short text and informally taken photos; YouTube viewers tend to discuss breaking news in the form of short videos with rough editing, and professional journalists usually produce standardized news stories with substantial text and well chosen images. Every day, a large number of videos, images, and text are shared on social media platforms. The cross-media data provide credible insights into public opinion, which are extremely beneficial to commercial firms, governments, and any organization concerned with societal opinions. As a result, assessing such voluminous data is beneficial. When cross-media information is combined, it can provide a more comprehensive and effective description of a given event or issue, as well as improve system performance. Also, in addition to being a significant source of information, cross-media information provided by various media is more robust, has a larger audience, and appears to be a more accurate reflection of real-world events. Consequently, robustly identifying topics from multimodal data from multiple media is a useful and sensible extension of cross-media topic identification. This paper presents a review of various cross-media topic detection approaches.

Keywords Cross-media topic detection · Topic detection approaches · Challenges · Applications

Supported by organization CSIR, New Delhi, India

S. Rani (✉) · M. Kumar
Computer Science and Engineering Department, University Institute of Engineering and Technology, Panjab University, Chandigarh 160014, India
e-mail: seemarani1307@yahoo.com

M. Tistarelli et al. (eds.), *Computer Vision and Machine Intelligence*,
Lecture Notes in Networks and Systems 586,
https://doi.org/10.1007/978-981-19-7867-8_45

1 Introduction

The social media world has evolved into a real-time repository of information in recent years. The advancement of video compression techniques has made it easier to share videos over the Internet. Thus, users post more videos than textual posts, which drive up the information load on the Web. The solution is to identify topics from video data in order to address this issue. Researchers can use this data for many purposes, such as tracking opinions about new products and services, looking for trends in popular culture, examining the effects of prescription drugs, investigating fraud and other types of criminal activity, examining opinions about political candidates, motor vehicle defects, and studying the consumption habits of different groups [36]. The general framework for cross-media topic detection has been shown in Fig. 1.

The Topic Detection and Tracking (TDT) method analyzes multimedia streams and identifies unknown topics automatically. There have been many different TDT techniques introduced in the past decade for dealing with different kinds of media (videos, images, and articles). The methods, however, are not suitable for identifying topics in social media videos because they are not generated by professionals. The vast majority of previous topic detection works have focused on data from a single medium; however, using data from multiple media can strengthen topic detection methods through their rich and complementary information. A single medium's intrinsic content of information is not as rich as complementary cross-media information delivered by multiple mediums. Additionally, cross-media information is often accepted by a broader demographic and reflects social reality from a wider spectrum. As a result, jointly detecting topics from the various data sources of multimodal data would be an interesting research area.

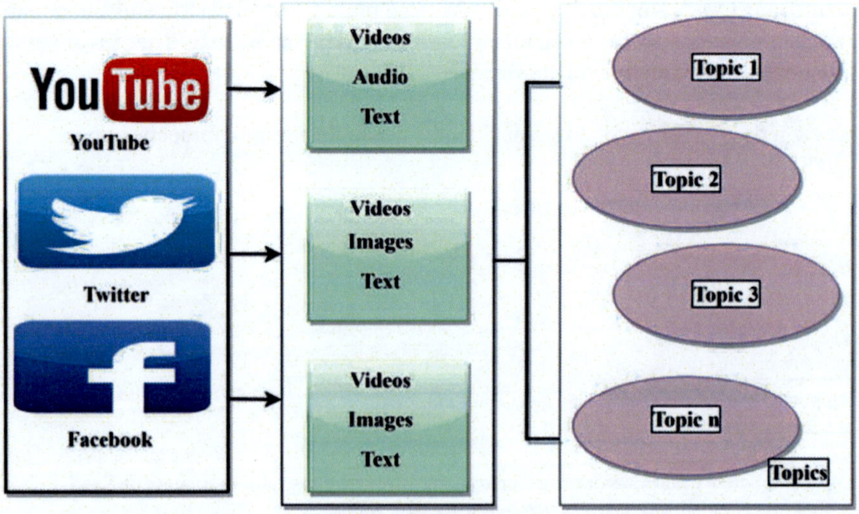

Fig. 1 Cross-media topics

2 Topic Detection Approaches

In this section, various approaches proposed for cross-media topic detection have been discussed. We have categorized the approaches into three categories: graph-based methods, machine learning-based methods, and deep learning-based methods.

2.1 Graph-Based Methods

Based on an evolution link graph, Cao et al. [3] presented a salient trajectory algorithm for topic discovery given a large set of videos recorded over months. Chen et al. [4] introduced a method for detecting Web video topics using a multi-clue fusion approach. They proposed a maximum average score and a burstiness score to extract dense-bursty tag groupings as a first step, employing video-related tag information. Second, near-duplicate keyframes (NDKs) from the videos are retrieved and merged with the retrieved tag groups. Following that, the top search phrases from the search engine are employed as topic detection guides. Finally, these clues are integrated to reveal the online video data's hidden topics. Shao (2012) presented a star-structured K-partite graph (SKG) for topic detection from Web videos that integrated multimodality data [37]. Gao et al. [8] proposed a method for detecting events, which generates a semantic entity called the microblog clique (MC), to assess highly linked information held by chaotic and transient microblogs. A hypergraph is constructed from the heterogeneous social media data, and the MCs are made up of the ones that are highly connected. In order to detect social events, these MCs are used to build a bipartite graph, which is then partitioned. A graph-based framework for visualizing real-world events has been proposed by Schinas et al. [34] to create visual summaries of the stream of posts related to those events. In addition to the posts, MGraph utilizes multiple signals and modalities.

2.2 Machine Learning-Based Methods

A method for detecting event structure with late fusion based on text/visual similarities within videos was presented by Wu et al. [43]. Detection of multimedia events was proposed with double fusion by Lan et al. [15]. An approach called topic recovery (TR) and multimodality fusion was proposed to accurately discover topics from cross-media data by Chu et al. [5]. This approach systematically blends heterogeneous multimodal data into a multimodality graph, which can further be used to identify useful topic candidates (TCs). Social event detection (SED) in huge multimedia archives can be detected with a scalable graph-based multimodal clustering technique [29]. Using example relevant clustering, the proposed approach learns a model of "same event" relationships between items in a multimodal domain and uses it to systematize them in a graph. Kim et al. [12] proposed analyzing trending multimedia content and summarizing it using a platform called Trends Summary. The Twitter Streaming API is utilized to retrieve trend keywords and their related keywords from raw Twitter data. Then, those keywords are annotated by adding information from Wikipedia and Google. Various Web sites are trawled for four different types of multimedia content according to expanded trend keywords. Naive Bayes was used to determine the best media type for this trend keyword. Based on the content of the selected media type, the most appropriate content was selected. Finally, we used TreeMap algorithms to display both the trend keywords and their multimedia contents on the screen. In 2015, Xue et al. [45] proposed a topic detection method that uses hot search queries as guidance. In 2013, Pang et al. [24] proposed a "constrained non-negative matrix factorization-based" semi-supervised co-clustering method for discovering cross-media topics. The mathematical rigorousness of this approach is demonstrated by the correctness and convergence of its approach.

Min et al. [22] presented a "cross-platform multimodal probabilistic model (CM3TM)" to handle the inter-platform recommendation challenge. Using this approach, topics on various platforms can be distinguished into shared and platform-specific topics, as well as aligned based on distinct modalities. Emerging topic detection and elaboration from multimedia streams across various online platforms had been introduced by Bao et al. [1]. A technique to build multimodal collection summaries based on latent topic analysis had been given by Camargo et al. [2]. This method models image semantics by fusing text information with visual information in the same latent space. Furthermore, Qian et al. [30] presented event topic model based on multimodal data. This model not only captures multimodal topics, but also acquires social event evolution trends and generates effective event summary details over time.

In 2017, Li et al. [16] proposed a "joint image-text news topic detection and tracking" technique. They proposed the "Multimodal Topic And-Or Graph," a structured topic representation that models image and text components of many topics simultaneously. The SWC-based cluster sampling method is used to detect topics. To deal with the constant updates of news streams, topics are also tracked throughout time. Pang et al. [27] proposed a method for extracting multimedia descriptions from noisy social media data. The devcovolution model is first employed to diminish similarities between non-informative words/images during Web topic detection. Second, the background-removed similarities are recreated during topic description to provide possible keywords and prototype images. By minimizing background similarities, this strategy creates a coherent and informative multimedia description for a topic.

To improve the interpretability of a Web video topic, Pang et al. [23] proposed a two-step technique that includes both "Sparse Poisson Deconvolution (SPD)" and "Prototypes from Submodular Function (PSF)." Rather to coming up with keywords to define a topic, analyzing a Web topic through its prototypes is a conceptually basic but effective method. The influences of incorrectly detected Webpages are decreased simply by incorporating sparse intratopic similarities into the classical Poisson deconvolution. The representative yet different prototypes are efficiently found from any symmetric similarities by presenting a broad prototype learning

Table 1 Summary of cross-media topic detection techniques

References	Paper title	Datasets	Technique used	Validation measures
[16]	"Joint image-text news topic detection and tracking by multimodal topic and-or graph"	Reuters-21578 dataset, UCLA Broadcast News Dataset	Swendsen-Wang Cuts (Swc), Multimodal Topic And-Or Graph (Mt-Aog)	Clustering accuracy, normalized mutual information, precision, recall
[19]	"Visual topic discovering, tracking and summarization from social media streams"	MCG-WEBV dataset, MicroblogV dataset	K-Partite Graph	Normalized mutual information (NMI), precision, average precision, average score
[18]	"Cross-media event extraction and recommendation"	Synthetic dataset	Convolutional neural network	Precision, recall, accuracy
[30]	"Multimodal event topic model for social event analysis"	MediaEval social event detection(SED)	Lda	MAP Score
[2]	"Multimodal latent topic analysis for image collection summarization"	MIR Flickr, Flickr4 Concepts	Convex non-negative matrix factorization (Convex-Nmf)	Reconstruction error, diversity score
[27]	"Robust latent poisson deconvolution from multiple features for Web topic detection"	MCG-WEBV, YKS	Latent poisson deconvolution, alternating direction method of multipliers (Admm)	Number of detected topics (NDT), false positive per topic (FPPT), accuracy
[5]	"Effective multimodality fusion framework for cross-media topic detection"	MCG-WEBV, YKS	Multimodality graph, Lda	Precision, recall, F-measure
[46]	"Fusing cross-media for topic detection by dense keyword groups"	MCG-WEBV dataset, CM-NV, CM-NV-N, CM-NV-V dataset	Keyword group extraction, keyword group refining and topic detecting	Hot degree of topics, precision, recall, F-measure
[1]	"Cross-platform emerging topic detection and elaboration from multimedia streams"	Synthetic dataset, Real-world data	Aging theory, co-clustering	Precision, normalized discount cumulative gain (NDCG), mAP score
[41]	"Cross-media topic detection: a multimodality fusion framework"	MCG-WEBV dataset	Convolutional neural network (CNN)	F-measure
[45]	"Topic detection in cross-media: a semi-supervised co-clustering approach"	YKSN dataset	Semi-supervised co-clustering	Precision, recall, F-measure
[44]	"Cross-media topic detection associated with hot search queries"	Synthetic dataset	Weighted co-clustering algorithm	Precision, recall, F-measure

(*continued*)

Table 1 (*continued*)

References	Paper title	Datasets	Technique used	Validation measures
[23]	"Increasing interpretation of web topic detection via prototype learning from sparse poisson deconvolution"	Sparse poisson deconvolution (SPD) and prototypes from submodular Function (PSF), k-nearest neighbor Hybrid similarity graph (HSG)	MCG-WEBV, YKS dataset	Accuracy, F1-score
[26]	"A two-step approach to describing web topics via probable keywords and prototype Images from Background-removed Similarities"	k-nearest neighbor Hybrid Similarity Graph (HSG), Text aided Poisson Deconvolution (TaPD) Alternating Direction Method of Multipliers (ADMM)	MCG-WEBV, YKS dataset	Accuracy, Domain Expert Evaluation (DEE) and Topic Description Intrusion (TDI)
[25]	"Two birds with one stone: A coupled poisson deconvolution for detecting and describing topics from multimodal web data"	Hybrid Similarity Graph, Coupled Poisson Deconvolution, Accelerated Proximal Gradient (APG), Affinity Propagation	MCG-WEBV, YKS dataset	Accuracy, F1-Score, Domain Expert Evaluation (DEE) and Topic Description Intrusion (TDI)
[38]	"A Multifeature Complementary Attention Mechanism for Image Topic Representation in Social Networks"	Attention Mechanism	Sina Weibo and Mir-Flickr 25K	Normalized Discounted Cumulative Gain (NDCG), and Mean Average Precision (MAP)

(PL) technique. Pang et al. [26] presented a two-step approach to describe Web topics using probable keywords and prototype images derived from background-removed similarities. The first step in understanding Web trends is to organize multimodal Webpages into popular topics. The less constrained social media, on the other hand, produce noisy user-generated content (UGC), making a detected topic less cohesive and interpretable This problem had been addressed by introducing a coupled Poisson deconvolution that simultaneously tackles topic detection and topic description. The interestingness of a topic is determined in topic detection by the similarities refined by topic description; the interestingness of topics is used to characterize topics in topic description. Two processes recognize interesting topics in a cyclical manner and provide a multimodal description of them [25] (Table 1).

2.3 Deep Learning-Based Methods

A novel image dominant topic model for topic detection had been introduced by Wang et al. [41]. This method creates a semantic simplex by combining text and visual modalities. Furthermore, an improved CNN feature is obtained by integrating the convolutional and fully connected layers to collect more visual information. "Complementary Attention Mechanism for Image Topic Representation (CATR)" has been proposed by Shi et al. [38]. To produce an accurate feature representation, the focused and unfocused features of the modeling image are first differentiated in this approach. Second, to generate a more concentrated image feature representation, the object feature is merged to construct the complementary attention mechanism. Finally, the whole depiction of the social network image topic is accomplished.

3 Challenges in Cross-media Topic Detection

In the past, there has been extensive research on single-media topic detection. The goal of cross-media topic detection is to comprehensively detect topics across multiple media channels with multimodal data [5]. Cross-media data handling challenges has also been shown in Fig. 2. The task of TDT for blog videos is much more challenging than for text streams or Web videos because of the following reasons:

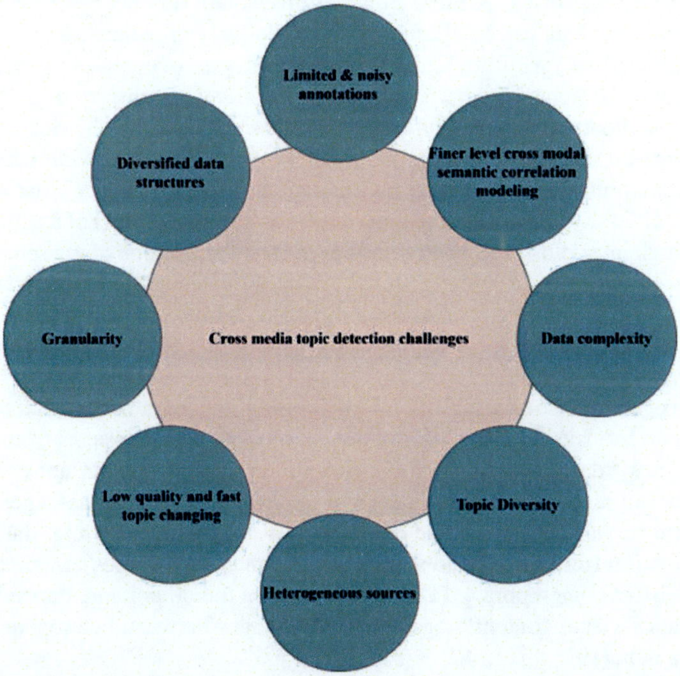

Fig. 2 Challenges in cross-media topic detection

1. **Sparse and noisy annotations**: In messages that include blog videos, there is a high degree of sparsity and noise. The message length limits of microblogging sites and textual annotations of videos make it impossible to retrieve effective textural information. Additionally, the texts are frequently short, ungrammarized, unstructured, and noisy [19].
2. **Low quality and Fast topic changing**: In social media, most video blogs are of poor quality and tend to drift from one topic to another. Since microblogging platforms are so popular, users post the topics in which they are interested, report events around them, and forward videos related to those topics. Furthermore, a video may contain fragments that address various topics, a phenomenon often called topic drifting [19].
3. **Length of text data**: It is not uniform how long text data is, as the length varies considerably between different mediums. Variation in length of text proclaims challenges in topic detection from cross-media data.
4. **Finer-level cross-modal semantic correlation modeling**: Compared to other data modes, there is relatively little research produced on the relationship between them. It is necessary to design a new model in order to analyze the complex relationships between different types of data.
5. **Scalability on large-scale data**: The emergence of large storage devices, mobile devices, and fast networks has contributed to the generation and distribution of more and more multimedia resources. Multimodal data growth requires the creation of scalable, effective algorithms that can withstand the fast growth of distributed platforms. A more in-depth investigation is needed to analyze how modalities of data can be efficiently and effectively organized.
6. **Data complexity**: Multimodal data has a high degree of complexity due to the significantly different characteristics, such as information capacity, noise levels, data mode, and data structure. Consequently, it is hard to utilize the rich cross-media information comprehensively, since different modalities provide incomparable data and multimodal data from multiple media are inefficiently structured [37].
7. **Topic diversity**: Topics are composed largely of dense clusters of multimodal data with high intra-cluster similarity; however, their distribution of granularity, topics, and noise levels differ widely. In light of such considerable topic diversity, topic detection is becoming increasingly difficult. Traditional clustering techniques, such as k-means and spectral clustering, cannot sort out the unknown number of diversified dense clusters from highly noisy multimodal data [5].
8. **Heterogeneous sources**: Both microblogging networks and associated sharing sites can be used to gather information on blog videos. Messages from blogs and their forwarding, according to the first source, might contain not only a set of words, but also information in several dimensions, such as messages, forwarding and forwarding times, and comments. We can extract textural data or visual information from the second source (for instance, key frames contained in video data). Most contemporary TDT approaches, on the other hand, depend solely on textual or visual content, neglecting the linkages between heterogeneous social media data [19].

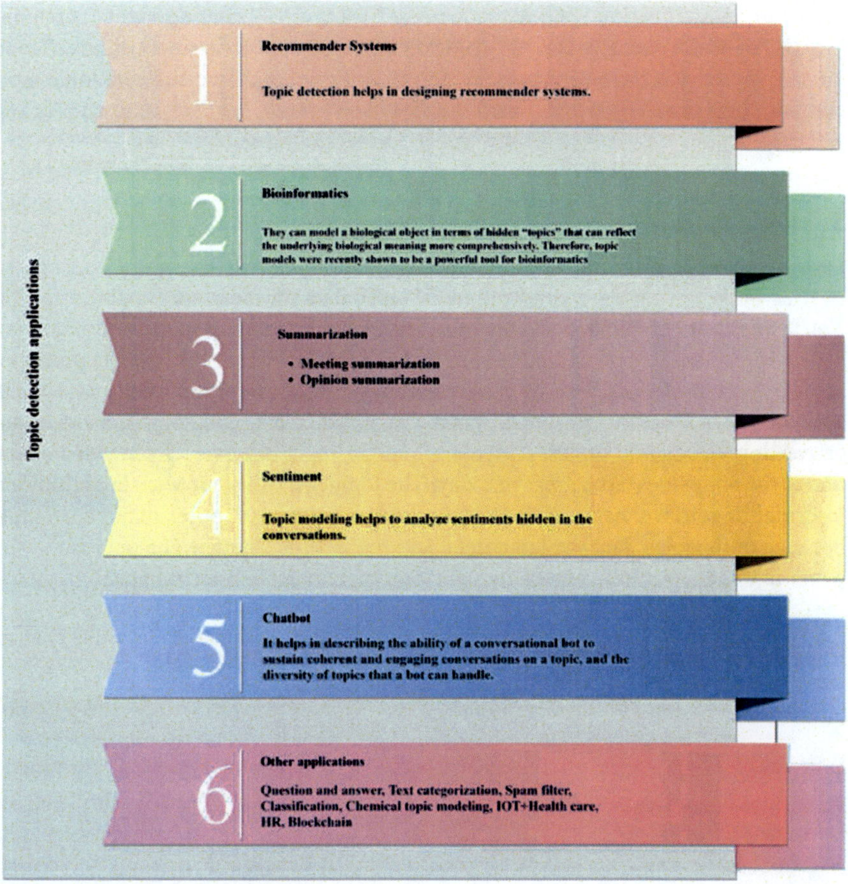

Fig. 3 Topic detection applications

9. **Diversified data structures**: Different mediums' data structures are frequently incomplete because no single medium has all of the available modalities at the same time [5].
10. **Granularity**: Topic granularity (i.e., time duration) varies widely, which is a common problem in topic detection [5].

4 Topic Detection Applications

Topic detection has been useful in various applications such as living lab [42], bioinformatics [17], summarization [11, 14], sentiment analysis [32], chatbot [9], topic tracking [40], question and answer [13],text categorization [10], similarity [39], spam filter [20], classification [21], recommender System [31], chemical topic modeling [35], IOT+health care [7], HR [28], blockchain [6]. Some other applications of topic modeling methods have been presented in [33]. Figure 3 shows different topic modeling applications.

5 Conclusion

In light of the revolution in the volume of video data on social media platforms like YouTube, Twitter, Flickr, and others, it has become very difficult to monitor and access important topics on the Internet. Among the different types of data posted on social media platforms, Web videos are gaining popularity for their rich audio-visual content. All of these factors emphasize the importance of organizing Web videos and converting videos into topics. As a consequence, topic discovery from video data is one of the hottest research areas of today. The topic discovery problem in multimodal information retrieval has been extensively studied, yet it still poses major challenges because of its noise, large volume data, and rapid topic change. The purpose of this paper is to review various cross-media video topic detection techniques.

Acknowledgements This work is supported by the Council of Scientific and Industrial Research (CSIR), Govt. of India, fellowship under grant no. 09/135(0745)/2016-EMR-I.

References

1. Bao, B.K., Xu, C., Min, W., Hossain, M.S.: Cross-platform emerging topic detection and elaboration from multimedia streams. ACM Trans. Multimed. Comput. Commun. Appl. (TOMM) **11**(4), 1–21 (2015)
2. Camargo, J.E., González, F.A.: Multimodal latent topic analysis for image collection summarization. Inform. Sci. **328**, 270–287 (2016)
3. Cao, J., Ngo, C.W., Zhang, Y.D., Li, J.T.: Tracking web video topics: Discovery, visualization, and monitoring. IEEE Trans. Circuit. Syst. Video Technol. **21**(12), 1835–1846 (2011)
4. Chen, T., Liu, C., Huang, Q.: An effective multi-clue fusion approach for web video topic detection. In: Proceedings of the 20th ACM International Conference on Multimedia, pp. 781–784 (2012)
5. Chu, L., Zhang, Y., Li, G., Wang, S., Zhang, W., Huang, Q.: Effective multimodality fusion framework for cross-media topic detection. IEEE Trans. Circuit. Syst. Video Technol. **26**(3), 556–569 (2014)
6. Chung, K., Yoo, H., Choe, D., Jung, H.: Blockchain network based topic mining process for cognitive manufacturing. Wirel. Pers. Commun. **105**(2), 583–597 (2019)
7. Dantu, R., Dissanayake, I., Nerur, S.: Exploratory analysis of internet of things (iot) in healthcare: a topic modelling & co-citation approaches. Inform. Syst. Manage. **38**(1), 62–78 (2021)
8. Gao, Y., Zhao, S., Yang, Y., Chua, T.S.: Multimedia social event detection in microblog. In: International Conference on Multimedia Modeling, pp. 269–281. Springer (2015)
9. Guo, F., Metallinou, A., Khatri, C., Raju, A., Venkatesh, A., Ram, A.: Topic-based evaluation for conversational bots. arXiv preprint arXiv:1801.03622 (2018)

10. Haribhakta, Y., Malgaonkar, A., Kulkarni, P.: Unsupervised topic detection model and its application in text categorization. In: Proceedings of the CUBE International Information Technology Conference, pp. 314–319 (2012)
11. Huang, T.C., Hsieh, C.H., Wang, H.C.: Automatic meeting summarization and topic detection system. Data Technol. Appl. (2018)
12. Kim, D., Kim, D., Jun, S., Rho, S., Hwang, E.: Trendssummary: a platform for retrieving and summarizing trendy multimedia contents. Multimed. Tools Appl. **73**(2), 857–872 (2014)
13. Kim, K., Song, H.J., Moon, N.: Topic modeling for learner question and answer analytics. In: Advanced Multimedia and Ubiquitous Engineering, pp. 652–655. Springer (2017)
14. Ku, L.W., Lee, L.Y., Wu, T.H., Chen, H.H.: Major topic detection and its application to opinion summarization. In: Proceedings of the 28th Annual International ACM SIGIR Conference on Research and Development in Information Retrieval, pp. 627–628 (2005)
15. Lan, Z.Z., Bao, L., Yu, S.I., Liu, W., Hauptmann, A.G.: Double fusion for multimedia event detection. In: International Conference on Multimedia Modeling, pp. 173–185. Springer (2012)
16. Li, W., Joo, J., Qi, H., Zhu, S.C.: Joint image-text news topic detection and tracking by multimodal topic and-or graph. IEEE Trans. Multimed. **19**(2), 367–381 (2016)
17. Liu, L., Tang, L., Dong, W., Yao, S., Zhou, W.: An overview of topic modeling and its current applications in bioinformatics. SpringerPlus **5**(1), 1–22 (2016)
18. Lu, D., Voss, C., Tao, F., Ren, X., Guan, R., Korolov, R., Zhang, T., Wang, D., Li, H., Cassidy, T., et al.: Cross-media event extraction and recommendation. In: Proceedings of the 2016 Conference of the North American Chapter of the Association for Computational Linguistics: Demonstrations, pp. 72–76 (2016)
19. Lu, Z., Lin, Y.R., Huang, X., Xiong, N., Fang, Z.: Visual topic discovering, tracking and summarization from social media streams. Multimed. Tools Appl. **76**(8), 10855–10879 (2017)
20. Ma, J., Zhang, Y., Wang, Z., Yu, K.: A message topic model for multi-grain sms spam filtering. Int. J. Technol. Human Interact. (IJTHI) **12**(2), 83–95 (2016)
21. Ma, J., Zhang, Y., Zhang, L., Yu, K., Liu, J.: Bi-term topic model for sms classification. Int. J. Business Data Commun. Netw. (IJBDCN) **13**(2), 28–40 (2017)
22. Min, W., Bao, B.K., Xu, C., Hossain, M.S.: Cross-platform multi-modal topic modeling for personalized inter-platform recommendation. IEEE Trans. Multimed. **17**(10), 1787–1801 (2015)
23. Pang, J., Hu, A., Huang, Q., Tian, Q., Yin, B.: Increasing interpretation of web topic detection via prototype learning from sparse poisson deconvolution. IEEE Trans. Cybern. **49**(3), 1072–1083 (2018)
24. Pang, J., Jia, F., Zhang, C., Zhang, W., Huang, Q., Yin, B.: Unsupervised web topic detection using a ranked clustering-like pattern across similarity cascades. IEEE Trans. Multimed. **17**(6), 843–853 (2015)
25. Pang, J., Tao, F., Huang, Q., Tian, Q., Yin, B.: Two birds with one stone: A coupled poisson deconvolution for detecting and describing topics from multimodal web data. IEEE Trans. Neural Netw. Learn. Syst. **30**(8), 2397–2409 (2018)
26. Pang, J., Tao, F., Li, L., Huang, Q., Yin, B., Tian, Q.: A two-step approach to describing web topics via probable keywords and prototype images from background-removed similarities. Neurocomputing **275**, 478–487 (2018)
27. Pang, J., Tao, F., Zhang, C., Zhang, W., Huang, Q., Yin, B.: Robust latent poisson deconvolution from multiple features for web topic detection. IEEE Trans. Multimed. **18**(12), 2482–2493 (2016)
28. Peters, N.S., Bradley, G.C., Marshall-Bradley, T.: Task boundary inference via topic modeling to predict interruption timings for human-machine teaming. In: International Conference on Intelligent Human Systems Integration, pp. 783–788. Springer (2019)
29. Petkos, G., Papadopoulos, S., Schinas, E., Kompatsiaris, Y.: Graph-based multimodal clustering for social event detection in large collections of images. In: International Conference on Multimedia Modeling, pp. 146–158. Springer (2014)
30. Qian, S., Zhang, T., Xu, C., Shao, J.: Multi-modal event topic model for social event analysis. IEEE Trans. Multimed. **18**(2), 233–246 (2015)

31. Qiu, J., Liao, L., Li, P.: News recommender system based on topic detection and tracking. In: International Conference on Rough Sets and Knowledge Technology, pp. 690–697. Springer (2009)
32. Rana, T.A., Cheah, Y.N., Letchmunan, S.: Topic modeling in sentiment analysis: A systematic review. J. ICT Res. Appl. **10**(1) (2016)
33. Rani, S., Kumar, M.: Topic modeling and its applications in materials science and engineering. Mater. Today: Proc. **45**, 5591–5596 (2021)
34. Schinas, M., Papadopoulos, S., Kompatsiaris, Y., Mitkas, P.A.: Mgraph: multimodal event summarization in social media using topic models and graph-based ranking. Int. J. Multimed. Inform. Retrieval **5**(1), 51–69 (2016)
35. Schneider, N., Fechner, N., Landrum, G.A., Stiefl, N.: Chemical topic modeling: Exploring molecular data sets using a common text-mining approach. J. Chem. Inform. Model. **57**(8), 1816–1831 (2017)
36. Schreck, T., Keim, D.: Visual analysis of social media data. Computer **46**(5), 68–75 (2012)
37. Shao, J., Ma, S., Lu, W., Zhuang, Y.: A unified framework for web video topic discovery and visualization. Pattern Recogn. Lett. **33**(4), 410–419 (2012)
38. Shi, L., Luo, J., Cheng, G., Liu, X., Xie, G.: A multifeature complementary attention mechanism for image topic representation in social networks. Sci. Program. **2021** (2021)
39. Spina, D., Gonzalo, J., Amigó, E.: Learning similarity functions for topic detection in online reputation monitoring. In: Proceedings of the 37th International ACM SIGIR Conference on Research & Development in Information Retrieval, pp. 527–536 (2014)
40. Tu, D., Chen, L., Lv, M., Shi, H., Chen, G.: Hierarchical online nmf for detecting and tracking topic hierarchies in a text stream. Pattern Recogn. **76**, 203–214 (2018)
41. Wang, Z., Li, L., Huang, Q.: Cross-media topic detection with refined cnn based image-dominant topic model. In: Proceedings of the 23rd ACM International Conference on Multimedia, pp. 1171–1174 (2015)
42. Westerlund, M., Leminen, S., Rajahonka, M.: A topic modelling analysis of living labs research. Technol. Innov. Manage. Rev. **8**(7) (2018)
43. Wu, X., Lu, Y.J., Peng, Q., Ngo, C.W.: Mining event structures from web videos. IEEE Multimed. **18**(1), 38–51 (2011)
44. Xue, Z., Jiang, S., Li, G., Huang, Q., Zhang, W.: Cross-media topic detection associated with hot search queries. In: Proceedings of the Fifth International Conference on Internet Multimedia Computing and Service, pp. 403–406 (2013)
45. Xue, Z., Li, G., Zhang, W., Pang, J., Huang, Q.: Topic detection in cross-media: a semi-supervised co-clustering approach. Int. J. Multimed. Inform. Retriev. **3**(3), 193–205 (2014)
46. Zhang, W., Chen, T., Li, G., Pang, J., Huang, Q., Gao, W.: Fusing cross-media for topic detection by dense keyword groups. Neurocomputing **169**, 169–179 (2015)

Water Salinity Assessment Using Remotely Sensed Images—A Comprehensive Survey

R. Priyadarshini, B. Sudhakara, S. Sowmya Kamath, Shrutilipi Bhattacharjee, U. Pruthviraj, and K. V. Gangadharan

Abstract In the past few years, the problem of growing salinity in river estuaries has directly impacted living and health conditions, as well as agricultural activities globally, especially for those rivers which are the sources of daily water consumption for the surrounding community. Key contributing factors include hazardous industrial wastes, residential and urban wastewater, fish hatchery, hospital sewage, and high tidal levels. Conventional survey and sampling-based approaches for water quality assessment are often difficult to undertake on a large-scale basis and are also labor and cost-intensive. On the other hand, remote sensing-based techniques can be a good alternative to cost-prohibitive traditional practices. In this article, an attempt is made to comprehensively assess various approaches, datasets, and models for determining water salinity using remote sensing-based approaches and in situ observations. Our work revealed that remote sensing techniques coupled with other techniques for

R. Priyadarshini and B. Sudhakara—Equal contribution.

R. Priyadarshini (✉) · B. Sudhakara · S. Sowmya Kamath · S. Bhattacharjee
Department of Information Technology, National Institute of Technology Karnataka, Surathkal, Mangalore 575025, India
e-mail: priya.217it003@nitk.edu.in

B. Sudhakara
e-mail: sudhakarab.217it006@nitk.edu.in

S. Sowmya Kamath
e-mail: sowmyakamath@nitk.edu.in

S. Bhattacharjee
e-mail: shrutilipi@nitk.edu.in

U. Pruthviraj
Department of Water and Ocean Engineering, National Institute of Technology Karnataka, Surathkal, Mangalore 575025, India
e-mail: pruthviu@nitk.edu.in

K. V. Gangadharan
Department of Mechanical Engineering, National Institute of Technology Karnataka, Surathkal, Mangalore 575025, India
e-mail: kvganga@nitk.edu.in

M. Tistarelli et al. (eds.), *Computer Vision and Machine Intelligence*,
Lecture Notes in Networks and Systems 586,
https://doi.org/10.1007/978-981-19-7867-8_46

estimating the salinity of water offer a clear advantage over traditional practices and also is very cost-effective. We also highlight several observations and gaps that can be beneficial for the research community to contribute further in this significant research domain.

Keywords Water salinity · Remote sensing · Regression · Data analytics · Machine learning

1 Introduction

Due to extensive encroachment of human settlements near water bodies across the world, water pollution and related adverse climatic conditions are already rampant. Majorly caused by urbanization, these effects can also be attributed to rapid increase in industrial establishments like product-related industries, fuel/power generation industries, etc., inefficient waste disposal practices, accidental oil spills, etc. Non-sustainable and unscientific water management practices have caused considerable changes in water characteristics, resulting in critical challenges in water quality and biodiversity management. Actively monitoring water quality can help in understanding natural causative processes in the target ecosystem, for assessing the effect of human activities for setting up restorative measures for addressing the degradation.

Water quality is affected by several physical, chemical, and biological parameters, such as the extent of suspended solids, sedimentation, and microbial presence. A critical parameter of interest is salinity, a measure of dissolved salts in the water, typically expressed in parts per million (ppm). Increase in salt levels in the water body can have severe consequences on farming activities, quality of potable water, and other undertakings. Increased concentration of salt requires extensive filtration and processing to make it suitable for human and animal consumption, which is often cost-prohibitive. Variation in salinity levels also affects soil quality and is also known to alter the features of sea water [16]. Many natural parameters like sea surface temperature, power of hydrogen (pH) value of water, chlorophyll, etc., also majorly affect water quality.

The spatio-temporal coverage of large bodies of water has inherent limitations, as measurement mechanisms such as specialized vessels and buoys, are often costly. Remote sensing images from satellites offer a cost-effective alternative for assessing surface salinity, with advantages like temporal/spatial imaging, large spatial and more frequent temporal coverage, etc. Understandably, assessing salinity from remote sensing images has seen increasing research interest over the years. However, the resolution is often too coarse to monitor sea surface salinity in coastal and estuary zones. Sea surface salinity (SSS) can be tracked with the help of optical remote sensing, using the linear relationships between Landsat Multispectral Scanner (MSS) bands and SSS. Recent Landsat satellites, such as Landsat 8 data, provide several new capabilities compared to other datasets, such as adding a few shorter wavelengths like blue band, ultra-blue, and narrower near-infrared (NIR) band, 12-bit radiometric resolution, and more significant signal-to-noise ratios. This can help to monitor high clarity water bodies, with low reflectance in bands such as blue and red. Similar improvements in the OLI sensor of Landsat 8 also help improve measurement of other water quality parameters, like colored dissolved organic matter, which plays an essential role in the significant fraction of dissolved organic matter (DOM) in many natural waters.

Detecting and tracking salinity levels both in the short and long term is a challenging task, making it a slow-onset threat [25]. In contrast to remote sensing approaches, in situ methods for determining water salinity take much time and are also highly expensive. Several other challenges persist as the salinity assessment is to be simulated using remote sensing, and relevant satellite, water salinity-sensitive spectral band parameters, and appropriate modeling algorithms should be chosen [1]. In this article, an exhaustive survey of different approaches such as empirical analysis, machine learning, and deep learning techniques for calculating salinity of water

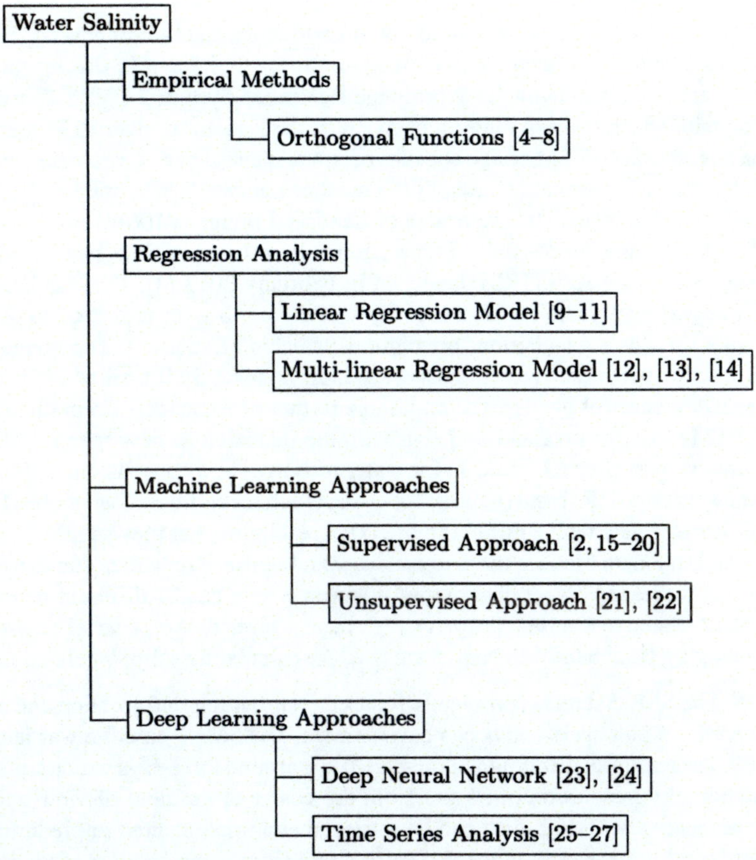

Fig. 1 Summary of water salinity assessment survey

from remotely sensed images is presented and a summary of the work is illustrated in the following Fig. 1.

The remainder of this manuscript is presented as follows: Section 2 presents a detailed view of the availability of remote sensing datasets and other data sources relevant to the salinity assessment. In Sect. 3, we discuss various water salinity prediction techniques that have been commonly used. Section 4 discusses the challenges in retrieving water salinity from remotely sensed images, followed by conclusions.

2 Data Sources

The data sources used for salinity assessment can be of wide variety, including satellite images from Landsat, Soil Moisture and Ocean Salinity (SMOS), and Soil Moisture Active Passive (SMAP) missions, each of which are discussed in detail in this section.

Landsat. Some of the most commonly used Landsat datasets are Landsat 4, 5, 7, and Landsat 8. Landsat 5 collects surface images of the earth; it has a 16-day repeat cycle, and data are collected using the Worldwide Reference System-2 (WRS-2) path/row system, with swath overlap ranging from 7% at the Equator to over 85% at extreme latitudes. Landsat 4–5 TM image consists of seven spectral bands with a resolution of 30 m. The band 6, thermal infrared (TIR) band of Landsat 5, was collected at 120 m and resampled to 30 m. The scene size of Landsat 5 image is 106 mi × 114 mi [20]. The Landsat 8 mission consists of Operational Land Imager (OLI) instruments and thermal infrared sensor (TIRS), launched in February 2013 [4]. This satellite has a 16-day repeat cycle, and it collects images of the earth with reference to WRS-2. Landsat 8 satellite's acquisitions are eight days behind Landsat 7. The image spans 170 km north-south and 183 km east-west as in Landsat 5 (106 mi × 115 mi). The OLI sensor's spectral bands are comparable to that of Enhanced Thematic Mapper Plus (ETM+) sensor on Landsat 7, but with the inclusion of new spectral bands: a deep blue visible channel (band 1) for water resources and coastal zone exploration and a new infrared (IR) band (band 9) for cirrus cloud recognition. Two thermal bands (TIRS) acquire data with a minimum of 100 m resolution, but they supply 30 m OLI data. Landsat 8 data files are more extensive than Landsat 7 data files due to the extra bands and increased 16-bit data output. Landsat 8 data has 11 different bands, and each band related to a certain property, e.g., band 1 gives data on coastal aerosol, and bands ranging from band 1 to band 5 are used for calculating salinity-related indices.

SMOS. The SMOS data is completely focused on providing information and insight about earth's water cycle and ocean environment. The SMOS satellite was launched in 2009. The temporal resolution of this satellite is around three days which has spatial and radiometric resolution of 35 km from the center of the field of view and 0.8–2.2 k, respectively. SMOS data can be retrieved and implemented in predicting the climatic conditions, based on the self-retrieving values using forward modeling and neural network models [11]. Existing approaches include SMOS data for forecasting the soil variations, retrieving salinity, and other oceanic parameters based on latitude and longitude values [15].

SMAP. SMAP is primarily used to record soil moisture and to find water-related properties of water bodies and soil information about the land area covered [10]. SMAP has a spatial resolution of 1–3 km; a data range from level 1 (L1) to level 4 (L4) and a temporal resolution of 49 min. SMAP has been used extensively for assessing Sea Surface Salinity (SSS) along with in situ data for validation [31].

In Situ Measurements. In situ data are commonly used for estimating water quality parameters, including salinity, or for validating satellite measurements and model output. The salinity is measured using physical instruments like a refractometer, which is placed in the selected physical location and based on its recorded value with respect to the water quality; the parameters are measured. So far, most existing techniques considered respective in situ measurements to either compare original data or test results. These help in retrieving actual salinity values of the ocean directly using manual experiments. Several works have been proposed, especially for salinity, where validating of satellite retrieved data is of primary concern. The satellite data of SMAP have been used for salinity assessment along with the in situ measurements, revealing that SMAP L band data provide high accuracy [30].

Preprocessing. Preprocessing techniques also play a major role in improving performance of approaches. As the images downloaded from any satellite source exhibit high deformation, cloud cover, etc., preprocessing is highly recommended before ecological experiments. One of the significant steps involved is geometric correction that includes pixel location correction and direction correction of elements present in the image [12]. Next comes the radiometric correction, which includes surface reflectance, brightness correction, sensor corrections, etc. This radiometric correction is a must for cases where images are required to be compared for any results or predictions. Topographic correction, atmospheric correction, solar correction, and preprocessing for spectral indices based on digital number (DN) are some additional techniques that are popular. Before applying preprocessing techniques, it is essential to know the type of data to be worked upon, like whether it is a shortwave infrared or thermal image, as preprocessing requirements vary significantly based on the type of data.

3 Water Salinity Prediction Approaches

Water salinity assessment problem has been addressed using techniques like physical experiments, survey data-based hydrological modeling, and data-driven prediction models. However, most methods require a large amount of input data and have a high computing demand, hence, are not always desirable. As an alternative, empirical, data analytics-based, and AI-based methods have been explored, since they demand less data and less processing time while producing equivalent results. They are broadly categorized into—(1) Empirical methods (2) Regression-based methods (3) Machine learning-based methods, and (4) Deep learning-based methods. We discuss state-of-the-art works in these areas in the subsequent sections.

3.1 Empirical Methods

Empirical methods are used when direct or indirect data points with reference to the target environment are available through observation or experimentation. Methods which use empirical evidences like in situ measurements in the case of the salinity prediction task fall in this category. These approaches help validate the results achieved with the measurements. We discuss existing research work that can be classified as empirical methods in this section.

Alvera et al. [3] used data interpolating empirical orthogonal functions (DINEOFs) to reconstruct missing salinity values in SMOS data. DINEOF is applied for calculating daily sea surface salinity values with very less noise and error; hence, reconstruction of any missing data in SMOS can be achieved. With EOF, the data mean is calculated, and the missing values are replaced, which can help in detailed analysis with augmented data. Zhang et al. [28] experimented with Ocean Reanalysis System (ORAS4) datasets for finding the El Nino events that occur in the tropical pacific with the help of Sea Surface Temperature (SST) and the Sea Surface Salinity (SSS) values. Basically, a comparison between two events in the Central Pacific and eastern Pacific regions of the Pacific ocean was undertaken, using index-based anomaly analysis of SST and SSS.

Agarwal et al. [2] proposed a method for retrieving salinity parameters with the help of SMOS data in the Indian Ocean. They formulated a retrieval algorithm with mean salinity, and three other principal component values and Argo values were used for comparing the values retrieved. Finally, root mean square error (RMSE), coefficient of determination R^2 is calculated, and their correctness is measured. Casagrande et al. [8] proposed an approach based on empirical orthogonal functions (EOFs) to model wave effects, where salinity and sea surface temperature act as important parameters. Decomposition of data variability in the pattern of the data is also attempted. The authors used the Medwin formula with salinity, sea surface temperature and depth as input for their assessment. Ouyang et al. [27] measured SSS using the ORAS5 dataset, using the space-time variations in SSS. EOF techniques were used for determining mixed-layer salinity, and regional differences in SSS for the Tropical Pacific Ocean were predicted. The results are plotted on an annual basis from 1980 to 2015 considering the changes that have occurred in SSS.

3.2 Regression Analysis for Salinity Prediction

In recent studies, regression models have been used for the task of salinity prediction. Regression analysis encompasses a set of statistical techniques for estimating a continuous result variable (y) depending on the values of one or more observed variables (x). It has been extensively used for identifying the independent variable that is highly related to the dependent variable. The salinity parameter acts as dependent variable, while a few salinity index-based variables are used as independent variables, for finding the most related parameter.

Ansari and Akhoondzadeh [4] conducted a study of the Karun river basin of Iran for mapping water salinity using Landsat 8 satellite images and electrical conductivity (EC) samples. After applying radiometric and atmosphere corrections, the water body pixels are extracted using an NDVI mask image. Next, Sobel's sensitivity analysis is to used to determine the best combination of bands in the Landsat image. They used a regression model to predict the salinity of water by establishing a relationship between the reflectance of the Landsat 8 images and in situ measured data. Also, methods such as ordinary least square (OLS), multilayer perceptron (MLP) with genetic algorithm, and support vector regression (SVR) methods were applied to predict the salinity. An empirical model developed by Ferdous and Rahman [13] measures the salinity of water in coastal Bangladesh from Landsat 5 TM, Landsat 8 OLI, and sample EC values. Landsat images are preprocessed using radiometric and atmospheric corrections. The salinity EC data of surface water from 74 sampling locations is added to the ArcGIS platform's attribute table. A total of 13 band compositions are formed using red, green, and blue bands, and the coefficient of determination R^2 is calculated using EC data. Next, multiple regression analysis is performed for both Landsat sensors image to get the equation for finding water salinity. This study successfully detects salinity with 82 and 76% accuracy using Landsat 5 Thematic Mapper and Landsat 8 OLI images, respectively.

Wang and Xu [32] designed a regression model for measuring the close relationship between salinity and water reflectance from different wavelengths, using eight cloud-free images of Landsat 5 TM and also in situ measurements. First, the noise and cloud cover was removed using Fourier transforms; then, the radiometric corrections were applied. Later, OLS regression is applied for in situ data and reflectance data derived from Landsat image. Analysis of spatial and temporal change in salinity is also carried out with the help of a prediction model in the whole area of Pontchartrain Lake on the eight dates. Analysis of variance (ANOVA) was used to analyze the change in average salinity values over the whole lake between the eight different dates for which images are acquired. Their model showed that the reflectance of 1st, 2nd, and 4th bands' is positively correlated with salinity levels, while that of the 3rd and 5th bands are correlated negatively. Yang et al. [33] present a method to retrieve salinity profiles from SMOS data using linear, nonlinear transfer functions and neural networks. Root mean square error (RMSE) is calculated for both; two types of validation are performed here, one using monthly grid data and another using real-time data.

Zhao et al. [34] retrieved the salinity of the Arabian gulf using an empirical algorithm built on Landsat 8 data. The algorithm extracts the salinity of water using in situ measurements and achieved a 70% coefficient of determination R^2 using a multivariate regression model. Maliki et al. [19] estimated total dissolved solids in river water body using spectral indices of Landsat 8 satellite data. Six sampling locations were chosen along the river to assess the total dissolved solids (TDSs) concentration. Three Landsat 8 images were downloaded by matching the temporal resolution of gathered samples to determine the river's salinity. Atmospheric corrections and extraction of spectral characteristics were applied to the satellite images as preprocessing operations. Regression and correlation analysis was performed to identify the correct model to detect water salinity based on the spectral reflectance of Landsat 8 OLI.

3.3 Machine Learning-Based Approaches

The study on remote sensing techniques to predict salinity intrusion in an attempt to build a numerical model to find salinity intrusion has been explored by many researchers. Due to wide applicability of learning-based models, ML algorithms have been adopted for varied problems like predicting the levels of groundwater, mapping potential recharge of groundwater, assessing vulnerability of groundwater, and predicting quality of groundwater.

Nguyen et al. [25] proposed an approach for assessing the correlation between Landsat 8 OLI image reflectance and in situ measurements. A total of 103 recorded samples are split into two halves for training and testing. Techniques such as decision trees (DTs), random forest (RF), and multiple linear regression (MLR) approaches were used in this study. The reflectance and salinity data at the location are merged as stepwise model factors, revealing that salinity had a favorable association. However, the limited size of the data poses a substantial difficulty, adversely affecting the performance. Melesse et al. [22] proposed a model for predicting river water salinity for the Babol-Rood River in Iran. ML algorithms like M5Prime, random forest, and a hybrid of eight different combinations were used for the same. The input variables were also of different varieties, and various water parameters are considered in which total dissolved solids (TDSs) are regarded as prime input. Bayati et al. [5] used both sentinel 2 and Landsat images for finding or tracking the changes in the water surface of Lake Urmia. Artificial neural network (ANN) and adaptive network-based fuzzy inference system (ANFIS) are used for finding the actual relationship between the water surface and reflectance, followed by the salinity parameter. For ANFIS, a Takagi-Sugeno-Kang (TSK) system with various inputs and rules is designed, producing a nonlinear relationship between the variables. Then, a multilinear regression model is derived for the dependent variables. The best band selection process for Landsat and Sentinel satellite data is also observed.

Muller et al. [24] proposed an algorithm for mapping surface water across Australia with the help of Landsat time series images over a period of 25 years. They used the regression tree classifier for finding the intermittent water bodies, while also testing it using the spectral bands, normalized difference ratio, and combination of both. The difference in water is calculated based on quality indices, including salinity as one of its parameters. Therefore, for each band, different spectral indices are calculated, and based on its value, they are classified. Ranhotra et al. [29] experimented with a different dataset called Blackbridge. Specific preprocessing techniques like background subtraction, object separation, rudimentary morphological process, and texture analysis are applied. Texture analysis is performed to detect the idiosyncratic characteristics of the images; the luminance threshold is fixed, and RGB color is converted into YCbCr. Several parameters like pH value, chemical factors, temperature, etc., in the water bodies were checked by comparing the color of the resultant image.

Liu et al. [18] proposed an integration technique for predicting salinity in coastal regions with the help of in situ measurements. They used Pearson's correlation for retrieving the parameters like pH value, sea surface temperature, total inorganic nitrogen, etc., which contributes to salinity levels. Then, spatial interpolation between the values is carried out using the ordinary kriging technique, to enable the sea surface salinity prediction using the random forest algorithm. Olmedo et al. [26] attempted to improve retrieved SMOS salinity values with the help of multifractals. They showed that salinity and temperature are related, and the changes that can happen as one of these parameters vary. They focused on improving the temporal and spatial resolution of the already retrieved salinity indices from SMOS. Biguino et al. [6] presented a method for evaluating the salinity indices from SMOS L4 data with recorded in situ measurements. They performed statistical analysis where RMSE, absolute percentage difference, relative percentage difference between the physically measured value (in situ), and satellite retrieved value are calculated. Next, a matchup analysis is carried out between the in situ and satellite data. Several factors related to error are calculated, and any links between biochemical variables and salinity are checked. Daniels et al. [9] used k-nearest neighbors, neural networks, and decision trees for predicting water quality parameters. Here, the input parameters like pH, salinity, temperature, conductivity, dissolved oxygen, and time are used for training and testing the model.

3.4 Deep Learning Approaches

In the last decade, deep neural models have garnered significant research interest and have been adapted for water quality studies. They are well-suited for exploring latent patterns of interest in the data and other unknown relationships between variables to be modeled, as they can automatically learn such insights from input data. They also offer significant advantages in terms of learnability and domain adaption when large quantities of remote sensing data are available, while being better attuned to handle any missing data. Furthermore, DL models are capable of handling unstructured data across different data sources and can achieve better accuracy.

Liu et al. [17] proposed a model for retrieving sea surface salinity utilizing sea parameters, such as pH values, total inorganic nitrogen, chlorophyll, and sea surface temperature. They implemented a backpropagation model with four input layers and one output layer for predicting salinity values. The random forest algorithm with the sigmoid tangent activation function with a learning rate of 0.2 to 0.15 performed well. Nardelli et al. [7] proposed a method for interpolation of SMOS data with in situ measurements using optimal interpolation techniques, specifically for sea surface salinity. This approach is further validated with in situ data and wavenumber spectral analysis, based on which they reported promising results.

Matsuoka et al. [21] formulated a new algorithm for detecting water bodies from space using MODIS and SMOS L2 salinity data. The salinity parameter is retrieved using the surface model, a two-scale model. A mass balance metric is applied between salinity and other satellite retrieved parameters for discrimination. The variability of sea surface salinity in SMOS is also retrieved. Meng et al. [23] proposed a convolutional neural network (CNN) model for reconstruction of anomalies such as salinity, temperature, and other subsurface parameters. They adapted single shot (SS) and ensemble (EN) models trained on SMOS data for comparing their prediction performance with subsurface salinity anomaly (SSA) and temperature. Jin et al. [14] attempted to predict three different ocean parameters such as salinity, temperature, and flow fields using deep learning models. The convolutional long short-term memory (LSTM) model trains the optimization algorithm to predict the three required parameters.

4 Discussion

Based on the comprehensive review to understand the scope of water salinity assessment problem using remote sensing techniques, several insights and challenges are observed. A significant volume of research work focuses on retrieval of salinity and dependent parameters like temperature, pH values, etc., to utilize them to predict salinity levels. Existing approaches also employed a wide variety of data, spanning physical, chemical, biological, spatial and other modalities of observatory data and have adapted both parametric and non-parametric methods for the task. Approaches include pure hydrology-based modeling, empirical methods, and regression modeled on Landsat data. Recent advancements have been in the adaptation of ML and DL models for exploiting greater learning and generalizability.

We noted that most researchers employed machine learning models and regression analysis for the salinity prediction, whereas deep learning models have helped achieve improved results. However, the major challenge is the limited data availability as DL models require larger datasets for better generalizability. The possible solution would be to either use more time series data or use augmentation techniques to create more samples.

Some observed gaps and potential solutions for the observed challenges are listed below.

- Most predictions are validated on in situ data, but it is not easy to acquire ground truth for all regions. Thus, available salinity data can be used for training ML classifiers to predict the salinity of water bodies, along with the knowledge captured from relevant satellite data. Thus, a combination of multiple such data sources can be used for validation, even when in situ data are not available.
- Global salinity predictions can be performed by leveraging data obtained from different satellites and designing techniques for comparative assessments.
- Choosing specific regions or stations for the prediction instead of entire areas has been reported to be more advantageous, as empirical analysis-based techniques can be applied. In most existing works, authors considered either the whole Indian Ocean or the Pacific ocean. This is a very vast region, and local characteristics will be overlooked entirely; hence, the accuracy of salinity prediction is significantly affected.
- For data sources that provide pre-computed salinity values for a specific region of interest, data validations can be explored with available ground truth/in situ measurements with different applications.
- Recently, other techniques like texture analysis have been explored to assess salinity parameters, but the results are not as expected. Very few works have used such methods; thus, a detailed study of different patterns and surface textures of satellite images can be undertaken for analyzing other water salinity parameters.

5 Concluding Remarks

A detailed review of techniques that leverage remote sensing data for the task of water salinity assessment was presented in this article. The Landsat 8 OLI provides a simple way to enable the prediction of salinity intrusion, and many studies reported that Band 1–6 plays an important role in designing an accurate salinity prediction model. The most important requirement is to increase the target area sample size by covering a larger area of interest with the help of remote sensing techniques to understand better the association between salinity levels and reflectance wavelength from satellite images. Several challenges exist, as very few measurements are taken during OLI acquisition times due to the satellite's 16-day revisit frequency and the requirement for cloud-free conditions for satellite measurements, especially in tropical areas. Also, seasonal and temperature changes in the Landsat 8 OLI image reflectance need to be addressed, by integrating findings from extensive statistical analysis with the limited data resources for geographically diverse locations.

Acknowledgements The authors gratefully acknowledge the computational resources made available as part of the AI for Earth Grant funded by Microsoft.

References

1. Abdelmalik, K.: Role of statistical remote sensing for inland water quality parameters prediction. Egypt. J. Rem. Sens. Space Sci. **21**(2), 193–200 (2018)
2. Agarwal, N., Sharma, R., Basu, S., Agarwal, V.K.: Derivation of salinity profiles in the indian ocean from satellite surface observations. IEEE Geosci. Remote Sens. Lett. **4**(2) (2007)
3. Alvera-Azcrate, A., Barth, A., Parard, G., Beckers, J.M.: Analysis of smos sea surface salinity data using dineof. Rem. Sens. Environ. 180 (2016)
4. Ansari, M., Akhoondzadeh, M.: Mapping water salinity using landsat-8 oli satellite images (case study: Karun basin located in iran). Adv. Space Res. **65**(5), 1490–1502 (2020)
5. Bayati, M., Danesh-Yazdi, M.: Mapping the spatiotemporal variability of salinity in the hypersaline lake urmia using sentinel-2 and landsat-8 imagery. J. Hydrol. **595**, 126032 (2021)
6. Biguino, B., Olmedo, E., Ferreira, A., Zacarias, N., et al.: Evaluation of smos l4 sea surface salinity product in the western iberian coast. Remote Sens. **14**(2) (2022)
7. Buongiorno Nardelli, B., Droghei, R., Santoleri, R.: Multi-dimensional interpolation of smos sea surface salinity with surface temperature and in situ salinity data. Remote Sens. Environ. **180**, 392–402 (2016)
8. Casagrande, G., Stephan, Y., Warn Varnas, A.C., Folegot, T.: A novel empirical orthogonal function (eof)-based methodology to study the internal wave effects on acoustic propagation. IEEE J. Oceanic Eng. **36**(4), 745–759 (2011)
9. Daniels, A., Koutsougeras, C.: Predicting Water Quality Parameters in Lake Pontchartrain Using Machine Learning: A Comparison on K-Nearest Neighbors, Decision Trees, and Neural Networks to Predict Water Quality. ACM (2021)
10. Das, N.N., Entekhabi, D., Njoku, E.G.: An algorithm for merging smap radiometer and radar data for high-resolution soil-moisture retrieval. IEEE Trans. Geosci. Remote Sens. **49**(5), 1504–1512 (2011)
11. De Rosnay, P., Calvet, J.C., Kerr, Y., et al.: Smosrex: a long term field campaign experiment for soil moisture and land surface processes remote sensing. Remote Sens. Environ. 102(3-4) (2006)
12. Devaraj, C., Shah, C.A.: Automated geometric correction of landsat mss l1g imagery. IEEE Geosci. Remote Sens. Lett. **11**(1), 347–351 (2014)
13. Ferdous, J., Rahman, M.T.U.: Developing an empirical model from landsat data series for monitoring water salinity in coastal bangladesh. J. Environ. Manage. **255**, 109861 (2020)
14. Jin, Q., Tian, Y., Sang, Q., Liu, S., et al.: A deep learning model for joint prediction of three-dimensional ocean temperature, salinity and flow fields. In: 2021 6th International Conference on Automation, Control and Robotics Engineering (CACRE), pp. 573–577 (2021)
15. Kerr, Y., Philippe, W., Wigneron, J.P., et al.: The smos mission: new tool for monitoring key elements of the global water cycle. Proc. IEEE **98** (2010)
16. Lagerloef, G.S., Swift, C.T., Le Vine, D.M.: Sea surface salinity: the next remote sensing challenge. Oceanography **8**(2), 44–50 (1995)
17. Liu, M., Liu, X., Jiang, J., Xia, X.: Artificial neural network and random forest approaches for modeling of sea surface salinity. Int. J. Remote Sens. Appl. **3** (2013)
18. Liu, M., Liu, X., Liu, D., Ding, C., Jiang, J.: Multivariable integration method for estimating sea surface salinity in coastal waters from in situ data and remotely sensed data using random forest algorithm. Comput. Geosci. **75** (2015)
19. Maliki, A.A., Chabuk, A., Sultan, M.A., et al.: Estimation of total dissolved solids in water bodies by spectral indices case study: Shatt al-arab river. Water Air Soil Pollut. **231**(9) (2020)
20. Markham, B.L., Storey, J.C., Williams, D.L., Irons, J.R.: Landsat sensor performance: history and current status. IEEE Trans. Geosci. Remote Sens. **42**(12), 2691–2694 (2004)
21. Matsuoka, A., Babin, M., Devred, E.C.: A new algorithm for discriminating water sources from space: a case study for the southern beaufort sea using modis ocean color and smos salinity data. Remote Sens. Environ. **184**, 124–138 (2016)
22. Melesse, A.M., Khosravi, K., Tiefenbacher, J.P., et al.: River water salinity prediction using hybrid machine learning models. Water **12**(10) (2020)

23. Meng, L., Yan, C., Zhuang, W., et al.: Reconstructing high-resolution ocean subsurface and interior temperature and salinity anomalies from satellite observations. IEEE Trans. Geosci. Remote Sens. **60**, 1–14 (2022)
24. Mueller, N., Lewis, A., Roberts, D., Ring, S., et al.: Water observations from space: Mapping surface water from 25 years of landsat imagery across australia. Remote Sens. Environ. **174** (2016)
25. Nguyen, P.T., Koedsin, W., McNeil, D., Van, T.P.: Remote sensing techniques to predict salinity intrusion: application for a data-poor area of the coastal mekong delta, vietnam. Int. J. Rem. Sens. **39**(20), 6676–6691 (2018)
26. Olmedo, E., Martnez, J., Umbert, M., Hoareau, N., et al.: Improving time and space resolution of smos salinity maps using multifractal fusion. Remote Sens. Environ. **180**, 246–263 (2016)
27. Ouyang, Y., Zhang, Y., Chi, J., Sun, Q., Du, Y.: Regional difference of sea surface salinity variations in the western tropical pacific. J. Oceanogr. **77**, 647–657 (2021)
28. Qi, J., Zhang, L., Qu, T., et al.: Salinity variability in the tropical pacific during the central-pacific and eastern-pacific el nio events. J. Mar. Syst. **199**, 103225 (2019)
29. Ranhotra, S.S.: Detection of Salinity of Sea Water Using Image Processing Techniques, pp. 76–81 (2014)
30. Tang, W., Fore, A., Yueh, S., Lee, T., Hayashi, A., Sanchez-Franks, A., Baranowski, D.: Validating smap sss with in situ measurements. In: 2017 IEEE International Geoscience and Remote Sensing Symposium (IGARSS), pp. 2561–2564 (2017)
31. Tang, W., Fore, A., Yueh, S., Lee, T., et al.: Validating smap sss with in situ measurements. Remote Sens. Environ. **200**, 326–340 (2017)
32. Wang, F., Xu, Y.J.: Development and application of a remote sensing-based salinity prediction model for a large estuarine lake in the us gulf of mexico coast. J. Hydrol. **360**(1–4), 184–194 (2008)
33. Yang, T., Chen, Z.Z., He, Y.: A new method to retrieve salinity profiles from sea surface salinity observed by smos satellite. Acta Oceanologica Sinica 34 (2015)
34. Zhao, J., Temimi, M.: An empirical algorithm for retreiving salinity in the arabian gulf: application to landsat-8 data. In: 2016 IEEE International Geoscience and Remote Sensing Symposium (IGARSS), pp. 4645–4648. IEEE (2016)

Domain Adaptation: A Survey

Ashly Ajith and G. Gopakumar

Abstract In computer vision, domain shifts are a typical issue. A classifier that has been trained on a source domain will not be able to perform well on a target domain. As a result, a source classifier taught to discriminate based on a particular distribution will struggle to classify new data from a different distribution. Domain adaptation is a hot area of research due to the plethora of applications available from this technique. Many developments have been made in this direction in recent decades. In light of this, we have compiled a summary of domain adaptation research, concentrating on work done in the last few years (2015–2022) for the benefit of the research community. We have categorically placed the important research works in DA under the chosen methodologies and have critically assessed the performances of these techniques. The study covers these features at length, and thorough descriptions of representative methods for each group are provided.

Keywords Domain adaptation · Computer vision · Transfer learning

1 Introduction

Machine learning is not the same as human learning. Humans can learn from a small number of labeled instances and apply what they have learned to new examples in unique situations. On the other hand, supervised machine learning approaches only work well when the test data is from the same distribution as the training data [22, 26]. They perform poorly when the testing dataset is from a non-identical distribution [13]. This happens due to the shift between the domain distributions.

Domain adaptation has found many applications in computer vision related to applying a trained network to real-world data. It can also label a synthetic dataset related to an earlier labeled dataset with less effort. Several works like [36] have utilized domain adaptation for segmentation problems in computer vision where the

A. Ajith (✉) · G. Gopakumar
Department of Computer Science and Engineering, Amrita Vishwa Vidyapeetham, Amritapuri, Kerala 690525, India
e-mail: ashlyajith619@yahoo.com

M. Tistarelli et al. (eds.), *Computer Vision and Machine Intelligence*,
Lecture Notes in Networks and Systems 586,
https://doi.org/10.1007/978-981-19-7867-8_47

testing set of data is from a distribution dissimilar to the training set. Under such cases, traditionally trained models will perform poorly. It has also found several applications under image captioning as well. Many domain adaptation techniques were devised to neutralize the performance reduction caused by domain shift. They can be broadly categorized into:

- **Supervised**: Under supervised domain adaptation, most of the samples in the target domain are labeled.
- **Semi-Supervised**: In the target domain, a few labeled samples are supplied to learn the suitable model adaption. This form of semi-supervised learning can extract invariant features from both domains. However, it also requires a small sample of target images to be labeled.
- **Unsupervised**: Unsupervised domain adaptation (UDA) reduces the shift between domains through unlabeled target datasets while seeking to maximize the classifier's performance on them. The target images are passed simultaneously with the source images. The network tries to classify the target images on the labels provided by the source domain images [3, 14].

This study aims to assess current progress in domain adaptation techniques and to offer some inferences from research directions.

2 Datasets Used for Domain Adaptation

Datasets used for domain adaptation simulate the condition where the data is from a different but related distribution. Datasets are therefore developed to generalize models across different domains. We have listed a few of the popular datasets used in domain adaptation. These datasets are used as the benchmark for calculating the domain adaptation of a technique.

2.1 Office 31

Office 31 [25] is a benchmark dataset with 4110 images with 31 categories and 3 domains. It contains Amazon, which has images extracted from Amazon, DSLR has images taken by a DSLR camera, and Webcam consists of images taken using a Web camera under various photographic settings. Office dataset was created in 2010 and has since been the benchmark dataset for DA problems. Sample images of the datasets are given in Fig. 1.

Fig. 1 Office 31 dataset samples

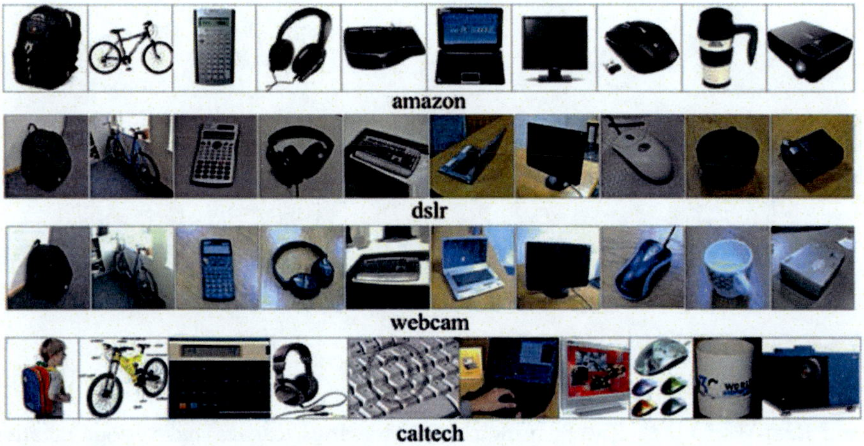

Fig. 2 Office-Caltech-256 dataset samples

2.2 *Caltech*

Caltech-256 [15] is a 257-class object recognition dataset that contains 30,607 real-world photos of various sizes. There are at least 80 photos for each class. The Caltech-101 dataset is a subset of this dataset. Caltech-256 is more intricate and demanding due to the more significant variability in size, backdrop, and other factors. We have added some samples from Office and Caltech, which share overlapping classes in Fig. 2. However, recently, more extensive datasets (Sects. 2.3 and 2.4) have been created to explore the scope of adaptation learning.

2.3 *Office Home*

Office Home [32] has four domains and 15,500 pictures or snapshots from 65 distinct categories. The four domains are Art, i.e., Ar, Clipart, i.e., Cl, Product, i.e., Pr,

Fig. 3 Office Home dataset samples

and Real-World, i.e., Rw. Snapshots from drawings, canvas and various creative renderings of images are included in the art domain. The Clipart domain is a collection of clipart pictures. Real-World is made up of typical photographs acquired with a camera, while Product comprises images without a background (refer Fig. 3).

2.4 *MNIST and MNIST-M*

MNIST [18] dataset is made up of 70,000 hand-written digits images in gray-scale. MNIST-M [10] is a dataset made from the MNIST dataset with the background containing varied patches of color from colored photos. It contains around 59,001 training and 90,001 test images. The MNIST/MNIST-M dataset samples have been given in Fig. 4.

3 Methods of Deep Domain Adaptation

We have divided distinct domain adaption techniques into categories depending on their methodology in this section.

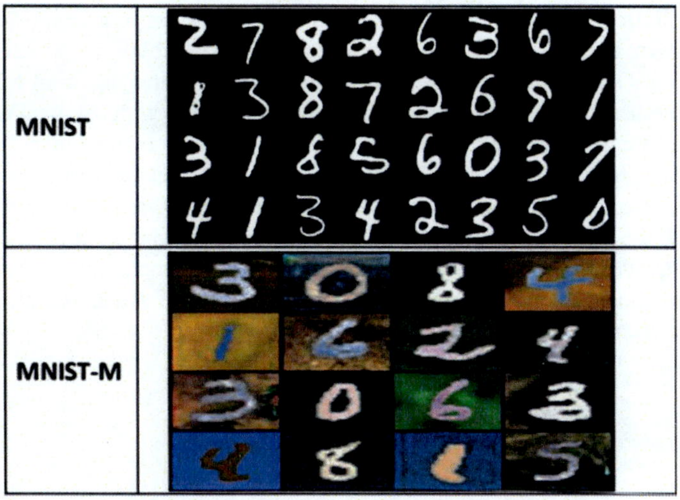

Fig. 4 MNIST/MNIST-M dataset samples

3.1 Discrepancy Based

Recent works show that deep domain adaptation networks, which perform discrepancy-based domain adaptation, produce better results than earlier multi-step works [29, 30]. Under discrepancy-based works, the commonly seen criteria to increase domain adaptation are.

Maximum Discrepancy Loss (MMD) MMD [21, 35] used a residual block attached to the end of the source network. The loss function was the sum of minimum mean discrepancy (MMD) loss [20] and entropy loss. The high representational features were extracted using the residual block, while the domain alignment was improved using MMD loss. The principle behind maximum mean discrepancy loss (MMD loss) is that distances between distributions are represented as distances between mean embeddings of features. The difference between the two alternative projections of the mean is the MMD. The maximum mean discrepancy (MMD) loss uses the kernel trick to find infinite moments across the distributions and reduce the distance between the expectation of distribution from the samples.

Correlational Alignment (CORAL) Similar to MMD, DeepCORAL [27, 28] was developed to improve domain adaptation using second-order distributions statistics. CORAL [27] explores the second-order statistics between the source and target domains using the higher representational features to align them. It was used within a DA network, reducing the domains' correlational alignment (CORAL) by minimizing the covariance between the higher representational tensors within the fully connected layers. This was done by using the CORAL loss obtained from the network layers. The domains were aligned by reducing the coral loss within the network.

Optimal Transport (OT) Optimal transport (OT) was proposed as a discrepancy technique to improve domain adaptation in joint distribution optimal transport (JDOT) [7]. Optimal transport is unique because it operates as a tool for converting one (continuous) probability distribution into another with minimal effort. The OT solution determines the most efficient technique to transform one distribution into another. The solution may be used to interpolate between them and acquire intermediate transformations seamlessly. The source data was transformed to a subspace with the shortest Wasserstein distance [2] across domains using OT. However, it scaled quadratically in computational cost with the sample size. This drawback was overcome in DeepJDOT [8] by implementing the OT stochastically in a deep adaptation network (DAN).

3.2 *Adversarial Based*

Adversarial network-based domain adaptation has recently attracted much attention since it provides state-of-the-art accuracy that outperforms discrepancy-based approaches. These networks follow a generator-discriminator architecture where the generator will learn to produce domain invariant and discriminative characteristics from the domains. A discriminator, which functions as a domain classifier, will predict the source of the image from the generator. This creates domain confusion within the network. Thus, working adversarially, these networks will produce domain invariant features which can be used to perform UDA.

Domain-adversarial neural network (DANN) [9] model was developed based on this principle of using GANs to reduce domain shift between the domains. Unsupervised domain adaptation is achieved by linking the feature extractor to a domain classifier through a gradient reversal layer that, during backpropagation-based training, doubles the gradient by a negative constant. The gradient reversal feature within the network was responsible for extracting the domain invariant features from both the domains within the networks.

Coupled generative adversarial networks (CoGANs) [19] use two GANs through which the domains are simultaneously passed, and the weights are shared between the networks. This encourages learning the joint distribution from both domains without using target domain labels.

PixelDA [4] uses adversarial architecture to reduce the shift between source and target distribution, to enable the classification of the target distribution samples. Instead of attempting to capture the domain's invariant features, this approach shifts the source domain to the target domain using GAN. Once trained, the model was able to produce and classify samples in the target domain. However, they are computationally expensive and limited by the dimensions of the input sample.

Adversarial discriminative domain adaptation (ADDA) [31] model uses a pretrained source encoder which is pretrained using labeled source domain images. The model uses the target and source images such that the discriminator cannot identify the origin of the domain from the images. The joint distribution of domains is learned from this and used for feature transformation and testing. Unsupervised classification of the target domain images is conducted based on parameters learned from these domain invariant features.

Selective adversarial networks (SANs) [6] are a type of deep adversarial adaptation proposed to perform partial transfer learning. It is used in applications where the target label is a subset of the source label category. Consequently, all the source domain labels are not present within the target domain. SAN can match source and target data distributions in a common subspace while segregating the outlier classes in the source domain. It is accomplished by increasing the alignment between similar data distributions in the latent space to the maximum degree possible.

3.3 Reconstruction Based

Deep reconstruction domain adaptation [12] is a method to improve adaptation which creates domain invariant features from each of the domains using an external reconstruction. It uses an encoder-decoder network architecture to classify and reconstruct the target distribution images. The higher representation of the target distribution is learned by the model trained on the source distribution and is used to categorize the target. After training, the reconstructed images from the original distribution show characteristics similar to the target distribution samples. Thus, it shows that the network has learned the joint distribution from both domains. This is then used to classify the target samples.

In the work [11] using an MTAE (Multitask autoencoder), the model reconstructs the images in different domains. This enables us to find the naturally occurring inter-domain variability when images are present across different domains. Such a process enables to find invariant features from both distributions. These features are then extracted from the different domains.

Domain separation networks (DSNs) [5] learn the higher-order representations and partition them into two subspaces. One of the subspaces is private to that particular domain, while the other subspace is shared across the domains. The partial representation is used to reproduce the image from both domains. The private and source representations are used in a shared decoder which learns to replicate the input domain distribution. The discriminability of the classes is increased by increasing the orthogonality constraints between the private and the shared subspaces components. The alignment of the domains is improved using the similarity loss between the shared subspace loss components.

3.4 Combination Based

Minimum discrepancy estimation [24] had used a combination of discrepancy-based losses of MMD and CORAL within the neural net to reduce the domain shift jointly.

Table 1 Table showing the summary view of all works

Method	Title	Year	Dataset	Acc%
Discrepancy-based	Learning transferable features with deep adaptation networks [20]	2015	Office 31	72.9
	Unsupervised domain adaptation with residual transfer networks [21]	2016	Office 31	73.7
	Mind the class weight bias: weighted maximum mean discrepancy for unsupervised domain adaptation [35]	2017	Office 31	72.1
	Return of frustatingly easy domain adaptation [27]	2016	Office 31	69.4
	Deep coral: Correlation alignment for deep domain adaptation [28]	2016	Office 31	72.1
	Joint distribution optimal transportation for domain adaptation [7]	2017	Caltech-office	80.04
	Deepjdot: Deep joint distribution optimal transport for unsupervised domain adaptation [8]	2018	MNIST MNIST-M	92.4
Adversarial-based	Unsupervised domain adaptation by backpropagation [9]	2015	MNIST MNIST-M	89.01
	Coupled generative adversarial networks [19]	2016	MNIST USPS	91.2
	Adversarial Discriminative Domain Adaptation [31]	2017	MNIST USPS	89.4
	Partial transfer learning with selective adversarial networks [6]	2017	Office 31	87.27
	Unsupervised pixel-level domain adaptation with generative adversarial networks [4]	2017	MNIST MNIST-M	98.2
Reconstruction-based	Deep reconstruction-classification networks for unsupervised domain adaptation [12]	2016	MNIST USPS	91.8
	Domain generalization for object recognition with multi-task autoencoders [11]	2015	Office Caltech	86.29
	Domain separation networks [5]	2016	MNIST MNIST-M	83.2
Combination-based	On minimum discrepancy estimation for deep domain adaptation [24]	2020	Office 31	74.6
	Dynamic Weighted Learning for Unsupervised Domain Adaptation [34]	2021	Office 31	87.1
	The domain shift problem of medical image segmentation and vendor-adaptation by unet-gan [36]	2019	Philips Siemens	83.5
	Cross-domain contrastive learning for unsupervised domain adaptation [33]	2022	Office 31	90.6
	Category contrast for unsupervised domain adaptation in visual tasks [17]	2022	Office 31	87.6
	Online unsupervised domain adaptation via reducing inter- and intradomain discrepancies adaptation [37]	2022	Office 31	85.3
Transformation-based	Direct domain adaptation through reciprocal linear transformations [1]	2021	MNIST MNIST-M	70
	Learning transferable parameters for unsupervised domain adaptation [16]	2022	Office 31	90.9

The source and target samples are sent simultaneously into the network. The encoded outputs from the network are taken from the higher representational layer and passed to the discrepancy-based adaptation losses. The losses from the MMD and CORAL loss are combined and backpropagated within the network for improved alignment within the domains.

Weighted DWL [34] utilizes a combination of criteria inspired by earlier works of adversarial and discrepancy-based models to build a DAN for improving the alignment. The network model focuses on two fronts: improving domain alignment and improving class discriminability within the domains. The class imbalance problem is also considered here by reweighing the samples before passing them into the network for training. This reduces the model bias due to the class imbalance of the training samples. The MMD is used to improve the alignment, and the linear discriminate analysis (LDA) loss improves the class discriminability.

Application of domain adaptation in medical research has been implemented by combining U-Net and adversarial-based GAN networks [36] to perform domain adaptation and image segmentation. The GAN is used for domain adaptation of the images (MRI scans) from different vendors, and U-Net is used to image segmentation.

Another application of the UDA combining technique is followed in OUDA [37], wherein the target domain is taken from the online streaming source. The methodology followed has two parts: The initial method involves reducing the discrepancy-based difference between the domains. A trained subspace is obtained from the initial phase, which captures domain invariant features through feature-level, sample-level, and domain-level adaptation. In the second part, the online classification uses the lower-dimensional alignment of the incoming target samples to the trained subspace. This is then used to reduce the intradomain distance between the online samples (target domain), which is then classified.

Unsupervised methods have also been developed and used with the earlier mentioned losses to overcome the target domain's label-free and source-free classifications. The work in [33] used a self-supervised method of implementation where the target was initially clustered based on KNN [23] and given pseudo-labels based on the source domain. These were then taken to perform contrastive learning by reducing the distance between the same classes from the target and increasing the distance between different classes. This was performed by decreasing intra-class and improving the inter-class distance, respectively.

A similar method that uses contrastive learning is followed in [17]. This approach creates a dictionary-like structure consisting of samples from the labeled source and unlabeled target domains. The samples from the unlabeled target domain are given pseudo-labels based on the source domain categories. The category contrast (CaCo) approach utilizes contrastive learning on the dictionary to reduce the distance between the same classes. The dictionary created for this purpose will also focus on class balance and class discriminability between the categories to minimize bias. Thus, the technique also accounts for class imbalance while training.

3.5 Transformation Based

Other than discrepancy-based loss functions, reconstruction, and GAN-based end-to-end architectures discussed earlier, another significant contribution to domain adaptation is the transformation applied as preprocessing on the input domain samples. A seminal work in this domain is the direct domain adaptation (DDA [1]). DDA using reciprocal linear transformation is a recent method based on prepossessing the input data to reduce the domain shift before the samples are passed through the network. The technique matches the signal-to-noise ratio between domains. The samples from the target domain are convoluted with the source domain and vice versa to reduce the shift. Thus, the domain shift is reduced outside the network before the training and testing.

A different form of transformation technique is explored under TransPar [16]. This work follows a method of identifying parameters in a network that learns the domain invariant features while training. Based on the lottery ticket hypothesis, the approach finds a network's transferable and untransferable parameters. The ratio of transferable parameters is inversely related to the domain shift distance between the distributions. The backpropagated weights from the loss identify the transferable parameters. The model then updates both parameters separately, focusing on the parameters that can generalize across the domains better. It is focused on reducing the domain-specific information learned by the network and can be integrated into current UDA-based models.

A summary of all works and their results has been given in Table 1. We have categorized them based on their approaches to aligning the domains. The performance comparison of these models is also compared to each other in Table 1. Their performance shows that adversarial methods produce the best results from all the models. Combination-based models also have results close to adversarial methods. These combinational models are often a combination of techniques involving adversarial and discrepancy-based methods. Reconstruction networks based on encoder-decoder architecture are also producing good results on diverse datasets.

4 Conclusions

This work has surveyed several papers on different approaches to domain adaptation techniques. In most real-world cases, it is likely that the target domain is not labeled and shares only a few classes similar to the source domain. This promises extensive research in unsupervised domain adaptation in the coming years.

5 Future Works

A promising direction in domain adaptation is adversarial networks and combinational models of different domain adaptation techniques. These types of approaches produce better results than most techniques in DA. Due to the more significant number of applications available for domain adaptation, much work in this direction is expected in the future.

References

1. Alkhalifah, T., Ovcharenko, O.: Direct Domain Adaptation Through Reciprocal Linear Transformations (2021)
2. Arjovsky, M., Chintala, S., Bottou, L.: Wasserstein generative adversarial networks. In International Conference on Machine Learning, pp. 214–223. PMLR (2017)
3. Ashokkumar, P., Don, S.: High dimensional data visualization: a survey. J. Adv. Res. Dyn. Control Syst. **9**(12), 851–866 (2017)
4. Bousmalis, K., Silberman, N., Dohan, D., Erhan, D., Krishnan, D.: Unsupervised pixel-level domain adaptation with generative adversarial networks. In: Proceedings of the IEEE Conference on Computer Vision and Pattern Recognition, pp. 3722–3731 (2017)
5. Bousmalis, K., Trigeorgis, G., Silberman, N., Krishnan, D., Erhan, D.: Domain separation networks. Adv. Neural Inform. Process. Syst. **29** (2016)
6. Cao, Z., Long, M., Wang, J., Jordan, M.I.: Partial transfer learning with selective adversarial networks. In: Proceedings of the IEEE Conference on Computer Vision and Pattern Recognition, pp. 2724–2732 (2018)
7. Courty, N., Flamary, R., Habrard, A., Rakotomamonjy, A.: Joint Distribution Optimal Transportation for Domain Adaptation (2017). arXiv preprint arXiv:1705.08848
8. Damodaran, B. B., Kellenberger, B., Flamary, R., Tuia, D., Courty, N.: Deepjdot: deep joint distribution optimal transport for unsupervised domain adaptation. In: Proceedings of the European Conference on Computer Vision (ECCV), pp. 447–463 (2018)
9. Ganin, Y., Lempitsky, V.: Unsupervised domain adaptation by backpropagation. In: International Conference on Machine Learning, pp. 1180–1189. PMLR (2015)
10. Ganin, Y., Ustinova, E., Ajakan, H., Germain, P., Larochelle, H., Laviolette, F., Marchand, M., Lempitsky, V.: Domain-adversarial training of neural networks. J. Mach. Learn. Res. **17**(1), 2096–2030 (2016)
11. Ghifary, M., Kleijn, W.B., Zhang, M., Balduzzi, D.: Domain generalization for object recognition with multi-task autoencoders. In: Proceedings of the IEEE International Conference on Computer Vision, pp. 2551–2559 (2015)
12. Ghifary, M., Kleijn, W.B., Zhang, M., Balduzzi, D., Li, W.: Deep reconstruction-classification networks for unsupervised domain adaptation. In: European Conference on Computer Vision, pp. 597–613. Springer (2016)
13. Gopika, P., Sowmya, V., Gopalakrishnan, E.A., Soman, K.P.: Transferable approach for cardiac disease classification using deep learning. In: Deep Learning Techniques for Biomedical and Health Informatics, pp. 285–303. Elsevier (2020)
14. Gressel, G., Hrudya, P., Surendran, K., Thara, S., Aravind, A., Prabaharan, P.: Ensemble learning approach for author profiling. In: Notebook for PAN at CLEF, pp. 401–412 (2014)
15. Griffin, G., Holub, A., Perona, P.: Caltech-256 Object Category Dataset (2007)
16. Han, Z., Sun, H., Yin, Y.: Learning transferable parameters for unsupervised domain adaptation. IEEE Trans, Image Proces (2022)

17. Huang, J., Guan, D., Xiao, A., Lu, S., Shao, L.: Category contrast for unsupervised domain adaptation in visual tasks. In: Proceedings of the IEEE/CVF Conference on Computer Vision and Pattern Recognition, pp. 1203–1214 (2022)
18. LeCun, Y., Bottou, L., Bengio, Y., Haffner, P.: Gradient-based learning applied to document recognition. Proc. IEEE **86**(11), 2278–2324 (1998)
19. Liu, M.-Y., Tuzel, O.: Coupled generative adversarial networks. Adv. Neural Inform. Proces. Syst. **29** (2016)
20. Long, M., Cao, Y., Wang, J., Jordan, M.: Learning transferable features with deep adaptation networks. In: International Conference on Machine Learning, pp. 97–105. PMLR (2015)
21. Long, M., Zhu, H., Wang, J., Jordan, M.I.: Unsupervised Domain Adaptation with Residual Transfer Networks (2016). arXiv preprint arXiv:1602.04433
22. Murugaraj, B., Amudha, J.: Performance assessment framework for computational models of visual attention. In: The International Symposium on Intelligent Systems Technologies and Applications, pp. 345–355. Springer (2017)
23. Peterson, L.E.: K-nearest neighbor. Scholarpedia **4**(2), 1883 (2009)
24. Rahman, M.M., Fookes, C., Baktashmotlagh, M., Sridharan, S.: On minimum discrepancy estimation for deep domain adaptation. In: Domain Adaptation for Visual Understanding, pp. 81–94. Springer (2020)
25. Saenko, K., Kulis, B., Fritz, M., Darrell, T.: Adapting visual category models to new domains. In: European Conference on Computer Vision, pp. 213–226. Springer (2010)
26. Sai, B.N.K., Sasikala, T.: Object detection and count of objects in image using tensor flow object detection api. In: 2019 International Conference on Smart Systems and Inventive Technology (ICSSIT), pp. 542–546. IEEE (2019)
27. Sun, B., Feng, J., Saenko, K.: Return of frustratingly easy domain adaptation. In: Proceedings of the AAAI Conference on Artificial Intelligence, vol. 30 (2016)
28. Sun, B., Saenko, K.: Deep coral: Correlation alignment for deep domain adaptation. In: European Conference on Computer Vision, pp. 443–450. Springer (2016)
29. Tamuly, S., Jyotsna, C., Amudha, J.: Deep learning model for image classification. In: International Conference On Computational Vision and Bio Inspired Computing, pp. 312–320. Springer (2019)
30. Thampi, S.M., Piramuthu, S., Li, K.-C., Berretti, S., Wozniak, M., Singh, D.: Machine Learning and Metaheuristics Algorithms, and Applications: Second Symposium, SoMMA 2020, Chennai, India, 14–17 Oct 2020, Revised Selected Papers, vol. 1366. Springer Nature (2021)
31. Tzeng, E., Hoffman, J., Saenko, K., Darrell, T.: Adversarial discriminative domain adaptation. In: Proceedings of the IEEE Conference on Computer Vision and Pattern Recognition, pp. 7167–7176 (2017)
32. Venkateswara, H., Eusebio, J., Chakraborty, S., Panchanathan, S.: Deep hashing network for unsupervised domain adaptation. In: Proceedings of the IEEE Conference on Computer Vision and Pattern Recognition, pp. 5018–5027 (2017)
33. Wang, R., Wu, Z., Weng, Z., Chen, J., Qi, G.-J., Jiang, Y.-G.: Cross-domain contrastive learning for unsupervised domain adaptation. IEEE Trans, Multimed (2022)
34. Xiao, N., Zhang, L.: Dynamic weighted learning for unsupervised domain adaptation. In: Proceedings of the IEEE/CVF Conference on Computer Vision and Pattern Recognition, pp. 15242–15251 (2021)
35. Yan, H., Ding, Y., Li, P., Wang, Q., Xu, Y., Zuo, W.: Mind the class weight bias: Weighted maximum mean discrepancy for unsupervised domain adaptation. In: Proceedings of the IEEE Conference on Computer Vision and Pattern Recognition, pp. 2272–2281 (2017)
36. Yan, W., Wang, Y., Gu, S., Huang, L., Yan, F., Xia, L., Tao, Q.: The domain shift problem of medical image segmentation and vendor-adaptation by unet-gan. In: International Conference on Medical Image Computing and Computer-Assisted Intervention, pp. 623–631. Springer (2019)
37. Ye, Y., Pan, T., Meng, Q., Li, J., Tao Shen, H.: Online unsupervised domain adaptation via reducing inter-and intra-domain discrepancies. IEEE Trans. Neural Netw. Learn. Syst. (2022)

Multi-branch Deep Neural Model for Natural Language-Based Vehicle Retrieval

N. Shankaranarayan and S. Sowmya Kamath

Abstract Natural language interfaces (NLIs) have seen tremendous popularity in recent times. The utility of natural language descriptions for identifying vehicles in city-scale smart traffic systems is an emerging problem that has received significant research interest. NL-based vehicle identification/retrieval can significantly improve existing systems' usability and user-friendliness. In this paper, the problem of NL-based vehicle retrieval is explored, which focuses on the retrieval/identification of a unique vehicle from a single-view video given the vehicle's natural language description. Natural language descriptions are leveraged to identify a specific target vehicle based on its visual features and environmental features such as trajectory and neighbours. We propose a multi-branch model that learns the target vehicle's visual features, environmental features, and direction and uses the concatenated feature vector to calculate a similarity score by comparing it with the feature vector of the given natural language description, thus identifying the vehicle of interest. The Cityflow-NL dataset was used for the purpose of training/validation, and the performance was measured using MRR (Mean Reciprocal Rank). The proposed model achieved a standardised MRR score of 0.15, which is on par with state-of-the-art models.

Keywords Natural language processing · Vehicle retrieval/identification · Vision-based transformers

1 Introduction

The growing number of applications connected to smart cities and autonomous driving systems has presented difficult challenges such as vehicle tracking, re-identification and retrieval. Recent research works focused on exploring the utility

N. Shankaranarayan (✉) · S. Sowmya Kamath
Department of Information Technology, National Institute of Technology Karnataka, Surathkal, Mangalore 575025, India
e-mail: n.shankaranarayan@gmail.com

S. Sowmya Kamath
e-mail: sowmyakamath@nitk.edu.in

M. Tistarelli et al. (eds.), *Computer Vision and Machine Intelligence*,
Lecture Notes in Networks and Systems 586,
https://doi.org/10.1007/978-981-19-7867-8_48

of natural language descriptions in identifying vehicles in a city-scale traffic system. Natural language-based vehicle retrieval can be formally defined as the process of retrieving a unique vehicle from a single-view video given a natural language description of the vehicle. The goal is to adopt natural language descriptions to identify a specific target vehicle based on its visual features and environmental features such as trajectory and neighbours. Natural language-based vehicle retrieval utilises not only general vehicle features such as type, colour and size but also environmental information such as neighbouring cars and the target vehicle's trajectory.

The problem of natural language-based vehicle retrieval entails several challenges. The problem of noisy detection occurs when the bounding box covering the vehicle of interest has some extra unnecessary information that does not correspond to the vehicle. Variations in illumination can affect the performance of automated systems, as the colour factor of the vehicle is affected by the illumination, due to which the same colour may look different. Viewpoint variance can occur when different viewpoints of the same vehicle are unrelatable, i.e., the same car from different angles might look different. When an object blocks the target vehicle, it causes issues due to occlusion, i.e., hidden features of the vehicle. People's ambiguous descriptions of observed events can also adversely affect a model's performance.

Most existing models achieve the task of natural language-based vehicle retrieval by following a two-branch approach, as shown in Fig. 1. One branch extracts visual features of the target vehicle and the environment, and the other extracts features from the given natural language query. The feature vectors generated by these two branches are later used to calculate a similarity score to retrieve the best matching vehicle corresponding to the natural language query. In this paper, we adopt a similar approach as [17], additionally, different visual and text embedding models are also considered for experiments and benchmarking with state-of-the-art models. For text embedding models, BERT and RoBERTa were considered, while ResNet50, ResNext50, VGGnet and ViT (Vision-Based Transformers) were selected for image feature extraction.

The rest of this article is structured as follows. In Sect. 2, we present a detailed discussion on existing work in the domain of NL-based vehicle retrieval. Section 3

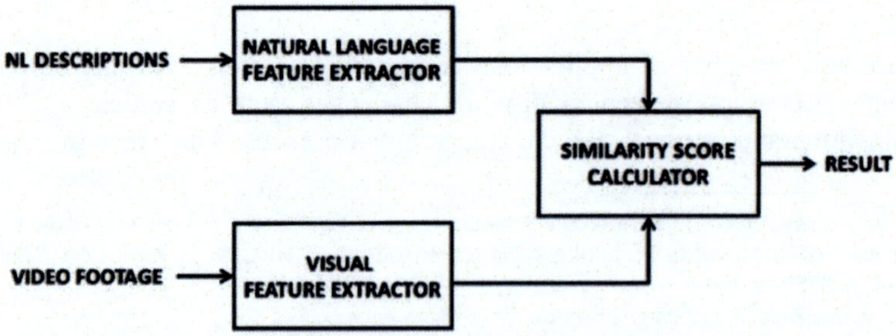

Fig. 1 General workflow of NL-based vehicle retrieval models

details the defined methodology and models designed for addressing the task. In Sect. 4, the experimental results and discussion are presented, followed by conclusion and directions for future work.

2 Related Work

Though NL-based vehicle retrieval has a wide range of applications, it has seen limited exploration due to the lack of well-annotated datasets. Research on NL-based vehicle retrieval picked up with the introduction of the CityFlowNL dataset [5] as part of the AI City Challenge 2021, where a baseline for NL-based vehicle retrieval was established for the first time. The dataset contains more than 5000 unique natural language descriptions of target vehicles, provided by at least three different people capturing realistic variation and ambiguities. Feng et al. [5] designed a two-branched model consisting of a pre-trained ResNct-50 to extract visual features and a pre-trained BERT to extract features from the given NL query. A similarity score was used to retrieve the best matching vehicle corresponding to the NL-based query. This baseline model achieved a MRR of 0.0269 and was later adopted by multiple researchers for further improvements on the baseline. Nguyen et al. [17] proposed a model which is structurally similar to the baseline model [5]. Their model also uses two main branches, a text encoder accepting the natural language description as input for text feature generation and a second branch for visual feature extraction. They experimented with Bi-directional InfoNCE and Marginal triplet loss and observed that the Bi-directional InfoNCE outperformed the marginal triplet loss.

Bai et al. [1] proposed a model that utilised language models such as BERT and RoBERTa to extract query features. They adopted two methods to increase the robustness of the text feature extraction model, backtranslation to generate semantic invariants of the training data, thus increasing the robustness as well as training data and Spacy, which extracts the subject from the given sentence and puts them in the beginning to strengthen the subject (i.e., the vehicle) and avoid inference caused by the environment. They followed a two streamed architecture for extracting visual features, one takes just the vehicle as input and covers the local features such as vehicle type, colour and size, and the other focused on global information such as environment and trajectory by constructing an augmented motion image that preserves the background and motion data. The model uses Instance loss for visual feature extraction and Symmetric InfoNCE loss for jointly training textual and visual features, which are finally projected into a single space.

Park et al. [19] proposed a model that combined the target vehicle's features with that of the vehicles around it using features such as colour, type, movement, neighbour car colour/type, to which weight is assigned. The model used a ResNet50 pre-trained on ImageNet dataset to extract the visual features such as colour and type. For movement analysis, GPS is used to decide the direction in which the vehicle is moving, and Kalman filter is used to estimate the vehicle's position and velocity from GPS data. The use of GPS eliminates the issue where a car captured near a camera, a car captured afar from the same camera, makes a smaller change in the images. Finally, a variable weight is assigned to the features as each of them has a different error rate and correlation. This method demonstrated that the variable weighing technique is a simple yet powerful method. Khorramshahi et al. [8] proposed a model which utilised feature averaging. The average natural language feature is obtained by extracting features from all-natural language descriptions of the same event and averaging them. The track features are obtained using frame-wise feature extraction using the Contrastive Language Image Pre-trained (CLIP) model [22] and averaging them. Even though the proposed model outperforms the baseline with a MRR of 0.1364, it can be observed that the model averages every single frame to obtain track features. The model can be improved by adopting better feature extraction methods and by selectively picking frames, as not all the frames are significant for tracking.

Certain models tried a segmentation-based approach to the problem. Lee et al.'s [9] model consisted of three parts, the *Natural language Module (NLM)* which utilised ELECTRA [2] for extracting features from the natural language description, the *Image Processing Module (IPM)* which utilised a ResNet50 backbone for extracting visual features and a *multi-modal module* to interpret the image and NL features, to achieve co-attention by combining them. They utilised a substitution module where the natural language features are used to generate image features and the image features are used to generate natural language features to exploit the fact that an image and its natural language description are semantically the same and exchangeable. Their proposed model also used a mask prediction module for mask prediction followed by a future prediction module which predicts the next frame. Finally, the model computes the probability of matching between the natural language description and vehicle based on mask prediction ratio, substitution similarity and colour and type matching probability. It is worth noting that this model outperformed the baseline model without utilising any post-processing method to increase the performance.

Some researchers utilised re-ranking methods to achieve better accuracy. Sun et al. [27] proposed a Dual-path Temporal Matching Network which used a convolutional neural network called ResNet-IBN [18] to extract visual features from vehicles and GloVe [20] extract textual features from the given retrieval query. Further, the model uses bidirectional GRUs to understand the temporal relation between given video and NL query. This was followed by the implementation of circle-loss and K-reciprocal-based re-ranking for post-processing.

3 Proposed Methodology

The proposed methodology is depicted in Fig. 2. The model consists of two main branches. One branch is used to extract text-based features from the given natural language query, while the other branch is used to extract vision-based features from the target video.

Dataset Specifics. The Cityflow-NL dataset [5] is currently the only dataset for NL-based vehicle retrieval tasks, which extends the older CityFlow benchmark by

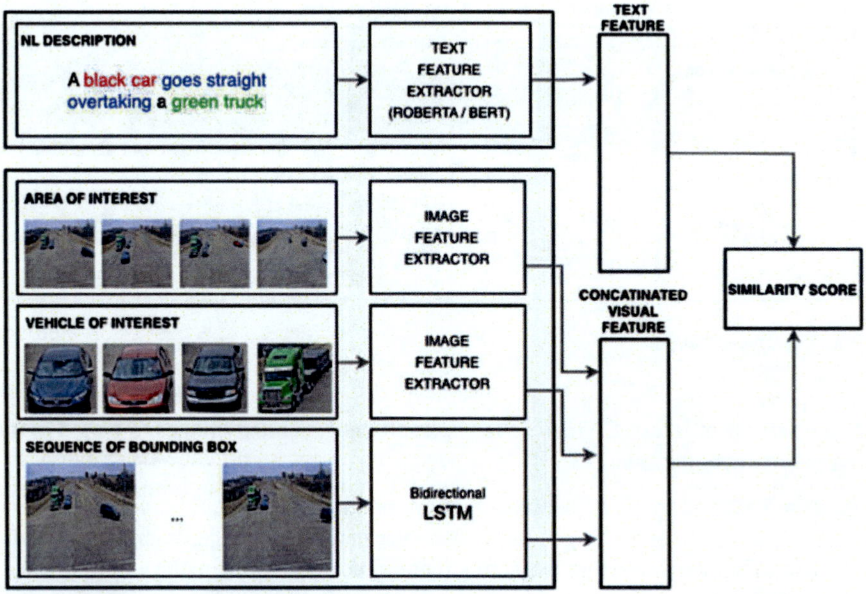

Fig. 2 Proposed methodology

including NL descriptions for vehicle targets. It consists of almost 5000 unique natural language descriptions of vehicle targets, making it the first dataset containing NL descriptions for multi-target, multi-camera tracking. Each track contains three distinct natural language descriptions which describe the vehicle of interest, as shown in Fig. 3. The raw traffic videos are annotated with each vehicle track's frames, bounding boxes and three natural language descriptions of the target.

Preprocessing. The raw traffic video as input and are split into individual frames, which are later used for extracting environmental features. Also, we generate cropped images of the vehicle of interest, which are used for extracting features corresponding

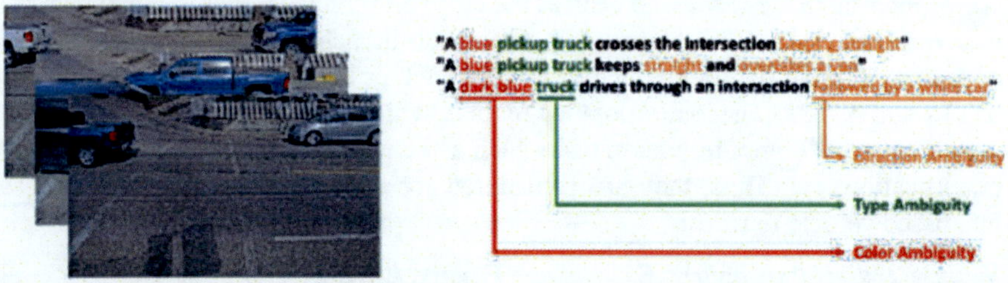

Fig. 3 Cityflow-NL: tracks and descriptions

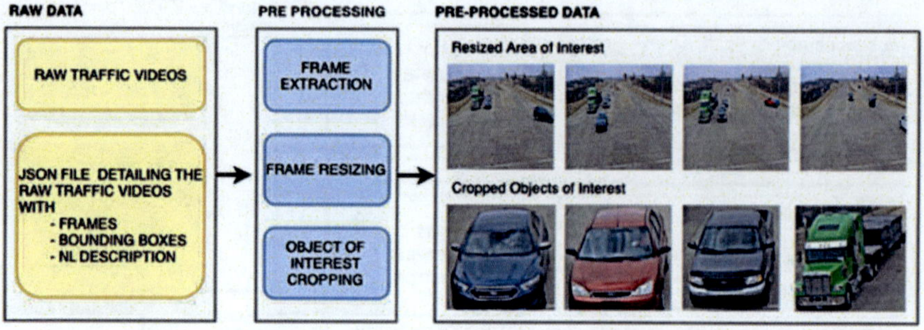

Fig. 4 Preprocessing

to the vehicle of interest. Finally, we resize all the generated images to be fed to the feature extraction model (Fig. 4).

Feature Extraction. The feature extraction model has two main branches. The first branch focuses on extracting features from the natural language descriptions, while the second branch focuses on extracting features corresponding to the environment as well as the vehicle of interest. The particular models adopted for each are discussed in detail below.

Text Feature Extraction: The first branch of the proposed approach focuses on text feature extraction. Two different pre-trained models were experimented with, namely, *BERT base uncased* [3] and *RoBERTa* [11]. A total of 768 features were extracted and later converted into 256 features by utilising a single layer feedforward network that takes 768 features as input and generates 256 features as output. The extracted feature vector (T) represents characteristics such as the vehicle of interest's colour, type and trajectory/direction of motion. It may also contain information regarding the neighbouring vehicles to the vehicle of interest.

Image Feature Extraction: The second branch consists of three sub-branches—the first sub-branch extracts features corresponding to the environment of the vehicle of interest by utilising the resized frames from the raw traffic video. The feature extractor generates a feature vector of size 512. The second sub-branch extracts features corresponding to the vehicle. It utilises the cropped vehicle images to generate a feature vector of size 512. In the first and second sub-branches, experimentation is done with five different feature extraction models such as ResNet50, ResNeXt50, VGG, ViT16 and ViT32. These are popular models which were successfully adopted in a wide range of computer vision tasks such as anomaly detection, re-identification, classification, etc. Thus, they are considered for experimentation for the task for NL-based vehicle retrieval.

Vehicle Trajectory Feature Extraction: Finally, the third sub-branch utilises a bi-directional LSTM to learn the vehicle's trajectory/direction of motion. The bounding box information is fed to the BiLSTM to generate a feature vector of size 128. The feature vectors of all the three sub-branches are concatenated to form a combined feature vector (V) of size 1152, and it is later converted to 256 features. The feature vectors (T) and (V) obtained from the branches are used to find the similarity score.

Metric Learning. For evaluation, Nguyen et al. [17]'s metric learning approach is utilised, it is a combination of image-to-text and text-to-vehicle loss. The model training involves two loss functions. The first one is an image-to-text contrastive loss Eq. 2, and the other is the text-to-image contrastive loss Eq. 1. The final loss is then computed as a summation of these two losses over a coefficient λ which acts as a factor for the bidirectional loss calculation.

1. *Text to Image loss*: Utilises the cosine distance between the anchor image, anchor text/description and negative vehicles to compute the loss over a mini-batch M.
2. *Image to Text loss*: Utilises the cosine distance between the anchor image, anchor text and negative text/description to compute the loss over a mini-batch M.

For the purpose of training, K positive and negative pairs are constructed for each mini-batch. It is to be noted that the dataset contains multiple vehicles' track information from the same camera. So, making use of this to construct the positive and negative pairs required for training. Pairing of the anchor text s_i with negative examples of the vehicle tracks from the same video as the anchor is done. This enables the model to learn harder to detect true positive pairs.

$$\text{Text to Image loss } l^{(s\to v)} = -\frac{1}{M}\sum_{i=1}^{M}\log\left(\frac{e^{g(s_i,v_i)/\tau}}{\sum_{k=1}^{K} e^{g(s_i,v_k)/\tau}}\right) \tag{1}$$

$$\text{Image to Text loss } l^{(v\to s)} = -\frac{1}{M}\sum_{i=1}^{M}\log\left(\frac{e^{g(v_i,s_i)/\tau}}{\sum_{k=1}^{K} e^{g(v_i,s_k)/\tau}}\right) \tag{2}$$

$$\text{Combined Bidirectional loss } l = \lambda l^{(s\to v)} + (1-\lambda) l^{(v\to s)} \tag{3}$$

$$\text{Cosine similarity } g(s, v) = \frac{sv}{\|s\|\,\|v\|} \tag{4}$$

The loss functions are computed as per Eqs. (1)–(4), where v_i is the anchor vehicle and s_i represents the anchor text/description. Also, $v_{k's}$ represents the negative vehicles, while $s_{k'v}$ provides the negative text/description. M indicates the mini-batch size, and λ is the coefficient used to weigh the directional loss.

4 Experimental Results and Discussion

The proposed method is implemented using PyTorch, and the experiments are carried out on a CUDA-based high-performance GPU cluster. We use Mean Reciprocal Rank

(MRR) and Recall to measure the performance of the model. MRR is a metric for evaluating any process that generates a list of possible results to a set of queries, arranged by likelihood of correctness. The reciprocal rank of a query response is the multiplicative inverse of the rank of the first correct answer (e.g., 1 point for first place, 1/2 point for second place, 1/3 point for third place and so on.) The mean reciprocal rank for a sample of queries Q is the MRR of the reciprocal ranks of results, where rank(i) refers to the rank position of the first relevant document for the i_{th} query. Recall is determined by dividing the total number of positive samples by the number of positive samples accurately categorised as positive. The recall is a metric that evaluates how well a model can detect positive samples. The higher the recall, the greater the number of positive samples found. Here, TP stands for True Positive, and FN stands for False Negative.

$$\text{MRR} = \frac{1}{|Q|}\sum_{i=1}^{|Q|} \frac{1}{\text{rank}(i)} \tag{5}$$

$$\text{Recall} = \frac{TP}{TP + FN} \tag{6}$$

We experimented with various visual feature extractors and text feature extractors. A total of ten different experimental setups were considered for this work. Table 1 showcases the observation in Recall@5, Recall@10 and Mean Reciprocal Rank for each of these experimental setups. Each row represents a varied experimental setup, and every column represents its corresponding performance. Here, recall@x represents the recall value for the top x retrievals. We experiment with a total of ten variations adapted from the base model in terms of the type of visual/text feature

Table 1 Performance of various models in terms of Recall and MRR

Feature extractor models		Recall@5	Recall@10	MRR
Visual	Text			
Baseline		0.1081	0.2039	0.0910
ResNet50	BERT	0.1154	0.1941	0.0977
ResNeXt50	BERT	0.1375	0.2162	0.0954
VGG	BERT	0.1425	0.2285	0.1042
ViT_b_16	BERT	0.1867	0.3218	0.1462
ViT_l_32	BERT	0.1916	0.3267	0.1336
ResNet50	RoBERTa	0.1425	0.2530	0.1059
ResNeXt50	RoBERTa	0.1351	0.2334	0.1004
VGG	RoBERTa	0.1597	0.2358	0.1063
ViT_b_16	RoBERTa	**0.2088**	**0.3341**	**0.1515**
ViT_l_32	RoBERTa	0.1818	0.3194	0.1411

extractors used. Out of the ten variations, four utilise BERT model for text embedding, and the others use RoBERTa for text embedding. We trained the models for 20 epochs with a learning rate of 1e−5.

Experiments revealed that the models that adopted RoBERTa for text embedding outperformed models that adopted BERT. This could be because, unlike BERT in RoBERTa, masking is done each time a natural language description is incorporated into the mini-batch. Thus, it contains different masked versions of the same sentence. Also, it can be noticed that vision-based transformers tend to outperform models which adopted conventional CNN-based feature extractors like VGGnet, ResNet and ResNext. In general, when it comes to NLP models, transformers have a high success rate, and they're now being used to recognise photos. ViT separates the images into visual tokens by separating a picture into fixed-size patches, embeds each one appropriately, and passes positional embedding to the transformer encoder as an input, whereas CNNs employ pixel arrays. Additionally, vision-based transformer has a self-attention layer that allows users to embed information over the entire image. These characteristics make ViT perform better when compared to conventional CNNs

5 Conclusion and Future Work

In this paper, a multi-branch approach to natural language-based vehicle retrieval is presented. The proposed model adopted bidirectional InfoNEC loss for metric learning. A total of ten different experimental setups were considered for the study. Each of these experimental setups utilised different visual as well as text embedding models. Through this, we demonstrated the effectiveness of vision-based transformers compared to conventional CNN-based feature extractors for image embedding and RoBERTa when compared to BERT for text embedding. As part of future work, we intend to further improve the model by adopting preprocessing steps such as backtranslation and multi-branched text embedding technique, which could extract more accurate text-based features from the natural language description.

References

1. Bai, S., Zheng, Z., Wang, X., Lin, J., Zhang, Z., Zhou, C., Yang, H., Yang, Y.: Connecting language and vision for natural language-based vehicle retrieval. In: Proceedings of the IEEE/CVF Conference on Computer Vision and Pattern Recognition, pp. 4034–4043 (2021)
2. Clark, K., Luong, M., Le, Q., Manning, C.: Electra: Pre-training Text Encoders as Discriminators Rather than Generators (2020). ArXiv Preprint ArXiv:2003.10555
3. Devlin, J., Chang, M.-W., Lee, K., Toutanova, K.: Bert: Pre-training of Deep Bidirectional Transformers for Language Understanding (2018). arXiv preprint arXiv:1810.04805
4. Dosovitskiy, A., Beyer, L., Kolesnikov, A., Weissenborn, D., Zhai, X., Unterthiner, T., Dehghani, M., Minderer, M., Heigold, G., Gelly, S., et al.: An Image Is Worth 16 × 16 Words:

Transformers for Image Recognition at Scale (2020). ArXiv Preprint ArXiv:2010.11929
5. Feng, Q., Ablavsky, V., Sclaroff, S.: CityFlow-NL: Tracking and Retrieval of Vehicles at City Scale by Natural Language Descriptions (2021). ArXiv Preprint ArXiv:2101.04741
6. Feng, Q., Ablavsky, V., Sclaroff, S.: CityFlow-NL: Tracking and Retrieval of Vehicles at City Scale by Natural Language Descriptions (2021). arXiv:2101.04741,2021
7. He, K., Zhang, X., Ren, S., Sun, J.: Deep residual learning for image recognition. In: Proceedings of the IEEE Conference on Computer Vision and Pattern Recognition, pp. 770–778 (2016)
8. Khorramshahi, P., Rambhatla, S., Chellappa, R.: Towards accurate visual and natural language-based vehicle retrieval systems. In: Proceedings of the IEEE/CVF Conference on Computer Vision and Pattern Recognition, pp. 4183–4192 (2021)
9. Lee, S., Woo, T., Lee, S.: SBNet: Segmentation-based network for natural language-based vehicle search. In: Proceedings of the IEEE/CVF Conference on Computer Vision and Pattern Recognition, pp. 4054-4060 (2021)
10. Leviathan, Y., Matias, Y.: An AI system for accomplishing real-world tasks over the phone. Google Duplex (2018)
11. Liu, Y., Ott, M., Goyal, N., Du, J., Joshi, M., Chen, D., Levy, O., Lewis, M., Zettlemoyer, L., Stoyanov, V.: Roberta: A Robustly Optimised Bert Pretraining Approach. arXiv preprint arXiv:1907.11692 (2019)
12. Naphade, M., Anastasiu, D., Sharma, A., Jagrlamudi, V., Jeon, H., Liu, K., Chang, M., Lyu, S., Gao, Z.: The NVIDIA AI city challenge. In: Prof, SmartWorld (2017)
13. Naphade, M., Chang, M., Sharma, A., Anastasiu, D., Jagarlamudi, V., Chakraborty, P., Huang, T., Wang, S., Liu, M., Chellappa, R., Hwang, J., Lyu, S.: The 2018 NVIDIA AI city challenge. In: Proceedings of CVPR Workshops, pp. 53–60 (2018)
14. Naphade, M., Tang, Z., Chang, M., Anastasiu, D., Sharma, A., Chellappa, R., Wang, S., Chakraborty, P., Huang, T., Hwang, J., Lyu, S.: The 2019 AI city challenge. In: The IEEE Conference On Computer Vision And Pattern Recognition (CVPR) Workshops, pp. 452–460 (2019)
15. Naphade, M., Wang, S., Anastasiu, D., Tang, Z., Chang, M., Yang, X., Yao, Y., Zheng, L., Chakraborty, P., Lopez, C., Sharma, A., Feng, Q., Ablavsky, V., Sclaroff, S.: The 5th AI city challenge. In: The IEEE Conference on Computer Vision and Pattern Recognition (CVPR) Workshops (2021)
16. Naphade, M., Wang, S., Anastasiu, D., Tang, Z., Chang, M., Yang, X., Zheng, L., Sharma, A., Chellappa, R., Chakraborty, P.: The 4th AI city challenge. In: The IEEE Conference on Computer Vision and Pattern Recognition (CVPR) Workshops, pp. 2665–2674 (2020)
17. Nguyen, T., Pham, Q., Doan, L., Trinh, H., Nguyen, V., Phan, V.: Contrastive learning for natural language-based vehicle retrieval. In: Proceedings of the IEEE/CVF Conference on Computer Vision and Pattern Recognition, pp. 4245–4252 (2021)
18. Pan, X., Luo, P., Shi, J., Tang, X. Two at once: enhancing learning and generalisation capacities via ibn-net. In: Proceedings of the European Conference on Computer Vision (ECCV), pp. 464–479 (2018)
19. Park, E., Kim, H., Jeong, S., Kang, B., Kwon, Y.: Keyword-based vehicle retrieval. In: Proceedings of the IEEE/CVF Conference on Computer Vision and Pattern Recognition, pp. 4220–4227 (2021)
20. Pennington, J., Socher, R., Manning, Glove, C.: Global vectors for word representation. In: Proceedings of the 2014 Conference on Empirical Methods in Natural Language Processing (EMNLP), pp. 1532–1543 (2014)
21. Perez, E., Strub, F., De Vries, H., Dumoulin, V., Courville, A.: Film: visual reasoning with a general conditioning layer. In: Proceedings of the AAAI Conference on Artificial Intelligence **32** (2018)
22. Radford, A., Kim, J., Hallacy, C., Ramesh, A., Goh, G., Agarwal, S., Sastry, G., Askell, A., Mishkin, P., Clark, J.: Learning Transferable Visual Models from Natural Language Supervision (2021). ArXiv Preprint ArXiv:2103.00020

23. Santoro, A., Raposo, D., Barrett, D., Malinowski, M., Pascanu, R., Battaglia, P., Lillicrap, T.: A Simple Neural Network Module for Relational Reasoning (2017). ArXiv Preprint ArXiv:1706.01427
24. Scribano, C., Sapienza, D., Franchini, G., Verucchi, M., Bertogna, M.: All you can embed: natural language based vehicle retrieval with spatio-temporal transformers. In: Proceedings of the IEEE/CVF Conference on Computer Vision and Pattern Recognition, pp. 4253–4262 (2021)
25. Sebastian, C., Imbriaco, R., Meletis, P., Dubbelman, G., Bondarev, E., et al.: TIED: a cycle consistent encoder-decoder model for text-to-image retrieval. In: Proceedings of the IEEE/CVF Conference on Computer Vision and Pattern Recognition, pp. 4138–4146 (2021)
26. Simonyan, K., Zisserman, A.: Very Deep Convolutional Networks for Large-scale Image Recognition (2014). ArXiv Preprint ArXiv:1409.1556
27. Sun, Z., Liu, X., Bi, X., Nie, X., Yin, Y.: DUN: Dual-path temporal matching network for natural language-based vehicle retrieval. In: Proceedings of the IEEE/CVF Conference on Computer Vision and Pattern Recognition, pp. 4061–4067 (2021)
28. Tan, M., Le, Q.: Efficientnet: rethinking model scaling for convolutional neural networks. In: International Conference on Machine Learning, pp. 6105–6114 (2019)
29. Tang, Z., Naphade, M., Liu, M., Yang, X., Birchfield, S., Wang, S., Kumar, R., Anastasiu, D., Hwang, J.: CityFlow: a city-scale benchmark for multi-target multi-camera vehicle tracking and re-identification. In: The IEEE Conference on Computer Vision and Pattern Recognition (CVPR), pp. 8797–8806 (2019)
30. Wang, H., Hou, J., Chen, N.: A survey of vehicle re-identification based on deep learning. IEEE Access. 7, 172443–172469 (2019)
31. Xie, S., Girshick, R., Dollár, P., Tu, Z., He, K.: Aggregated residual transformations for deep neural networks. In: Proceedings of the IEEE Conference on Computer Vision and Pattern Recognition, pp. 1492–1500 (2017)

Kullback–Leibler Distance-Based Fuzzy K-Plane Clustering Approach for Noisy Human Brain MRI Image Segmentation

Puneet Kumar, R. K. Agrawal, and Dhirendra Kumar

Abstract MR images are complex as the data distribution of tissues in MR images is non-spherical and overlapping in nature. The fuzzy k-plane clustering method (FkPC) is the most suitable soft clustering method to cluster non-spherical shape data and solve the overlapping problem of MR tissues. Generally, MR images are corrupted with imaging artifacts such as noise and intensity inhomogeneity. The FkPC method performs poor in the presence of noise. In this work, a new objective function for handling noise in the plane-based clustering approach is discussed with application to the MR image segmentation problem. We have utilised Kullback–Leibler (KL) distance measure to dampen the effect of noise in the k-plane clustering approach for the image segmentation problem. The proposed method is termed as KL distance-based fuzzy k-plane clustering (KLFkPC) method. Three publicly available MRI image datasets were utilised in the study to assess the efficiency of the proposed KLFkPC approach. The results are evaluated using a variety of performance metrics that demonstrate the efficacy of the proposed method over the ten existing related methods.

Keywords Fuzzy k-plane clustering · KL distance measure · Fuzzy local membership information · MRI image segmentation

1 Introduction

The human brain is an essential part of our body which is responsible for coordinating and controlling many day-to-day activities, including all voluntary and involuntary actions. The human brain is made up of three tissues: cerebrospinal fluid (CSF), white matter (WM), and grey matter (GM). Brain tissues degenerate with time due

P. Kumar (✉) · R. K. Agrawal
School of Computer & Systems Sciences, Jawaharlal Nehru University, Delhi, India
e-mail: me.puneetkumar94@gmail.com

D. Kumar
Department of Applied Mathematics, Delhi Technological University, Delhi, India

M. Tistarelli et al. (eds.), *Computer Vision and Machine Intelligence*,
Lecture Notes in Networks and Systems 586,
https://doi.org/10.1007/978-981-19-7867-8_49

to unhealthy lifestyles and ageing and effect the proportion of these three brain tissues. The improper proportion of tissue content in brain leads to many neuro-degenerative diseases such as Ataxia, Stroke, Alzheimer, brain tumour, and other brain disorders. which affects the lifestyle of a person suffering from these diseases. Hence, the analysis of the brain tissues is an essential step in diagnosing neurological ailments. For the analysis of brain diseases, two types of imaging techniques are found in the medical literature, namely invasive and non-invasive imaging techniques. Invasive imaging techniques are considered a gold standard in medical research. These techniques are used to identify the chemical constituent of the organs in the human body. Optical coherence tomography (OCT), intravascular MRI (IV-MRI), and intravascular ultrasound (IVUS), etc. are few examples of invasive techniques. Non-invasive imaging techniques capture the tissue structures of the organs such as computed tomography (CT), positron emission tomography (PET), and magnetic resonance imaging (MRI), etc. These imaging techniques are widely used to view the internal organs of the human body. For the primary diagnosis, non-invasive techniques are preferred to invasive techniques, as they are painless, risk-free, and do not harm the human body's soft tissues. Amongst all, MRI is the most popular non-invasive technique. MRI uncovers the crucial aspect of the human brain that accurately determines the soft tissue structure in biomedical research. As a result, MRI image segmentation is a critical stage in medical image processing. Segmentation using clustering techniques is a prevalent topic in the field of computer vision, machine intelligence, and image processing.

In the literature, many clustering approaches are proposed based on distribution of the data and the representation of the cluster prototype. Based on distribution of the data, i.e. overlapping or non-overlapping nature of the dataset, there are two basic groups of clustering algorithms, namely hard clustering methods [13] and soft clustering methods [2, 14, 17]. Each data point in the hard clustering approach is assigned to just one cluster, whereas in the soft clustering method, each data point might belong to many clusters. Real-world datasets such as MRI data contain soft boundaries between tissues, i.e. boundary pixels, may belong to more than one cluster. In such a scenario, the soft clustering methods perform well [3]. Fuzzy clustering is a soft clustering approach that is based on Fuzzy Set Theory. Fuzzy Set Theory was introduced by Zadeh et al. [18] which allows a given element to belong to multiple clusters with the degree of belongingness. The literature suggests an another category of clustering method which is based on the representation of the cluster prototype, i.e. point-based clustering method [10] and plane-based clustering method [4, 11, 12, 14]. In the point-based clustering method, the assumption is that the data points are distributed around a point or centroid. Hence, the point-based clustering method's performance is good when the distribution of data points is spherical in nature. The popular k-means clustering [10] method is the first point-based clustering method, which gathers the data points around cluster centroid by minimising the sum of square error between each data point to the cluster centroids. Plane-based clustering approaches are suggested to represent cluster prototypes for non-spherical data distribution. Plane-based clustering method has been proven to be more efficient and give better clustering results for non-spherical shaped data than point-based clustering methods [4, 12]. The first plane-based clustering technique, k-plane clustering (kPC) [4], may efficiently cluster non-spherically scattered data points by minimising the total sum of the square of each data point's distance from cluster planes. However, the cluster planes obtained from the kPC method are not well approximated as it considers only within-cluster information. Liu et al. [12] devised an expanded version of kPC called k-proximal plane clustering (kPPC), which utilises both within and between clusters information in the objective function. Hence, it is more robust than kPC.

Nevertheless, the kPC method does not handle the overlapping cluster problem. Zhu et al. [14] proposed a fuzzy k-plane clustering (FkPC) [14] approach to solve overlapping cluster issues in non-spherical shape clusters to overcome the limitations of kPC. In FkPC, a fuzzier exponent parameter (m) controls the degree of fuzziness in cluster overlapping regions. The clusters represented in terms of fuzzy sets are associated with uncertainty in defining their membership value which is difficult to measure. Entropy can be considered as an alternative way to quantify the uncertainty involved in defining the fuzzy sets. It's computed as the average of element non-membership in fuzzy sets [17]. Motivated by the concept of entropy in fuzzy sets, in this work [11], fuzzy entropy k-plane clustering (FEkPC) method is suggested which is based on the maximum entropy principle. The objective function of the FEkPC method consists of a fuzzy partition entropy term with a fuzzy entropy parameter in the conventional kPC. The fuzzy entropy parameter controls the degree of fuzziness, the same as the fuzzier parameter in the fuzzy clustering method. The FEkPC clustering method is a suitable application in the segmentation of brain MRI images. However, its performance is poor when the MR images are corrupted with imaging artefacts [7], i.e. partial volume effect, low-resolution, intensity in-homogeneity, noise, etc.

This research is the first attempt to explore the plane-based clustering method with Kullback–Leibler (KL) distance to incorporate local information, referred to as KL distance-based fuzzy k-plane clustering (KLFkPC) method. The proposed KLFkPC method objective function consists of (i) kPC objective function and (ii) KL distance term. The first term produces a cluster plane prototype same as the kPC method. The second term is the KL distance, which measures the relative distance between the membership of pixels and the average membership of immediate neighbourhood pixels. As a result, minimising the KL distance increases the pixel cluster membership towards the smoothed membership function of the immediate neighbourhood of the pixel. This suppresses noise and produces a clustered image with piecewise homogeneous regions.The parameter associated with KL term in the KLFkPC controls both the degree of fuzziness of overlapping clusters and local information in the KLFkPC clustering method. The proposed KLFkPC method solves the optimisation problem by utilising the underlined constraints with the help of Lagrange's method. Thus the KLFkPC clustering method produces an optimised fuzzy partition matrix and cluster plane by solving a series of eigenvalue problems. The highlight of the proposed work can be summarised as follows:

- We developed a new variant of the kPC to segment noisy images, referred to as the KLFkPC method.
- KL distance incorporates the local information by minimising the relative entropy between the membership of a pixel and the average local membership of the pixel's neighbourhood.
- KLFkPC produces an optimised fuzzy partition matrix, which improves the segmentation result when noise is present in MRI.
- We show the effectiveness and applicability of the suggested methods on three neuroimaging datasets and compared its performance with 10 related methods.

We used three freely accessible neuroimaging datasets to validate the efficacy and application of the suggested KLFkPC method: BrainWeb, IBSR, and MRbrainS18 datasets. To assess the efficacy of the proposed KLFkPC approach over other related methods such as kPC [4], kPPC [12], FCM [2], FEC [17], FkPC [14], FEkPC [11], FCM_S [1], EnFCM [16], FGFCM [5], LMKLFCM [8], the average segmentation accuracy (ASA) and Dice score (DS) are utilised.

The rest of this article is organised in the following manner; Preliminaries and related work were described in Sect. 2. In Section 3, the suggested KLFkPC method's optimisation formulation is described. Section 4 presents a summary of the datasets and experimental outcomes. Section 5 presents the conclusion and future directions.

2 Preliminaries and Related Work

In this section, some of the preliminaries and related work to the proposed method are presented.

2.1 Concept of Relative Entropy (a.k.a. Kullback–Leibler Distance) for Fuzzy Sets

Fuzzy set P and Q, over a universal set X can be represented as [18]

$$P = \{(x, \mu_P(x)) : x \in X\}, \mu_P(x) \in [0, 1] \tag{1}$$

$$Q = \{(x, \mu_Q(x)) : x \in X\}, \mu_Q(x) \in [0, 1] \tag{2}$$

where $\mu_P(x)$ and $\mu_Q(x)$ are membership function, which quantifies the degree of belongingness of the element x in fuzzy set P and Q respectively.

Kullback–Leibler Distance or Relative Entropy of Fuzzy Set: The relative entropy, also known as the Kullback–Leibler distance, is a measure of the difference between two fuzzy sets P's and Q's membership functions [19]. $D_{KL}(P||Q)$ is defined as relative entropy or KL distance as [9]:

$$D_{KL}(P||Q) = \sum_{i=1}^{N} \mu_P(x_i) \log \frac{\mu_P(x_i)}{\mu_Q(x_i)} \tag{3}$$

where $\mu_P(x_i)$ and $\mu_Q(x_i)$ represent membership value of the element $x_i \in X, i = 1, 2, \ldots, N$ in fuzzy set P and Q.

2.2 kPC Method

This research work [4] suggested a k-plane clustering (kPC) where a cluster prototype is represented with the help of a plane. This strategy can efficiently cluster the data points which are non-spherical in nature. The goal of the kPC method's optimisation problem is to reduce the total sum of the squares of distances between ith datapoints and the jth cluster plane prototype. The optimisation formulation of the kPC method can be given as [4]:

$$\begin{gathered} \min_{\{\mathbf{W},\mathbf{b}\}} \sum_{i=1}^{N} \sum_{j=1}^{k} \|x_i^T \mathbf{w_j} + b_j\|^2 \\ \text{subject to} \|\mathbf{w_j}\|^2 = 1, \ 1 \le j \le k \end{gathered} \tag{4}$$

where x_i is a datapoint in $\mathbf{X}$, $\mathbf{w_j}$ is the plane parameter for representing the cluster prototype, and b_j is the plane parameter for representing the cluster prototype.

3 Proposed Kullback–Leibler Distance-Based Fuzzy K-Plane Clustering Approach

This research work presents a noise robust plane-based clustering method for the image segmentation problem. To dampen the effect of noise, KL divergence-based distance measure is utilised on fuzzy partition matrix that incorporates the local similarity information along with regularisation parameter. The proposed objective function for clustering the data points includes the advantage of KL distance measure to incorporate local membership information in the conventional k-plane clustering (kPC) method termed as Kullback–Leibler distance-based fuzzy k-plane clustering (KLFkPC) method. The optimisation problem of the proposed KLFkPC involves the minimisation of the sum of two terms. The first term of the KLFkPC method is the kPC objective function term; this term minimises the sum of distances of ith pixel to the jth cluster plane. The second term represents sum of KL relative distance between the membership μ_{ij} and an average membership π_{ij} of its immediate neighbourhood of ith pixel to jth cluster. As a result, minimising the KL distance refines the membership of noisy pixels with its local membership information and helps in correctly label the pixel to appropriate cluster. The proposed KLFkPC method suppresses noise and produces a piecewise homogeneous segmented regions. The regularisation parameter α associated with the KL divergence term regulates both fuzziness index between the clusters and local information simultaneously. The optimisation formulation of the proposed KLFkPC can be given as:

$$\min_{\{\mathbf{U},\mathbf{W},\mathbf{b}\}} \sum_{i=1}^{N}\sum_{j=1}^{k} \mu_{ij}\|\mathbf{x}_i^T\mathbf{w}_j + b_j\|_2^2 + \alpha\sum_{i=1}^{N}\sum_{j=1}^{k}\mu_{ij}\log\frac{\mu_{ij}}{\pi_{ij}}$$
$$\text{subject to} \sum_{j=1}^{k}\mu_{ij} = 1,\ \|\mathbf{w}_j\|^2 = 1,\ \mu_{ij}\in[0,1] \qquad (5)$$
$$\forall j \in \{1,2,\ldots k\}, \forall i \in \{1,2,\ldots N\}$$

$$\pi_{ij} = \frac{1}{N_r}\sum_{r\in N_i}\mu_{rj} \qquad (6)$$

The iterative closed-form solution for cluster plane prototype and fuzzy membership can be obtained using the Lagrange method of the undetermined multiplier. The Lagrangian function $\mathbf{L}\{\mu_{ij}, \mathbf{w_j}, b_j, \lambda_i, \xi_j\}$ of the proposed method is represented as:

$$\mathbf{L}\{\mu_{ij}, \mathbf{w_j}, b_j, \lambda_i, \mathbf{\xi_j}\} = \sum_{i=1}^{N}\sum_{j=1}^{k}\mu_{ij}\|\mathbf{x}_i^T\mathbf{w}_j + b_j\|_2^2 + \alpha\sum_{i=1}^{N}\sum_{j=1}^{k}\mu_{ij}\log\frac{\mu_{ij}}{\pi_{ij}}$$
$$+\sum_{i=1}^{N}\lambda_i\left(1-\sum_{j=1}^{k}\mu_{ij}\right) + \sum_{j=1}^{k}\mathbf{\xi_j}(\|\mathbf{w_j}\|_2^2 - 1) \qquad (7)$$

where λ_i and $\mathbf{\xi_j}$ are the Lagrange multiplier.

The partial derivatives of the Lagrangian function $\mathbf{L}\{\mu_{ij}, \mathbf{w_j}, b_j, \lambda_i, \xi_j\}$ w.r.t μ_{ij}, $\mathbf{w}_j, b_j, \lambda_i, \mathbf{\xi_j}$ equate to zero leads to the following equations:

$$\frac{\partial L}{\partial \mu_{ij}} = \|\mathbf{x}_i^T\mathbf{w}_j + b_j\|_2^2 + \alpha\left(1+\log\frac{\mu_{ij}}{\pi_{ij}}\right) - \lambda_i = 0 \qquad (8)$$

$$\frac{\partial L}{\partial \mathbf{w}_j} = \sum_{i=1}^{N}\mu_{ij}(\mathbf{x}_i^T\mathbf{w}_j + b_j)\mathbf{x}_i + \mathbf{\xi_j}\mathbf{w}_j = 0 \qquad (9)$$

$$\frac{\partial L}{\partial b_j} = \sum_{i=1}^{N}\mu_{ij}(\mathbf{x}_i^T\mathbf{w}_j + b_j) = 0 \qquad (10)$$

$$\frac{\partial L}{\partial \lambda_i} = 1 - \sum_{i=1}^{k}\mu_{ij} = 0 \qquad (11)$$

$$\frac{\partial L}{\partial \xi_j} = ||\mathbf{w}_j||_2^2 - 1 = 0 \tag{12}$$

Equation (8) can be rewritten as:

$$\mu_{ij} = \pi_{ij} \exp\left(\frac{-(\|\mathbf{x}_i^T \mathbf{w}_j + b_j\|_2^2)}{\alpha}\right) \exp\left(\frac{\lambda_i - \alpha}{\alpha}\right) \tag{13}$$

Using Eqs. (11) and (13), we have

$$\mu_{ij} = \frac{\pi_{ij} \exp\left(\frac{-(\|\mathbf{x}_i^T \mathbf{w}_j + b_j\|_2^2))}{\alpha}\right)}{\sum_{p=1}^{k} \pi_{ip} \exp\left(\frac{-(\|\mathbf{x}_i{}^T \mathbf{w}_p + b_p\|_2^2)}{\alpha}\right)} \tag{14}$$

Solving Eq. (10), we have

$$b_j = \frac{-\sum_{i=1}^{N} \mu_{ij}(x_i^T \mathbf{w}_j)}{\sum_{i=1}^{N} \mu_{ij}^m} \tag{15}$$

To obtain iterative formula for $\mathbf{w}_j$, we substitute b_j into Eq. (9) and have

$$\left[\frac{(\sum_{i=1}^{N} \mu_{ij}\mathbf{x}_i)(\sum_{i=1}^{N} \mu_{ij}\mathbf{x}_i^T)}{\sum_{i=1}^{N} \mu_{ij}} - \sum_{i=1}^{N} \mu_{ij}\mathbf{x}_i\mathbf{x}_i^T\right]\mathbf{w}_j = \xi_j \mathbf{w}_j \tag{16}$$

The preceding equation may be re-written as an eigenvalue problem.

$$D_j \mathbf{w}_j = \xi_j \mathbf{w}_j \tag{17}$$

where

$$D_j = \left[\frac{(\sum_{i=1}^{N} \mu_{ij}\mathbf{x}_i)(\sum_{i=1}^{N} \mu_{ij}\mathbf{x}_i^T)}{\sum_{i=1}^{N} \mu_{ij}} - \sum_{i=1}^{N} \mu_{ij}\mathbf{x}_i\mathbf{x}_i^T\right] \tag{18}$$

We can simply determine that $\mathbf{w}_j$ is the eigenvector of D_j corresponding to the smallest eigenvalue ξ_j of D_j. After determining $\mathbf{w}_j$, Eq. (15) can be used to determine b_j. The KLFkPC method alternatively computes the fuzzy partition matrix and cluster plane prototype until convergence. The Algorithm 1 summarises the suggested method's outline.

4 Dataset and Experimental Results

The datasets, evaluation measures, and outcomes are all described in this section.

Algorithm 1 The proposed KLFkPC algorithm

Input: Set the number of clusters k, the smoothing parameter $\alpha > 0$, and the threshold value ϵ to their respective values.
1. Fuzzy membership matrix should be randomly initialised. $\mathbf{U}^1$
2.t$\leftarrow$ 1
3.**Repeat**
4.Update the plane parameters $\mathbf{w}_\mathbf{j}^\mathbf{t}$ using equation (17)$\forall j \in \{1, 2, \ldots k\}$
5.Update the plane parameters $\mathbf{b}_\mathbf{j}^\mathbf{t}$ using equation (15) $\forall j \in \{1, 2, \ldots k\}$
6.Update the fuzzy membership matrix $U^{t+1} = \{\mu_{ij}^{t+1}\}_{N\times k}$ using Eq. (14)
7. t $\leftarrow$ t+1;
8.**Until** $||\mathbf{U}^{t+1} - \mathbf{U}^t|| < \epsilon$
9.**Return**: $\mathbf{U} = \{\mu_{ij}\}_{N\times k}$, $\mathbf{W} = \{\mathbf{w_j}\}_{d\times k}$ and $\mathbf{b} = \{b_j\}_{1\times k}$

4.1 Datasets

To compare the efficacy of the proposed KLFkPC technique to the state of the art, we utilised three freely accessible T1-weighted MRI imaging datasets of the human brain with ground truth. The first is a simulated Brainweb dataset [6], while the second is a real brain MRI dataset obtained from the Internet Brain Segmentation Repository (IBSR) (available: https://www.nitrc.org/projects/ibsr), and the third dataset is a real brain MRI dataset obtained from MICCAI 2018 grand challenge on MR Brain Segmentation (MRBrainS18) (available: https://mrbrains18.isi.uu.nl/). For scull stripping, we used the brain extraction tool [15]. Table 1 provides a detailed description of the datasets.

4.2 Performance Metrics

To assess the efficacy of the suggested KLFkPC method, we have converted fuzzy partition matrix into crisp set with the maximum membership aka defuzzification process. The average segmentation accuracy (ASA) and Dice score (DS) were used to evaluate the performance of MRI imaging data, which are mathematically given as:

$$ASA = \sum_{i=1}^{k} \frac{|X_i \cap Y_i|}{\sum_{j=1}^{k} |X_j|} \quad DS = \frac{2|X_i \cap Y_i|}{|X_i| + |Y_i|}$$

Table 1 Dataset description

3D MRI Dataset	Volume dimension	Voxel size
Simulated Brainweb MRI dataset	181 × 217 × 181	1 mm × 1 mm × 1 mm
Real brain IBSR dataset	256 × 256 × [49–59]	1 mm × 1 mm × 1 mm
Real brain MRBrainS18 dataset	240 × 240 × 56	0.958 mm × 0.958 mm × 3.0 mm

where X_i and Y_i signify the collection of pixels in the ith ground truth image and ith cluster of the segmented image respectively. The cardinality of pixels in X_i corresponding to the ith class of the ground truth image is denoted by $|X_i|$, Similarly, the cardinality of pixels in the Y_i corresponding to the ith cluster of the segmented image is denoted by $|Yi|$. The higher the DS and ASA values for a algorithm, the greater the method's performance.

4.3 *Results on BrainWeb: Simulated Brain MRI Database*

We conduct experiments on a simulated BrainWeb MRI dataset to evaluate the effectiveness and utility of the proposed KLFkPC technique for MRI image segmentation. The proposed KLFkPC approach yields quantitative results, shown on the BrainWeb MRI dataset and compared its performance with kPC, kPPC, FCM, FEC, FkPC, FEKPC, FCM_S, EnFCM, FGFCM, and LMKLFCM methods for 7 and 9% noise in Table 2. From Table 2, the following conclusion may be drawn:

- The suggested KLFkPC approach performs better than other relevant methods in terms of ASA on BrainWeb MRI dataset.
- The suggested KLFkPC approach performs better than other relevant methods in terms of DS of GM, WM and CSF on BrainWeb MRI dataset.
- The performance of all segmentation methods deteriorates as the percentage of noise increases.

We compare the proposed KLFkPC approach to the kPC, kPPC, FCM, FEC, FkPC, FEkPC, FCM_S, EnFCM, FGFCM, and LMKLFCM methods in terms of average time of 100 runs on a the BrainWeb MRI image of size 181 × 217 in Table 3. The suggested KLFkPC method takes longer compute computation time than all other state-of-the-art methods, as indicated in Table 3.

4.4 *Results on IBSR: Real Clinical Brain MRI Database*

To assess the effectiveness and application of the suggested KLFkPC approach on a real MRI dataset, we consider MRI data with case no. '111_2', '112_2', '13_3', '16_3', '1_24', '205_3', '4_8', '5_8', and '7_8'. We present quantitative results in

Table 2 Comparison of performance in terms of ASA and DS on the simulated BrainWeb MRI images for 7 and 9% noise

Method	ASA		DS (GM)		DS (WM)		DS (CSF)	
	7% Noise	9% Noise	7% Noise	9% Noise	7% Noise	9% Noise	7% Noise	9% Noise
kPC	0.8787	0.8312	0.8663	0.8055	0.9328	0.8936	0.7966	0.7566
kPPC	0.8787	0.8312	0.8663	0.8055	0.9328	0.8936	0.7966	0.7566
FCM	0.8742	0.8207	0.8619	0.7950	0.9300	0.8860	0.7911	0.7458
FEC	0.8785	0.8312	0.8662	0.8056	0.9326	0.8936	0.7965	0.7566
FkPC	0.8784	0.8314	0.8661	0.8057	0.9326	0.8935	0.7963	0.7575
FEkPC	0.8803	0.8376	0.8686	0.8145	0.9326	0.8934	0.8001	0.7714
FCM_S	0.9029	0.8921	0.8971	0.8778	0.9559	0.9437	0.8074	0.8082
EnFCM	0.8766	0.8641	0.8636	0.8429	0.9435	0.9285	0.7548	0.7592
FGFCM	0.8898	0.8729	0.8807	0.8541	0.9505	0.9321	0.7789	0.7792
LMKLFCM	0.9060	0.9025	0.9074	0.8934	0.9559	0.9493	0.7981	0.8098
KLFkPC	**0.9099**	**0.9045**	**0.9079**	**0.8979**	**0.9581**	**0.9504**	**0.8183**	**0.8178**

terms of mean along with standard deviation of the performance of 9 MRI of the IBSR Brain MRI dataset. From Table 4, the following conclusion may be drawn:

- The proposed KLFkPC method performs better than other related methods in terms of average ASA over all MRI considered from the real IBSR Brain dataset.
- The proposed KLFkPC method performs better than other related methods in terms of average DS for CSF, WM, and GM over all MRIs considered from the real IBSR Brain dataset.

4.5 *Results on MRBrainS18: Real Brain Challenge MRI Database*

To determine the KLFkPC method's efficacy and applicability on a real MRI dataset, we consider MRI data with case no. '14', '148', '4' and '5' of the MRBrainS18 Brain MRI dataset. From Table 5, the following conclusion may be drawn:

- The proposed KLFkPC method outperforms other related methods in terms of average ASA over all MRI considered from the real brain MRBrainS18 dataset.
- The proposed KLFkPC method outperforms other related methods in terms of average DS for CSF, WM, and GM over all MRI considered from the real brain MRBrainS18 dataset.

Tables 2, 4, and 5 demonstrate that the proposed KLFkPC approach performs better than existing related methods on all the three neuroimaging MRI datasets corrupted with noise. In the proposed KLFkPC approach, the KL term measures

Table 3 Comparison of computation time of the proposed method with related work

Method	kPC	kPPC	FCM	FEC	FkPC	FEkPC	FCM_S	EnFCM	FGFCM	LMKLFCM	KLFkPC
Time (s)	0.0587	0.0722	0.4302	0.6421	0.5145	0.8772	1.8764	0.1036	0.2683	1.9004	2.2160

Table 4 Results obtained on the real brain IBSR MRI dataset for ASA and DS measure [(mean (standard deviation)]

Method	ASA		DS (GM)		DS (WM)		DS (CSF)	
	5% Noise	10% Noise	5% Noise	10% Noise	5% Noise	10% Noise	5% Noise	10% Noise
kPC	0.62(0.07)	0.59 (0.07)	0.62 (0.07)	0.60 (0.07)	0.77 (0.09)	0.74 (0.10)	0.10 (0.04)	0.10 (0.03)
kPPC	0.63 (0.07)	0.61 (0.07)	0.63 (0.08)	0.61 (0.07)	0.78 (0.09)	0.74 (0.09)	0.11 (0.04)	0.12 (0.06)
FCM	0.59 (0.07)	0.56 (0.07)	0.60 (0.07)	0.58 (0.05)	0.77 (0.09)	0.73 (0.10)	0.09 (0.03)	0.09 (0.03)
FEC	0.62 (0.07)	0.58 (0.08)	0.62 (0.07)	0.60 (0.06)	0.77 (0.09)	0.73 (0.10)	0.10 (0.04)	0.09 (0.03)
FkPC	0.62 (0.07)	0.58 (0.07)	0.62 (0.07)	0.60 (0.06)	0.77 (0.09)	0.74 (0.10)	0.10 (0.04)	0.09 (0.03)
FEkPC	0.62 (0.07)	0.59 (0.07)	0.62 (0.08)	0.60 (0.07)	0.77 (0.09)	0.73 (0.10)	0.10 (0.04)	0.10 (0.03)
FCM_S	0.62 (0.07)	0.60 (0.07)	0.62 (0.08)	0.61 (0.07)	**0.78 (0.08)**	0.76 (0.09)	0.11 (0.04)	0.11 (0.04)
EnFCM	0.62 (0.07)	0.60 (0.07)	0.62 (0.08)	0.60 (0.07)	**0.78 (0.08)**	0.75 (0.09)	0.11 (0.04)	0.11 (0.04)
FGFCM	0.63 (0.07)	0.60 (0.07)	0.62 (0.08)	0.61 (0.07)	**0.78 (0.08)**	0.75 (0.09)	0.11 (0.04)	0.11 (0.04)
LMKLFCM	0.64 (0.07)	0.62 (0.07)	0.64 (0.08)	0.63 (0.07)	**0.78 (0.08)**	0.75 (0.09)	0.11 (0.04)	0.11 (0.04)
KLFkPC	**0.65 (0.06)**	**0.64 (0.07)**	**0.65 (0.07)**	**0.66 (0.07)**	**0.78 (0.08)**	**0.76 (0.07)**	**0.13 (0.05)**	**0.13 (0.06)**

Table 5 Results obtained on the real brain MRbrain_S18 MRI dataset for ASA and DS measure [mean (standard deviation)]

Method	ASA		DS (GM)		DS (WM)		DS(CSF)	
	5% Noise	10% Noise	5% Noise	10% Noise	5% Noise	10% Noise	5% Noise	10% Noise
kPC	0.72 (0.04)	0.71 (0.04)	0.70 (0.05)	0.69 (0.04)	0.78(0.08)	0.78 (0.08)	0.71 (0.05)	0.70 (0.06)
kPPC	0.72 (0.04)	0.71 (0.04)	0.70 (0.05)	0.69 (0.04)	0.78 (0.08)	0.78 (0.08)	0.71 (0.05)	0.70 (0.06)
FCM	0.72 (0.04)	0.71 (0.03)	0.70 (0.05)	0.70 (0.04)	0.79 (0.08)	**0.78 (0.07)**	0.70 (0.05)	0.69 (0.06)
FEC	0.72 (0.04)	0.71 (0.03)	0.70 (0.05)	0.69 (0.04)	0.78 (0.08)	0.78 (0.08)	0.71 (0.05)	0.69 (0.06)
FkPC	0.72 (0.04)	0.71 (0.03)	0.70 (0.05)	0.70 (0.04)	**0.79 (0.07)**	**0.78 (0.07)**	0.71 (0.05)	0.69 (0.06)
FEkPC	0.72 (0.04)	0.71 (0.04)	0.70 (0.05)	0.69 (0.04)	0.78 (0.08)	0.78 (0.08)	0.71 (0.05)	0.70 (0.06)
FCM_S	0.73 (0.04)	0.73 (0.04)	0.69 (0.05)	0.69 (0.05)	0.78 (0.08)	0.78 (0.08)	0.75 (0.06)	0.75 (0.06)
EnFCM	0.71 (0.03)	0.71 (0.03)	0.68 (0.04)	0.68 (0.04)	0.78 (0.07)	**0.78 (0.07)**	0.71 (0.05)	0.71 (0.05)
FGFCM	0.72 (0.04)	0.72 (0.04)	0.69 (0.05)	0.69 (0.05)	0.78 (0.08)	**0.78 (0.07)**	0.72 (0.05)	0.72 (0.05)
LMKLFCM	0.74 (0.05)	0.74 (0.04)	0.70 (0.06)	0.69 (0.06)	0.78 (0.08)	0.78 (0.08)	0.77 (0.05)	0.78 (0.05)
KLFkPC	**0.75 (0.05)**	**0.75 (0.05)**	**0.71 (0.06)**	**0.71 (0.06)**	0.78 (0.08)	**0.78 (0.07**	**0.78 (0.05)**	**0.79 (0.05)**

relative distance between the pixel membership and the average membership of its immediate neighbourhood pixels. As a result, minimising the KL distance increases the pixel cluster membership towards the smoothed membership function of the immediate neighbourhood of the pixel. This suppresses noisy pixels and produces a clustered image with piecewise homogeneous regions. Hence, the proposed KLFkPC method suppresses noise by refining its membership with neighbourhood information and also provides a type of noise smoothing technique for noisy image. The proposed KLFkPC method produces piecewise homogeneous regions in the presence of noise shown in Fig. 1.

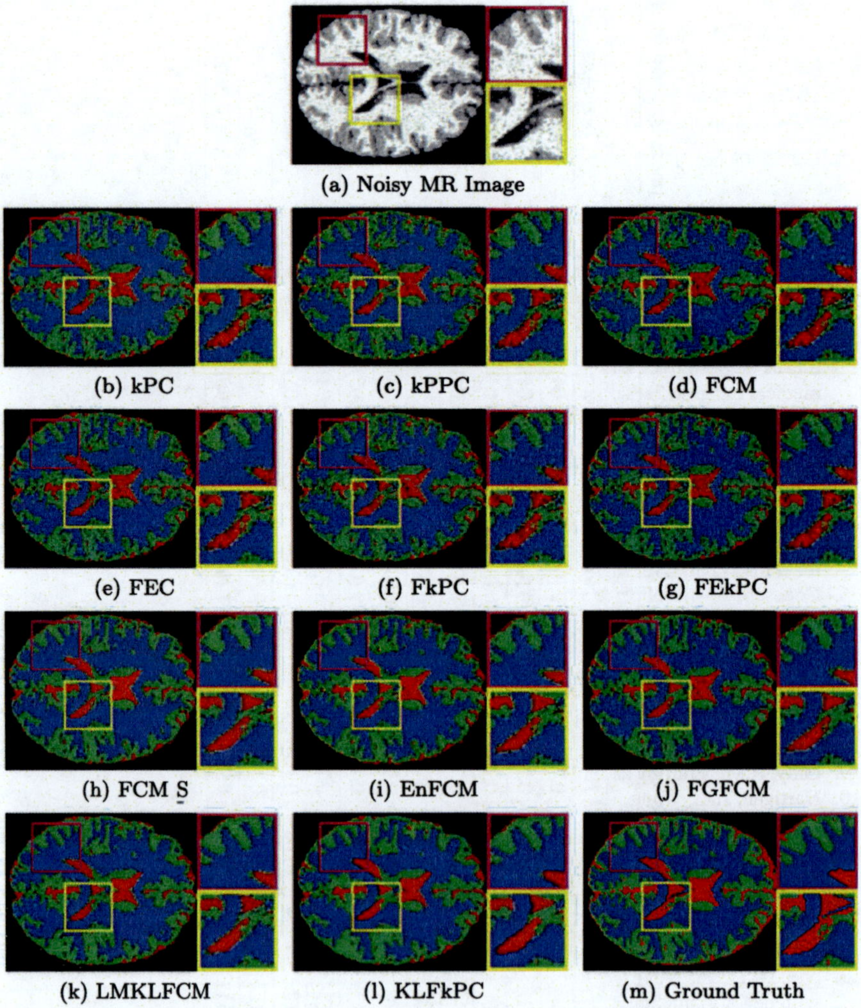

Fig. 1 Qualitative segmentation result on a simulated BrainWeb MRI dataset image corrupted with 9% Rician noise

5 Conclusion and Future Direction

This paper proposes a fuzzy Kullback–Leibler distance-based fuzzy k-plane clustering (KLFkPC) method. The method is formulated by the incorporation of KL distance-based local membership information in the objective function of conventional k-plane clustering method. The KL term in the proposed KLFkPC method measures information between the membership of pixels and the average membership of its immediate neighbourhood pixels. Hence, the proposed KLFkPC method smooths the membership of noisy pixels and is immune to the effect of noise in the segmentation process. After defuzzification of the fuzzy partition matrix, piecewise homogeneous segmented regions are obtained. The experimental results are compared using two quantitative metrics on the three publicly available MRI brain imaging datasets. The proposed KLFkPC approach outperforms other related methods in the presence of noise, as demonstrated by experimental results. In the future, we'll concentrate on expanding the proposed method for complex MRI datasets, i.e. multi-modal MR imaging dataset.

References

1. Ahmed, M.N., Yamany, S.M., Mohamed, N., Farag, A.A., Moriarty, T.: A modified fuzzy c-means algorithm for bias field estimation and segmentation of mri data. IEEE transactions on medical imaging **21**(3), 193–199 (2002)
2. Bezdek, J.C., Ehrlich, R., Full, W.: Fcm: the fuzzy c-means clustering algorithm. Comput. Geosci. **10**(2–3), 191–203 (1984)
3. Bora, D.J., Gupta, D., Kumar, A.: A Comparative Study Between Fuzzy Clustering Algorithm and Hard Clustering Algorithm (2014). arXiv preprint arXiv:1404.6059
4. Bradley, P.S., Mangasarian, O.L.: K-plane clustering. J. Glob. Optim. **16**(1), 23–32 (2000)
5. Cai, W., Chen, S., Zhang, D.: Fast and robust fuzzy c-means clustering algorithms incorporating local information for image segmentation. Pattern Recogn. **40**(3), 825–838 (2007)
6. Cocosco, C.A., Kollokian, V., Kwan, R.K.S., Pike, G.B., Evans, A.C.: Brainweb: online interface to a 3d mri simulated brain database. In: NeuroImage. CiteSeer (1997)
7. Doi, K.: Computer-aided diagnosis in medical imaging: historical review, current status and future potential. Comput. Med. Imag. Graph. **31**(4–5), 198–211 (2007)
8. Gharieb, R., Gendy, G.: Fuzzy c-means with a local membership kl distance for medical image segmentation. In: 2014 Cairo International Biomedical Engineering Conference (CIBEC), pp. 47–50. IEEE (2014)
9. Gray, R.M.: Entropy and information theory. Springer Science & Business Media (2011)
10. Hartigan, J.A., Wong, M.A.: Algorithm as 136: a k-means clustering algorithm. J. Roy. Stat. Soc. Ser. C (Appl. Stat.) **28**(1), 100–108 (1979)
11. Kumar, P., Kumar, D., Agrawal, R.K.: Fuzzy entropy k-plane clustering method and its application to medical image segmentation. In: 6th IAPR International Conference on Computer Vision and Image Processing (CVIP-2021), pp. 1–12. Springer (2022)
12. Liu, L.M., Guo, Y.R., Wang, Z., Yang, Z.M., Shao, Y.H.: k-proximal plane clustering. Int. J. Mach. Learn. Cybern. **8**(5), 1537–1554 (2017)
13. Nie, F., Li, Z., Wang, R., Li, X.: An effective and efficient algorithm for k-means clustering with new formulation. IEEE Trans. Knowl. Data Eng. (2022)
14. Pan, Z.L.W.S.t., Bin, Y.H.H.: Improved fuzzy partitions for k-plane clustering algorithm and its robustness research. J. Electron. Inform. Technol. **8** (2008)

15. Smith, S.M.: Bet: Brain Extraction Tool
16. Szilagyi, L., Benyo, Z., Szilágyi, S.M., Adam, H.: Mr brain image segmentation using an enhanced fuzzy c-means algorithm. In: Proceedings of the 25th Annual International Conference of the IEEE Engineering in Medicine and Biology Society (IEEE Cat. No. 03CH37439), vol. 1, pp. 724–726. IEEE (2003)
17. Tran, D., Wagner, M.: Fuzzy entropy clustering. In: Ninth IEEE International Conference on Fuzzy Systems. FUZZ-IEEE 2000 (Cat. No. 00CH37063), vol. 1, pp. 152–157. IEEE (2000)
18. Zadeh, L.A., Klir, G.J., Yuan, B.: Fuzzy Sets, Fuzzy Logic, and Fuzzy Systems: Selected Papers, vol. 6. World Scientific (1996)
19. Zarinbal, M., Zarandi, M.F., Turksen, I.: Relative entropy fuzzy c-means clustering. Inform. Sci. **260**, 74–97 (2014)

Performance Comparison of HC-SR04 Ultrasonic Sensor and TF-Luna LIDAR for Obstacle Detection

Upma Jain, Vipashi Kansal, Ram Dewangan, Gulshan Dhasmana, and Arnav Kotiyal

Abstract Obstacle detection is one of the most challenging fields due to the different shapes, sizes, and materials of obstacles. This work provides insight into commonly used obstacle avoidance sensors. The main contribution of this paper is the comparison of the performance of TF-Luna (LIDAR) with an ultrasonic sensor (HC-SR04) to detect various different obstacles. The performance of obstacle avoidance sensors has been evaluated in two different cases. At first, a single object is placed in the vicinity of the sensor, and readings have been taken. In the case of a single object, four different obstacle materials have been considered. The behavior of sensors with respect to multiple objects is also analyzed. An Arduino UNO microcontroller unit is used to collect the data from the sensor. The difference between actual and measured values is used to analyze the data. This analysis will help to select the right sensor for handling the obstacle detection problem.

Keywords Obstacle detection sensor · Ultrasonic sensor · TF-Luna

1 Introduction

Sensors are being used more frequently as the demand for autonomous projects increases [17, 18]. These are sophisticated devices that turn physical parameters (such as temperature, pressure, humidity, speed, and so on) into an electrical signal that can be measured. Robots have been used in many areas including agriculture [27], medical [16], industrial [19], planetary exploration [2], search and rescue [4], localization of hazardous odor sources [13], intelligent vehicles [8], etc.

To perform a particular task, a robot needs to collect information about its environment. This information is provided by various kinds of sensors depending on the application. One of the most common tasks of robots is to move from one location

U. Jain (✉) · V. Kansal · G. Dhasmana · A. Kotiyal
Graphic Era Deemed to be University, Dehradun, India
e-mail: upma.jain88@gmail.com

R. Dewangan
Thapar Institute of Technology, Patiyala, India

M. Tistarelli et al. (eds.), *Computer Vision and Machine Intelligence*,
Lecture Notes in Networks and Systems 586,
https://doi.org/10.1007/978-981-19-7867-8_50

to another. During this, a robot may encounter an obstacle when moving toward a specified location. A range of sensors is available for obstacle detection, such as infrared sensors [12, 23, 30], ultrasonic sensors [9, 30], vision systems [29], and light detection and ranging (LIDAR) [11, 28] (laser-based sensor system). LIDAR has been considered one of the most promising sensors to provide information about the shape and size of the object. Several different surveys had been published on the obstacle detection sensors [1, 5, 15, 25] that mainly focuses on vision-based system, ultrasonic sensor, and IR sensor.

This paper mainly compares the performance of TF-Luna (LIDAR) with an ultrasonic sensor (HC-SR04). This paper is arranged as follows: Sect. 2 provides the technical specifications as well as the characteristics of different sensors. Section 3 provides the details of the experimental setup for observing the performance of HC-SR04 and TF-Luna. Results and discussion are given in Sect. 4 followed by a conclusion in Sect. 5.

2 Sensor Characteristic and Technical Specification

With so many sensors on the market, it is important to pick the right one with attributes like accuracy, environmental conditions, range, calibration, resolution, cost, and repeatability. Therefore, we carried out a study of the following sensors, and their technical specifications are given in Table 1.

IR Sensor: It is an electrical device that detects and analyzes infrared radiation in its surroundings. IR sensors are commonly used to measure distances, and they can be employed in robotics to avoid obstacles. The power consumption of an infrared sensor is lower than that of ultrasonic sensors [10], but the response time is faster [24]. It has some drawbacks, such as the color of the obstacle can also affect the performance [1].

Table 1 Technical specification of IR sensor, ultrasonic sensor, and LIDAR

S. No.	Properties	IR sensor	Ultrasonic sensor	LIDAR
1	Cost	Low	Low	Higher cost
2	Fast	No	No	Yes
3	Measurement range	Up to 30 cm	Up to 400 cm	Up to 800 cm
4	3D imaging compatible	No	No	Yes
5	Sensitive to external condition	Yes	No	No
6	FOV	7.6–7.6°	Up to 160°	2°
7	Supply voltage	5–20 V	+5 V	+5 V

Environmental factors such as rain, fog, dust, smog, and transmission can have an impact on the performance of the sensor. The data transfer rate is slow [7] in the IR sensor.

Ultrasonic sensor: An ultrasonic sensor works on the basis of the time of flight method. It considers the time taken by a pulse to travel back and forth from the transmitter to the obstacle [14]. Furthermore, an ultrasonic sensor can identify any sort of impediment (metal, concrete wall, wooden-based object, plastics, transparent objects, rubber-based goods, etc.) and is unaffected by low lighting [22]. However, due to their sensitivity to mirror-like surfaces and large beam-width, US sensors have some limits [21, 26].

LIDAR: LIDAR has been considered as one of the most promising sensors to provide information about the shape and size of the object [3, 6, 20]. It is a form of laser distance sensor that detects the distance between two points. They operate by launching pulses in all directions and measuring the timing of how long it takes for them to return. They function very much like ultrasonic sensors. LIDARs use laser beams instead of sound waves to measure distances and analyze objects, and their operational distinction is the frequency at which they work. With very good accuracy, they do have the ability to measure 3D structures. We can use these sensors in the fields of robotics for obstacle avoidance, environment mapping, etc. The performance of LIDAR is much better in comparison with ultrasonic or IR sensors, but it has a higher cost, and it is harmful to the naked human eye.

3 Experimental Setup

An Arduino UNO microcontroller unit is used to collect data from the ultrasonic HC-SR04 sensor and TF-Luna LIDAR sensors for various types of obstructions. This was afterward logged to the PC and interpreted. The sensors were connected directly to the controller, without using any other interface.

3.1 Interfacing of Ultrasonic Sensor with Arduino UNO

An ultrasonic sensor uses ultrasonic sound waves to detect the distance to an object. The ultrasonic sensor's TRIG pin permits sound waves to be sent, and we may receive these sound waves by turning off the TRIG pin and turning on the echo pin.

The HC-SR04 ultrasonic sensor contains four pins: TRIG, ECHO, VCC, and GND. The sensor's TRIG pin is connected to the microcontroller's pin 9, the sensor's ECHO pin is connected to pin 10, the sensor's GND pin is connected to the GND pin, and the sensor's VCC pin is connected to the microcontroller's 5 V pin as shown in Fig. 1.

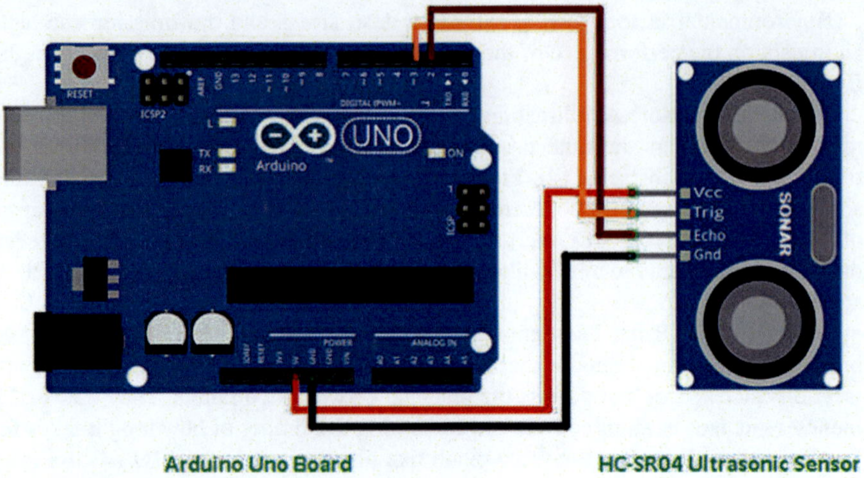

Fig. 1 Interfacing of ultrasonic sensor with Arduino UNO

3.2 Interfacing TF-Luna with Arduino UNO

The time of flight (TOF) principle is used by TF-Luna to calculate distance, and it regularly emits near-infrared-modulated pulses. The phase difference between the incident and reflected waves is used by TF-Luna to estimate time, which is subsequently used to calculate relative distance.

Pin 1 is the +5 V power supply, pin 2 is SDA/RXD receiving/data, pin 3 is SCL/TDX transmitting/clock, pin 4 is ground, and pin 5 is configuration input. In I2C mode, pin 6 is the communications mode multiplexing output. When pin 5 is unplugged or connected to 3.3 volts, serial port communication begins. The TF-Luna will be configured to receive RXD from pin 2 and send TXD from pin 3. The TF-Luna enters I2C mode when pin 5 is grounded, with pin 2 acting as SDA data and pin 3 acting as the SCL clock.

In this experiment, the serial port data communication mode is used for collecting the data with TF-Luna. Pin 1 is connected to +5 V of the UNO, pin 2 is connected to pin 10, pin 3 is connected to pin 11, and pin 4 is connected to the ground pin of the Arduino as shown in Fig. 2.

4 Results and Discussion

The performance of HC-SR04 and TF-Luna is analyzed in different cases. In case 1, a single object is placed within the detection range of the sensor, and readings were measured. Four distinct objects have been taken into consideration in this case:

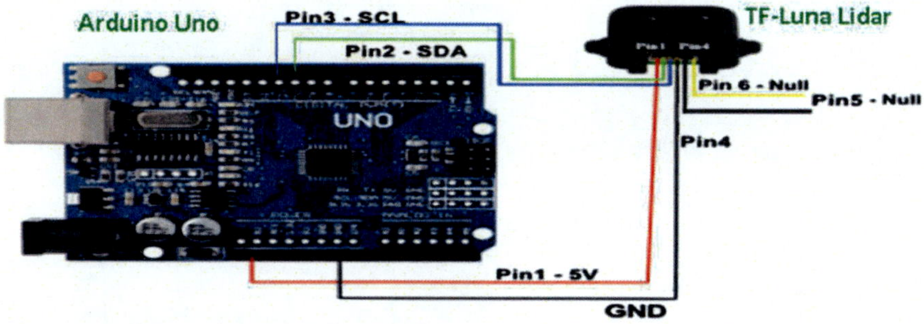

Fig. 2 Interfacing of TF-Luna with Arduino UNO

a cardboard box, a plastic box, a rubber box, and a metal bottle. The behavior of sensors with respect to several objects is also analyzed in case 2. Two objects are positioned differently to generate two distinct scenarios. Overall, six different setups have been created for evaluating the performance of the sensors.

Performance comparison in case 1: Figs. 3, 4, 5, and 6 provide the difference in the measured and actual value of sensors with obstacle materials cardboard, plastic, rubber, and metal, respectively.

Figure 3 constitutes the graph of actual versus measured distance values for a cardboard box that has been placed within the range of the TF-Luna LIDAR and ultrasonic sensors at different positions to collect the readings. It can be observed from the graph

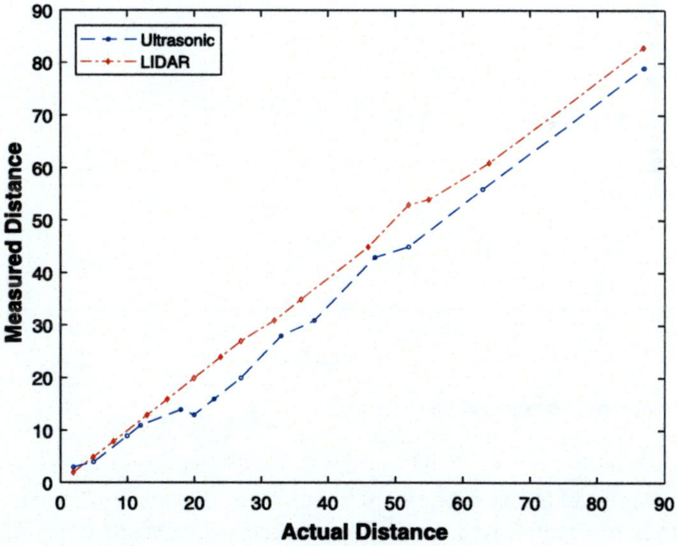

Fig. 3 Actual versus measured distance for cardboard

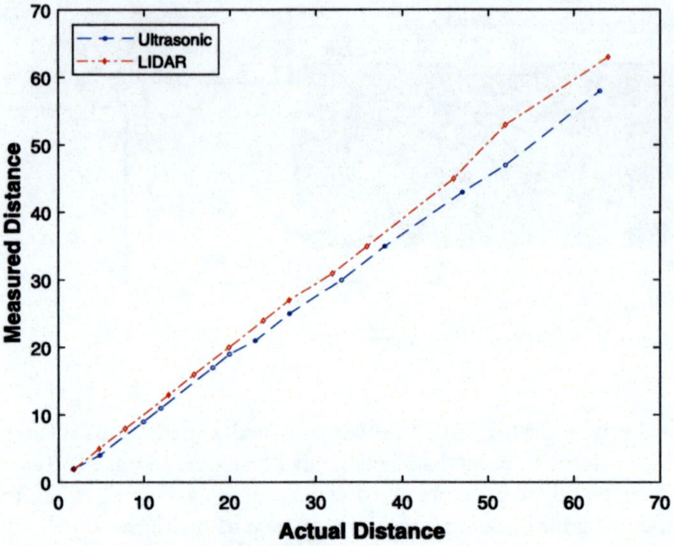

Fig. 4 Actual versus measured distance for plastic

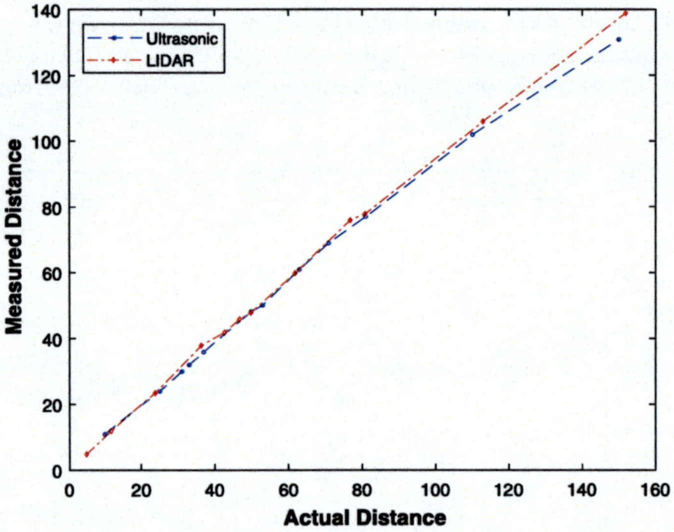

Fig. 5 Actual versus measured distance for rubber

that the readings of TF-Luna are more accurate, as the ultrasonic sensor works on the principle of sound waves. Due to the surface of cardboard, some of the sound waves were absorbed, which in turn affected the reading of the ultrasonic sensor.

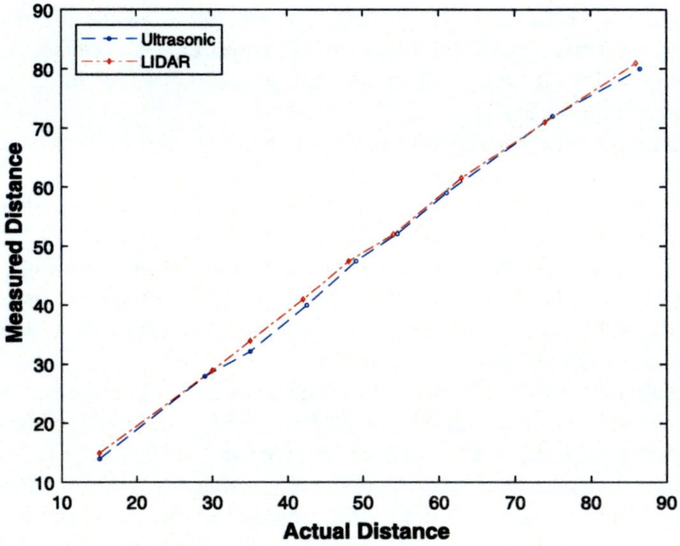

Fig. 6 Actual versus measured distance for metal

Figure 4 provides the graph of actual versus measured distance values for a plastic box that is being placed within the range of the TF-Luna LIDAR and ultrasonic sensors at different positions to collect the readings. It can be observed from the graph that the variation in ultrasonic sensor readings as compared to cardboard material is less, due to the smooth surface of the plastic that helps in the reflection of most of the sound waves falling on it.

TF-Luna works by repeatedly emitting infrared-modulated waves. These waves are not absorbed as sound waves, so the readings provided by TF-Luna are more accurate.

Figures 5 and 6 depict the comparison of TF-Luna LIDAR and ultrasonic sensor readings for rubber and metal obstacles, respectively. It can be seen that the readings of both sensors are almost overlapped as the rubber and metal surface reflects most of the sound waves falling on it.

It can be observed from Figs. 4, 5, 6 that the readings achieved by TF-Luna are more accurate as compared to the ultrasonic sensor. TF-Luna works by repeatedly emitting infrared-modulated waves. These waves are not absorbed as sound waves, so the readings provided by TF-Luna are more accurate.

As the type of material changes, the behavior of both sensors gradually changes. Additionally, as object distance increases, both sensors' performance degrades.

Performance comparison in case 2 (multiple objects): Two different scenarios are created by placing obstacles at different positions.

In scenario 1, obstacles are placed parallel to each other but at a different distance. The ultrasonic sensor HC-SR04 returns the distance of a nearer object. In TF-Luna, field of view (FOV) is very small, at only 2 degrees, so it returns the distance of the object, i.e., within its FOV.

In scenario 2, when an obstacle is placed one after another, both of the sensors provide the distance of the object, i.e., nearer.

Key findings: The readings provided by TF-Luna are more accurate and have a detection range of 8 m, while HC-SR04 has a detection range of 4 m. Although no major differences have been noticed in the actual and measured readings, the higher detection range of TF-luna makes it more suitable for applications like self-driving cars. The HC-SR04 comes at a low cost and is suitable for applications where a lower detection range is suitable.

Shape and size of the obstacles: For evaluating the performance of the sensors, not only the different materials of objects have been considered, but the shape and size of each object also vary. It can be observed that the readings with respect to objects considered do not have much effect on the shape and size of the object. But, if the object is too small, then the readings might become more erroneous as they depend on the number of waves falling on its surface for both of the sensors.

5 Conclusion

This work provides a brief discussion on IR, ultrasonic, and LIDAR sensors for obstacle detection along with the pros and cons associated with the respective sensors. Furthermore, the performance of the HC-SR04 ultrasonic sensor and TF-Luna is evaluated by collecting the readings from the sensors for different obstacles. The difference between measured and actual distance helps to understand the performance of both sensors. The readings of TF-Luna are more accurate as compared to the HC-SR04; on the other hand, HC-SR04 is cheaper as compared to the TF-Luna. As well as the difference in their FOV and maximum detection range makes one suitable over another depending on the application.

References

1. Adarsh, S., Kaleemuddin, S.M., Bose, D., Ramachandran, K.: Performance comparison of infrared and ultrasonic sensors for obstacles of different materials in vehicle/robot navigation applications. In: IOP Conference Series: Materials Science and Engineering. vol. 149, p. 012141. IOP Publishing (2016)
2. Apostolopoulos, D.S., Pedersen, L., Shamah, B.N., Shillcutt, K., Wagner, M.D., Whittaker, W.L.: Robotic antarctic meteorite search: Outcomes. In: Proceedings 2001 ICRA. IEEE International Conference on Robotics and Automation (Cat. No. 01CH37164). vol. 4, pp. 4174–4179. IEEE (2001)

3. Asvadi, A., Premebida, C., Peixoto, P., Nunes, U.: 3d lidar-based static and moving obstacle detection in driving environments: An approach based on voxels and multi-region ground planes. Robot. Auton. Syst. **83**, 299–311 (2016)
4. Baxter, J.L., Burke, E., Garibaldi, J.M., Norman, M.: Multi-robot search and rescue: a potential field based approach. In: Autonomous Robots and Agents, pp. 9–16. Springer (2007)
5. Chang, S., Zhang, Y., Zhang, F., Zhao, X., Huang, S., Feng, Z., Wei, Z.: Spatial attention fusion for obstacle detection using mmwave radar and vision sensor. Sensors **20**(4), 956 (2020)
6. Dewan, A., Caselitz, T., Tipaldi, G.D., Burgard, W.: Motion-based detection and tracking in 3d lidar scans. In: 2016 IEEE International Conference on Robotics and Automation (ICRA), pp. 4508–4513. IEEE (2016)
7. Discant, A., Rogozan, A., Rusu, C., Bensrhair, A.: Sensors for obstacle detection-a survey. In: 2007 30th International Spring Seminar on Electronics Technology (ISSE). pp. 100–105. IEEE (2007)
8. Duchoň, F., Hubinský, P., Hanzel, J., Babinec, A., Tölgyessy, M.: Intelligent vehicles as the robotic applications. Procedia Eng. **48**, 105–114 (2012)
9. Gageik, N., Müller, T., Montenegro, S.: Obstacle detection and collision avoidance using ultrasonic distance sensors for an autonomous quadrocopter, pp. 3–23. Aerospace Information Technologhy, University of Wurzburg, Wurzburg, germany (2012)
10. Grubb, G., Zelinsky, A., Nilsson, L., Rilbe, M.: 3d vision sensing for improved pedestrian safety. In: IEEE Intelligent Vehicles Symposium, 2004. pp. 19–24. IEEE (2004)
11. Hutabarat, D., Rivai, M., Purwanto, D., Hutomo, H.: Lidar-based obstacle avoidance for the autonomous mobile robot. In: 2019 12th International Conference on Information & Communication Technology and System (ICTS). pp. 197–202. IEEE (2019)
12. Ismail, R., Omar, Z., Suaibun, S.: Obstacle-avoiding robot with IR and PIR motion sensors. In: IOP Conference Series: Materials Science and Engineering. vol. 152, p. 012064. IOP Publishing (2016)
13. Jain, U., Tiwari, R., Godfrey, W.W.: Multiple odor source localization using diverse-PSO and group-based strategies in an unknown environment. J. Comput. Sci. **34**, 33–47 (2019)
14. Jansen, N.: Short range object detection and avoidance. Traineeship Report p. 17 (2010)
15. Karasulu, B.: Review and evaluation of well-known methods for moving object detection and tracking in videos. J. Aeronaut. Space Technol. **4**(4), 11–22 (2010)
16. Khan, Z.H., Siddique, A., Lee, C.W.: Robotics utilization for healthcare digitization in global covid-19 management. Int. J. Environ. Res. Public Health **17**(11), 3819 (2020)
17. Kumar, A., Patil, P.: Graphic Era University, Dehradun, India (2017)
18. Kumar, A., Jassal, B.: Remote sensing through millimeter wave radiometer sensor. J. Graph. Era Univ. 47–52 (2019)
19. Leggieri, S., Canali, C., Caldwell, D.G.: Design of the crawler units: toward the development of a novel hybrid platform for infrastructure inspection. Appl. Sci. **12**(11), 5579 (2022)
20. Li, Q., Dai, B., Fu, H.: Lidar-based dynamic environment modeling and tracking using particles based occupancy grid. In: 2016 IEEE International Conference on Mechatronics and Automation. pp. 238–243. IEEE (2016)
21. Majchrzak, J., Michalski, M., Wiczynski, G.: Distance estimation with a long-range ultrasonic sensor system. IEEE Sens. J. **9**(7), 767–773 (2009)
22. Mustapha, B., Zayegh, A., Begg, R.K.: Ultrasonic and infrared sensors performance in a wireless obstacle detection system. In: 2013 1st International Conference on Artificial Intelligence, Modelling and Simulation. pp. 487–492. IEEE (2013)
23. Nanditta, R., Venkatesan, A., Rajkumar, G., .B, N., Das, N.: Autonomous obstacle avoidance robot using IR sensors programmed in Arduino UNO. Int. J. Eng. Res. **8**, 2394–6849 (2021)
24. Qiu, Z., An, D., Yao, D., Zhou, D., Ran, B.: An adaptive Kalman predictor applied to tracking vehicles in the traffic monitoring system. In: IEEE Proceedings. Intelligent Vehicles Symposium, 2005. pp. 230–235. IEEE (2005)
25. Ristić-Durrant, D., Franke, M., Michels, K.: A review of vision-based on-board obstacle detection and distance estimation in railways. Sensors **21**(10), 3452 (2021)

26. Shrivastava, A., Verma, A., Singh, S.: Distance measurement of an object or obstacle by ultrasound sensors using p89c51rd2. Int. J. Comput. Theory Eng. **2**(1), 64–68 (2010)
27. Ulloa, C.C., Krus, A., Barrientos, A., Del Cerro, J., Valero, C.: Trend technologies for robotic fertilization process in row crops. Front. Robot. AI **9** (2022)
28. Villa, J., Aaltonen, J., Koskinen, K.T.: Path-following with lidar-based obstacle avoidance of an unmanned surface vehicle in harbor conditions. IEEE/ASME Trans. Mechatron. **25**(4), 1812–1820 (2020)
29. Xie, L., Wang, S., Markham, A., Trigoni, N.: Towards monocular vision based obstacle avoidance through deep reinforcement learning. arXiv:1706.09829 (2017)
30. Yılmaz, E., Tarıyan Özyer, S.: Remote and autonomous controlled robotic car based on Arduino with real time obstacle detection and avoidance (2019)

Infrared and Visible Image Fusion Using Morphological Reconstruction Filters and Refined Toggle-Contrast Edge Features

Manali Roy and Susanta Mukhopadhyay

Abstract In this paper, the authors have discussed a simple scheme for infrared-visible image fusion to combine the complementary information captured using multiple sensors with different properties. It employs a pair of morphological connected filters, i.e., opening and closing by reconstruction applied at each scale, brings out categorical bright and dark features from the source images. At each specific scale, the bright (or dark) features at a given spatial location from the constituting images are mutually compared in terms of clarity or prominence. Additionally, the cumulative gradient information is extracted using multiscale toggle-contrast operator which is further refined using a guided filter with respect to the source images. The final fusion is achieved by combining the best bright (or dark) features along with the refined edge information onto a base image. Experiments are conducted on pre-registered infrared-visible image pairs from the TNO IR-visible image dataset along with subjective and objective performance evaluation with due comparison against other state-of-the-art methods. The proposed method yields realistic fusion outputs and exhibits appreciable levels of performance with a better representation of complementary information which could facilitate decision-making in higher levels of image processing tasks.

Keywords IR-visible fusion · Guided filter · Multiscale morphology · Filters by reconstruction · Toggle-contrast operator

1 Introduction

A multi-sensor data acquisition system acquires an image using multiple sensory devices with varying perspectives and possibly at multiple resolutions. The features expressed in an imaging modality differ with the sensor used for its capture. Infrared cameras (mostly used in night-time imaging systems) capture the thermal contrast

M. Roy (✉) · S. Mukhopadhyay
Indian Institute of Technology (Indian School of Mines) Dhanbad, Jharkhand 826004, India
e-mail: manalir66@gmail.com

M. Tistarelli et al. (eds.), *Computer Vision and Machine Intelligence*,
Lecture Notes in Networks and Systems 586,
https://doi.org/10.1007/978-981-19-7867-8_51

in a scene in the form of the long-wave infrared wavelength emitted by the object whereas visible light cameras capture the light rays reflected by an object in presence of an additional light source. Consequently, infrared images typically have low resolution but are resistant to poor illumination and severe weather conditions (rain, fog, snow, smoke). On the contrary, visible images have high resolution, and rich detail with distinct tonal contrast, but are prone to low-light conditions. Multi-sensor images have complementary properties and are insufficient in terms of individual information content. Therefore, multi-sensor data can be optimally utilized if the features best represented in individual imaging modalities are synergistically integrated or fused. Such fused image pairs fusion has vast applications in the field of medical diagnosis, defect inspection, concealed weapon detection, remote sensing and military surveillance [1, 16, 19].

With gradual development in multi-sensor data acquisition and fusion research, several efficient algorithms and fusion rules are actively being proposed [10]. By and large, they can be grouped into pixel-level, feature-level and decision-level fusion. In pixel-level (or image-level) fusion, the pixels in the fused image are generated by colligating the targeted properties of the pixels from the candidate images. Being the lowest level of fusion, it is commonly used due to its low computational complexity, easy implementation, robustness and higher efficiency [12]. Keeping up with the context of the paper, this section briefly reviews the literature on IR-visible image fusion carried out at pixel-level either in the spatial, spectral or hybrid domain. Multiscale (or resolution) analysis in image fusion gives us a complete idea about the varying size and resolution of the details existing at different locations within an image to choose from [3, 5]. A few of the earlier multiscale techniques include the Gaussian pyramid, filter-subtract-decimate (FSD) pyramid, gradient pyramid, Laplacian pyramid, morphological pyramid, etc. Fusion algorithms using wavelets and their related transforms form another group under the umbrella of such techniques, where an image is decomposed into a series of coarse and fine resolution sub-bands using an appropriate wavelet transform [2]. A detailed overview and comparison of some of the early research on image fusion in connection with wavelets are presented in [13]. Multiscale decomposition is also being used to develop visual saliency-based fusion approaches that target the regions of interest (salient regions) by simulating the perception of the human visual system. Typically, the source images are decomposed into one or more layers followed by the application of a saliency detection method to generate saliency weight maps [6, 15]. The fused image is constructed back by combining the individual layers. Multiscale decomposition also serves as a crucial pre-processing step for fusion algorithms that utilizes deep neural networks (CNN, GAN, Encoder-Decoder) and their improved models. In [11], prior to the application of pre-trained VGG-19, the authors have decomposed the source images into base and detail components. The detail component is fused using the deep features whereas the base is combined using weighted averaging. A variety of end-to-end learning models introduced in [4, 23, 24] facilitates IR-visible image fusion, preserving color information, edge features and adaptive similarity between the source images and fusion result.

The key highlights of this paper are as follows:

- The proposed method employs morphological reconstruction filters to locate, identify and select the targeted features across a pair of constituting images corresponding to a scale which are mutually compared to retain the prominent ones.
- Also for every scale, the gradient features from the source images are evaluated using erosion (and dilation) like operations devised from toggle-contrast operator followed by refinement using guided filter with respect to the source images.
- The novelty in our work lies at the summation of scale-specific edge features which takes all the edges of varying scales into account followed by refinement using guided filter. The enhanced edges thus improve the perceptual quality of the fused image.

The remaining paper is structured as follows: Sect. 2 gives a brief introduction to preliminaries followed by the proposed method in Sect. 3; experimental results along with subjective and objective evaluation are presented in Sect. 4, and lastly, Sect. 5 provides the concluding remarks.

2 Preliminaries

2.1 Grayscale Morphological Reconstruction Filters

The basic morphological filtering operations, i.e., erosion (or dilation) and opening (or closing) expand and enhance the darker (or brighter) regions, respectively, in a grayscale image. However, these conventional filters fail to assure perfect preservation of edge information due to which morphological filters with geodesic reconstruction have been introduced [21]. These filters are centered around geodesic dilation and erosion operators. The elementary geodesic dilation (erosion) of size one of an image u with respect to the reference image v is defined as the minimum (maximum) between the dilation (erosion) of u with a structuring element, S of size one and v.

$$\delta_S^1(u, v) = \min(u \oplus S, v), \quad \epsilon_S^1(u, v) = \max(u \ominus S, v) \tag{1}$$

Now, iterating the elementary operations in Eq. 1, geodesic operations of arbitrary size can be obtained as,

$$\delta_S^i(u, v) = \min(\delta_S^{i-1}(u, v) \oplus S, v), \quad \epsilon_S^i(u, v) = \max(\delta_S^{i-1}(u, v) \ominus S, v)$$

Practically, the iteration stops at an integer n (i=n) when there is no change in the resultant image, i.e., $\delta_S^{(n)}(u, v) = \delta_S^{(n-1)}(u, v)$ and $\epsilon_S^{(n)}(u, v) = \epsilon_S^{(n-1)}(u, v)$. The stable reconstructed outputs after final (n^{th}) iteration are termed as reconstruction by dilation ($\delta_{\text{rec}}(u, v)$) and erosion ($\epsilon_{\text{rec}}(u, v)$), respectively.

$$\delta_{\text{rec}}(u, v) = \delta_S^{(n)}(u, v), \quad \epsilon_{\text{rec}}(u, v) = \epsilon_S^{(n)}(u, v) \tag{2}$$

Based on Eq. 2, opening by reconstruction of opening (or opening by reconstruction) and closing by reconstruction, denoted by $u\bar{\circ}S$ and $u\bar{\bullet}S$, respectively, are two important filtering operations.

$$u\bar{\circ}S = \delta_{\text{rec}}(u \circ S, u), \quad u\bar{\bullet}S = \epsilon_{\text{rec}}(u \bullet S, u)$$

Opening (Closing) by reconstruction retains the whole of the feature even if a part of it can be accommodated within the structuring element thereby avoiding abrupt truncation of edges.

2.2 *Toggle-Contrast Filter*

Morphological toggle mappings were first introduced in [17] for contrast enhancement in images. The two-state toggle mapping (TM) based on morphological dilation ($\oplus$) and erosion ($\ominus$) operations is expressed as below:

$$TM_S(x, y) = \begin{cases} f \oplus S(x, y), \text{ if} \\ (f \oplus S - f)(x, y) < (f - f \ominus S)(x, y) \\ f \ominus S(x, y), \text{ if} \\ (f \oplus S - f)(x, y) > (f - f \ominus S)(x, y) \end{cases} \tag{3}$$

where f is the image and S denotes the structuring element. Each pixel value in a toggle-mapped output image switches between the eroded and dilated versions of the source image depending on which is closer to the input pixel value. This operator along with its multiscale variants are widely explored in image fusion (Fig. 1).

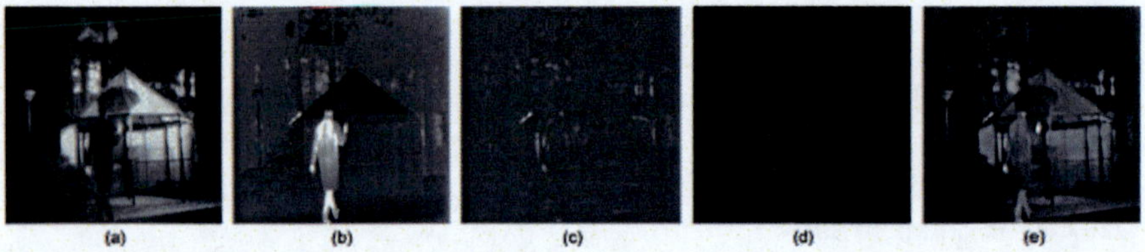

Fig. 1 **a** Ir image; **b** Visible image; **c** Resultant feature image ($F^c_{\text{op}} - F^c_{\text{cl}}$); **d** Resultant edge image (f_{e_r}); **e** Fused image using the proposed method

3 Proposed Fusion Scheme

The proposed method comprises four steps, (a) feature extraction, (b) feature comparison, (c) feature cumulation and refinement and (d) final fusion. This section discusses the proposed method in detail along with the block diagram presented in Fig. 2.

3.1 Feature Extraction Using Open (Close) and Toggle-Contrast Filters

A 2D grayscale image comprises salient features exhibited by peaks and valleys which vary in height (or depth), width and respective spatial locations. Application of conventional opening (or closing) to a grayscale image with respect to a structuring element (SE) shortens those peaks (or valleys) which are narrower than SE. However, for nonlinear 2D signals (i.e., image), conventional open-close operators

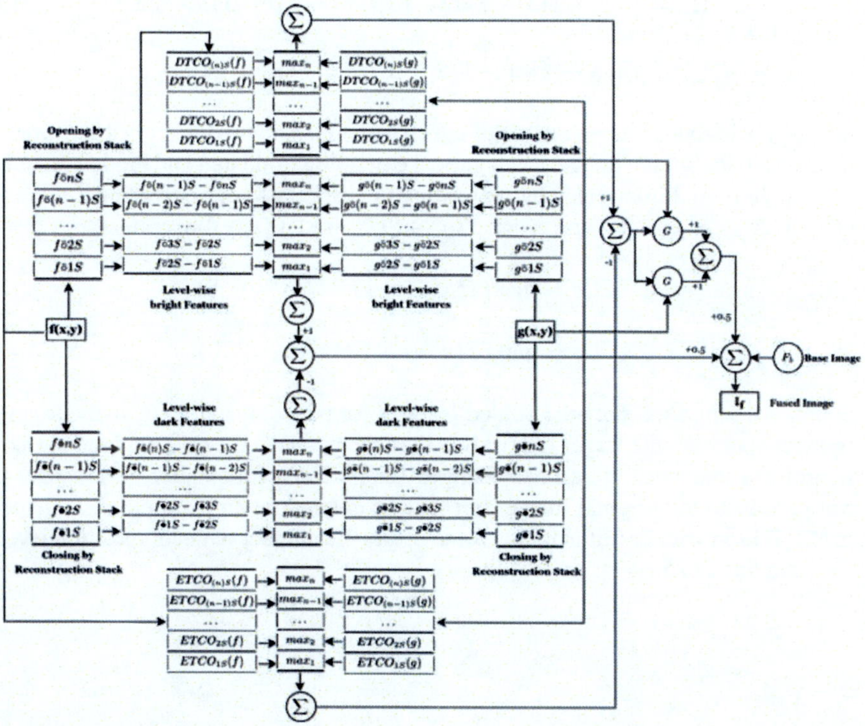

Fig. 2 Block diagram of the proposed method

cause deletion and drifting of edge features. Morphological filters by reconstruction solve this issue in two steps, a) firstly, the bright and dark features are narrower than SE are eliminated using simple opening-closing followed by b) performing repetitive geodesic dilation to reconstruct those features partially retained as a result of (a). As a result, bright (or dark) features are either completely removed or wholly retained. So, opening (or closing) by reconstruction of a grayscale image with an isotropic SE of increasing radii will generate a group of reconstructed opened (or closed) versions of input images. The collection of images is stored in separate stacks for further processing. So, for a pair of input images (say f and g), we obtain four such stacks.

$$f\bar{\circ}iS = \delta_{\text{rec}}(f \circ iS, f), \quad g\bar{\circ}iS = \delta_{\text{rec}}(g \circ iS, g) \tag{4}$$

$$f\bar{\bullet}iS = \delta_{\text{rec}}(f \bullet iS, f), \quad g\bar{\bullet}iS = \delta_{\text{rec}}(g \bullet iS, g) \tag{5}$$

where $i = 1, 2, 3, \ldots (n-1)$, n stands for the scale of SE. Carrying out a difference operation between two successive levels (say i and $i+1$) in the opening (closing) stack, bright (dark) features of the input image at a scale greater than or equal to i, but less than $i+1$ can be obtained.

$$d_f^{\bar{\circ}}(i) = f\bar{\circ}(i-1)S - f\bar{\circ}iS, \quad d_g^{\bar{\circ}}(i) = g\bar{\circ}(i-1)S - g\bar{\circ}iS \tag{6}$$

$$d_f^{\bar{\bullet}}(i) = f\bar{\bullet}iS - f\bar{\bullet}(i-1)S, \quad d_g^{\bar{\bullet}}(i) = g\bar{\bullet}iS - g\bar{\bullet}(i-1)S \tag{7}$$

As a toggle-mapped image consists of selective output between the eroded and dilated pixels from the source image, a categorical extraction of dilated and eroded features is possible at multiple scales by replicating the properties of conventional dilation ($f \oplus S(x, y) \geq f(x, y)$) and erosion ($f \ominus S(x, y) \leq f(x, y)$) operators.

$$\text{DTCO}_i = \max(\text{TM}(f(x, y) - f(x, y)), 0) \tag{8}$$

$$\text{ETCO}_i = \max(f(x, y) - \text{TM}(f(x, y)), 0) \tag{9}$$

Equation 8 brings out only the dilated features (or pixels) resulting from the toggle-contrast operator with larger gray values than the pixels in the original image. Likewise, Eq. 9 generates the eroded features (or pixels) with smaller gray values in comparison to the original image. For feature extraction, an isotropic disk-shaped flat SE is used with its initial radius set to 2, and the number of decomposition levels has been fixed to 5.

3.2 Feature Comparison

The feature images obtained in Eqs. 6 and 7 either consist of only peaks or only valleys over the entirety. When identical components of the participant images are compared, these peaks and valleys play crucial roles in identifying, selecting and locating the most prominent ones for constructing the final fused image. Here, the mutual comparison between bright (or dark) features extracted at every level is performed by taking a pixel-wise maximum. A similar approach is adopted at each scale to keep the highest dilation and erosion features obtained from Eqs. 8 and 9 using the toggle-contrast operator. Hence, at every level of the feature stack, we evaluate,

$$F^i_{\mathrm{op}} = \max(d^{\bar{\circ}}_f, d^{\bar{\circ}}_g), \quad F^i_{\mathrm{cl}} = \max(d^{\bar{\bullet}}_f, d^{\bar{\bullet}}_g) \tag{10}$$

$$F^i_{\mathrm{ETCO}} = \max(\mathrm{DTCO}_f, \mathrm{DTCO}_g), \quad F^i_{\mathrm{DTCO}} = \max(\mathrm{ETCO}_f, \mathrm{ETCO}_g) \tag{11}$$

Therefore, we obtain two intermediate stacks for each type which holds the best bright (dark) and dilated (eroded) features for further processing in the subsequent steps.

3.3 Feature Cumulation and Refinement

The prominent bright and dark features captured at each level in the intermediate stack are integrated to obtain a cumulative feature image. Alongside, the scale-specific gradients from toggle-contrast operators are combined as well. Mathematically, it is expressed as,

$$F^c_{\mathrm{op}} = \sum_{i=1}^{n} F^i_{\mathrm{op}}, \quad F^c_{\mathrm{cl}} = \sum_{i=1}^{n} F^i_{\mathrm{cl}}, \quad F^c_{\mathrm{DTCO}} = \sum_{i=1}^{n} F^i_{\mathrm{DTCO}}, \quad F^c_{\mathrm{ETCO}} = \sum_{i=1}^{n} F^i_{\mathrm{ETCO}} \tag{12}$$

As F^c_{DTCO} and F^c_{ETCO} contains bright and dark edge features from all scales, respectively, a resultant edge image is generated using Eq. 13 to achieve a perceptual balance between them.

$$f_{e_r} = F^c_{\mathrm{DTCO}} - F^c_{\mathrm{ETCO}} \tag{13}$$

The edge image thus obtained is subjected to a refinement process using the edge-preserving guided filter (G), taking the source images as a guidance images. Guided filter can be expressed as a local linear model between an input image (p), a sharp guidance image (I) and filter output (q) [9].

$$f^{'}_{e_r} = G(f_{e_r}, f(x, y), r, \epsilon) + G(f_{e_r}, g(x, y), r, \epsilon) \tag{14}$$

3.4 Final Fusion

The final fusion is achieved by integrating the salient features along with the improved edges obtained earlier with the base image. For constructing the base image, opening and closing by reconstruction are performed on the source image with the SE at the highest scale (n). As, opening (closing) by reconstruction enhances the bright (dark) features for both the source images, choosing the minimum (maximum) value among them would keep the bright (dark) features in proportion thereby not increasing the overall brightness (darkness) in the result.

$$m_x = \min(f \bar{\circ} nS, g \bar{\circ} nS), \quad m_y = \max(f \bar{\bullet} nS, f \bar{\bullet} nS); \tag{15}$$

The base image is obtained by averaging the opened and closed versions of the input images as follows:

$$F_b = [m_x + m_y] \star 0.5 \tag{16}$$

The final fused image is constructed as follows:

$$I_f = F_b + 0.5 \times (F^c_{\mathrm{op}} - F^c_{\mathrm{cl}}) + 0.5 \times f^{'}_{e_r} \tag{17}$$

4 Experimental Analysis and Discussion

This section presents a qualitative and quantitative judgment of the approach with other recently developed fusion algorithms. Experiments performed on multi-sensor image pairs from TNO IR-visible dataset [22] confirm the efficacy of the said algorithm. The executions are performed on MATLAB R2017b with a 64-bit windows operating system, Intel 2.60 Hz Core i7 CPU and 16 GB RAM. It is compared with six standard IR-visible fusion approaches such as FFIF [26], SA [14], MISF [25], IFCNN [28], PMGI [27] and U2 [24]. Seven image fusion metrics, namely feature mutual information (FMI) [7], average correlation coefficient (CC_{avg}), average similarity index ($SSIM_{\mathrm{avg}}$), naturalness image quality evaluator($NIQE$) [18], Piella's metric (Q, Q_w) [20] and visual information fidelty (VIFF) [8], are adopted to compare the results. In guided filter, the value for r and ϵ is fixed at 8 and 0.01, respectively. The average values of quantitative metrics are presented in Table 1.

Table 1 Average objective evaluation on IR-visible image pairs

Images	Methods							
	Metrics	FFIF [26]	SA [14]	MISF [25]	IFCNN [28]	U2 [24]	PMGI [27]	Proposed
IR-visible images from TNO dataset [22] (512 × 512)	Q	0.7569	0.8257	0.8309	0.8251	0.7652	0.7736	**0.8377**
	Q_w	0.7090	0.8474	0.8574	0.8336	0.7905	0.7681	**0.8598**
	FMI	0.8824	0.8816	0.9054	0.8894	0.8813	**0.8979**	0.8962
	$NIQE$	4.6701	4.3234	4.4591	4.8064	5.6877	4.6742	**4.2267**
	$VIFF$	0.3256	0.4503	0.4298	0.4121	0.5156	**0.5421**	0.4683
	CC_{avg}	0.6741	0.6716	0.6620	0.7266	0.7367	0.7403	**0.7408**
	$SSIM_{avg}$	0.6779	0.6863	0.6906	0.7225	0.6817	0.6972	**0.7385**

4.1 Subjective Evaluation

A good image fusion algorithm should be efficient enough to pick up crucial features selectively from the source image depending on the purpose of fusion application. The fusion result should not introduce (or enhance) extra features or distortion beyond the source images. This section contains visual quality comparison of the fused results with other similar methods. For all the source pairs, results in FFIF, SA and MISF methods have produced irregular blotches in their fused results (Figs. 3, 4, 5, 6c, d, e). It can be attributed to improper capture of dark pixel information from the infrared images. As observed from the results, IFCNN and U2-based methods have introduced grainy effect in the fused images (Figs. 3, 5, 6f, g). Furthermore, in Fig. 4f, the clouds are not visibly distinct whereas in Fig. 4g, the fused results suffer from enhanced contrast. Results from PMGI (Figs. 3, 4, 5, 6h) show comparable performance in terms of visual quality. The results from our method (Figs. 3, 4, 5, 6i) have outperformed other approaches in terms of perceptual quality. It scores better in almost all the objective fusion metrics with natural and realistic appearance validated by the lowest value of $NIQE$ as presented in Table 1. The features have been proportionately captured without missing out on edge clarity, brightness or contrast. Unlike other methods, it has also effectively eliminated the grain effect from the fused results (Fig. 7).

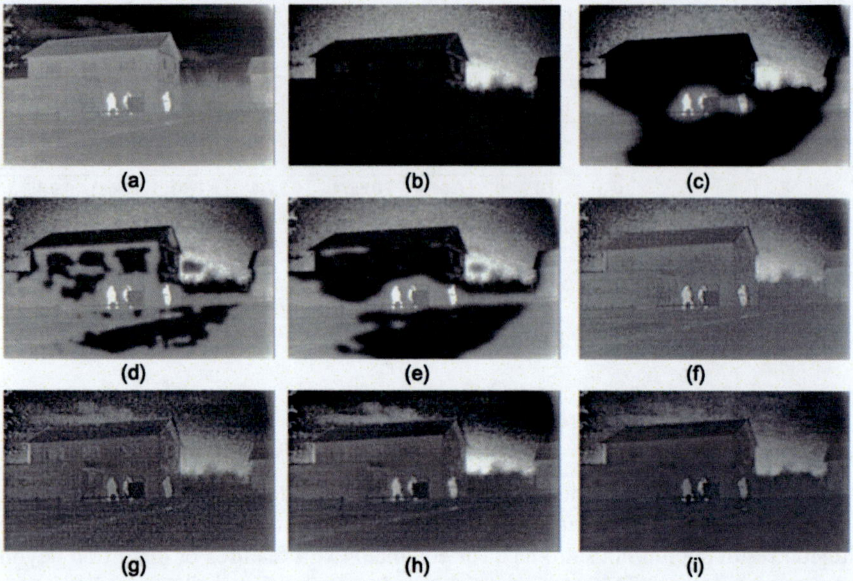

Fig. 3 Source pair from TNO dataset [22] and fused results: **a** IR image; **b** Visible image; **c** FFIF result; **d** SA result; **e** MISF result; **f** IFCNN result; **g** U2 result; **h** PMGI result; **i** Result from proposed method

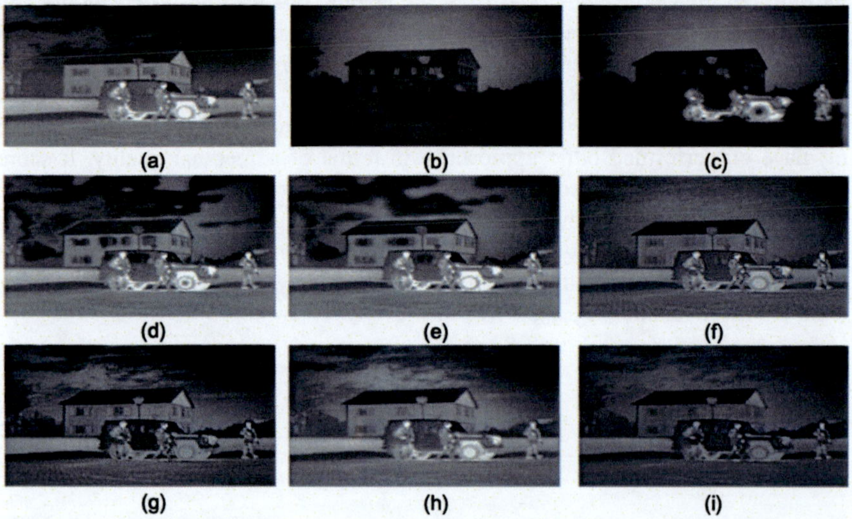

Fig. 4 Source pair from TNO dataset [22] and fused results: **a** IR image; **b** Visible image; **c** FFIF result; **d** SA result; **e** MISF result; **f** IFCNN result; **g** U2 result; **h** PMGI result; **i** Result from proposed method

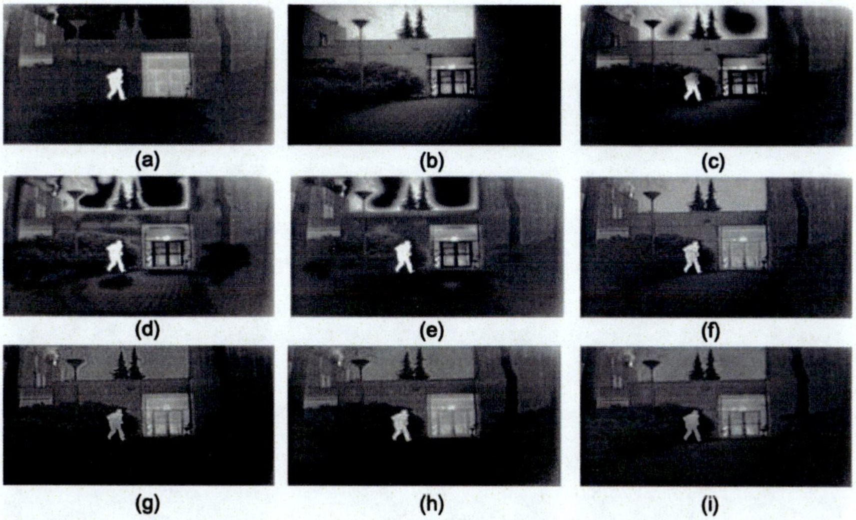

Fig. 5 Source pair from TNO dataset [22] and fused results: **a** IR image; **b** Visible image; **c** FFIF result; **d** SA result; **e** MISF result; **f** IFCNN result; **g** U2 result; **h** PMGI result; **i** Result from proposed method

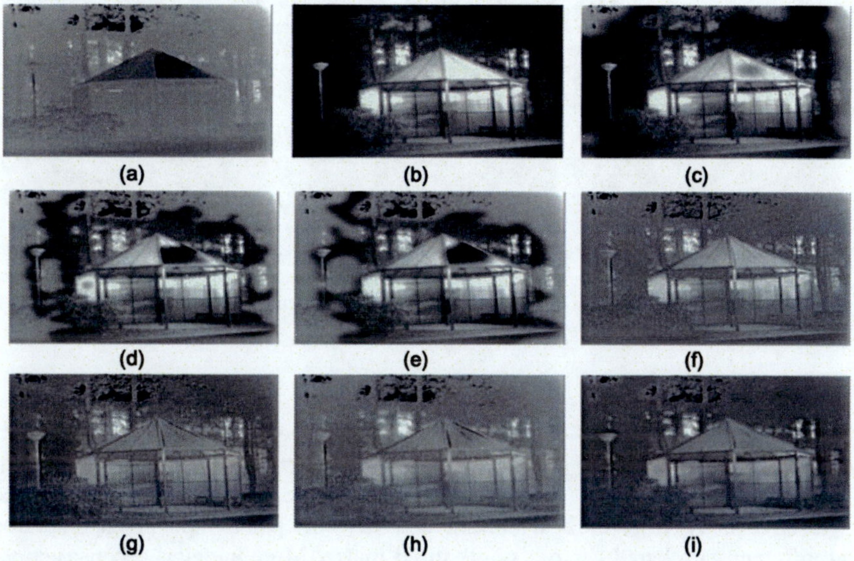

Fig. 6 Source pair from TNO dataset [22] and fused results: **a** IR image; **b** Visible image; **c** FFIF result; **d** SA result; **e** MISF result; **f** IFCNN result; **g** U2 result; **h** PMGI result; **i** Result from proposed method

Fig. 7 Other source pairs from TNO dataset [22] and fused results: **a**, **d**, **g** Visible image; **b**, **e**, **h** IR image; **c**, **f**, **i** Result from proposed method

5 Conclusion

In this paper, a scheme for a fusion of infrared-visible images has been introduced with an objective to get the best relevant features from a pair of input images captured at different wavelengths into a single fused image. Morphological reconstruction-based filters are utilized to bring out the bright and dark features of varying scales which are conjointly compared to select the best of them. As a novel step, multiscale toggle-contrast operators are employed to extract the cumulative gradient features which are refined with respect to the source images using the popular edge-preserving guided filter. The final fused image with improved perceptual quality is composed by superimposing the cumulative feature images along with the refined edges onto a base image. The performance of the proposed method has been compared using seven fusion quality metrics with six other relevant methods on a standard dataset. Experimental results show that our method achieves superior performance in the assessment criteria and performs considerably well.

References

1. Blum, R.S., Liu, Z.: Multi-sensor Image Fusion and Its Applications. CRC Press (2018)
2. Cai, H., Zhuo, L., Chen, X., Zhang, W.: Infrared and visible image fusion based on bemsd and improved fuzzy set. Infrared Phys. Technol. **98**, 201–211 (2019)
3. Chen, J., Li, X., Luo, L., Mei, X., Ma, J.: Infrared and visible image fusion based on target-enhanced multiscale transform decomposition. Inform. Sci. **508**, 64–78 (2020)
4. Ciprián-Sánchez, J.F., Ochoa-Ruiz, G., Gonzalez-Mendoza, M., Rossi, L.: Fire-gan: a novel deep learning-based infrared-visible fusion method for wildfire imagery. Neural Comput. Appl. 1–13 (2021)
5. Dogra, A., Goyal, B., Agrawal, S.: From multi-scale decomposition to non-multi-scale decomposition methods: a comprehensive survey of image fusion techniques and its applications. IEEE Access **5**, 16040–16067 (2017)
6. Guo, Z., Yu, X., Du, Q.: Infrared and visible image fusion based on saliency and fast guided filtering. Infrared Phys. Technol. 104178 (2022)
7. Haghighat, M.B.A., Aghagolzadeh, A., Seyedarabi, H.: A non-reference image fusion metric based on mutual information of image features. Comput. Electr. Eng. **37**(5), 744–756 (2011)
8. Han, Y., Cai, Y., Cao, Y., Xu, X.: A new image fusion performance metric based on visual information fidelity. Inform. Fusion **14**(2), 127–135 (2013)
9. He, K., Sun, J., Tang, X.: Guided image filtering. IEEE Trans. Pattern Anal. Mach. Intel. **35**(6), 1397–1409 (2012)
10. Jin, X., Jiang, Q., Yao, S., Zhou, D., Nie, R., Hai, J., He, K.: A survey of infrared and visual image fusion methods. Infrared Phys. Technol. **85**, 478–501 (2017)
11. Li, H., Wu, X.J., Kittler, J.: Infrared and visible image fusion using a deep learning framework. In: 2018 24th International Conference on Pattern Recognition (ICPR), pp. 2705–2710. IEEE (2018)
12. Li, S., Kang, X., Fang, L., Hu, J., Yin, H.: Pixel-level image fusion: a survey of the state of the art. Inform. Fusion **33**, 100–112 (2017)
13. Li, S., Yang, B., Hu, J.: Performance comparison of different multi-resolution transforms for image fusion. Inform. Fusion **12**(2), 74–84 (2011)
14. Li, W., Xie, Y., Zhou, H., Han, Y., Zhan, K.: Structure-aware image fusion. Optik **172**, 1–11 (2018)
15. Lin, Y., Cao, D., et al.: Adaptive Infrared and Visible Image Fusion Method by Using Rolling Guidance Filter and Saliency Detection. Optik, p. 169218 (2022)
16. Ma, J., Ma, Y., Li, C.: Infrared and visible image fusion methods and applications: a survey. Inform. Fusion **45**, 153–178 (2019)
17. Meyer, F., Serra, J.: Contrasts and activity lattice. Signal Proces. **16**(4), 303–317 (1989)
18. Mittal, A., Soundararajan, R., Bovik, A.C.: Making a "completely blind" image quality analyzer. IEEE Signal Proces. Lett. **20**(3), 209–212 (2012)
19. Patel, A., Chaudhary, J.: A review on infrared and visible image fusion techniques. In: Intelligent Communication Technologies and Virtual Mobile Networks, pp. 127–144. Springer (2019)
20. Piella, G., Heijmans, H.: A new quality metric for image fusion. In: Proceedings 2003 International Conference on Image Processing (Cat. No. 03CH37429), Vol. 3, pp. III–173. IEEE (2003)

21. Salembier, P., Serra, J.: Flat zones filtering, connected operators, and filters by reconstruction. IEEE Trans. Image Proces. **4**(8), 1153–1160 (1995)
22. Toet, A.: The tno multiband image data collection. Data Brief **15**, 249 (2017)
23. Wang, B., Zou, Y., Zhang, L., Li, Y., Chen, Q., Zuo, C.: Multimodal super-resolution reconstruction of infrared and visible images via deep learning. Opt. Lasers Eng. **156**, 107078 (2022)
24. Xu, H., Ma, J., Jiang, J., Guo, X., Ling, H.: U2fusion: a unified unsupervised image fusion network. IEEE Trans. Pattern Anal. Mach. Intell. **44**(1), 502–518 (2020)
25. Zhan, K., Kong, L., Liu, B., He, Y.: Multimodal image seamless fusion. J. Electron. Imaging **28**(2), 023027 (2019)
26. Zhan, K., Xie, Y., Wang, H., Min, Y.: Fast filtering image fusion. J. Electron. Imaging **26**(6), 063004 (2017)
27. Zhang, H., Xu, H., Xiao, Y., Guo, X., Ma, J.: Rethinking the image fusion: a fast unified image fusion network based on proportional maintenance of gradient and intensity. In: Proceedings of the AAAI Conference on Artificial Intelligence, vol. 34, pp. 12797–12804 (2020)
28. Zhang, Y., Liu, Y., Sun, P., Yan, H., Zhao, X., Zhang, L.: Ifcnn: a general image fusion framework based on convolutional neural network. Inform. Fusion **54**, 99–118 (2020)

Extractive Text Summarization Using Statistical Approach

Kartikey Tewari, Arun Kumar Yadav, Mohit Kumar, and Divakar Yadav

Abstract Nowadays, text summarization is an important research area because of its large requirements in society. The usage is increasing in different areas such as education, health care, and legal. In the past, researchers have carried a lot of work on different text summarization techniques. Also, they propose machine learning/deep learning-based approaches for text summarization. This study proposed a novel statistical approach for extractive multi-document text summarization. The proposed approach is based on weight assignment of keywords, in local files and global files. We evaluate the approach of the publicly available "Multi-news" dataset. Results show that the proposed approach outperforms 1.08% in the ROUGE-L $F1$ score on the state-of-the-art method.

Keywords Text summarization · Extractive summarization · Statistical approach · Multi-news · ROUGE · BLEU

1 Introduction

In the era of social media, large number of information is streaming on the Internet. It becomes challenging to retrieve relevant information for this corpus. Regarding text data, it is quit complex and thorough process to gather and identify basic meaning from huge data. Text summarization is the way of retrieving quality information from in brief from documents. Its manual process is time-consuming and difficult for identifying human experts in different areas. Automatic text summarization mechanism can help to resolve the issues of manual text summarization. In the last decade, researchers work on automatic text summarization, and a summary of text is selected from original text followed by the number of documents it summarizes. They demonstrated that broadly classification of documents as single-document text summarization and multi-document text summarization [1–3]. In single-document text summarization, the algorithm summarizes a single document. Thus, each sum-

K. Tewari · A. K. Yadav · M. Kumar (✉) · D. Yadav
National Institute of Technology Hamirpur, Hamirpur, H.P. 177005, India
e-mail: mohit@nith.ac.in

M. Tistarelli et al. (eds.), *Computer Vision and Machine Intelligence*, Lecture Notes in Networks and Systems 586,
https://doi.org/10.1007/978-981-19-7867-8_52

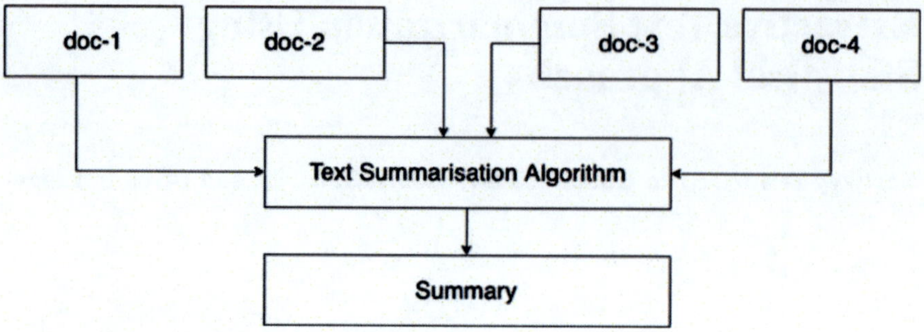

Fig. 1 Multi-document text summarization overview

mary is mapped to a single original text. In multi-document text summarization, the summarizing algorithm generates a summary from a set of documents. Thus, a single summary is mapped to a group of original documents [4]. Each type of text summarization can be further classified based on whether the summary text is generated by selecting the phrases of original text. In extractive text summarization, the generated text contains the same content/phrases as the original text. It is made by selecting some phrases from the original text and concatenating them to make the summary. In abstractive text summarization, the summarization algorithm can may or may not contain the same phrases as the original text. Thus, it can add more phrases or words than the original text [1]. The main applications areas where text summarization are used such as news summarization [5], opinion/sentiment summarization [2], e-mail summarization [4], literature summarization [6], and story/novel summarization [7]. It is also used in domain-specific use cases like biomedical text summarization [8], legal document summarization [1], and scientific paper summarization [9]. In this study, we proposed multi-document text summarization using statistical approach as shown in Fig. 1.

Most summarization process are used machine learning and deep learning approaches or its variants. The main Achilles heel of extractive multi-document text summarization algorithms is that they do not contain the context by which the summary text is being generated. Thus, they generate context-free summary which does not reflect the true information in the text. To resolve the mentioned issues, this study involves following contribution:

- A novel context-dependent multi-document text summarization algorithm is proposed.
- The evaluation of the proposed methodology, along with the competing approaches, is done using a publicly available dataset—*multi news*.
- The results are evaluated on standard evaluation metrics and found 1.08% better than state-of-the-art method on same dataset.

Rest of the paper is organized as follows. Section 2 explains the literature review on text summarization, methodologies used and their effects during summarization. Section 3 describes the proposed algorithm and dataset description. Section 4 discusses the experiments performed on multi-news dataset on different local file weights. It also elucidates on the ROUGE Scores and BLUE scores obtained by the experiments. Section 5 discusses the conclusion of the paper and the future scope of work.

2 Literature Survey

Text summarization is the way of creating a summary of a document. Manual/human experts' text summarization is time taken and difficult to validate. Also, human experts can give a different opinion on same documents, and it is not possible to search human experts for different dimensions. To resolve the mentioned issues, automatic text summarization (ATS) can play an important role. In the past, researchers discuss various classification of text summarization and its implementation methods [10, 11]. Initially, ATS was proposed in the paper [12] for generating abstract summary. Broadly, text summarization is classified as abstractive, extractive, and hybrid on single/multi-document. Extractive text summarization (ETS) provides the result based on given documents. Abstractive text summarization is used generate summary that is not completely dependent on the input documents, and it may include some sentences in the summary that are not present in the input document. Abstractive text summarization is much complex than ETS for multi-document, while the hybrid approach is combination of ETS and abstractive text summarization. The researchers commented that hybrid provides lower quality results as compared to ETS [13]. This section is presented to discuss the different text summarization methods used by the researchers in different dimensions.

In the paper [14], authors work on style specific news stories based on spoken. They evaluated the features of stylic and content-based contribution and concluded that for containing content information, both are significant. A single-document extractive summarization is proposed in the paper [15]. Authors used deep learning-based auto-encoder for extractive query-oriented summarization using term frequency. They add random noise to local term frequency to input sentence to know effect of noise in auto-encoder. In the paper [16], authors proposed sequence-to-sequence (seq2seq) model using deep learning to generate and improve the summary. A reduced redundancy model is proposed in the paper [17]. They used DUC-2005 on proposed solo content outline model for summarization. A fuzzy logic-based text summarization is proposed in the paper [18]. They used DUC-2002 dataset and calculated ROUGE-R values on the proposal of single report summarization of sentences. They used evaluation of scores of words to find the weight of sentences. In current scenario, most of the research is going on application of machine learning and deep learning for text summarization [18, 19]. An extractive text summarization method is proposed in the paper [20]. In this paper, they try to reduce the redundancy of summary as two different documents produce same summery. To meet the mentioned objective, authors used vector space model in topic modeling for extractive text summarization. Again, in the paper [21], authors used multi-document text summarization generation. It describes about PRIMERA (Pyramid-based masked sentence pre-training for Multi-document summarization). It uses a novel pre-trained objective to connect information across documents and teach the model. It also uses an efficient encoder-decoder transformer to simplify the processing of input documents. It has the best result on multi-news dataset, with a ROUGE-L $F1$-score of 25.9. The other methodologies used for extractive summarization are statistical-based method [22, 23], concept-based method [24], topic-based method, clustering-based method [25] and graph-based method [26]. Methods such as semantic-based method [27], machine learning-based method [28, 29], deep learning-based method [30, 31], optimization-based method [32], and fuzzy logic-based methods [33] are also used for extractive text summarization algorithms.

On the study of literature, it can be concluded that extractive summarization is more fast and simpler than abstractive approach. This also generally leads to high quality summary documents than abstractive approaches. The generated summary contains the same tokens as the original text; hence, the information content is similar to the original text [34]. Again, its major disadvantages include redundancy in summary sentences and longer length sentences than average [35]. There is also lack of semantics and cohesion in summary because of erroneous links between sentences and dangling amphora. Sometimes, output summary can also be wrong for text that contain information about several topics [36]. The schematic diagram of extractive text summarization is in Fig. 2.

On the basis of discussion of past work, it can be observed that most of the text summarization has been performed using machine/deep learning models or other approaches that take more time to generate summary but it also does not provide much

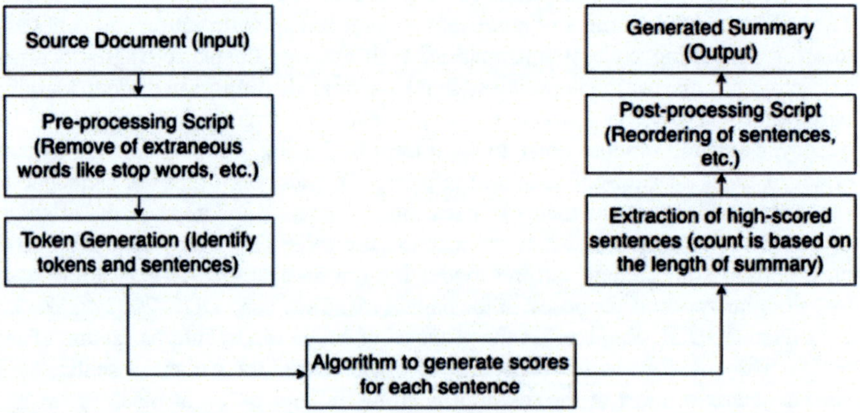

Fig. 2 Schematic diagram of extractive text summarization [37]

validated summary. To resolve the same, this article proposes a statistical approach-based extractive text summarization algorithm that takes less time to generate the summary. As discussed in literature review, it is clear that there is a lack of context-dependent multi-document text summarization algorithms. To resolve the same, we propose a novel approach for context-based extractive text summarization.

3 Proposed Work

This section describe the proposed work based on issued identified in the literature. Past study shows that summarization process follows the context-free multi-document text summarization, that lagging from semantic meaning of the summary. To resolve the mentioned issues, we propose an algorithm for context-based extractive text summarization.

The proposed algorithm performs the following steps to generate the summary: (1) Preprocessing of text, (2) Token generation of documents, (3) Token weight formation for each document (local file weight), (4) Cumulative token weight formation (global file weight), (5) Generation of complex token weight per document, (6) Scoring of sentences, and (7) Generation of summary. Algorithms 1, 2, and 3 are defined as follows:

Algorithm 1 Generate LFTFD for a file

```
procedure GET_LFTFD(file)
    map<string,int> lftfd
    for word in file do
        if word ≠ stopword then
            lftfd[word]+=1
        end if
    end for
    return lftfd
end procedure
```

Algorithm 1 is used to generate local file term frequency document (LFTFD). In this regard, it calls the function GET. LFTFD() and pre-processes the text and removes all stop words from the documents followed by conversion of all texts to tokens of words and sentences. At the end, it generates the frequency of each word in a document.

Algorithm 2 Generate GFTFD for a file

```
procedure GET_GFTFD(vector < map < string, int >> list_lftfd)
    map<string,int> gftfd
    for file_lftfd in list_lftfd do
        for word in file_lftfd do
            gftfd[word] += file_lftfd[word]
        end for
    end for
    return gftfd
end procedure
```

Algorithm 2 is used to generate global file term frequency(GFTFD) of each token in the document. All the files along with its LFTFD is passed to the gen_gftfd function. It generates the global word frequency of each word in the whole set. Algorithm 3 is used to generate the summary of documents. It creates local file weight and global file weight along with LFTFD and GFTFD of the documents.

Algorithm 3 Generate summary for a file

```
procedure GET_SUMMARY(lfw, gfw, lgtfd, gftfd)
    map<sentence,int> sentence_score_all
    for sentence in sentences do
        sentence_score=0
        for word in sentence do
            if word ≠ stop_words then
                word_score = (lfw*lftfd[word]) + (gfw*gftfd[word])
                sentence_score += word_score
            end if
        end for
        sentence_score_all[sentence]=sentence_score
    end for
    summary_sentence = get_top_frequency(sentence_score_all)
    summary = catenate(summary_sentence)
    return summary
end procedure
```

It generated the summary by iterating through all the documents and its words. It generates the score of each word as (LFW * LFTFD (word)) + (GFW * GFTFD (word)). The local score of a sentence is generated by summing all the scores of individual words in it. This process is continued for all sentences in a document-set. The sentences with the highest score are selected for generating the summary. The schematic diagram of the complete process of summary generation is shown in Fig. 3.

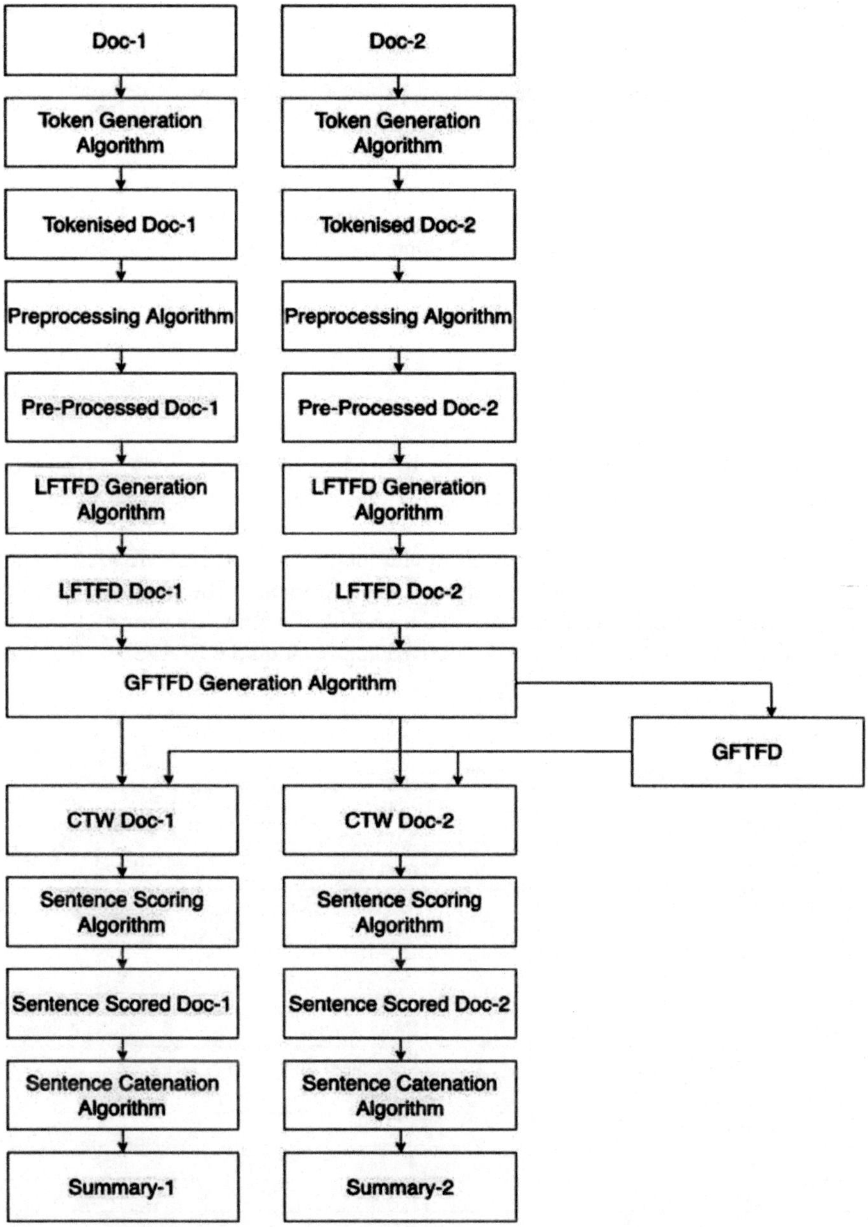

Fig. 3 Block diagram of proposed work

3.1 Dataset

To evaluate the proposed model, this study uses publicly available multi_news dataset [38] (version 1.0.0) from TensorFlow. It contains news articles from the website [39] including human-generated summary of all the articles. Thus, it acts as a good dataset for benchmarking the proposed work, against the previous work done on text summarization. The dataset is divided into three parts—Train Split (44,972 document-set, 80%), Test Split (5622 document-set, 10%), and Validation Split (5622 document-set, 10%).

4 Experiments and Results

To evaluate the proposed work, this study uses MacOS operating system. The tools used, along with their version, to program and run the code is depicted in Table 1. The configuration is shown in Table 2.

Based on the proposed method, the popular metrics such as ROUGE (1, 2, L) and BLEU (1, 2, 3, 4) are calculated to estimate the performance. The results on ROUGE metrics are shown in Table 3, and the results on BLEU metrics are shown in Table 4. Figure 4 depicts the performance of proposed approach on the ROUGE-L $F1$ score metric.

Table 1 Tools used along with their versions

Tool	Version
Python	3.9.13
Makefile	GNU make 3.81
Bash script	5.1.16(1)-release (x86_64-apple-darwin21.1.0)

Table 2 Configuration of the system used in experiments

Type	Specification
Processor name	Quad-Core Intel Core i5
Processor speed	1.1 GHz
Number of processors	1
Total number of cores	4
L2 cache (per core)	512 KB
L3 cache	6 MB
Hyper-threading technology	Enabled
Memory	8 GB

Table 3 ROUGE scores obtained by the proposed approach for different LFW values

LFW	ROUGE-1			ROUGE-2			ROUGE-L		
	Rec.	Prec.	*F*1	Rec.	Prec.	*F*1	Rec.	Prec.	F1
91	49.0335	21.9592	28.5167	21.1759	8.1180	10.8579	45.0624	20.1346	26.1624
92	49.0311	21.9567	28.5142	21.1719	8.1159	10.8553	45.0578	20.1321	26.1595
93	49.0278	21.9579	28.5133	21.1702	8.1176	10.8552	45.0543	20.1336	26.1590
94	49.0324	21.9561	28.5138	21.1734	8.1166	10.8553	45.0593	20.1326	26.1603
95	49.0341	21.9552	28.5139	21.1753	8.1149	10.8537	45.0589	20.1303	26.1585
96	49.0337	21.9536	28.5137	21.1769	8.1158	10.8550	45.0524	20.1270	26.1557
97	49.0394	21.9504	28.5140	21.1837	8.1120	10.8545	45.0569	20.1236	26.1558
98	49.0442	21.9441	28.5090	21.1944	8.1065	10.8500	45.0622	20.1193	26.1526
99	49.1014	21.9534	28.5340	21.2434	8.1058	10.8572	45.1206	20.1325	26.1802
100	49.1241	21.7348	28.4279	21.1009	7.7421	10.5347	45.0585	19.8906	26.0309

Here, LFW is in %

Table 4 BLEU scores achieved by the proposed approach at different LFW values

LFW (%)	BLEU-1	BLEU-2	BLEU-3	BLEU-4
91	22.7421	8.8985	5.0520	3.7362
92	22.7367	8.8952	5.0501	3.7347
93	22.7321	8.8932	5.0489	3.7338
94	22.7280	8.8927	5.0484	3.7330
95	22.7210	8.8904	5.0462	3.7308
96	22.7113	8.8887	5.0463	3.7311
97	22.6926	8.8845	5.0439	3.7286
98	22.6527	8.8702	5.0337	3.7191
99	22.5822	8.8575	5.0252	3.7083
100	21.5297	8.3728	4.5559	3.2323

Here LFW is in %

4.1 Range of LFW and GFW Used

The constant sum of LFW and GFW is taken to be 100 as it is a gradual gradient which gives a good weight distribution to both the local file weight (LFW) and global file weight (GFW).

Further, since local file weight (LFW) has to give more weight in summary generation, the range of LFW-GFW split in the experiment was from LFW-GFW split of 91–9% split (local file weight is 91% and global file weight is 9%) to 100–0% split (local file weight is 100% and global file weight 0%).

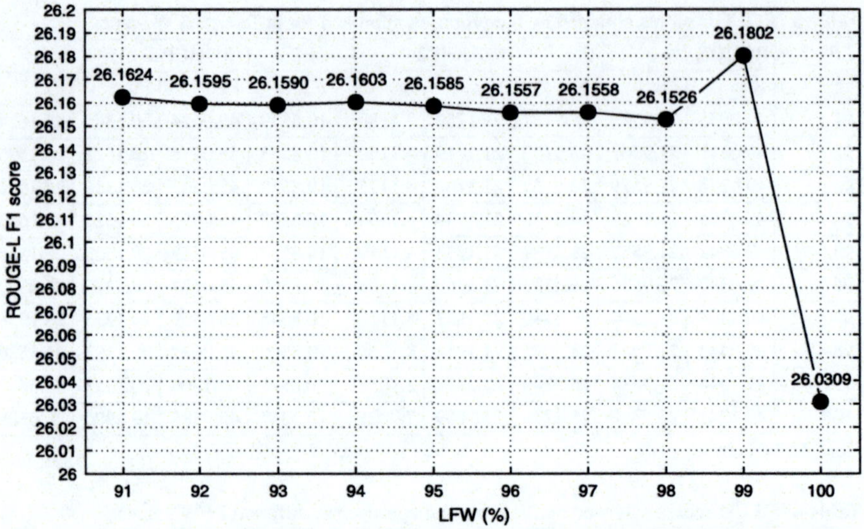

Fig. 4 ROUGE-L $F1$ score of the proposed approach with different LFW values

4.2 Length of Generated Summary

The length of generated summary was taken to be around 40% of original text. Also, all results were run on the test split of the multi-news dataset.

The following observations can be made from the above results. Firstly, if the generated token is weighted solely from local file and the weight calculation from global file weight is given no value, then the generated summary has a lower quality. As can be seen from the graphs and tables, all measured metrics have a low point at LFW of 100 except recall for ROUGE-1. Therefore, it is important that a proportion of weight is generated from global file weight to generate context-based summaries that result in better performance. Secondly, the proposed method provides the highest ROUGE-L $F1$-score (26.1802) at local file weight and global file weight split of 99% and 1%, respectively. Thus, this split seems to be the most optimal weight distribution for generating summaries of a document-set, while incorporating both local and global context from the whole set of documents.

4.3 Comparison with State-of-the-art Literature

In the paper [21], the authors propose an algorithm, "PRIMERA" for multi-document text summarization. The algorithm was also run on multi-news dataset. A comparison of results of PRIMERA and the proposed algorithm is shown in Table 5.

Table 5 Comparison with state-of-the-art on ROUGE-L $F1$ metric

S. No.	Evaluation metric	PRIMERA	Proposed model
1	ROUGE-1 $F1$-score	**49.9**	28.534
2	ROUGE-2 $F1$-score	**21.1**	10.8572
3	ROUGE-L $F1$-score	25.9	**26.1802**

Here, proposed approach is used with 0.99 LFW

5 Conclusion and Future Work

The paper proposes a novel approach to context-dependent multi-document extractive text summarization. The proposed approach is evaluated on *multi_news* dataset on popular metrics like ROUGE (1, 2, L) scores and BLEU scores. The proposed algorithm provides a ROUGE-L $F1$-score of 26.18, that outperforms the state-of-the-art method by 1.08%. It also achieves comparable results to state-of-the-art approaches in other metrics such as precision and recall. Furthermore, the proposed offers some significant advantages. It does not suffer from the standard disadvantages of machine and deep learning-based approaches such as the requirement of a large dataset for training, large amount of training time, and inherent data biases.

References

1. Anand, D., Wagh, R.: Effective deep learning approaches for summarization of legal texts. J. King Saud. Univ. Comput. Inf., Sci (2019)
2. Bhargava, R., Sharma, Y.: Deep extractive text summarization. Procedia Comput. Sci. **167**, 138–146 (2020)
3. Lovinger, J., Valova, I., Clough, C.: Gist: general integrated summarization of text and reviews. Soft Comput. **23**(5), 1589–1601 (2019)
4. Muresan, S., Tzoukermann, E., Klavans, J.L.: Combining linguistic and machine learning techniques for email summarization. In: Proceedings of the ACL 2001 Workshop on Computational Natural Language Learning (ConLL) (2001)
5. Radev, D., Hovy, E., McKeown, K.: Introduction to the special issue on summarization. Comput. linguist. **28**(4), 399–408 (2002)
6. Mihalcea, R., Ceylan, H.: Explorations in automatic book summarization. In: Proceedings of the 2007 Joint Conference on Empirical Methods in Natural Language Processing and Computational Natural Language Learning (EMNLP-CoNLL), pp. 380–389 (2007)
7. Kazantseva, A., Szpakowicz, S.: Summarizing short stories. Comput. Linguist. **36**(1), 71–109 (2010)
8. Menéndez, H.D., Plaza, L., Camacho, D.: Combining graph connectivity and genetic clustering to improve biomedical summarization. In: 2014 IEEE Congress on Evolutionary Computation (CEC), pp. 2740–2747. IEEE (2014)
9. Alampalli Ramu, N., Bandarupalli, M.S., Nekkanti, M.S.S., Ramesh, G.: Summarization of research publications using automatic extraction. In: International Conference on Intelligent Data Communication Technologies and Internet of Things, pp. 1–10. Springer (2019)
10. Meshram, S., Anand Kumar, M.: Long short-term memory network for learning sentences similarity using deep contextual embeddings. Int. J. Inf. Technol. **13**(4), 1633–1641 (2021)

11. Sintayehu, H., Lehal, G.: Named entity recognition: a semi-supervised learning approach. Int. J. Inf. Technol. **13**(4), 1659–1665 (2021)
12. Luhn, H.P.: A statistical approach to mechanized encoding and searching of literary information. IBM J. Res. Dev. **1**(4), 309–317 (1957)
13. Nallapati, R., Zhou, B., Gulcehre, C., Xiang, B., et al.: Abstractive text summarization using sequence-to-sequence rnns and beyond. arXiv preprint arXiv:1602.06023 (2016)
14. Abolhassani, M., Fuhr, N.: Applying the divergence from randomness approach for content-only search in xml documents. In: European Conference on Information Retrieval, pp. 409–419. Springer (2004)
15. Text summarization using unsupervised deep learning: Expert Syst. Appl. **68**, 93–105 (2017)
16. Li, P., Lam, W., Bing, L., Wang, Z.: Deep recurrent generative decoder for abstractive text summarization. arXiv preprint arXiv:1708.00625 (2017)
17. Alguliev, R.M., Aliguliyev, R.M., Mehdiyev, C.A.: psum-sade: a modified p-median problem and self-adaptive differential evolution algorithm for text summarization. Appl. Comput. Intell. Soft Comput. **2011** (2011)
18. Patel, D., Chhinkaniwala, H.: Fuzzy logic-based single document summarisation with improved sentence scoring technique. Int. J. Knowl. Eng. Data Mining **5**(1–2), 125–138 (2018)
19. Yang, M., Li, C., Shen, Y., Wu, Q., Zhao, Z., Chen, X.: Hierarchical human-like deep neural networks for abstractive text summarization. IEEE Trans. Neural Netw. Learn. Syst. **32**(6), 2744–2757 (2020)
20. Belwal, R.C., Rai, S., Gupta, A.: Text summarization using topic-based vector space model and semantic measure. Inf. Process. Manage. **58**(3), 102536 (2021)
21. Xiao, W., Beltagy, I., Carenini, G., Cohan, A.: Primer: pyramid-based masked sentence pre-training for multi-document summarization. arXiv preprint arXiv:2110.08499 (2021)
22. Afsharizadeh, M., Ebrahimpour-Komleh, H., Bagheri, A.: Query-oriented text summarization using sentence extraction technique. In: 2018 4th International Conference on Web Research (ICWR), pp. 128–132. IEEE (2018)
23. Gambhir, M., Gupta, V.: Recent automatic text summarization techniques: a survey. Artif. Intell. Rev. **47**(1), 1–66 (2017)
24. Sankarasubramaniam, Y., Ramanathan, K., Ghosh, S.: Text summarization using wikipedia. Inf. Process. Manage. **50**(3), 443–461 (2014)
25. Nazari, N., Mahdavi, M.: A survey on automatic text summarization. J. AI Data Mining **7**(1), 121–135 (2019)
26. Baralis, E., Cagliero, L., Mahoto, N., Fiori, A.: Graphsum: Discovering correlations among multiple terms for graph-based summarization. Inf. Sci. **249**, 96–109 (2013)
27. Mashechkin, I., Petrovskiy, M., Popov, D., Tsarev, D.V.: Automatic text summarization using latent semantic analysis. Program. Comput. Softw. **37**(6), 299–305 (2011)
28. Alguliyev, R.M., Aliguliyev, R.M., Isazade, N.R., Abdi, A., Idris, N.: Cosum: Text summarization based on clustering and optimization. Expert Syst. **36**(1), e12340 (2019)
29. John, A., Premjith, P., Wilscy, M.: Extractive multi-document summarization using population-based multicriteria optimization. Expert Syst. Appl. **86**, 385–397 (2017)
30. Cheng, J., Lapata, M.: Neural summarization by extracting sentences and words. arXiv preprint arXiv:1603.07252 (2016)
31. Kobayashi, H., Noguchi, M., Yatsuka, T.: Summarization based on embedding distributions. In: Proceedings of the 2015 Conference on Empirical Methods in Natural Language Processing, pp. 1984–1989 (2015)
32. Sanchez-Gomez, J.M., Vega-Rodriguez, M.A., Perez, C.J.: Experimental analysis of multiple criteria for extractive multi-document text summarization. Expert Syst. Appl. **140**, 112904 (2020)
33. Kumar, A., Sharma, A.: Systematic literature review of fuzzy logic based text summarization. Iran. J. Fuzzy Syst. **16**(5), 45–59 (2019)
34. Tandel, A., Modi, B., Gupta, P., Wagle, S., Khedkar, S.: Multi-document text summarization-a survey. In: 2016 International Conference on Data Mining and Advanced Computing (SAPIENCE), pp. 331–334. IEEE (2016)

35. Gupta, V., Lehal, G.S.: A survey of text summarization extractive techniques. J. Emerg. Technol. Web Intell. **2**(3), 258–268 (2010)
36. Moratanch, N., Chitrakala, S.: A survey on extractive text summarization. In: 2017 International Conference on Computer, Communication and Signal Processing (ICCCSP), pp. 1–6. IEEE (2017)
37. El-Kassas, W.S., Salama, C.R., Rafea, A.A., Mohamed, H.K.: Automatic text summarization: A comprehensive survey. Expert Syst. Appl. **165**, 113679 (2021)
38. Fabbri, A.R., Li, I., She, T., Li, S., Radev, D.R.: Multi-news: a large-scale multi-document summarization dataset and abstractive hierarchical model (2019)
39. Newser.com. www.newser.com

Near-Infrared Hyperspectral Imaging in Tandem with Machine Learning Techniques to Identify the Near Geographical Origins of Barley Seeds

Tarandeep Singh, Apurva Sharma, Neerja Mittal Garg, and S. R. S. Iyengar

Abstract The nondestructive identification of the geographical origins of the seeds is a crucial step in the food industry. The seeds from near geographical origins are challenging to be separated due to identical climatic and agronomic conditions. The current study implemented the idea of combining near-infrared hyperspectral imaging (NIR-HSI) with machine learning to distinguish barley seeds concerning their geographical origins. Hyperspectral images of barley seeds from four near geographical origins were captured within the range of 900–1700 nm. Sample-wise spectra were extracted from the hyperspectral images and pretreated with different spectral preprocessing techniques, viz., standard normal variate (SNV), multiplicative scatter correction (MSC), Savitzky–Golay smoothing (SGS), Savitzky–Golay first derivative (SG1), Savitzky–Golay second derivative (SG2), and detrending. Unprocessed and preprocessed sample-wise spectra were given as input to four different machine learning models. Support vector machines (SVMs), *K*-nearest neighbors (KNNs), random forest (RF), and partial least squares discriminant analysis (PLS-DA) were used for the classification based on the 1D spectral features. Among these classifiers, SVM showed the best classification accuracy of 93.66% when applied with the SG2 preprocessing technique. The results revealed the significance of using hyperspectral and machine learning to make a clear, fast, and accurate difference among barley varieties based on their geographical origins.

Keywords Barley · Geographical origins · Near-infrared hyperspectral imaging

T. Singh · A. Sharma · N. M. Garg (✉)
CSIR-Central Scientific Instruments Organisation, Chandigarh 160030, India
e-mail: neerjamittal@csio.res.in

Academy of Scientific and Innovative Research (AcSIR), Ghaziabad 201002, India

S. R. S. Iyengar
Indian Institute of Technology Ropar, Ropar, Punjab 140001, India

M. Tistarelli et al. (eds.), *Computer Vision and Machine Intelligence*,
Lecture Notes in Networks and Systems 586,
https://doi.org/10.1007/978-981-19-7867-8_53

1 Introduction

Barley is one of the oldest cultivated grains, ranked fourth (i.e., after wheat, maize, and rice) among most widely grown crops worldwide. It is known to be developed in extreme climatic conditions and is enriched with various kinds of vitamins, minerals, and protein [1]. It is widely used in different applications, such as food, feed, and the malting industry for preparing beer [2]. Assuring the quality of barley has become one of the essential steps in the food and agriculture industry. Geographical origin is one of the factors that can affect the quality of seed and its commercial price. Differences in soil, climate, seasons, and environment depend on geographical origin. These environmental variations affect the chemical and physical properties of the same kind of the seed [3]. With the variation in the region, the barley seeds show diversity in features and attributes such as shape, size, texture, nutritional value, and taste [4]. The seeds from near geographical origins are challenging to be separated due to identical climatic and agronomic conditions. Traditionally, manual segregation and chemical-based methods (e.g., mineral and nutrient composition, high-performance liquid chromatography, etc.) were applied to differentiate the seeds' geographical origins. These methods are time-consuming, expensive, and destructive. Recently, the authors measured the nutrients and mineral elements of the barley samples to identify five geographical origins using linear discriminant analysis (LDA). The study exhibited a significant difference in nutritional value and minerals in barley from various areas [5]. However, there is a need to develop a quick and nondestructive approach for determining barley geographical origin.

Several optical-based nondestructive techniques, including digital imaging, near-infrared spectroscopy (NIRS), and hyperspectral imaging (HSI), have been developed. Among these, spectroscopy is the most commonly used method, which provides spectral information for fast and nondestructive analysis. Recently, several studies have widely employed spectroscopy combined with machine learning techniques for seed discrimination based on geographical origin. Giraudo at el. [6] used NIRS to obtain spectral features and developed a PLS-DA model for classifying coffee beans cultivated from different origins. Zhao at el. [7] determined the geographical origin of the wheat seeds using NIRS coupled with LDA and PLS-DA. Richter at el. [8] employed SVM to classify white asparagus based on features obtained from NIRS. However, NIRS suffers from some drawbacks like the position of the samples and gathering spectral information in a relatively small region of the samples.

HSI is another famous optical technique researchers adopt for nondestructive seed quality evaluation. It has the ability to combine spectral as well as spatial features in a single system. Spectral and morphological features can be attained by capturing images in 3D format (also known as hypercube). Gao at el. [9] classified Jatropha curcas seeds using the features obtained from the HSI system combined with the least squared support vector machines. Wang at el. [3] successfully identified the origins of two maize varieties using HSI and PLS-DA. Mo at el. [10] discriminated white rice from different origins using the HSI system and establishing the PLS-DA model. Sun at el. [11] employed HSI to identify the four origins of the rice seeds with the help of SVM.

Nevertheless, no reports are available to identify the geographical origins of barley seeds using the HSI system. The main objective of this study is to examine the effectiveness of using NIR-HSI combined with machine learning techniques to discriminate barley seeds from different geographical origins.

2 Material and Methods

2.1 Barley Samples

The barley seeds from four near geographical origins were provided by certified seed growers in India, namely, National Bureau of Plant Genetic Resources (Delhi), Indian Institute of Wheat and Barley Research (Karnal, Haryana), University Seed Farm (Ladhowal, Ludhiana, Punjab), Rajasthan Agricultural Research Institute (Durgapura, Jaipur, Rajasthan). The seeds comprised 29 hulled and 6 naked barley varieties harvested in four different years (2016–2019). The seeds with uniform morphology and without visual defects were chosen for the experimental work. In total, 5544 seeds were chosen from each geographical origin. Hence, the total number of seeds included in the study was 22,176 (4 geographical origins × 5544 seeds).

2.2 Hyperspectral Image Acquisition and Calibration

The seeds from different geographical origins were placed onto a seed holding plate (Fig. 1a). The plate was designed with 72 wells to carry seeds and painted with matt black color. The hyperspectral images of both the sides of the seeds were obtained by a push-broom HSI system in the NIR range (900–1700 nm). The typical HSI system is constituted by a camera, spectrograph, halogen light source, linear translation stage to carry samples, focusing lens, a system covering box to obstruct stray light, and a computer with data acquisition software (Fig. 1b). The acquired hypercube was corrected by using standard black and white reference. The camera used in the current study captured the 2D images at 168 wavelengths with 4.9 nm spectra resolution. Only the wavelengths from 955.62 nm to 1668.87 nm (total of 147 wavelengths) were taken into account to remove the noisy images present at both the edges of the wavelength range.

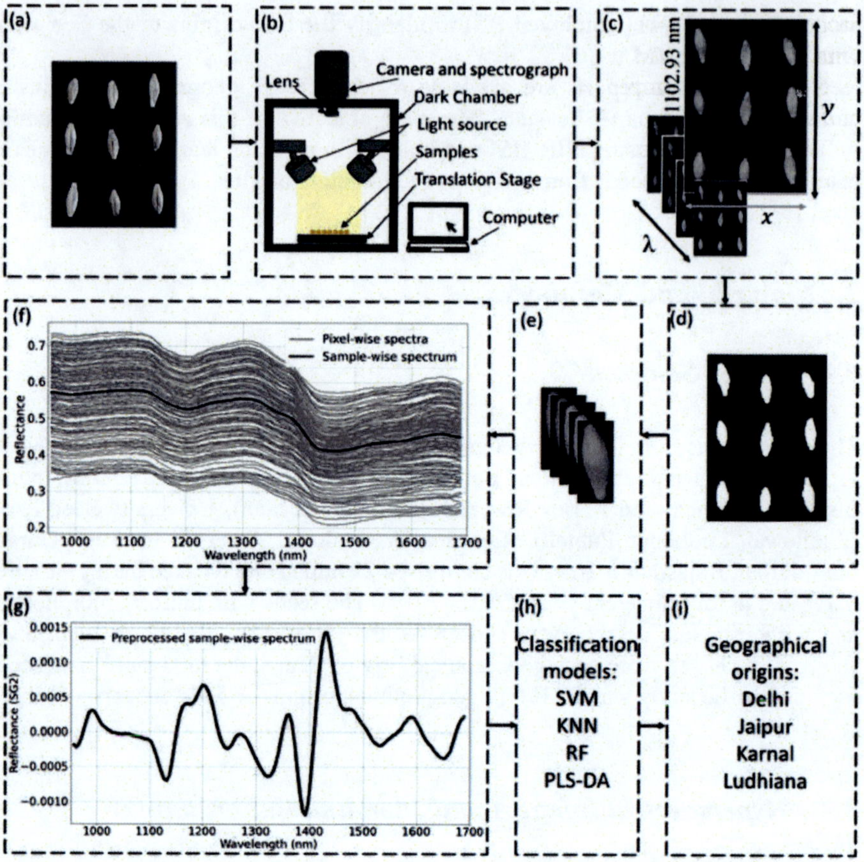

Fig. 1 Hyperspectral data analysis: **a** seed samples, **b** hyperspectral imaging system, **c** hypercube $I\ (x \times y \times \lambda)$, **d** binary mask, **e** region of interest, **f** pixel-wise spectra and sample-wise spectrum, **g** preprocessed sample-wise spectrum, **h** application of different machine learning models, and **i** final output of the different models

2.3 Image Processing and Extraction of Sample-Wise Spectra

Figure 1c illustrates the hypercube $I\ (x \times y \times \lambda)$ stored in the computer system acquired by the camera. The hypercube contains the 2D images $(x \times y)$ at different wavelengths (λ). Firstly, the dead pixels present with the hypercube were removed by applying a median filter. Next, the image at 1102.93 nm was picked by observing the high contrast between the reflectance values of seeds and the plate. This image was used to separate the seeds from the background plate by applying the threshold segmentation technique. The threshold value of 0.27 was nominated to select the seeds' area and discard the background area belonging to the plate. All the pixels' values greater than 0.27 were considered as pixels corresponding to the seeds and background otherwise. This step resulted in a binary mask which is shown in Fig. 1d. This binary mask was used to select the region of interest (ROI) belonging to each seed. Figure 1e exemplifies the ROI of a seed that can be considered as a hypercube of a single seed.

Two types of spectral data are extracted from the ROIs: pixel-wise and sample-wise spectra [12]. Pixel-wise spectra comprise the spectra corresponding to each pixel present within the ROI. A sample-wise spectrum is computed by averaging the pixel-wise spectra. Figure 1f depicts the pixel-wise spectra and sample-wise spectrum extracted from the ROI of a seed. The NIR spectra suffer from undesirable effects, e.g., scattering of the radiation [13]. In this study, the sample-wise spectra were pretreated by six widely used spectral preprocessing techniques to remove the undesirable effects. The sample-wise spectrum preprocessed by the SG2 method is shown in Fig. 1g.

2.4 Classification Models' Development and Validation

The ultimate goal of hyperspectral image analysis is to draw useful information from spectral data. The raw and the preprocessed sample-wise spectra were used as inputs to the different machine learning models (Fig. 1h). The final output of the models was the performance in classifying the spectra belonging to one of the four geographical origins (Fig. 1i). In our study, four classification models were implemented: SVM, KNN, RF, and PLS-DA. Among them, SVM, KNN, and RF are nonlinear classifiers, whereas PLS-DA is a linear classifier. SVM model forms a decision boundary to segregate the *n*-dimensional input data into the desired number of classes [9]. In this study, the SVM model was involved with a radial basis function (RBF) to obtain the higher dimensional features. KNN model considers the similarity between the unknown and known samples to classify them into a class well-suited category [12]. RF is constituted by several decision trees developed on different subsets of the dataset that operates on the ensemble technique. PLS-DA is a supervised type of principal component analysis that attains the dimensionality reduction by projecting the input data to a lesser number of latent vectors (LVs) [3].

The barley spectral dataset was divided into 80% training and 20% testing sets. The hyperparameters of the models and the best preprocessing technique were selected by applying five-fold cross-validation on the training set. The hyperparameters tuned in the current study were: the penalty parameter and kernel function parameter for SVM, number of neighbors for KNN, number of the decision tree in RF, and number of LVs for PLS-DA. The performance of the models was compared with the different metrics retrieved from the confusion matrix. The confusion matrix is incorporated to compute true positives, true negatives, false positives, and false negatives. Subsequently, accuracy, precision, recall, and F1-score are obtained. The methodology adopted in the current study was implemented using Python language. Mainly, 'Spectral', 'OpenCV', 'Skimage', 'Sklearn', and 'Numpy' libraries were explored to analyze the hyperspectral data.

3 Results and Discussion

The mean of sample-wise spectra extracted from barley seeds of four near geographical origins (Delhi, Jaipur, Karnal, and Ludhiana) is shown in Fig. 2. All the spectral curves of different geographical origins have similar trends with the difference in the reflectance intensity. These similarities and differences indicate that seeds of different geographical origins have identical internal chemical composition, but the individual constituents are different in content. As illustrated in Fig. 2, prominent absorption peaks were witnessed at around 980, 1200, and 1450 nm. These peaks are related to different constituents, including moisture (980, 1450 nm), carbohydrates or starch (980, 1200 nm), and protein (1450 nm) [12].

The sample-wise spectra extracted from 22,176 (4 geographical origins × 5544 seeds) barley seeds were used to develop different classification models. The sample-wise spectra were extracted from the crease up and crease down side of the seeds, resulting in a dataset of 44,352 sample-wise spectra. The dataset was divided into 35,398 spectra as the training set and 8954 spectra as the testing set. The five-fold cross-validation was applied to the training set, where 28,244 spectra were used to train, and the remaining 7154 spectra were kept to validate the model.

The performance comparison of the four machine learning models (SVM, KNN, RF, and PLS-DA) with the best combination of the hyperparameter(s) and preprocessing technique is shown in Fig. 3. The optimal hyperparameters for the models were: a penalty parameter of 32 and kernel function parameter of 0.0078125 for SVM, 7 neighbors for KNN, 900 decision trees for RF, and 40 LVs for PLS-DA. SG2 spectral preprocessing technique was found to be the best for the SVM, KNN, and SVM, whereas PLS-DA was well suited for unprocessed data. The validation accuracy of 93.41 ± 0.48, 83.24 ± 0.22, 80.93 ± 0.38, and 70.21 ± 0.27 was achieved in the case of SVM, KNN, RF, and PLS-DA, respectively.

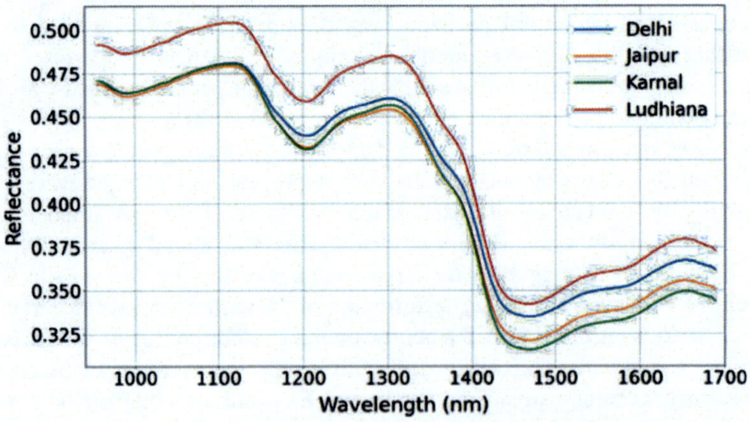

Fig. 2 Mean sample-wise spectrum for four geographical origins of barley seeds

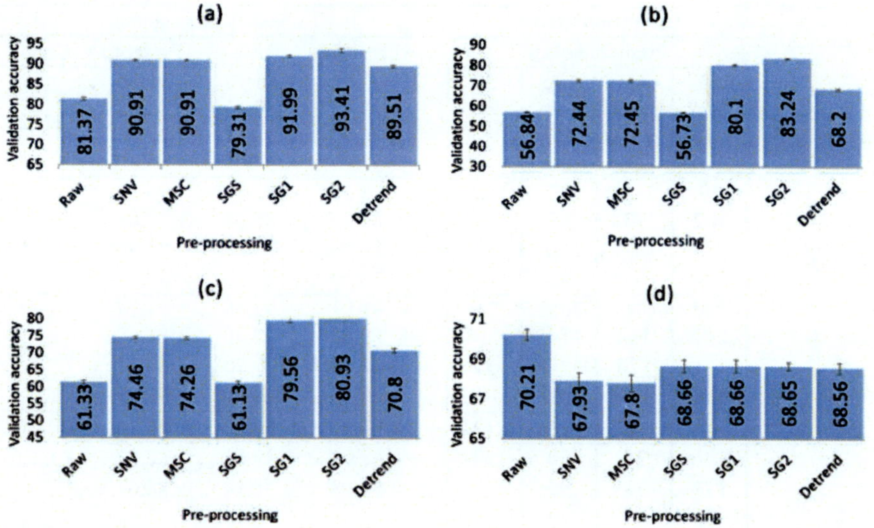

Fig. 3 Validation accuracy of **a** SVM, **b** KNN, **c** RF, and **d** PLS-DA model combined with spectral preprocessing methods

The final comparison of all the models was carried out on the same testing set (8954 spectra) after training them with the same training set (35,398 spectra). SVM model outperformed other classifiers with a testing set accuracy of 93.66% (Fig. 4). The model achieved more than 93% accuracy for both sides (i.e., crease up and crease down side) of the seeds. This advocated the model's suitability for scanning the seeds from either side in an online detection setup to identify their geographical origins. The second highest accuracy of 84.70% was achieved by KNN, followed by 82.09% in the case of RF, followed by 70.05% for PLS-DA. The SVM outperformed the other models because higher-dimensional features retrieved by RBF were more capable of discriminating the spectra of seeds from different geographical origins. Moreover, spectral preprocessing treatment was needed to improve the model's performance.

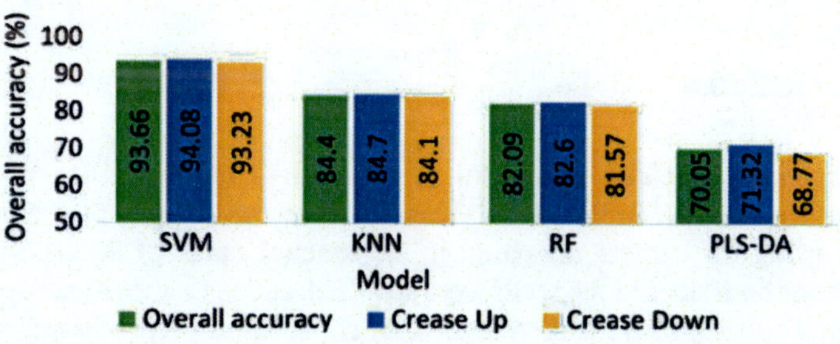

Fig. 4 Classification accuracy of different models on the testing set

Actual class	Confusion matrix – Predicted class: D	J	K	L	Classification report – Precision	Recall	F1-score	Count
D	2075	45	106	36	90.97	91.73	91.35	2262
	1887	106	151	118	77.46	83.42	80.33	
	1783	154	213	112	76.92	78.82	77.86	
	1448	326	282	206	67.95	64.01	65.92	
J	59	2122	39	6	95.37	95.33	95.35	2226
	162	1933	92	39	87.19	86.84	87.01	
	173	1875	105	73	83.59	84.23	83.91	
	174	1897	82	73	71.42	85.22	77.71	
K	107	46	2055	36	91.91	91.58	91.74	2244
	241	144	1745	114	85.62	77.76	81.50	
	191	176	1778	99	81.00	79.23	80.11	
	414	401	985	444	65.58	43.89	52.59	
L	40	12	36	2134	96.47	96.04	96.26	2222
	146	34	50	1992	88.02	89.65	88.83	
	171	38	99	1914	87.08	86.14	86.61	
	95	32	153	1942	72.87	87.40	79.48	

Legend – D: Delhi, J: Jaipur, K: Karnal, L: Ludhiana. Color coding: SVM, KNN, RF, PLS-DA.

Fig. 5 Confusion matrix and classification report of different machine learning models

The confusion matrix and the classification report (precision, recall, and $F1$-score) of all the models for each geographical origin are presented in Fig. 5. The SVM model achieved the best performance on the seeds collected from Ludhiana origin ($F1$-score = 96.26%), followed by KNN ($F1$-score = 88.83%), RF ($F1$-score = 86.61%), and PLS-DA ($F1$-score = 79.48%). In the case of the SVM model, most of the predicted and actual spectra match. Out of 2222 spectra belonging to Ludhiana origin, 96.04% were correctly classified. The lowest value of the $F1$-score was 91.35% for the spectra extracted from the Delhi origin seeds. For seeds from this geographical origin, out of 2262, 91.73% of the spectra were correctly classified.

4 Conclusions

This study was conducted to nondestructively identify barley seeds near geographical origins using the NIR-HSI system. The hyperspectral images of the seeds were captured using the camera operating in the spectral range of 900–1700 nm. The hypercube of both sides of the seeds were acquired, and the sample-wise spectra were extracted. The spectra were pretreated with six spectral preprocessing techniques and given as input to the four machine learning models. The SVM outperformed the rest of the models and achieved an accuracy of 93.66%. Furthermore, the model achieved more than 93% accuracy for crease up and crease down side of the seeds. The model was found to be suitable for scanning the seeds from either side in an online detection setup. The classification accuracy of other models was observed to be 84.70%, 82.90%, and 70.05% for KNN, RF, and PLS-DA, respectively. The outcomes of the study advocated that NIR-HSI, in tandem with an SVM, has excellent potential to identify the near geographical origins of the barley seeds. In future, barley seeds from far geographical origins can be included in the study. Further studies can employ the deep learning models (e.g., the CNN model) that are well suited for spectral data analysis. Identical research can be performed on the different types of seeds, e.g., wheat, rice, maize, etc. Furthermore, the current study can be further extended to assess the quality parameters of the seeds, e.g., protein and starch.

Acknowledgements The authors are grateful to National Bureau of Plant Genetic Resources (Delhi), Indian Institute of Wheat and Barley Research (Karnal, Haryana), University Seed Farm (Ladhowal, Ludhiana, Punjab), Rajasthan Agricultural Research Institute (Durgapura, Jaipur, Rajasthan) for providing the seeds.

References

1. Hussain, A., Ali, S., Hussain, A., et al.: Compositional profile of barley landlines grown in different regions of Gilgit-Baltistan. Food Sci Nutr **9**, 2605–2611 (2021)
2. Sohn, M., Himmelsbach, D.S., Ii, F.E.B., et al.: Near-infrared analysis of ground barley for use as a feedstock for fuel ethanol production. Appl. Spectrosc. **61**, 1178–1183 (2007)
3. Wang, Q., Huang, M., Zhu, Q.: Characteristics of maize endosperm and germ in the geographical origins and years identification using hyperspectral imaging. In: 2014 ASABE Annual International Meeting. American Society of Agricultural and Biological Engineers, pp. 1–6 (2014)
4. Gordon, R., Chapman, J., Power, A., et al.: Mid-infrared spectroscopy coupled with chemometrics to identify spectral variability in Australian barley samples from different production regions. J. Cereal. Sci. **85**, 41–47 (2019). https://doi.org/10.1016/j.jcs.2018.11.004
5. Zhang, T., Wang, Q., Li, J., et al.: Study on the origin traceability of Tibet highland barley (Hordeum vulgare L.) based on its nutrients and mineral elements. Food Chem. **346**, 128928 (2021). https://doi.org/10.1016/j.foodchem.2020.128928
6. Giraudo, A., Grassi, S., Savorani, F., et al.: Determination of the geographical origin of green coffee beans using NIR spectroscopy and multivariate data analysis. Food Control **99**, 137–145 (2019). https://doi.org/10.1016/j.foodcont.2018.12.033
7. Zhao, H., Guo, B., Wei, Y., Zhang, B.: Near infrared reflectance spectroscopy for determination of the geographical origin of wheat. Food Chem. **138**, 1902–1907 (2013). https://doi.org/10.1016/j.foodchem.2012.11.037
8. Richter, B., Rurik, M., Gurk, S., et al.: Food monitoring: Screening of the geographical origin of white asparagus using FT-NIR and machine learning. Food Control **104**, 318–325 (2019). https://doi.org/10.1016/j.foodcont.2019.04.032
9. Gao, J., Li, X., Zhu, F., He, Y.: Application of hyperspectral imaging technology to discriminate different geographical origins of Jatropha curcas L. seeds. Comput. Electron. Agric. **99**, 186–193 (2013). https://doi.org/10.1016/j.compag.2013.09.011
10. Mo, C., Lim, J., Kwon, S.W., et al.: Hyperspectral imaging and partial least square discriminant analysis for geographical origin discrimination of white rice. J. Biosyst. Eng. **42**, 293–300 (2017). https://doi.org/10.5307/JBE.2017.42.4.293

11. Sun, J., Lu, X., Mao, H., et al.: A method for rapid identification of rice origin by hyperspectral imaging technology. J. Food Process. Eng. **40**, e12297 (2017). https://doi.org/10.1111/jfpe.12297
12. Singh, T., Garg, N.M., Iyengar, S.R.S.: Non-destructive identification of barley seeds variety using near-infrared hyperspectral imaging coupled with convolutional neural network. J. Food Process. Eng. **44**, 1–13 (2021). https://doi.org/10.1111/jfpe.13821
13. Dai, Q., Sun, D.-W., Cheng, J.-H., et al.: Recent Advances in de-noising methods and their applications in hyperspectral image processing for the food industry. Compr. Rev. Food Sci. Food Saf. **13**, 1207–1218 (2014). https://doi.org/10.1111/1541-4337.12110

MultiNet: A Multimodal Approach for Biometric Verification

Poorti Sagar and Anamika Jain

Abstract These days, the safety of personal information has become a matter of great concern for everyone. In this matter, the concept of Multimodal Biometrics has attracted the interest of the researchers because to the ability to solve a number of limitation of uni-modal biometric system. In this paper, we have presented multimodal biometric-based verification system, which is based on convolutional neural network to verify a individual using multi traits biometric modalities, i.e., fingerprint iris by score level fusion. We have achieved 98.8% accuracy over CASIA V fingerprint and iris dataset. Obtained results shows that using two different biometric trait in proposed biometric verification systems achieved better result than single biometric trait.

Keywords Biometric · Feature extraction · Score fusion · CNN

1 Introduction

In the modern social, political, and legal systems, the real-time identification of an individual is crucial in many situations such as during migrations, while crossing international borders, or accessing a software system on a computer. Many applications use tokens-based mechanisms such as a passport or Id card. On the other hand, some applications are using knowledge-based methods like a pin or password as a method of authorizing the individuals [1]. These two methods, i.e., token-based and knowledge-based methods can be stolen, forgotten, or forged. To overcome this limitation, the biometric identification process came into the picture [2].

P. Sagar (✉) · A. Jain
Centre for Advanced Studies, AKTU, Lucknow, India
e-mail: Poortisgr25@gmail.com

A. Jain
MIT-WPU, Pune, India

M. Tistarelli et al. (eds.), *Computer Vision and Machine Intelligence*,
Lecture Notes in Networks and Systems 586,
https://doi.org/10.1007/978-981-19-7867-8_54

Biometric is one of the recent solutions for overcoming the limitations of previous security methods [2]. With a biometric approach, one can identify the individual by their physical appearance. Biometrics has broadly explained into different parts, i.e., physiological biometrics and behavioral biometric. Physiological biometrics includes face, iris, fingerprint, etc. Behavioral biometric includes, signature, gait, keyboard dynamics, etc. Biometrics has its application in various fields such as healthcare systems, airport security, social networks applications, cloud computing and homeland security. Besides this, billions of smartphone users unlock their phones with their fingerprint or face for security purposes.

With the increasing demand for better human authentication, the use of a single biometrics system that runs on a single trait feature is mostly not able to obtain satisfactory results and sometimes also leads to fraudulent access. In addition, high level security applications and large civilian recognition systems place stringent accuracy necessities that cannot be met with the help of unibiometric systems [3]. To provide better authentication, multimodal biometric authentication is proposed. If the biometric system uses at least two biometric trait, then that biometric system is known as a multi-biometric system or multimodal biometric system. Multi-biometric authentication gives higher verification rates as different independent biometric features are used. The main part of a multimodal biometric system is the fusion of various levels of biometric features [1]. Four levels are used which include a sensor, feature extraction, matching score and decision level. In the research, we have used score-level fusion.

1.1 Biometric Review

The increasing incidents of fraud and criminal activities in today's society are creating chaotic situations. These matters have created doubts and confusion over the existing systems. Earlier, the systems relied on methods that require users to remember their passwords and pins to prove their identities. However, users faced challenges such as remembering pins and passwords. As a solution, the researchers proposed an approach to user identification based on their behavioral and physiological attributes termed Biometrics. Biometrics is described as the recognition of an individual using their unique attributes such as the face, iris, fingerprint, gait, signature, etc [4]. Biometrics provides an advantage over previous approaches in that you don't need to remember any pin or password. Biometrics offers better security as biometric attributes cannot be lost, stolen, or transferred.

To overcome these problems biometric system came into picture. The Biometric system is a recognition system that use the information about a user to identify that user. For identifying user, every biometric system consists of these four steps, (i) Image acquisition: In this the image of biometric trait is acquired and submitted to the system for further process. (ii) Feature Extraction: This process the acquired images by extracting their salient features. (iii) Feature matcher: This matches the

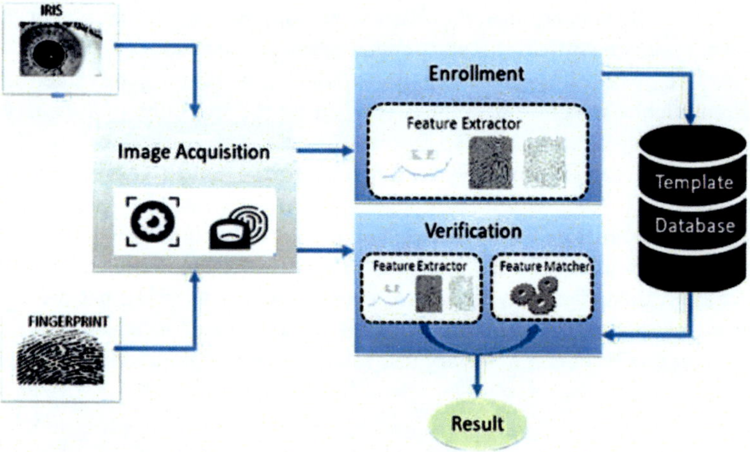

Fig. 1 General verification process of biometric system

extracted features of images with those stored in system's database images to generate the match score. (iv) Decision Module: This rejects or verify the user based on the matching score. Figure 1 illustrates the biometric system.

1.2 Multimodal Biometric System

Instead of enormous advances in technology, the recognition system based on a single biometric cannot guarantee comprehensive security. This is because of several factors. For example, attackers can easily forge a single biometric to gain unauthorized access. To resist these kinds of frauds, the multimodal biometric system came into frame [5]. Such systems uses multiple biometric. For e.g., the face, iris and palm print of a person are combined to establish the identity. These systems provide better security.

Multibiometrics systems uses information from different traits for verification. There can be different types of approaches for fusion:

1. Sensor Level Fusion: The raw information gained from the multiple sensors are fused before extracting feature. This can be only possible when no. of images are fused, when they are captured using similar sensor [6]. If different sensors are being used then the image from different sources must be compatible [7]. There is lot information in raw data, but at the same time it can be corrupted.
2. Feature Level Fusion: The information from different traits are separately processed, then feature extraction is done. After a merged feature vector is calculated to match it against the stored template. The fusion of traits can be easily achieved, if feature extraction is done using similar algorithm otherwise it will become

tedious [7]. This is because the feature sets may not be compatible, or the relationship between different feature sets was not properly known.

3. Score level fusion: This is combination of matching score which is taken as output from particular matcher. Matching score shows the proximity of the query image to the template stored. It is also refer as fusion at confidence level [6]. Matching score is very high in info above feature level fusion and also easy to combine the value [8].
4. Decision level fusion: Fusion at decision level is only possible when output of the particular biometric matcher is present. Firstly every biometric trait is processed separately, then fuses the outputs of every modality to give final decision, i.e., reject or accept [3]. Maximum no. of votes can be used to make final decision. Fusion at this level contains very less information as compared to other fusions.

2 Literature Review

There are many methods [6, 9–14] that have worked on the biomertic verification. Some of them uses multiple modalities and some uses single modalities. The verification rate of multi modal systems relies on different factor, for example: fusion technique, selected features, extraction techniques and the modalities used. The following section shows a brief overview of related work of multimodal biometric systems.

In Khuwaja et al. [9] presented a bimodal method that take face and finger biometric as input for identification of an individual. Authors have performed fusion at the sensor level. The input images has been fused together before feature extraction process. The integrated feature is then taken from the fused image and the features has been fused to an adaptive artificial neural network [9].

He et al. in [6] have used Gabor filters with particle swarm optimization. Later authors have trained a model for final decision. They have combined Facial Recognition database, i.e., FAFB, CASIA-V and FVC2005 to form a dataset of iris,fingerprint and face biometric [6].

In [10], authors have shown that the recognition rate of multimodal systems depends on multiple factors, i.e., fusion scheme, fusion technique, features extraction and selection techniques. They have calculated the compatibility of the feature vectors extracted from various traits [10] . In [11], authors presented a fusion approach for fingerprint and iris. In their method, authors have extracted the features by log Gabor filter from both the traits, fingerprint and iris and these feature vectors are joined together. The output feature vectors are sub divide to generate the distinctive pattern of selected fingerprint and iris to bitwise biometric stored template. Using Hamming distance they resulted the final match score. Authors have performed experimentation on a dataset of 50 persons giving FAR = 0% and FRR = 4.3% [11]. In [12] authors have proposed rank level method, that combines the info from different traits. They have used principal component Fisher's linear methods to extract features from the traits. Rank of every single matchers are joined by calculating the highest rank, Borda count [12].

Authors in [13] proposed a work for biometric fusion which is based on individual trait matching implementation for chosen modalities, i.e., fingerprint and iris. For the experiment they have used the WVS biometric database. Database contains 400 images (4 enrolment images × 100 users). They have used single hamming distance for matching to provide high accuracy than the uni-modal systems [13].
In [15] Minaee et al. proposed Face Recognition biometric system which is based on scattering Conv Architecture. For feature extraction they have used scattering transform method and for classification they have used SVM. However, invariant scattered feature is used to improve the accuracy [15].

In [14] authors have proposed the primary working procedure on iris recognition which shapes the reason for the majority of the formative exercises in iris biometrics till date. They procured a human eye with the assistance of a camera and distinguishes it. The operator search for circular way, where there is greatest change in the pixel values. By differing focus and range position. This is applied in a manner such that the smoothing logically decreased to accomplish exact limitation. The integro-differential should be visible as a variety of the Hough transform, since it is also a utilization of first subsidiaries of the picture and plays out an inquiry to track down mathematical boundaries. Since it works with raw subordinate data, it doesn't experience the problem effects of threshold issues of the Hough transform [14].

3 Dataset

In the experiment, two datasets are used to implement the proposed system. One dataset is for the iris images and other is for the fingerprint images. Both the datasets are explained in next section.

3.1 Fingerprint Dataset

For the fingerprint input images CASIA V fingerprint dataset is used. CASIA Fingerprint V contains 2000 fingerprint images. The fingerprint image of dataset were captured using URU400 sensor. Here we are using total of 50 images of fingerprints from dataset [16]. Figure 2 shows the images of the fingerprint dataset.

Fig. 2 Sample images of fingerprint dataset

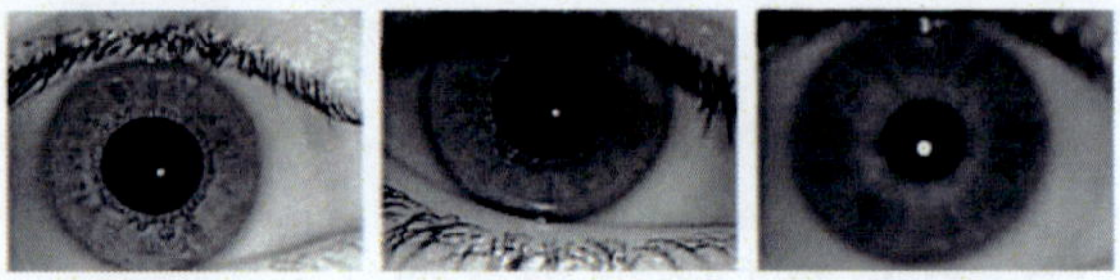

Fig. 3 Sample images of iris dataset

3.2 Iris Dataset

For iris input images we are using the CASIA V iris datast. CASIA-Iris V contain a total number of 5400 images from approx genuine 800 subjects and virtual 400 subjects. All images are of eight bit gray-level. We are using total of 50 images of iris from dataset [17]. In Fig. 3, sample images of iris dataset has been shown.

4 Methodology

In this work, we have proposed a CNN-based (MultiNet), score level fusion approach for verification of an individual user. The system uses iris and fingerprint traits. The workflow of the proposed method has been shown in Fig. 4. MultiNet is used for feature extraction and classification. Score level fusion was chosen because of the strong trade-off between the simplicity in merging the biometric traits data and better information [18]. The workflow of the proposed method is divided into 5 parts (1) Data acquisition, (2) Preprocessing, (3) Feature Extraction, (4) Classification, (5) Score fusion. Each module is discussed in the following sections.

4.1 Network Architecture

Convolutional neural network has proved its superiority in the field of feature extraction and classification. CNN architecture is composed of three different layers, i.e., convolutional layer, maxpooling layer and fully connected layers. Our MultiNet is 5 layered architecture. Out of five layers 3 are conv layers and remaining two are fully connected layer. Figure 5 shows the architecture of the proposed MultiNet. The first

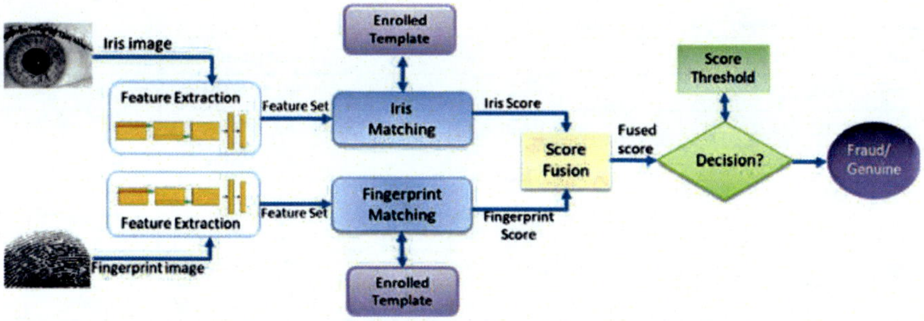

Fig. 4 Workflow of multimodal biometric verification

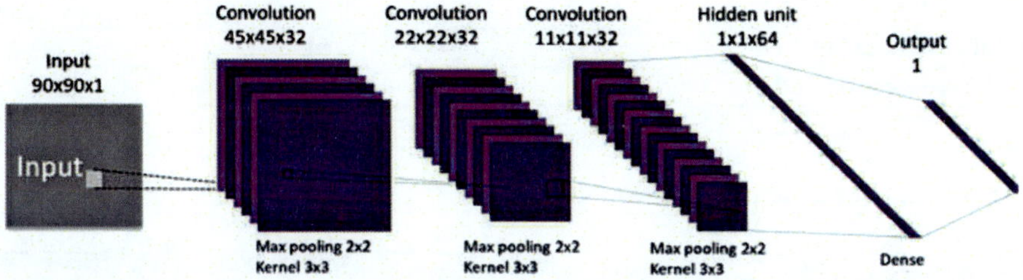

Fig. 5 Architecture of proposed MultiNet

Conv layer (C_1) has 32 filters of size 3 × 3. Initial convolutional layer extracts the generic feature like edges, boundary, etc. and the last layers are responsible to detect more complex pattern and specific shape in the images. In all the conv layers, we have used ReLU activation. With the activation function, we can introduce non-linearity in the network. This non-linearity lead the network to learn complex shapes in the data. The output feature map of C_1 fed to maxpool layer M_1. Max pool layer helps in reducing the CNNs complexity by reducing the size of the feature maps generated by conv layers. In our work we have used maxpool. The output of the M_1 has been given to the second convolutional layer (C_2). This layer have 32 filters of size 3 × 3. The ReLU activation has been applied on C_2 and the resultant feature map has been sent to the next maxpool layer (M_2). The resultant reduced feature maps of M_2 has been fed to the third convolutional layer (C_3). C_3 have 32 filters of size 3 × 3. Activation function has been applied on C_3 and the resultant feature map has been sent to the next maxpool layer (M_3).The result of the M_3 has been flattened and given to the fully connected layers. In our work we have stacked two fully connected layers. Fully connected layers FC_1, FC_2 has 64, 1 neurons, respectively. Sigmoid activation function has been applied on the last fully connected layer (FC_2).

In the proposed work, We are using proposed CNN model for identifying the iris and fingerprint. Proposed model receives an input of 90 × 90, contains 3 convolutional layers, 3 maxpooling layers and 2 fully connected layers (FCL) as shown in Table 1.

Table 1 Architecture of the MultiNet

Layers	Shape of the output feature map
Input layer	90 × 90 × 1
Convolution layer ($C1$) [$S = 1$, $F = 3$, Padding = 'Same']	90 × 90 × 32
ReLU	90 × 90 × 32
Pool ($M1$) [$S = 2$, $F = 2$]	45 × 45 × 32
Convolution layer ($C2$) [$S = 1$, $F = 3$, Padding = 'Same']	45 × 45 × 32
ReLU	45 × 45 × 32
Pool ($M2$) [$S = 2$, $F = 2$]	22 × 22 × 32
Convolution layer ($C3$) [$S = 1$, $F = 3$, Padding = 'Same']	22 × 22 × 32
ReLU	22 × 22 × 32
Pool ($M3$)[$S = 2$, $F = 2$]	11 × 11 × 32
Fully connected ($FC1$)	64
Fully connected ($FC2$)	1

Table 2 Hyper-parameter of the MAMMO-Net

Parameters	Value
Optimizer	Adam
Batch size	32
No of epochs	10

4.2 Data Preprocessing

In the proposed work, we have applied data augmentation in both iris and fingerprint images and we have also performed image resizing. With data augmentation there can be less changes of over-fitting. This augmentation approach is used to increase the amount of training data [19]. The augmentation technique which was used to increase the number of input images of fingerprint and iris, were rotation, zoom, flipping. All the images of iris and fingerprint were resized to 90 × 90 size.

4.3 Fusion Approach

In proposed work, we have used feature and score level fusion approach. The feature extracted from the images of iris and fingerprint has been subtracted and fed to the classifier for generation of the scores. In score level fusion, we get the resultant score from both the modalities, i.e., iris and fingerprint. The resultant scores from both the modalities, i.e., iris and fingerprint, are then merged to generate one score. In the proposed work we have performed score fusion with arithmetic mean. The arithmetic

Table 3 Comparison with the state of the art

Author	Approach	Accuracy (%)
Shamil Mustafa [18]	Decision level	95
Sarhan [21]	Feature level	93.5
Gawande [4]	Score level	91
Vishi [17]	Decision level	90.2
Proposed	Score level	98.8

Fig. 6 Loss curve

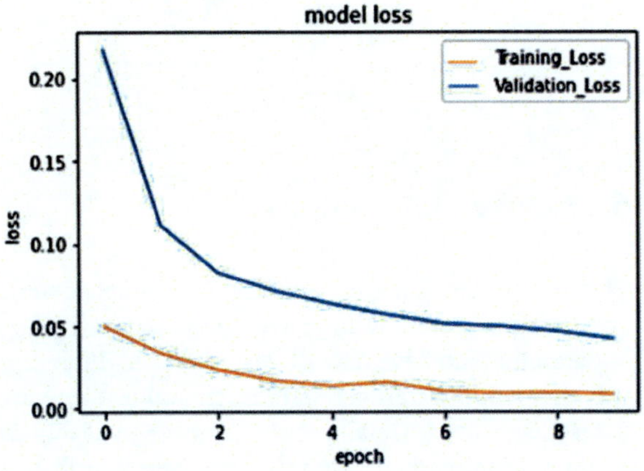

mean is calculate by using the formula given in Eq. (1):

$$\text{Mean}(S) = \frac{\text{Score}(i) + \text{Score}(f)}{2} \tag{1}$$

Where Mean (S) is the fused score, Score (i) is the score of iris, Score (f) is the score of the fingerprint.
If fused score value of the input fingerprint and iris images,is above or equal to set threshold value. Then, the person is considered genuine, else fraud [20].

5 Experimental Setup

The experiment was performed using Intel core i7 processor and 8 GB RAM. We used the validation set for model performance evaluation. The training of the proposed CNN model was done using a batch size equal 32. Table 2 shows the hyperparameter used in the experiment performed.

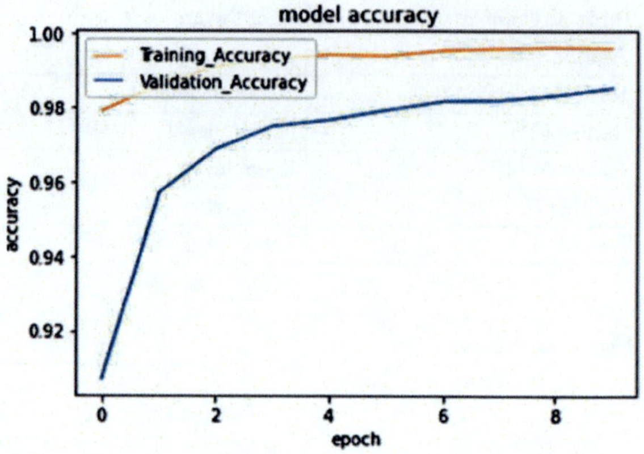

Fig. 7 Accuracy curve

6 Results and Analysis

The experiment is set for multimodal biometric verification. It is pre-configured to compare the results with real-world situations. The results were tested on CASIA V dataset which have data of iris images and fingerprint images. The no. of images was divided into training, validation and test sets. For this experiment, the CASIA dataset is utilized. Details about the storage of this datasets are provided.
In this, 50 individuals are randomly selected from the database and for every 50 individuals one fingerprint input and one iris input is selected as the input data and different pictures are chosen at different times as enrolled id images. Each input data, i.e., iris and fingerprint is compared against the selected enrolled id image stored, i.e., against 50 users images. When the fused match score of iris and fingerprint is below a certain fixed value, the user is declared a scammer. To determine authenticity, the maximum matching score for each individual of all registered images is calculated. Then a threshold is calculated. The threshold is set at 80. If the outcome of the match is equal to or greater than the selected limit that person is considered real. When the match score is below a certain fixed value, the user is declared a scammer. To facilitate this analysis, results obtained from an experiment are presented below in Figs. 6 and 7.

Table 3 shows the comparison of the state of art results with the the proposed MultiNet.

7 Conclusion

There are numerous issues introduced in single biometric verification frameworks like loud information, absence of uniqueness. As single modal biometric only use single biometric due to which sometimes it might be possible that it will be not available or readable. So here multimodal biometric verification may solve many problem which were coming in the single biometric systems. Our system works on the two biometrics that is iris and fingerprint for verification which describes the effective fingerprint and iris multimodal biometric system that are more secure toward frauds. The fusion of two features is performed using score fusion, approach obtained better accuracy.

References

1. Medjahed, C., Rahmoun, A., Charrier, C., Mezzoudj, F.: A deep learning-based multimodal biometric system using score fusion. IAES Int. J. Artif. Intell. **11**(1), 65 (2022)
2. Dargan, S., Kumar, M.: A comprehensive survey on the biometric recognition systems based on physiological and behavioral modalities. Expert Syst. Appl. **143**, 113114 (2020)
3. Sudhamani, M., Venkatesha, M., Radhika, K.: Fusion at decision level in multimodal biometric authentication system using iris and finger vein with novel feature extraction. In: Annual IEEE India Conference (INDICON), pp. 1–6. IEEE (2014)
4. Gawande, U., Zaveri, M., Kapur, A.: Fingerprint and iris fusion based recognition using RBF neural network. J. Signal Image Process. **4**(1), 142 (2013)
5. Joseph, T., Kalaiselvan, S., Aswathy, S., Radhakrishnan, R., Shamna, A.: A multimodal biometric authentication scheme based on feature fusion for improving security in cloud environment. J. Ambient Intell. Humanized Comput. **12**(6), 6141–6149 (2021)
6. He, F., Liu, Y., Zhu, X., Huang, C., Han, Y., Chen, Y.: Score level fusion scheme based on adaptive local Gabor features for face-iris-fingerprint multimodal biometric. J. Electron. Imaging **23**(3), 033019 (2014)
7. Ryu, R., Yeom, S., Kim, S.-H., Herbert, D.: Continuous multimodal biometric authentication schemes: a systematic review. IEEE Access **9**, 34541–34557 (2021)
8. Jain, A.K., Nandakumar, K., Ross, A.: 50 years of biometric research: accomplishments, challenges, and opportunities. Pattern Recogn. Lett. **79**, 80–105 (2016)
9. Khuwaja, G.A.: Merging face and finger images for human identification. Pattern Anal. Appl. **8**(1), 188–198 (2005)
10. Ammour, B., Boubchir, L., Bouden, T., Ramdani, M.: Face-iris multimodal biometric identification system. Electronics **9**(1), 85 (2020)
11. Radha, N., Kavitha, A.: Rank level fusion using fingerprint and iris biometrics. Indian J. Comput. Sci. Eng. **2**(6), 917–923 (2012)
12. Monwar, M.M., Gavrilova, M.L.: Multimodal biometric system using rank-level fusion approach. IEEE Trans. Syst. Man Cybern. B (Cybern.) **39**(4), 867–878 (2009)
13. Baig, A., Bouridane, A., Kurugollu, F., Qu, G.: Fingerprint-iris fusion based identification system using a single hamming distance matcher. In: Symposium on Bio-inspired Learning and Intelligent Systems for Security, pp. 9–12. IEEE (2009)
14. Nsaef, A.K., Jaafar, A., Jassim, K.N.: Enhancement segmentation technique for iris recognition system based on daugman's integro-differential operator. In: International Symposium on Instrumentation & Measurement, Sensor Network and Automation (IMSNA), vol. 1, pp. 71–75. IEEE (2012)
15. Minaee, S., Abdolrashidi, A., Wang, Y.: Face recognition using scattering convolutional network. In: IEEE Signal Processing in Medicine and Biology Symposium (SPMB), pp. 1–6. IEEE (2017)
16. Shekhar, S., Patel, V.M., Nasrabadi, N.M., Chellappa, R.: Joint sparse representation for robust multimodal biometrics recognition. IEEE Trans. Pattern Anal. Mach. Intell. **36**(1), 113–126 (2013)

17. Vishi, K., Yayilgan, S.Y.: Multimodal biometric authentication using fingerprint and iris recognition in identity management. In: 2013 9th International Conference on Intelligent Information Hiding and Multimedia Signal Processing, pp. 334–341. IEEE (2013)
18. Mustafa, A.S., Abdulelah, A.J., Ahmed, A.K.: Multimodal biometric system iris and fingerprint recognition based on fusion technique. Int. J. Adv. Sci. Technol. **29**, 7423–7432 (2020)
19. AbuAlghanam, O., Albdour, L., Adwan, O.: Multimodal biometric fusion online handwritten signature verification using neural network and support vector machine. In: Transactions, vol. 7, p. 8 (2021)
20. Punyani, P., Gupta, R., Kumar, A.: A multimodal biometric system using match score and decision level fusion. Int. J. Inf. Technol. **14**(2), 725–730 (2022)
21. Sarhan, S., Alhassan, S., Elmougy, S.: Multimodal biometric systems: a comparative study. Arab. J. Sci. Eng. **42**(2), 443–457 (2017)

Synthesis of Human-Inspired Intelligent Fonts Using Conditional-DCGAN

Ranjith Kalingeri, Vandana Kushwaha, Rahul Kala, and G. C. Nandi

Abstract Despite numerous fonts already being designed and easily available online, the desire for new fonts seems to be endless. Previous methods focused on extracting style, shape, and stroke information from a large set of fonts, or transforming and interpolating existing fonts to create new fonts. The drawback of these methods is that generated fonts look-alike fonts of training data. As fonts are created from human handwriting documents, they have uncertainty and randomness incorporated into them, giving them a more authentic feel than standard fonts. Handwriting, like a fingerprint, is unique to each individual. In this paper, we have proposed GAN-based models that automate the entire font generation process, removing the labor involved in manually creating a new font. We extracted data from single-author handwritten documents and developed and trained class-conditioned DCGAN models to generate fonts that mimic the author's handwriting style.

Keywords DCGAN · Intelligent fonts · Random fonts · cGAN · FID · WCFID · BCFID

Supported by I-Hub foundation for Cobotics.

R. Kalingeri (✉) · V. Kushwaha · R. Kala · G. C. Nandi
Center of Intelligent Robotics Indian Institute of Information Technology Allahabad Prayagraj, 211015 UP, India
e-mail: ranji.iitb@gmail.com

V. Kushwaha
e-mail: kush.vandu@gmail.com

R. Kala
e-mail: rkala@iiita.ac.in

G. C. Nandi
e-mail: gcnandi@iiita.ac.in

M. Tistarelli et al. (eds.), *Computer Vision and Machine Intelligence*,
Lecture Notes in Networks and Systems 586,
https://doi.org/10.1007/978-981-19-7867-8_55

1 Introduction

The font is a graphical representation of a text. It is characterized by its design, weight, color, point size, and typeface. There are tons of use cases out of which some are listed below:

- Fonts add style and visual appeal to documents, web pages, newspapers, books, billboards, etc.
- Out of a vast number of fonts, specific fonts are selected based on specific medium, background, and context. Even in the same document, different fonts are used for different sections, authors, headings, titles, etc.

The fonts evolved into different classes like random fonts, variable fonts, handwriting style fonts, etc. are shown in Fig. 1.

1.1 Handwritten Fonts

Handwritten fonts, as shown in Fig. 1 resembles human handwriting style. Content created using these fonts look similar to content written with a pen or marker by hand. Handwritten fonts mirror the penmanship, so this kind of font can be integrated and used in various informal communications like:

- WhatsApp chats or any social messaging chat box.
- Personal letters and mails, greeting cards, invitation cards, and personalized notes preparations.

Fig. 1 Normal fonts, Handwritten fonts and Random font

Usually, these fonts are not incorporated with randomness (refer Sect. 1.2) or intelligence (refer Sect. 1.3). Font's appearance is the same every time they are displayed.

1.2 Random Fonts

Van Rossum has said, "A certain roughness or varying unevenness is quite pleasing to the eye". So, he and Van Blokland brought uncertainty and randomness to typography by co-designing the first random font, "Beowolf". In this font, the character appears differently every time it is displayed or printed, as shown in Fig. 1. This is achieved by randomly shifting the ragged edges each time the font is displayed. Random fonts are the first stepping stones toward intelligent fonts.

1.3 Intelligent Fonts

Intelligent fonts alter their appearance depending on its previous and succeeding character. Like random fonts, characters in intelligent fonts appears differently every time they are displayed or printed. The dependence of intelligent fonts on adjacent characters distinguishes them from random fonts, which has random appearance.

Intelligent font's concept is inspired from human handwriting style. In normal human hand writings, we found that the same character is written in different ways based on its adjacent characters. For example, in Fig. 2, the character 'r' (underlined in blue color) between 'e' and 's' is different from 'r' (underlined in green color) written between 'u' and '␣'. However, character 'r' between 'e' and 's' is similar to 'r' (underlined in red color) written between 't' and 'u' but not exactly the same.

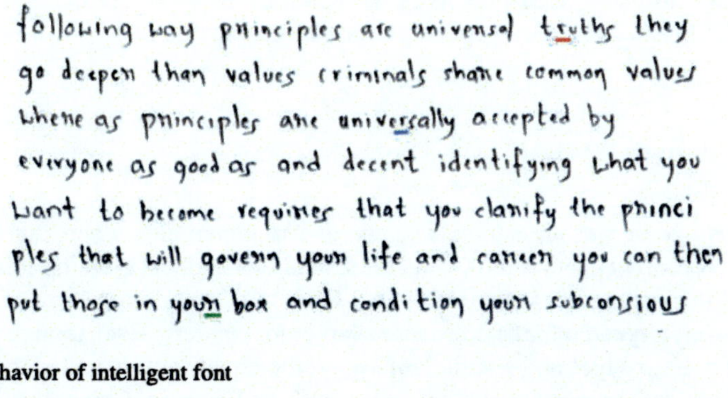

Fig. 2 Behavior of intelligent font

1.4 Motivation

This study aims to generate fonts from human handwritten TXT files to mimic the human handwritten style and computerize font generation instead of manually designing fonts due to the following reasons:

- Just as a fingerprint, handwriting is unique to each person. It is possible to create as many fonts as there are many humans with our model.
- In intelligent font, the appearance of each alphabet varies depending on its adjacent alphabets. So, a type designer needs to create large variations of alphabets to develop an intelligent font.
- As uncertainty and randomness is incorporated in intelligent fonts, it gives a more natural feeling than normal font.

1.5 Our Contribution

Here are the lists of contribution of our work:

1. We have proposed R-cDCGAN and I-cDCGAN models, two-class conditioned versions of DCGAN for generating random font and intelligent font, respectively.
2. We have collected key data from handwritten documents, named it as IIITA-ALPHABETS dataset, that is used for training our models. This method can also be used to extract key data from handwritten documents of other users.
3. We have proposed an algorithm to form sentences from generated alphabets.
4. The experimental results show that the sentences formed using synthetic fonts generated from proposed models are incorporated with uncertainty and randomness, giving them a more natural depiction than standard fonts.
5. We have also computed FID, WCFID, and BCFID for our proposed models R-cDCGAN and I-cDCGAN for its performance evaluation.

2 Literature Review

There have been various attempts lately to use generative adversarial networks (GANs) [1] for font generation. One of these methods is zi2zi [2] which combines the pix2pix [3], auxiliary classifier GAN (AC-GAN) [4], and domain transfer network [5] to convert a specific font pattern into a target font pattern. Even though generated fonts have sharp edges and a variety of styles, the target font is limited to alphabets with a significant number of character sets, such as Japanese, Chinese, and Korean; hence, applying this method to alphabets with few letters is challenging.

Several alternative approaches like example-based font generation using GANs are proposed. Chang and Gu [6] used U-net [7] architecture as generator for producing character patterns with the desired style, which is more efficient than zi2zi at balancing the loss functions. Azadi et al. [8] used conditional GAN (cGAN) [9] to generate fonts with limited examples.

In some studies, style and shape of the target alphabet are used as input for their GAN-based font creation network. Guo et al. [10] fed their GAN with skeleton vector of the target character and font style vector to generate fonts with target character. Later, Lin et al. [11] introduced a stroke-based font creation method in which two trained styles are interpolated by modulating a weight.

In the above GAN-based font generation methods, a new typeface is created by combining style information with character shape information retrieved from input character images. Since the generated font is influenced by the geometry of the input image, creating a novel font is difficult. We attempted to overcome this problem through our proposed models, intelligent font-conditional deep convolutional GAN (I-cDCGAN) and random font-cDCGAN (R-cDCGAN), to generate fonts inspired by human handwritten style. First, we extracted shape and location information related to all alphabets from the author's handwritten documents and used this extracted information to train class-conditioned DCGAN to produce fonts that mimic the author's handwriting style.

3 Dataset

3.1 About Dataset

We created a dataset consisting of, 12552 lower case alphabetical character images of size 64 × 64 pixel from handwritten documents of a single author and named it IIITA-ALPHABETS dataset as shown in Fig. 3.

The dataset is in the form of a CSV file, as shown in Table 1. It has five columns and 12552 rows, where each row represents an alphabet of the dataset and column represents crucial information related to alphabet as described below:

(a) The first column and the third column refer to previous and next alphabets of the current alphabet (second column).

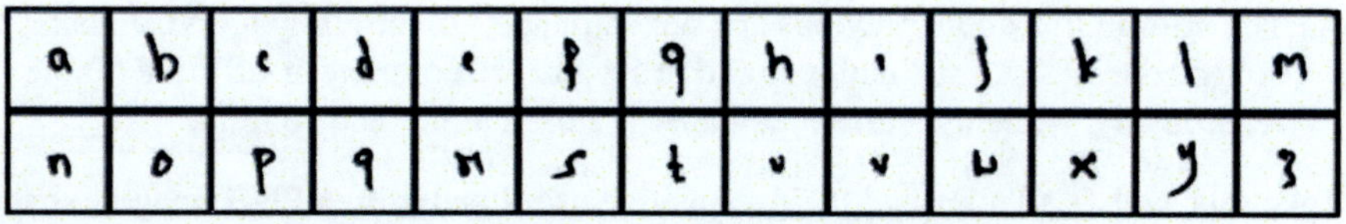

Fig. 3 Sample of a to z alphabets collected from IIITA-ALPHABETS data set

Table 1 First three rows in CSV file of IIITA-ALPHABETS data set

Previous alphabet	Current alphabet	Next alphabet	Current alphabet position	Current alphabet shape
␣	f	o	[[0, 0], [39.14, 67.000]]	[[255, 255, ..., 255], ...]
f	o	1	[[0, 1], [44.83, 84.265]]	[[255, 255, ..., 255], ...]
o	1	1	[[0, 2], [40.78, 100.57]]	[[255, 255, ..., 255], ...]

(b) The fourth column contains spacial information of the current alphabet in the handwritten PNG image file. It is in the form of Python 2D list like [[ele1, ele2], [ele3, ele4]]. ele1 refers to the line number of the current alphabet in a handwritten PNG image file, ele2 refers to after position of the current alphabet in the ele1 line. ele3 and ele4 represent Y-mean and X-mean of the current alphabet in the handwritten PNG image file in pixel units.

(c) The fifth column represents pixel values of the current alphabet's image of size 64×64.

Note that preceding and succeeding character of the current alphabet need not be only alphabets they can be spaces ('␣') too, as the alphabet might be present at the end or start of the word in the handwritten PNG image file.

3.2 Dataset Collection Process

1. Twelve TXT files (Fig. 4a) of text data are collected. From this text data, all non-alphabetic characters are removed except spaces and all upper case alphabets are converted into lowercase.
2. Then, the author is asked to manually write content in each TXT file on an A4 sheet. So, an equivalent handwritten PNG image (Fig. 4b) corresponding to each TXT file is created.
3. We traversed left to right on every handwritten line in the handwritten PNG images (Fig. 4c). During traversal, if we found any alphabet that is not encountered yet, we collected and stored the spatial and shape information related to that alphabet in the 4th and 5th column of CSV file (Table 1) respectively.
4. The information about previous and next alphabets of currently found alphabet is collected from TXT file and is stored in 1st and 3rd columns of CSV file (Table 1) respectively, while the current alphabet is stored in the 2nd column.

Like this, with TXT files and their respective user handwritten PNG images, shape and spatial information of each alphabet is collected and stored in CSV file.

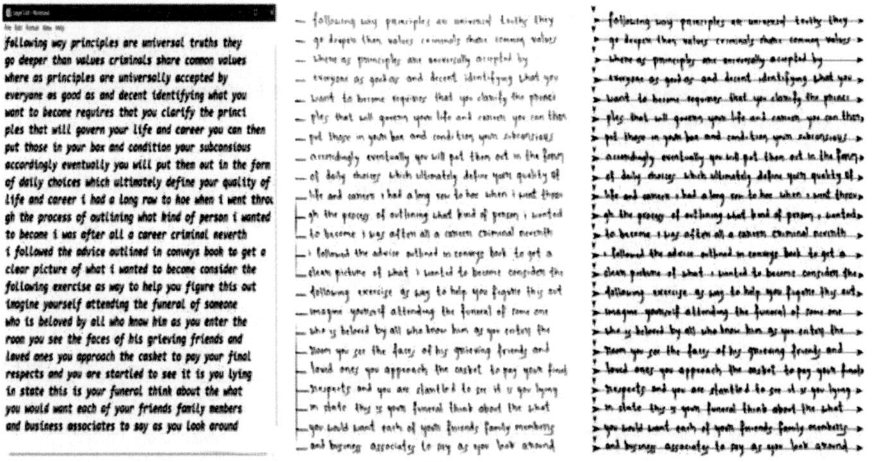

(a) (b) (c)

Fig. 4 **a** TXT file **b** handwritten PNG image of TXT file **c** direction of traversal on Handwritten PNG images, arrows indicate top to bottom and left to right traversal

4 Preliminary Knowledge

4.1 *Generative Adversarial Network (GAN)*

GAN is proposed by Ian Goodfellow et al. in [1] consists of two neural networks: the generator G and the discriminator D. The generator accepts a z-dimensional vector of random numbers as input and outputs data with the same dimensions as the training data. The discriminator, on the other hand, distinguishes between samples from genuine data and data created by the generator.

G and D play the minimax game in training with the value function $\mathcal{L}_{\text{GAN}}$ is represented by (1):

$$\min_{G}\max_{D}\mathcal{L}_{\text{GAN}}(G,D)=\mathbb{E}_{x\sim P_{data}(x)}\big[\log D(x)\big] + \mathbb{E}_{z\sim P_z(z)}\big[\log\big(1-D\big(G(z)\big)\big)\big] \quad (1)$$

where $P_{\text{data}}(x)$ and $P_z(z)$ denote the training data and z distributions, respectively. The $D(x)$ refers to discriminator output, probability of x belongs to the real data distribution, while mapping from z to data space is denoted by $G(z)$.

4.2 Conditional GAN (cGAN)

In traditional GANs, predicting what sort of pattern would be created from a given input z via G is tricky. Mirza et al. [9] demonstrated a conditional GAN that controls the class of the output image by adding class information (y) encoded as a one-hot vector to the generator's input and a channel encoding the class to the discriminator's input. The loss function of cGAN is represented by (2):

$$\begin{aligned}\min_G \max_D \mathcal{L}_{\text{cGAN}}(G, D) &= \mathbb{E}_{x\sim P_{\text{data}}(x)}\big[\log D(x \mid y)\big] \\ &+ \mathbb{E}_{x\sim P(z)}\big[\log\big(1 - D\big(G(z \mid y)\big)\big)\big]\end{aligned} \tag{2}$$

4.3 Deep Convolutional GAN (DCGAN)

DCGAN proposed by Metz et al. [12] is a variant of GAN architecture based on convolutional neural networks (CNNs). The DCGAN architectural guidelines are given below:

- Strided convolutions in D and fractional-strided convolutions in G are used instead of any pooling layers.
- Both G and D uses batchnorm.
- For deeper architectures, remove fully connected hidden layers.
- Tanh activation is used for the last layer, while other layers uses ReLU activation in G.
- For all layers, use LeakyReLU activation in D.

However, the loss function of DCGAN is the same as GAN.

5 Proposed Methodology

We implemented class-conditional DCGAN (cDCGAN), where class label (y) is initially fed into the embedding layer to obtain dense vector representation. This dense vector is concatenated with the noise (z) and input image (x) of the generator (G) and discriminator (D) of DCGAN, respectively.

5.1 Random Font-cDCGAN (R-cDCGAN)

R-cDCGAN is proposed to generate synthetic handwritten alphabets (Random font) of a given class. For training of R-cDCGAN, IIITA-ALPHABETS dataset consisting of 26 classes is used. Alphabet 'a' is labeled as class 0, 'b' as class 1, and so on.

The class label is passed into the embedding layer to generate 100-dimensional one-hot encoded vector, which is further concatenated with 100-dimensional uniformly distributed noise. Then, the resultant vector is fed as input into the generator of R-cDCGAN.

Similarly, in the discriminator, the class label is projected into 64×64-dimensional vector space when passed through the embedding layer. This resultant encoded vector and 64×64 pixel image input are concatenated before being fed as input into the discriminator of R-cDCGAN.

5.2 Intelligent Font-cDCGAN (I-cDCGAN)

I-cDCGAN (Fig. 5) is proposed to generate synthetic handwritten alphabets of a given class whose appearance depends on the previous and next alphabets (Intelligent font). As shown in Fig. 5a, previous, current, and next class labels are passed into the embedding layer to produce 20, 60, and 20 dimensional class encoded vectors, respectively. This resultant class encoded vectors and 100-dimensional uniformly distributed noise are concatenated before being fed as input into the generator (Fig. 5a) of I-cDCGAN.

Similarly, in I-cDCGAN discriminator (Fig. 5b), current, previous, and next class labels are passed into the embedding layer to produce $(3 \times 64 \times 64)$(channel $\times$ height $\times$ width), $(1 \times 64 \times 64)$ and $(1 \times 64 \times 64)$ class encoded dense vectors, respectively. The resultant class encoded vector and 64×64 input pixel image are concatenated before being fed as input into the discriminator of I-cDCGAN.

5.3 Sentence Formation from Generated Alphabets

1. The generated alphabets from proposed models are projected on a new white image at specific target pixel locations to form a sentence. The target pixel location of the first alphabet is [64, 64] by default. In Fig. 6 the alphabet ‘h’ is projected at pixel location [64, 64].
2. Above target pixel location is obtained by adding displacement values (In Fig. 6, ‘o’ is projected 4 rows below and 21 columns right to its previously projected alphabet ‘h’. In this case, [4, 21] is considered as displacement values to the target pixel location of the previously projected alphabet.
3. From the data set, all instances where the current alphabet is followed by the previous alphabet are taken into consideration. The mean of all displacement values collected from those instances give an estimated displacement at which the current alphabet should be projected from the previous alphabet.

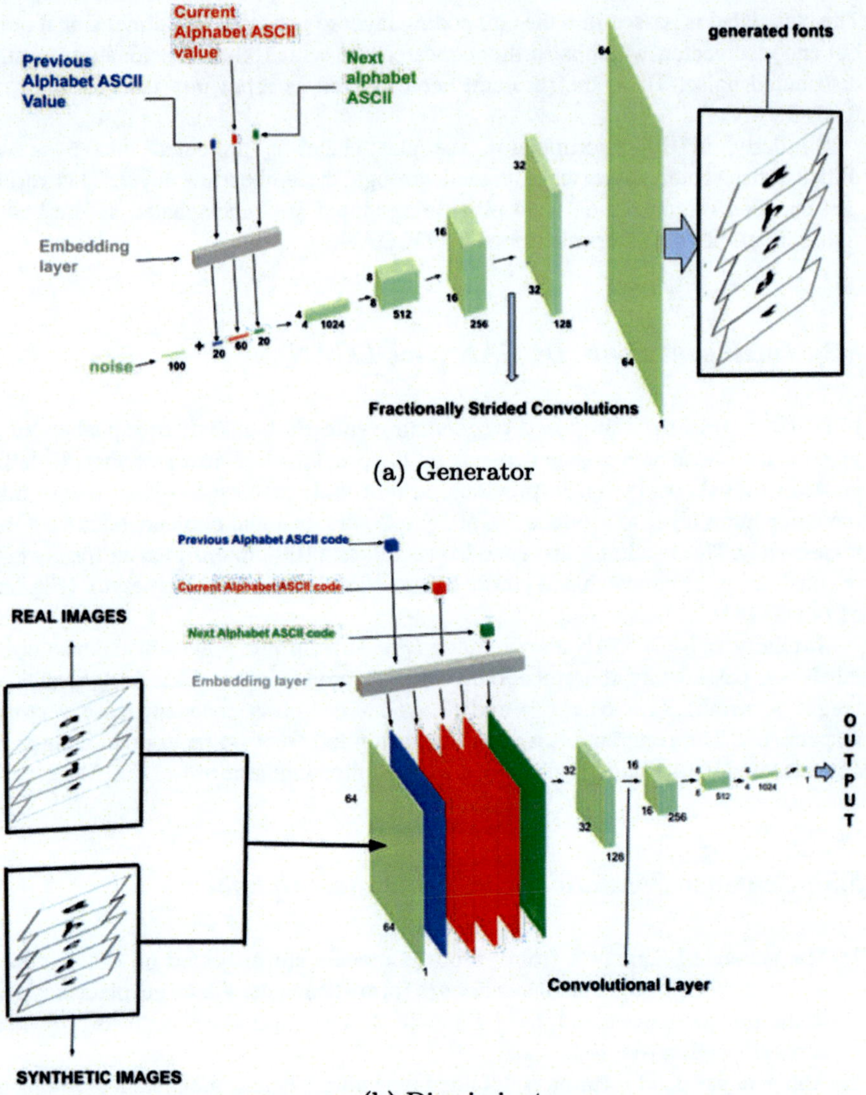

(a) Generator

(b) Discriminator

Fig. 5 Structure of a generator and **b** discriminator in I-cDCGAN. R-cDCGAN structure is similar to I-cDCGAN except only current class label information is fed into the generator and the discriminator

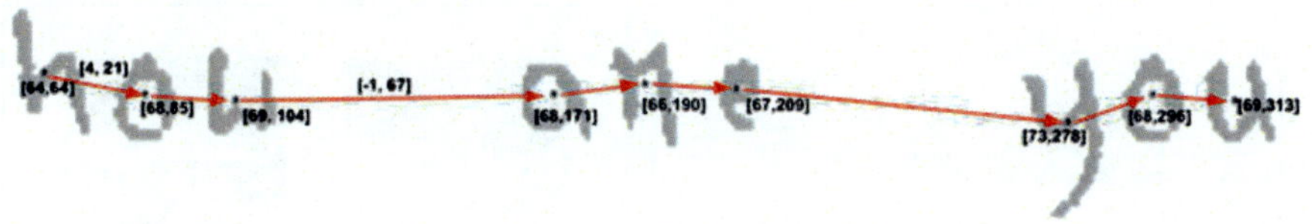

Fig. 6 Sentence formation with generated fonts using our proposed models

4. If there is a space between the previous alphabet and the current alphabet, then we added an extra 30 pixels to the column displacement value. In Fig. 6 the mean displacement of 'a' from 'w' is $[-1, 37]$ and by adding an extra 30 pixels to the column displacement value, the resultant mean displacement values become $[-1, 67]$.

Steps 2–4 are repeated until all generated alphabets are projected onto a new image to form a sentence.

6 Results and Analysis

6.1 Model Training and Parameters of R-cDCGAN and I-cDCGAN

We have tried some combination of parametric values for training our proposed models and out of those the best ones are mentioned here. For weight updation, Adam optimizer[13] is used with the parameter, $\beta_1 = 0$ and $\beta_2 = 0.9$ and learning rate $= 2e - 4$. We have initialized all the weights using Gaussian distribution with mean $= 0$ and standard deviation $= 0.02$. For the LeakyReLU, the slope is 0.2 in all models. Per generator iteration, the number of discriminator iterations is set to 5. The number of learning iterations per epoch $= 12552$, batch size $= 1$. The proposed models are trained for 35 epochs each.

6.2 Results

Figure 7 shows four variations of alphabets 'a', 'e', 'o', and 's' generated using DCGAN, R-cDCGAN, and I-cDCGAN. For comparison purposes, samples from real data are also added. Alphabets generated by DCGAN are poor in quality, and there is less variation in generated data as only a few classes of alphabets are generated.

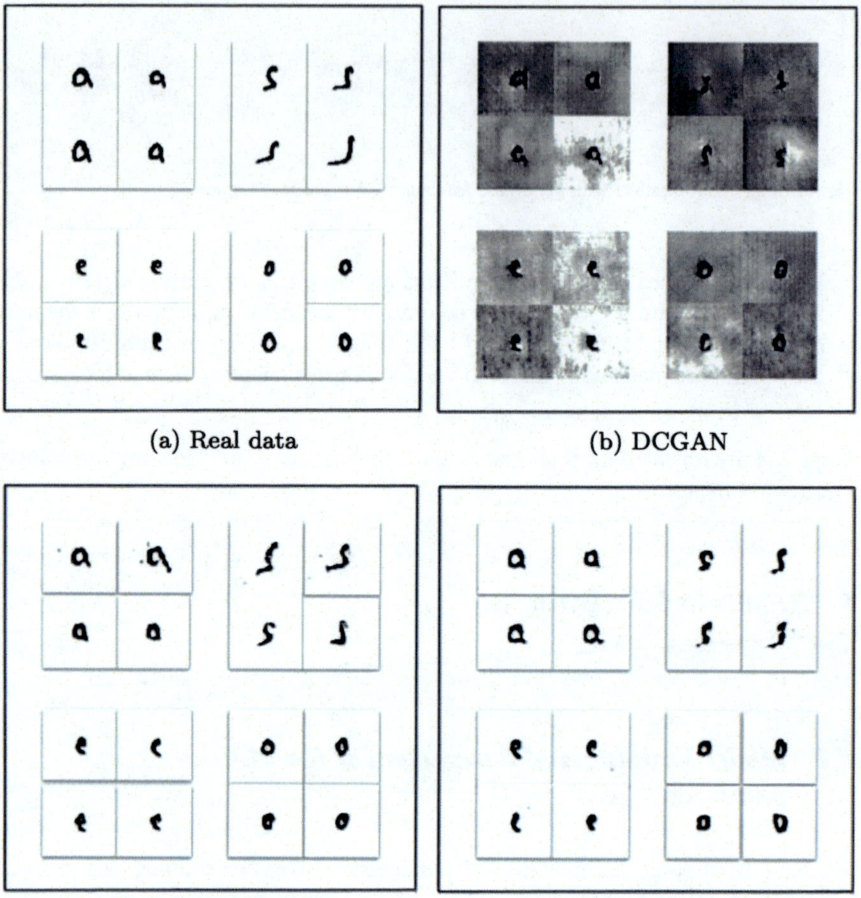

(a) Real data (b) DCGAN

(c) R-cDCGAN (d) I-cDCGAN

Fig. 7 Alphabets 'a', 'e', 'o', and 's' collected/synthesized using **a** real data **b** DCGAN **c** R-cDCGAN **d** I-cDCGAN

R-cDCGAN and I-cDCGAN generated good quality images and results showed variations not only across the classes but also within each class.

Figure 8a shows three sentences generated using commonly used normal fonts. The same sentences are generated using alphabets synthesized by R-cDCGAN (Fig. 8c) and I-cDCGAN (Fig. 8d). In Fig. 8b, sentences are generated using alphabets taken from real data; this gives a visual estimation of the author's handwriting style. Sentences produced by alphabets generated using I-cDCGAN give a more realistic look than R-cDCGAN as it considered neighboring alphabets along with the current alphabet in the training process.

(a) Normal Font

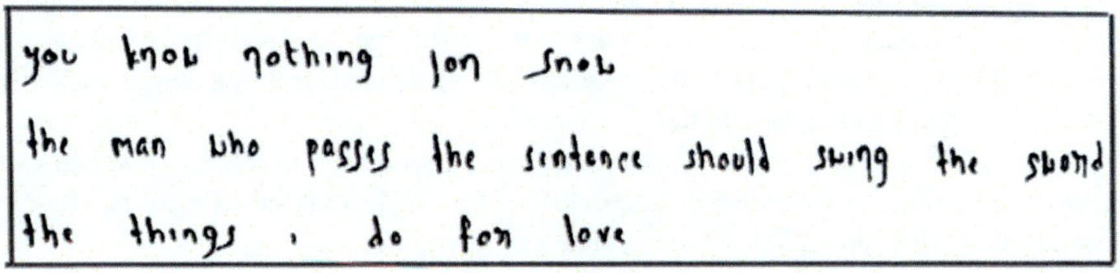

(b) Real data

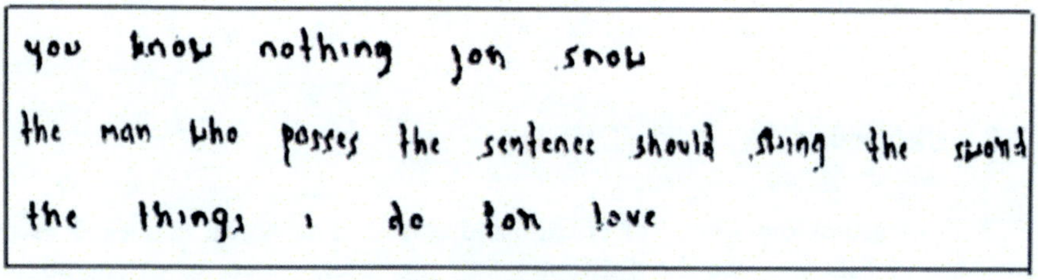

(c) R-cDCGAN

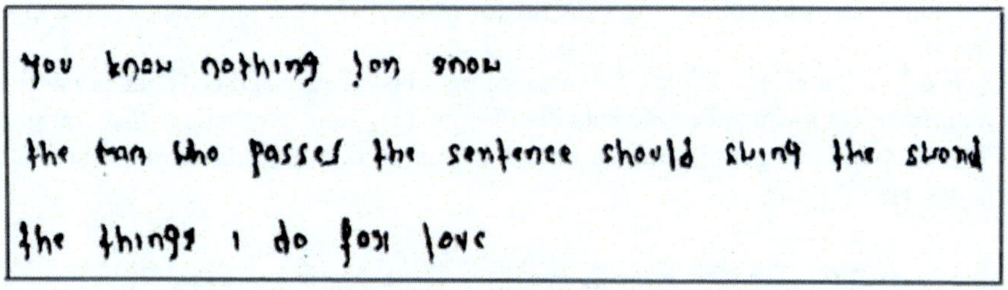

(d) I-cDCGAN

Fig. 8 Text generated with alphabets collected/synthesized using **a** normal fonts **b** real data **c** R-cDCGAN **d** I-cDCGAN

7 Performance Evaluation

To measure the diversity and quality of synthetic images produced by the generator, Fréchet Inception Distance (FID) [14] score is used. Benny et al. [15] generalized FID score for evaluating generative models in the class-conditional image generation setting. They proposed Within Class FID (WCFID) and Between Class FID (BCFID) for comparing the performances of cGANs. FID, WCFID, and BCFID scores are computed to measure the performance of our proposed models.

7.1 Fréchet Inception Distance (FID)

FID score computes resemblance between synthetic (S) data distribution and actual (A) data distribution. The lower FID score indicates that the diversity and quality of synthetic images generated by the generator are more similar to the actual ones. The lower FID score indicates a better GAN model.

If μ^A, Σ^A and μ^S, Σ^S are the centers and covariance matrices of actual data distribution ($\mathcal{D}_A$) and synthetic data distribution ($\mathcal{D}_S$) obtained using a pre-trained feature extractor, then FID score is calculated using (3).

$$\text{FID}\left(\mathcal{D}_A, \mathcal{D}_S\right) = \left\|\mu^A - \mu^S\right\|^2 + \text{Tr}\left(\Sigma^A + \Sigma^S - 2\left(\Sigma^A\Sigma^S\right)^{\frac{1}{2}}\right) \tag{3}$$

7.2 Between Class FID (BCFID)

BCFID measures the spread of conditioned classes over actual classes. It is the measure of FID between the distributions of mean feature vector of conditioned classes in synthetic (S) data distribution and mean feature vector of actual (A) classes in actual data distribution. Like FID, the lower BCFID score indicates a better GAN model.

If μ_B^A, Σ_B^A and μ_B^S, Σ_B^S are the mean of per class means and covariance matrices of all per class means of actual data distribution ($\mathcal{D}_A$) and synthetic data distribution ($\mathcal{D}_S$) obtained using a pre-trained feature extractor, then BCFID score is calculated using (4).

$$\text{BCFID}\left(\mathcal{D}_A, \mathcal{D}_S\right) = \|\mu_B^A - \mu_B^S\|^2 + \text{Tr}\left(\Sigma_B^A + \Sigma_B^S - 2\left(\Sigma_B^A\Sigma_B^S\right)^{\frac{1}{2}}\right) \tag{4}$$

7.3 Within Class FID (WCFID)

WCFID measures the resemblance between each conditioned class to its corresponding real class. WCFID is the average of all FID scores within all the classes in synthetic (S) data distribution and actual data distribution. The lower WCFID score, similar to FID and BCFID, denotes a superior GAN model.

If μ_c^A, $\Sigma_{W,c}^A$ and μ_c^S, $\Sigma_{W,c}^S$ are within class means and within class covariance matrices of actual data distribution ($\mathcal{D}_A$) and synthetic data distribution ($\mathcal{D}_S$) of class

(C) obtained using a pre-trained feature extractor, then WCFID score is calculated using (5).

$$\text{WCFID}\left(\mathcal{D}_A, \mathcal{D}_S\right) = \mathbb{E}_{c\sim\mathcal{D}_C}\left[\left\|\mu_c^A - \mu_c^S\right\|^2 + \text{Tr}\left(\Sigma_{W,c}^A + \Sigma_{W,c}^S - 2\left(\Sigma_{W,c}^A \Sigma_{W,c}^S\right)^{\frac{1}{2}}\right)\right] \tag{5}$$

7.4 *Experiments*

For the feature extractor model, we developed a simple CNN classifier (with 2 convolutional layers and 2 fully connected layers), trained it on IIITA-ALPHABETS data set, and got a test accuracy of 99.35%. This pre-trained CNN classifier is used as a feature extractor to calculate FID score. From real and generated distributions, 50 samples are randomly collected from each class for measuring FID score. The activation of the penultimate layer is employed as the extracted features. The extracted feature dimension is 4096.

In Table 2, real data FID scores are obtained not from synthetic data, but from real data itself, two distributions used for FID score calculations are collected from real data itself. Real data FID scores serve as upper bound to evaluate the performance of GANs. Since, DCGAN generated only a few classes of alphabets, so there is high diversity between actual data distribution and synthetic data distribution, resulting in poor FID score.

WCFID score of I-cDCGAN is better than R-cDCGAN because, in real data, the appearance of the alphabet depends on its previous and next alphabet. I-cDCGAN mimics this behavior by considering previous and next alphabet class label along with current alphabet class label as input while training and generating alphabet. So within each class, it produces more diverse data than R-cDCGAN which considers only the current alphabet class label.

BCFID score of R-cDCGAN is better than I-cDCGAN because it is the measure of FID between the distributions of the mean feature vector of conditioned classes in synthetic data distribution and the mean feature vector of actual classes in actual data distribution. As discussed earlier, R-cDCGAN within each class produces less diverse data, but more mean-centric data compared to I-cDCGAN as it considers only the current alphabet class label while training and generating the alphabet.

However, the overall performance of I-cDCGAN is better than R-cDCGAN, as it has a better FID score, indicating that the resemblance between synthetic data distribution and actual data distribution is more in the case of I-cDCGAN than R-cDCGAN.

Table 2 FID, WCFID, and BCFID metrics on IIITA-ALPHABETS data set for DCGAN, R-cDCGAN, and I-cDCGAN

Model	FID	WCFID	BCFID
Real data	−0.8097	−0.0002064	−9.26e−10
DCGAN	137692	N/A	N/A
R-cDCGAN	46.1970	1.1398	0.0213
I-cDCGAN	43.4696	1.0302	0.0327

8 Conclusion and Future Work

R-cDCGAN and I-cDCGAN are proposed, two-class conditioned versions of DCGAN for generating random font and intelligent font, respectively. The alphabet class is passed in the R-cDCGAN generator to generate random font, whereas the previous, current, and next alphabet classes are passed in the I-cDCGAN generator to generate intelligent font. The experimental results showed that the sentences formed using synthetic fonts generated from proposed models are incorporated with uncertainty and randomness, giving them a more natural depiction than standard fonts. Instead of random TXT documents, lexicons can be created based on the most frequently used words in English to reduce the number of handwritten documents that the author needs to write. Numbers, special characters, and uppercase alphabets can also be added to training data.

Acknowledgements The present research is partially funded by the I-Hub foundation for Cobotics (Technology Innovation Hub of IIT-Delhi set up by the Department of Science and Technology, Govt. of India).

References

1. Goodfellow, I.J.: Pouget-Abadie, Jean, Mirza, Mehdi, Xu, Bing, Warde-Farley, David, Ozair, Sherjil. Aaron C., and Bengio, Yoshua. Generative adversarial nets. NIPS, Courville (2014)
2. Tian, Y.: zi2zi: Master Chinese calligraphy with conditional adversarial networks (2017). https://kaonashi-tyc.github.io/2017/04/06/zi2zi.html. Accessed 16 Apr 2019
3. Isola, P., Zhu, J.-Y., Zhou, T., Efros, A.A.: Image-to-image translation with conditional adversarial networks. In: Proceedings of the IEEE Conference on Computer Vision and Pattern Recognition, pp. 1125–1134. CVPR (2017)
4. Odena, A., Olah, C., Shlens, J.: Conditional image synthesis with auxiliary classifier GANs. In: Proceedings of the 34th International Conference on Machine Learning, pp. 2642–2651. ICML (2017)
5. Taigman, Y., Polyak, A., Wolf, L.: Unsupervised cross-domain image generation. arXiv preprint arXiv:1611.02200
6. Chang, J., Gu, Y.: Chinese typography transfer. arXiv preprint arXiv:1707.04904
7. Ronneberger, O., Fischer, P., Brox, T.: U-net: convolutional networks for biomedical image segmentation. In: International Conference on Medical image Computing and Computer-Assisted Intervention, pp. 234–241. Springer, Cham (2015)

8. Azadi, S., Fisher, M., Kim, V., Wang, Z., Shechtman, E., Darrell, T.: Multi-content GAN for few-shot font style transfer. In: Proceedings of the IEEE Conference on Computer Vision and Pattern Recognition (CVPR), pp. 7564–7573 (2018)
9. Mirza, M., Osindero, S.: Conditional generative adversarial nets (2014). arXiv preprint arXiv:1411.1784
10. Guo, Y., Lian, Z., Tang, Y., Xiao, J.: Creating new Chinese fonts based on manifold learning and adversarial networks. In: Diamanti, O., Vaxman, A. (eds) Proceedings of the Eurographics—Short Papers. The Eurographics Association (2018)
11. Lin, X., Li, J., Zeng, H., Ji, R.: Font generation based on least squares conditional generative adversarial nets. Multimedia Tools Appl. 1–15 (2018)
12. Radford, A., Metz, L., Chintala, S.: Unsupervised representation learning with deep convolutional generative adversarial networks. arXiv preprint http://arxiv.org/abs/1511.06434 (2015)
13. Kingma Diederik, P., Adam, J.B.: A method for stochastic optimization. arXiv preprint arXiv:1412.6980 (2014)
14. Heusel, M., Ramsauer, H., Unterthiner, T., Nessler, B., & Hochreiter, S. (2017). Gans trained by a two time-scale update rule converge to a local nash equilibrium. In Advances in neural information processing systems (pp. 6626-6637)
15. Benny, Y., Galanti, T., Benaim, S. et al. Evaluation Metrics for Conditional Image Generation. Int J Comput Vis 129, 1712-1731 (2021). https://doi.org/10.1007/s11263-020-01424-w

Analysis and Application of Multispectral Data for Water Segmentation Using Machine Learning

Shubham Gupta, D. Uma, and R. Hebbar

Abstract Monitoring water is a complex task due to its dynamic nature, added pollutants, and land build-up. The availability of high-resolution data by Sentinel-2 multispectral products makes implementing remote sensing applications feasible. However, overutilizing or underutilizing multispectral bands of the product can lead to inferior performance. In this work, we compare the performances of ten out of the thirteen bands available in a Sentinel-2 product for water segmentation using eight machine learning algorithms. We find that the shortwave-infrared bands (B11 and B12) are the most superior for segmenting water bodies. B11 achieves an overall accuracy of 71% while B12 achieves 69% across all algorithms on the test site. We also find that the Support Vector Machine (SVM) algorithm is the most favorable for single-band water segmentation. The SVM achieves an overall accuracy of 69% across the tested bands over the given test site. Finally, to demonstrate the effectiveness of choosing the right amount of data, we use only B11 reflectance data to train an artificial neural network, BandNet. Even with a basic architecture, BandNet is proportionate to known architectures for semantic and water segmentation, achieving a 92.47 mIOU on the test site. BandNet requires only a fraction of the time and resources to train and run inference, making it suitable to be deployed on web applications to run and monitor water bodies in localized regions. Our codebase is available at https://github.com/IamShubhamGupto/BandNet.

Keywords Water · Sentinel-2 · Machine-learning · Artificial neural network · BandNet

Work done partially while interning at CDSAML, PES University and RRSC-S, ISRO.

S. Gupta (✉) · D. Uma
PES University, Bangaluru, KA 560085, India
e-mail: shubhamgupto@gmail.com

R. Hebbar
Regional Remote Sensing Centre–South, ISRO, Bengaluru, KA 560037, India

M. Tistarelli et al. (eds.), *Computer Vision and Machine Intelligence*,
Lecture Notes in Networks and Systems 586,
https://doi.org/10.1007/978-981-19-7867-8_56

1 Introduction

Water is one of the most essential resources on our planet. With the rise in population in cities, water bodies have shrunk or disappeared entirely. In the coastal regions, due to global warming, the increase in water levels has begun to submerge land in low-lying areas [13]. To take action and prevent irreparable damages, spatio-temporal awareness of water is a necessity.

The recent advancements in Convolutional Neural Network (CNN) models show promising results in object classification [14, 22] and segmentation [5, 12, 20] from images and reflectance values [11, 15, 21]. The reflectance models rely on the combination of either or all of the visible spectrum, near-infrared, and shortwave-infrared spectrum bands. To our knowledge, there is no ranking of the multispectral bands for their usability for water segmentation. In this work, we study the relationship between water segmentation performance and multispectral bands using several machine learning algorithms [1, 2, 6, 8, 9, 17, 18, 23]. The goal is to find the multispectral band and machine learning algorithm that is best suited to segmenting water bodies.

By utilizing the best multispectral band to delineate water bodies, we show that a single band is adequate to perform similar to large segmentation models [5] trained on multispectral false-infrared images that are composed of B8, B4, and B3. We achieve this result on a simple Artificial Neural Network, BandNet. Unlike existing literature, BandNet does not treat multispectral data as images but rather sees each pixel as a unique data-point. Furthermore, it requires a fraction of the training time and resources to produce respectable results. To summarize our contributions to this paper:

1. We compare the water segmentation performance of the multispectral bands present in the Sentinel-2 product using multiple machine learning algorithms.
2. We compare the performances of various machine learning algorithms for single-band water segmentation.
3. We demonstrate the importance of data selection by training a simple ANN, BandNet, that performs similar to other segmentation models - trained on images and relfectance data.

2 Satellite Data and Study Site

This study uses the multispectral products from the European Space Agency satellite constellation Sentinel-2. All the products are made available in L2A mode, giving us direct access to Bottom of Atmosphere (BOA) reflectance [3]. We study bands B2, B3, B4, B5, B6, B7, B8, B8A, B11, and B12 which are re-sampled to $10m$ spatial resolution using SNAP, SeNtinel Applications Platform.

Since one of the requirements of training Deeplabv3+ [5] is an extensively annotated dataset, we required images of water bodies from more than one Sentinel-2

Table 1 Metadata of the Sentinel-2 data used

Tile	Date	Type	Images
T43PGQ	2019–01–04	Lakes	186*
T44RQN	2019–01–21	Rivers	64
T43PFS	2019–02–16	Lakes, rivers	47
T43QGU	2019–03–18	Lakes, rivers	172*
T43PFQ	2019–03–28	Lakes	310*
T44QLE	2019–03–30	Rivers	175
T43RFJ	2019–04–20	Rivers	242*
T43QHV	2019–04–24	Lakes	250*

* Rotation is applied to images as part of data augmentation

product. For this reason, we acquired a total of eight Sentinel-2 products. As for the image from the products themselves, we utilize false infrared images generated from B8, B4, and B3 of each product. The entire image is 10980 × 10980 pixels. Since we cannot process an image of this scale at once, we break down the false-infrared images into 549 × 549 pixel tiles.

We use the scene classification map available within each Sentinel-2 product to annotate water bodies. We discard tiles and the corresponding annotation if they do not contain any water bodies. The resulting data acquired from these products is summarized in Table 1. We obtain 1446 images with annotations after applying a simple rotation augmentation to five of the eight products.

On the other hand, a single Sentinel-2 product can provide 120 million data points as the values are based on each pixel. Due to hardware limitations, we can only take a subset of the product at a time. Figure 1 shows the false-infrared images of the two subsets we will be using to evaluate our findings. Image (a) in Fig. 1 represents a subset with North latitude 12.963, West longitude 77.630, South latitude 12.889, East longitude 77.703. This subset contains Bellandur lake and has a good representation of land built-up, water bodies, and vegetation. Image (c) in Fig. 1 represents a subset with North latitude 12.655, West longitude 77.131, South latitude 12.581, East longitude 77.205. This subset contains vegetated land and lakes at close proximity. We will be referring to them as subset1 and subset2, respectively.

3 Methodology Used

3.1 Data Processing

From subset1, we consider an equal number of water and non-water data points. For reproducibility, we set a seed value [16] so that the same data points will be selected when the experiment is re-run. The dataset is split into test, train, and validation

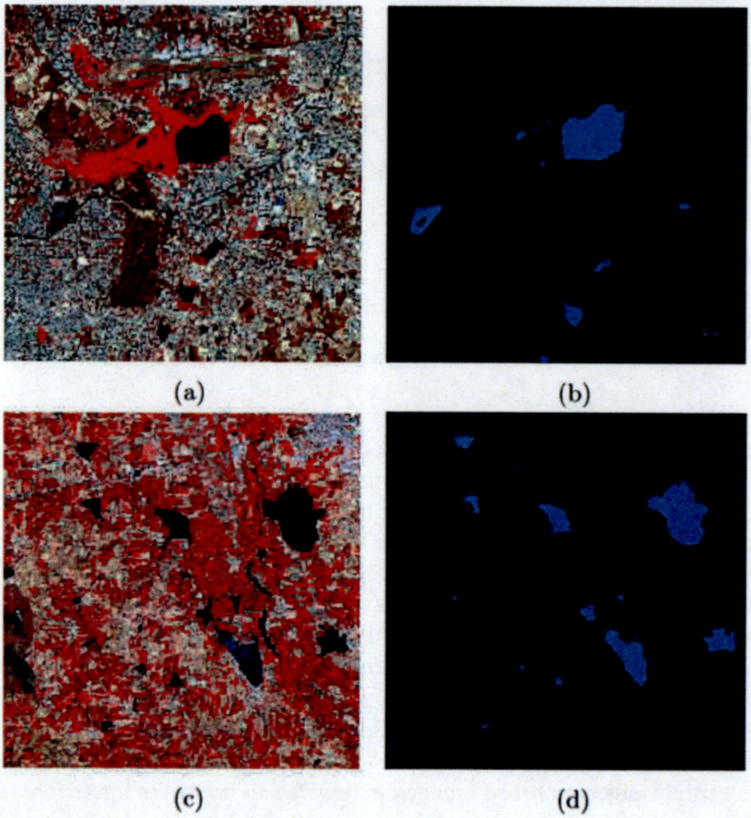

Fig. 1 False-infrared image generated from the Sentinel-2 product T43PGQ subset dated 2019-01-04. **a** Preview of subset1. **b, d** Corresponding water body annotation generated from scene classification map. **c** Preview of subset2

sets, each being mutually exclusive. We evaluate and report the performance of the machine learning algorithms on the validation set. The validation set has never been seen before by the algorithms. Data points from subset2 are entirely used to evaluate the algorithms and models.

3.2 Band Reflectance Analysis

We employ eight statistical machine learning algorithms on the reflectance data of individual bands to gauge their performance for water segmentation, namely: Logistic Regression (LR) [9], Gaussian Naive Bayes (GNB) [2], K-neighbors (KN) [23], Decision Tree (DT) [17], Random Forest (RF) [1], XGBoost (XGB) [6], Stochastic Gradient Descent (SGD) [18], and Support Vector Machine (SVM) [8]. All eight

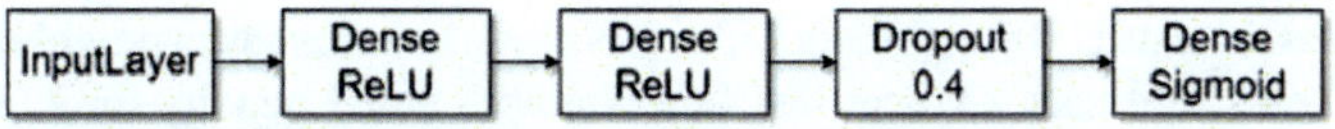

Fig. 2 Proposed architecture of BandNet for segmentation of water bodies

algorithms are first made to fit the training set, followed by testing their performance on the validation set.

3.3 BandNet

Based on the performance of individual bands, further discussed in Sect. 5, we designed a simple ANN architecture, BandNet, to create segmentation maps of water bodies using raw reflectance data as input.
BandNet consists of two Dense layers with ReLU activation functions. Before the classification head, a Dropout layer of 0.4 is in place to deal with the class imbalance problem and improve generalization. The classification head is a Dense layer with Sigmoid activation. The architecture of BandNet is shown in Fig. 2.

3.4 Multispectral Image Analysis

We generate false color infrared images from B8, B4, and B3 to train Deeplabv3+ [5], a deep neural network architecture that uses Atrous Spatial Pyramid Pooling (ASPP) [4] and a encoder-decoder [12] structure. The ASPP is used to encode the contextual information with the help of varying atrous rates r.

$$y[i] = \sum_k x[i + r.k]w[k] \tag{1}$$

The decoder then uses the contextual information to generate fine object boundaries. Finally, depthwise separable convolution [7] is applied to both ASPP and the decoder module for reducing computational complexity while preserving the performance of the model.

4 Implementation Details

Here we define the details used to train the various models and algorithms. We use an Nvidia GTX 1050 graphics card, 16GB of memory, and an Intel i5 8300H processor for training and inference. The SVM classifier uses a radial basis function (rbf) kernel.

The RF classifier uses 100 estimators. The KN classifier uses seven neighbors to vote. The SGD classifier uses a Huber loss function with an L1 penalty for 25 iterations. The remaining machine learning algorithms ran with no extra parameters.

For each multispectral band and model, we generate a confusion matrix, which then helps us calculate the mean intersection over union (mIoU). mIoU for N classes can be defined in terms of true positives (TP) as:

$$\mathrm{IoU_i} = \frac{\mathrm{TP_i}}{\mathrm{Sum\ Of\ Rows_i} + \mathrm{Sum\ Of\ Columns_i} - \mathrm{TP_i}} \tag{2}$$

$$\mathrm{mIoU} = \frac{\sum_{n=1}^{N} \mathrm{IoU_i}}{N} \tag{3}$$

We train BandNet with a BinaryCrossEntropy loss function and BinaryAccuracy as the metric. The optimizer Adam has a learning rate of $1e-4$. We use EarlyStopping to avoid overfitting with a patience value of 5.

We run the DeepwatermapV2 [11] and WatNet [15] directly using pretrained weights for inference on B2, B3, B4, B8A, B11, and B12 based on the guides provided by the authors.

The Deeplabv3+ uses a learning rate of 0.001, weight decay of 0.00004, atrous rates at (6,12,18), output stride of 16, momentum of 0.9, decoder output stride at 4, batch size of 1 for 30K iterations. We use Xception65 [7] and MobileNetv2 [19] as the network backbones pretrained on the PASCALVOC-2012 dataset [10].

5 Results and Discussion

In Table 2, we present our results of generating water segmentation from each of the ten bands paired with each of the eight machine learning algorithms. We calculate the percent column for a band as a fraction of the sum of the achieved mIoU across all algorithms, to the maximum attainable mIoU. The percent row for each algorithm is calculated as the sum of the achieved mIoU across all bands divided by the maximum attainable mIoU (Table 2).

In that order, B11, B12, and B8A are the best performing bands across all tested algorithms. B11 outperforms B2 by an absolute value of 0.36 which makes B11 102% better than B2 for water segmentation. We also notice a trend of SWIR bands (B11 and B12), followed by NIR bands (B8, B8A) and finally visible spectrum bands (B2, B3, B4) in terms of water segmentation performance. However, the algorithms used to test these bands do not present such large variations in performance. The best performing algorithm, Support Vector Machines [8], outperforms Linear Regression [9] by an absolute value of 0.08 or 13.11%.

In Table 3, we compare the performance of BandNet on different band combinations to existing solutions for water segmentation. This comparison is carried out on subset1. We directly use Deepwatermapv2 and WatNet using their existing weights

Table 2 Comparison of single-band water segmentation performance (mIoU) using various machine learning models on validation set of subset1

Band	LR	GNB	RF	KN	DT	SGD	XGB	SVM	Percent
B02	36.93	34.23	42.07	42.74	42.45	35.26	42.62	40.76	0.35
B03	36.79	32.04	41.58	41.69	41.89	40.99	41.45	42.06	0.35
B04	44.37	42.66	51.34	51.58	51.82	48.33	51.43	61.06	0.43
B05	49.17	48.23	55.02	55.3	55.86	58.11	56.51	56.48	0.47
B06	72.26	77.27	76.02	72.68	76.75	79.44	79.41	78.92	0.67
B07	73.66	79.53	76.52	73.88	77.16	78.5	77.62	78.18	0.67
B08	70.15	75.08	74.4	78.02	75.72	73.16	79.05	79.78	0.66
B8A	75.81	82.77	77.16	72.82	77.72	78.19	78.03	79.69	0.68
B11	80.75	87.71	81.71	78.61	82.3	78	82.2	84.65	0.71
B12	71.33	86.21	78.01	78.21	80.06	77.36	84.68	84.69	0.69
Percent	0.61	0.65	0.65	0.65	0.66	0.65	0.67	0.69	

The mIoU is color coded from low to high as brick-red to blue. The percent row and column describes the accuracy of a band across various algorithms and the accuracy of an algorithm across various bands, respectively. The color coding from low to high follows as red to teal

Table 3 Comparison of BandNet to other models on subset1 in terms of performance, training time, parameters. mIOU: mean Intersection over Union

Model	Data	Parameters ↓	Time ↓	mIOU ↑
DeepWaterMapv2 [11]	B2-B3-B4-B8A-B11-B12	37.2M	–	89.01
WatNet [15]	B2-B3-B4-B8A-B11-B12	3.4M	–	89.74
BandNet (Ours)	B8-B4-B3	1217	0.41	89.92
MobileNetv2 [5, 19]	Image	2.1M	12	90.27
Xception65 [5, 7]	Image	41M	16	92.44
BandNet (Ours)	**B11**	**1153**	**0.45**	**92.47**
BandNet (Ours)	B12-B11	1185	0.27	92.96
BandNet (Ours)	B12-B11-B8A	1217	0.65	92.99

Time is measured in Hours. ↓: The lower the better. ↑: The higher the better

Table 4 Comparison of BandNet to reflectance models on subset1 in terms of performance, training time, parameters. mIOU: mean Intersection over Union

Model	Data	mIOU ↑
WatNet [15]	B2-B3-B4-B8A-B11-B12	86.92
BandNet (Ours)	B12-B11-B8A	87.00
BandNet (Ours)	B12-B11	87.33
BandNet (Ours)	B11	87.42
DeepWaterMapv2 [11]	B2-B3-B4-B8A-B11-B12	89.33

Time is measured in Hours. ↑: The higher the better

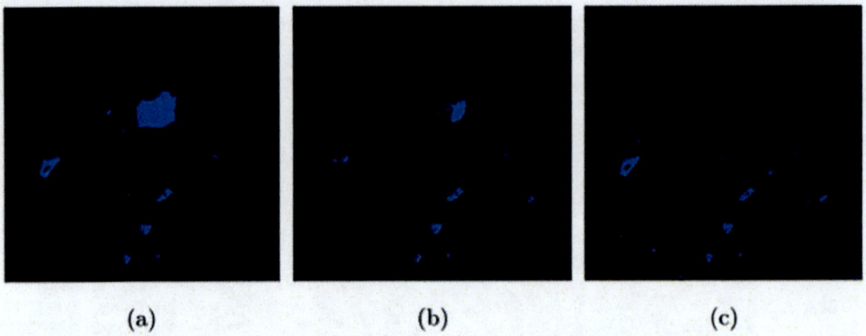

Fig. 3 Application of BandNet: Monitoring of Bellandur lake using BandNet in subset1. The maps are generated on the T43PGQ subset dated: **a** 2019–01–04, **b** 2020–03–29 and **c** 2021–03–04

provided by the authors. Both models perform respectably considering they have never seen the data before in training or testing phases.

To make a fair comparison of images to reflectance data, we compared Deeplab V3+ to BandNet trained on the reflectance of B8-B4-B3. We notice BandNet performs similarly to Deeplabv3+ with MobileNetv2 as the backbone. However, BandNet trains at a fraction of the parameters and time required by Deeplabv3+ for the same performance. This performance gap is bridged when we train BandNet on B11, the best performing multispectral band. We observe that BandNet with B11 outperforms Deeplabv3+ by an absolute value of 2.2 and 0.03 mIOU on MobileNetv2 and Xception65 backbones, respectively. BandNet further surpasses Deeplab3+ if we pass along additional high-performing bands with B11.

In Table 4, we compare the inference of models on subset2. BandNet sees a slight drop in performance but is still comparable to Deepwatermapv2 and WatNet. These results underscore the idea behind this work that choosing the right set of data is important when building predictive models. We believe this drop in performance is due to the simplistic nature of BandNet and that it was originally designed for one multispectral band only. However, this makes it convenient to re-train and deploy BandNet over a very localized region using minimal resources and time. The small size of BandNet allows it to be hosted on a web application and run inferences at a very low cost. This is extremely helpful if someone were to host their own water body monitoring solution (Fig. 3).

6 Conclusion

In this paper, we have compared the performances of individual bands for water segmentation, using machine learning algorithms. We create a hierarchy of multispectral bands based on their performance, placing SWIR bands (B11, B12) on the top, followed by NIR bands (B8, B8A) and finally visible spectrum bands (B2, B3, B4). We also observed that the Support Vector Machine followed by XG Boost algorithm are favorable is single-band water segmentation compared to other algorithms.

Using the best performing band, B11, we developed a simple ANN that is able to compete with specialized segmentation architectures in performance while requiring only a fraction of the time and resources to train. This lightweight nature of BandNet makes it suitable for deploying it on web applications to monitor water bodies in localized regions.

The objective we want to underscore in this study is that using the right amount and type of data is sufficient to compete with existing solutions with no improvements to the architecture. Now that we have established a hierarchy of multispectral bands, we can further look into designing a novel architecture or training regime for water segmentation. Our process can also be extended to other features such as vegetation or built-up land and can be used in combinations for multi-class segmentation.

Acknowledgements This work has been supported by Center of Data Science and Applied Machine Learning, Computer Science and Engineering Department of PES University, and Regional Remote Sensing Centre—south. We would like to thank Dr. Shylaja, S. S. of PES University and Dr. K. Ganesha Raj of Regional Remote Sensing Centre - south for the opportunity to carry out this work.

References

1. Breiman, L.: Random forests. Mach. Learn. **45**(1), 5–32 (2001). https://doi.org/10.1023/A:1010933404324
2. Chan, Golub, L.: Updating formulae and a pairwise algorithm for computing sample variances. http://i.stanford.edu/pub/cstr/reports/cs/tr/79/773/CS-TR-79-773.pdf (1979)
3. Chen, J., Li, Y., Ma, Q., Shen, X., Zhao, A., Li, J.: Preliminary evaluation of sentinel-2 bottom of atmosphere reflectance using the 6sv code in Beijing area. In: IGARSS 2018—2018 IEEE International Geoscience and Remote Sensing Symposium. pp. 7760–7763 (2018). https://doi.org/10.1109/IGARSS.2018.8517598
4. Chen, L.C., Papandreou, G., Schroff, F., Adam, H.: Rethinking atrous convolution for semantic image segmentation (2017)
5. Chen, L.C., Zhu, Y., Papandreou, G., Schroff, F., Adam, H.: Encoder-decoder with atrous separable convolution for semantic image segmentation (2018)
6. Chen, T., Guestrin, C.: XGBoost: a scalable tree boosting system. In: Proceedings of the 22nd ACM SIGKDD International Conference on Knowledge Discovery and Data Mining. pp. 785–794. KDD '16, ACM, New York, NY, USA (2016). https://doi.org/10.1145/2939672.2939785. http://doi.acm.org/10.1145/2939672.2939785
7. Chollet, F.: Xception: deep learning with depthwise separable convolutions. In: 2017 IEEE Conference on Computer Vision and Pattern Recognition (CVPR). pp. 1800–1807 (2017)
8. Cortes, C., Vapnik, V.: Support-vector networks. Mach. Learn. **20**(3), 273–297 (1995)
9. Cox, D.R.: The regression analysis of binary sequences. J. Roy. Stat. Soc. Ser. B (Methodol) **20**(2), 215–232 (1958)
10. Everingham, M., Van Gool, L., Williams, C.K.I., Winn, J., Zisserman, A.: The PASCAL visual object classes challenge 2012 (VOC2012) results. http://www.pascal-network.org/challenges/VOC/voc2012/workshop/inde-x.html
11. Isikdogan, L., Bovik, A., Passalacqua, P.: Seeing through the clouds with deepwatermap. IEEE Geosci. Remote Sens. Lett. **PP**, 1–5 (2019). https://doi.org/10.1109/LGRS.2019.2953261

12. Islam, M., Rochan, M., Bruce, N., Wang, Y.: Gated feedback refinement network for dense image labeling. pp. 4877–4885 (07 2017). https://doi.org/10.1109/CVPR.2017.518
13. Kazama, S., Oki, T.: The effects of climate change on water resources (2006)
14. Liu, Z., Hu, H., Lin, Y., Yao, Z., Xie, Z., Wei, Y., Ning, J., Cao, Y., Zhang, Z., Dong, L., et al.: Swin Transformer V2: Scaling Up Capacity and Resolution. arXiv preprint arXiv:2111.09883 (2021)
15. Luo, X., Tong, X., Hu, Z.: An applicable and automatic method for earth surface water mapping based on multispectral images. Int. J. Appl. Earth Obs. Geoinf. **103**, 102472 (2021)
16. Picard, D.: Torch. manual_seed (3407) is all you need: On the influence of random seeds in deep learning architectures for computer vision. arXiv preprint arXiv:2109.08203 (2021)
17. Quinlan, J.R.: Induction of decision trees. Mach. Learn. **1**, 81–106 (1986)
18. Ruder, S.: An overview of gradient descent optimization algorithms. arXiv preprint arXiv:1609.04747 (2016)
19. Sandler, M., Howard, A., Zhu, M., Zhmoginov, A., Chen, L.C.: Mobilenetv2: inverted residuals and linear bottlenecks. pp. 4510–4520 (2018). 10.1109/CVPR.2018.00474
20. Wei, Y., Hu, H., Xie, Z., Zhang, Z., Cao, Y., Bao, J., Chen, D., Guo, B.: Contrastive learning rivals masked image modeling in fine-tuning via feature distillation. arXiv preprint arXiv:2205.14141 (2022)
21. Yang, L., Driscol, J., Sarigai, S., Wu, Q., Lippitt, C.D., Morgan, M.: Towards synoptic water monitoring systems: a review of ai methods for automating water body detection and water quality monitoring using remote sensing. Sensors **22**(6) (2022). https://doi.org/10.3390/s22062416, https://www.mdpi.com/1424-8220/22/6/2416
22. Zhang, H., Li, F., Liu, S., Zhang, L., Su, H., Zhu, J., Ni, L.M., Shum, H.Y.: Dino: Detr with improved denoising anchor boxes for end-to-end object detection. arXiv preprint arXiv:2203.03605 (2022)
23. Zhang, M.L., Zhou, Z.H.: A k-nearest neighbor based algorithm for multi-label classification. In: Hu, X., Liu, Q., Skowron, A., Lin, T.Y., Yager, R.R., Zhang, B. (eds.) GrC. pp. 718–721. IEEE (2005). http://cs.nju.edu.cn/zhouzh/zhouzh.files/publication/grc05.pdf

MangoYOLO5: A Fast and Compact YOLOv5 Model for Mango Detection

Pichhika Hari Chandana, Priyambada Subudhi, and Raja Vara Prasad Yerra

Abstract Detection of fruits in orchards is crucial for agricultural applications like automatic estimation and mapping of yield. The state-of-the-art approaches for this task are based on hand-crafted features and therefore are liable to variations in a real orchard environment. However, the current deep learning-based one-stage object detection methods like the YOLO provide excellent detection accuracy at the cost of increased computational complexity. So this paper presents an improved, fast, and compact YOLOv5s model named as MangoYOLO5 for detecting mangoes in the images of open mango orchards given by the MangoNet-Semantic dataset. The proposed MangoYOLO5 has adopted a few improvements over YOLOv5s. Firstly, the feedback convolutional layer is removed from the BottleneckCSP module of the original YOLOv5s model, reducing the convolutional layers by 11. Secondly, two convolutional layers, one from the focus module and another just after the focus module, are removed to reduce the overall weight of the architecture. It is observed from precision, recall, and mAP metrics of the experimental results that MangoYOLO5 detection performance is 3.0% better than the YOLOv5s, addressing several factors such as occlusion, distance, and lighting variations. In addition, the realized lighter model requires 66.67% less training time as compared to original YOLOv5s, which can significantly affect its real-time implementations.

Keywords Deep learning · Lightweight · Mango detection · YOLOv5

P. Hari Chandana · P. Subudhi · R. Vara Prasad Yerra (✉)
Indian Institute of Information Technology, Sri City, Chittoor, Sri City, Chittoor, AP, India
e-mail: yrv.prasad@iiits.in

P. Hari Chandana
e-mail: harichandana.p@iiits.in

P. Subudhi
e-mail: priyambada.s@iiits.in

M. Tistarelli et al. (eds.), *Computer Vision and Machine Intelligence*,
Lecture Notes in Networks and Systems 586,
https://doi.org/10.1007/978-981-19-7867-8_57

1 Introduction

Various fruits, vegetables, and crops are grown by Indian farmers every year [1]. Among fruits, mangoes are produced on a large scale in India (54.2% of total world production) and are having a high demand worldwide because of their aroma, exotic taste and high nutritional value. India has the richest collection of mango cultivators, and among other states, Andhra Pradesh (AP) has occupied the second position in the amount of mango production with an area of 4.31 *lakh* hectares and annual production of 43.5 *lakh* metric tons [2]. As our research group is in AP, we aim at automatic pre-harvest yield estimation of mangoes in nearby mango orchards. This will be a valuable guide for the farmers in making harvesting and marketing plans.

The first step in yield estimation is the detection of mangoes. Manual detection of mangoes is expensive in terms of both time and labor. Moreover, the use of hand-crafted features like color, shape [3], texture [4], for the detection of mangoes suffers from the problem of illumination changes, indistinguishable backgrounds, high overlap, occlusion conditions, and low resolutions [5]. Nevertheless, significant research works have been carried in the application of deep convolutional neural networks for fruit detection and recognition [6]. Deep learning-based systems recognize objects by implicitly finding their unique features as opposed to the traditional image processing methods. Research works in literature such as model trained using MobileNet [7], randomized Hough transform, and back propagation NN [8], R-CNN [9], and Faster R-CNN [10] were effectively used for mango detection. However, these methods involve two-stage object detection, where the region of interest is generated in the first stage. In the second stage, features are extracted for the bounding box and classification regression task. Despite reasonable detection accuracy, two-stage object detection methods suffer low detection speed. As a solution to this problem, one-stage object detection models were introduced. Such a model generates class probabilities and bounding boxes as a single regression task and hence is faster as compared to the two-stage detection models [11]. Among others, You Only Look Once (YOLO) [12] is the most popular one-stage object detection model, having its different versions used in diverse application areas. An improved YOLOv3 model, namely MangoYOLO [13], is also used for mango detection. However, it is specifically designed for images taken in artificial lighting conditions and hence is not suitable for images taken in daylight with natural illumination variations. Also, the complexity of the model is high, and for its use in a real-time environment, the model needs to be lightweight. In this regard, our contribution to the current work is twofold, as specified below.

- We have used the faster, lightweight, and more accurate (as compared to other YOLO variants) YOLOv5s [14] model for mango detection, which has not been reported previously.
- The YOLOv5s model is improved in terms of making it more lightweight by removing nearly 13 convolutional layers in a way that, in turn, results in high accuracy in the detection of mangoes in lesser time. The resulting model is named $MangoYOLO5$.

Experiments are conducted on the publicly available MangoNet-Semantic dataset [6], and results are also compared with those obtained through MangoNet and original YOLOv5s. MangoYOLO5 shows robustness in detecting the mangoes with higher accuracy under different conditions of lighting, occlusion, differences in scale, and mango density.

The remaining of the manuscript is structured as follows. Literatures related to mango detection and the YOLO model are reviewed in Sect. 2. The proposed methodology and the experimental result analysis are discussed in Sects. 3 and 4, respectively. The conclusions and future scopes are presented in Sect. 5.

2 Literature Review

Over the previous years, deep learning-based methods have shown remarkable improvement in the automatic detection of mangoes in mango orchards as compared to the state-of-the-art approaches. In this regard, Qiaokang Liang et al. [15] have proposed a real-time on-tree mango detection framework based on a single-shot Multibox Detector (SSD) network and claimed an $F1$ score of 91.1 at 35 FPS. However, the model does not identify the mangoes when there is a large overlap between the mangoes or mangoes and leaves. In [16], the authors have proposed an improved faster R-CNN architecture for locating and identifying mangoes. They reported an $F1$ score of 90% for mango detection which dropped to 56% for cultivar identification. However, it faced challenges in identifying the various fruit cultivars in the segmented tree. A precise method uses MangoNet [6], a model based on deep convolutional neural network which performs semantic segmentation for detection of mangoes. The experiments exhibit that it performs superior compared to Fully Convolutional Networks (FCN) architecture. The results obtained in terms of accuracy are 73.6%, with an $F1$ score of 84.4%. They have created the $MangoNet\ semantic\ dataset$ [6], which is publicly available. The work presented in this paper adopted this dataset for detection of mangoes.

The one-stage object detection approaches are gaining enormous attention in object detection and semantic segmentation due to their speed and accurate detection capability [11]. Two representative methods in this category are SSD [17] and YOLO [12]. The methods based on SSD are highly accurate in segmenting large-sized objects, but fail to show the desired performance for small-sized objects. On the contrary, the YOLO model can detect smaller objects with higher accuracy and lesser time than SSD. There are many variants of YOLO, such as YOLOv2, YOLOv3, YOLOv4, and YOLOv5, which are widely used for different fruit detection [18]. In [19], the authors have given a method based on YOLOv2 for detection of mangoes in images of mango orchards taken by unmanned aerial vehicle (UAV). A Light-YOLOv3 [20] was proposed as a fast detection approach for green mangoes. Changes were made in ResNet unit in YOLOv3, soft-NMS and added a multiscale context aggregation module (MSCA) and image enhancement algorithm (CLAHE-Mango). Koirala et al. [13] have given an architecture named as MangoYOLO for real-time

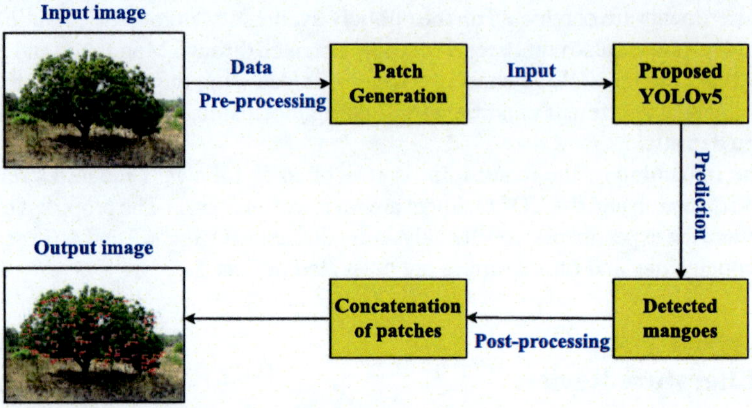

Fig. 1 Steps followed in the proposed method

mango detection. However, this model has not been tested with different lighting conditions. A lightweight apple target detection method [21] was proposed for pick and place robotic applications using improved YOLOv5s. In [18], the authors presented an automatic kiwifruit defect detection method using YOLOv5.

3 Methodology

The proposed methodology is discussed precisely in this section. The various steps involved in our method are shown in Fig. 1. The MangoNet semantic dataset is used for the experimental analysis in this paper. Data is pre-processed before applying the MangoYOLO5 model to this dataset. The mango detection results obtained from MangoYOLO5 are further post-processed to get the final output. The details involved in each of these steps are given in the following subsections.

3.1 Data Pre-processing

MangoNet semantic dataset [6] contains 45 open orchard mango tree images taken in daylight, each of size 4000×3000. Each of these images are divided into paths of size 200×200 and $11,096$ such patches are used for training and 1500 patches for testing which are generated from 40 and 4 original images, respectively. As the original dataset is not labeled in the required YOLOv5 format, the entire image dataset is labeled manually using makesense.ai online labeling tool by drawing rectangular boxes around the mango objects. The text format files are generated, and the annotations are saved in the required YOLOv5 format. After labeling, the dataset is split into training, test, and validation sets with ratios of 70%, 20%, and 10%, respectively.

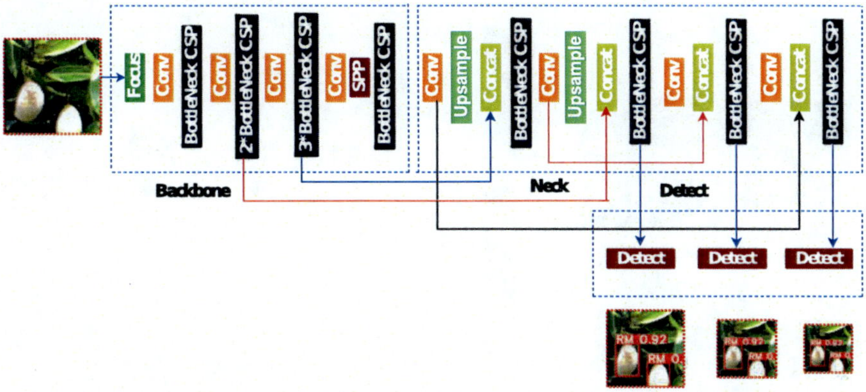

Fig. 2 Architecture of original YOLOv5s network

3.2 *MangoYOLO5 Network Architecture*

The proposed MangoYOLO5 is an improved YOLOv5s model for mango detection. Four different architectures exist for YOLOv5, specifically named YOLOv5s, YOLOv5m, YOLOv5l, and YOLOv5x, respectively [14]. The key difference among these architectures lies in the number of modules for feature extraction and convolution kernels. As the size of the weight file and the number of model parameters of the YOLOv5s model is small, so the MangoYOLO5 model is based on the YOLOv5s architecture. YOLOv5s network comprises three main modules: backbone, neck, and detect as shown in Fig. 2.

3.2.1 Backbone Network

The backbone network of YOLOv5s comprises four main modules namely focus, convolution, BottleneckCSP, and SPP modules. The focus module is used to lower the computation of the overall model and expedite the training. The architecture of this module can be visualized in Fig. 3. The working of this module is explained with an example given as follows. If the original image of size $3 \times 640 \times 640$ is given as the input, a slicing operation is performed to form four slices of size $3 \times 320 \times 320$ as a first step, which are then concatenated to form a feature map of size $12 \times 320 \times 320$. This feature map is then passed through a convolutional layer of 32 convolutional kernels, resulting in the feature map of size $32 \times 320 \times 320$, which is given as the input to the Batch Normalization (BN) and Hard swish function. The output obtained from this function is then passed to the next layer. The convolution module comprises a convolution layer, a BN layer, and a LeakyReLU layer. The deep features of the image are obtained using the BottleneckCSP module which comprises the bottleneck and CSP modules. There are four BottleneckCSP modules in the backbone network

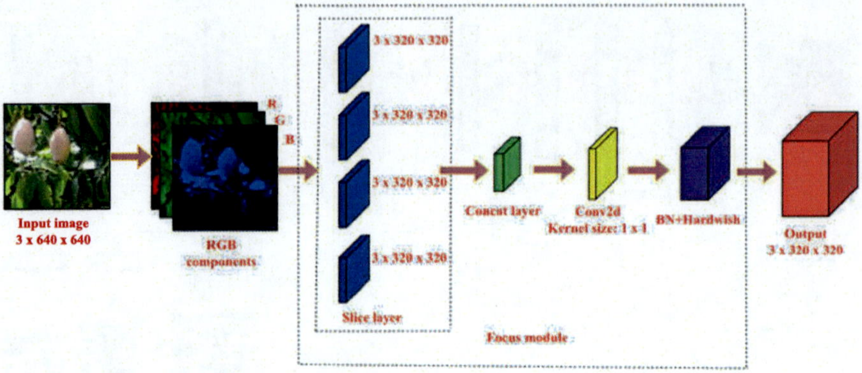

Fig. 3 Structure of focus module

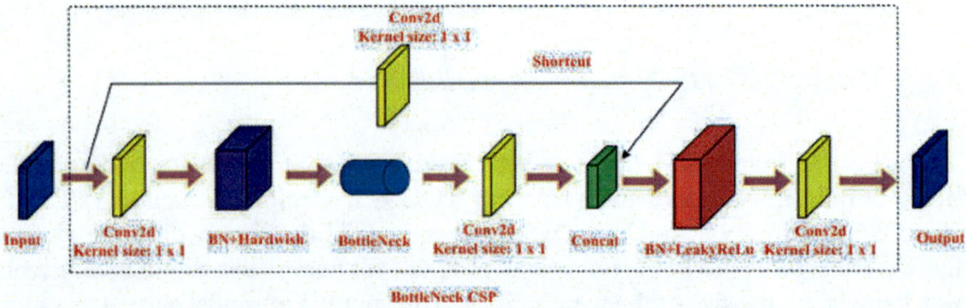

Fig. 4 Structure of BottleneckCSP module

which contain overall 42 convolutional layers. It makes two divisions of the input; the first one is given to the bottleneck module, and the other one is directly passed through a 1×1 convolution layer bypassing the bottleneck module as shown in Fig. 4. The bottleneck module consists of a 1×1 convolution kernel followed by a 3×3 convolution kernel and a 1×1 convolution kernel. The Spatial pyramid pooling (SPP) module is used to remove the fixed size constraint of the network. It executes the maximum pooling with different kernel sizes and fuses the features by concatenating them.

The improved backbone network replaces the original YOLOv5s architecture's focus and BottleneckCSP modules with Focus1 and BottleneckCSP1. Convolution operation is effective in extracting the image features; however, it results in more number of parameters in the detection model. In the focus network, there is a convolutional layer having kernel size 1×1 and 64 such kernels. As it is known that 1×1 convolutions are used to increase or decrease the depth of the feature map and not to extract any new features, removal of the 1×1 convolutional layer will not affect the accuracy of the result. So we have removed this 1×1 convolutional layer from the focus module and redesigned it as $Focus1$, as shown in Fig. 5, and we are able to reduce 64 parameters here. We have also removed the convolutional layer after the

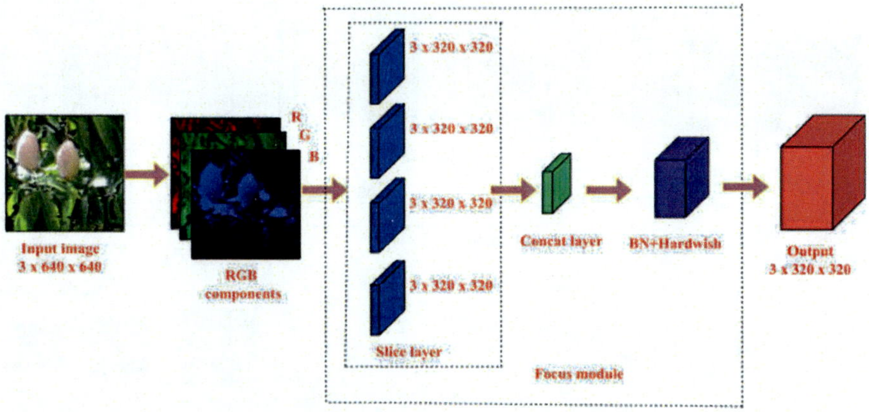

Fig. 5 Architecture of Focus1 module

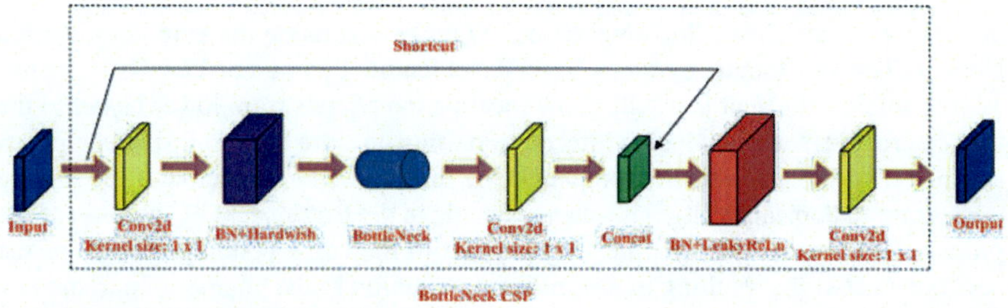

Fig. 6 Architecture of BottleNeckCSP1 module

focus module which has the kernel size 3 × 3, and there are 128 such kernels. This is useful in extracting the features; however, after each BottleneckCSP module, we have a convolutional layer of kernel size 3 × 3. So we tried removing the first 3 × 3 convolutional layer and passed the input through the BottleneckCSP directly. The BottleneckCSP module in the proposed model is also modified. We have removed a convolutional layer on the shortcut path of the original module, which is also a 1 × 1 convolutional layer. Thus, now, the input feature map of the BottleneckCSP module is directly connected with the output feature map of another branch in depth. This effectively reduces the number of parameters as we have 11 BottleneckCSP modules in the architecture. The improved BottleneckCSP architecture is depicted in Fig. 6 and named as BottleneckCSP1.

3.2.2 Neck

The neck network is used to create feature pyramids to generalize for different dimensions of the objects. Images of the given object in different sizes and scales can be

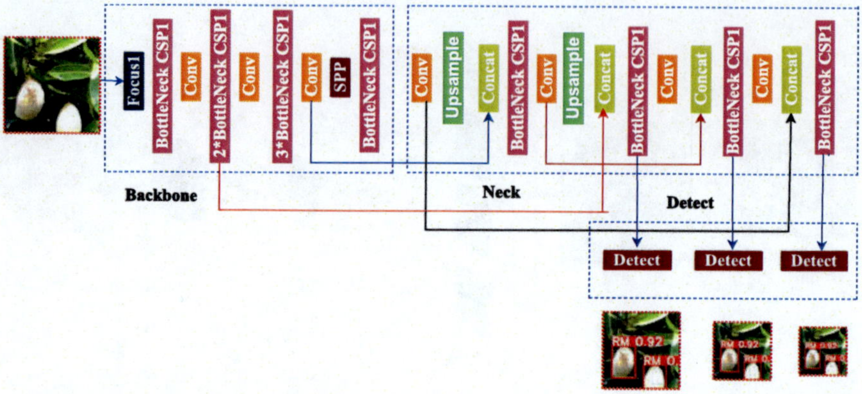

Fig. 7 Architecture of MangoYOLO5 network

detected with the above implementation. YOLOv5s is using the Path Aggregation Network (PANet) feature pyramid. The PANet feature pyramid of YOLOv5s creates an information shortcut that will allow localization signals from lower layers to the top feature layers without losing information using an extra bottom-up path augmentation. YOLOv5s also use the BottleneckCSP module at the neck module to enhance the feature fusion capability. The changes made in the BottleneckCSP module of the proposed MangoYOLO5 are the same as that in the corresponding module of the backbone network, resulting in a significant reduction in the number of parameters.

3.2.3 Detect

The detect module is primarily responsible for the final detection result of the network. It uses anchor boxes to construct the feature map output from the previous layer and generate vectors containing class probability, objectness score, and bounding box coordinates. YOLOv5s is having three detect layers, where each input is a feature map of size 80 × 80, 40 × 40, and 20 × 20, respectively. These layers are used to detect objects of various sizes in the image. The output of every detect layer is a 21-dimensional vector with 2 classes, 1 class probability, 4 surrounding box position coordinates, and 3 anchor boxes. The architecture of the lightweight MangoYOLO5 model for mango detection is as shown in Fig. 7.

The detection results we are getting are in the form of patches of size 200 × 200, which are then concatenated together to form the complete image of size 4000 × 3000 in the post-processing stage.

Fig. 8 Patch outputs for randomly selected test images using MangoYOLO5

4 Result Discussion and Experimental Analysis

Experiments were conducted on a Google Colab Tesla T4 12 GB GPU system with 16 GB RAM for the validation of the proposed lightweight MangoYOLO5 model. Training of the model is carried out using 11096 image patches of size 200 × 200 with Learning Rate (LR), batch size, and number of epochs as 0.01, 16, and 25, respectively. We validated and tested the trained model with 3000 and 1500 image patches of size 200 × 200, respectively. The model's output for a few arbitrarily considered test image patches is illustrated in Fig. 8. From the figure, it can be noticed that the MangoYOLO5 model is capable of detecting all the mangoes even when the mangoes were cut into halves or small portions in the process of making patches. The results of mango detection on three complete test images using the MangoNet, YOLOv5, and MangoYOLO5 models are shown in Fig. 9. It can be visualized from the figure that more mangoes are detected in the proposed MangoYOLO5 model as compared to both MangoNet and YOLOv5s.

4.1 Analysis Using Performance Evaluation Metrics

Evaluation metrics like precision, recall, mAP@IoU=0.5 (IoU: Intersection over Union), and Estimation time are used to evaluate the performance of the models. IoU is a basic metric employed to compare object detection systems, and the threshold for the IoU metric depends on the ground truth bounding box annotated in this work. The threshold helps the model in predicting the bounding box during the testing process. True positive (TP) is computed if the IoU is larger than the defined threshold value, and similarly, false positive (FP) is also calculated. Precision, recall, and mAP performance metrics can be computed using Eqs. (1), (2), and (3), respectively, with the obtained TP, true negative (TN), FP, and false negative (FN) values. The threshold value is fixed at 0.5.

Fig. 9 Mango detection results using **a** MangoNet model **b** original YOLOv5s model **c** MangoYOLO5 model on 3 test images

$$\text{Precision} = \frac{\text{TP}}{\text{TP} + \text{FP}} \quad (1)$$

$$\text{Recall} = \frac{\text{TP}}{\text{TP} + \text{FN}} \quad (2)$$

$$\text{mAP} = \frac{1}{N}\sum_{i=1}^{N} \text{AP}_i \quad (3)$$

N: Number of queries, AP: Average precision

The performance of the MangoYOLO5 model is evaluated with respect to the existing YOLOv5s and MangoNet models based on the precision, recall, and mAP metrics, and the values are given in Table 1. The mAP value for the MangoYOLO5

Table 1 Comparison of performance metrics for MangoYOLO5 and its variants

Model name	Precision (%)	Recall (%)	mAP (%)	Estimation time
MangoNet	88	51	73.6	–
YOLOv5s	86.9	86.9	91.4	~30 min
MangoYOLO5	91.6	90	**94.4**	~10 min

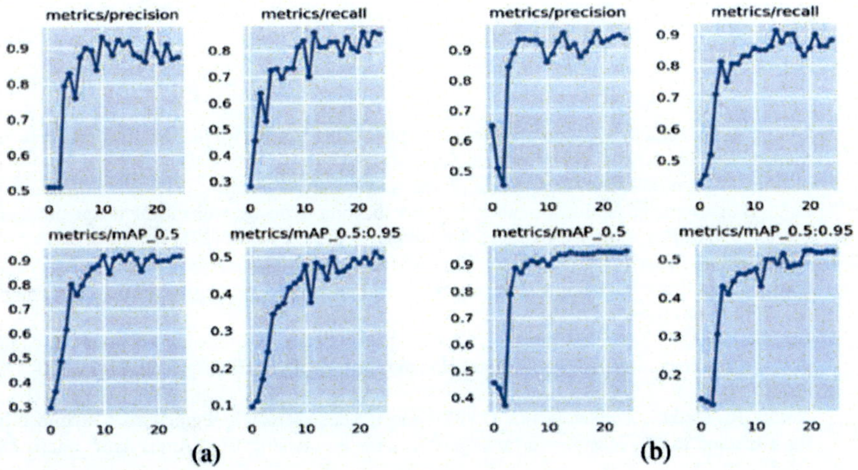

Fig. 10 Plots of different performance metrics against number of epochs for **a** YOLOv5s model. **b** MangoYOLO5 model

recognition network is 3.0 and 20.6% higher than that of the original YOLOv5s network and MangoNet, respectively. This indicates that the MangoYOLO5 model is better for mango detection, which can also be verified from Fig. 10 based on the graph plotted for the four-evaluation metrics against the number of epochs. The figure illustrates that metrics in the MangoYOLO5 model are stable, but there are fluctuates in case of the YOLOv5s model. The storage space for MangoYOLO5 is 13.7 MB, which is 9.89 MB lower than the original YOLOv5s and also the training time of our model is observed to be 66.67% lesser as compared to the later, by which we can conclude that the proposed model can be adopted for real-time mango detection.

5 Conclusion and Future Scope

This paper presents a fast and lightweight approach for detecting mangoes, which can be used for value-added applications like yield estimation in an open field environment. Focus and BottleneckCSP modules in the original YOLOv5s model are modified in the proposed MangoYOLO5 architecture to make it lighter by reducing the total parameters by a factor of almost 30, 000. The observed mAP in case of MangoYOLO5 for detection of mangoes is 94.4%. Comparisons are made with the other models like YOLOv5s and MangoNet, which clearly shows that the MangoYOLO5 gives better performance to changes in scale, lighting, contrast, and occlusion. In the future, we will extend this work for estimating the yield and train the model for detection of multiple varieties of mangoes.

References

1. Ullagaddi, S., Raju, S.V.: Automatic robust segmentation scheme for pathological problems in mango crop. Int. J. Mod. Edu. Comput. Sci. **9**(1) (2017)
2. National mango data base. https://mangifera.res.in/indianstatus.php
3. Sahu, D., Dewangan, C.: Identification and classification of mango fruits using image processing. Int. J. Sci. Res. Comput. Sci. Eng. Inf. Technol. **2**(2), 203–210 (2017)
4. Kadir, M.F.A., Yusri, N.A.N., Rizon, M., bin Mamat, A.R., Makhtar, M., Jamal, A.A.: Automatic mango detection using texture analysis and randomised hough transform. Appl. Math. Sci. **9**(129), 6427–6436 (2015)
5. Lin, G., Tang, Y., Zou, X., Cheng, J., Xiong, J.: Fruit detection in natural environment using partial shape matching and probabilistic Hough transform. Precision Agricult. **21**(1), 160–177 (2020)
6. Kestur, R., Meduri, A., Narasipura, O.: Mangonet: a deep semantic segmentation architecture for a method to detect and count mangoes in an open orchard. Eng. Appl. Artif. Intell. **77**, 59–69 (2019)
7. Liwag, R.J.H., Cepria, K.J.T., Rapio, A., Cabatuan, K., Calilung, E.: Single shot multi-box detector with multi task convolutional network for carabao mango detection and classification using tensorflow. In: Proceedings of the 5th DLSU Innovation and Technology, pp. 1–8 (2017)
8. Nanaa, K., Rizon, M., Abd Rahman, M.N., Ibrahim, Y., Abd Aziz, A.Z.: Detecting mango fruits by using randomized hough transform and backpropagation neural network. In: 2014 18th International Conference on Information Visualisation, pp. 388–391. IEEE (2014)
9. Stein, M., Bargoti, S., Underwood, J.: Image based mango fruit detection, localisation and yield estimation using multiple view geometry. Sensors **16**(11), 1915 (2016)
10. Basri, H., Syarif, I., Sukaridhoto, S.: Faster r-cnn implementation method for multi-fruit detection using tensorflow platform. In: International Electronics Symposium on Knowledge Creation and Intelligent Computing (IES-KCIC), pp. 337–340. IEEE (2018)
11. Junos, M.H., Mohd Khairuddin, A.S., Thannirmalai, S., Dahari, M.: An optimized yolo-based object detection model for crop harvesting system. IET Image Process. **15**(9), 2112–2125 (2021)
12. Redmon, J., Divvala, S., Girshick, R., Farhadi, A.: You only look once: unified, real-time object detection. In: Proceedings of the IEEE Conference on Computer Vision and Pattern Recognition, pp. 779–788 (2016)
13. Koirala, A., Walsh, K.B., Wang, Z., McCarthy, C.: Deep learning-method overview and review of use for fruit detection and yield estimation. Comput. Electron. Agricult. **162**, 219–234 (2019)
14. ultralytics. yolov5. https://github.com/ultralytics/yolov5
15. Liang, Q., Zhu, W., Long, J., Wang, Y., Sun, W., Wu, W.: A real-time detection framework for on-tree mango based on SSD network. In: International Conference on Intelligent Robotics and Applications, pp. 423–436. Springer (2018)
16. Borianne, P., Borne, F., Sarron, J., Faye, E.: Deep mangoes: from fruit detection to cultivar identification in colour images of mango trees. (2019) *arXiv preprint* arXiv:1909.10939
17. Liu, W., Anguelov, D., Erhan, D., Szegedy, C., Reed, S., Fu, C.-Y., Berg, A.C.: Ssd: single shot multibox detector. In: European Conference on Computer Vision, pp. 21–37. Springer (2016)

18. Yao, J., Qi, J., Zhang, J., Shao, H., Yang, J., Li, X.: A real-time detection algorithm for kiwifruit defects based on yolov5. Electronics **10**(14), 1711 (2021)
19. Xiong, J., Liu, Z., Chen, S., Liu, B., Zheng, Z., Zhong, Z., Yang, Z., Peng, H.: Visual detection of green mangoes by an unmanned aerial vehicle in orchards based on a deep learning method. Biosyst. Eng. **194**, 261–272 (2020)
20. Xu, Z.-F., Jia, R.-S., Sun, H.-M., Liu, Q.-M., Cui, Z.: Light-yolov3: fast method for detecting green mangoes in complex scenes using picking robots. Appl. Intell. **50**(12), 4670–4687 (2020)
21. Yan, B., Fan, P., Lei, X., Liu, Z., Yang, F.: A real-time apple targets detection method for picking robot based on improved yolov5. Remote Sensing **13**(9), 1619 (2021)

Resolution Invariant Face Recognition

Priyank Makwana, Satish Kumar Singh, and Shiv Ram Dubey

Abstract Face images recorded from security cameras and other similar sources are generally of low-resolution and bad quality. There are many recent face recognition models which extract face features/encodings using deep neural networks (DNNs) and give very good results when tested against images with higher resolution (HR). Moreover, the performance of these types of algorithms deteriorates to a great extent for images with low-resolution (LR). To reduce the shortcoming, we used convolution neural network (CNN) architecture along with the combination of super-resolution (SR) technique during the pre-processing steps to achieve the comparable results on the state-of-the-art techniques. The proposed method can be outlined in 4 steps: Face retrieval, image pre-processing and super-resolution, training the model, and face detection/classification. The dataset used for this study is a publicly available with name Face Scrub Dataset, subset of this dataset is used containing 20050 images of 229 people for the experiments.

Keywords Convolutional neural network (CNN) · Super-resolution (SR) · Face recognition · Low-resolution

1 Introduction

Face recognition (FR) is a way of identifying or verifying a subject's identity using their face image as a query. These face recognition systems are used to identify people in various pictures, videos, or real-time frames. It is a process that is applied in the physical world on a regular basis in multiple scenarios; including monitor-

P. Makwana (✉) · S. Kumar Singh · S. Ram Dubey
Department of Information Technology, IIIT Allahabad, Prayagraj, India
e-mail: mit2020045@iiita.ac.in

S. Kumar Singh
e-mail: sk.singh@iiita.ac.in

S. Ram Dubey
e-mail: srdubey@iiita.ac.in

M. Tistarelli et al. (eds.), *Computer Vision and Machine Intelligence*,
Lecture Notes in Networks and Systems 586,
https://doi.org/10.1007/978-981-19-7867-8_58

ing cameras, criminological studies, and controlling passage to sensitive areas. Face recognition (FR) has always been a well-liked topic of research and therefore the developments in deep neural networks (DNN) have increased the accuracy of such tasks [1, 2]. Many researchers have made a great progress in developing novel techniques for FR [3–6], and in other areas such as occluded thermal recognition [7]. The overall performance at the traditional benchmarks like LFW [8] has been progressed to a great extent via the means of latest face recognition methods; however, as soon as they are tested against low-resolution images the results degrade with an enormous margin.

The use of surveillance cameras has become popular in many areas/places, and the images captured by such devices usually have low-resolution (LR) and the images might have different poses and illumination conditions, while the faces used for the recognition are of high-resolution, so that poses a challenging recognition task. Since the gap between the LR images and HR images is huge, it makes the overall problem statement more complex. The LR images have a different domain in comparison to HR images which causes feature dimension mismatch problems. Due to these reasons, recognition in the lower resolution domain is an understudied field when we compare it with recognition in the high-resolution domain.

There are two major approaches in the literature and previous works to deal with this taxing task;

1. Super-resolution techniques try to get the images from the lower resolution domain to the higher resolution feature domain and try to perform the recognition in the HR domain only [9–12].
2. Resolution robust techniques try to reduce the difference of features extracted from images from both low and high-resolution domains into some common feature space and perform recognition in that space [13–16].

Both the mentioned techniques work well and give good results; however, the difference is in the complexity of both the approaches [9, 14]. One of the methods to reduce the gap between the facial features from images of both the resolution domains is to use images from both the domains; i.e., HR, and their corresponding down-sampled LR domain for training the network [13]. It turns out to be a good technique for recognition in the LR domain, but when the recognition is done in the HR domain, a drop in the accuracy score is observed. That is because, in the process of getting the features from both the HR and LR images close to each other, the neural network loses some of the high-resolution information. In order to evade that obstacle, we propose to utilize super-resolution for lower dimensional images along with the backbone CNN network [17] for recognition. As Fig. 1a shows that when raw HR and LR images are trained through network, they have higher distance in common space [13] where as in Fig. 1b when the LR images are fed through SR network before training, the distance reduces considerably.

Generally, the information available in high-resolution images is higher when compared with low-resolution images; therefore the differentiation of low-resolution images is overlooked by some of the existing cross-resolution face recognition

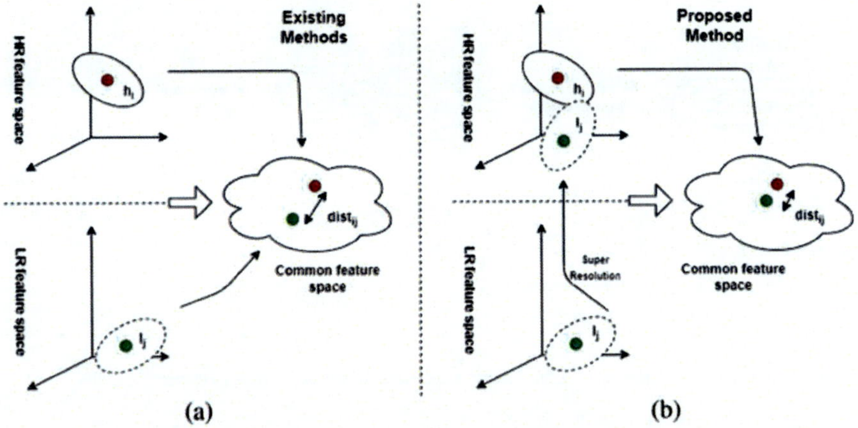

Fig. 1 Overview for the feature space: **a** Existing methods have higher distance b/w HR and LR features in common space [13]. **b** Proposed method uses SR to reduce the gap b/w images from high- and low-resolution in feature plane

techniques. The proposed work considers both the high and low-resolution images equally, which results in better utilization of information (Fig. 1b). In order to achieve the comparison, the dimension of LR images is raised using the super-resolution method. Super-resolution (SR) is the method that is used to fabricate HR images from the LR domain and try to make information available as much as possible. SR techniques require higher computational cost than interpolation but the results are much cleaner and crisp. In order to achieve the resolution robustness, the real HR images are mixed with pre-processed LR images, and the combined data is trained in a supervised way using a deep convolutional network.

Our contribution to the solution of this problem are as follows:

1. We propose an architecture as depicted in Fig. 2 which makes use of EDSR [18] as super-resolution network in combination with the backbone CNN network which helps in learning of resolution invariant features.
2. We introduced the "Three Resolution" training setting. The results obtained from this setting beats the generic "Two resolution" settings in most of the testing resolution domains as presented in the results section.

2 Related Works

The problem of resolution invariant face recognition is to recognize face identity with good accuracy given facial images of any resolution. Low-resolution face recognition (LR FR) has relatively fewer studies if compared to recognition of faces in HR domain (HR FR). Due to lack of availability of extensive datasets for both training and testing,

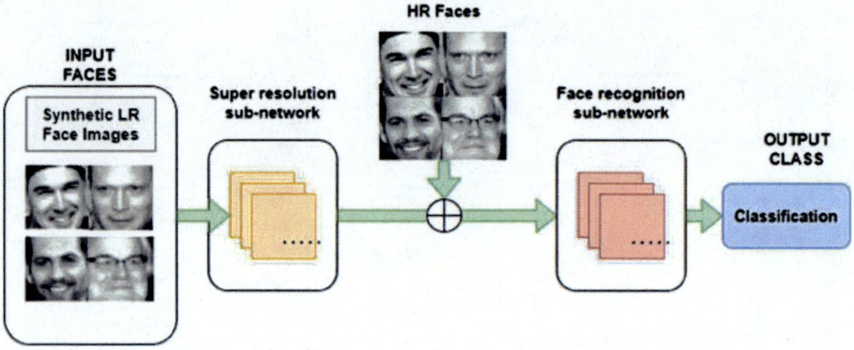

Fig. 2 Overview of proposed architecture : super-resolution with multi-resolution CNN

majority of the previous researches focus on the synthetic dataset generated by down sampling the high-resolution images. The techniques that try to solve this problem are categorized mainly into two groups, i.e., super-resolution (SR) technique [9, 10, 19] and resolution robust technique [13, 15, 20].

2.1 *Super-Resolution Techniques*

SR techniques have undergone recent development, and have been enhanced wildly [9, 10, 19] in order to be used in the recent studies and developments. SR techniques are majorly used to synthesize high-resolution (HR) images from the corresponding low-resolution (LR) images. Xu et al. [10] use multiple shallow feature super-resolution modules based on the dimension of input LR images. Gunturk et al. [19] use the SR reconstruction on the down-sampled low-resolution image space.

2.2 *Resolution Invariant Techniques*

These resolution robust techniques have been popular for low-resolution recognition tasks because of their relatively less complexity as compared to SR algorithms. These algorithms rely on the search of the common feature space where the recognition is done. There are various techniques depicted in the previous works [13, 15, 20] which try to achieve this unified space of features with different training algorithms. Even though these methods are computationally less expensive but when the difference between the high and low-resolution images increases these methods tend to degrade their performance.

3 Methodology and Workflow

Training the model using only high-resolution images is not sufficient to learn the resolution invariant features. The models that are trained only in high-resolution images work very well with the high-resolution test images but when they are tested with very low-resolution images, the results come out to be very poor and the accuracy degrades to a great extent. Therefore, in order to build the resolution robust model, low-resolution features are also necessary. We are using a multi-resolution convolutional neural network as our backbone model; this model is fed with the pre-processed LR images of multiple resolutions to achieve the aim of learning features for resolution robustness.

We propose architecture to recognize the facial images with high accuracy across all the multiple resolutions. At first, we extract the facial images of the higher resolution domain from the dataset based on the bounding box and form the HR image dataset. Now, for the process of learning features from the low-resolution domain, HR images are down-sampled to multiple low resolutions, i.e., 30×30, 40×40, 60×60 and 100×100. These images are then shifted from low-resolution domain to high-resolution domain using the super-resolution technique [18]. The formulated dataset is then used to train the backbone network along with different classifiers.

The methodology includes the steps breakdown which is used in the implementation of the project.

3.1 Data Pre-processing

The images from the dataset [21] are 3-channel images, for us to make the proposed architecture simpler, these images are converted into gray-scale which are used for further processing. The HR images are down-sampled and the image pair is created. To match the resolutions of HR and LR images for training purpose, the LR images are up-sampled using a pre-trained super-resolution network which is Enhanced Deep Residual Networks (EDSR) [18]. EDSR is one of the eminent SR model which is based on SRResNet architecture. It has the ability to upscale the images upto 4 times the original size. According to [22] EDSR is among the top performing SR methods in context to PSNR score (approx 34db) and higher the PSNR score, better is the reconstructive power of that particular SR algorithm.

Figure 3 shows the comparison of images from multiple resolution domains; i.e., HR, LR, and super-resolution. Identity in first, second and third columns belong to HR, LR, and SR domains, respectively.

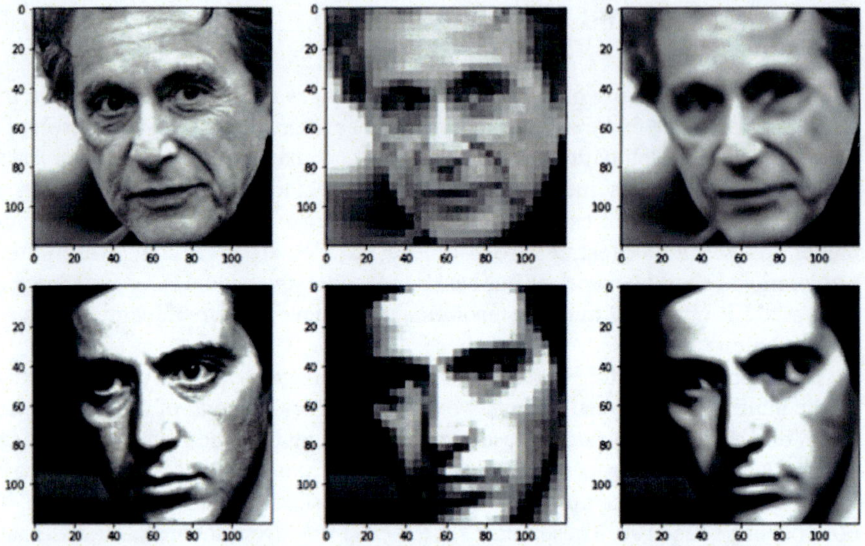

Fig. 3 Comparison of sample HR, LR, and SR images from FaceScrub dataset [21]

3.2 Network Architecture

The network architecture model is based on convolutional neural network (CNN) which is very effective for feature extraction and pattern recognition from images [17]. Neural networks need to see data from all the classes. The network architecture consists mainly of an input layer, followed by several hidden layers and then the final output layer. The hidden layers are constructed using multiple convolution layers for feature extraction. The features extracted by CNN are used by multiple dense layers or classifiers for classification purposes.

We train the multi-resolution convolutional neural network (MR-CNN) on multiple proposed settings, i.e., combination of two and three resolutions in the discussed environment in order to recognize the face images in the test set. Figure 4, depicts the architecture diagram of the network. The proposed network takes 120×120 dimension images as the input. The output to the network is 229 classes which depicts the identity of the image.

As depicted in Fig. 4; the model that is proposed is a combination of 2 sections; the first section contains six convolution layers that are distributed in the group of two convolution layers with each group followed by a batch normalization layer, max-pooling layer, and a dropout layer. In the second part of the section, we use three convolution layers, with each layer is followed by batch normalization, max-pooling, and a dropout layer. The dropout layer prevents the network from overfitting by dropping out units randomly with the help of probability. This results in the simplification of the computation for the network.

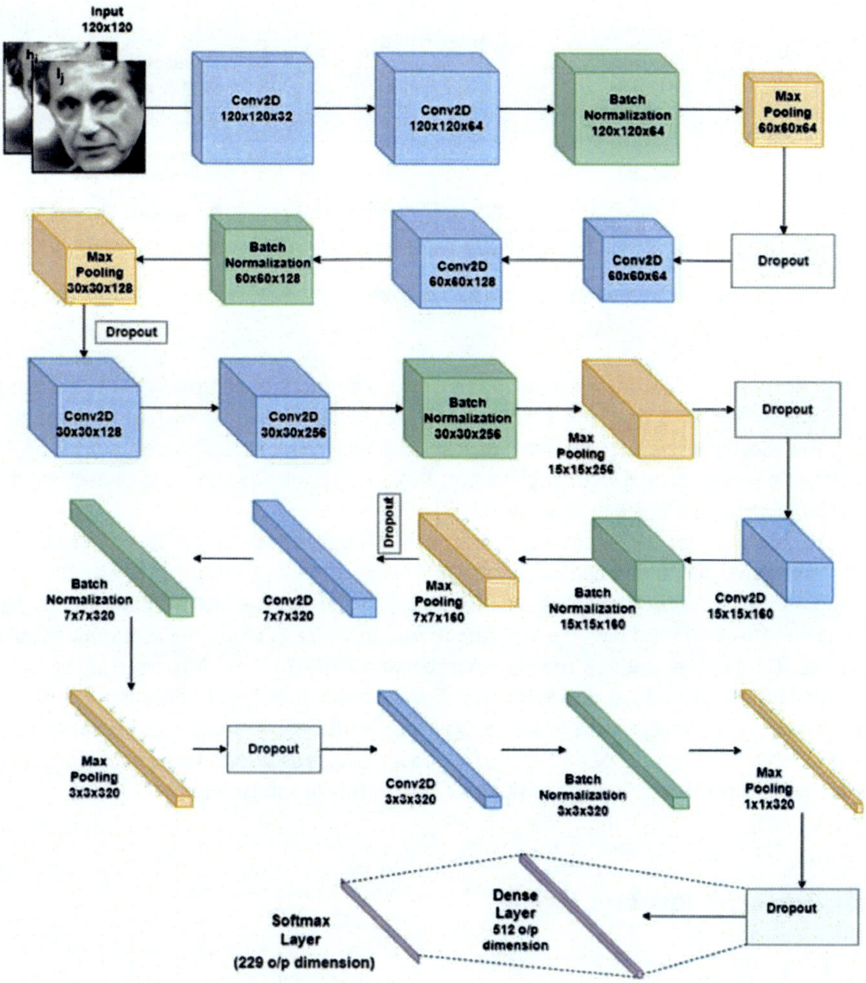

Fig. 4 Proposed deep CNN architecture

We have fixed the output dimension of the last dense layer as 512, so the features extracted from the last dense layer could be used for further processing. This helps us in comparing the model by removing the last layer (softmax layer) and training different classifiers on top of the trained CNN. In that scenario, the trained CNN without the softmax layer acts as the feature extractor network and classifier helps in the multi-class classification of the identities.

When the softmax layer is not used for classification, it is discarded when the training of the model is complete. This makes the trained deep CNN as the feature extractor. Then for each test image, we obtain a feature vector of dimension 512 by passing the images through the model. These extracted feature vectors are used to

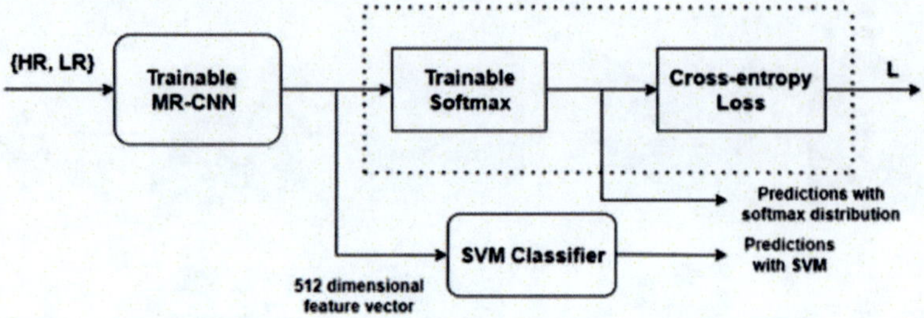

Fig. 5 Block diagram of feature extractor with classifier

train the classifier for the recognition purpose. The mentioned flow is depicted using the Fig. 5. We have used the SVM and KNN classifiers on top of feature extractor for the multi-class classification purpose and the performance is compared. SVM performs better than KNN (depicted in Table 1), so empirically it is chosen as the feature extractor for the architecture [23].

To summarize, we accomplish the task of resolution robust face recognition by using two mechanisms, i.e., firstly by using super-resolution [18] for the pre-processing step which enables the LR images to attain the HR domain. Secondly, training the combined data through the model in order to attain the common feature space. These steps reduce the distance between the LR and HR images as compared to interpolated LR and HR images as visualized in Fig. 1. Training the model by reducing the categorical crossentropy loss in the above mentioned environment helps low-resolution features to get close to the high-resolution features. Hence, this mechanism serve efficiently for the task of resolution robustness.

4 Experiments and Result

4.1 Dataset

In this study, we have used a subset of FaceScrub dataset [21]. This is an extensive dataset in which carry 106,863 high-resolution images of 530 different people with high difference in posture and lighting conditions across face images of the same identity. Figure 6 is the visualization of the sample images from the dataset. In this research work, we are working with a large number of various resolutions, it was computationally very demanding to use all the images; therefore we are working with 20,050 high-resolution images of 229 identities.

The metadata for each image is as follows:

1. Name (of identity)
2. image_id
3. face_id

Fig. 6 Sample images from FaceScrub dataset [21] of dimension 120 × 120 with four different images of four different identities. Each row depicts a single identity and four images of each identity is presented

4. url (of the perticular image)
5. bbox (bounding box co-ordinate for the face in the image)

4.2 Implementation Details

Training the deep model on the mixture of low-resolution (LR) and high-resolution (HR) data is the process of looking for a common feature sub-space. The data is extracted from the URL present in the FaceScrub [21] dataset, the brightness of the image is adjusted if the mean brightness is less than 0.1. Along with that the images are converted into gray-scale and cropped into 120 × 120 dimensions. We use a deep

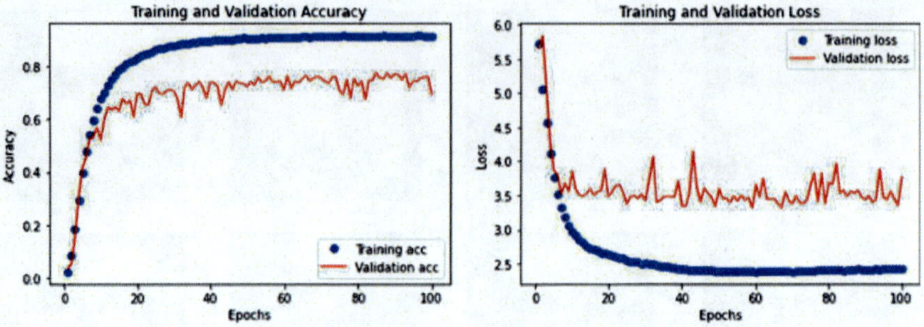

Fig. 7 Accuracy and loss plot—three res. (MR-CNN + SR)

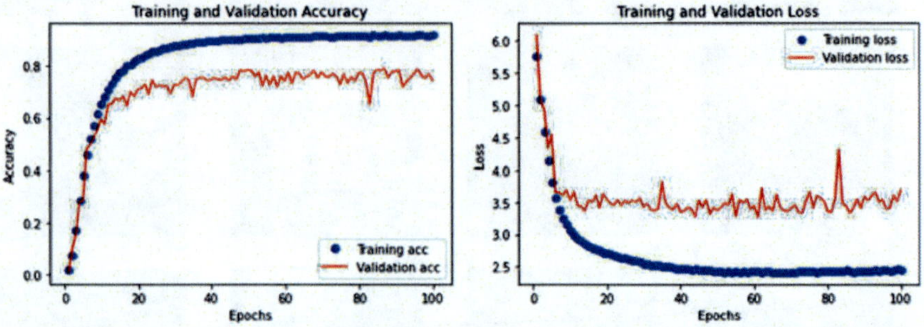

Fig. 8 Plot—three res. (MR-CNN + Bi-cubic int.)

convolutional network trained on the dataset as a feature extractor. We use Root Mean Squared Propagation (RMSProp) [24] technique for model optimization.

The network is trained using Keras/Tensorflow [25] having batch size of 40. The learning rate (LR) of the network to be trained is set to 0.001 initially and then the validation loss is monitored by learning rate reduction with the patience of 8. The LR is reduced by the component of 0.8 if the validation loss is not improved after 8 consecutive epochs. The minimum LR is set to 0.0000000001. The model is trained till 100 epochs.

For the optimizer RMSProp [24], the rho is set to 0.9, epsilon to 1e-08, and decay as 0. Once the model training is finished, the model is tested with the softmax distribution and then the last layer is popped out. Then for each facial picture, feature vector of dimension 512 is extracted. These feature vectors are obtained by passing the images to the network. For the comparative recognition purposes, the SVM classifier [26] is trained with the extracted features and the recognition is done by that classifier. Then we're evaluating out results on the obtained accuracy scores for different resolution testing settings (30×30, 40×40, 60×60, and 100×100). The training accuracy and loss are depicted in the graphs (Figs. 7 and 8).

Table 1 Comparison with different training settings on FaceScrub [21]

Method	Accuracy LR images (%)			HR images (%)
	30×30	40×40	60×60	100×100
DeepCNN + Bi-cubic int. [20] (Distillation approach with fixed HR network)	78	80	82	84
MR-CNN + Bi-cubic int. (Two resolution (40×40 and 120×120))	73.87	82.59	83.24	83.47
MR-CNN + SR (Two resolution (40×40 and 120×120))	76.02	80.93	83.04	83.40
MR-CNN + Bi-cubic int. (Three res. (30×30, 60×60 and 120×120))	79.33	81.53	81.80	81.70
MR-CNN + SR + KNN (k=11) (Three res. (30×30, 60×60 and 120×120))	76.17	78.48	80.56	80.68
Proposed: MR-CNN + SR (Three res. (30×30, 60×60 and 120×120))	**79.07**	**80.37**	**82.79**	**82.89**
Proposed: MR-CNN + SR + SVM (Three res. (30×30, 60×60 and 120×120))	**79.83**	**81.21**	**83.19**	**83.15**

4.3 Results

The resolution robust face recognition task is evaluated using FaceScrub [21], which is a large-scale dataset. This dataset contains facial images from the public domain with large variations in pose and illumination across the images of the same identities.

We are showing the results on these methods for given four low-resolution constraints which are $30 \times 30, 40 \times 40, 60 \times 60$, and 100×100. As we can see in Table 1, that the two resolution setting achieves the accuracy of 76.02%, 80.93%, 83.04% and 83.40% in 30×30, 40×40, 60×60 and 100×100 resolution domain respectively, where as the model in three resolution setting achieves the accuracy of 79.83%, 81.21%, 83.19% and 83.15% in 30×30, 40×40, 60×60 and 100×100 resolution domain, respectively. The proposed method work well in generalization and improve the accuracy for low-resolution images.

The comparative analysis with different training settings and architecture methods is shown. From Table 1, we can depict that the three resolution setting works better than the two resolutions setting in lower resolution test domains.

5 Conclusion

The problem of robustness in resolution for face recognition is addressed in this work, with a major focus on low-resolution face recognition. We proposed an improved version of convolution neural network (CNN) which is assisted using the super-resolution technique (EDSR [18]) along with different training settings. The backbone model trained on combined images from both the resolution domain (i.e., HR & LR) is then dissected into a feature extractor and classifier. The feature extractor helps in gathering features from test images and the classifier helps with recognition purposes. The resultant architecture is tested on multiple resolutions from the FaceScrub [21] dataset and results are recorded.

References

1. Taigman, Y., Yang, M., Ranzato, M.A., Wolf, L.: Deepface: closing the gap to human-level performance in face verification. In: 2014 IEEE Conference on Computer Vision and Pattern Recognition, pp. 1701–1708 (2014)
2. Wang, H., Wang, Y., Zhou, Z., Ji, X., Gong, D., Zhou, J., Li, Z., Liu, W.: Cosface: large margin cosine loss for deep face recognition (2018)
3. Chakraborty, S., Singh, S.K., Chakraborty, P.: Local gradient hexa pattern: a descriptor for face recognition and retrieval. IEEE Trans. Circ. Syst. Video Technol. **28**(1), 171–180 (2018)
4. Chakraborty, S., Singh, S.K., Chakraborty, P.: Local quadruple pattern: a novel descriptor for facial image recognition and retrieval. Comput. Electr. Eng. **62**, 92–104 (2017)
5. Dubey, S.R., Singh, S.K., Singh, R.K.: Rotation and illumination invariant interleaved intensity order-based local descriptor. IEEE Trans. Image Process. **23**(12), 5323–5333 (2014)
6. Chakraborty, S., Singh, S.K., Chakraborty, P.: Centre symmetric quadruple pattern: a novel descriptor for facial image recognition and retrieval. Pattern Recogn. Lett. **115**, 50–58 (2018). (Multimodal Fusion for Pattern Recognition)
7. Kumar, S., Singh, S.K.: Occluded thermal face recognition using bag of CNN (*bo*CNN). IEEE Sig. Process. Lett. **27**, 975–979 (2020)
8. Ramesh, M., Berg, T., Learned-Miller, E., Huang, G.B.: Labeled faces in the wild: a database for studying face recognition in unconstrained environments. In: Technical Report 07-49 University of Massachusetts, Amherst (2007)
9. Yin, X., Tai, Y., Huang, Y., Liu, X.: Fan: feature adaptation network for surveillance face recognition and normalization (2019)
10. Xu, L.Y., Gajic, Z.: Improved network for face recognition based on feature super resolution method. Int. J. Autom. Comput. **18**, 10 (2021)
11. Biswas, S., Bowyer, K.W., Flynn, P.J.: Multidimensional scaling for matching low-resolution face images. IEEE Trans. Pattern Anal. Mach. Intell. **34**(10), 2019–2030 (2012)
12. Massoli, F.V., Amato, G., Falchi, F.: Cross-resolution learning for face recognition. Image Vision Comput. **99**, 103927 (2020)
13. Zeng, D., Chen, H., Zhao, Q.: Towards resolution invariant face recognition in uncontrolled scenarios. In: 2016 International Conference on Biometrics (ICB), pp. 1–8 (2016)
14. Lu, Z., Jiang, X., Kot, A.: Deep coupled resnet for low-resolution face recognition. IEEE Sig. Process. Lett. **25**(4), 526–530 (2018)
15. Mishra, N.K., Dutta, M., Singh, S.K.: Multiscale parallel deep CNN (mpdCNN) architecture for the real low-resolution face recognition for surveillance. Image Vision Comput. **115**, 104290 (2021)

16. Talreja, V., Taherkhani, F., Valenti, M.C., Nasrabadi, N.M.: Attribute-guided coupled GAN for cross-resolution face recognition. In: 2019 IEEE 10th International Conference on Biometrics Theory, Applications and Systems (BTAS), pp. 1–10 (2019)
17. Albawi, S., Mohammed, T.A., Al-Zawi, S.: Understanding of a convolutional neural network. In: 2017 International Conference on Engineering and Technology (ICET), pp. 1–6 (2017)
18. Lim, B., Son, S., Kim, H., Nah, S., Mu Lee, K.: Enhanced deep residual networks for single image super-resolution (2017)
19. Gunturk, B.K., Batur, A.U., Altunbasak, Y., Hayes, M.H., Mersereau, R.M.: Eigenface-domain super-resolution for face recognition. IEEE Trans. Image Process. **12**(5), 597–606 (2003)
20. Khalid, S.S., Awais, M., Feng, Z.H., Chan, C.H., Farooq, A., Akbari, A., Kittler, J.: Resolution invariant face recognition using a distillation approach. IEEE Trans. Biometrics Behav. Identity Sci. **2**(4), 410–420 (2020)
21. Facescrub Dataset.: Available at https://www.vintage.winklerbros.net/facescrub.html
22. Bashir, S.M.A., Wang, Y., Khan, M., Niu, Y.: A comprehensive review of deep learning-based single image super-resolution. Peer J. Comput. Sci. **7**, e621 (2021)
23. Borah, P., Gupta, D.: Review: support vector machines in pattern recognition. Parashjyoti Borah et al. / Int. J. Eng. Technol. (IJET) **9** (2017)
24. Mukkamala, M.C.,. Hein, M.: Variants of rmsprop and adagrad with logarithmic regret bounds (2017)
25. Chollet, F., et al.: Keras (2015)
26. Zhang, Y.: Support vector machine classification algorithm and its application. In: Liu, C., Wang, L., Yang, A. (eds.) Information Computing and Applications, pp. 179–186. Springer, Berlin, Heidelberg (2012)

Target Detection Using Transformer: A Study Using DETR

Akhilesh Kumar, Satish Kumar Singh, and Shiv Ram Dubey

Abstract Transformer has been proposed to augment the attention mechanism in neural networks without using recurrence and convolutions. Starting with machine translation, it graduated to vision transformer. Among the vision transformers, we explore the DEtection TRansformer (DETR) model proposed in the End-to-end Object Detection with Transformers paper by the team at Facebook AI. The authors have demonstrated interesting object detection results from the DETR model. That triggered the curiosity to use the model for detection of custom objects. Here, we are presenting the way to fine-tune the pre-trained DETR model over custom dataset. The fine-tuning results demonstrate significant improvement with respect to number of training epochs, both visibly as well as statistically.

Keywords Object Detection · Transformer · Bipartite matching

1 Introduction

Object detection is one the most researched domain in Computer Vision. It involves localization of objects in the given image and then classifying them by providing labels. While CNN-based techniques for image classification matured to the level of human and even surpassing them in certain scenes, these techniques have been extensively used for object detection tasks as well.

A. Kumar (✉)
Defence Institute of Psychological Research (DIPR), DRDO, Delhi, India
e-mail: akhilesh.dipr@gov.in

S. K. Singh · S. R. Dubey
Computer Vision and Biometrics Laboratory, Indian Institute of Information Technology, Allahabad, Prayagraj, India
e-mail: sk.singh@iiita.ac.in

S. R. Dubey
e-mail: srdubey@iiita.ac.in

M. Tistarelli et al. (eds.), *Computer Vision and Machine Intelligence*,
Lecture Notes in Networks and Systems 586,
https://doi.org/10.1007/978-981-19-7867-8_59

The object detection pipelines have been evolved into two main categories, i.e., multi-stage pipeline and single-stage pipeline. The CNN-based object detection algorithms started with Region-based CNN (RCNN) [1], going through Fast-RCNN [2] and Faster-RCNN [3]. Later many secondary tasks like semantic-segmentation were also baselined over object detection (e.g., Mask RCNN [4] using Faster-RCNN as baseline). These are multi-stage networks where the candidate-proposals having possible objects are either generated outside the network through separate mechanism (e.g., Selective Search [5] in case of Fast-RCNN) or through an in-build network (e.g., RPN [3] in case of Faster-RCNN) generating candidate-proposals and then passing them for detection. Later research explored single-stage networks, where the object detection task is modeled as an end-to-end task. YOLO series (version 1,2 and 3) [6–8] and SSD [9] are best examples of single-stage object detection networks. Results from both the networks are comparable. While Faster-RCNN still enjoys the benefit of better detection alignments, it's the YOLOv3 which gives the flexibility to scale the detection network based on small, medium or large object detection tasks. No doubt, the inference time of single-stage networks are less as compared to multi-stage ones. Today, while one community is a strong follower of Faster-RCNN pipeline, the other one believes on YOLOv3 pipeline. Both the flavors are evolving parallelly.

Recently, Transformer has emerged as an alternative to the convolution and recurrent networks. Initially the transformer was designed and proposed for the task of natural language processing [10]. Later, the DETR [11] model adapted the transformer architecture for object recognition. Since then, transformer architecture is extended to Vision Transformer (ViT) for visual recognition tasks. Deformable DETR [12] has been proposed to overcome the slow convergence and limited feature spatial resolution observed in the DETR model. Further improvement to DETR resulted in XCiT [13], where the authors have used a "transposed" version of the self-attention mechanism [10]. More new improvements include LocalViT [14], Pyramid Vision Transformer [15], and Multi-Scale Vision Longformer [16]. Recently, Vision transformer has been used for image retrieval [17] also.

2 Related Work

With still ongoing R&D based on Faster-RCNN and YOLO pipeline, these two designs are somewhat saturating with very less significant increments. With an aim to explore some different architectures for possible improvements over detection results and inference time, researchers at Facebook AI [11] have started exploring the Transformer network [8, 10], initially proposed by Google for machine translation. The team at Facebook AI designed a new framework called Detection transformer or DETR [11], especially for object detection. This framework, being a streamlined one, removed the use of non-maximal suppression and anchor-box generation components which were essential ingredients of the earlier models [11]. Still the model demonstrates performance which is at par with the Faster-RCNN baseline on the challenging COCO [19] object detection dataset. Later improvements were proposed in [12, 13].

The DETR model can be used to detected custom targets in two ways–whether training the model from scratch with the custom dataset or fine-tune the pre-trained model for the custom dataset. To train the model from scratch, the data requirements should meet that of the model. Training a huge model with less data will not give good performance results. Moreover, training from scratch may not be possible in general, due to heavy compute requirements. In this work, the size of our dataset is significant enough with very limited number of targets. But we have utilized the pre-trained weights to initialize the model, rather than random weights. Moreover, some changes were also carried out so that the pre-trained model can be re-trained on custom dataset with different number of target classes than COCO. Details are presented in subsequent sections. We shall start with some discussion of basic transformer model and then elaborate the customization and training followed by results.

2.1 The Transformer Model

The Google research team came up with this novel architecture for machine translation. This model incorporated the attention mechanism also. It was claimed in the paper that it is more parallel than the existing RNN-based models for the same task, thereby taking comparatively less time in training as well as inference [10]. Subsequently, the transformer model has been used in many NLP tasks.

Broadly, the transformer model architecture is divided into two block–encoder and decoder. Inside each block there are multiple units of Multi-Head Attention (MHA) network and Feed Forward Network (FFN) as shown in Fig. 1. The MHA consists of multiple heads of self-attention block. The self-attention relates different positions of a single sequence and hence computes intra-attention [10]. The self-attention is implemented using *Scaled dot-product attention* blocks which is briefly explained in subsequent paragraphs along with diagram in Fig. 2.

The inputs to the module are set of queries $Q = [q_1, q_2, \ldots q_m]$, set of keys $K = [k_1, k_2, \ldots k_n]$, and set of values $V = [v_1, v_2, \ldots v_n]$. All the individual vectors inside each set have common dimension d. The attention on a set of queries is obtained as

$$F = A(Q, K, V) = \text{softmax}\left(\frac{QK^T}{\sqrt{d}}\right)V \tag{1}$$

where F will be a $m \times d$ dimension output.

Multiple independent heads of this attention module are concatenated in parallel to make MHA. Thus, the attention obtained by MHA is

$$F = MA(Q, K, V) = \text{Concat}(h_1, h_2, \ldots h_h)W^o \tag{2}$$

Fig. 1 The Transformer architecture in an encode-decoder fashion. N denote the number of times each block is stacked in each of encoder and decoder module. Image credit: Attention is all you need [10]

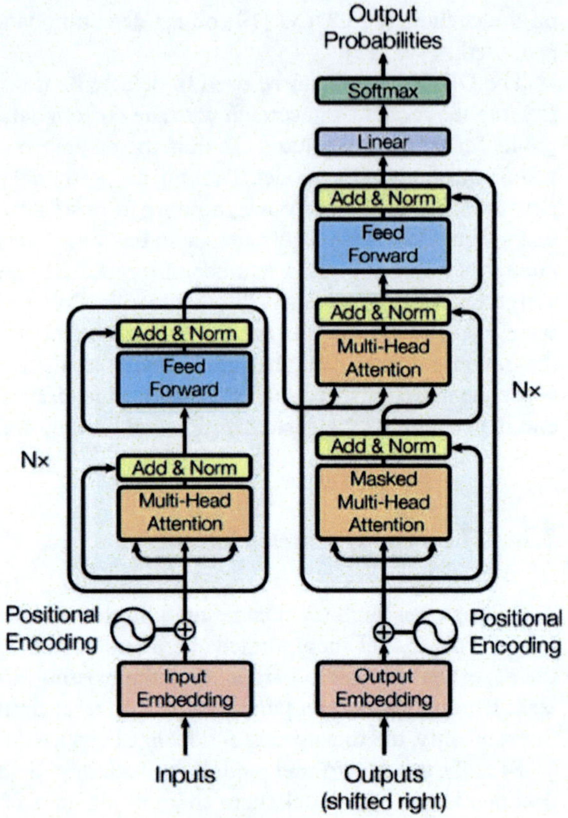

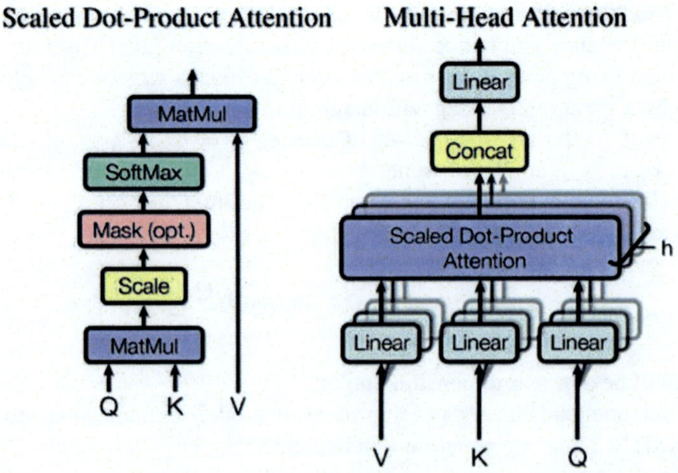

Fig. 2 (left) Scaled dot-product attention. (right) Multi-head attention. Image Credit: Attention is all you need [10]

where

$$h_i = A\left(QW_i^Q, KW_i^K, VW_i^V\right) \quad (3)$$

W_i^Q, W_i^K, and W_i^V are the matrices to project Q, K and V of the *i*th attention head. W^o is the output projection matrix to combine the outputs of all the heads. After the MHA, the output is passed to FFN which is a two-layer network with ReLU activation and Dropout in between.

$$\text{FFN}(x) = \text{FC}(\text{Dropout}(\text{ReLU}(\text{FC}(x)))) \quad (4)$$

These two modules are the basic ingredients of both the encoder and decoder modules.

While encoder calculates self-attention using MHA, where each of query, key, and values are obtained from same input, the decoder uses the encoded output as the guided attention along with its own self-attention [20]. Also, shortcut connection and layer normalizations are applied after each module of MHA and FFN, to simplify optimization. More details can be sought from the paper [10].

2.2 *The DETR Model [11]*

The team at Facebook AI formulated the object detection model as direct set prediction problem. The basic idea of encoder-decoder architecture based on transformers has been adopted for this task.

The DETR predicts all objects at once. The complete network is trained end-to-end. The error is calculated by Bipartite matching between predicted and ground truth objects. This results in a set loss function. The bipartite matching is carried out using Hungarian loss. This guarantees that each object has unique matching and enforces permutation invariance. A basic flow depicting all the functionalities is shown in Fig. 3.

The bipartite matching is obtained as

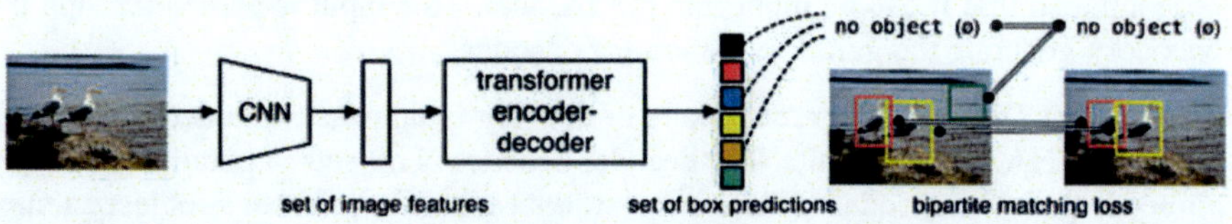

Fig. 3 DETR flow (image credit: End-to-end object detection with transformers [11])

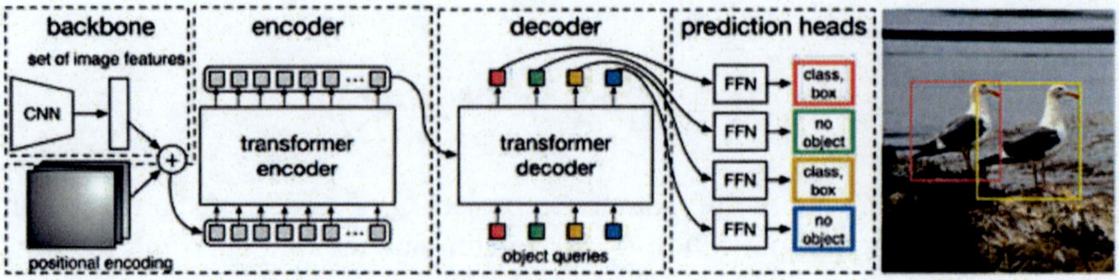

Fig. 4 DETR architecture (image credit: End-to-end object detection with transformers [11])

$$\hat{\sigma} = \underset{\sigma \in \mathfrak{S}_N}{\arg\min} \sum_{i}^{N} \mathcal{L}_{\text{match}}\left(y_i, \hat{y}_{\sigma(i)}\right) \tag{5}$$

where $\mathcal{L}_{\text{match}}$ is a pair-wise matching cost between ground truth and prediction with index $\sigma(i)$. $\mathfrak{S}_N$ is one permutation of N predictions.

Hungarian algorithm is used to compute above assignment. The similarity for class as well bounding box is taken into account. The Hungarian loss for all pairs matched as per Eq. (5) is obtained. This loss takes into account both the bipartite matching as per Eq. (5) as well as the bounding box loss. The box predictions are done directly rather than predicting offsets as done in Faster-RCNN and other such algorithms. A linear combination of L_1 loss and generalized IoU loss makes the Bounding Box loss.

DETR Architecture

The DETR architecture is shown in Fig. 4. The whole architecture is divided into three main components: the backbone, an encoder-decoder transformer and the prediction heads. Now let's understand these components one by one.

Backbone. It is a CNN-based backbone to generate a low-resolution feature map of the input image. While the input image is of dimension $3 \times H_o \times W_o$, the feature map dimension is $C \times H \times W$. The authors have used $C = 2048$ and $H, W = \frac{H_o}{32}, \frac{W_o}{32}$.

Transformer Encoder. The feature map from backbone is subject to 1×1 convolution to reduce the channel dimension from C to d. The spatial dimensions of the resulting feature map are further collapsed into one dimension ($d \times HW$) to make the input sequential so that it can be input into the encoder. This input is passed through the encoder t generate the guided attention for decoder.

Transformer Decoder. The architecture of decoder is same as explained earlier with multiple MHA and FFN units. The decoder decodes N objects in parallel. The input embeddings in the decoder are learned positional encoding referred as object queries. Output embedding are generated using these N object queries. Box coordinates and class labels are obtained by decoding these embeddings by the decoder. Thus, N final predictions are obtained.

FFN. It is a three-layer perceptron with ReLU activations. The hidden dimension is d with linear projections. The FFN is used to predict the normalized coordinates of bounding Box (center coordinates, height, width) with respect to the input image. Class labels are predicted by softmax over the linear layer. Additional class label ϕ is predicted in place of no-object.

Prediction FFNs and Hungarian losses are added after each decoder layer. This proved helpful during training. Parameter are shared among FFNs.

3 Target Detection with Transformer

The available pre-trained DETR model is trained on COCO object detection dataset. Because of this it can only detect the classes present in COCO dataset, labeling the missing classes with similar class label from the COCO dataset as shown in Fig. 5. Here the class "weapon" is labeled as "skis" in one image and "motorbike" in the other. Hence, the pre-trained model cannot be used as such for detecting custom objects. In this study, we explore the training of the DETR model on our custom dataset with pre-trained DETR weights as the initial ones.

We have used the code from the GitHub repository of the Facebook AI research team [11]. We shall be explaining the dataset details, training details, and the results obtained in subsequent texts.

3.1 Dataset

We have generated our dataset for four target classes–People, Building, Weapon, and Vehicle. Images were collected from open domain and object annotation was done in-house using MATLAB's Image labeler tool.

The annotated dataset has 27,983 train images and 6995 validation images. Some of the images in thumbnail are shown in Fig. 6. The dataset has been prepared

Fig. 5 DETR inference result on image having different object (weapon) than COCO dataset

Fig. 6 Images in the dataset

following the guidelines as per the standard object detection datasets like PASCAL [21] and COCO. All the annotations were done manually. Since the train-code of DETR restricts us to input annotations in COCO format, so we converted the MATLAB annotations into JSON format shown in Fig. 7. Images were arranged in *train2017* and *val2017* folders. Annotations are kept under *annotations* folder. This arrangement has been done as per COCO format.

3.2 Customizations

Since the DETR was trained on COCO with 90 object classes, so the classification head of the pre-trained model needs to be customized before loading the state-dictionary. The sample code in python is shown below.

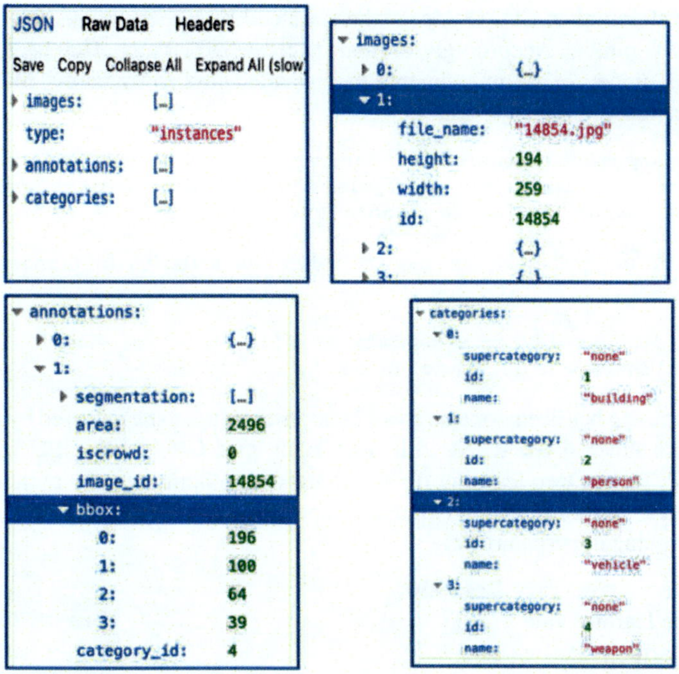

Fig. 7 Annotation description (.JSON file)

```
from hubconf import detr_resnet50
model = detr_resnet50(pretrained=False)
# Get pretrained weights
checkpoint = torch.load('../model_zoo/detr-r50-e632da11.pth')
model.load_state_dict(checkpoint["model"])
# Remove class weights
del checkpoint["model"]["class_embed.weight"]
del checkpoint["model"]["class_embed.bias"]
# Save the customized model
torch.save(checkpoint,'detr-r50_no-class-head.pth')
```

Now we can load the DETR model with above saved checkpoint and start training on our custom dataset with different number of target classes.

3.3 Training

DETR is originally trained on COCO object detection dataset which has 90 classes. To train the pre-trained model over different target categories, the existing code needs small amendment. The current Github repository [22] contains the code to train/val the DETR over COCO dataset. There is no provision to pass custom number of classes different than COCO. Hence the main code has been amended with number of classes argument. Accordingly, the code to create DETR model has been amended. The code snippets of the amendments are presented next. In "main.py" file following code was added:

```
# add argument parameter for number of classes
parser.add_argument('--num_classes',    type=int,    default=5,
help="Number of classes in dataset+1")
```

Also, in "detr.py" under the "models" folder amend the line for number of classes as follows:

```
#num_classes = 20 if args.dataset_file != 'coco' else 91
num_classes = args.num_classes
```

The training has been done on a workstation-class machine powered by dual Xeon processors and 256 GB RAM with Dual Titan RTX GPU (each GPU with 24 GB RAM). PyTorch Deep learning library along with other necessary python libraries were used for training. Average time taken per epoch was ~ 25 min. Training is done with following hyperparameters.

Backbone	Resnet50
Backbone learning rate	1e-6
DRER learning rate	1e-5
Batch_size	2
Learning rate drop	10 epochs
Number of queries	20
Epochs	50
Distributed	True
Resume	True

The original DETR supports two backbones ResNet50 and ResNet101. Since the number of classes as well as the dataset size are quits less as compared to COCO, so small backbone, i.e., ResNet50 is considered. Also, the learning rate is kept at smaller levels so that the machine can learn over the pre-trained weights. Default augmentations with scale and random-crop are used during training.

4 Results

The snapshots show the detection of custom objects. The validation results as shown in Figs. 8 and 9 depicts the improvement in learning from epoch 10 to epoch 50.

Both Average Recall as well as Average Precision have increased with increase in number of epochs of training. We hope that more training will result better performance. Further, some of the result snapshots after epoch 10 and epoch 50 are shown in Figs. 10 and 11, respectively.

```
IoU metric: bbox
 Average Precision  (AP) @[ IoU=0.50:0.95 | area=   all | maxDets=100 ] = 0.371
 Average Precision  (AP) @[ IoU=0.50      | area=   all | maxDets=100 ] = 0.652
 Average Precision  (AP) @[ IoU=0.75      | area=   all | maxDets=100 ] = 0.356
 Average Precision  (AP) @[ IoU=0.50:0.95 | area= small | maxDets=100 ] = 0.170
 Average Precision  (AP) @[ IoU=0.50:0.95 | area=medium | maxDets=100 ] = 0.360
 Average Precision  (AP) @[ IoU=0.50:0.95 | area= large | maxDets=100 ] = 0.528
 Average Recall     (AR) @[ IoU=0.50:0.95 | area=   all | maxDets=  1 ] = 0.294
 Average Recall     (AR) @[ IoU=0.50:0.95 | area=   all | maxDets= 10 ] = 0.541
 Average Recall     (AR) @[ IoU=0.50:0.95 | area=   all | maxDets=100 ] = 0.607
 Average Recall     (AR) @[ IoU=0.50:0.95 | area= small | maxDets=100 ] = 0.391
 Average Recall     (AR) @[ IoU=0.50:0.95 | area=medium | maxDets=100 ] = 0.605
 Average Recall     (AR) @[ IoU=0.50:0.95 | area= large | maxDets=100 ] = 0.778
```

Fig. 8 Val results after epoch 10

```
IoU metric: bbox
 Average Precision  (AP) @[ IoU=0.50:0.95 | area=   all | maxDets=100 ] = 0.422
 Average Precision  (AP) @[ IoU=0.50      | area=   all | maxDets=100 ] = 0.728
 Average Precision  (AP) @[ IoU=0.75      | area=   all | maxDets=100 ] = 0.411
 Average Precision  (AP) @[ IoU=0.50:0.95 | area= small | maxDets=100 ] = 0.217
 Average Precision  (AP) @[ IoU=0.50:0.95 | area=medium | maxDets=100 ] = 0.417
 Average Precision  (AP) @[ IoU=0.50:0.95 | area= large | maxDets=100 ] = 0.591
 Average Recall     (AR) @[ IoU=0.50:0.95 | area=   all | maxDets=  1 ] = 0.321
 Average Recall     (AR) @[ IoU=0.50:0.95 | area=   all | maxDets= 10 ] = 0.580
 Average Recall     (AR) @[ IoU=0.50:0.95 | area=   all | maxDets=100 ] = 0.647
 Average Recall     (AR) @[ IoU=0.50:0.95 | area= small | maxDets=100 ] = 0.434
 Average Recall     (AR) @[ IoU=0.50:0.95 | area=medium | maxDets=100 ] = 0.648
 Average Recall     (AR) @[ IoU=0.50:0.95 | area= large | maxDets=100 ] = 0.813
```

Fig. 9 Val results after epoch 50

Fig. 10 Prediction after epoch 10

We can see that the results after epoch 50 are better as compared to epoch 10. In fact, the amount of redundant bounding box predictions is significantly minimized after epoch 50. Further training may reduce the redundant predictions. Also, the bounding boxes are better aligned after epoch 50 as evident from Figs. 8 and 9.

The improvement in average recall suggests its strong surveillance related applications, where we cannot afford missing any targets.

Fig. 11 Prediction after epoch 50

5 Conclusion

We explored DETR to detect custom targets with an aim for surveillance related applications. We used the available code from the author's repository and customized that to train our custom dataset. The results, both statistically and qualitatively, show the applicability of DETR architecture in custom-target detection. We hope that more training epochs with some changes / additions at architecture level of DETR shall provide better results. The results for small and medium target detections are still under promising and requires significant improvement. In fact, automatically detecting small targets will be the real augmentation to the mankind. Future work shall look into this research.

Acknowledgements We thank DIPR, DRDO for providing the R&D environment to carry out the research work. We also thank IIIT Allahabad for providing the opportunity to carry out the PhD course under the Working Professional Scheme.

References

1. Girshick, R., Donahue, J., Darrell, T., Malik, J.: Rich feature hierarchies for accurate object detection and semantic segmentation (2013). (Online). Available: http://arxiv.org/abs/1311.2524
2. Girshick, R.: Fast R-CNN (2015). (Online). Available: http://arxiv.org/abs/1504.08083
3. Ren, S., He, K., Girshick, R., Sun, J.: Faster R-CNN: towards real-time object detection with region proposal networks. IEEE Trans. Pattern Anal. Mach. Intell. **39**(6), 1137–1149 (2017). https://doi.org/10.1109/TPAMI.2016.2577031
4. He, K., Gkioxari, G., Dollár, P., Girshick, R.: Mask R-CNN. IEEE Trans. Pattern Anal. Mach. Intell. **42**(2), 386–397 (2020). https://doi.org/10.1109/TPAMI.2018.2844175
5. Uijlings, J.R.R., van de Sande, K.E.A., Gevers, T., Smeulders, A.W.M.: Selective search for object recognition. Int. J. Comput. Vision **104**(2), 154–171 (2013). https://doi.org/10.1007/s11263-013-0620-5
6. Redmon, J., Divvala, S., Girshick, R., Farhadi, A.: You Only Look Once: Unified, Real-Time Object Detection (2015) (Online). Available: http://arxiv.org/abs/1506.02640

7. Redmon, J., Farhadi, A.: YOLOv3: An incremental improvement. *arXiv* (2018)
8. Redmon, J., Farhadi, A.: YOLO9000: better, faster, stronger. In: Proceedings—30th IEEE Conference on Computer Vision and Pattern Recognition, CVPR 2017, vol. 2017-Janua, pp. 6517–6525 (2017). https://doi.org/10.1109/CVPR.2017.690.
9. Liu, W., et al.: SSD: single shot multibox detector. In: Lecture Notes in Computer Science (including subseries Lecture Notes in Artificial Intelligence and Lecture Notes in Bioinformatics), vol. 9905 LNCS, pp. 21–37 (2016). https://doi.org/10.1007/978-3-319-46448-0_2
10. Vaswani, A., et al.: Attention Is All You Need (2017). (Online). Available: http://arxiv.org/abs/1706.03762
11. Carion, N., Massa, F., Synnaeve, G., Usunier, N., Kirillov, A., Zagoruyko, S.: End-to-End Object Detection with Transformers (2020). (Online). Available: http://arxiv.org/abs/2005.12872
12. Zhu, X., Su, W., Lu, L., Li, B., Wang, X., Dai, J.: Deformable DETR: Deformable Transformers for End-to-End Object Detection (2020). (Online). Available: http://arxiv.org/abs/2010.04159
13. El-Nouby, A., et al.: XCiT: Cross-Covariance Image Transformers. (2021). (Online). Available: http://arxiv.org/abs/2106.09681
14. Li, Y., Zhang, K., Cao, J., Timofte, R., van Gool, L.: LocalViT: Bringing Locality to Vision Transformers (2021). (Online). Available: http://arxiv.org/abs/2104.05707
15. Wang, W., et al.: Pyramid Vision Transformer: A Versatile Backbone for Dense Prediction without Convolutions (2021). (Online). Available: http://arxiv.org/abs/2102.12122
16. Zhang, P., et al.: Multi-Scale Vision Longformer: A New Vision Transformer for High-Resolution Image Encoding (2021). (Online). Available: http://arxiv.org/abs/2103.15358
17. Dubey, S.R., Singh, S.K., Chu, W.-T.: Vision Transformer Hashing for Image Retrieval (2021). (Online). Available: http://arxiv.org/abs/2109.12564
18. Muñoz, E.: Attention is all you need: Discovering the Transformer paper (2020)
19. Lin, T.Y., et al.: Microsoft COCO: common objects in context. In: Lecture Notes in Computer Science (including subseries Lecture Notes in Artificial Intelligence and Lecture Notes in Bioinformatics), vol. 8693 LNCS, no. PART 5, pp. 740–755 (2014). https://doi.org/10.1007/978-3-319-10602-1_48.
20. Yu, J., Li, J., Yu, Z., Huang, Q.: Multimodal Transformer with Multi-View Visual Representation for Image Captioning (2019). (Online). Available: http://arxiv.org/abs/1905.07841
21. Everingham, M., van Gool, L., Williams, C.K.I., Winn, J., Zisserman, A.: The pascal visual object classes (VOC) challenge. Int. J. Comput. Vision **88**(2), 303–338 (2010). https://doi.org/10.1007/s11263-009-0275-4
22. fmassa et al.: facebookresearch/detr (2020)

Document Image Binarization in JPEG Compressed Domain Using Dual Discriminator Generative Adversarial Networks

Bulla Rajesh, Manav Kamlesh Agrawal, Milan Bhuva, Kisalaya Kishore, and Mohammed Javed

Abstract Image binarization techniques are being popularly used in enhancement of noisy and/or degraded images catering different Document Image Analysis (DIA) applications like word spotting, document retrieval, and OCR. Most of the existing techniques focus on feeding pixel images into the convolution neural networks to accomplish document binarization, which may not produce effective results when working with compressed images that need to be processed without full decompression. Therefore in this research paper, the idea of document image binarization directly using JPEG compressed stream of document images is proposed by employing Dual Discriminator Generative Adversarial Networks (DD-GANs). Here the two discriminator networks—Global and Local work on different image ratios and use focal loss as generator loss. The proposed model has been thoroughly tested with different versions of DIBCO dataset having challenges like holes, erased or smudged ink, dust, and misplaced fibers. The model proved to be highly robust, efficient both in terms of time and space complexities, and also resulted in state-of-the-art performance in JPEG compressed domain.

Keywords Compressed domain · Deep learning · DCT · JPEG · CNN · Adversarial network · Handwritten · DD-GAN

B. Rajesh (✉) · M. K. Agrawal · M. Bhuva · K. Kishore · M. Javed
Department of IT, IIIT Allahabad, Prayagraj, U.P 211015, India
e-mail: rsi2018007@iiita.ac.in; br_cse@vignan.ac.in

M. K. Agrawal
e-mail: iit2018178@iiita.ac.in

M. Bhuva
e-mail: iit2018176@iiita.ac.in

K. Kishore
e-mail: iit2018079@iiita.ac.in

M. Javed
e-mail: javed@iiita.ac.in

B. Rajesh
Department of CSE, Vignan University, Guntur, A.P 522213, India

M. Tistarelli et al. (eds.), *Computer Vision and Machine Intelligence*,
Lecture Notes in Networks and Systems 586,
https://doi.org/10.1007/978-981-19-7867-8_60

1 Introduction

Document image binarization is a critical stage in any image analysis task, where eventually the image pixels are classified into text and background as shown in Fig. 1. This dominant stage can hamper recognition tasks in the later stages [11]. The need for this stage arises due to natural degradation of historical documents, such as aging effects, ink stains, bleed through, stamps, and faded ink [12]. Moreover, digitized documents themselves might be compromised due to bad camera quality, disturbances, non-uniform illumination, watermarks, etc. These documents constitute plethora of information which could prove beneficial for us humans. Therefore, in the literature, there are plenty of research work focused on image binarization using both the handcrafted feature-based methods [2, 11] and deep learning-based methods [1, 4, 20, 29]. These methods are pixel image driven, which may not be feasible when working with compressed document images that need to be processed without full decompression. This is because full decompression becomes an expensive task when huge volume of document images are to be processed. Therefore, in this research paper, the novel idea of image binarization using compressed document images is proposed that trains the deep learning model directly with the compressed stream of data.

In digitization's early stages, document binarization meant using single or hybrid thresholding techniques, such as Otsu method, Nick method, multi-level Otsu's method, and CLAHE algorithm [2]. However, with the advent of deep learning, CNNs' have been continually used to solve this rather taxing problem. Its superior performance over standard thresholding approaches can be due to its ability to capture the spatial dependence among the pixels [7]. The GANs', deep learning tech-

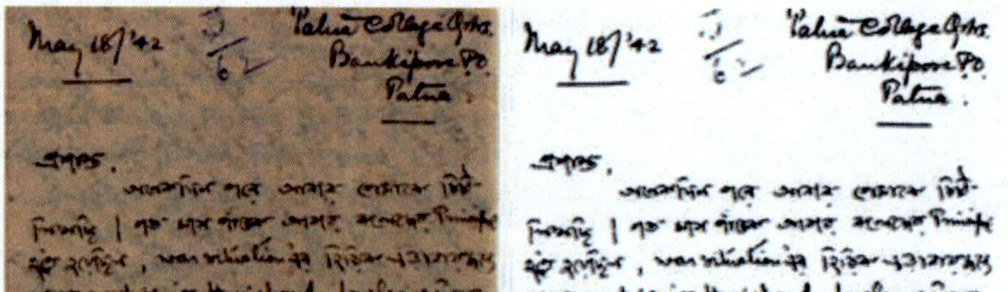

Fig. 1 Problem of document binarization in case of a historical document

nique, have been more successful than CNNs' in the domain of image generation, manipulation, and semantic segmentation [27]. GANs' such as Cycle GAN [25], ATANet coupled with UDBNet [14], conditional GAN [29], and GAN with U-Net [20] can also combat the issue of limited data. Even conditional GANs' have been quite successful working on watermark removal from documents [6]. Dual Discriminator GANs', are a relatively new concept, have been used to remove degradation, and they have proved to be robust and have performed on par with the current techniques. These deep learning networks have not only performed well on English-based documents [18], but also on Arabic- and Hindi-based documents [17]. GANs' can also perform on Persian heritage documents [1], and on handwritten Sudanese palm leaf documents [28].

However, Dual Discriminator Generative Adversarial Networks [4] are quite slow despite being shallow, owing to the presence of a heavy U-Net Architecture and two convolution neural networks. This issue can be solved to a certain extent by decreasing the size of the images, and we intend to use JPEG compression algorithm here. This algorithm intelligently employs Discrete Cosine transformation, quantization, and serialization to remove redundancy of redundant features from an image. The human psychovisual system discards high-frequency information such as abrupt intensity changes and color hue in this type of compression. The accuracy of current techniques is frequently assessed through ICDAR conference [19].

From the above context, the major contributions of this paper are given as follows:

- The idea of accomplishing document image binarization directly using JPEG compressed stream.
- The modified DD-GANs architecture with two discriminator networks—Global and Local working on different image ratios and using focal loss as generator loss, in order to accommodate JPEG compressed documents.
- Sate-of-the-art performance in JPEG compressed domain with reduced computation time and low memory requirements

The rest of the paper is organized in three sections. Section 2 discusses the related literature, Sect. 3 briefs the proposed model and deep learning architecture. Section 4 reports the experimental results and presents analysis. Finally, Sect. 5 concludes the work with a brief summary and future work.

2 Related Literature

In this section, we present some of the prominent document binarization techniques that are reported in the literature. Boudraa et al. [2] have introduced a data pre-processing algorithm called CLAHE. This algorithm enhances visual contrast while preventing over-amplification of sound. However, applying the algorithm to all images is not a good strategy because it can modify the object boundaries and add distortion, affecting vital info in some circumstances. As a result, the contrast value is used. CLAHE is only done when the contrast value lies below a specific threshold. In [26], the research goal is to improve the document images by tackling several types of degeneration. Some of these are removal of watermark(s), cleaning up of documents, and binarization. The purpose is to segment the image with watermark and the text in the foreground and background, respectively. This is achieved using primarily detecting the watermark and then passing it through a designated model.

The research work by [4] proposes dual discriminator GAN. The architecture consists of U-Net as a generator and a self-designed local and global discriminator. Unlike normal GANs', in dual discriminator GAN, local discriminator works on lower level features, whereas global discriminator works on background features. Discriminators in normal GAN usually learn higher level or lower level features, but they cannot focus on both. Dual discriminators work better here as they can learn to recognize both. Since the input documents contain more background noise, the focal loss function is used as the generator loss function to avoid the imbalance in the dataset, as suggested in the literature [15], and binary cross-entropy is used for the discriminator networks. Here, they use a total loss function to reduce over-fitting in generator and discriminator. The local discriminator contributes more than the global discriminator to total loss as the local discriminator learns low level features which is to be reproduced by generator and thus is more important.

The work by [8] revisits the formulation of JPEG algorithm. At the time of its creation, compression techniques such as predictive coding, block coding, cosine transformation, vector quantization, and combination of these were proposed. The Karhunen–Loeve Transform (KLT) introduced at the time was the most optimum compression technique; however, it was the most computationally intensive. DWT, other such technique, is also a more optimum compression technique; however, it too was not feasible with the hardware. Thus, Discrete Cosine transform, which could be calculated very fast by using Fourier transforms, along with vector quantization and serialization with zig-zag encoding was decided as the major component of the algorithm. Also, the block size for compression was a major topic of discussion. Neither could it be so small that pixel-to-pixel correlation be missed, nor could it be large that block tries to take advantage of a correlation that might not exist. Thus, the block size of 8×8 was decided. Finally, the image was encoded using two algorithms simultaneously. The first pixel of each block is considered to be the AC component that has Digital Pulse Code Modeling applied on it, whereas for the other 63 pixels, which are considered to be the DC component have run length encoding applied on them.

There are some recent efforts to accomplish different DIA operations like text segmentation [22, 23], word recognition [21] and classification, and retrieval [16], etc., directly using the JPEG compressed stream. To the best of our knowledge, presently there is no image binarization technique available for JPEG compressed document images. Also JPEG is the most popular compression algorithm supported worldwide, and more than 90% of images in the Internet world are in JPEG format [22]. Therefore, the image binarization method in this paper is focused only on addressing JPEG compressed document images.

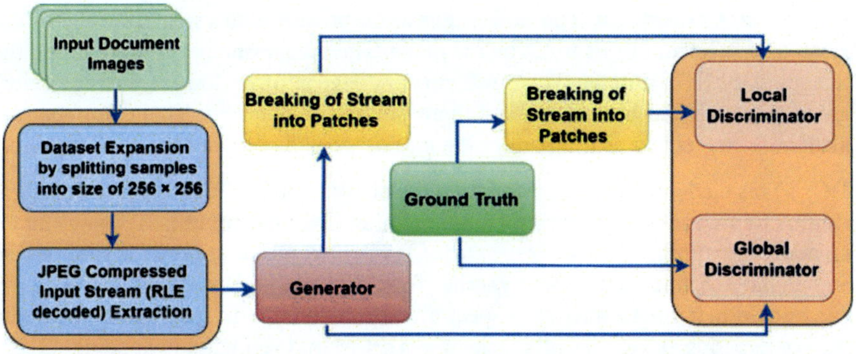

Fig. 2 Proposed model for binarization of JPEG compressed document images. The deep learning architecture here uses average instead of max-pooling layer, and it has one additional convolution layer in the local discriminator in comparison with the base model [4]

3 Proposed Model

Our introduced model comprises two major parts: pre-processing of pixel images and DD-GAN, as shown in Fig. 2. It is very important to note that the pre-processing stage is required only for those documents that are not directly available in the JPEG compressed form. The input to the generator is always the JPEG compressed stream of document image to be binarized.

3.1 Image Pre-processing

We have used JPEG compression algorithm for pre-processing the images that are not made available in the JPEG compressed form. The JPEG algorithm takes advantage of the anatomical characteristics of the human eye. This method considers that humans are more sensitive to color illumination than chromatic value of an image, and that we are more sensitive to low frequency content in any picture than high frequency content. JPEG algorithm consists of steps such as splitting of each image into blocks of size 8×8, color-space transform, DCT, quantization, serialization, vectoring, encoding, and decoding. The step by step procedure of the algorithm is explained in [22, 30].

Splitting of Image: The process of choosing the right block size, though might seem less significant, is one of the most significant parts of the JPEG algorithm. If we choose a small block size, then we will not be able to find any relevant correlation between the pixels of the image; however, a rather large size would lead to unnecessary advantage of a correlation that is not present. After careful consideration, the JPEG came up with the block size of 8×8 for images of size 720×575 or less. Since, each of the image that we have are of the size 256×256, we use this block size.

Color-Space Transform: The name suggests the process of this step in the algorithm. Here, a switch from RGB to YCbCr is made. The transformation is done for all the images in the RGB domain. We transform our image into this domain, as these colors are less sensitive to human eye and therefore can be removed. This color space is also more convenient as it separates the luminance and chrominance of the image.

DCT: This mathematical transformation is a crucial step in JPEG compression algorithm, known as Discrete Cosine Transformation (DCT); it comprises various mathematical algorithms, such as Fast-Fourier Transform, which are used in turn to take any signal and transform it into another form. Since image is a type of signal, we can transform it into frequency or spectral information, so that it can be manipulated for compression using algorithms such as DPCM and run length encoding (RLE). This transformation basically expresses each of those 8×8 block pixels as sum of cosine waves. This helps us calculate the contribution of each cosine wave. As the high frequency contents of the image will have a less coefficient of cosine wave, we can remove those and only retain the low frequency ones.

Quantization: Quantization is the process where we remove the high frequency cosine waves while we retain the low frequency ones. For this, we have used the standard chrominance and luminance quantization tables. These tables can be edited to change the JPEG compression ratio.

Serialization: In serialization, we reduce the redundancy in the image, by zig-zag pattern scanning and serialize this data. Also, this groups the low frequency coefficients in the top of the vector.

Vectoring: After applying DCT on the image, we are left 64 cosine waves, in which the first pixel is DC value, whereas other 63 values are AC values. These DC values are large and can be varied, but they will be similar to previous 8×8 block, and thus is vectorized using Digital Pulse Code Modeling (DPCM). We also use run length encoding to encode the AC components of the image.

Encoding: We use Huffman encoding technique to shrink the file size down further, and then reverse the process to decode the image. However, full decoding is not necessary here in this research work, but partial decoding is needed to extract JPEG compressed DCT coefficients to be fed into the deep learning model.

3.2 Network Architecture

The CNN used by us is Dual Discriminator-based Generative Adversarial Network (DD-GAN) [4]. This model embraces a generator and a global and a local discriminator. A basic GAN model, vanilla GAN model, makes use of one of each generator and discriminator, where the generator attempts to develop images that are plausible enough to fool the discriminator. These GANs' coupled with CNNs' for feature extraction have made it plausible to rehabilitate ancient murals to a degree [3]. The main restriction in a single discriminator GAN model will be its limitation of having to choose between high level features or low level features, whichever gives a better plausibility to the model. And, since we ignore either one of them, it is plausible that the model is ignoring a rather significant feature of the image. Dual discriminator will allow each discriminator to extract both high level and low level features. The global discriminator is fed the entire image so that it extracts high level features like image background and texture, and the local discriminator, which is fed in patches of the entire image, i.e., 32×32 size of images extracted from 256×256 image size, will extract low level features like the text strokes, edges, blobs, etc.

Generator: We use the U-Net architecture for the generator, proposed in the paper [24]. This generator architecture proposes down-sampling of the image, followed by up-sampling of same image. The down-sampling uses a typical convolution architecture of two 3×3 convolution layers followed my a max pool having a stride of 2. The up-sampling of the image is done with a similar architecture as that of down-sampling architecture with the exception that the max-pooling layer is replaced with a up-convolution layer. Also, the generator uses the focal loss function. A major disadvantage of DIBCO dataset is the existence of pixels that belong to the background rather than text such as colored or white background, than the foreground pixels such as text strokes. This will create a largely imbalanced generator which will focus more on the background. This in turn will lead to a bad generator model. The focal loss function can treat this issue to some extent. This function petitions a modulating cross-entropy loss term in order to focus learning on hard miss classified instances and thus deals with class imbalance problem aptly.

Discriminator: We use two discriminators: Global and Local. We have used binary cross-entropy (BCE) loss as a local loss function for both discriminators. The global discriminator is made up of two convolution layers with batch normalization and the Leaky Relu activation function, followed by one average pooling layer and three fully connected layers. The global has a lesser number of layers than the local discriminator as it is supposed to deal with a background in the images, which consists of fewer features in ground truth images. The local discriminator is made up of five convolution layers coupled to batch normalization layers, and the activation function used is Leaky Relu. In addition, four fully connected layers are added to the discriminator. There is no average pooling layer as we consider patches of images of size 32×32, and the use of a pooling layer might lead to loss of spatial information, which will be unfavorable as the discriminator is supposed to learn intricate features of the image.

3.3 *Total GAN Loss*

The total GAN loss tells how well the model is being trained. We use the focal loss from the discriminator and BCE loss from the discriminators in the following formula. The λ value is very high when compared to other values to signify that the generator is the main model being trained.

- The total loss function:

$$\boxed{L_{\text{total}} = \mu(L_{\text{global}} + \sigma L_{\text{local}}) + \lambda_{\text{Lgen}}} \quad (1)$$

- L_{total} = Total loss, L_{global} = global loss, L_{local} = local loss averaged over all patches of images, Lgen = generator loss.
- The value of μ given in the paper [4] is 0.5, 5, and 75.
- The value of σ is higher than 1 to indicate that the global discriminator contributes lesser to the loss function than local .

4 Experiment and Results

4.1 DIBCO Dataset

We have used the DIBCO 2014 [13], 2016 [9], 2017 [10] datasets to perform the experiment on the proposed model. DIBCO is a standardized dataset primarily used for document models which represent the challenges of binarization of historic handwritten manuscripts. Some of the sample document images are shown in Fig. 3.

There are ten document images in the 2014 H-DIBCO, while H-DIBCO 2016 and 2017 consist of 20 document images each. DIBCO dataset consists of document images as well as ground truth images which were built manually. This dataset poses the following challenges: holes, erased and smudged ink, dust, and misplaced

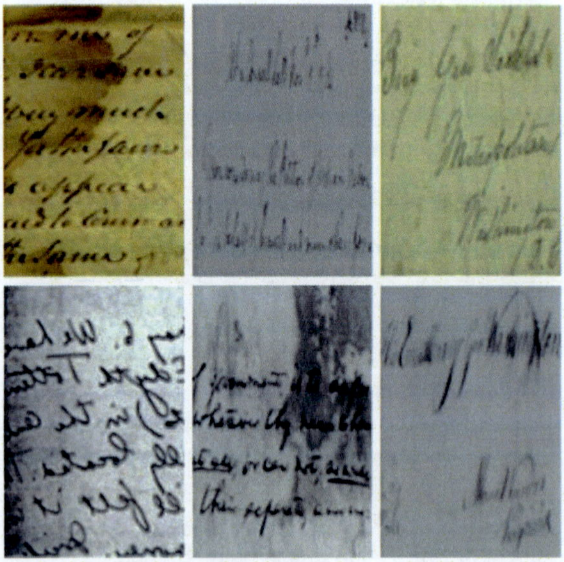

Fig. 3 Some document images from DIBCO 09, 10, 11 datasets [17]

fibers, owing to seventeenth-century historical cloth documents. Furthermore, the digitization of these documents was done by institutions that own these documents and thus add non-uniformity in lighting, resolution, etc., to the digitized documents. To escalate the size of the training and testing dataset, we pad each image with 128 black spaces on all sides and divide the image into blocks of 256 × 256. This also allows us to train the model on less-resolution images, making the training faster. The train and test images of the entire dataset were passed through the JPEG algorithm. We got an average compression ratio of 20:1.

Firstly, we pre-process the DIBCO dataset images obtained from UCI machine learning repository [5] which are not directly available in the JPEG compressed format. The dataset consists of 20 documents and ground truth images. However, we cannot train a deep learning model with few images. Thus to expand the dataset, we divide each image into segments of 256 × 256. This expansion creates sufficiently large training data of size 2352 images. After this, we pass the images through the JPEG compression algorithm to compress the images and feed them to the proposed DD-GAN model. Primarily, We train a GAN by feeding the document images and ground truth of the document images. The document images are fed to the generator while the ground truth images are relegated to the discriminator. We feed the entire image (256 × 256) to the global discriminator and patches of size 32 × 32 into the local discriminator. During testing also, we need to convert the testing image into compressed form, same as in training, before passing it through the generator. The generator will produce an image devoid of bleed through, stain marks, uneven pen strokes, etc. Finally, we apply global thresholding of 127 to the generated image, giving us a binarized image.

4.2 Results

The experimental results of the proposed model tested on the standard dataset H-DIBCO have been tabulated in Tables 1, 2, and 3. We have employed the Peak Signal-to-Noise Ratio (PSNR) metric as the performance measurement as given in Eq. (2).

$$\text{PSNR} = 10\log_{10}\frac{255^2}{\text{MSE}} \tag{2}$$

The performance of the proposed model has been compared with the performance of the existing pixel domain model in the literature. In all the experiments, the proposed model has achieved better performance directly in the compressed domain, as shown in the tables. Similarly, we have calculated the performance in terms of pixel domain, where the PSNR value of the generated output of the proposed model is better when it is fully decompressed and compared with pixel domain output, as shown in the tables. Some of the output images in the compressed domain for the compressed input streams fed to the model are shown in Fig. 4. The middle two columns are the input stream and output stream in the compressed domain. In the figure, the uncompressed

Table 1 Test results on 2014 dataset of H-DIBCO with the proposed model

Procedure	PSNR (%)
Base model [4]	22.60
Proposed model with compressed JPEG images	23.57
Proposed model with fully decompressed images	24.73

Table 2 Test results on 2016 dataset of H-DIBCO with the proposed model

Procedure	PSNR (%)
Base model [4]	18.83
Proposed model with compressed JPEG images	19.64
Proposed model with fully decompressed images	20.51

Table 3 Test results on 2017 dataset of H-DIBCO with the proposed model

Procedure	PSNR(%)
Base model [4]	18.34
Proposed model with compressed JPEG images	18.76
Proposed model with fully decompressed images	19.79

image of the compressed input stream of the model and the fully decompressed image of the compressed output stream computed by the proposed model are shown in the first and last columns, as shown in Fig. 4, for human visual perception.

Further, the advantage of the proposed model applying to direct compressed stream is in two folds. One is computational gain, and the second one is storage efficiency. We have conducted an experiment on the proposed model to verify these two advantages. The details are tabulated in Table 4. The model takes an average time of 708 s to run one epoch for pixel images and 333 s for compressed images, as shown in the table. Similarly, the storage cost of the compressed input stream is 48 Kb, which is very low compared to the uncompressed stream with a size of 3072 Kb, as shown in the table. Similarly, we have extended this experiment to the entire dataset, where the computational analysis for n number of images vs the computational time it requires in both compressed and pixel domains is graphically shown in Fig. 5. In both cases, the compressed input has shown reduced costs and improved performance in both computational and storage costs.

The proposed model has been tested on different challenging input cases such as erased or smudged ink, uneven lighting, holes, and dust, present in the H-DIBCO dataset. The output images of such cases are shown in Fig. 6. In the figure, the first row shows the output for the smudged ink. The second row shows the holes, and the third row shows the dust, and finally, the final row shows the results of uneven lighting. In all the cases, it can be observed that the proposed model has achieved significant performance directly in the compressed domain. During training, since the proposed

Fig. 4 Output images generated by the proposed model for the sample input document images in the compressed domain

Table 4 Performance details of the proposed model in terms of computational time and storage costs

Procedure	Time/epoch	Space/batch
Base model [4]	708 s	3072 kb
Proposed model	333 s	48 kb

model was only trained on small patches of the input document images, we tested the model by verifying it on entire document images. Therefore, the proposed model may be ensured that it is not just limited to the small patches of the document image but also applicable to processing the entire document images. The experimental results on the entire document images are shown in Fig. 7. In the context of all the experiments discussed above, the overall observation and conclusion are that the proposed model in the compressed domain has proved to be an effective solution for binarizing document images directly in the compressed domain.

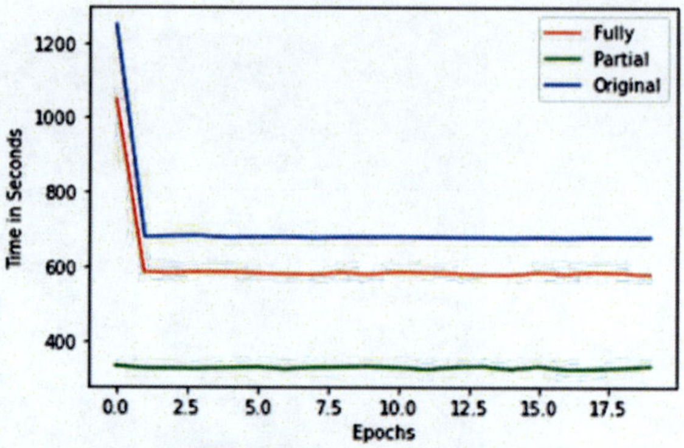

Fig. 5 Time analysis of proposed model based on raw images (original), JPEG compressed and fully decompressed images

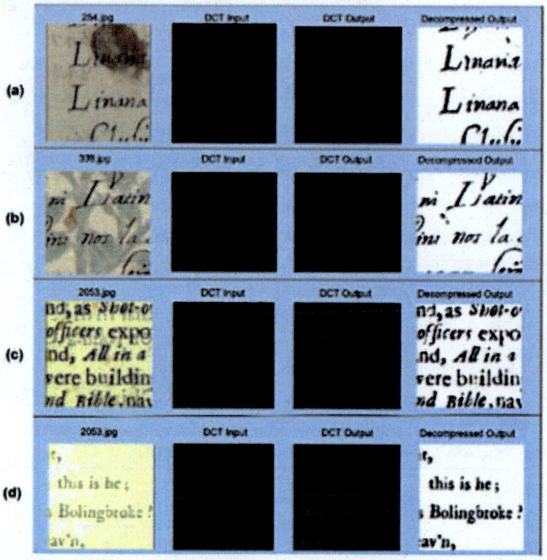

Fig. 6 Experimental results of the proposed model tested on different challenging cases like **a** smudged ink, **b** hole, **c** dust, and **d** uneven lighting present in the dataset

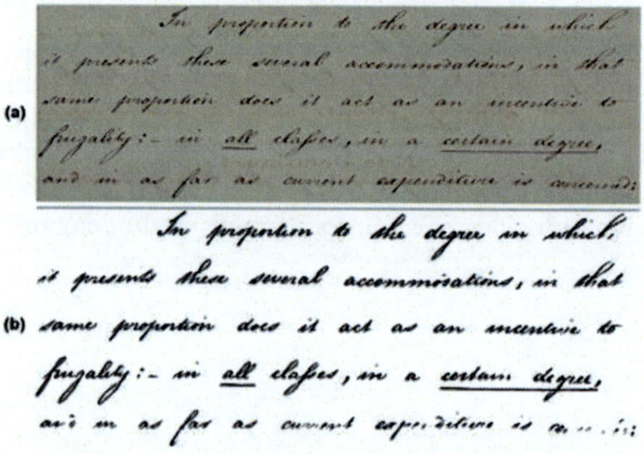

Fig. 7 Experimental of the model tested on the entire document image showed in the decompressed domain (row-wise), **a** sample input image with noise and **b** predicted binarized image

5 Conclusion

The present research paper proposed a model for performing the document image binarization using a dual discriminate generative adversarial network. The contribution of this research work is that the direct compressed stream of document images is fed to the proposed model to perform the binarization task directly in the compressed representation without applying decompression. The model has been tested on the benchmark dataset DIBCO, and the experimental results of the proposed model have shown the promising and stat-of-the-art performance directly in the compressed domain.

References

1. Ayatollahi, S., Nafchi, H.: Persian heritage image binarization competition (2012). pp. 1–4 (03 2013). https://doi.org/10.1109/PRIA.2013.6528442
2. Boudraa, O., Hidouci, W., Michelucci, D.: Degraded historical documents images binarization using a combination of enhanced techniques (2019)
3. Cao, J., Zhang, Z., Zhao, A., Cui, H., Zhang, Q.: Ancient mural restoration based on a modified generative adversarial network. Heritage Sci. **8**, 7 (2020). https://doi.org/10.1186/s40494-020-0355-x
4. De, R., Chakraborty, A., Sarkar, R.: Document image binarization using dual discriminator generative adversarial networks. IEEE Sig. Process. Lett. 1–1 (2020). https://doi.org/10.1109/LSP.2020.3003828
5. Dua, D., Graff, C.: UCI machine learning repository (2017). http://archive.ics.uci.edu/ml
6. Dumpala, V., Kurupathi, S., Bukhari, S., Dengel, A.: Removal of historical document degradations using conditional GANs. pp. 145–154 (2019). https://doi.org/10.5220/0007367701450154
7. Ehrlich, M., Davis, L.S.: Deep residual learning in the JPEG transform domain. CoRR **abs/1812.11690** (2018). http://arxiv.org/abs/1812.11690
8. Hudson, G., Léger, A., Niss, B., Sebestyén, I.: Jpeg at 25: still going strong. IEEE MultiMedia **24**(2), 96–103 (2017). https://doi.org/10.1109/MMUL.2017.38
9. Ioannis, P., Konstantinos, Z., George, B., Basilis, G.: Icfhr2016 competition on handwritten document image binarization. In: Proceedings of the International Conference on Document Analysis and Recognition, ICDAR, pp. 619–623 (2016)
10. Ioannis, P., Konstantinos, Z., George, B., Basilis, G.: Icdar2017 competition on handwritten document image binarization. In: Proceedings of the International Conference on Document Analysis and Recognition, ICDAR, pp. 1395–1403 (2017)
11. Javed, M., Bhattacharjee, T., Nagabhushan, P.: Enhancement of variably illuminated document images through noise-induced stochastic resonance. IET Image Process. **13**(13), 2562–2571 (2019)
12. Khamekhem Jemni, S., Souibgui, M.A., Kessentini, Y., Fornés, A.: Enhance to read better: a multi-task adversarial network for handwritten document image enhancement. Pattern Recogn. **123**, 108370 (2022)
13. Konstantinos, N., Basilis, G., Ioannis, P.: Icfhr2014 competition on handwritten document image binarization. In: Proceedings of the International Conference on Document Analysis and Recognition, ICDAR, pp. 809–813 (2014)
14. Kumar, A., Ghose, S., Chowdhury, P.N., Roy, P.P., Pal, U.: UDBNET: unsupervised document binarization network via adversarial game. CoRR **abs/2007.07075** (2020). https://arxiv.org/abs/2007.07075

15. Lin, T.Y., Goyal, P., Girshick, R., He, K., Dollár, P.: Focal loss for dense object detection. In: Proceedings of the IEEE International Conference on Computer Vision, pp. 2980–2988 (2017)
16. Lu, Y., Tan, C.L.: Document retrieval from compressed images. Pattern Recogn. **36**(4), 987–996 (2003)
17. Mahmoud, S.A., Ahmad, I., Alshayeb, M., Al-Khatib, W.G., Parvez, M.T., Fink, G.A., Märgner, V., Abed, H.E.: Khatt: Arabic offline handwritten text database. In: 2012 International Conference on Frontiers in Handwriting Recognition, pp. 449–454 (2012). https://doi.org/10.1109/ICFHR.2012.224
18. Marti, U.V., Bunke, H.: A full english sentence database for off-line handwriting recognition. In: Proceedings of the 5th International Conference on Document Analysis and Recognition. ICDAR '99 (Cat. No.PR00318), pp. 705–708 (1999). https://doi.org/10.1109/ICDAR.1999.791885
19. Pratikakis, I., Gatos, B., Ntirogiannis, K.: Icdar 2013 document image binarization contest (DIBCO 2013), pp. 1506–1510 (2011). https://doi.org/10.1109/ICDAR.2011.299
20. Quang-Vinh, D., Guee-Sang, L.: Document image binarization by GAN with unpaired data training. Int. J. Contents **16**(2), 1738–6764 (2020)
21. Rajesh, B., Jain, P., Javed, M., Doermann, D.: Hh-compwordnet: holistic handwritten word recognition in the compressed domain. In: 2021 Data Compression Conference (DCC), pp. 362–362 (2021). https://doi.org/10.1109/DCC50243.2021.00081
22. Rajesh, B., Javed, M., Nagabhushan, P.: Automatic tracing and extraction of text-line and word segments directly in jpeg compressed document images. IET Image Processing (2020)
23. Rajesh, B., Javed, M., Nagabhushan, P.: Fastss: fast and smooth segmentation of jpeg compressed printed text documents using dc and ac signal analysis. Multimedia Tools Appl. 1–27 (2022)
24. Ronneberger, O., Fischer, P., Brox, T.: U-net: Convolutional networks for biomedical image segmentation. CoRR **abs/1505.04597** (2015). http://arxiv.org/abs/1505.04597
25. Sharma, M., Verma, A., Vig, L.: Learning to Clean: A GAN Perspective, pp. 174–185 (2019). https://doi.org/10.1007/978-3-030-21074-8_14
26. Souibgui, M.A., Kessentini, Y.: DE-GAN: A conditional generative adversarial network for document enhancement. CoRR **abs/2010.08764** (2020). https://arxiv.org/abs/2010.08764
27. Sungho, S., Jihun, K., Paul, L., Yong., L.: Two-stage generative adversarial networks for document image binarization with color noise and background removal (2020)
28. Suryani, M., Paulus, E., Hadi, S., Darsa, U.A., Burie, J.C.: The handwritten sundanese palm leaf manuscript dataset from 15th century. In: 2017 14th IAPR International Conference on Document Analysis and Recognition (ICDAR). vol. 01, pp. 796–800 (2017). https://doi.org/10.1109/ICDAR.2017.135
29. Tensmeyer, C., Martinez, T.: Document image binarization with fully convolutional neural networks, pp. 99–104 (11 2017). https://doi.org/10.1109/ICDAR.2017.25
30. Wallace, G.K.: The JPEG still picture compression standard. IEEE Trans. Consum. Electron. **38**(1), 18–34 (1992)

Author Index

M. Tistarelli et al. (eds.), *Computer Vision and Machine Intelligence*, Lecture Notes in Networks and Systems 586,
https://doi.org/10.1007/978-981-19-7867-8